AF861677

SCIENCE OF HARD MATERIALS — 3

SCIENCE OF HARD MATERIALS — 3

Proceedings of the 3rd International Conference on the Science of Hard Materials, Nassau, The Bahamas, November 9–13, 1987

EDITOR: V. K. SARIN

Reprinted from the journal
Materials Science and Engineering
Vol. A105/106

ELSEVIER APPLIED SCIENCE
LONDON and NEW YORK

ELSEVIER APPLIED SCIENCE PUBLISHERS LTD
Crown House, Linton Road, Barking, Essex IG11 8JU, England

Sole Distributor in the USA and Canada
ELSEVIER SCIENCE PUBLISHING CO., INC.
655 Avenue of the Americas, New York, NY 10010, U.S.A.

British Library Cataloguing in Publication Data

International Conference on the Science of
Hard Materials (*3rd: Nassau, Bahamas*)
Science of hard materials 3.
1. Hard materials
I. Title II. Sarin Vinod K.
III. Materials science and engineering
620.1'1

ISBN 1-85166-982-5

Library of Congress Cataloging in Publication Data

International Conference on the Science of Hard Materials
(*3rd : 1987 : Nassau, Bahamas*)
Science of hard materials.
Reprinted from the journal *Materials Science and Engineering* vol. A105/106
Bibliography: p.
Includes index.
1. Hard materials—Congresses. I. Sarin, V. K.
II. Title.
TA418.45.I57 1987 620.1'12 88-33447
ISBN 1-85166-982-5

© ELSEVIER SEQUOIA S.A.. 1988

The selection and presentation of material and the opinions expressed in this publication are the sole responsibility of the authors concerned

All rights reserved. No part of this publication may be reproduced, stored in a retrieval system, or transmitted in any form or by any means, electronic, mechanical, photocopying, recording, or otherwise, without the prior written permission of the copyright owner.

Printed in The Netherlands by Krips Repro B.V.

Preface

This volume is a collection of papers that were presented at the Third International Conference on the Science of Hard Materials held in Nassau, Bahamas, in November 1987. As with the two previous conferences, our objective was to provide an interdisciplinary forum for addressing the fundamental role of the physical, chemical and mechanical properties of hard materials, and for exploring the theoretical and applied possibilities for their advancement. To this end, a delegation of 125 scientists and engineers from 18 countries made over 90 presentations, highlighted by five invited talks by Professors W. S. Williams, A. G. Evans, R. M. German and Drs. J. M. Blocher and S. Yazu. Of these, 73 were selected and reviewed to present a balanced and comprehensive survey of the state of the art in the "science of hard materials". Thematically, the book is divided into six sections, namely (1) Fundamental properties, (2) Ceramics (structure/properties), (3) Cemented carbides (process/structure), (4) Cemented carbides (properties), (5) Surface modification and (6) Diamond (synthesis/properties).

The objective was to provide a volume of high quality which could serve as an invaluable reference for researchers and technologists concerned with hard materials and their characteristics. I am grateful to all the authors for their patience, understanding and cooperation, and to the referees for their invaluable assistance in helping me achieve this goal. I would also like to acknowledge the staff of Elsevier Sequoia (Oxford, England), without whom the collation and timely publication of these proceedings would not have been possible. Last, but not least, I would like to acknowledge with special thanks the dedication and work of the members of the Organizing Committee (A. G. Evans, J. Gurland, A. Krawitz, M. Lee, L. Toth and W. Williams) and the International Liaison Committee.

An undertaking of this magnitude for a non-profitmaking organization such as the "Science of Hard Materials" group is impossible without sponsorship. Financial assistance provided by GTE Laboratories, Kennametal Inc., Mitsubishi Metal Corporation, Sumitomo Electric Industries Ltd., Stellman Societe Anonyme, Toshiba Tungaloy Co. Ltd. and GTE Valenite is gratefully acknowledged.

Current research in hard materials, such as composite ceramics, ultrahard materials, surface modified systems and cemented carbides, suggests that the full potential of these materials has yet to be realized. Continued investigation is essential in order to develop new and improved hard materials to meet the technological challenges of the future. The fulfillment of these multifaceted objectives will require the close co-operation of, and dialogue between, scientists, technologists and manufacturers. It is our hope that these conferences will continue to provide this interdisciplinary forum for the presentation, discussion and exchange of theoretical and applied concepts and ideas to aid in the development of novel hard materials.

VINOD K. SARIN

Dr. Eric A. Almond (1940–1987)

This volume is dedicated to the memory of Dr. Eric A. Almond, whose presence and contributions in the hard materials community will be greatly missed. Even during his illness, he maintained an extremely active role in the organization of these conferences, no doubt as a result of his keen desire to enhance our understanding of this class of materials. His work and achievements will always be an inspiration to those of us who knew him.

CONTENTS

CEMENTED CARBIDES (PROCESS/STRUCTURE)

CEMENTED CARBIDES (PROPERTIES)

SURFACE MODIFICATION

Materials Science and Engineering, *A105/106* (1988) 1–10

Physics of Transition Metal Carbides*

WENDELL S. WILLIAMS

Department of Materials Science and Engineering, Case Western Reserve University, Cleveland, OH 44106 (U.S.A.)

(Received January 25, 1988)

Abstract

The refractory compounds TiC, ZrC, HfC, VC, NbC and TaC comprise a family of materials characterized by high melting point, great hardness and low chemical reactivity. These properties, which are exploited in tribological applications, are reviewed and discussed in terms of models taken from solid state physics. Particular attention is given to the important role of non-stoichiometry in these compounds and its influence on bonding and on transport properties. It is argued that the non-stoichiometry is a metastable condition, however, in view of the presence of ordered phases at certain values of carbon-to-metal ratio. The substantial impact of order–disorder transformations on transport properties is analyzed. The partially covalent bonding is discussed in terms of mixing of C 2p and metal 3d or 4d atomic states, and the high Peierls stress resisting dislocation motion is thereby justified.

1. Introduction

As this volume is on the science of hard materials, it seems appropriate to begin with a discussion of how the complex and useful properties of some hard materials can be described and interpreted from the perspective of solid state physics. This approach will also allow us to identify interrelationships between various properties, with the ultimate goal being a comprehensive, coherent and yet simple picture which will be helpful in designing new materials and expanding the applications of existing ones.

2. Transition metal carbides

The materials which will be the focus of this discussion are the monocarbides of the six group IV and V transition metals: TiC, ZrC, HfC and VC, NbC, TaC. These six compounds are isomorphic; they all crystallize in the NaCl structure. In addition, they all share the properties of exceptional hardness, high melting point, low thermal conductivity and semimetallic electrical conductivity. Despite these similarities, which are sufficient to establish these six materials as a class, some of the members have interesting and practically significant differences. For example, two of the group V carbides are fairly good superconductors—NbC and TaC—and TaC has a much higher brittle-to-ductile transition temperature than TiC. These various properties suggest numerous practical applications.

Having seen that the transition metal monocarbides present numerous application possibilities, including the strengthening of cemented WC–Co tool materials and the establishment of a class of TiC–(Ni, Mo) cemented carbide tools, we turn to an examination of the properties of the transition metal carbides from groups IV and V in more detail. For brevity and consistency, the results to be reviewed are drawn primarily from the present author's laboratory; they are not meant to exhaust the field. Longer, although earlier, reviews have been given by Toth [1], Storms [2] and Williams [3].

3. Non-stoichiometry

The composition of the transition metal carbides can be represented by the non-stoichiometric chemical formula MC_x, where M represents one of the six transition metals and x represents the carbon-to-metal ratio. The value of x has the astonishing range 0.5–0.97 without change in crystal structure. The carbon deficiency is accounted for by carbon atom vacancies in the f.c.c. carbon sublattice in the NaCl structure. This high concentration of vacancies (with the minimum value being 2 or 3%) far exceeds the level

*Paper presented at the 3rd International Conference on the Science of Hard Materials, Nassau, The Bahamas, November 9–13, 1987.

attained in metals, semiconductors or ionic solids by quenching, cold work or radiation damage.

Although at one time the non-stoichiometric NaCl-structure form of the transition metal carbides was regarded as their equilibrium state, the presence of ordered phases at specific values of carbon-to-metal ratio x (*e.g.* $0.833 = 3/8$ and $0.875 = 7/8$) in vanadium carbide and niobium carbide suggests that the familiar form is only metastable. Nevertheless, this is the form which generally develops when transition metal carbides are synthesized by reducing the oxide and carburizing the metal; so it is still of interest to examine the non-stoichiometric form in detail.

How do the vacancies affect the properties of these non-stoichiometric compounds? From the perspective of solid state physics, vacancies are recognized as scattering centers for electrons and phonons, and as potentially mobile defects for atomic self-diffusion. Thus, we would expect the transition metal carbides to exhibit several defect-controlled properties, and, indeed, experiment bears out this expectation.

The carbon deficiency expressed by the chemical formula MC_x can be viewed in a different light, namely as indicating a reduction in the number of C–M bonds relative to the stoichiometric limit of $x = 1$. Material properties which depend on the number of such bonds, *e.g.* melting point, elastic constants and cohesive energy, would thus be expected to vary systematically with variations in x. Again, experiment supports this expectation.

Thus the extraordinary deviation from stoichiometry exhibited by the transition metal monocarbides of groups IV and V can affect both defect-related transport properties and bonding-related bulk properties. Examples of both effects are given below.

4. Thermal conductivity

The thermal conductivity of the transition metal monocarbides is surprisingly low for crystalline metallic solids. Indeed, it is low enough that TiC has been proposed as a thermal barrier liner for the cylinder walls of the adiabatic diesel engine currently under development. Figure 1 indicates how much the thermal conductivity has been suppressed by various scattering processes by illustrating how much it increases when the dominant scattering process is somehow eliminated; below a certain critical temperature, the

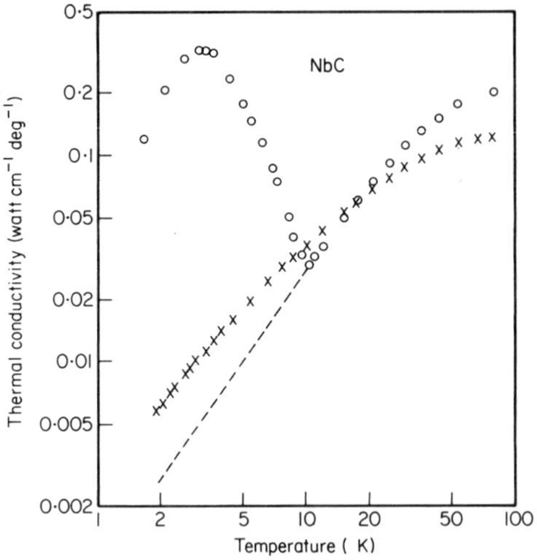

Fig. 1. Experiment demonstrating the suppression of thermal conductivity in niobium carbide at temperatures above 10 K and the reason for it by showing a 200-fold increase below 10 K, where the specimen becomes superconducting. The scattering of phonons by conduction electrons is progressively eliminated by the developing energy gap at lower temperatures. (From ref. 4.)

thermal conductivity of $NbC_{0.97}$ begins to increase rapidly, reaching a maximum which is over two orders of magnitude higher than the corresponding value for another specimen of niobium carbide which differs only in having a carbon-to-metal ratio (value of x) equal to 0.78 but which shows no abrupt change at this or any other temperature within the range of the experiment [4].

The explanation for this dramatic increase in thermal conductivity below 10 K for $NbC_{0.97}$ lies in the fact that the material becomes superconducting at this critical temperature T_c. This explanation is not obvious, however, since for most superconducting metals the thermal conductivity decreases rather than increases below T_c because the superconducting electrons cannot carry heat, *i.e.* energy associated with disorder; in the normal state, most of the heat is carried by the conduction electrons. However, in niobium carbide, electrons in the normal state cannot carry much heat because they are severely scattered by carbon atom vacancies and have mean free paths of the order of only tens of ångströms even at low temperatures [5]. Thus the lattice component of the thermal conductivity is a substantial fraction of the total, as in alloys. Therefore, anything which affects the lattice component will make a large impact on the total thermal conductivity of niobium carbide.

In related experiments, it was shown that the lattice component of the thermal conductivity of transition metal carbides is severely limited by scattering of lattice waves (or their quanta, phonons) by conduction electrons [5]. Although the scattering of phonons by carbon vacancies was expected to be dominant, the theoretical modelling of the scattering process by the Callaway formalism, taking into account the importance of the short electron mean free path, led to the conclusion that this scattering mechanism is not primarily responsible for the extremely low thermal conductivity of the carbides at low temperatures; rather, the dominant process is phonon–electron scattering (Figs. 2 and 3) [5]. This conclusion, based initially on curve fitting

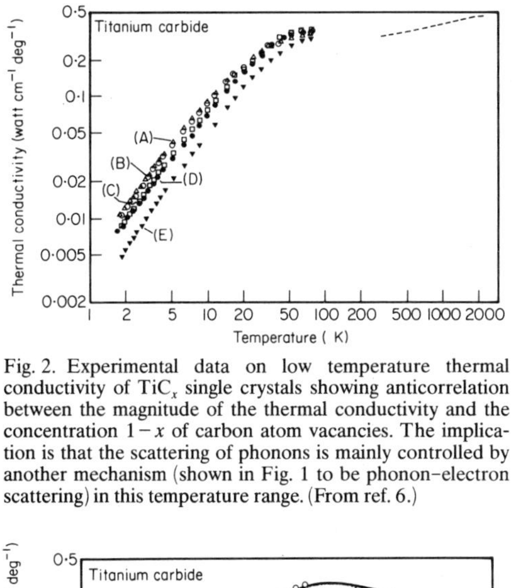

Fig. 2. Experimental data on low temperature thermal conductivity of TiC_x single crystals showing anticorrelation between the magnitude of the thermal conductivity and the concentration $1-x$ of carbon atom vacancies. The implication is that the scattering of phonons is mainly controlled by another mechanism (shown in Fig. 1 to be phonon–electron scattering) in this temperature range. (From ref. 6.)

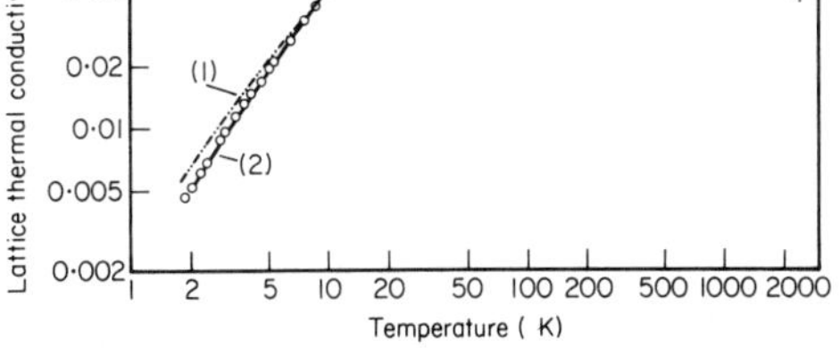

Fig. 3. Experimental points and curve fit from Callaway analysis using Debye approximation and summing individual scattering probabilities for four different phonon scattering mechanisms. (From ref. 6.)

using appropriate scattering cross-sections for the various processes, was then experimentally confirmed by results on the superconducting niobium carbide [4].

As the temperature falls below the superconducting transition temperature, a small energy gap begins to open at the Fermi energy (the energy of the most energetic electrons in the conduction electron gas). As there are no energy states accessible to electrons in the gap, the phonon–electron scattering probability, which is proportional to the density of final states, decreases. As a result, the phonons are increasingly able to pass through the solid, carrying thermal energy without being impeded [4].

This effect in NbC is the largest "phonon bump" ever observed. It could be viewed as a "heat switch" and may have important practical applications in cryogenic heat transfer systems, an unusual application for a material usually considered a high temperature refractory compound and noted for its hardness. Similar but much smaller effects have been seen in the new high T_c superconductors.

At high temperatures the unexpectedly low thermal conductivity of the carbides is dominated by a significantly different mechanism: the scattering of phonons as well as electrons by carbon atom vacancies [6]. The thermal conductivity increases rapidly as the number of point defects (vacancies) decreases near the stoichiometric limit. Clearly, for a thermal barrier coating, a value of x far removed from this limit would be advantageous.

5. Electrical resistivity

Although the transition metal carbides are hard enough to qualify as ceramics, the transport of electric charge is via conduction electrons rather than by ions. In fact, ion migration is extremely difficult in these materials, as we shall see in Section 6.

These conduction electrons are not very numerous in the group IV carbides, however, having a concentration of only about 0.08 electrons atom^{-1} in TiC, compared with 1 electron atom^{-1} for copper, according to Hall coefficient measurements [7, 8]. Materials of this description are sometimes called semimetals.

The electrical resistivity of the carbides is strongly affected by the presence of the numerous point defects (carbon vacancies) in the lattice.

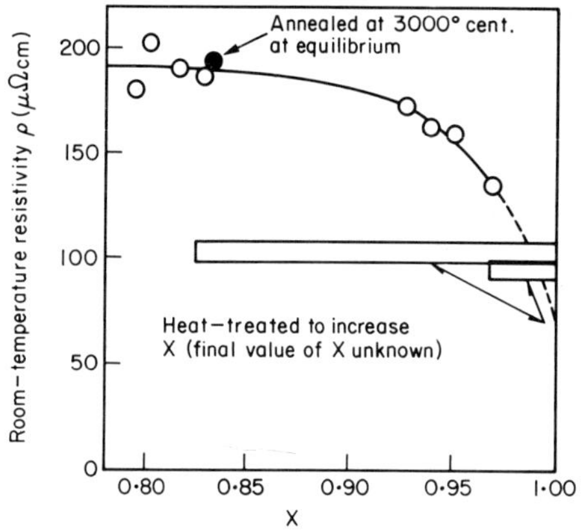

Fig. 4. Electrical resistivity of TiC_x single crystals of differing vacancy concentrations $1-x$ at room temperature. The correlation supports identifying vacancies (point defects) as the cause of the large residual resistivity of the carbides. (From ref. 8.)

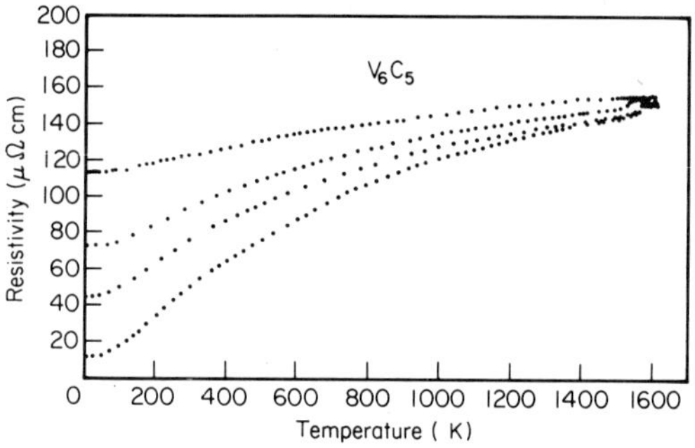

Fig. 5. Electrical resistivity of vanadium carbide single crystals with a carbon-to-metal ratio equal to 5/6. The resistivity of the disordered NaCl-structure $VC_{0.83}$ specimen is high down to very low temperatures (high residual resistivity), whereas a similar specimen in the ordered phase V_6C_5 shows a low residual resistivity and a sigmoidal resistivity *vs.* temperature curve. The results confirm the importance of vacancy scattering of conduction electrons. (From ref. 10.)

Indeed, at room temperature and below, defect scattering of electrons can dominate the resistivity (Fig. 4) [8]. Such scattering in simple metals is generally temperature independent (Matthiessen's rule) but in the carbides the apparent influence of the defects lessens as the temperature increases [9, 10]. This behavior may be a result of interband scattering.

A consequence of the decreasing influence of point defects with increasing temperature is a relatively small temperature coefficient of resistivity for many of the transition metal compounds. Tantalum nitride is an extreme example and finds practical application in high temperature microelectronics as a temperature-insensitive thin film resistor.

A vivid demonstration of the importance of point defect scattering of conduction electrons in establishing the measured resistivity of carbides is provided by experiments which exploit transformations to ordered phases [9–11]. These transformations can occur for specimens with certain values of x corresponding to ratios of integers (*e.g.* 5/6 = 0.833 and 7/8 = 0.875). On slow cooling through the transformation temperature (of the order of 1000 °C), the randomly arranged carbon vacancies in the NaCl structure of the transition metal carbide become ordered, defining new crystalline phases and structures. As a result, the vacant atomic sites become part of the periodic potential of the solid and do not scatter conduction electrons. The low temperature resistivity is greatly affected; the high residual resistivity characteristic of these materials is essentially eliminated, revealing a sigmoidal curve of resistivity *vs.* temperature (Fig. 5) resulting from intrinsic scattering of electrons by phonons [10].

The shape of this curve is somewhat peculiar, however, as simple metals show a linear dependence of resistivity on temperature. The sigmoidal shape is found for a number of superconductors and other metals and has been attributed to a "leakage" current from interband transitions [12, 13].

6. Diffusion

As hard materials experiencing elevated temperatures in cutting and grinding applications, the transition metal carbides are subjected to chemical and thermal stresses which would be expected to be relieved by diffusion. In view of the abundance of vacancies already present in these non-stoichiometric compounds, diffusion of carbon requires only sufficient energy for migration of a carbon atom into an adjacent vacant lattice site. However, experiment shows that the energy required is actually quite large: 4.5 eV atom^{-1} for TiC_x [14–16]. It is interesting to note that this energy is approximately that required to disorder an ordered specimen by electron bombardment in the electron microscope, presumably because single migration jumps by carbon atoms are sufficient to disrupt the long-range periodicity.

In view of this large migration energy, carbon diffusion is slow even at elevated temperatures, except where short-circuit paths, *e.g.* grain boundaries, are present. Diffusion of the metal species takes even more energy (approximately 8 eV atom^{-1}) because vacancies must be created on the metal f.c.c. lattice as well as moved. Thus, practical applications of carbides where diffusion might prove to be a problem turn out to be less threatened than might be expected. For example, the chemical vapor deposition TiC coatings commonly used on cemented carbide cutting tools provide diffusion barriers against cobalt moving out and iron moving in.

7. Bonding

A single-phase hard material must perforce exhibit strong interatomic bonding. In the transition metal carbides, bonding is a complex mixture of covalent, ionic and metallic types. We shall not attempt to separate them numerically but try to identify the fundamental origin of each.

The covalent component arises from the mixing of metal d states with C 2p states near the Fermi energy, according to numerous electron energy band calculations (Fig. 6) [17]. It is primarily responsible for the large values of the elastic constants of the carbides and their high Debye temperatures.

The ionic contribution results from partial electron transfer from metal to carbon on forming the compound. This result is predicted by energy band calculations and supported by most of the relevant experiments, although there is some debate about interpretation. The most direct evidence comes from electron spectroscopy for chemical analysis (ESCA) (X-ray photoelectron scattering) studies, which show a slight shift in the binding energies of the electrons close to the nucleus in metal and carbon in the compound compared with the elements (Fig. 7) [18]. This energy shift is explained as resulting

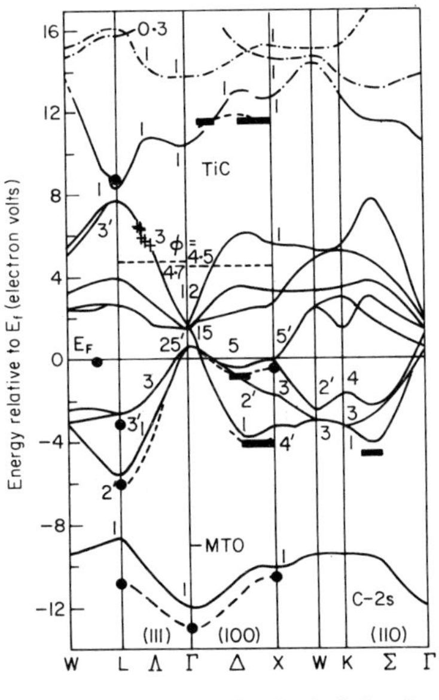

Fig. 6. Electron energy band calculation for TiC, showing features typical of other such calculations: mixing of Ti 3d and C 2p near Fermi energy to account for strong covalent bond; pushing up of Ti 4s state to leave it unpopulated and to provide for electron transfer to carbon and partial ionicity of bond. The Fermi energy occurs at a minimum in the density of states. (From ref. 17.)

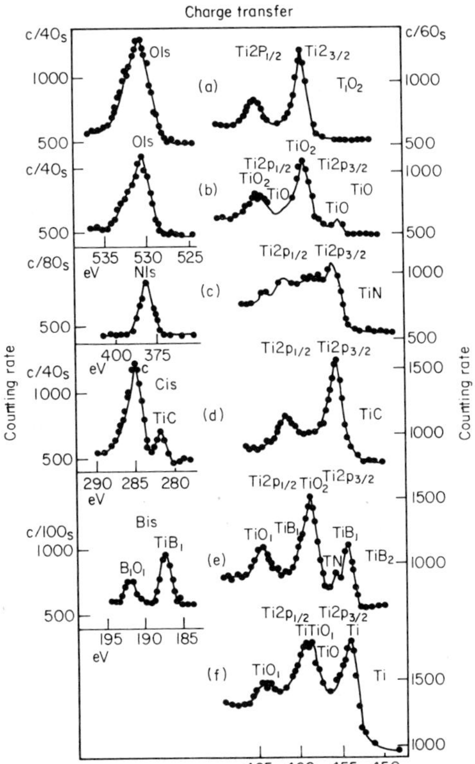

Fig. 7. Shifts in titanium peaks in the ESCA spectrum to higher binding energies compared with titanium metal on formation of TiC and other compounds, implying partial electronic charge transfer from titanium to a non-metal atom. (From ref. 18.)

from a redistribution of electronic charge in the direction of the carbon particle. One interpretation holds that the amount of this charge transfer is sufficient to account for the heat of formation of the compound from the elements through the Madelung energy [18].

The metallic component of the bonding is difficult to separate from the others but probably includes d band bonding typical of the transition metals, resulting from partial filling of atomic d states when the solid is formed.

Theoretical justification for the bonding model above comes not only from electron energy band calculations for the infinite solid but also from two cluster calculations, which are not widely known.

A relatively simple approach, using modified Hückel theory, treats clusters of increasing complexity, starting with one titanium atom surrounded by six carbon atoms. Even this simple model delivers several important features which are found in the results from the infinite solid; when the titanium carbide compound is formed from the elements, the Ti 4s states are pushed upwards in energy and unpopulated; both bonding and antibonding states are formed between C 2p and Ti 3d states; there is some electron transfer from the titanium towards the carbon. These results are unchanged as larger clusters are investigated (Figs. 8 and 9) [19].

A more detailed and rigorous cluster calculation, using the unrestricted Hartree–Fock approach, produces a similar set of findings for a cluster of one titanium and five carbon atoms, with bonds tied off with hydrogen. This model is meant to simulate the surface of a titanium carbide crystal with the {111} plane exposed (Fig. 10) [20].

8. Hardness

The strong bonding of the carbides accounts for their hardness through the creation of a high Peierls stress (intrinsic lattice resistance to dislocation motion) [21]. Although the carbon atom vacancies might be expected to act as pinning points for dislocations, the decrease in hardness with increasing defect concentration argues against this interpretation and in favor of an average increase in the strength of the chemical bond and hence in the Peierls stress (Fig. 11) [22, 23]. This situation is one in which both aspects of the non-stoichiometry of the compound—defects and bonding—might be implicated, but one is seen to dominate.

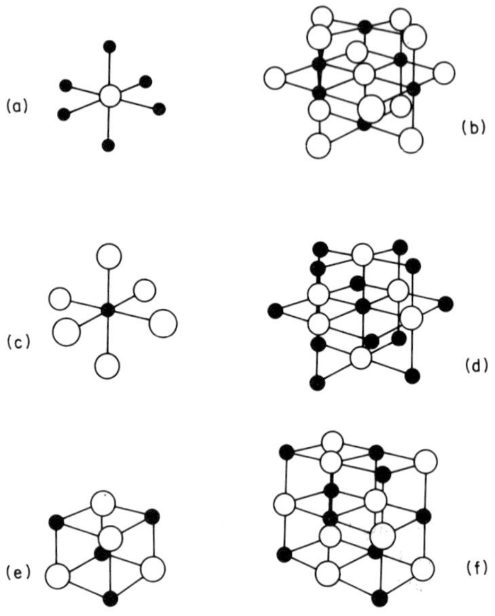

Fig. 8. Various clusters of titanium (○) and carbon (●) atoms investigated using modified Hückel theory. (From ref. 19.)

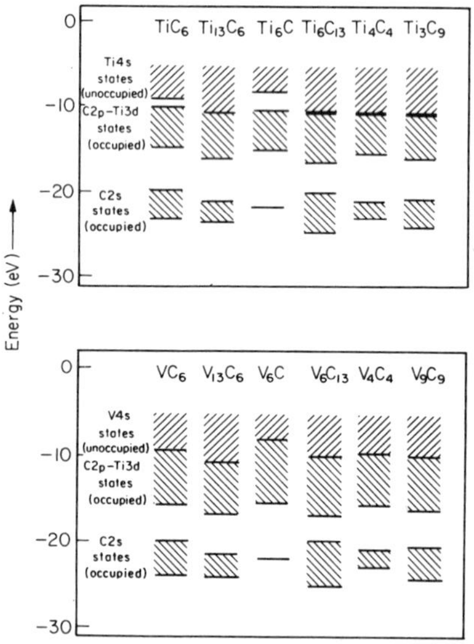

Fig. 9. Energy states calculated for clusters shown in Fig. 8. (From ref. 19.)

Measurements of the hardness of the carbides, as with any very hard material, require diamond

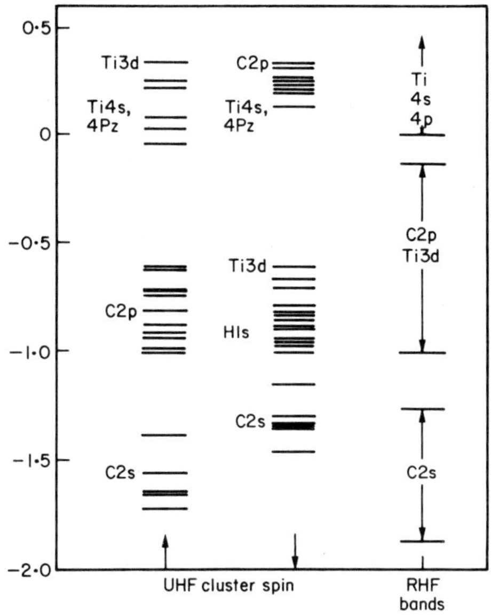

Fig. 10. More exact cluster calculation for TiC using the unrestricted Hartree–Fock approach with a cluster of one titanium and five carbon atoms, simulating a (100) surface of the NaCl-structure solid. (From ref. 20.)

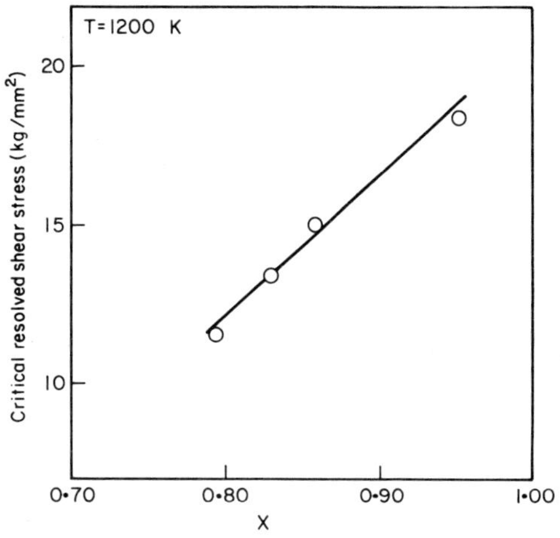

Fig. 11. Dependence of microhardness of TiC_x on value of x, or $1-N_v$, where x is the carbon-to-metal ratio, and N_v the number of carbon vacancies. The hardness increases as the number of vacancies decreases, implying that vacancy pinning of dislocations is less important than the Peierls stress. (From ref. 23.)

indenters and deliver microhardness values. Dislocation production under the indenter is evidently very localized, as etching after room temperature indenting produces only a large cavity around the previous impression; a high temperature anneal preceding the etching allows the typical rosette pattern to develop, however. Evidently the elastic stress associated with the impression is relieved at high temperatures as dislocations receive enough thermal energy to move away [22].

Measurements of orientation dependence of the microhardness values obtained with a Knoop indenter on single crystals of carbides have produced periodic variations. These curves are clear indications of the participation of crystallographic defects (dislocations) in the formation of the impression (Fig. 12) [22]. With the aid of theoretical models, the operative slip system can be inferred from the results. The result obtained in this way for TiC is $\{110\}\langle 1\bar{1}0\rangle$, although a recent transmission electron microscopy study has raised a question about this conclusion and suggested that the slip plane is $\{111\}$ at room temperature, as it is at high temperatures.

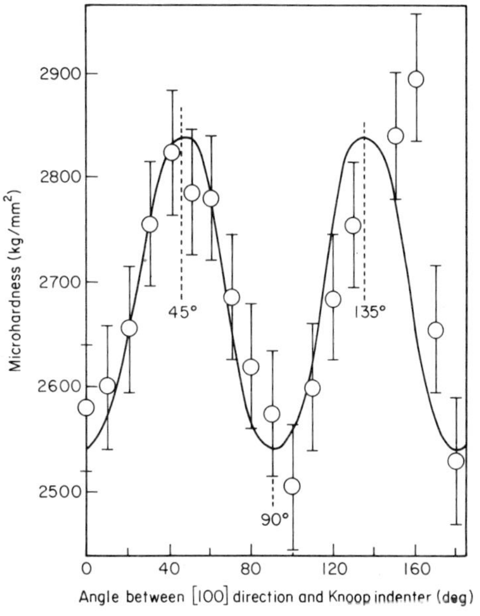

Fig. 12. Periodic variation in microhardness with orientation of Knoop indenter in TiC single crystals on the (100) plane. (From ref. 22.)

9. High temperature plastic deformation

At temperatures above approximately 800 °C, TiC crystals become sufficiently soft to accommodate macroscopic plastic deformation in uniaxial compression ([23]; the slip system is then {111}⟨1$\bar{1}$0⟩ [22], although the {110} planes and even {100} can also be activated in ZrC and presumably in TiC). The hardness drops by two orders of magnitude from room temperature, following an exponential temperature dependence (Fig. 13) [23]. Thermal activation over the Peierls barrier is no doubt one of the mechanisms involved in this softening, but the experimental results suggest that more than one mechanism is active in reducing the hardness at high temperatures [23].

This loss of the very quality that makes the material valuable to hard materials producers and users is a serious limitation on the applicability of this class of materials for structural purposes. Attempts to raise the yield stress of the light-weight transition metal carbides, *e.g.* TiC and VC, at high temperatures have produced some interesting results.

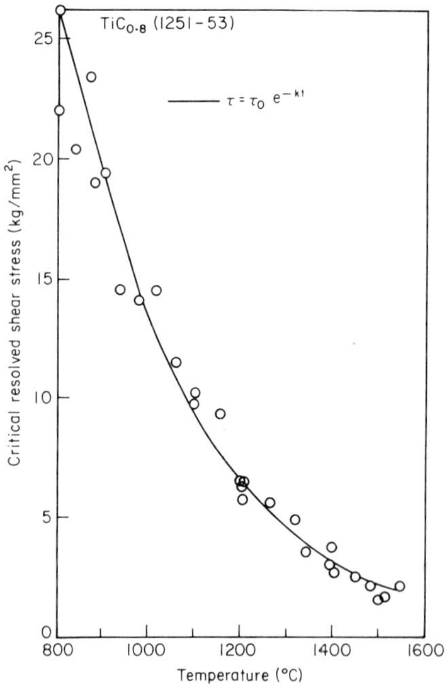

Fig. 13. Temperature dependence of high temperature critical resolved shear stress of TiC single crystals measured in uniaxial compression. $\tau_0 = 1.32 \times 10^3$ kg mm^{-2}, $k = 3.61 \times 10^{-3}$ deg K^{-1}.

It has been found that the addition of a mere fraction of a percentage of boron can increase the critical resolved shear stress of TiC single crystals by a factor of 6 at 1600 °C (Fig. 14) [24]. The strengthening mechanism is the formation of precipitates of TiB_2 (another refractory compound) on the active slip plane, {111} [24]. This plane in the NaCl structure is the plane of like atoms in a close-packed array. Such an arrangement of titanium atoms is also like the basal plane of the hexagonal TiB_2 structure. Thus a precipitate forms, with the next layer beyond the titanium being boron atoms in an hexagonal net. The boundaries of the microscopic precipitate are Shockley partial dislocations. For plastic deformation to occur, these defects must move, but the precipitate introduces an extreme form of Suzuki hardening, and motion of the defect is limited by the ability of boron to diffuse through the lattice. As this diffusion is expected to be energetically expensive, the precipitates increase the yield stress of the doped carbide [25, 26].

An analogous effect is found in the case of TaC which is considerably harder than TiC at high temperatures. This fact is exploited by the cemented carbide industry, but the reason has not been known until recently. It has been pointed out that stacking faults between partial dislocations in TaC must have a different symmetry from

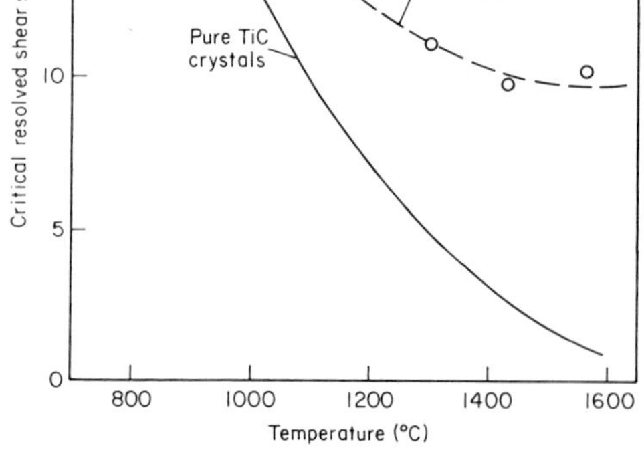

Fig. 14. Increase in yield stress of TiC with the addition of a fraction of per cent of boron, which forms TiB_2 precipitates on the slip plane. (From ref. 24.)

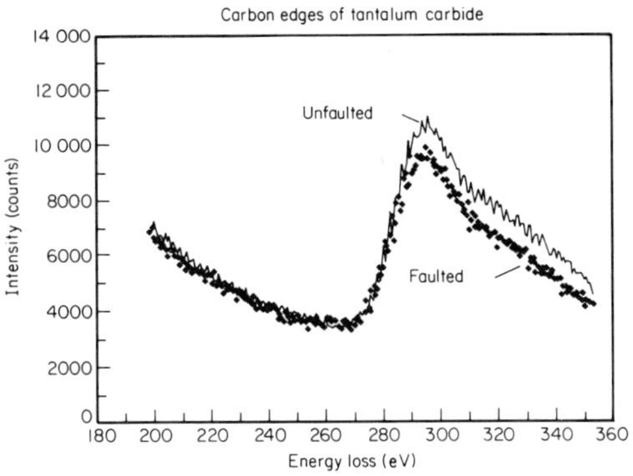

Fig. 15. Study of microchemistry of stacking fault between Shockley partial dislocations in TaC_x using EELS. (From ref. 26.)

the host lattice—in this case, hexagonal rather than cubic. However, there is no hexagonal phase of tantalum carbide with a 1:1 carbon-to-metal ratio, so the stacking fault takes on the hexagonal Ta_2C symmetry and chemistry. The equivalent carbon-to-metal ratio is 1:2, or 0.5, which is less than the x value of the host TaC_x lattice [25].

This difference in carbon-to-metal ratio between stacking fault, a few nanometers in thickness, and host lattice has been detected experimentally by electron energy loss spectroscopy (EELS) (Fig. 15) [26]. (The method also confirmed the chemical identity of TiC precipitates in TiB_2 [27]. EELS is particularly helpful in studying microscopic precipitates containing light elements.)

In view of the difference between the chemical composition of the bulk and that of the stacking fault in TaC, again a kind of Suzuki hardening is established, and again plastic deformation will be diffusion limited [25]. Similar processes do not occur in TiC because the dislocations do not appear to dissociate.

Another attempt to strengthen carbides against plastic deformation at high temperatures showed that solid solutions of VC and TiC had substantially higher yield stresses than either of the pure materials. Considering the utility of the preceding argument, it is likely that the solid solution hardening in the carbides depends on stacking fault formation, as with the noble metals.

10. Summary

This review of transition metal carbides has pointed out some interesting and practical properties of this class of hard materials and has attempted to interpret them in the light of models of solid state physics.

References

1 L. E. Toth, *Transition Metal Carbides and Nitrides*, Academic Press, New York, 1971.
2 E. K. Storms, *The Refractory Carbides*, Vol. II, Academic Press, New York, 1968.
3 W. S. Williams, Transition metal carbides, *Prog. Solid State Chem., 6* (1971) 57.
4 L. G. Radosevich and W. S. Williams, Lattice thermal conductivity of superconducting niobium carbide, *Phys. Rev., 118* (1969) 770–773.
5 L. G. Radosevich and W. S. Williams, Phonon scattering by conduction electrons and by lattice vacancies in carbides of the transition metals, *Phys. Rev., 181* (1969) 1110–1117.
6 L. G. Radosevich and W. S. Williams, Thermal conductivity of transition metal carbides, *J. Am. Ceram. Soc., 53* (1970) 30–33.
7 J. Piper, *J. Appl. Phys., 33* (1962) 2394.
8 W. S. Williams, Scattering of electrons in nonstoichiometric crystals of titanium carbide, *Phys. Rev., 135* (1964) A505–A510.
9 W. S. Williams and L. W. Shacklette, Scattering of electrons by vacancies through an order–disorder transition in vanadium carbide, *J. Appl. Phys., 42* (1971) 4698–4703.
10 L. W. Shacklette and W. S. Williams, Influence of order–disorder transformations on the electrical resistivity of vanadium carbide, *Phys. Rev. B, 7* (1973) 5041–5053.
11 L. C. Dy and W. S. Williams, Resistivity, superconductivity and order–disorder transformations in transition metal carbides and carbonitrides, *J. Appl. Phys., 53* (1982) 8915–8927.
12 B. Chakraborty and P. B. Allen, *Phys. Rev. Lett., 42* (1979) 736.
13 M. Gurvitch, in H. Suhl and M. B. Maple (eds.), *Superconductivity in d- and f-band Metals*, Academic Press, New York, 1980, pp. 317–334.
14 S. Sarian and J. C. Criscione, *J. Appl. Phys., 38* (1967) 1794.
15 S. Sarian, *J. Appl. Phys., 39* (1968) 5036; *40* (1969) 3305.
16 D. L. Kohlstedt, W. S. Williams and J. B. Woodhouse, Chemical diffusion in titanium carbide crystals, *J. Appl. Phys., 41* (1970) 4476–4484.
17 A. Neckel, *Int. J. Quant. Chem., 23* (1983) 1317–1353.
18 L. Ramquist, K. Hamrin, G. Johansson, A. Fahlman and C. Nordling, Charge transfer in transition metal carbides and related compounds studied by ESCA, *J. Phys. Chem. Solids, 30* (1969) 1835.
19 K. L. Yip, A. B. Kunz and W. S. Williams, Electronic structure of TiC and VC from iterative Hückel theory, *Phys. Status Solidi B, 75* (1976) 533–541.
20 D. R. Jennison, D. L. Klein, A. B. Kunz and W. S. Williams, Relationship between electronic structure and catalytic behavior of transition metal carbides, *Electrocatalysis on Non-Metallic Surfaces, Proc. NBS Workshop, Gaithersburg, MD, December 9–12, 1975*, in *NBS Spec.*

Publ. 455, 1976, pp. 87–92 (National Bureau of Standards).
21 W. S. Williams and R. D. Schaal, Elastic deformation, plastic flow and dislocations in single crystals of titanium carbide, *J. Appl. Phys., 33* (1962) 955–962.
22 W. S. Williams, Brittle–ductile behavior of the transition metal carbides, in *Propriétés Thermodynamiques, Physiques et Structurales des Dérivés Semimétalliques*, Editions du Centre National de la Recherche Scientifique, Paris, 1967, pp. 181–189.
23 W. S. Williams, Influence of temperature, strain rate, surface condition and composition on plasticity of transition metal carbides crystals, *J. Appl. Phys., 35* (1964) 1329–1338.
24 W. S. Williams, Dispersion hardening of titanium carbide by boron doping, *Trans. AIME, 236* (1966) 211–216.
25 M. Hoffman and W. S. Williams, A simple model for the deformation behavior of tantalum carbide, *J. Am. Ceram. Soc., 69* (1986) 612–614.
26 A. Allison, M. Hoffman and W. S. Williams, Electron energy loss spectroscopy of carbon in dissociated dislocations in tantalum carbide, *J. Appl. Phys., 53* (1982) 6757–6761.
27 P. Mochel, C. Allison and W. S. Williams, Study of titanium carbide precipitates in titanium diboride by electron energy loss spectroscopy, *J. Am. Ceram. Soc., 64* (1981) 185–187.

Materials Science and Engineering, A105/106 (1988) 11–18

Theory of Bulk Moduli of Hard Solids*

MARVIN L. COHEN

Department of Physics, University of California, and Materials and Chemical Sciences Division, Lawrence Berkeley Laboratory, Berkeley, CA 94720 (U.S.A.)

(Received December 12, 1987)

Abstract

Two theoretical approaches are described to examine the bulk moduli of solids. The first is based on an ab initio method requiring only the atomic numbers of the constituent atoms as input. The second is a much simpler empirical approach which is applicable to diamond and zinc blende solids and uses the nearest-neighbor bond length as input. Both methods give comparable results, but the ab initio approach can be used to study the details of the structural, bonding and electronic properties of solids. This aspect of the ab initio method is illustrated by a description of the bonding changes during a graphite-to-diamond pressure-induced transition and by a study of the structural and bonding properties of BN, BP, BAs and SiC.

1. Introduction

The focus of this conference is on studies of hard materials. Although the hardness of a solid depends on temperature, impurities, dislocations and defects of almost all types, it is still instructive in a study of hardness to use theory to examine the fundamental structural and electronic properties of ideal solids at zero temperature. Of particular interest is the bulk modulus which for ideal solids scales roughly with hardness. An *ab initio* approach was described [1] at our previous conference with applications to some covalent and ionic solids. This approach is based on a calculation of the total energy of a solid for a given structural configuration using a pseudopotential scheme [2] and requires only the atomic numbers of the constituent atoms to determine the potentials. When total energies for different structural arrangements and volumes are compared, a lowest energy structure can be chosen from a group of candidates. Pressure-induced solid–solid transformations can be simulated by changing the volume and comparing the total energies at the reduced volume. This technique yields a determination of the transition volumes and the transition pressure for the solid–solid structural phase transition. The cohesive energies of each structural phase at a given volume can be determined by comparing the total energy of the phase with that of a model system where the atomic distances are large enough to simulate a system of isolated atoms.

The method described above can also be used to compute vibrational spectra [1, 2] by comparing the total energy of a system distorted by a lattice vibrational wave with an ideal undistorted structure. When a similar calculation is done for a compressed volume, the phonon mode Gruneisen constants also can be evaluated. Recently, it has become possible to compute electron–phonon coupling constants. The only additional input necessary for determining phonon spectra and electron–phonon couplings is the atomic masses of the atoms in the solid.

One of the most dramatic achievements of the pseudopotential–total energy scheme is the successful prediction [3] of superconductivity in compressed simple hexagonal silicon. At our last conference, it was suggested [1] that simple hexagonal silicon would be superconducting because of the covalent nature of the electronic charge distribution. Earlier theoretical studies [4] pointed out the advantages of covalent-like bonds in metal systems for superconductivity. It was this observation that stimulated the *ab initio* calculations of the electron–phonon coupling constants which in turn motivated the experimental study.

*Paper presented at the 3rd International Conference on the Science of Hard Materials, Nassau, The Bahamas, November 9–13, 1987.

0921-5093/88/$3.50

© Elsevier Sequoia/Printed in The Netherlands

Although the *ab initio* scheme provides many of the important properties needed for a variety of studies of solids, the method is computationally complex and requires large computers. Hence, there has been considerable effort to develop simpler schemes [5, 6] for describing solid state effects. Often these approaches sacrifice accuracy for simplicity and are most useful for exploring general trends in groups of materials.

The method of Phillips [5] used spectral data and dielectric constants to establish a scale of ionicity. This approach is capable of separating fourfold- and sixfold-coordinated structures according to the spectral parameters. It was this approach which was used as a starting point for an empirical scaling theory of bulk moduli [7]. Some aspects of this approach will be described here, and applications will be given.

In the next few sections of this paper the results for the *ab initio* and empirical approaches will be discussed and some perspectives about these methods will be provided.

2. Total energy pseudopotential method

As described above and in ref. 1, the total energy pseudopotential method is an *ab initio* approach requiring only the atomic numbers and atomic masses of the constituent atoms as input. Many applications have been made using a free-electron or plane wave basis set for the wavefunction calculation. This basis has the advantage of starting the calculation without a bias because the initial electronic charge density is constant everywhere in a unit cell. However, for localized charge configurations such as those expected for the 2p states of carbon, many plane waves are required in the basis set. This results in large-sized matrices to be diagonalized to obtain the electronic energy bands and wavefunctions. Although a localized basis set reduces the matrix size and computer time required, often these schemes are more biased and less accurate. An exception is the accurate approach developed by Chelikowsky and Louie [8]. This local-orbital approach is used together with *ab initio* pseudopotentials and is particularly suitable for treating localized p orbitals such as those expected for atoms of the first row of the periodic table.

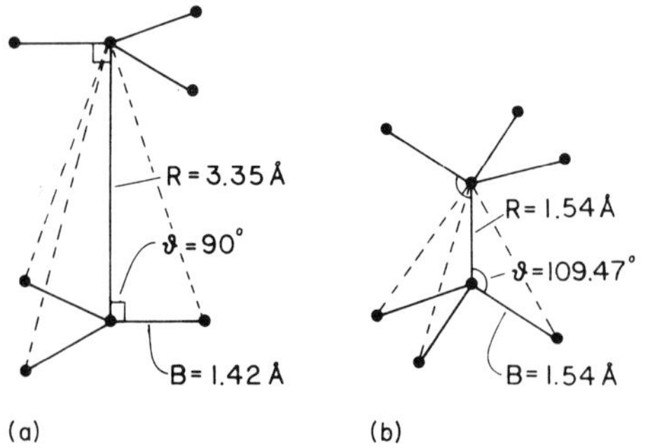

Fig. 1. Coordinates and geometry of the local structure of (a) rhombohedral graphite and (b) diamond (the [111] direction of the conventional cubic coordinates of diamond is vertical): ---, the basis vectors of the rhombohedral lattice.

An application [9] of the local-orbital method to various structural forms of carbon yields results consistent with previous plane wave calculations and extends these calculations. In addition, the reduced computational complexity of the local-orbital method has made possible detailed studies of the rhombohedral-graphite-to-diamond [10] and hexagonal-graphite-to-hexagonal-diamond [11] transitions. These calculations yield a detailed description of the energetics, electronic structure and bonding changes occurring during the transformations. Some results for the rhombohedral-graphite-to-diamond transition are described below.

About 5%–15% of a given sample of graphite contains the rhombohedral structure intermixed in a mosaic combination with the more common hexagonal form. The rhombohedral structure has ABC stacking in contrast with the AB stacking of the hexagonal structure. A convenient set of coordinates for examining the transition is the bond length R between layers, the bond length B within the layers and the buckling angle θ. As shown in Fig. 1, the transition from rhombohedral graphite to diamond involves a change in the structural parameters from $R = 0.335$ nm, $B = 0.142$ nm and $\theta = 90°$ to $R = B = 0.154$ nm and $\theta = 109.47°$. A continuous transformation involves the reduction in R while B and θ are increased. The path which minimizes the total energy $E(R, B, \theta)$ is determined by treating R as an independent variable and varying B and θ to minimize $E(R, B, \theta)$. A similar approach is used in an extended Huckel band calculation [12].

The electronic charge density for points on the path which minimizes the energy as a function of R is shown in Fig. 2. In Fig. 2(a), the double-peak

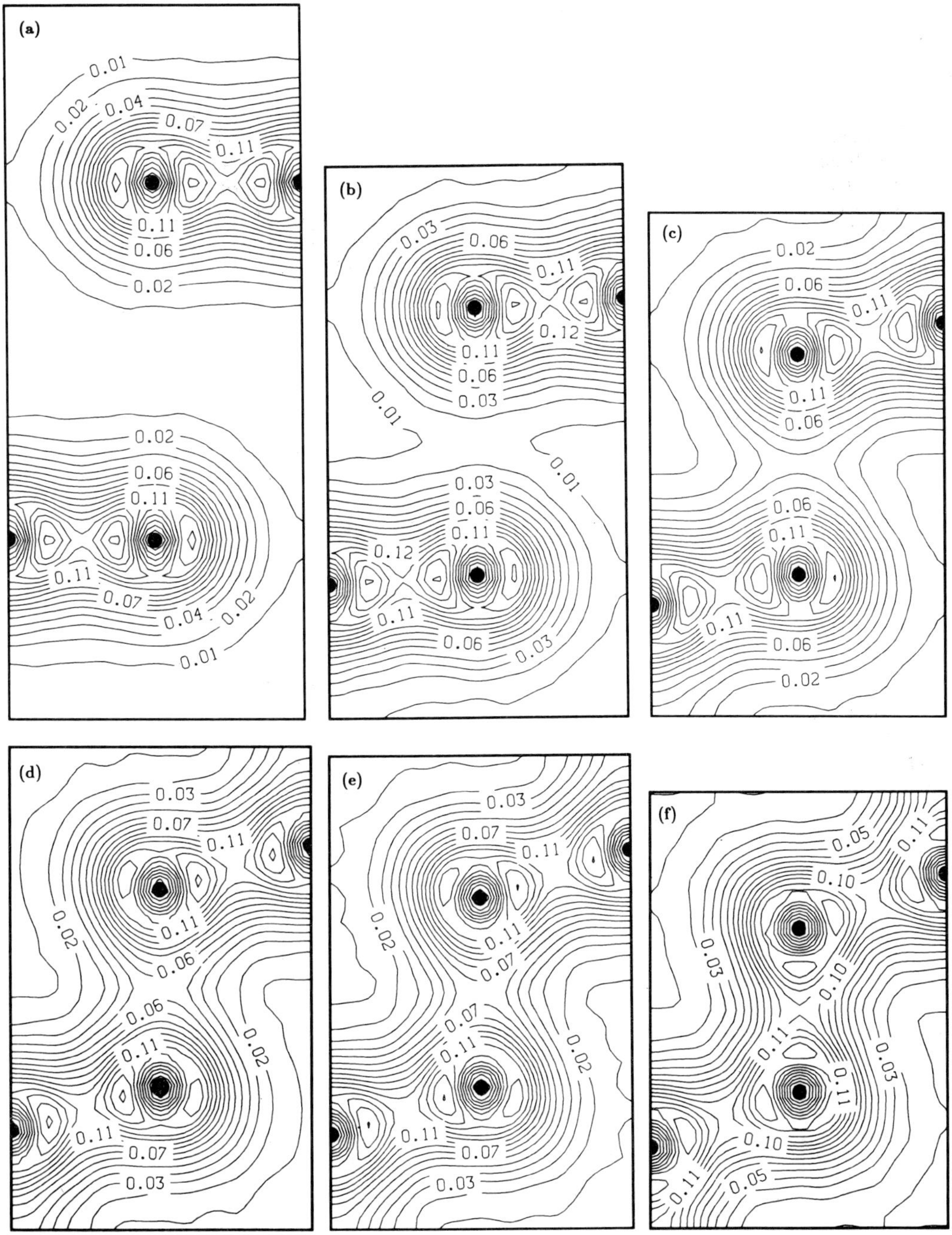

Fig. 2. Valence electron charge density for carbon displayed in a plane containing a bond located between layers and a bond lying within layers for various points on a path chosen to minimize the energy for each value of the interlayer bond length R: ●, atomic positions. The values of R are (a) 0.335 nm, (b) 0.25 nm, (c) 0.207 nm, (d) 0.188 nm, (e) 0.18 nm and (f) 0.154 nm.

charge density characteristic of C–C bonds [1] is displayed within the layers, but the charge density is small and homogeneous between the layers characteristic of non-interacting layers. As R decreases, charge builds up between the layers, and a bond begins to develop when R is reduced to 0.207 nm as shown in Fig. 2(c). At this point, the bonds in the plane begin to buckle. In Fig. 2(f), the buckling is complete and $R = B = 0.154$ nm. Once the peak in the charge density between planes develops, a gap in the electronic density of states between the valence and conduction bands also develops characteristic of insulating diamond. However, there is no metallic behavior found for any of the structure studies along the transition path.

The energy barrier for the path chosen is 0.33 eV atom^{-1}. When a hydrostatic pressure condition is modeled by reducing the volume and maintaining rhombohedral symmetry, a graphite-like structure is found to be metastable up to 80 GPa, at which point the density is 60% greater than graphite at zero pressure and 2% greater than diamond at zero pressure. For pressures above 80 GPa, diamond is the only stable or metastable structure. Although the path chosen has a large energy barrier and very high pressures are required for this ideal graphite-to-diamond transition, the results are suggestive about how this interesting class of transitions occurs. Cross-linking of hexagonal rings leads to tetrahedral coordination when the interlayer distance between the ring is between 0.21 and 0.23 nm. Hopefully, these studies can be useful in studies of the role of catalysts in the graphite-to-diamond transition.

In contrast with carbon, earlier calculations [1] demonstrated that the graphite structure lies at a higher energy [1] for silicon and hence is not expected to be stable at any pressure. In fact, an expansion would be required to form this structure. Much of the difference between silicon and carbon can be attributed [1, 2] to the single *vs.* double-peak structure of their respective bonds. This feature has its origins in the differences between the silicon and carbon cores. Because there are no p states in the carbon core, the valence 2p electrons lie near their positions in the isolated atoms and are not pushed out to form a single peaked bond as in silicon and germanium.

Other manifestations of the lack of core p states are also evident in the compounds BN, BP and BAs. For III–V compounds formed from elements not in the first row of the periodic table, a typical charge density plot for the valence electrons shows a partially covalent partially ionic bond configuration. The ionic component occurs because the anion accumulates more valence electron charge because of its more positive core potential. A similar result is obtained for BN but not for BP or BAs. Studies [13–15] of these compounds illustrate the role of the p electron core potential in determining the bonding, lattice constant and bulk modulus. Since BN has the highest bulk modulus of all ionic solids, this bonding study is instructive for understanding hardness.

As shown earlier, the strong attractive p potential for carbon is not canceled by a repulsive Pauli potential from inner p electrons as it is for higher rows in the periodic table. Hence, the p electron peak in the charge density occurs close to the carbon atoms. A similar situation occurs for

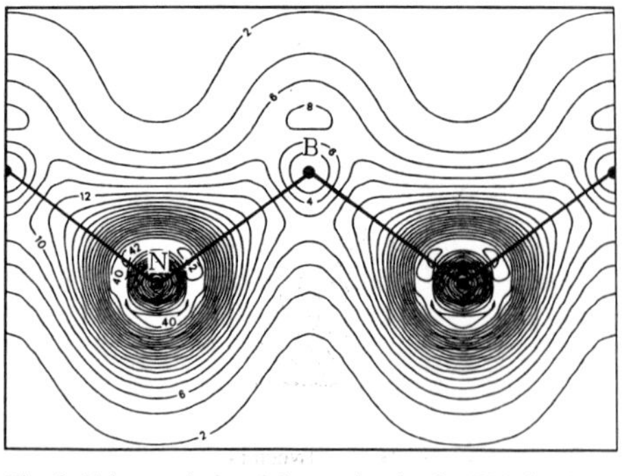

Fig. 3. Valence electron charge density for BN displayed in the (110) plane in units of electrons per cell volume.

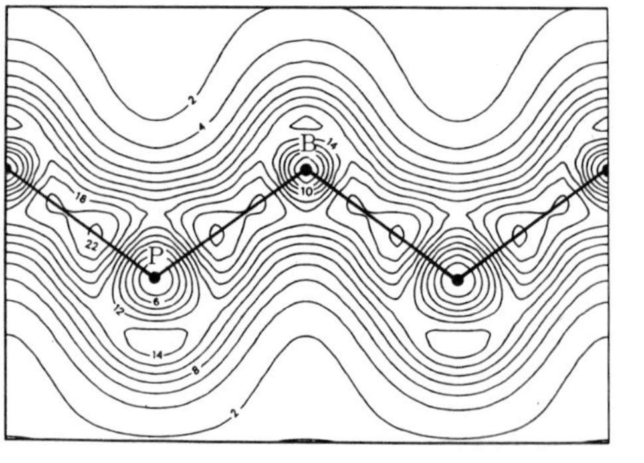

Fig. 4. Valence electron charge density for BP displayed in the (110) plane in units of electrons per cell volume.

boron and nitrogen. When BN is formed, the additional attractive ionic potential of nitrogen results in an ionic charge distribution as shown in Fig. 3. The valence electron charge density is localized in the region close to the nitrogen atom, and one local maximum in the charge density is found. In contrast, for BP (Fig. 4), the charge density is more evenly distributed between the cation and the anion. The bond has a small ionic component and is double peaked in a manner similar to that found in diamond. This distribution of charge has its origins in the fact that, unlike nitrogen, phosphorus does not have a strong attractive p potential. Hence, the ionic attraction of the phosphorus is compensated by the strong B p potential, and the charge is shared in a covalent-like configuration.

The above features are accentuated in BAs (Fig. 5) where a covalent-like charge distribution is also found with a slightly higher charge density near the boron sites. This unusual situation is the reverse of what is found in other III–V compounds, and boron may be more accurately described as an anion rather than a cation.

The strong ionic–covalent bonding in BN results in a short bond length and high bulk modulus while the weaker bonds of BP and BAs yield longer bond lengths and lower bulk moduli even though both are highly covalent systems. The results for the lattice constants a and bulk moduli B_0 are compared with measured values [16–19] in Table 1. A large discrepancy exists between the calculated value of B_0 and the experimental estimate [17] for BN. A similar situation exists for one of the two reported values [19] for BP. We are not aware of measurements of B_0 for BAs. Other properties such as the frequencies of the zone center transverse optical phonon mode, cohesive energies and high pressure properties and structures have been studied. Most of the results are given in refs. 13–15.

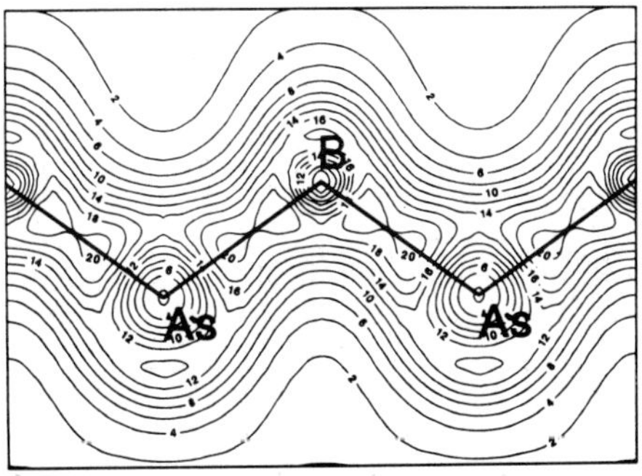

Fig. 5. Valence electron charge density for BAs displayed in the (110) plane in units of electrons per cell volume.

An interesting and unusual zinc blende compound involving a first-row element is SiC. Figure 6 contains both the calculated [20] valence charge densities for a superposition of carbon and silicon atomic charge distributions and the valence charge density in the zinc blende crystal. In both cases, the strong C 2p potential localizes the charge density near the carbon atom. For the superposition of atomic charges, the charge distribution is spherical around each atom. In the zinc blende crystal, the charge redistributes. The charge density at the atomic sites decreases, and there is an increase in the bonding region. For silicon, the charge moves from the region near the atom and in the antibonding region into the bond.

The transfer of charge into the bonding region close to carbon causes the ionic character of the SiC. However, unlike the case for III–V zinc blendes where the ionicity is caused by the valency difference, the charge redistribution for SiC rises from the stronger carbon potential and, in particular, from the C 2p potential. The overall result is an ionic-like bonding of modest strength with a charge configuration resembling that of BN in Fig. 3.

Results for the calculated [20] and measured

TABLE 1 Lattice constants *a* and bulk moduli B_0 for BN, BP and BAs

Material	*Method*	*a* (nm)	B_0 (GPa)
BN	Experiment	0.3615[a]	465[b]
BN	Calculated	0.3606[c]	367[c]
BN	Calculated		367[d]
BP	Experiment	0.4538[a]	173[e]
BP	Experiment		267[f]
BP	Calculated	0.4558[c]	165[c]
BP	Calculated		166[d]
BAs	Experiment	0.4777[a]	—
BAs	Calculated	0.4777[g]	145[g]
BAs	Calculated		138[c]

[a]Ref. 16.
[b]Ref. 17.
[c]*Ab initio* calculation, ref. 13.
[d]Empirical calculation, ref. 7.
[e]Ref. 18.
[f]Ref. 19.
[g]Ref. 14.

[16, 21] values of lattice constant and bulk modulus are given in Table 2 together with results for carbon and silicon for comparison. The results are in good agreement with the measured values. A comparison with the properties of silicon and carbon shows that the lattice constant of zinc blende SiC is 0.0138 nm smaller than an average of the lattice constants of silicon and carbon. This exception to Vegard's rule arises from the charge redistribution. The SiC bulk modulus is 28% smaller than the average of carbon and silicon, but it is larger than that of its isoelectronic pair BP (Table 1).

The studies of the charge distributions and the associated calculations of the lattice constants and bulk moduli yield useful information and physical pictures related to the role of bonding in determining the mechanical properties of solids. These calculations provide the basis for a simpler scheme which does not require complete band structures or total energy calculations for determining the material properties. This scheme is discussed in Section 3.

3. Empirical theory of bulk moduli

For specific models of the attractive and repulsive potentials, the bulk modulus B_0 can be estimated. Examples include inert gas solids [22] and alkali halide crystals [22, 23]. In both cases, the potential and electronic charge distributions are mostly spherical around the atomic sites. Charge redistribution effects are small in these cases and not accounted for. In the opposite extreme, a free-electron gas model for a metal [22, 24] is based on a constant charge density. For this case, the atomic potentials and charge redistributions do not play a role, and the system is characterized by parameters such as the electron concentration n, the electron gas parameter r_s and the Fermi energy E_F. Dimensional analysis suggests that B_0 should scale with E_F divided by the volume per electron. The direct calculation [22, 24] gives

$$B_0 = \tfrac{2}{3} n E_F = \left(\frac{6.13}{r_s}\right)^3 \text{ GPa} \qquad (1)$$

Since covalent solids are among those with largest B_0, it would be useful to have a scheme similar to those described above for this system. However, spherical potentials and spherical charge densities or constant charge densities are not appropriate since the electrons are concentrated in bonds and the charge distributions are

TABLE 2 Lattice constants *a* and bulk moduli B_0 for carbon, silicon and SiC

Material	*Method*	*a* (nm)	B_0 (GPa)
C	Experiment	0.3567[a]	443[a]
C	Calculated	0.3561[b]	438[b]
C	Calculated		435[c]
Si	Experiment	0.5429[a]	99[a]
Si	Calculated	0.5433[b]	92[b]
Si	Calculated		98[c]
SiC	Experiment	0.4360[a]	224[d]
SiC	Calculated	0.4361[b]	212[b]
SiC	Calculated		211[c]

[a]Ref. 16.
[b]Ref. 20.
[c]Ref. 7.
[d]Ref. 21.

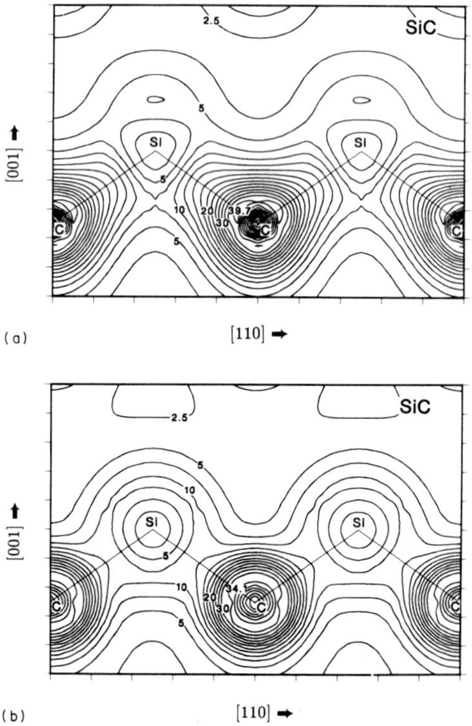

Fig. 6. Valence electron charge density in the (110) plane for zinc blende SiC: (a) self-consistent charge density for the crystal; (b) superposition of atomic charge densities.

not spherical about the atoms. However, some of the scaling arguments described above should hold true for these systems. In particular, B_0 should scale as the covalent bond energy divided by the bond volume, and this is the approach in ref. 7.

It is possible to determine [5] an average covalent gap E_h from spectral data and/or measurements of the dielectric constant ε using

$$\varepsilon = 1 + \frac{E_p^{\,2}}{E_h^{\,2} + C^2} \tag{2}$$

where E_p is the plasma frequency and C is the ionic contribution to the average gap. Typical numbers for the series germanium, GaAs and ZnSe are $E_p \approx 15.6$ eV and $E_h \approx 4.3$ eV. The change in ε for these three crystals therefore arises from the change in ionic character given by $C = 0$ eV, 2.9 eV and 5.6 eV for the three solids respectively. Since the lattice constants are about the same for these three solids, it is E_h which determines the bond length and is dominant in fixing the value of B_0.

Using the scaling law discussed before and the observation that for these systems the bonds are roughly cylindrical in shape with volumes of about $\pi(2a_B)^2 d$ where a_B is the Bohr radius and d is the bond length, then [7]

$$B_0 = 45.6 E_h d^{-1} \tag{3}$$

for B_0 in gigapascals, E_h in electronvolts and d in ångströms. If standard values of E_h are used, then the B_0 values obtained [7] are in good agreement with the measured data. However, since E_h scales [5] as $d^{-2.5}$, a simpler expression depending only on d can be obtained

$$B_0 = 1761 d^{-3.5} \tag{4}$$

where again B_0 is in gigapascals and d is in ångströms.

It is possible to include the effects of ionicity to lowest order using a parameter λ which distinguishes between the purely covalent elements of group IV of the periodic table ($\lambda = 0$), the partially ionic III–V compounds ($\lambda = 1$) and the more ionic II–VI compounds ($\lambda = 2$). Including [7] the effects arising from ionicity gives

$$B_0 = (1971 - 220\lambda) d^{-3.5} \tag{5}$$

Equation (5) gives better results than eqn. (4) does for the II–VI compounds. Some results for III–V and II–VI compounds are given in Table 3. The results for BN, BP, BAs, carbon, silicon and SiC are given in Tables 1 and 2.

The microscopic origins of the scaling relations of eqns. (3)–(5) derived more semiempirically in ref. 7 were re-examined [25] using analytic expressions based on the total energy pseudopotential scheme. Expressions are derived [25] for the purely covalent case in terms of the equilibrium Wigner–Seitz radius R_0. Although the power laws for the scaling differ from eqn. (4), this equation does represent a good compromise to the R_0^{-1}, R_0^{-2} and R_0^{-3} terms. When the ionic contribution is included, the results for the III–V compounds is similar to those obtained from eqn. (5). However, the results for the II–VI materials are not satisfactory. Adjustments can be made [25], but the analytic microscopic calculation does not give overall agreement with experiment on the level of that obtained from eqn. (5).

4. Discussion and summary

The results of the empirical approach in ref. 7 are surprisingly good and partially derivable from microscopic theory. The results clearly show that this expression gives accurate estimates for B_0 and that the highest bulk moduli are obtained for this class of crystals when the ionicity is low and the bond length is short. Diamond is an excellent

TABLE 3 Comparison of calculated III–V and II–VI bulk moduli B_0 using the empirical theory in ref. 7 and the measured bond lengths d (the experimental values are taken from ref. 16)

Material	*d* (nm)	*Experimental* B_0 (GPa)	*Calculated* B_0 (GPa)
III–V			
AlP	0.236	86.0	86.7
AlAs	0.243	77.0	78.3
AlSb	0.266	58.2	57.0
GaP	0.236	88.7	86.7
GaAs	0.245	74.8	76.1
GaSb	0.265	57.0	57.8
InP	0.254	71.0	67.0
InAs	0.261	60.0	61.0
InSb	0.281	47.4	47.1
II–VI			
ZnS	0.234	77.1	78.1
ZnSe	0.245	62.4	66.5
ZnTe	0.264	51.0	51.2
CdS	0.252	62.0	60.3
CdSe	0.262	53.0	52.6
CdTe	0.281	42.4	41.2
HgSe	0.263	50.0	51.9
HgTe	0.278	42.3	42.7

example. New suggestions for materials with large B_0 are systems with C–N bonds and B–N–C mixtures. Unfortunately, the empirical approach does not predict bond lengths or structures, but ionic radii can be used to make estimates.

In principle, the *ab initio* approach is capable of predicting structures and bond lengths, once candidate structures are proposed. It is interesting to note the results for the bulk moduli for the compounds BN, BP, BAs and SiC which were recently computed using the *ab initio* approach are almost identical with those obtained using the empirical approach. Perhaps the most fruitful direction to pursue now is a combined approach using the *ab initio* and empirical calculational schemes.

Acknowledgments

This work was supported by National Science Foundation under Grant DMR-8319024 and by the Director, Office of Energy Research, Office of Basic Energy Sciences, Materials Sciences Division, U.S. Department of Energy, under Contract DE-AC03-76SF00098.

References

1 M. L. Cohen, in E. A. Almond, C. A. Brookes and R. Warren (eds.), *Proc. 2nd Int. Conf. on Science of Hard Materials*, Institute of Physics, Bristol, 1986, p. 1.
2 M. L. Cohen, *Science, 234* (1986) 549.
3 K. J. Chang, M. M. Dacorogona, M. L. Cohen, J. M. Mignot, G. Chouteau and G. Martinez, *Phys. Rev. Lett., 54* (1985) 2375.
4 M. L. Cohen and P. W. Anderson, in D. H. Douglass (ed.), *Superconductivity in d- and f-Band Metals*, American Institute of Physics, New York, 1972, p. 17.
5 J. C. Phillips, *Bonds and Bands in Semiconductors*, Academic Press, New York, 1973.
6 W. A. Harrison, *Electronic Structure and Properties of Solids*, Freeman, San Francisco, CA, 1980.
7 M. L. Cohen, *Phys. Rev. B, 32* (1985) 7988.
8 J. R. Chelikowsky and S. G. Louie, *Phys. Rev. B, 29* (1984) 3470.
9 S. Fahy and S. G. Louie, *Phys. Rev. B, 36* (1987) 3373.
10 S. Fahy, S. G. Louie and M. L. Cohen, *Phys. Rev. B, 34* (1986) 1191.
11 S. Fahy, S. G. Louie and M. L. Cohen, *Phys. Rev. B, 35* (1987) 7623.
12 M. Kertesz and R. Hoffmann, *J. Solid State Chem., 54* (1984) 313.
13 R. M. Wentzcovitch, K. J. Chang and M. L. Cohen, *Phys. Rev. B, 34* (1986) 1071.
14 R. M. Wentzcovitch and M. L. Cohen, *J. Phys. C, 19* (1986) 6791.
15 R. M. Wentzcovitch, M. L. Cohen and P. K. Lam, *Phys. Rev. B, 36* (1987) 6058.
16 *Landolt–Bornstein, New Series*, Vol. 17a, Springer, Berlin, 1982, and references cited therein.
17 J. A. Sanjurjo, E. Lopez-Cruz, P. Vogl and M. Cardona, *Phys. Rev., 28* (1983) 4579.
18 W. Wettling and J. Windscheif, *Solid State Commun., 50* (1984) 33.
19 T. Suzuki, T. Yagi and S. Akimoto, *J. Appl. Phys., 54* (1983) 748k.
20 K. J. Chang and M. L. Cohen, *Phys. Rev. B, 35* (1987) 8196.
21 D. H. Yean and J. R. Riter, Jr., *J. Phys. Chem. Solids, 32* (1971) 653.
22 C. Kittel, *Introduction to Solid State Physics*, 6th edn., Wiley, New York, 1986.
23 M. P. Tosi, *Solid State Physics., 16* (1964) 1.
24 N. W. Ashcroft and N. D. Mermin, *Solid State Physics*, Holt, Rinehart, and Winston, New York, 1976.
25 P. K. Lam, M. L. Cohen and G. Martinez, *Phys. Rev. B, 35* (1987) 9190.

Materials Science and Engineering, A105/106 (1988) 19–28

Hardness Anisotropies: A New Approach*

S. G. ROBERTS, P. D. WARREN and P. B. HIRSCH

Department of Metallurgy and Science of Materials, University of Oxford, Parks Road, Oxford OX1 3PH (U.K.)

(Received November 9, 1987)

Abstract

Hardness testing is often the only mechanical property test that can be applied to hard solids. The apparent simplicity of hardness tests belies their true complexity; however, understanding the fundamentals of the responses of the materials to indentation allows considerable information to be extracted from such tests. In particular, anisotropies in hardness, for example the Knoop hardness anisotropy on a single-crystal plane, or differences in hardness between different test planes, have been commonly used to investigate the active slip systems in a wide variety of materials. We describe here a new approach to the analysis of hardness anisotropy results, in which the whole slip pattern around indentations is modelled. The hardness anisotropy is then described in terms of the stresses on the active dislocations, of their efficiency as mechanisms for material transport, and of the possible interactions between them. It will be shown that this approach leads to a full description of the slip patterns and hardness behaviour of a wide variety of crystals and leads to a better understanding of other indentation phenomena, such as plastic recovery and the initiation and propagation of fracture.

1. Introduction

(Micro)hardness testing is commonly used to investigate the flow and fracture properties of very hard materials. It is the only simple technique that can be used at low temperatures, and specialized instruments are capable of tests up to 1600 °C. The apparent simplicity of hardness tests belies their true complexity [1]. While quantitative analyses of crack length data can be used to obtain values of fracture toughness [2], the correlation via the micromechanisms of plasticity between "hardness" and conventional yield strength is not well understood, especially for hard brittle solids.

However, microhardness testing has been used extensively to study the nature of the active slip systems in hard single crystals. An indenter of relatively low symmetry (usually a "Knoop" profile) is used to study the variation in hardness with indenter orientation on various crystal planes. It is assumed that orientations of low hardness are those where the orientation of the indenter to some slip systems in the crystal is such that these slip systems are subject to a high resolved shear stress (RSS). Early work in this area, generally based on resolving a single stress state associated with each indenter facet onto the possible active slip systems, was reviewed by Brookes *et al.* [3]. Brookes *et al.* refined this type of model further to take into account likely interactions between the active slip systems, adjacent material and the indenter facets, producing the "effective resolved shear stress" (ERSS) model. The ERSS on a slip system is the RSS from the assumed stress state, multiplied by a geometric "constraint factor". The constraint factor is intended to model restrictions on slip directed into the indenter facets or into neighbouring slip zones.

The ERSS model has been used to determine, for example, the change in slip systems with temperature, in various transition metal carbides [4], and the slip systems in diamond [5] and cubic boron nitride [6]. In this last case, the distinction between $\{111\}\langle 1\bar{1}0\rangle$ and $\{110\}\langle 1\bar{1}0\rangle$ slip systems, for which the same type of hardness variation (on the $\{001\}$ plane studied) is predicted by the ERSS model, was made by examining the slip steps visible around high temperature indentations. This approach, combining hardness variation with indenter orientation and the observation of slip traces and dislocation arrays, has now been

*Paper presented at the 3rd International Conference on the Science of Hard Materials, Nassau, The Bahamas, November 9–13, 1987.

0921-5093/88/$3.50

© Elsevier Sequoia/Printed in The Netherlands

considerably extended [7–10] and forms the basis of the new model summarized in this paper.

2. Models for prediction of slip patterns

The essential difference between the new approach and the ERSS model of Brookes *et al.* lies in the choice of indenter stress field. The ERSS model assumes that the slip systems controlling the Knoop hardness anisotropy (KHA) are those immediately adjacent to the indenter facets; the new model uses "extended" stress fields, so as to predict the likely slip patterns and the stresses on them in the region around the indenter. Constraint factors are not used, as the use of more realistic stress fields should not predict material to flow in unrealistic directions. This type of approach was suggested by initial results for the KHA of germanium [8]. Here, it was found that hardness anisotropy on the {001} surfaces was detectable only above the temperature at which dislocation activity was apparent at some distance away from the indenter, either as slip steps or by etch pitting. Thus it was thought that the use of an extended stress field should be the basis of an improved model.

The stress fields used are those for the elastic contact of flat punches with a flat, elastically isotropic surface. These are, of course, extreme simplifications of the very complex stress fields operating around a real indentation. Ideally, the stress field used should be "elastic–plastic", incorporating not only the stresses from the indenter–surface contact but also all the stresses from the dislocations (and other defects) in the "plastic zone" generated during indentation. This plastic zone is itself highly structured in a way that varies with the relative orientations of the indenter and specimen crystallography. However, the very simple stress fields used here might at least be expected to give an insight into the types of slip that initiated during the first stages of indentation, and which might then be expected to continue in operation. It will be shown that in fact there is a good qualitative fit between the predictions of such models and observed slip patterns.

Two computational methods are used.

(a) The most activated slip systems at an array of points on a plane section through the centre of an "indenter facet" are considered (see Fig. 1). A computer program evaluates the stress field components at each point, transforms these to principal stresses and then resolves these principal stresses onto the slip systems available. A map is then produced showing the most stressed slip system at each point, together with contours of this maximum RSS, with the option of RSS contours for specific slip systems. Maps can then be produced for any desired orientation of the indenter on a given test plane. It is assumed that the most stressed systems are those likely to be active in a particular region, especially if they are on slip planes which intersect the crystal surface under or near to the indentation (since we would expect slip to be nucleated there). Thus the changes in slip pattern with orientation can be predicted.

(b) The variation in stress with indenter orientation is considered on all slip systems at one particular point (relative to the indenter). This produces similar plots to the ERSS model, except that no constraint factors are used, and also in that several "typical" points must be studied rather than the implicit single point of the ERSS model.

Two types of stress field have been used: (a) that of an infinitely long flat punch of finite width subjected to uniform loading [11] (hereafter referred to as the IP field); (b) that of a finite rectangular, uniformly loaded punch [12] (hereafter referred to as the FP field). The IP field was initially used because of its mathematical simplicity. The stress components are

$$\sigma_{xx} = -\frac{P}{2\pi}\{2(\phi_1 - \phi_2) + \sin(2\phi_1) - \sin(2\phi_2)\}$$

$$\sigma_{zz} = -\frac{P}{2\pi}\{2(\phi_1 - \phi_2) - \sin(2\phi_1) + \sin(2\phi_2)\}$$

$$\sigma_{yy} = \nu(\sigma_{xx} + \sigma_{zz})$$

$$\sigma_{xz} = \frac{P}{2\pi}\{\cos(2\phi_1) - \cos(2\phi_2)\}$$

The notation for the stress directions and ϕ_1 and ϕ_2 are shown in Fig. 1; this also shows how for Knoop indentations the two facet edges meeting at an obtuse angle are replaced by a single punch edge*.

*Although glass is isotropic, it provides an interesting check on the applicability of this elastic stress field to hardness testing. The shear bands beneath a Vickers indentation in glass, investigated by Hagan [13], follow closely the lines of maximum shear (*i.e.* those at 45° to the principal stress directions) in the IP field [11].

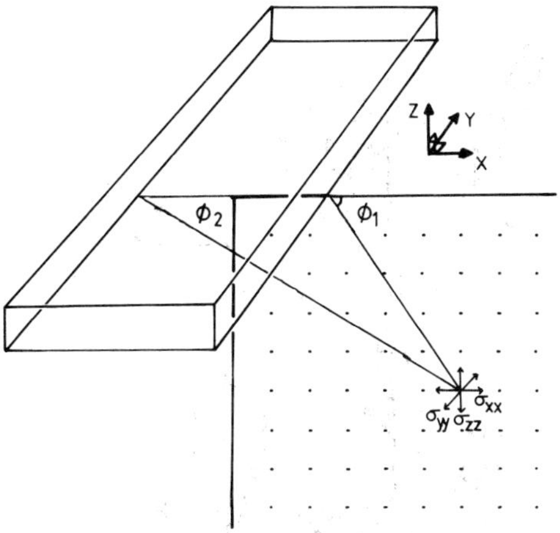

Fig. 1. The IP model. The angles used in computing the stresses at various points relative to the indenter are shown.

It was found that the IP field could be used to give reasonable solutions for real indenters of more complex shape, considered as being built up from sets of IP edges (see for example ref. 7). The more complicated FP stress field was used to check the validity of this approach, approximating a Knoop indenter by a rectangle of 7:1 aspect ratio, and a Vickers indenter by a uniformly loaded square. It was found that the FP solutions gave essentially the same results (near the centre of the long edge) as the IP solutions for aspect ratios of 2:1 upwards (the main difference being that the IP gives no stresses at the surface, whereas the FP gives in-plane stresses).

For the purposes of comparison, the ERSS model used was a simplified version of that used by Brookes *et al.* [3]. The stress state used was a simple tensile stress at 22° to the specimen surface. Only the ERSS values resulting from a single facet (considered, as for the IP field, to be the "average" of the two facets on one side of the Knoop indenter) were computed.

Analyses of hardness anisotropy in a variety of cubic materials are now presented. It will be shown that the changes in hardness with angle may be due either to the variation in resolved shear stress on the available slip systems or to the effectiveness of the operating slip systems as routes for removal of material from the indentation site. Models of this type can also give insights into mechanisms of crack nucleation and post-indentation plastic recovery [7–9].

3. Cubic semiconductors

For diamond cubic materials, indented on {001}, the new model predicts different flow patterns and stress variations from those of the ERSS model. Figure 2 compares plots of type (b) above, showing the variation in RSS (using the IP field) with indenter angle at various points relative to one indenter edge, with the variation in ERSS with angle for the simplified version of the ERSS model (since the {001} face has fourfold symmetry, the opposite edge produces the same stresses on symmetry-related slip systems).

The ERSS model ascribes the KHA in these materials to the variation, with indenter angle, in the total effective resolved shear stress on the {111}⟨110⟩ slip systems. Modelling using the IP field showed that, in the whole region surrounding the "punch", the resolved shear stress on the most highly stressed slip systems varies very little with angle, although the particular slip system with the highest RSS changes frequently.

However, the new model predicts the complete three-dimensional pattern of slip; the types of slip system activated vary considerably with orientation. Figures 3(a) and 3(b) show maps of the most highly stressed slip systems around IP "indenters" in [100] and [110] orientations respectively on the (001) surface. It is useful to distinguish two types of slip system: (i) those with the slip direction parallel to the surface; (ii) those with the slip direction inclined to the surface. Slip of the first type ("rosette" slip) is only expected when the indenter is near ⟨100⟩ (the hardest orientation) and then only in the near-surface regions; this is as observed experimentally (see below). Inclined slip (type (ii)) is predicted beneath the indenter in this orientation. Where slip planes of this latter type intersect, locking dislocations are formed. In ⟨100⟩ orientations, only this lock-forming type of inclined slip is predicted (see Fig. 3(a)).

However, in the ⟨110⟩ (softest) orientations, the inclined slip is of two types: under the indenter, slip occurs on planes which converge beneath the indenter, with the extra half-plane lying above the slip plane, as for the ⟨100⟩ orientation; outside the indenter, slip occurs on diverging slip planes with the extra half-plane of the dislocations involved lying beneath the slip plane (see Figs. 3(b) and 3(c)). It should be noted that, since the ERSS model is effectively confined to the stress state at one point, while it predicts the convergent slip

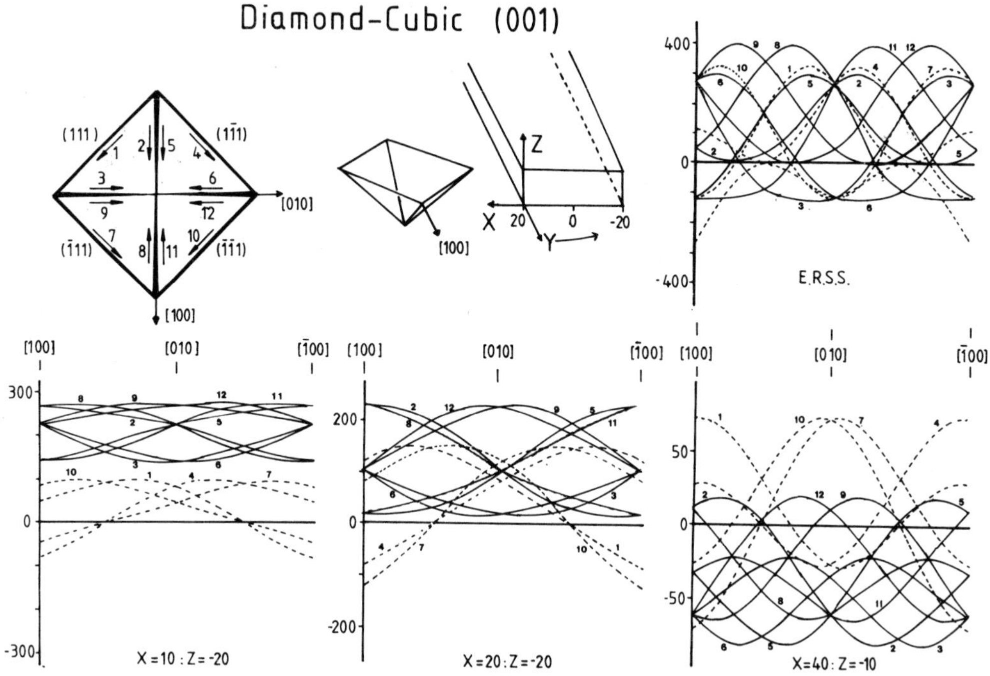

Fig. 2. Comparison of the ERSS and the IP models for diamond cubic materials with (001) indentation. The variation, with indenter angle, in ERSS and RSS on the possible slip systems at various points around an IP indenter is shown. The indenter orientation is given by the orientation of its long axis *Y*. The slip systems are labelled so that, if the RSS is positive, the side of the slip plane nearer the crystal surface slips in the direction shown. The stresses on "rosette" systems (those with slip vectors parallel to the surface, systems 1, 4, 7 and 10), are shown by broken lines. The reversal of the sense of RSS on the slip systems with slip vectors inclined to the surface (——) in going from underneath to outside the indenter profile should be noted.

mode, it does not predict this diverging slip. This type of slip generates no locks, and its extent is limited only by the mobility of the dislocations on it.

Thus, consideration of the activated slip systems shows that the indenter orientation with the highest hardness ⟨100⟩ corresponds to a maximum in lock-forming dislocation intersections, *i.e.* KHA in these materials is associated with a variation in work hardening with indenter orientation. In the ⟨110⟩ (soft) orientation, a pair of non-work-hardening slip planes are activated, on which slip can penetrate deep into the crystal. Confirmation of the slip pattern predictions of this model has been provided by cross-sectioning and etching experiments in germanium [14].

So as to test the validity of this model, further work was carried out on GaAs [8]. In this non-centrosymmetric solid, otherwise identical dislocations with opposite senses of slip can have very different velocities. Because of this asymmetry, hardness anisotropy is observed between [110] and [1$\bar{1}$0]*. It was found that the softer of the two ⟨110⟩ indenter orientations corresponds to the motion of the faster dislocation type on the divergent slip planes; this slip mode is not predicted by the ERSS models. Cross-sectioning and etching experiments and transmission electron microscopy (TEM) studies [15] confirm these slip patterns.

It should be noted that both cubic boron nitride [6] and diamond [5] show considerable hardness anisotropy at room temperature. Since the extent of slip would be expected to be very limited in these cases, the causes of the aniso-

*There is also an asymmetry in Vickers hardness between the (111) and ($\bar{1}\bar{1}\bar{1}$) faces in GaAs [7], InSb [9] and other non-centrosymmetric crystals. This effect can also be explained in terms of the different mobilities of the opposite senses of dislocations on the active slip planes [7]. The softer face is that for which the faster dislocation type moves on the diverging inclined slip planes.

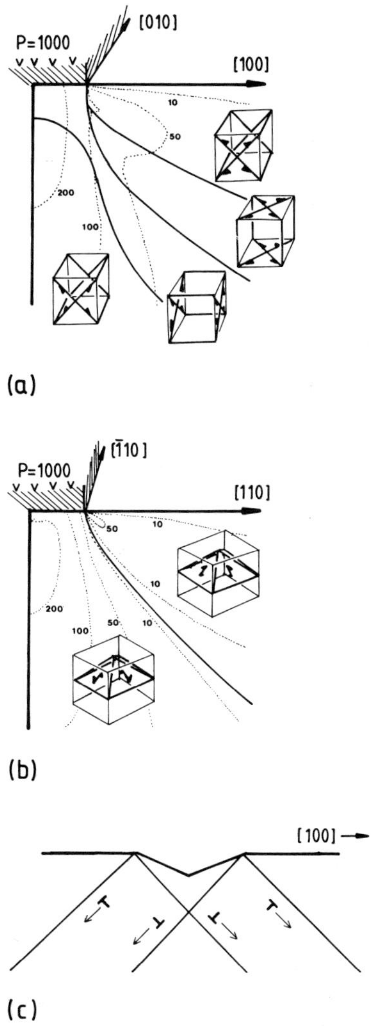

Fig. 3. Diamond cubic materials with {001} indentation. (a), (b) Maps of the most highly stressed slip systems, using the IP field for (a) the indenter lying along [100] (hard orientation) and (b) the indenter lying along [110] (soft orientation). The slip planes active in a particular region are shown by thicker lines, and the sense of shear is indicated by arrows on the active slip direction. The broken lines show contours of equal RSS (relative to the indentation pressure of 1000). (c) Dislocations active around a ⟨110⟩ indentation, assuming that slip nucleates near the surface.

tropy are unclear; possibly some variant of the ERSS approach might be appropriate here. Further work, involving TEM of subsurface deformation patterns, would be needed to resolve the problem.

4. Magnesium oxide

This material slips on {110}⟨1$\bar{1}$0⟩; its KHA behaviour has been studied on {001} and {110} surfaces [3, 10]. On {001} faces, there is a large variation in hardness between the ⟨100⟩ directions (soft) and the ⟨110⟩ directions (hard). On the {110} faces, the ⟨001⟩ direction is softest, with the hardness increasing to a peak near ⟨1$\bar{1}$1⟩ and then dropping slightly to a local minimum at ⟨1$\bar{1}$0⟩.

The ERSS analysis fits these results reasonably well [3]. The envelope of maximum values of ERSS shows the correct variation with indenter angle (being a minimum when hardness is a maximum, and vice versa). Figure 4 compares, for the (001) indentation plane, the predictions of the active slip systems from the simplified version of the ERSS model with the predictions of the FP model. It can be seen that the ERSS model gives very little activation of any slip systems in the hard ⟨110⟩ directions, and activation of one inclined slip system and both "rosette" slip systems in each of the soft ⟨100⟩ orientations*.

In contrast, the FP model predicts that, under the indenter, all inclined slip systems should be roughly equally activated, with very little stress on the "rosette" systems; there is very little variation in the stress on any of the slip systems with angle.

Outside the indenter, the stresses vary strongly with angle. The rosette systems are strongly activated in the ⟨100⟩ orientations but have no stress on them in the ⟨110⟩ orientations. The inclined slip systems have a high stress on them in the ⟨100⟩ orientations (of the opposite sign to that on those under the indenter and in the ERSS model), which falls to a minimum in the ⟨110⟩ orientations. Figures 5(a) and 5(b) show maps of the most highly stressed slip systems around IP indenters in [100] and [110] orientations respectively on the (001) surface; Fig. 5(c) shows a section normal to an indenter along ⟨100⟩, showing the predicted slip patterns (it should be noted that strong locks are not expected to form under the indenter as the dislocations on the intersecting slip systems have perpendicular Burgers vectors).

The actual slip patterns beneath indentations made in ⟨100⟩ and ⟨110⟩ orientations on MgO can be revealed by chemical stripping-polishing and etch pitting. Results for indentations made on an {001} face are shown in Fig. 6.

*For convenience in this paper, "rosette" slip is used to mean slip outside the indenter profile, with the slip vector parallel to the indented surface, so as to allow contrast with slip systems with the slip vector inclined to the surface.

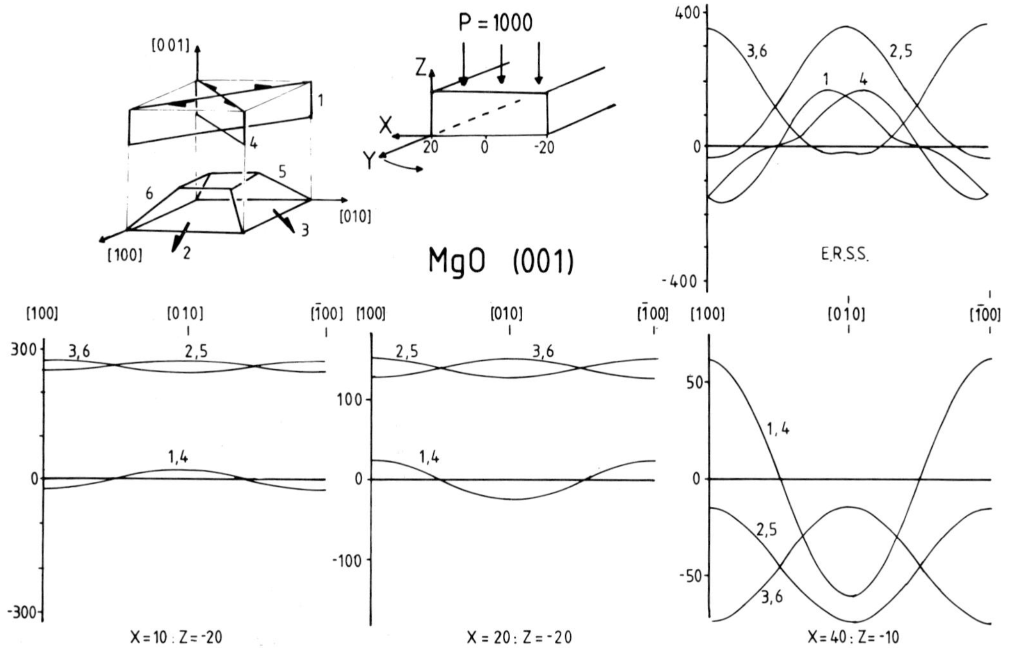

Fig. 4. Comparison of ERSS and FP model predictions for MgO with (001) indentation. The variation, with indenter angle, in ERSS and RSS on the possible slip systems at various points around an IP indenter is shown. The indenter orientation is given by the orientation of its long axis *Y*.

The observed slip patterns correspond very well to those predicted by the FP model. In the soft ⟨100⟩ orientation, the inclined slip systems "parallel" to the indenter long axis are activated both beneath and outside the indenter; it is reasonable to assume that the dislocations on these slip planes are of opposite sign, with material from the indentation site effectively being punched into the bulk of the crystal between these two sets of planes (as in Fig. 5(c)). The IP model predicts the correct senses of slip on the inner and outer planes; the ERSS model gives the sense of slip corresponding to the inner set of planes only. It should be noted that the other inclined slip system is also activated, giving a square of dislocation etch pits at depth. Slip of this type is not predicted by the ERSS model. "Rosette"-type slip is also visible in the ⟨100⟩ orientation, but only close to the surface. This is as is predicted by the IP model.

In the hard ⟨110⟩ orientation, slip on all inclined systems is predicted by both models, and the observed slip patterns, as shown in Fig. 6, fit these predictions. However, slip is seen to penetrate to roughly equal depths for indenters in the hard and soft orientations. This corresponds more closely to the IP model, where the stresses on the inclined systems under the indenter do not vary greatly with orientation. Slip can also be seen on systems 1 and 4 (the "rosette" systems). This is because of the slight inclination of the edges of the real indenter from ⟨110⟩; while RSS on these systems is seen to be zero at ⟨110⟩ (for $X=40$, $Z=-10$), it rises very rapidly away from this orientation. The predicted RSS decreases very rapidly with depth. This tallies with the experiments (see Fig. 6), where the penetration of this type of slip is seen to be small. A similar effect has been noted by Brookes *et al.* [3] and is illustrated in their Figs. 6 and 9.

Thus, while both the ERSS and the IP model attribute the variation in hardness on the {001} planes to a variation in stress on the possible active slip planes, the ERSS model deals (implicitly) only with slip under the indenter. The IP model gives little variation in these stresses under the indenter, but a wide variation outside the indenter, and gives a full description, in accor-

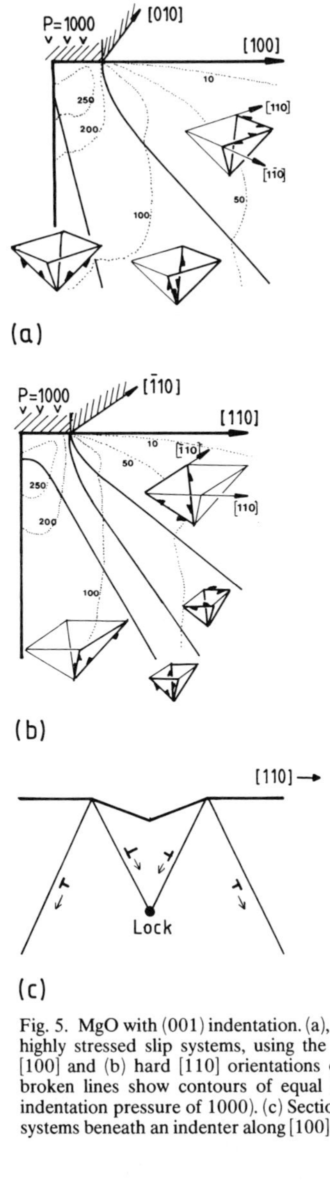

Fig. 5. MgO with (001) indentation. (a), (b) Maps of the most highly stressed slip systems, using the IP field, for (a) soft [100] and (b) hard [110] orientations of the indenter. The broken lines show contours of equal RSS (relative to the indentation pressure of 1000). (c) Section of the inclined slip systems beneath an indenter along [100].

dance with experimental results, of the active slip modes in all regions beneath and around the indenter.

In the ⟨100⟩ orientations, all the highly stressed slip systems can operate in ways which minimize the lock-forming dislocation interactions. In the ⟨110⟩ orientations, there is no activation of the "free-running" divergent slip mode outside the indentation, and all the slip modes activated interact strongly. Thus the hardness anisotropy on the {001} faces of MgO is interpreted as being due to a combination of stress variation on the possible slip systems (maximum stresses in the softest orientation) and an angular variation in lock-forming dislocation interactions (this form of work hardening being strongest in the hardest orientation).

Similar analyses have been carried out for the hardness anisotropy on {110} faces of MgO [10]. Here again, the IP model gives a full prediction of the experimentally observed slip patterns, and the hardness anisotropy is interpreted as being due to a combination of the variation in RSS on the available slip systems (especially in the regions outside the indentation) and the variation in work hardening of the activated slip systems.

5. Calcium fluoride

This material slips on {001}⟨110⟩ systems and has a hardness anisotropy on the {001} faces opposite to that of MgO, *i.e.* the ⟨110⟩ direction is soft compared with the ⟨100⟩ direction [3]. Analyses using the ERSS and the IP models give similar results (Fig. 7). Both models show peaks in the RSS curves in the ⟨110⟩ indenter orientations; for the IP model, this variation is roughly similar in magnitude both under and outside the indenter. In both cases, the dominant slip system in the [110] orientation on (001) is (001)[110] (slip system 1), and both models give the same sense of slip (upper side of the slip plane being sheared away from the indenter edge).

We now consider the models' predictions regarding the slip on the planes normal to the surface. In the [110] orientation, the ERSS model predicts slip on $(010)[\bar{1}0\bar{1}]$ (system 3) and $(100)[01\bar{1}]$ (system 6), with the side of the slip plane nearer the indenter edge shearing downwards into the bulk of the specimen. Using the IP stress field, stresses on these systems are predicted to be relatively high only at some depth beneath the indenter; close to the indenter, the stresses on $(010)[10\bar{1}]$ (system 4) and $(100)[0\bar{1}\bar{1}]$ (system 5) are predicted to be the higher. Indeed, near the surface outside the indentation profile, the stresses on these slip systems are higher than those on systems operating on the plane parallel to the crystal surface.

Thus the two models give opposite predictions as to the active slip directions on the planes normal to the surface. It would be very difficult to

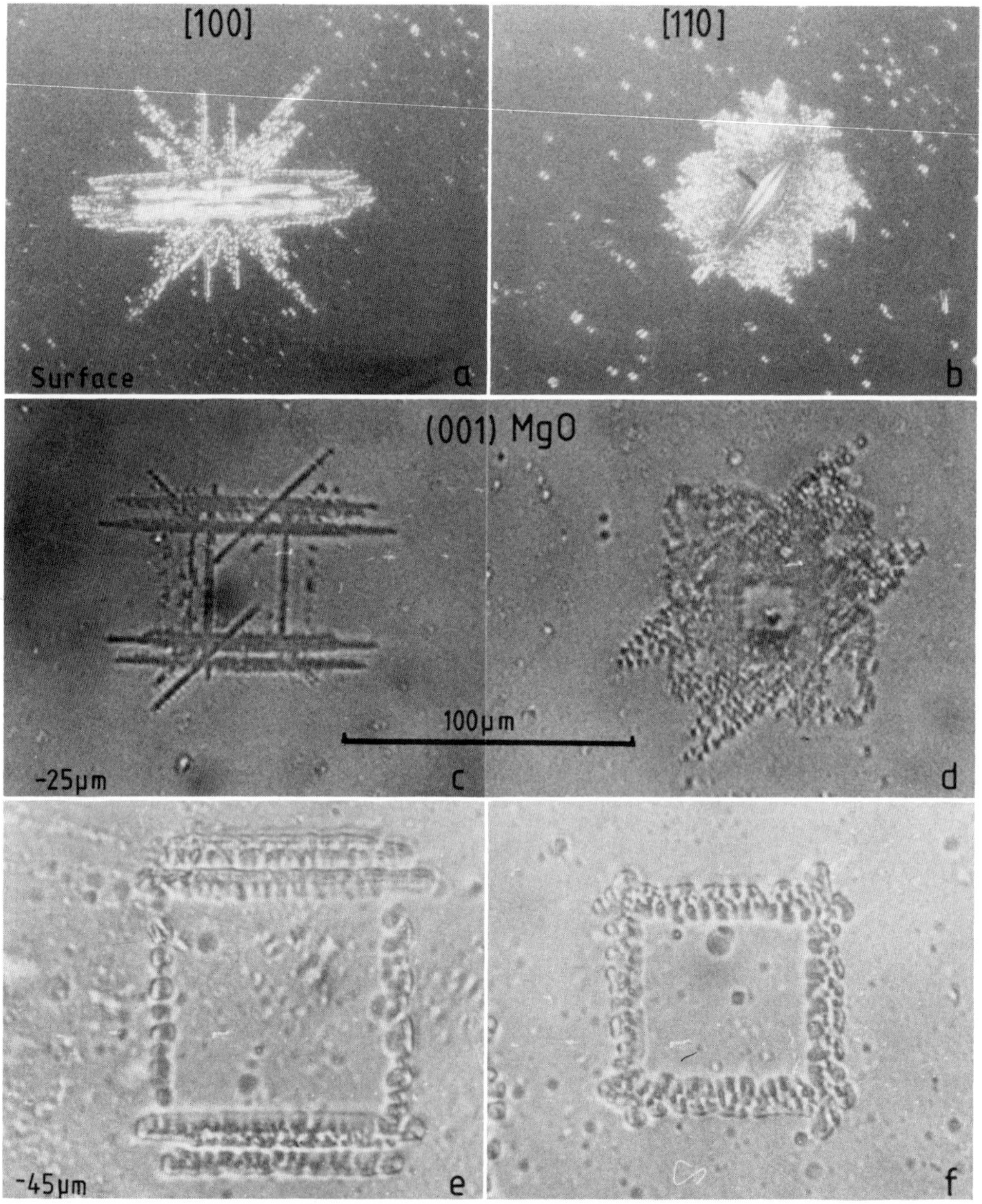

Fig. 6. Actual slip patterns beneath 100 gf Knoop indentations on an {001} surface: (a), (c), (e), indenter along ⟨100⟩; (b), (d), (f), indenter along ⟨110⟩. Dislocations are revealed by etch pitting, after chemical surface stripping to the indicated depths.

distinguish between the two possibilities by examinantion of slip lines or etch pit patterns; TEM or some other technique sensitive to the Burgers vector would be required.

6. Conclusions

The FP model and the ERSS model for KHA have been compared for cubic materials with

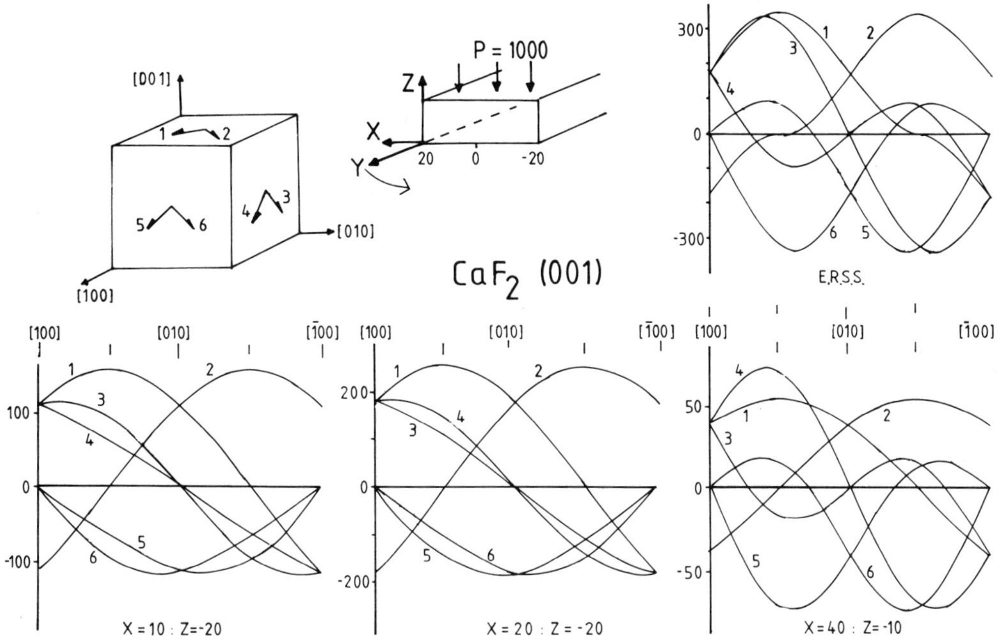

Fig. 7. Comparison of ERSS and FP model predictions for CaF_2 with (001) indentation. The variation, with indenter angle, in ERSS and RSS on the possible slip systems at various points around an IP indenter is shown. The indenter orientation is given by the orientation of its long axis *Y*.

three types of slip system. The FP model gives a more complete description of the slip patterns, and the KHA on {001} faces is described as arising differently in these three types of material.

(a) In diamond cubic and zinc blende semiconductors, slipping is on $\{111\}\langle 1\bar{1}0\rangle$; KHA is due to the availability of a non work-hardening mode of slip with the indenter aligned along ⟨110⟩ directions [8]. The model also describes the "hardness polarity" effect in zinc blende materials [7].

(b) In MgO, slipping is on $\{110\}\langle 1\bar{1}0\rangle$; KHA is due to the variation with indenter orientation both in RSS on the slip systems and in the availability of an "easy" slip mode in the ⟨100⟩ orientation. The model also describes the hardness anisotropy on {110}[10].

(c) In CaF_2, slipping is on {001}⟨110⟩; KHA is due to the variation in RSS on the available slip systems. The soft direction, ⟨110⟩, is where slip on the plane parallel to the indented surface is most strongly activated.

Further study is needed to establish the actual slip patterns around indentations in CaF_2. The model is currently being extended to hexagonal materials, where different types of slip system (basal, prismatic and pyramidal) may be active simultaneously, so as to interpret effects such as the hardness polarity of ZnO [16].

Acknowledgments

The work reported here was carried out with the support of the Science and Engineering Research Council of both S.G.R. (Postdoctoral Fellowship and Advanced Fellowship) and P.D.W. (Quota Studentship).

References

1 J. J. Gilman, in J. H. Westbrook and H. Conrad (eds.), *The Science of Hardness Testing and its Research Applications*, American Society for Metals, Metals Park, OH, 1973, p. 51.

2. B. R. Lawn and D. B. Marshall, *J. Am. Ceram. Soc., 62* (1978) 347.

3 C. A. Brookes, J. B. O'Neill and B. A. W. Redfern, *Proc. R. Soc. London, Ser. A*, *322* (1971) 73.

4 D. L. Kohlstedt, *J. Mater. Sci.*, *8* (1973) 777.

5 P. Humble and R. J. Hannink, *Nature, London, 234* (1978) 112.

6 C. A. Brookes, in E. A. Almond, C. A. Brookes and

R. Warren (eds.), *Science of Hard Materials, Proc. 2nd Int. Conf. on the Science of Hard Materials, Rhodes, September 23–28, 1984*, in *Inst. Physics Conf. Ser., 75* (1986) 207.
7 P. B. Hirsch, P. Pirouz, S. G. Roberts and P. D. Warren, *Philos. Mag. B, 52*(1985) 759.
8 S. G. Roberts, P. D. Warren and P. B. Hirsch, *J. Mater. Res., 1*(1986) 162.
9 S. G. Roberts, *Philos. Mag., 54*(1986) 37.
10 S. G. Roberts, *Philos. Mag.*, to be published.
11 A. Nadai, *Plasticity*, McGraw-Hill, New York, 1931, p. 247.
12 A. E. H. Love, *Proc. R. Soc. London, Ser. A, 228* (1929) 377.
13 J. T. Hagan, *J. Mater. Sci., 15*(1980) 1417.
14 S. G. Roberts, unpublished, 1984.
15 C. D. Qin and S. G. Roberts, to be published.
16 J. T. Czernuska, unpublished, 1986.

Materials Science and Engineering, *A105/106* (1988) 29–37

Vacancy Arrangements in NbC_x*

K. OHSHIMA† and J. HARADA

Department of Applied Physics, Nagoya University, Nagoya 464 (Japan)

M. MORINAGA

Toyohashi University of Technology, School of Production Systems Engineering, Tempaku-cho, Toyohashi-shi, Aichi 440 (Japan)

P. GEORGOPOULOS and J. B. COHEN

Midwest Associated Team for Research with Intense X-rays, and the Department of Materials Science and Engineering, The Technological Institute, Northwestern University, Evanston, IL 60208 (U.S.A.)

(Received January 25, 1988)

Abstract

The diffuse X-ray scattering from $NbC_{0.72}$ was measured on an absolute scale in a volume of reciprocal space. The measurements were repeated three times at two synchrotron sources (in the U.S.A. and Japan) and with a high intensity rotating-anode source (in Japan). This diffuse scattering, previously thought to be due to vacancy octahedra, is shown instead to be dominated by the scattering due to mean-square atomic displacements. From the sign and amplitude of the displacement parameters, the following model for the environment around a carbon vacancy is proposed: the niobium nearest neighbors to a vacancy move away from it, whereas second-neighbor carbon atoms collapse toward it. This appears to be caused by an enhancement of the strength of the Nb–C bond due to the presence of vacancies on the carbon sublattice. There is also evidence that these vacancies tend to be correlated along 211 vectors.

1. Introduction

Transition metal carbides have been studied extensively for both fundamental and practical reasons. They have extremely high melting points and hardnesses and exhibit metallic conductivity [1, 2]. The compound NbC_x with $x = 0.70–0.99$ has a wide single-phase region at high temperatures. It has an NaCl structure, with large numbers of carbon vacancies, as indicated by density measurements [2] and structure determination by X-ray diffraction [3]. A vacancy band below the Fermi level, occupied by Nb 4d electron states, is thought to be responsible for the stability of the NaCl structure [4]. These vacancies have a profound influence on its physical properties. For instance, its superconducting transition temperature varies with composition. Although the reasons for this are not yet clear [5–8], it has been suggested that vacancy ordering may play a role by modifying the electronic density of states [9]. NbC_x is extremely inert (boiling aqua regia does not attack it even in powder form), and so it finds applications in high energy batteries.

If this material is slowly cooled, long-range ordering of the vacancies occurs [10] whereas, in the disordered (quenched) condition, electron diffraction patterns exhibit strong diffuse scattering [11, 12] which has been attributed primaril to local ordering of the vacant sites. The shape (this diffuse scattering is depicted in Fig. 1. number of other metal carbides and nitrides giv quite similar diffraction patterns [13], attributed to the presence of octahedra in the crystal lattice, whose corners are a mixture of carbon atoms and vacancies, and with a niobium atom at the center; the arrangement of the carbon atoms and vacancies determines the modulations of the intensity. However, with one exception, all the previous work has considered only the position and shape of this scattering but not the actual intensity.

*Paper presented at the 3rd International Conference on the Science of Hard Materials, Nassau, The Bahamas, November 9–13, 1987.

†Present address: Institute of Applied Physics, University of Tsukuba, Sakura, Ibaraki 305, Japan.

0921-5093/88/$3.50

© Elsevier Sequoia/Printed in The Netherlands

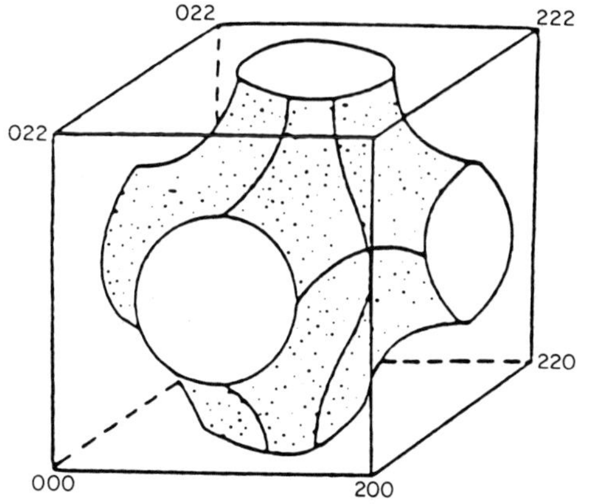

Fig. 1. Shape of diffuse scattering from NbC_x as observed in this study and in ref. 11.

There are strong indications that important contributions to the intensity arise from local atomic displacements. The static displacements of first-neighbor niobium atoms around a vacancy estimated from Debye–Waller factor determinations [14], channeling experiments and lattice statics calculations [15] are large, about 0.1 Å. The carbon atom displacements are considered to be much smaller but have not been reliably determined. Phonon dispersion curves also exhibit several anomalies [16].

A more complete study of the diffuse scattering due to atomic displacements is presented here, to provide information on the carbon atom displacements and more details on the niobium shifts. There has been one study of the actual intensity, using neutrons [17]. One plane in reciprocal space was sampled and, in attempting to separate the effects of local order, only the linear displacement terms (and not higher moments) were considered. From our results reported below, we estimate that such terms contribute only 1%–2% to the observed diffuse neutron scattering, so that the analysis techniques employed were satisfactory. However, our measurements with X-rays also indicate that the features of the diffuse scattering may be too sharp to record without distortions with the usual collimation employed in neutron diffraction.

There was one further motivation for this study. Even though the analysis of the diffuse scattering is by far the most complete and direct method of obtaining statistical information on the local atomic arrangements in a crystalline solid, such measurements are very time consuming with commonly available X-ray sources. Therefore, few such experiments have been conducted and there has never been a comparison of results on a single sample measured in different laboratories under different experimental conditions. The recent availability of synchrotron radiation and other very high intensity conventional sources makes it possible to complete diffuse scattering measurements in a matter of hours or at most a day or two, but the experimental conditions are sufficiently different from common practice to warrant careful evaluation of traditional methods and techniques. We have therefore repeated these measurements three times with the same specimen at different X-ray sources in the U.S.A. and Japan in an attempt to address these issues.

2. Experimental procedures

A $NbC_{0.72}$ single crystal was grown by a floating-zone technique in a radio frequency induction-heated furnace at 3650 K [18]. It was slowly cooled to 1400 K and then quenched to room temperature. The main impurities were 200–300 ppm oxygen and 160 ppm Ta. A disc 8 mm in diameter and 3 mm thick was cut and polished so that its faces were close to {111} planes. The lattice parameter was measured and found to be 4.436 Å.

The diffuse X-ray scattering in a volume in reciprocal space was obtained with monochromatic radiation (about 1.5 Å), in absolute units. The measurements were repeated three times, with the following experimental conditions.

(1) The rotating-anode source at Nagoya University, Japan, was operated at 55 kV and 1200 mA. It was equipped with a graphite monochromator. The beam collimation was approximately 0.5°, horizontal and vertical. The specimen environment was air, and a lead baffle was placed in front of the specimen to minimize air scattering.

(2) The bending-magnet line (10A) at the Photon Factory in Tsukuba, Japan, was operated at 2.5 GeV and 60–120 mA for the stored electron beam. An Si(111) elastically bent monochromator was used, producing a beam with divergence of approximately 0.01°×0.5° in and perpendicular to the scattering plane respectively. The specimen was in air.

(3) The six-pole wiggler line (A1) at Cornell University's CHESS facility was operated at 5.5

GeV and 5–15 mA for the stored electron beam. A Ge(111) elastically bent monochromator was used, followed by a flat float glass mirror, producing a beam with divergence of approximately $0.005° \times 0.25°$ in and perpendicular to the scattering plane respectively. The specimen was tested under an evacuated beryllium dome to eliminate air scattering.

After corrections for Compton and parasitic scattering (quite intense in all three cases), the data were placed on an absolute scale by characterizing the incident X-ray flux with integrated intensity measurements of peaks from an aluminum powder briquette and with large-angle scattering from amorphous polystyrene. After all the data manipulations, the agreement was quite good (within a few per cent), considering the widely different characteristics of each X-ray source. See ref. 19 for further details.

3. Analysis of the diffuse scattering

Any deviation from perfect periodicity in a crystal produces diffuse scattering (of X-rays, electrons or neutrons), in addition to the Bragg peaks from the average structure. This scattering can be caused by atomic level variations in composition (clustering or local order) or local deviations from ideal lattice positions, *i.e.* atomic displacements. When up to quadratic terms in an expansion of the phase factor $\exp(i\boldsymbol{k}\cdot\boldsymbol{u})$ due to displacements $\boldsymbol{u}$ are included, the diffuse scattering $I(\boldsymbol{k})$ is given as the sum of three terms:

$$I_{\mathrm{diff}}(\boldsymbol{k}) = I_{\mathrm{Laue}}\{I_{\mathrm{SRO}}(\boldsymbol{k}) + I_{\mathrm{SE}}(\boldsymbol{k}) + I_{\mathrm{TDS+H}}(\boldsymbol{k})\} \qquad (1)$$

where I_{Laue} is the Laue monotonic scattering from a random distribution of the species. The term $I_{\mathrm{SRO}}(\boldsymbol{k})$ is a function of the conditional pair probabilities of finding a B atom at the end of an interatomic vector if there is an A atom at the vector's origin. The second term $I_{\mathrm{SE}}(\boldsymbol{k})$ is a modulation consisting of terms proportional to $\boldsymbol{k}$ which provides information about the preferred interatomic distance of any atomic pair. The third term $I_{\mathrm{TDS+H}}(\boldsymbol{k})$ is due to the variances in these distances because of static (Huang scattering) and dynamic (thermal diffuse scattering). This effect is proportional to $\boldsymbol{k}\cdot\boldsymbol{k}$.

The total diffuse intensity may be written in terms of dimensionless coordinates h_i $(i = 1, 3)$ in reciprocal space and with $\eta = f_{\mathrm{Nb}}/f_{\mathrm{C}}$, where f_{Nb} and f_{C} are atomic scattering factors:

$$\begin{aligned}
\frac{I_{\mathrm{diff}}(h_i)}{I_{\mathrm{Laue}}} = {} & I_{\mathrm{SRO}}(h_i) + h_1 Q_x^{\mathrm{C\text{-}C}}(h_i) + h_2 Q_y^{\mathrm{C\text{-}C}}(h_i) \\
& + h_3 Q_z^{\mathrm{C\text{-}C}}(h_i) + h_1 \eta Q_x^{\mathrm{Nb\text{-}C}}(h_i) \\
& + h_2 \eta Q_y^{\mathrm{Nb\text{-}C}}(h_i) + h_3 \eta Q_z^{\mathrm{Nb\text{-}C}}(h_i) \\
& + h_1^2 \eta^2 R_x^{\mathrm{Nb\text{-}Nb}}(h_i) + h_1^2 \eta R_x^{\mathrm{Nb\text{-}C}}(h_i) \\
& + h_1^2 R_x^{\mathrm{C\text{-}C}}(h_i) + h_2^2 \eta^2 R_y^{\mathrm{Nb\text{-}Nb}}(h_i) \\
& + h_2^2 \eta R_y^{\mathrm{Nb\text{-}C}}(h_i) + h_2^2 R_y^{\mathrm{C\text{-}C}}(h_i) \\
& + h_3^2 \eta^2 R_z^{\mathrm{Nb\text{-}Nb}}(h_i) + h_3^2 \eta R_z^{\mathrm{Nb\text{-}C}}(h_i) \\
& + h_3^2 R_z^{\mathrm{C\text{-}C}}(h_i) + h_1 h_2 \eta^2 S_{xy}^{\mathrm{Nb\text{-}Nb}}(h_i) \\
& + h_1 h_2 \eta S_{xy}^{\mathrm{Nb\text{-}C}}(h_i) + h_1 h_2 S_{xy}^{\mathrm{C\text{-}C}}(h_i) \\
& + h_2 h_3 \eta^2 S_{yz}^{\mathrm{Nb\text{-}Nb}}(h_i) + h_2 h_3 \eta S_{yz}^{\mathrm{Nb\text{-}C}}(h_i) \\
& + h_2 h_3 S_{yz}^{\mathrm{C\text{-}C}}(h_i) + h_3 h_1 \eta^2 S_{zx}^{\mathrm{Nb\text{-}Nb}}(h_i) \\
& + h_3 h_1 \eta S_{zx}^{\mathrm{Nb\text{-}C}}(h_i) + h_3 h_1 S_{zx}^{\mathrm{C\text{-}C}}(h_i) \qquad (2)
\end{aligned}$$

Here

$$I_{\mathrm{Laue}} = N x_{\mathrm{C}}^{\mathrm{V}} x_{\mathrm{C}}^{\mathrm{C}} f_{\mathrm{C}}^{2} \qquad (3)$$

where N is the number of unit cells seen by the incident beam, $x_{\mathrm{i}}^{\mathrm{j}}$ is the sublattice fraction (with subscript indicating the sublattice and superscript the species). With the interatomic vector (la, ma, na),

$$\begin{aligned}
I_{\mathrm{SRO}}(h_i) = {} & \Sigma_l \Sigma_m \Sigma_n\, \alpha_{\mathrm{CC}}^{\mathrm{C\text{-}V}}(lmn) \cos(2\pi l h_1) \\
& \times \cos(2\pi m h_2) \cos(2\pi n h_3) \qquad (4)
\end{aligned}$$

Defining the conditional pair probability $P_{\mathrm{CC}}^{\mathrm{C\text{-}V}}$ of finding a vacancy at the end of an interatomic vector on the carbon sublattice, if there is a carbon atom at the vector's origin,

$$\alpha_{\mathrm{CC}}^{\mathrm{C\text{-}V}}(lmn) = 1 - \frac{P_{\mathrm{CC}}^{\mathrm{C\text{-}V}}(lmn)}{x_{\mathrm{C}}^{\mathrm{V}}} \qquad (5)$$

The short-range order parameter α has values only for $l + m + n = 2p/2$ (with p an integer).

Also, the Q terms involve the fractional displacements X, Y, Z from vectors of the average lattice:

$$\begin{aligned}
Q_x^{\mathrm{Nb\text{-}C}}(h_i) = {} & \Sigma_l \Sigma_m \Sigma_n\, \gamma_x^{\mathrm{Nb\text{-}C}}(lmn) \sin(2\pi l h_1) \\
& \times \cos(2\pi m h_2) \cos(2\pi n h_3) \qquad (6a)
\end{aligned}$$

$$\gamma_x^{\mathrm{Nb\text{-}C}}(lmn) = \frac{4\pi}{x_{\mathrm{C}}^{\mathrm{V}}} \langle X_{\mathrm{NbC}}^{\mathrm{Nb\text{-}C}}(lmn) \rangle \qquad (6b)$$

$$Q_x^{\text{C-C}}(h_i) = \Sigma\Sigma\Sigma\, \gamma_x^{\text{C-C}}(lmn)\sin(2\pi l h_1) \times \cos(2\pi m h_2)\cos(2\pi n h_3) \tag{6c}$$

$$\gamma_x^{\text{C-C}}(lmn) = 2\pi \left\{ \frac{x_{\text{C}}^{\text{C}}}{x_{\text{C}}^{\text{V}}} + \alpha_{\text{CC}}^{\text{C-V}}(lmn) \right\} \times \langle X_{\text{CC}}^{\text{C-C}}(lmn) \rangle \tag{6d}$$

with similar expressions for the Y and Z components, whereas the R and S terms involve variances in the displacements:

$$R_x^{\text{Nb-Nb}}(h_i) = \Sigma\Sigma\Sigma\, \delta_x^{\text{Nb-Nb}}(lmn)\cos(2\pi l h_1) \times \cos(2\pi m h_2)\cos(2\pi n h_3) \tag{7a}$$

$$\delta_x^{\text{Nb-Nb}}(lmn) = \frac{4\pi^2}{x_{\text{C}}^{\text{C}} x_{\text{C}}^{\text{V}}} \langle X_{\text{Nb}}^{\text{Nb}}(000) X_{\text{Nb}}^{\text{Nb}}(lmn) \rangle \tag{7b}$$

For $R_x^{\text{C-C}}(h_i)$,

$$\delta_x^{\text{C-C}}(lmn) = 4\pi^2 \left\{ \frac{x_{\text{C}}^{\text{C}}}{x_{\text{C}}^{\text{V}}} + \alpha_{\text{CC}}^{\text{C-V}}(lmn) \right\} \times \langle X_{\text{C}}^{\text{C}}(000) X_{\text{C}}^{\text{C}}(lmn) \rangle \tag{7c}$$

It should be noted that eqns. (7b) and (7c) differ from zero only for $l+m+n=2p/2$.

For $R_x^{\text{Nb-C}}(lmn)$,

$$\delta_x^{\text{Nb-C}}(lmn) = \frac{8\pi^2}{x_{\text{C}}^{\text{V}}} \langle X_{\text{Nb}}^{\text{Nb}}(000) X_{\text{C}}^{\text{C}}(lmn) \rangle \tag{7d}$$

which has a non-zero value only for $l+m+n=(2p+1)/2$. Finally,

$$S_x^{\text{Nb-Nb}}(h_i) = \Sigma_l \Sigma_m \Sigma_n\, \delta_{xy}^{\text{Nb-Nb}}(lmn)\sin(2\pi l h_1) \times \sin(2\pi m h_2)\cos(2\pi n h_3) \tag{8a}$$

where

$$\varepsilon_{xy}^{\text{Nb-Nb}}(lmn) = \frac{8\pi^2}{x_{\text{C}}^{\text{C}} x_{\text{C}}^{\text{V}}} \langle X_{\text{Nb}}^{\text{Nb}}(000)\, Y_{\text{Nb}}^{\text{Nb}}(lmn) \rangle \tag{8b}$$

and for $S_{xy}^{\text{C-C}}(h_i)$

$$\varepsilon_{xy}^{\text{C-C}}(lmn) = 8\pi^2 \left\{ \frac{x_{\text{C}}^{\text{C}}}{x_{\text{C}}^{\text{V}}} + \alpha_{\text{CC}}^{\text{C-V}}(lmn) \right\} \times \langle X_{\text{C}}^{\text{C}}(000)\, Y_{\text{C}}^{\text{C}}(lmn) \rangle \tag{8c}$$

Both eqn. (8b) and eqn. (8c) differ from zero only for $l+m+n=2p/2$.

Similarly, for $S_{xy}^{\text{Nb-C}}(lmn)$,

$$\varepsilon_{xy}^{\text{Nb-C}}(lmn) = \frac{16\pi^2}{x_{\text{C}}^{\text{V}}} \langle X_{\text{Nb}}^{\text{Nb}}(000)\, Y_{\text{C}}^{\text{C}}(lmn) \rangle \tag{9}$$

which differs from zero for $l+m+n=(2p+1)/2$.

The selection rules for the various coefficients arise from the fact that I_{SRO} has f.c.c. symmetry (involving only the carbon sublattice), whereas the Q, R and S terms have simple cubic symmetry and involve terms between sublattices, *i.e.* their repeat unit is the reciprocal of the smaller simple cube in the NaCl cell which is the proper unit cell when the identity of the species is ignored. Equation (2) results from the fact that the niobium atoms fill one sublattice and that the carbon atoms and vacancies are on the other sublattice.

On inspection of eqn. (2), it can be seen that each of I_{SRO}, Q, R, S is a Fourier series so that, if the total intensity can be separated into these 25 component series, we can obtain the α values and displacements by a Fourier transform. The following strategy is employed to achieve this separation: the intensity is measured at a set of points in reciprocal space related by the symmetries of the various components. Such points can always be found since each of I_{SRO}, Q, R and S is periodic. The separation is then accomplished by the least-squares method [20, 21]; a repeat volume in reciprocal space is filled for each component with the aid of its symmetry and this component is then Fourier transformed. We actually employed the repeat volume for R and S, which is twice the repeat volume for Q and I_{SRO}. This volume was sampled at intervals h of 0.1. For each such point, typically 50 other points were measured for a total of 9000–12 000 intensity measurements for each data set, which were combined for analysis. (Each individual data set was also analyzed individually, but only the results of the combination are discussed here.) A complete solution of eqn. (2) was attempted, as has been routinely done for binary alloys [22], but it was not successful; the I_{SRO} term is less than 1% of the total and, while the separated values had the shape expected from previous studies, the integrated total exceeded the required value of unity (see eqn. (2c) for $lmn=000$). In eqn. (2), in the case of X-rays or electrons, the R and S terms dominate (the Q terms contribute only approximately 10% of the total) and, furthermore, because η is only about 7% of η^2, all the terms involving carbon atoms make a small contribution. Stable solutions for the Q, R and S values were found only when I_{SRO} and the C–C and Nb–C terms in R and S were ignored, reducing eqn. (2) to nine terms. However, even with this simplification, the remaining separated intensity components and, in particular, the Q terms did not have the correct symmetries. This led to non-zero values for the displacements for interatomic vectors when the value should be zero, as mentioned earlier. Adding third-order terms did not improve the situation and, therefore, this method of analysis was abandoned.

An alternative method of analysis can be devised by substituting eqns. (4)–(9) into eqn. (2), *i.e.* expanding eqn. (2) further in the short-range order and displacement coefficients [23]. Each component series must of course be limited to a finite number of coefficients and each coefficient is then obtained by the least-squares method including all available measured intensities. Solutions excluding I_{SRO}, but including 83–113 of all the remaining terms gave reliability indices of 10%–12%. This analysis was performed with each of the three experiments separately.

4. Results

The Fourier coefficients from the least-squares solutions are given in Fig. 2 for the three sets of data. The agreement is good in all cases, and a comparison of the intensity simulation of one set and the measured intensities is presented in Fig. 3. The satisfactory nature of the solutions is apparent, as well as the fact that most of the X-ray scattering is due to the terms R and S, quadratic in the displacements. This must also be the case for the reported electron scattering.

Most of the diffuse scattering is concentrated near the Brillouin zone boundary. The contribution due to thermal effects is therefore small and the atomic displacements must be static in nature. The displacements are summarized in Table 1. Among the coefficients determined in this analysis, $\delta_x^{\text{Nb-Nb}}(000)$ has a special physical meaning. It is related to the mean-square x component of the total displacement of a niobium atom from the average lattice:

$$\delta_x^{\text{Nb-Nb}}(000)=\frac{4\pi^2}{x_C{}^C x_C{}^V}\langle[X_{\text{Nb}}{}^{\text{Nb}}(000)]^2\rangle \qquad (10)$$

where X is the displacement as a fraction of the lattice parameter. The Debye–Waller factor can

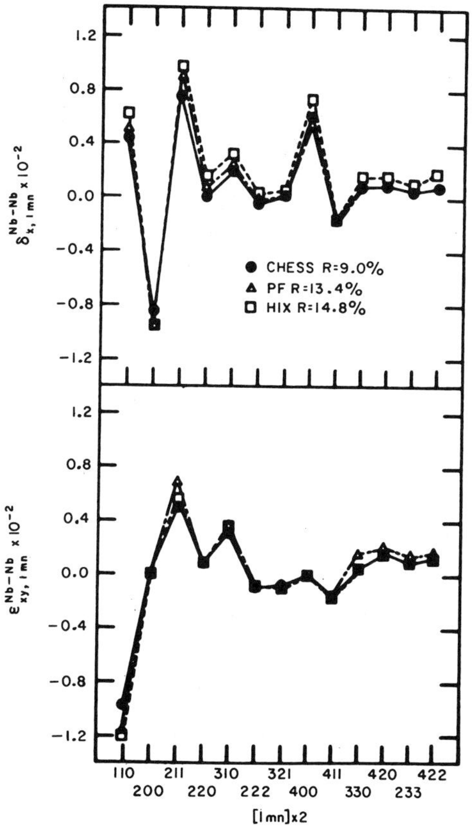

Fig. 2. Comparison of the Fourier coefficients for the three sets of data.

be calculated from this value, as it is $8\pi^2\langle(X^{\text{Nb}})^2\rangle a^2$. Using the values in Table 1, the result is 0.373 Å^2, in good agreement with the value of 0.32(6) Å^2 recently determined from an analysis of integrated intensities of Bragg peaks [3].

5. Discussion

It was found that displacements of the type illustrated in Fig. 4(a) around a carbon vacancy satisfy the following inequalities, which involve the dominant terms in Table 1:

$$\begin{aligned} &\gamma_x^{\text{Nb-C}}(100)<0, && \gamma_x^{\text{C-C}}(200)<0,\\ &\delta_x^{\text{Nb-C}}(100)>0, && \delta_x^{\text{Nb-Nb}}(200)<0,\\ &\delta_x^{\text{Nb-Nb}}(211)>0, && \varepsilon_{xy}^{\text{Nb-C}}(210)>0,\\ &\varepsilon_{xy}^{\text{Nb-Nb}}(110)<0, && \varepsilon_{xy}^{\text{Nb-Nb}}(211)>0, \end{aligned} \qquad (11)$$

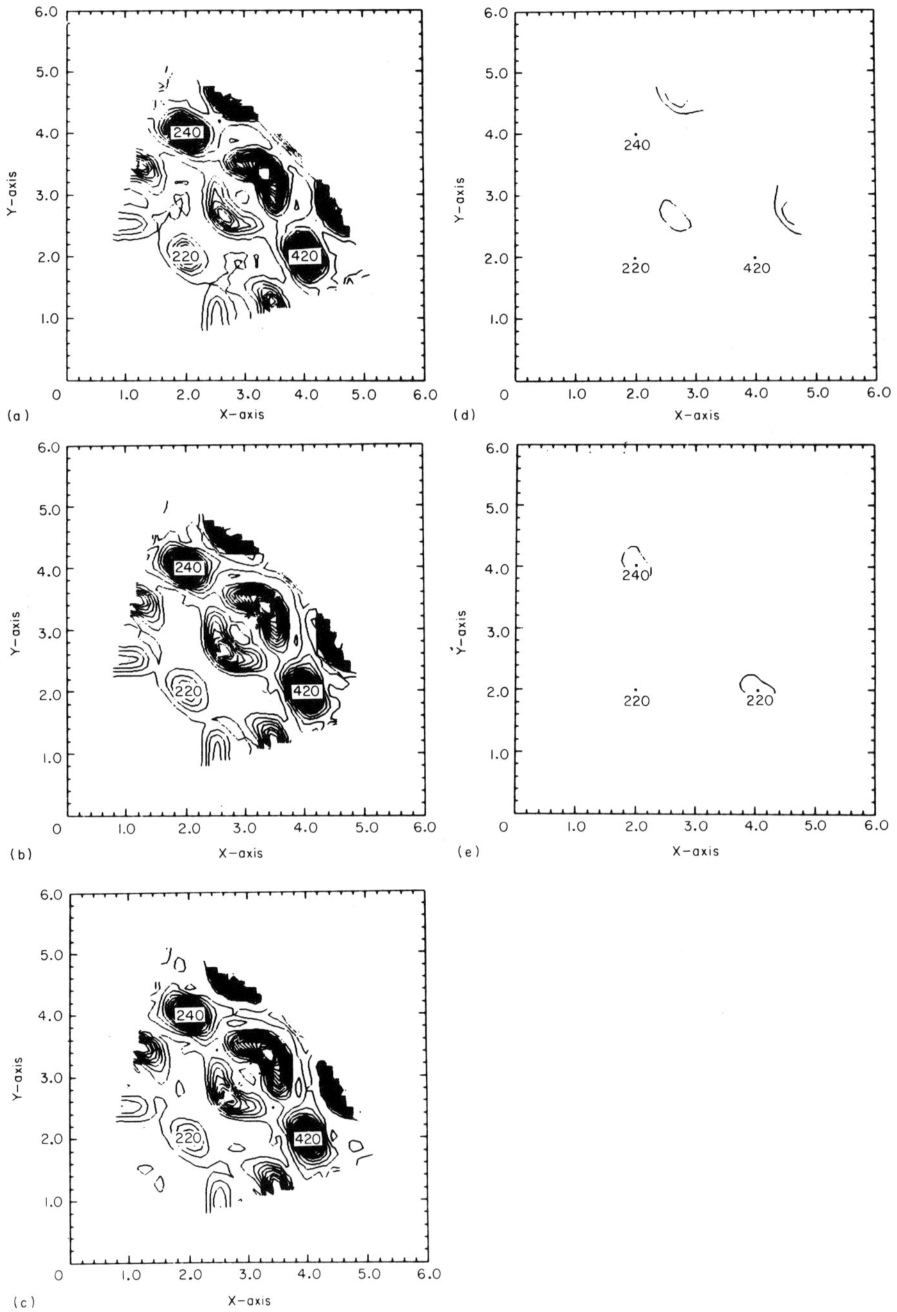

Fig. 3. Comparison of the (a) measured and (b)–(e) synthesized (CHESS) data in the (*hk*0) reciprocal lattice plane; (b) all terms; (c) only $\delta+$(Nb–Nb) terms; (d) only (Nb–C) terms; (e) only $\delta+$(Nb–C) terms. Equi-intensity contours are drawn every 40 Laue units.

TABLE 1 List of displacement parameters for $NbC_{0.72}$

$2x[lmn]$	$\langle X_{C\text{-}C}^{CC}\rangle$	$\langle Y_{C\text{-}C}^{CC}\rangle$	$\langle Z_{C\text{-}C}^{CC}\rangle$	$\langle X_A^{Nb}X_A^{Nb}\rangle$	$\langle Y_A^{Nb}Y_A^{Nb}\rangle$	$\langle Z_A^{Nb}Z_A^{Nb}\rangle$	$\langle X_A^{Nb}Y_A^{Nb}\rangle$	$\langle Y_A^{Nb}Z_A^{Nb}\rangle$	$\langle Z_A^{Nb}X_A^{Nb}\rangle$
Type 1		$(\times 10^{-2})$			$(\times 10^{-4})$			$(\times 10^{-4})$	
000	0	0	0	2.43[a]	2.43	2.43	0	0	0
110	−1.16	−1.16	0	0.21	0.21	−0.28	−0.25	0	0
200	−4.75	0	0	−0.48	0.18	0.18	0	0	0
211	1.02	−0.23	−0.23	0.40	0.09	0.09	0.13	0.00	0.13
220	−0.04	−0.04	0	−0.02	−0.02	0.10	−0.01	0	0
310	0.08	0.03	0	0.10	0.00	0.08	0.07	0	0
222	0.31	0.31	0.31	−0.03	−0.03	−0.03	−0.03	−0.03	−0.03
321	−0.40	−0.46	0.11	−0.01	−0.02	−0.01	−0.02	−0.01	−0.01
400	0.98	0	0	0.28	−0.06	−0.06	0	0	0
411	−0.28	0.12	0.12	−0.08	0.01	0.01	−0.04	0.02	−0.04
330	0.53	0.53	0	0.05	0.05	−0.01	0.01	0	0
420	−0.37	0.17	0	0.06	0.06	−0.00	0.03	0	0
233	0.02	0.24	0.24	0.03	0.03	0.03	0.02	0.02	0.02
422	0.19	0.04	0.04	0.05	0.01	0.01	0.02	0.00	0.02

$2x[lmn]$	$\langle X_{A\text{-}C}^{NbC}\rangle$	$\langle Y_{A\text{-}C}^{NbC}\rangle$	$\langle Z_{A\text{-}C}^{NbC}\rangle$	$\langle X_A^{Nb}X_C^{C}\rangle$	$\langle Y_A^{Nb}Y_C^{C}\rangle$	$\langle Z_A^{Nb}Z_C^{C}\rangle$	$\langle X_A^{Nb}Y_C^{C}\rangle$	$\langle Y_A^{Nb}Z_C^{C}\rangle$	$\langle Z_A^{Nb}Y_C^{C}\rangle$
Type 2		$(\times 10^{-2})$			$(\times 10^{-4})$			$(\times 10^{-4})$	
100	−0.46	0	0	0.71	0.20	0.20	0	0	0
111	0.14	0.14	0.14	0.14	0.14	0.14	−0.08	−0.08	−0.08
210	0.02	0.03	0	0.05	0.03	−0.03	0.18	0	0
300	0.17	0	0	0.20	0.07	0.07	0	0	0
122	−0.03	−0.02	−0.02	−0.03	0.07	0.07	0.03	0.02	0.03
311	−0.06	−0.04	−0.04	−0.02	−0.03	−0.03	0.00	0.07	0.00
320	−0.01	0.00	0	−0.04	−0.04	0.10	−0.05	0	0
410	−0.24	0.01	0	−0.00	0.01	−0.04	0.01	0	0
322	0.02	0.01	0.01	−0.05	−0.04	−0.04	−0.01	−0.04	−0.01
133	0.02	0.01	0.01	−0.05	0.01	0.01	−0.01	0.06	−0.01
421	0.02	0.00	0.01	0.05	−0.06	−0.02	0.01	−0.00	0.01
500	−0.02	0	0	−0.04	−0.02	−0.02	0	0	0
430	−0.01	−0.01	0	−0.01	−0.02	−0.02	0.05	0	0

[a] When $[lmn]$ is [000], $\langle X^{Nb}X^{Nb}\rangle$ becomes $\langle(X^{Nb})^2\rangle$. As the temperature parameter B_{Nb} is expressed by $8\pi^2\langle(X^{Nb})^2\rangle$, it was estimated to be about 0.37 Å^2. This value is comparable with the B_{Nb} value of 0.32 Å^2, determined by the measurements of integrated X-ray intensities of Bragg reflections.

The opposite displacement mode is shown in Fig. 4(b) and clearly does not satisfy the first two relationships. No other model satisfying these inequalities could be devised. The basic features of the proposed model are that the first nearest-neighbor niobium atoms move away from a carbon vacancy but that the second-neighbor carbon atoms move toward this vacancy. Kaufman and Meyer [15] have recently proposed a similar model for the local atomic displacements around a single vacancy (based on ion channeling and neutron diffraction). The displacement of the first niobium neighbors along [100] can be estimated from the results in Table 1 and the relation:

$$\langle X_{Nb}^{\ Nb}(000)X_{Nb}^{\ Nb}(200)\rangle = -x_C^{\ V}(1-x_C^{\ V})\langle(X^{Nb})^2\rangle \qquad (12)$$

which gives a value of 0.07 Å compared with the value of 0.09 Å for $NbC_{0.82}$ reported by Kaufman and Meyer.

Although the dominant coefficients in Table 1 are consistent with the displacement model given in Fig. 4(a), the magnitude of the displacements seems to be affected by the local vacancy arrangements. This observation is based on the fact that the coefficients $\delta_x^{Nb\text{-}Nb}(211)$ and $\delta_x^{Nb\text{-}Nb}(400)$ (which correspond to relatively long interatomic vectors) are large and positive. This could be due to vacancy–vacancy correlations along 211 vectors, because a vacancy causes large niobium displacements, as seen in Fig. 4(a). Intensity simulations (via eqn. (2)) on the $(hk0)$ reciprocal lattice plane, shown in Fig. 5 are helpful in establishing the importance of each coefficient. Simulations with coefficients up to 200 vectors (Fig. 5(b)) fail to describe the basic features of the observed diffuse scattering (Fig. 5(a)), whereas the inclusion of 211 coefficients (Fig. 5(c)) results in dramatic improvement. Furthermore, the diffuse scattering is concentrated near the superlattice reflections of Nb_6C_5. Although it is clear

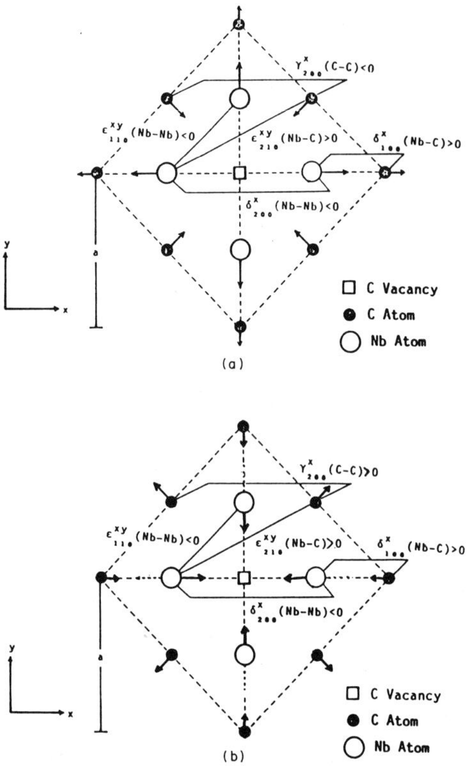

Fig. 4. Models for the atomic displacements: (a) those that satisfy the measured Fourier coefficients; (b) the opposite displacements.

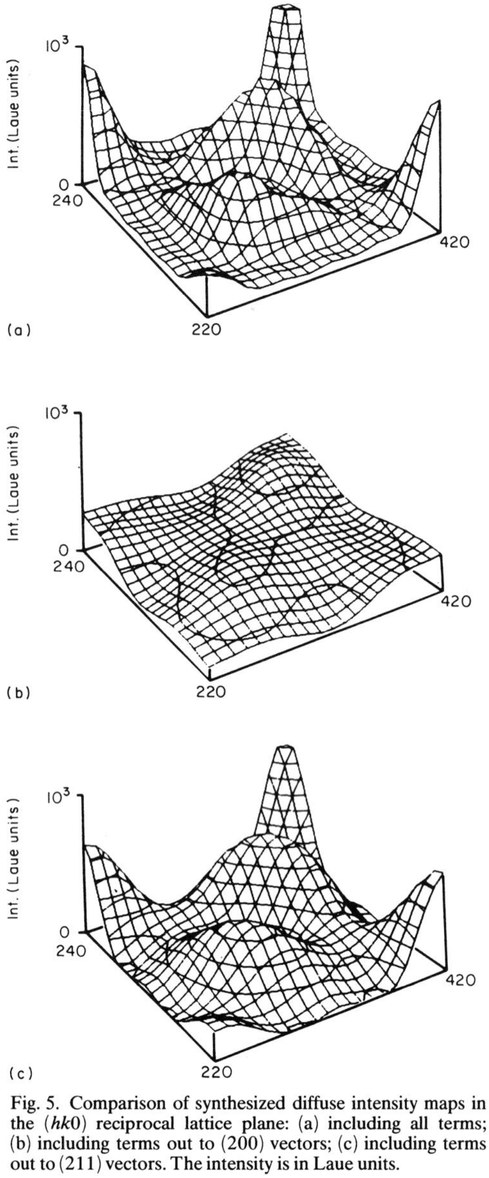

Fig. 5. Comparison of synthesized diffuse intensity maps in the (*hk*0) reciprocal lattice plane: (a) including all terms; (b) including terms out to (200) vectors; (c) including terms out to (211) vectors. The intensity is in Laue units.

that the atomic displacements are influenced by the local vacancy arrangements, the available information on short-range order parameters is simply inadequate at this point.

The electronic structure of NbC has recently been calculated by Aoki *et al.* [24] with the DV-Xα cluster method. Two clusters were examined: Nb_6C_{12} and Nb_6C_{13}, the former with a central vacancy. Their results indicate that the vacancy enhances the electronic bond strength between niobium and carbon in its vicinity. Such an interaction would shorten the Nb–C distance, moving niobium away (and carbon toward) the vacancy. This is what our analysis indicates. However, the presence of vacancies also enhances Nb–Nb bonding, which would shorten the Nb–Nb interatomic distance. The balance between these two forces determines the actual displacements and our results indicate that the Nb–C interaction is stronger.

6. Conclusions

(1) Diffuse X-ray scattering measurements can be reliably performed with synchrotron radiation in a matter of hours and methods of data reduction are well established.

(2) The diffuse X-ray (and electron) scattering that has been observed for NbC_x is largely due to

static displacements and not to local order as had previously been assumed.

(3) Niobium atoms are displaced away from a vacancy, whereas second-neighbor carbon atoms move toward the vacancy.

Acknowledgments

This joint U.S.–Japan study was supported in the U.S.A. by a travel grant from the National Science Foundation (Grant INT-8208308) and by research support from the U.S. Department of Energy (through a subcontract with Purdue University for the Midwest Associated Team for Research with Intense X-rays). For the Japanese investigators, support was provided by the Japan Society for the Promotion of Science. We thank Dr. E. Matsubara for his (considerable) advice and assistance during the measurements at CHESS, and the cooperation of Professor B. W. Batterman and Dr. D. Bildeback of the staff of this storage ring. We also thank Dr. S. Sasaki of the Photon Factory, KEK, and Professor M. Sakata of Nagoya University for their kind support in programming for the diffuse scattering measurements, and Dr. S. Otani of the National Institute for Research on Inorganic Materials for providing the crystal of $NbC_{0.72}$. The Japanese investigators are also grateful to the former directors of the Photon Factory, Professor K. Kohra and Professor T. Sasaki, for promoting this joint project. Finally, we thank Dr. C. H. deNovion for pointing out an early error in our calculations.

References

1 L. E. Toth, *Transition Metal Carbides and Nitrides*, Academic Press, New York, 1971.
2 E. K. Storms, *The Refractory Carbides*, Academic Press, New York, 1967.
3 M. Morinaga, K. Ohshima, J. Harada and S. Otani, *J. Appl. Crystallogr., 19* (1988) in the press.
4 M. Morinaga, K. Sato, A. Aoki, H. Adachi and J. Harada, *Philos. Mag. B, 47* (1983) 107–111.
5 A. L. Gigri, E. G. Szklarz, E. K. Storms, A. L. Bowman and B. T. Mattias, *Phys. Rev., 125* (1962) 837–838.
6 L. E. Toth, C. P. Wang and C. M. Yen, *Acta Metall., 14* (1966) 1403–1408.
7 L. E. Toth and J. Zbasnik, *Acta Metall., 16* (1968) 1177–1182.
8 L. E. Toth, M. Ichikawa and Y. A. Chang, *Acta Metall., 16* (1968) 1183–1187.
9 B. M. Klein, D. A. Papakonstantopoulos and L. L. Boyer, *Phys. Rev. B, 22* (1980) 1946–1966.
10 M. H. Lewis, J. Billingham and P. S. Bell, in E. Thomas, R. M. Fulrath and R. M. Fisher (eds.), *Electron Microscopy and Structure of Materials*, University of California Press, Berkeley, CA, 1972, p. 1084.
11 M. Sauvage and E. Parthè, *Acta Crystallogr., Sect. A, 28* (1972) 607–616.
12 M. Sauvage and E. Parthè, *Acta Crystallogr., Sect. A, 30* (1974) 239–249.
13 J. Billingham, P. S. Bell and M. H. Lewis, *Acta Crystallogr., Sect. A, 28* (1972) 602–606.
14 H. Metzger, J. Peisl and R. Kaufman, *J. Phys. F, 13* (1983) 1103–1113.
15 R. Kaufman and O. Meyer, *Phys. Rev. B, 28* (1983) 6216–6226.
16 W. Weber, *Phys. Rev. B, 8* (1973) 5082–5092.
17 V. Moisy-Maurice, C. H. deNovion, A. N. Christensen and W. Just, *Solid State Commun., 39* (1981) 661–665.
18 S. Otani, T. Tanaka and Y. Ichizawa, *J. Cryst. Growth, 62* (1983) 211–218.
19 K. Ohshima, J. Harada, M. Morinaga, P. Georgopoulos and J. B. Cohen, *J. Appl. Crystallogr., 19* (1986) 188–194.
20 P. Georgopoulos and J. B. Cohen, *J. Phys. (Paris), Colloq. C7, 38* (1977) 191–196.
21 T. B. Wu, E. Matsubara and J. B. Cohen, *J. Appl. Crystallogr., 16* (1983) 407–414.
22 J. B. Cohen, *Solid State Phys., 39* (1986) 131–206.
23 R. O. Williams, *ORNL Rep. 4828*, 1972 (Oak Ridge National Laboratory, Oak Ridge, TN).
24 A. Aoki, K. Sato, M. Morinaga, J. Harada and H. Adachi, to be published.

The Brittle-to-Ductile Transition in Silicon*

J. SAMUELS, S. G. ROBERTS and P. B. HIRSCH

Department of Metallurgy and Science of Materials, University of Oxford, Parks Road, Oxford OX1 3PH (U.K.)

(Received November 9, 1987)

Abstract

In order to study the brittle-to-ductile transition of a strong solid, silicon was chosen as a model material. Single crystals of silicon with semielliptical surface cracks parallel to the (111) cleavage plane, introduced by indentation, were deformed by four-point bending. The brittle-to-ductile transition is very sharp; brittle failure occurs by cleavage below a critical temperature T_c; in the transition region failure also occurs by cleavage, but at a stress intensity considerably greater than that in the low temperature brittle regime. Above the transition region macroscopic yielding of the whole specimen occurs. T_c depends on strain rate and this strain-rate dependence is found to be controlled by the activation energy for dislocation velocity. Dislocation mechanisms have been studied at T_c by the etch-pitting technique; the observations suggest that ductile behaviour is due to the shielding of the crack by dislocations emitted from a few dislocation sources at favourable sites along the crack front. A model has been developed in which dislocation sources are formed at the crack tip, followed by shielding of the whole of the crack by the emission and motion of dislocation loops from these sources. The dynamics of the shielding process has been simulated by a computer program which assumes mode III deformation. The theory explains the observed relation of the strain-rate dependence of T_c with dislocation velocity, accounts reasonably for the sharpness of the transition, and predicts T_c and a size effect (observed). The implications of these results on the criteria which control the brittle or ductile response of solids to mechanical stress are discussed.

1. Introduction

During the last few years there has been an upsurge of interest in the phenomena which control whether crystalline solids containing cracks deform in a brittle or ductile manner, and in the nature of the transition (at some characteristic temperature) from one regime of behaviour to the other. The current state of thinking in this area is described in the 1986 viewpoint set in *Scripta Metallurgica* [1] and in a review by Thomson [2]. Currently, interest is centred on the role of dislocations, either nucleated at the crack tip itself or moving from some nearby or distant sources.

The approach of Rice and Thomson [3], that a material would be ductile if dislocations could be spontaneously emitted from the crack tip at stresses less than the fracture stress, while correctly predicting the ductility of f.c.c. metals, suggests that the activation energy for the brittle-to-ductile transition (BDT) is that for the nucleation of small dislocation loops. However, experiments [4] on silicon showed that the activation energy for the BDT was that for dislocation motion. The exact processes controlling the BDT in silicon, however, were not clear. In this paper, we summarize our experimental work aimed at clarification of the nature of the BDT in silicon and describe a new model for the BDT, based on the dynamic behaviour of dislocations near the cracks.

2. Experimental work

Experiments were performed on differently doped silicon samples at different strain rates (equivalent to different stress intensity increase rates $\dot{K}$) to check whether the activation energy for the BDT would vary with doping in the same way as that for the dislocation velocity [5]. Semicircular cracks (on a {111} plane) were introduced by indentation into the tensile surfaces

*Paper presented at the 3rd International Conference on the Science of Hard Materials, Nassau, The Bahamas, November 9–13, 1987.

0921-5093/88/$3.50 © Elsevier Sequoia/Printed in The Netherlands

of miniature four-point bend specimens (25 mm × 3 mm × 1 mm); the specimens were annealed in vacuum at 800 °C to relieve the residual stresses associated with indentation. The results reported here are for crack radius of 13 μm; some specimens with a crack radius of 37 μm were also tested (see below). The specimen geometry is shown in Fig. 1. It should be noted that slip planes are tangent to the crack front at two points inside the specimen. Specimens were loaded under an inert atmosphere using an Instron machine, at a variety of strain rates and temperatures. Further experimental details are given in refs. 6 and 7.

Figure 2 shows typical results, consisting of fracture or bending stress as a function of temperature for a particular strain rate. The BDT was found to be characterized by a sharp increase in the stress to fracture; the range of temperature over which the transition occurs is less than 5 °C. The sharpness of the transition was emphasized by examination of the fracture surfaces, by etching to reveal dislocation positions. All specimens fractured below the BDT temperature T_c show no evidence of dislocation activity around the precursor crack (at least on the scale (about 1 μm) accessible by this technique). Above T_c, the whole specimen yields and the entire fracture surface (fracture was performed after cooling to room temperature) has a very high dislocation density. Specimens fractured exactly at T_c show clear "trains" of dislocations emanating from a few well-defined points on the precursor crack (in particular those where the crack was tangent to a slip plane) (Fig. 3); these specimens fracture at a higher value of the applied stress intensity factor K than room temperature K_{Ic}. Ohr [8] observed similar trains of dislocations being emitted from crack fronts in thin films of various metals.

Further study showed the following.

(a) The BDT depends on strain rate, giving a linear variation in log(strain rate) with $1/T_c$, enabling activation energies to be derived (Table 1). These activation energies were found to be the same as those for dislocation glide [5].

(b) There is an effect on the BDT from changing the size of the precursor flaw in the test pieces. Even after the effect of flaw size on K for a given stress is accounted for, specimens with larger precursor flaws have higher values of T_c.

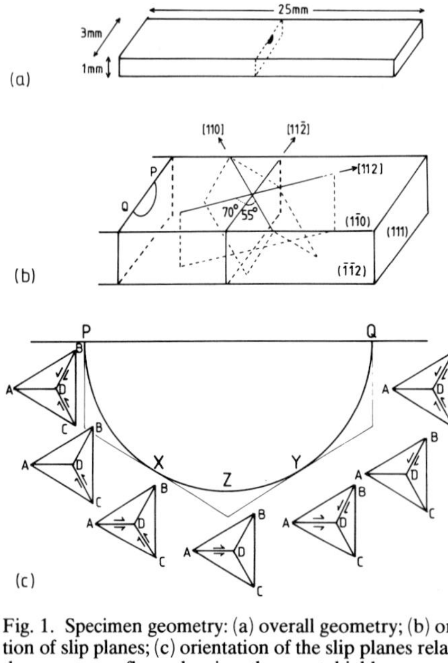

Fig. 1. Specimen geometry: (a) overall geometry; (b) orientation of slip planes; (c) orientation of the slip planes relative to the precursor flaw, showing the most highly stressed slip systems at various points. The apices D of the Thompson's tetrahedra in (c) are above the plane of the paper, and to the left in (b).

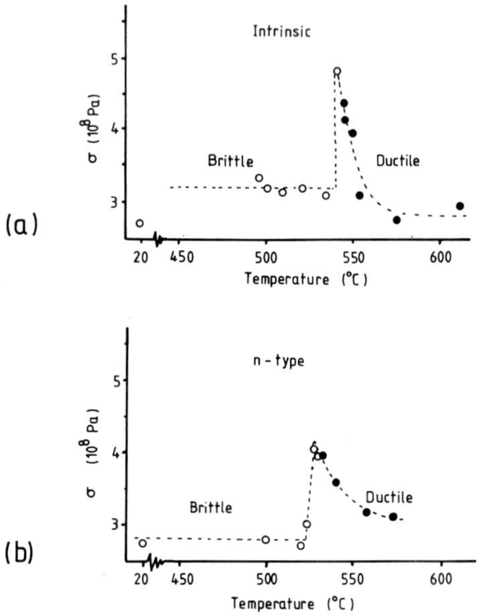

Fig. 2. Typical BDT curves, as stress to fracture or bending as a function of temperature, for (a) intrinsic and (b) n-type silicon. The sharpness of the transition should be noted. T_c for n-type silicon is about 15 °C lower than for intrinsic silicon.

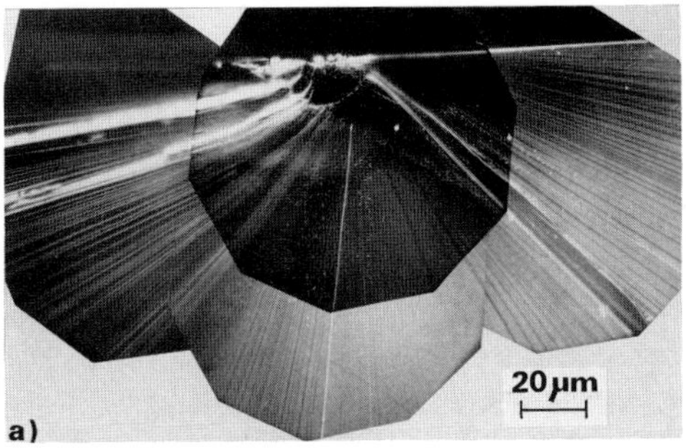

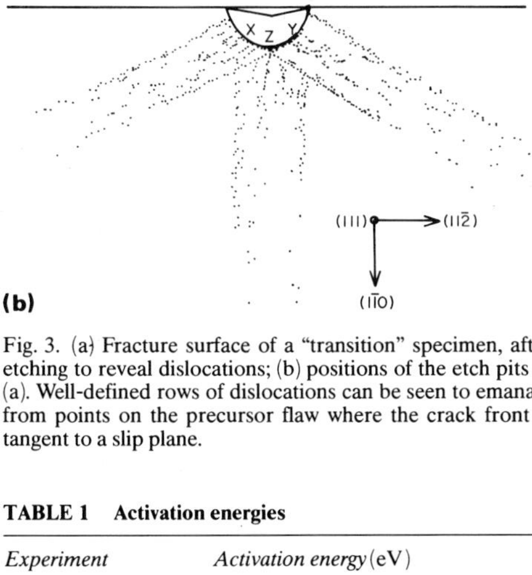

Fig. 3. (a) Fracture surface of a "transition" specimen, after etching to reveal dislocations; (b) positions of the etch pits in (a). Well-defined rows of dislocations can be seen to emanate from points on the precursor flaw where the crack front is tangent to a slip plane.

TABLE 1 Activation energies

Experiment	*Activation energy* (eV)	
	Intrinsic Si (2×10^{13} P cm^{-3})	*n-type Si* (2×10^{18} P cm^{-3})
BDT (Oxford [5])	2.1	1.6
BDT (St. John [4])	1.9	—
Dislocation velocity [6]	2.2	1.7

(c) The results of St. John [4], while showing the same temperature variation as ours, produced BDT temperatures about 100 °C higher.

All these effects are shown in Fig. 4. Obviously, any model for the BDT should be able to account for all these results and should also involve the observed behaviour of the dislocation trains near the cracks.

3. Dynamic modelling of dislocation arrays

Since it was found experimentally that the processes involved in controlling the BDT had pre-

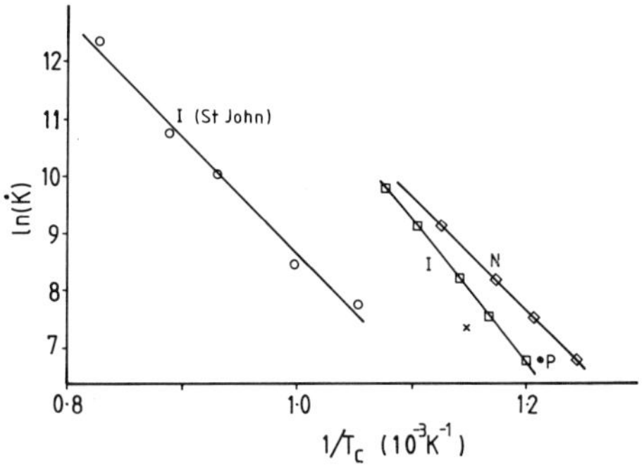

Fig. 4. Variation in T_c with strain rate, shown as an Arrhenius plot. The results for n-type and intrinsic silicon yield activation energies. The results for specimens with a larger precursor flaw X and for p-type silicon P are also shown, as are the results of St. John for intrinsic silicon (using a different crack geometry).

cisely the same activation energies as for dislocation glide, we conclude that it is the dynamic behaviour of dislocations in a configuration that evolves as stress is applied to the crack tip that controls the BDT. If some simplifications are made, this behaviour can be modelled. In particular, the mixed-mode crack is replaced by a mode III crack, emitting straight parallel screw dislocations. The stress on each dislocation (at positions x_i) is then given by the equation

$$\tau_{x_i} = \frac{K}{(2\pi_{x_i})^{1/2}} - \frac{\alpha\mu b}{x_i} - \frac{\mu b}{2\pi}\sum\left(\frac{x_j}{x_i}\right)^{1/2}\frac{1}{x_j - x_i} \quad (1)$$

where the first term is the crack tip stress from the applied stress field, the second term is the image stress and line tension and the third term is the stress from interactions with other dislocations (at positions x_j). Dislocations move with a velocity controlled by stress and temperature according to the equation

$$v = A\tau^m \exp\left(-\frac{U}{kT}\right)$$
$$= v_0\tau^m \quad (2)$$

where m is the stress exponent (which equals approximately 1.2 for silicon [5]), U is the activation energy referred to above and v_0 is the temperature-dependent part of the dislocation velocity function.

The model used is a computer simulation of the emission of dislocations from a crack tip, their subsequent motion and their influence on the stress state at the crack by "shielding" (rather than by blunting). Full details of the computer program used will be published elsewhere. Shielding reduces the stress intensity at the crack tip to an effective stress intensity $K_{\rm eff}$, where

$$K_{\rm eff} = K - \sum_j \frac{\mu b}{(2\pi x_j)^{1/2}} \tag{3}$$

The motion of each dislocation in a small time interval is calculated, and the process repeated, increasing the applied stress appropriately for the time increase. New dislocations are emitted by the crack tip when the net stress at the nucleation distance x_c (usually taken to be equal to b) becomes repulsive. Thus the first dislocation is emitted at $K = K_N$, found by setting the left-hand side of eqn. (1) to zero:

$$K_N = \alpha \mu b \left(\frac{2\pi}{x_c}\right)^{1/2} \tag{4}$$

At each stage, the shielding effect of the dislocations on the crack tip is calculated. It was assumed that fracture would occur if $K_{\rm eff}$ rose above the value of (room temperature) $K_{\rm Ic}$, the "brittle" critical stress intensity. If dislocations have been emitted, the applied stress intensity K will be above $K_{\rm eff}$, and the simulation will correspond to either a "ductile" test ($K_{\rm eff}$ does not reach $K_{\rm Ic}$ before the specimen yields plastically) or a "transition" test ($K_{\rm eff}$ eventually reaches $K_{\rm Ic}$, at $K = K^*$). All modelling discussed below is for intrinsic silicon, at the lowest experimental strain rate.

Combining eqns. (1) and (2), using $K = \dot{K}t$, gives

$$\left(\frac{\dot{K}}{v_0}\right)^{1/m}\left(\frac{{\rm d}x_i}{{\rm d}K}\right)^{1/m} = \frac{K}{(2\pi x_i)^{1/2}} - \frac{\alpha\mu b}{x_i} - \frac{\mu b}{2\pi}\sum\left(\frac{x_j}{x_i}\right)^{1/2}\frac{1}{x_j - x_i} \tag{5}$$

Thus, provided that K_N is independent of temperature (or is a function of $\dot{K}/v_0$), and neglecting the small variation in m with temperature, the dislocation configurations (and hence $K_{\rm eff}$ for a given K) will be identical for a particular value of $\dot{K}/v_0$. Hence the strain rate dependence of the BDT will have the same activation energy as dislocation glide, as is observed experimentally. If there were a nucleation process at the crack tip with a different activation energy than that for v_0 (*e.g.* as postulated in the original BDT model of Rice and Thomson), then the dislocation configuration at a given K would depend on this as well as on $\dot{K}/v_0$, and the activation energies derived from BDT temperature and K measurements would not be those for dislocation motion.

4. Model 1: $K_{\rm eff}$ at source

The model was used to produce graphs of the variation with time in the predicted dislocation positions and in the effective stress intensity $K_{\rm eff}$ at the crack tip. Schematic curves are shown in Fig. 5(a); it can be seen that, as K rises, disloca-

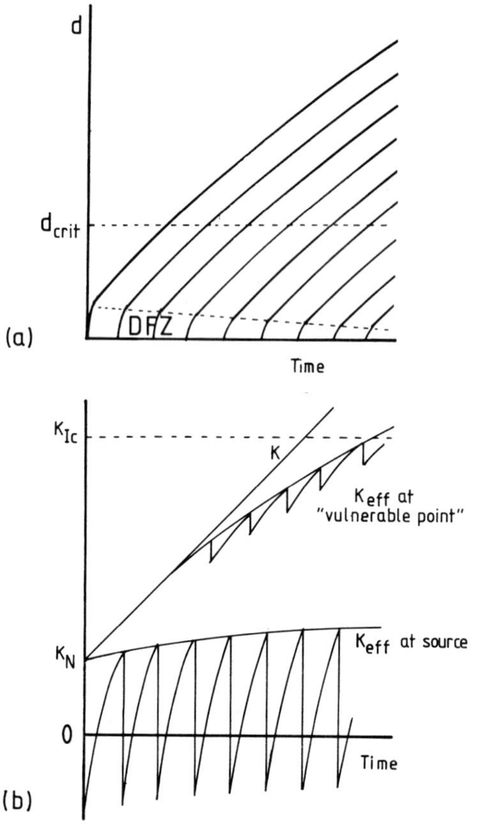

Fig. 5. Simple model for dislocation motion near a crack tip (schematic): (a) positions of dislocations; (b) effective K at the crack tip and at a "vulnerable point" (distance $d_{\rm crit}$). The dips in $K_{\rm eff}$ as each dislocation is emitted and crosses the DFZ should be noted.

tions are emitted and move away in a stable array, which for large numbers of dislocations corresponds roughly to an "inverted pile-up". The interactions between dislocations ensure that all dislocations in the array move under the same total stress (and hence the same velocity), with the exception of the dislocation nearest the crack tip. On emission, this dislocation rapidly crosses a "dislocation-free zone" (DFZ), whose width depends on temperature, where the crack tip stresses are much higher than the repulsive stresses from the dislocation array. Such arrays, and the DFZ, have been observed (in metals) in the transmission electron microscope [8] and more recently in silicon [9].

The stresses from the array of dislocations produce a K_{eff} at the crack tip which is below the applied K (see Fig. 5(b)). Each time that a dislocation is nucleated, K_{eff} drops sharply and then rises to the "base level" value as the dislocation crosses the DFZ, whereupon the process is repeated. The slowly varying trend in K_{eff} was used to predict ductile and brittle behaviour, as described above.

This model reproduces reasonably well the observed dislocation arrays and gives values of the transition temperatures reasonably close to those observed. However, the transitions predicted by this model are "soft", *i.e.* the BDT occurs over a wide temperature range, in two related ways: (a) much dislocation activity is predicted even in specimens which are predicted to fail eventually in a brittle manner; (b) the temperature range for which fracture occurs at a higher K than the room temperature K_{Ic} is similarly large. Thus the observed very sharp transition range (less than 5 °C on both the above criteria) is not reproduced by this version of the model.

5. Model 2: K_{eff} at a sensitive point on the crack

The model was then modified to take account of the fact that the point on the real curved crack front most vulnerable to fracture (the deepest part of the crack; see Z in Fig. 1(c)) is not where the dislocations are emitted (points where the slip plane is tangent to the crack front; see X and Y in Fig. 1(c); see also Fig. 3). Dislocations must travel some distance related to the crack size (d_{crit} in Fig. 5(a)), before they can act to shield the whole crack tip. This was simulated (roughly) in the model by requiring that only the dislocations beyond d_{crit} (typically 5–10 μm) have their shielding effect taken into account. Figure 5(b) shows K_{eff} at this vulnerable point, to be compared with K_{eff} at the dislocation source. This produces rather better simulation of the sharpness of the BDT in terms of K_{eff} and accounts to some extent for the "size effect" on the BDT but still predicts that long dislocation arrays should be seen even at temperatures considerably below the BDT; experimentally, this is not found. Although the model can only be an approximation to the real situation, no variation in parameters for dislocation nucleation (*e.g.* changing α or x_c, or requiring a non-zero nucleation stress) or motion (*e.g.* including a "work-hardening" term) could reproduce this sharp transition in dislocation behaviour. It was therefore thought that a sharp nucleation of the dislocation trains must occur at, or just below, K_{Ic}.

6. Experimental investigation of the nucleation event

Further experiments were carried out, holding specimens at temperatures very close to those for the BDT. Two test cycles were used. In both cases, the test temperature was 550 °C at a strain rate of $\dot{K} = 886$ N m$^{3/2}$ s^{-1}, *i.e.* within the transition range. One specimen was loaded to $0.91K_{Ic}$, unloaded, cooled to room temperature and fractured. It fractured at the normal K_{Ic} value and, on etching the fracture face, no arrays of dislocations were visible. The second specimen was loaded to $1.11K_{Ic}$, unloaded, cooled and fractured. This specimen fractured at $1.13K_{Ic}$, and etching of the fracture faces revealed well-developed arrays of dislocations. Thus, some nucleation event is operating at a value of K very close to K_{Ic}. (It should also be noted that the second experiment is an example of the "warm-prestressing" effect in silicon, whereby high temperature stressing can improve the low temperature fracture toughness; this is discussed further later in the paper.)

7. Model 3: with a nucleation event

Earlier versions of the computer model allowed dislocation nucleation to start at the "natural" stress intensity level K_N, defined in eqn. (4) and assuming that nucleation occurs at a critical distance x_c (normally taken to be equal to b, giving $K_N = 0.24$ MN m$^{3/2}$, equal to that observed experimentally by Michot and George [10]). In order to accommodate a nucleation event, dislocations in the computer model were "released" only above a level of K (K_0) between K_N and K_{Ic}; the other parameters controlling dislocation motion were left unchanged. Typical results are shown in Fig. 6. At low K_0, the length d of the dislocation array and the value of K^* increase slowly with temperature; as K_0 is increased, these both, initially, change over a somewhat wider temperature range, as the sudden "avalanche" of dislocations at K_0 causes K_{eff} to diverge more rapidly from K. As K_0 approaches K_{Ic}, however, K^* and d are predicted to change suddenly from one value to another at T_c.

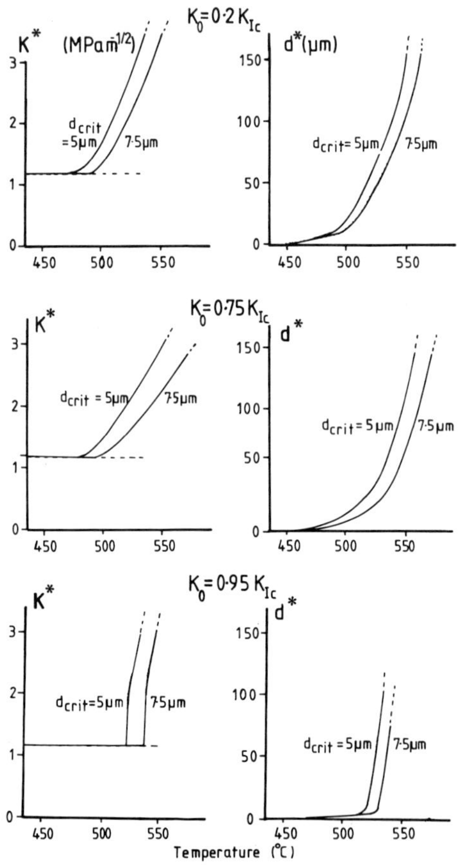

Fig. 6. Effect of changing K_0 on the predictions of the model; curves of d^* (the length of the dislocation array at fracture) and K^*, for two values of d_{crit} (5 and 7.5 μm). At low values of K_0, K^* and d^* increase at T_c over a wide temperature range. At intermediate values of K_0, the transition is broadened further. For K_0 just less than K_{Ic}, the transition involves a sudden initial jump in K^* and d^*.

Figure 7 shows the results of the simulation for $K_0 = 0.95K_{Ic}$. Below T_c, the dislocations would pass the critical distance for shielding the whole crack front only after K has exceeded K_{Ic}. Above T_c there is an initial "avalanche" of fast-moving dislocations, causing K_{eff}, both at the source and at the "sensitive points", to drop very rapidly. Precisely at T_c, there is still a small drop in K_{eff}, after which K_{eff} rises slowly to the level of K_{Ic}. The effect of altering the test temperature from just below to just above T_c should not be considered. Just below T_c, the "vulnerable point" will not be shielded before K (and K_{eff} at this point) reaches K_{Ic}; pure brittle fracture is therefore predicted. At a temperature just above T_c, the initial peak in K_{eff} will not quite reach K_{Ic}, but failure is predicted to occur some time later, with K^* considerably higher than the room temperature K_{Ic} and with considerable dislocation activity. At T_c, therefore, a sudden jump between the

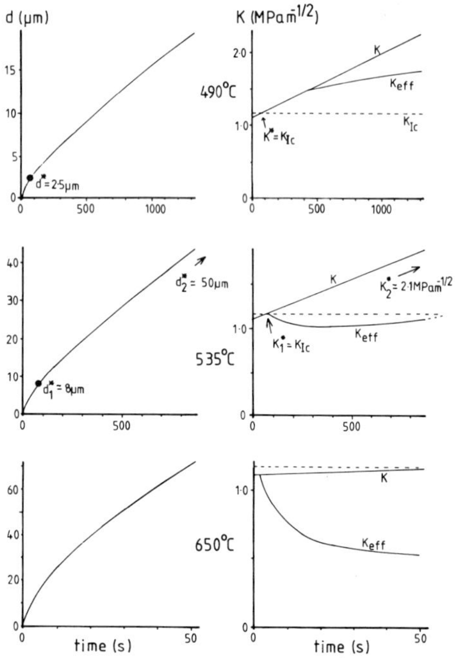

Fig. 7. Characteristics of the K_{eff} curves predicted by the model for high K_0. (a) Below T_c: K exceeds K_{Ic} before dislocations shield the whole crack profile. (b) At T_c: as K reaches K_{Ic}, dislocations shield the whole crack tip, and K_{eff} at the most vulnerable point drops. A small change in temperature from just below to just above T_c changes K^* from $K_1{}^*$ to $K_2{}^*$ and d^* from $d_1{}^*$ to $d_2{}^*$. (c) Above T_c: K_{eff} drops rapidly before K reaches K_{Ic}.

two values of K^* is predicted; such a jump is observed experimentally (Fig. 2(a)). It should be noted that the value of T_c predicted by the model using the "standard" parameters is very close to that observed experimentally.

The existence and magnitude of the "warm pre-stressing effect" referred to earlier are also reproduced by this model. If a test at the transition temperature is stopped before K reaches K^*, then $K_{\text{shielding}}$ is slightly greater than $K-K_{\text{Ic}}$, as the curve of K_{eff} is almost flat. Thus, if this dislocation configuration (and the associated shielding effect) is "frozen in" by cooling to room temperature, room temperature fracture would be expected to occur at a value of K very slightly above that used in the pre-stressing. This is what is observed experimentally.

Thus this type of model, based on the dynamic behaviour of dislocations, can reproduce the observed sharp BDT if there is a nucleation event causing dislocation sources to become active at a considerably higher stress intensity level than that needed for their subsequent operation. A possible mechanism for this is discussed below.

8. Formation of the crack tip dislocation sources

This nucleation event was modelled by assuming that the dislocation sources at the crack tips arise from pre-existing dislocations which move to the crack tip, cross slip and then operate to generate the dislocation trains. The obvious origin for such dislocations is the plastic zone of the indentation used to generate the precursor crack. Such sources can be generated by multiple cross-slip, allowing the active screw dislocations to follow the crack profile [11]. The mechanism is shown in Fig. 8. Analysis of the stress fields around the crack front [12] shows that the stresses on the primary and secondary cross-slip systems are consistent with the proposed mechanism. It should be noted that the loop crosses the plane of the crack as it expands towards A in Fig. 8 but is confined to one side of the crack in its extension towards the surface of the specimen (*i.e.* towards B in Fig. 8). This fits well with the observed etch pit patterns, where the dislocations are seen to run extensively into the bulk of the specimen but not towards the specimen surface (see Fig. 3).

The delayed nucleation at K_0 then occurs because of the time taken for the active dislocations to cross from the plastic zone to the crack tip. This time can be estimated [13] as follows:

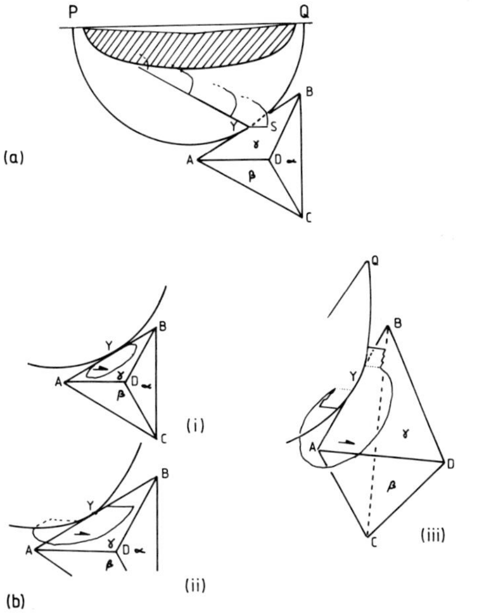

Fig. 8. (a) Cross-slip mechanism for the formation of a dislocation source at the crack tip, from a dislocation gliding from the indentation plastic zone; (b) as the loop expands from the source at Y on the primary glide plane (γ), the ends cross slip (on the β plane) to follow the crack profile.

$$v=\frac{\mathrm{d}r}{\mathrm{d}t}=\frac{fK^m}{(8\pi r)^{1/2}}\,v_0 \tag{6}$$

where f is an orientation factor and r is the distance from the crack tip. Integration of this expression (between K_d, at which dislocations start to move to the source, formed at K_0) leads to

$$K=\frac{v_0}{(8\pi)^{m/2}}\,\frac{f^m}{r_0^{\,1+m/2}}\,\frac{m+2}{2(m+1)}\left(K_0^{\,m+1}-K_d^{\,m+1}\right) \tag{7}$$

It should be noted that this criterion for nucleation of the dislocation source also has an activation energy equal to that for dislocation motion (since, neglecting the small temperature dependence of m, the temperature dependence is controlled by that of v_0).

If it is assumed that the sources are generated at $K=K_{\text{Ic}}$ and that K_d, the stress intensity at which the dislocation starts to move, is small, K_0 can be predicted. Hence the BDT temperatures can be predicted. Table 2 shows predicted temperatures, assuming $m=1$ and $f=\frac{1}{4}$. These pre-

TABLE 2 Transition temperatures (intrinsic silicon)

Crack size (μm)	*Predicted* T_c (°C)	*Experimental* T_c (°C)
10	≈575	550
30	≈640	600

dicted temperatures form an internally consistent set in good agreement with those observed. It can be seen that use of larger precursor cracks gives a higher BDT as it requires a longer time (and thus a higher applied K) to form the crack tip dislocation sources.

It also should be noted that this mechanism for formation of the dislocation source further enhances the sharpness of the BDT. At temperatures lower than the BDT, not only will dislocations from the sources take longer to pass the critical distance for shielding, but also the sources will take longer to form. Above the BDT, sources will form more rapidly, and dislocations will pass the critical distance more quickly.

Application of this model to St. John's results (see Fig. 4) gives an r_0 value of 60–450 μm (depending on the value of U chosen) [13]. The former value corresponds to a bulk source density of 10^6 cm^{-3}, consistent with the density of "swirl defects" in such material; the latter value would suggest operation of surface dislocation sources, as the specimens used in these experiments were about 1 mm thick.

9. Conclusions

Experimental work shows that the BDT in silicon is controlled by processes with the same activation energy as for dislocation motion. Modelling of the BDT based on the shielding effect of a train of moving dislocations indicates that the very sharp transition observed experimentally is caused by dislocation nucleation at the crack tip at a value of K (K_0) very close to K_{Ic}, with K_0 greater than K_N, the stress intensity needed to operate the dislocation source. This may be caused by motion of pre-existing dislocations to the crack tip, followed by cross-slip. A model incorporating this nucleation event reproduces well the experimentally observed behaviour, including the "warm-prestressing effect".

Acknowledgments

The work reported here was carried out with the support of the Venture Research Unit of BP plc and the Science and Engineering Research Council (S.G.R., Postdoctoral Fellowship and Advanced Fellowship; J.S., Quota Studentship).

References

1 *Scr. Metall., 20* (1986) 1465, 1467, 1473, 1477, 1483, 1489, 1495, 1501.
2 R. Thomson, *Solid State Phys., 39* (1986) 1.
3 J. R. Rice and R. Thomson, *Philos. Mag., 29* (1974) 73.
4 C. St. John, *Philos. Mag., 32* (1975) 1193.
5 A. George and G. Champier, *Phys. Status Solidi A, 53* (1979) 529.
6 J. Samuels and S. G. Roberts, *Proc. Brit. Ceram. Soc., 37* (1986) 267.
7 J. Samuels, *D.Phil. Thesis,* University of Oxford, 1987.
8 S. M. Ohr, *Mater. Sci. Eng., 72* (1985) 1.
9 Y. H. Chaio and D. R. Clarke, *Acta Metall.,* in the press.
10 G. Michot and A. George, *Scr. Metall., 20* (1986) 1495.
11 P. B. Hirsch, S. G. Roberts and J. Samuels, *Scr. Metall.,* to be published.
12 P. B. Hirsch, S. G. Roberts and J. Samuels, *Proc. R. Soc. London,* in the press.
13 P. B. Hirsch, S. G. Roberts and J. Samuels, *Proc. Conf. on Mechanisms and Mechanics of Plasticity, Aussois, 1987,* to be published.

Self-diffusion of Silicon-30 and Nitrogen-15 in α Phase Silicon Nitride*

KIRSTEN P. KUNZ

Department of Materials Science and Engineering, North Carolina State University, Box 7907, Raleigh, NC 27695-7907 (U.S.A.)

VINOD K. SARIN

GTE Laboratories, 40 Sylvan Road, Waltham, MA 02254 (U.S.A.)

ROBERT F. DAVIS

Department of Materials Science and Engineering, North Carolina State University, Box 7907, Raleigh, NC 27695-7907 (U.S.A.)

SCOTT R. BRYAN

Standard Oil Engineered Materials Co., 4440 Warrensville Center Road, Cleveland, OH 44128 (U.S.A.)

(Received November 8, 1987)

Abstract

Several methods for determining self-diffusion rates in α phase silicon nitride were explored. Techniques for creating solid–solid diffusion couples are described. The results of diffusion anneals in these samples are reported, together with results obtained from samples which were ion implanted with silicon-30.

The particular material selected for this study was a large-grained pyrolytically deposited α phase silicon nitride with no secondary grain boundary phase. Results of X-ray diffraction, transmission electron microscopy, scanning electron microscopy and secondary ion mass spectrometry (SIMS) analyses are reported. Isotopic diffusion profiles were analyzed by SIMS, where the issues of sample charging and depth profiling have been addressed.

1. Introduction

Silicon nitride, in several forms, is a promising refractory material. Occurring in α and β forms, grown or sintered with additives or secondary phases, the result is a wide variety of materials containing a range of grain sizes and grain boundary phases. These materials are candidates for high temperature applications such as turbine engines, bearings and corrosion-resistant refractories. In its hot-pressed form, silicon nitride exhibits good strength. The low coefficient of thermal expansion combined with the high thermal conductivity of silicon nitride give it outstanding thermal shock resistance as well. It is primarily for these reasons that it is an important engineering material. As such, a fundamental understanding of diffusion processes in this material is essential on several levels.

The nitridation of silicon is the usual procedure for the manufacture of silicon nitride powders. The reaction kinetics involved are not well understood. The diffusion rates of both silicon and nitrogen in the growing silicon nitride lattice must be known if the process is to be more fully understood; lattice diffusion may well be the rate-determining step [1].

Sintering of silicon nitride powders is crucial in many low cost forming techniques. It cannot be achieved, however, without the addition of sintering aids. The reasons lie in the covalent bonding of the lattice and the inherently slow lattice diffusion of both silicon and nitrogen. Thus, diffusion coefficients are necessary for better understanding and control [2].

Creep observed in silicon nitride materials also shows some dependence on simple lattice diffusion. For a more complete picture of these processes, data on lattice self-diffusion which are not currently available must be collected to determine the basic mechanisms involved.

*Paper presented at the 3rd International Conference on the Science of Hard Materials, Nassau, The Bahamas, November 9–13, 1987.

0921-5093/88/$3.50 © Elsevier Sequoia/Printed in The Netherlands

2. Past work

To date, self-diffusion studies on silicon nitride have produced discrepancies in calculated diffusion coefficients and activation energies owing to four major types of problem: (1) obtaining suitable silicon nitride samples for study, (2) providing a source of silicon and/or nitrogen isotopes as tracers, (3) finding a means of analyzing tracer profiles after diffusion anneals and (4) preventing sample decomposition during the anneals. The results of these diffusion studies are listed in Table 1.

The first issue was to choose a silicon nitride material for study. The available choices typically contained small grains (0.1–10 μm) of α and/or β phase silicon nitride with a large fraction of secondary phase such as magnesia, silicon oxynitride or elemental silicon. Diffusion measured in these materials represents a mixture of grain boundary and lattice diffusion. Cooper and Major [3], who used a hot-pressed material containing β phase silicon nitride and magnesia, were the only workers to attempt to separate the effects of lattice and grain boundary diffusion. This was done using the theoretical models of Fisher [4] and Suzuoka [5].

The second problem in self-diffusion studies of silicon nitride is providing a source of isotopes. Both silicon and nitrogen have stable isotopes: nitrogen-15 with a natural abundance of 0.36%, and silicon-29 and silicon-30 with natural abundances of 4.67% and 3.10% respectively. However, producing solid–solid silicon nitride diffusion couples requires a means of making isotopic silicon nitride and bonding it to the material for study. Batha and Whitney's [6] study, which involved interpreting thermal decomposition data, did not require isotopic tracers. Kijima and Shirasaki [7] and Cooper and Major [3] used a gas source of nitrogen-15 which is convenient but raises questions concerning phase boundary reaction rates. Wuensch and Vasilos [8] were the first to produce solid isotopic silicon nitride diffusion couples.

Sample decomposition during diffusion anneals represents the third major concern in these studies. With the exception of Batha and Whitney's [6] study (where thermal decomposition was the point of interest), investigators have confined their anneals to the temperature range 1200–1450 °C, where decomposition over the course of the diffusion annealing time was not an issue. However, diffusion data in the temperature range 1450–1800 °C are necessary for a better understanding of the kinetics associated with high temperature processes. Up to this point, diffusion coefficients in this range could only be extrapolated from existing data.

The fourth point mentioned was the analysis of diffusion profiles in annealed samples. The studies by Wuensch and Vasilos [8] and Cooper and Major [3] were the only studies based on measuring these profiles. Wuensch and Vasilos [8] proposed the use of secondary ion mass spectrometry (SIMS) to measure both silicon and nitrogen diffusion profiles. This was unsuccessful owing to charge build-up on the insulating samples. They used a serial sectioning technique as an

TABLE 1 Available self-diffusion data for silicon nitride

Temperature range (°C)	D_{Si} ($cm^2\ s^{-1}$)	D_N ($cm^2\ s^{-1}$)	*Q* ($kcal\ mol^{-1}$)	*Method*	Si_3N_4 *type*	*Reference*
1490–1750	—	$10^{10}\exp(-186000 \pm 10000/RT)$	N: 186 ± 10 Si: 93.0 ± 6.6	Thermal decomposition	β with some α, elemental Si	[6]
1200–1410	—	$1.2 \times 10^{-12}\exp(-55700/RT)$	N: 55.7	Isotope exchange gas source	α	[7]
1200–1410	—	$6.8 \times 10^{6}\exp(-185700/RT)$	N: 185.7	Isotope exchange gas source	β	[7]
1233–1450	—	$2.02 \times 10^{-11}\exp(-317000/RT)$ in grains $4.09\exp(-509000/RT)$ in grain boundaries	N: 31.7 in grains N: 50.9 in grain boundaries	Gas source nuclear microanalysis	Hot-pressed β, some Si_2N_2O and MgO glassy phase	[3]
1400	1.5×10^{-13}	—	—	Solid tracer spectrometry analysis	β	[8]

alternative. Silicon and nitrogen were microchemically separated from each sequential grinding and analyzed for isotopic ratios by mass spectrometry techniques. Owing to carbon contamination during sectioning, nitrogen isotopic profiles were lost. The diffusion profile for silicon and calculated diffusion coefficient were reported at a single temperature of 1400 °C.

The experimental procedures explored by the present authors were aimed at overcoming the problems encountered by past investigators so as to allow the clear delineation of lattice and grain boundary diffusion rates in the temperature range 1600–1800 °C.

3. Experimental details

α phase silicon nitride was deposited on a graphite susceptor by pyrolytic reaction of silicon tetrafluoride, ammonia and nitrogen. The purely α phase silicon nitride product contains large columnar grains with a preferred orientation along the (222) direction, as identified by X-ray diffraction. Scanning electron micrographs taken of thermally etched samples (Figs. 1(a) and 1(b)) were used to determine an average equivalent spherical diameter of 18.02 μm in the plane normal to diffusion. Column lengths are in excess of the expected diffusion zone. Transmission electron microscopy (TEM) shows clean sharp grain boundaries (Fig. 2). Energy-dispersive X-ray analysis (EDXA) and SIMS show minor impurities of copper and tin. The metal impurities are assumed to be a result of fluorine-induced transport from the walls of the reactor during deposition and appear in the form of small precipitates seen in transmission electron micrographs (Fig. 3). After the samples have been annealed for 3 days at 1800 °C, these precipitates are no longer visible, and the presence of copper or tin cannot be detected by EDXA or SIMS. Annealing did not result in any α-to-β-phase transition, and grain boundaries appear unchanged.

This large-grained material is well suited to the isolation of lattice diffusion coefficients alone. There is still a contribution to diffusion profiles from grain boundary transport to be considered. However, this effect can be separated in a theoretical treatment when the contribution is not overwhelming. The Suzuoka [5] model can be used in a treatment which follows from the work of Cooper and Major [3].

Fig. 1. Scanning electron micrographs of the microstructures (a) perpendicular to diffusion and (b) parallel to diffusion.

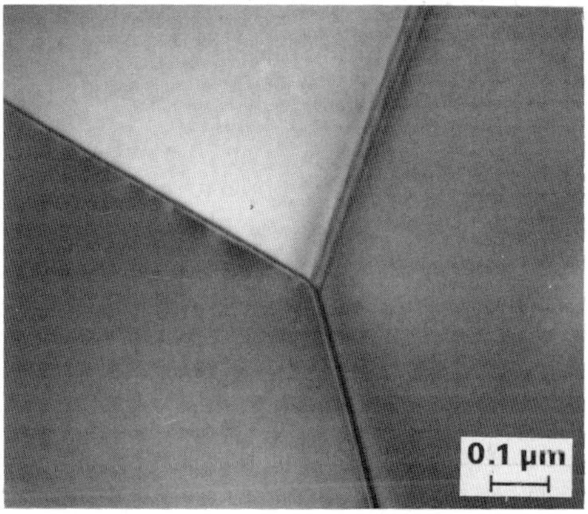

Fig. 2. Transmission electron micrograph of silicon nitride grain boundaries.

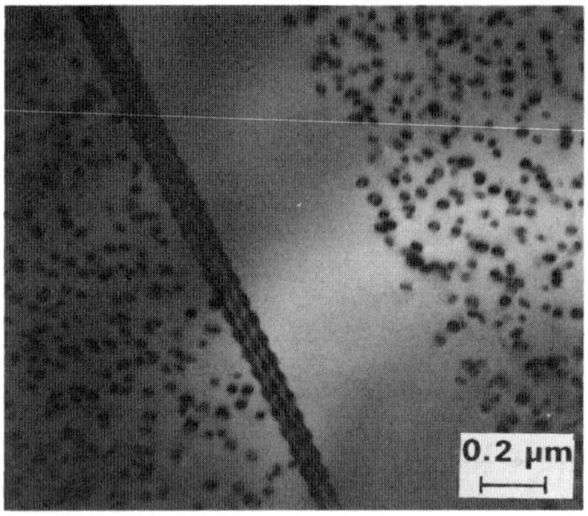

Fig. 3. Transmission electron micrograph of precipitates in silicon nitride.

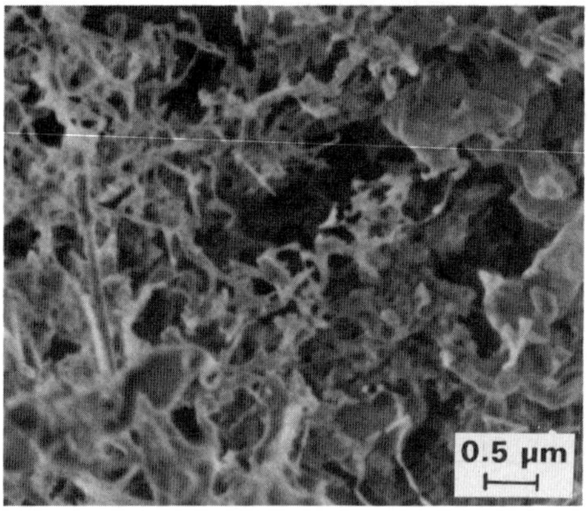

Fig. 4. Scanning electron micrograph of isotopic nitridation product.

Creating a solid isotopic diffusion couple with this material involved the nitridation of silicon-30 with nitrogen-15 and bonding the isotopic product to the pyrolytically deposited (α phase) silicon nitride (PDSN) samples. Nitridation of small amounts of silicon-30 was carried out in a high purity alumina tube closed at one end and attached to a vacuum pump. The nitrogen-14 and nitrogen-15 sources were at the other end. The alumina tube was placed in a tube furnace for heating. The system was pumped to a base pressure of 5×10^{-4} Torr at 300 °C and backfilled with nitrogen-14 five times to remove residual oxygen and water vapor. It was then heated to 1200 °C, at which temperature the nitrogen-14 was evacuated and replaced with nitrogen-15 at 1×10^{3} Torr. The temperature was raised 50 °C every 3 h until 1400 °C was reached. The slow rise in temperature was used to keep the highly exothermic reaction from melting the remaining silicon and lowering the surface area available for reaction. The temperature was held at 1400 °C for 14 h and then slowly cooled to room temperature. The product of this reaction was a mixture of α and β phase silicon nitride with no remaining free silicon, as seen in X-ray diffraction. Scanning electron micrographs (Fig. 4) show the needle-like α and platelet-type β mixture.

Several techniques for bonding the solid isotopic powder to the PDSN material were investigated. The first involved a variation of the traveling-solvent method (TSM) used by Davis *et al.* [9] in similar silicon carbide studies. The technique was introduced by Pfann [10] in 1955. The basic concept of TSM requires a molten metal solvent which will wet the material to be bonded and dissolve it but which is not soluble itself in the solid material. A further requirement is that it must not form, in this case, a metal silicide or nitride. It has been found that gadolinium metal fulfils these requirements.

Isotopic silicon nitride powder was painted on a PDSN "source" crystal by suspending it in glycerol and applying the suspension with a very fine camel-hair brush. The glycerol was allowed to evaporate. The source crystal and the "seed" crystal (that which is to become the diffusion couple) were then coated with 5 μm of high purity gadolinium by thermal evaporation. The two samples were then arranged in a sandwich configuration, source on the bottom, gadolinium in the middle, and seed on the top, and placed in a silicon-carbide-coated graphite boat for heating. The boat is heated by electrical resistance to a temperature such that the source–gadolinium interface reached 1328 °C, 15 °C above the melting point of gadolinium. The molten gadolinium will wet the isotopic silicon nitride and begin to dissolve it. To effect the transfer, the source–gadolinium interface was heated slowly to 1428 °C. Across the 10 μm metal layer, an 85–100 °C temperature drop was observed (by optical pyrometry). The temperature gradient results in mass transport: dissolution at the hotter interface and precipitation from a supersaturated solution at the cooler interface. After the assembly had been cooled to room temperature, the gadolinium was removed by etching for 2 h in concentrated HCl. EDXA showed that low levels of gadolinium (1–3 wt.%) remain in the deposited tracer. The gadolinium did not diffuse during the diffusion anneals, as confirmed by SIMS, and was judged to be innocuous. EDXA showed that there were no other metal impurity contents greater than 0.1 wt.% in the tracer layer. The isotopic layer deposited on the seed in this manner was approximately 2.5 μm thick. It was less distinctly crystalline but well bonded to the PDSN material. An extra advantage of using gadolinium as a solvent is that its high affinity for oxygen effectively removes the native oxide from the PDSN surface during transfer, allowing a true nitride–nitride couple.

The only problem encountered with this technique was the high solubility of silicon nitride in gadolinium at its melting point. The result of this high solubility was dissolution of the seed at the cooler interface and reprecipitation, resulting in mixing which effectively lowered the isotopic enrichment of the bonded tracer. SIMS analysis shows that the isotope transfer has occurred during the TSM process, but concentrations are much lower than in the isotopic powder. Silicon isotope ratios were monitored at each step of the process in order to confirm the validity of TSM. The isotopic silicon nitride powder was approximately 93 wt.% ^{30}Si and 7 wt.% ^{28}Si. The deposited tracer layer showed only 4–5 wt.% ^{30}Si. The isotope enrichments must be near 100% in the case of silicon, which has a high background so that subsequent diffusion profiles can be adequately measured.

In order to alleviate the mixing problem, a second technique was developed whereby the gadolinium layer is saturated with isotopic silicon nitride powder at 1428 °C before adding the PDSN seed to the arrangement. This was done by heating the PDSN source painted with isotopic powder and coated with gadolinium to 1428 °C first. This was then cooled as rapidly as possible (less than 1 min) to below 1200 °C and then to room temperature. The system was then opened to add the PDSN seed on top of the saturated gadolinium. The TSM process was then completed as previously described. However, it was found, on disassembling the source and seed and removing the gadolinium by acid etch, that only silicon had deposited on the seed. There are two possible explanations for this. First, the silicon nitride had visibly begun to precipitate on the top surface of the molten gadolinium even with no seed present. This resulted in a rough powdery surface which did not bond well with the seed once it had been introduced and heated back to 1428 °C. This left considerable porosity between gadolinium and the seed, where nitrogen gas evolving could have escaped, leaving only silicon in solution. Second, the gadolinium, being saturated with silicon nitride, may not have wet the seed, and the deposition may have been a result of vapor phase transport of silicon rather than liquid phase precipitation.

In a third scheme, a small amount of silicon-30 powder was painted directly on the silicon nitride seed and melted in an argon atmosphere to create a bond. The molten silicon tended to pool in droplets on the polished silicon nitride surface. However, the coated seed was then cooled and painted with a very thick layer of isotopic silicon nitride powder. When remelted, the silicon served to bond the powder to the seed. The large-surface-area powder dispersed the molten silicon evenly. When cooled, the silicon–silicon nitride tracer layer physically adhered to the seed. In addition, silicon nitride is slightly soluble in molten silicon, which could aid in effecting a chemical bond as well. This method was found to be most suitable for creating solid–solid diffusion couples.

As an alternative to solid–solid diffusion couples, some PDSN silicon nitride samples were ion implanted with silicon-30. Because silicon-30 has a high natural abundance, it is easily mass separated in an ion beam column in sufficient quantity to implant a high dose. A 180 keV implant at 773 K with a total dose of 1.08×10^{17} ions cm^{-2} was used. Using Lindhard–Scharft–Schiøtt statistics, the total depth of the implant region was estimated to be 0.3 μm. When these implanted samples are subjected to diffusion anneal, they will also yield diffusion coefficients which can be compared with results obtained for solid–solid diffusion couples.

To prevent sample loss during diffusion anneals, diffusion couples and implanted samples were covered with a second PDSN sample to protect the tracer layer and were packed in a powder mixture of 50 wt.% Si plus 50 wt.% α-Si_3N_4. High density carbon-impregnated graphite crucibles fitted with taper caps and set in the wells of a graphite block were used. Screw caps, which threaded into the top of the wells, applied pressure to the taper caps. This arrangement provided a tight seal within the graphite crucible.

Crucibles were loaded in a nitrogen atmosphere. During diffusion annealing, the evaporation of silicon results in the formation of silicon carbide along the edge of the taper cap, providing a complete seal. The overpressure of silicon and nitrogen resulting from powder evaporation greatly retards decomposition of diffusion couples. No sample loss was observed at 1600 °C after 6 days. Small sample losses observed at 1800 °C are believed to be only from the bottom and sides of the diffusion couple. TEM observations indicate no significant grain boundary voids in the diffusion zone.

SIMS analysis was chosen for collection of diffusion profiles in this study. When annealed solid–solid diffusion couples are beveled at an angle ϕ of 34° from the interface plane, the resulting magnification in depth is approximately 100:1. A primary ion beam 10 μm in diameter used to sputter the sample in SIMS stepped along the bevel results in a depth resolution of 1000 Å. The spreading of shallow implant profiles can be observed directly by depth profiling in the traditional sense. Although SIMS has been successfully performed on insulating samples for quite some time, owing to the high current density and sampling time necessary to collect nitrogen secondary ions which have a low ionization probability in this case, the sputtering parameters had to be carefully considered to avoid charging. It has been found that the use of a negatively charged oxygen beam for sample sputtering combined with adjustable positive sample bias and a 200 Å gold coating on the sample for charge dissipation has successfully eliminated sample charging. SIMS was performed on a Cameca IMS-3f instrument.

4. Results and discussion

It was possible during this study to conduct three diffusion anneals on silicon-30 ion-implanted samples at 1400, 1500 and 1600 °C, and one anneal at 1600 °C with a solid–solid diffusion couple. The diffusion couple was produced using molten silicon-30 to bond isotopic silicon nitride to the seed material.

The implanted samples were annealed for 2 h each. SIMS analyses were then performed on the silicon-30 concentration as a function of depth in each sample by sputtering through the implant zone (Fig. 5). To analyze the spreading of the initial implant profile, the data were fitted to a standard gaussian function for implant profiles [11]:

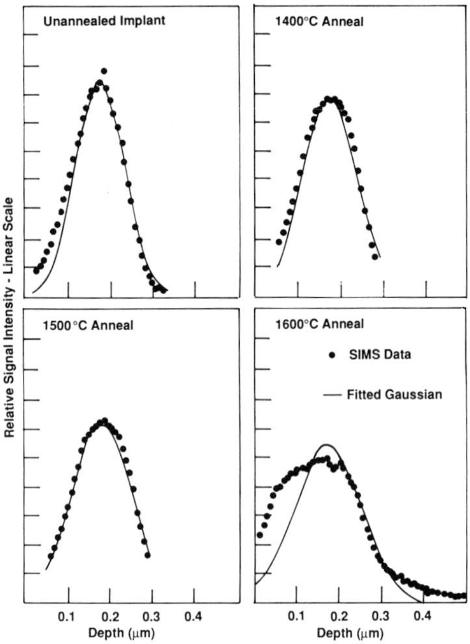

Fig. 5. SIMS analysis of annealed silicon-30 implants.

$$C(x)=\frac{\phi}{\pi^{1/2}\,\Delta R_p}\exp\left\{\frac{-(x-R_p)^2}{2\,\Delta R_p^{\,2}}\right\} \tag{1}$$

where R_p represents the projected range, ΔR_p is the straggle, and ϕ (ions cm^{-2}) is the dose. Using values of $R_p=0.182$ μm, $\Delta R_p=0.082$ μm and $\phi=1.08\times10^{17}$ ions cm^{-2}, the full curve for the unannealed implant was generated (Fig. 5). The SIMS signal intensities, obtained in arbitrary units, were then scaled to the calculated initial implant curve. The appropriate diffusion equation [11]

$$C(x)=\frac{\phi}{\pi(2\,\Delta R_p^{\,2}+4Dt)^{1/2}}\exp\left\{\frac{-(x-R_p)^2}{2\,\Delta R_p^{\,2}+4Dt}\right\} \tag{2}$$

where x is the distance and $C(x)$ is the concentration above the natural background was used to evaluate the diffusion coefficients which best fit the SIMS profiles at each temperature. The best fit in each case was determined by a least-squares analysis. The diffusion coefficients were determined to be 4.5×10^{-16} cm^2 s^{-1} at 1400 °C and 1.0×10^{-15} cm^2 s^{-1} at 1500 °C.

The experimental data at 1600 °C could not be fitted by the theoretical treatment. This is attributed to thermal grooving in the sample surface during the anneal. Although a second polished silicon nitride sample protected the implanted surface during the anneal, it was not actually bonded to the surface. This led to greater surface erosion in implanted samples than in solid–solid diffusion couples. The resulting topography invalidates the depth profiling done with SIMS. The SIMS signal at a given time during profiling results from the fact that various depths are sampled simultaneously. This would account for the flattening out of the gaussian profile. The theoretical curves plotted for 1600 °C show that the estimated diffusion coefficient of 2×10^{-15} cm^2 s^{-1} is within reason.

It is possible to conduct similar experiments for the analysis of nitrogen self-diffusion. However, since nitrogen-15 exists in low natural abundance (0.36%), an enriched source would have to be supplied for ion implantation.

It was not possible with these experiments to separate the effects of lattice and grain boundary diffusion. The analysis of grain boundary diffusion requires the clear discrimination of a diffusion curve at least $4(D_L t)^{1/2}$ cm deep, where D_L is the lattice diffusion coefficient. This is because of the assumption that diffusion in that region is due to the faster grain boundary contribution alone. Because of the high natural background of silicon-30 and limited concentration of the implant dose, deep diffusion of silicon-30 was not detected above background.

The issue of crystalline lattice damage caused by the fact that implanting ions affects the diffusion rates has not been addressed. The shallow implant profiles also fall in a region where subsurface polishing damage could have an effect. Both types of damage might artificially increase the observed diffusion rate owing to an increased concentration of vacancies and self-interstitials.

The diffusion coefficient reported for silicon at 1400 °C of 4.5×10^{-16} cm^2 s^{-1} is considerably lower than the only other reported data for silicon at 1400 °C. Wuensch and Vasilos reported $D = 1.5 \times 10^{-13}$ cm^2 s^{-1} in a small-grained β phase silicon nitride with considerable grain boundary phase. Even if the value at present reported is artificially high owing to implant and polishing damage, it indicates lattice diffusion rates which are exceedingly small. Lattice diffusion can be assumed to be the dominant factor in this PDSN material. An apparent activation energy can be calculated from the diffusion coefficients reported. On the assumption that the diffusion coefficients at 1400 and 1500 °C are more accurate than at 1600 °C, the activation energy calculated is 3.3×10^{-19} J atom^{-1} (2.06 eV atom^{-1}).

On the basis of an estimated diffusion coefficient of 2×10^{-15} cm^2 s^{-1} at 1600 °C from the implant studies, the solid–solid diffusion couple was annealed for 4 days in order to obtain a diffusion zone of at least 1 μm. This sample was then beveled to reveal the isotope–seed interface and to profile the silicon-30 concentration in the diffusion zone. The SIMS data are presented in Fig. 6. The semi-infinite solution to Fick's second law,

$$C(x) - C_2 = \frac{C_1 - C_2}{2} \operatorname{erfc}\left\{\frac{x}{2(Dt)^{1/2}}\right\} \quad (3)$$

where C_1 is the initial background in the seed and C_2 is the initial concentration of isotope in the tracer layer, was used to fit the data. Since the depth of the diffusion zone is only a few times greater than the SIMS depth resolution, the number of datum points was small. As a result, an accurate curve fit was not possible, and an exact diffusion coefficient cannot be reported. However, it is clear that the diffusion coefficient is of the order of 2×10^{-15} cm^2 s^{-1} at 1600 °C, as estimated in the implant case. It was not possible to separate clearly a contribution from grain boundary diffusion in this case either, for the reasons stated previously.

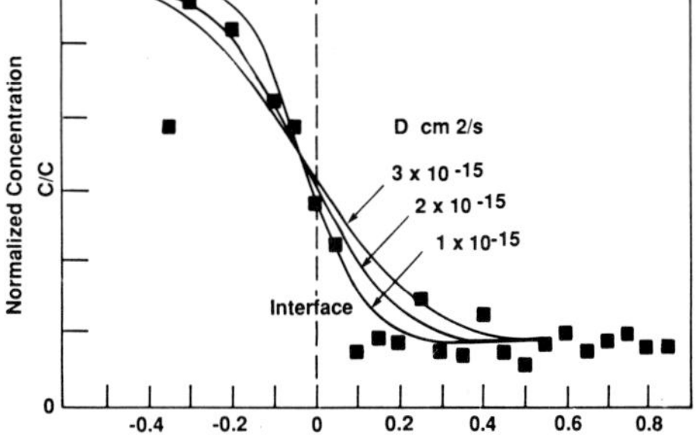

Fig. 6. SIMS analysis of silicon-30 profile in an annealed solid–solid diffusion couple.

Nitrogen self-diffusion is expected to be slower than that of silicon on the basis of previous reports. Therefore a longer diffusion anneal and consequently a larger diffusion zone would be required for the accurate determination of nitrogen diffusion coefficients by this technique.

5. Conclusions

Experimental difficulties encountered in past studies have been addressed. Results of material characterization show that a pure α material is well suited to the study of self-diffusion. The size and shape of the grains will be considered in applying the theoretical Suzuoka [5] model for separation of lattice and grain boundary diffusion rates if current limitations on obtaining deep diffusion profiles can be overcome. The present data suggest diffusion coefficients for silicon of $(0.45-2)\times10^{-15}$ cm^2 s^{-1} in the temperature range 1400–1600 °C and an apparent activation energy of 3.3×10^{-19} J atom^{-1} (2.06 eV atom^{-1}).

The TSM for tracer deposition has not produced suitable solid–solid diffusion couples, but using molten isotopic silicon to bond isotopic powder to the host material has produced the characteristic error function diffusion profile associated with a semi-infinite diffusion couple. Further work is needed to collect diffusion coefficients for both silicon and nitrogen using solid–solid diffusion couples.

Sample decomposition during diffusion anneals has been satisfactorily controlled in the solid–solid diffusion couple case, but it remains a problem for implant studies. SIMS has been shown to be a successful technique for diffusion profile analysis with excellent depth resolution. SIMS has been shown to be capable of resolving nitrogen as well as silicon profiles. Theoretically, longer diffusion anneals of solid–solid diffusion couples will yield diffusion profiles for nitrogen similar to those obtained for silicon.

Acknowledgments

The authors are pleased to acknowledge the support of the National Science Foundation (under Grant DMR-8512405), without whose help this work could not have been accomplished.

Appreciation is also extended to Dr. Calvin Carter for the design and building of the annealing furnace and for many useful discussions, Mr. James Gibson at Standard Oil Engineered Materials Co. for performing the SIMS analysis, Dr. Sopa Chevacharoenkel for TEM analysis, and Dr. Roderick Kunz for scanning electron microscopy and EDXA.

References

1 H. M. Jennings, *J. Mater. Sci., 18* (1983) 951.
2 F. Thummler, *Mater. Sci. Res., 13* (1980) 247–277.
3 A. R. Cooper and L. D. Major, Use of nuclear microanalysis, Part II, Nuclear microanalysis of materials, nitrogen transport in Si_3N_4, *NTIS Rep. AD-A-069004*, 1979 (National Technical Information Service, U.S. Department of Commerce).
4 J. C. Fisher, *J. Appl. Phys., 22* (1951) 74.
5 T. Suzuoka, *J. Phys. Soc. Jpn., 19* (1964) 839.
6 H. D. Batha and E. D. Whitney, *J. Am. Ceram. Soc., 56* (7) (1973) 365.
7 K. Kijima and S. Shirasaki, *J. Chem. Phys., 65* (7) (1976) 2668.
8 B. J. Wuensch and T. Vasilos, Self-diffusion in silicon nitride, *NTIS Final Rep. AD-A-021175*, 1975 (National Technical Information Service, U.S. Department of Commerce).
9 R. F. Davis, J. D. Hong and M. H. Hon, *Mater. Sci. Res., 2* (1978) 653–661.
10 W. G. Pfann, *Trans. AIME, 203* (1955) 961.
11 S. Wolf and R. N. Tauber, *Silicon Processing for the VLSI Era*, Lattice Press, Sunset Beach, CA, 1986, p. 248.

Materials Science and Engineering, A105/106 (1988)55–63

Characterization of Carbides by Scanning Tunneling Microscopy*

DAWN A. BONNELL

IBM Research, Thomas Watson Research Center, Yorktown Heights, NY (U.S.A.)

(Received January 13, 1988)

Abstract

Scanning tunneling microscopy and scanning tunneling spectroscopy were used to characterize several carbide surfaces. A single-crystal (6H) SiC growth surface parallel to the basal plane exhibited macroscopic growth steps and facets of the order of 1.4–3.0 nm. Three midgap electronic states were identified on this surface, which allowed imaging at relatively low biases for a large-band-gap semiconductor. Facets were observed on a $TiC_{0.95}$(100) fracture surface with dimensions of five to ten times the unit-cell size. Surface reactions were observed by characterizing the SiC surfaces before and after exposure to the atmosphere and by in situ monitoring of a surface adsorbate on TiC. The first topographic images of a phase boundary were obtained in a polycrystalline Al_2O_3–TiC composite.

1. Introduction

Scanning tunneling microscopy (STM) is a new method of imaging the surfaces of materials, including those classified as hard materials. The microscope can be operated as a simple profilometer to obtain topographic images of surfaces which would be of use, for example, in the characterization of wear surfaces and machining damage. In operating the microscope under conditions for higher resolution, more detailed geometric and electronic structural information can be obtained. As a result of recent advances, it is now possible to determine surface atomic structure and detailed electron density of state structure as a function of position by combining STM imaging with scanning tunneling spectroscopy (STS) [1, 2]. The technique has been applied to metals and conventional semiconductors but has not yet been extended to the characterization of a wide variety of materials. However, a recent investigation of STM of conductive ceramic systems indicates that the technique has wider applicability than previously thought possible [3].

The purpose of the present study was to determine what useful information can be gained by the application of STM and STS to single-crystal carbides and to carbides as part of composite systems. This work represents an extension of the technique in that STM and STS characterization of carbides had not been attempted before. As STM is a recently developed characterization tool, a short description of the technique will first be given. Then, topographic and spectroscopic analyses of an SiC single-crystal growth surface and a TiC single-crystal fracture surface will be presented. Preliminary analysis of an Al_2O_3–TiC composite will be described including the first STM imaging of a phase boundary intersecting the surface. In summarizing, the implications of these results for use of STM and STS in materials science will be discussed.

2. Scanning tunneling microscopy

STM relies on tunneling electrons to produce a current which is used to gain structural and spatially localized electronic information about materials [4, 5]. The potential information gained as well as the basis for the simple design of the microscope become evident in the consideration of the fundamental current relation which follows.

A metallic probe tip is brought close enough to the sample surface (within 1–2 nm) that high energy electrons near the Fermi level can tunnel through the potential barrier between the tip and sample surface. The current I produced by the tunneling electrons is given by:

$$I=\frac{2\pi e}{\hbar}\sum_{s,t} f(E_s)\{1-f(E_t+eV)\}|M_{s,t}|^2\delta(E_s-E_t) \qquad (1)$$

*Paper presented at the 3rd International Conference on the Science of Hard Materials, Nassau, The Bahamas, November 9–13, 1987.

0921-5093/88/$3.50

© Elsevier Sequoia/Printed in The Netherlands

where M is the tunneling matrix, $f(E)$ is the Fermi function, e is the charge on an electron, $\hbar$ is Planck's constant, and s and t refer to the sample and tip respectively. Tersoff and Hamann [6] evaluated the tunneling matrix and, in the limit of a low applied bias V and a low temperature, the formula reduces to

$$I \propto D_s(E) \exp(-2\kappa Z) \qquad (2)$$

where Z is the sample-to-tip separation, D_s is the sample density of states near the Fermi level and κ, which is the inverse decay length of the state with energy E, equals $(2m\phi/\hbar^2)^{1/2}$ where m is the mass of an electron and the effective barrier potential ϕ is $(\phi_t + \phi_s)/\{2 - E + (eV/2)\}$. Equation (2) is sufficient to specify the experimental variables in the microscope and indicates that, by manipulating appropriate parameters, we can directly measure the surface topography, *i.e.* Z, the local density D_s of states and the tunnel barrier height ϕ_t.

The tunneling configuration in the STM is shown schematically in Fig. 1 which illustrates the simplicity of the microscope design. A bias voltage is applied between the sample and tip, and the resulting tunneling current is monitored in a feedback controller. As the tip is scanned across the sample surface, the feedback controller alters the sample-to-tip separation to maintain a constant current. The exponential dependence of tunneling current on sample-tip separation indicated in eqn. (2) provides for high sensitivity which, under the best circumstances, allows subnanometer spatial resolution. The resulting image represents a map of constant charge density which can, at times, be related to atom positions. Further consideration of eqn. (2) indicates that, if the feedback controller is disengaged while the bias voltage is ramped (implying a constant sample-to-tip separation Z), the resulting changes in current will be related to the electronic structure of the sample surface, specifically to the density of states at that position. The dependence of the measured density of states on applied bias voltage is the basis for STS, normally presented as current–voltage (I–V) characteristics or as voltage derivatives of current which are proportional to the density of states. The scanning tunneling microscope with modifications for materials science applications that was used in this study has been described in detail elsewhere [3].

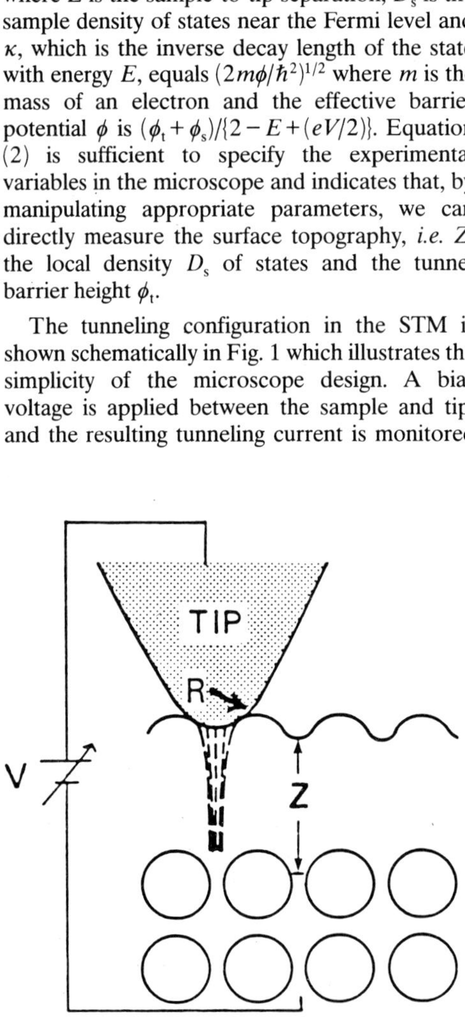

Fig. 1. Schematic representation of tunneling as it occurs in the STM configuration. The bias V applied between the sample and tip results in a tunneling current which is used to control the sample-to-tip separation Z. Both image and spectroscopy resolution depend on the radius R of the probe tip.

3. Results

STM, including topographic and spectroscopic analyses, was used to characterize a series of carbide surfaces. Results from an SiC single-crystal growth surface, a TiC single-crystal growth surface, and an interface in an Al_2O_3–TiC composite are described here.

3.1. SiC single-crystal growth surface

The SiC single crystals produced by the Acheson process were confirmed by X-ray diffraction and Laue backscattering techniques to be of the 6H modification with the basal plane parallel to the surface. Samples analyzed in air were first rinsed in acetone and then ethanol to remove any contamination accumulated during handling. Those analyzed in vacuum (about 10^{-8} Torr) were subsequently soaked in an aqueous 45% HF solution for 24 h to remove surface reactants and transferred to the vacuum within 3 min. It is assumed that this prolonged exposure to aqueous HF solution removes any surface oxide and that the oxygen replacement reaction is slow enough that any resulting oxide contains only a couple of atomic layers. Imaging and spectroscopic analyses were performed with an electrochemically etched tungsten tip.

The images of 25 nm × 25 nm areas of the SiC surface shown in Fig. 2 illustrate the topographical structure typical of this material. Facets are clearly observed with step heights varying from 1.4 to 3.0 nm. These facets are consistent with macroscopic growth steps observed on this surface by optical microscopy and might be expected from the crystal growth process. The electronic structures of the crystal as determined by STS in vacuum and in air are compared in Fig. 3 which shows the $I-V$ characteristics as a function of sample-to-tip separation. A striking difference in electronic behavior is illustrated by the asymmetry in the $I-V$ characteristics in the sample analyzed in air (Fig. 3(a)) which is not present in the sample analyzed in vacuum (Fig. 3(b)). For example, a positive applied bias of +2.0 V produces a much larger tunneling current than does an applied bias of −2.0 V in the sample analyzed

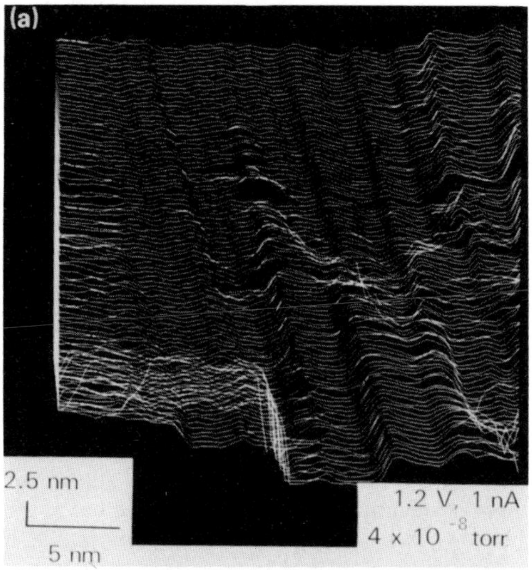

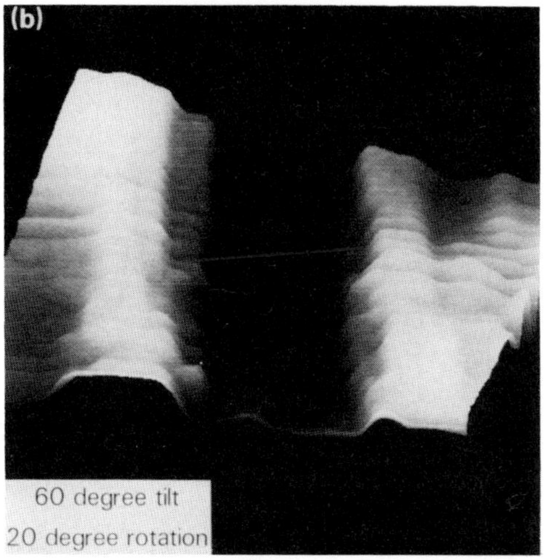

Fig. 2. Constant-current images of (a) 25 nm × 25 nm and (b) 20 nm × 20 nm areas of the SiC basal plane etched in HF acid, observed in a vacuum of 10^{-8} torr. The smallest steps observed were about 0.4 nm in height and spaced about 1.45 nm apart. The marker gives lateral (x, y) and vertical (z) scales.

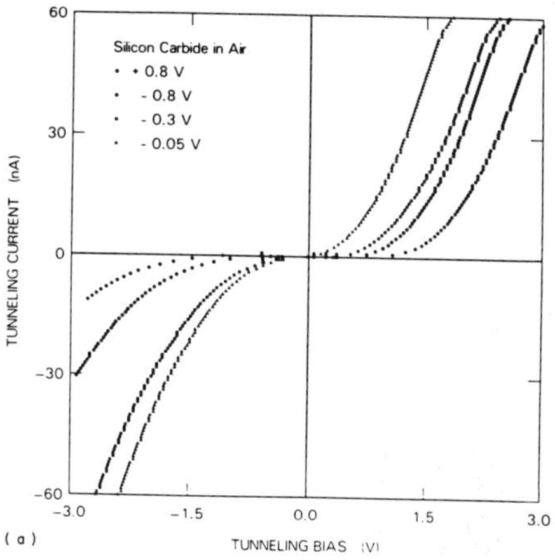

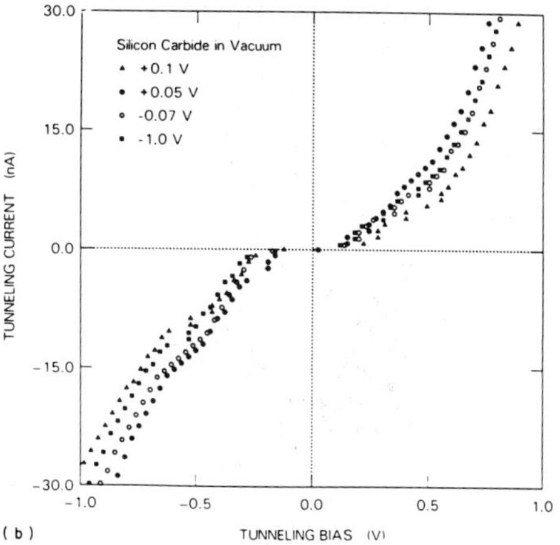

Fig. 3. STS of the SiC basal plane expressed as $I-V$ characteristics comparing the electronic structure of the surface (a) in air and (b) in vacuum. Each curve was acquired at a different sample-to-tip separation in the range 0.5–2 nm, which is indicated by the applied bias in the upper left of the plots.

in air. Although such rectification can be caused by field intensification due to tip geometry [7], a more probable cause is suggested by the image in Fig. 4. The topographic image of a 50 nm × 50 nm area of SiC acquired in air exhibits features indicative of surface adsorption of environmental contaminants. This reaction of the SiC surface with the atmosphere must alter the tunnel barrier height which would also result in the observed rectification.

Complete spectroscopic analysis of the SiC in vacuum indicates the semiconducting nature of the materials by exhibiting an energy band gap and clearly identifying three states between − 1 and + 1 eV, as seen in the logarithmic derivative spectrum in Fig. 5. As the band gap of SiC is 2.89 eV, these are midgap states [8]. It was not possible in these initial measurements to determine whether the states resulted from surface reaction with fluorine from the cleaning procedure or are bulk states induced by the doping acquired during the crystal growth process. Nevertheless, it is clearly the presence of these states that allows tunneling into, and consequently imaging of, SiC at low bias voltages. The tip radius implied by the resolution of the images in Fig. 2 indicates that the areas analyzed in STS measurements were about 4 nm^2.

3.2. *TiC fracture surface*

Cleaved surfaces of a 5 at.% C deficient TiC single crystal ($TiC_{0.95}$) grown by the floating-zone method [9] were characterized. Surfaces parallel to the (100) and (112) planes were soaked in HF acid for 1 h to remove any thick oxide film which may have developed before analysis. Both topographic and spectroscopic analysis were performed in air with mechanically ground PtIr tips. These tips proved to be electronically more stable than the tungsten tips, presumably owing to the absence of a surface oxide known to grow on the tungsten tips. However, the PtIr tips used in this analysis happened to be larger than the tungsten tips, which necessarily implies lower resolution imaging conditions.

The topography of a TiC(100) fracture surface is shown in the wide-range line scan image in Fig. 6 in which 3–5 nm steps are observed on the plane. The low resolution of the image indicated by the roundness of the steps precludes precise measurement of their sharpness which may be of atomic dimensions. The surface is inclined with increasing steps from the bottom left to the upper right of the image. The electronic structure of the (100) surface is given by the STS results in Fig. 7.

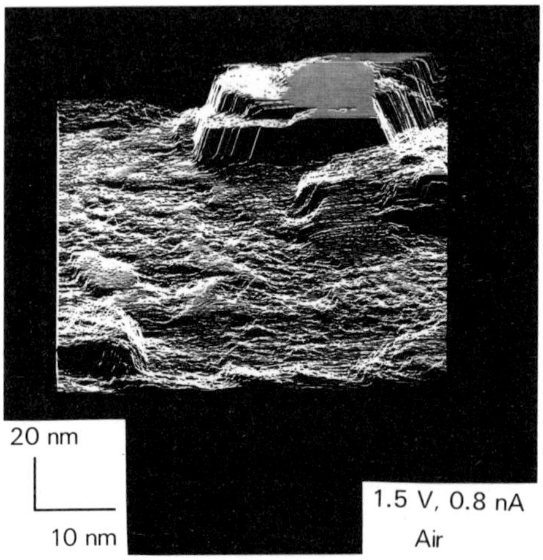

Fig. 4. Constant-current image of the SiC basal plane rinsed in acetone and ethanol and observed in air with a tungsten tip. The large feature in the upper right corner is crystallographic in nature, as indicated by the sharpness of the facet. The smaller rolling features in the center are indicative of the presence of surface contaminants and represent a convolution of their geometric structure and induced change in surface work function. The marker gives lateral (x, y) and vertical (z) scales.

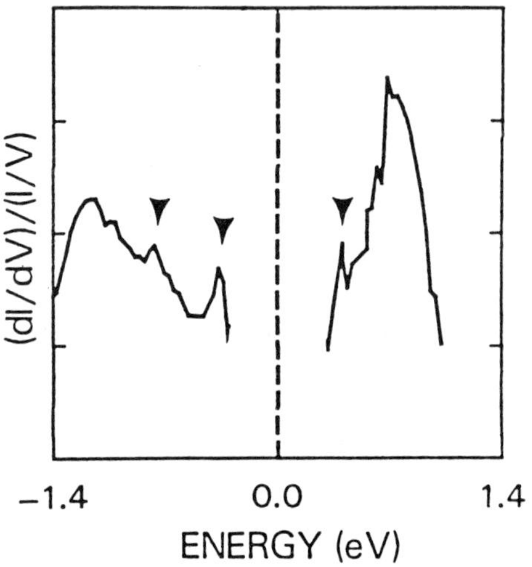

Fig. 5. STS of SiC analyzed in vacuum expressed as a logarithmic derivative spectrum. The spectrum is proportional to the density of states and identifies three midgap states in this material, as indicated by the arrows.

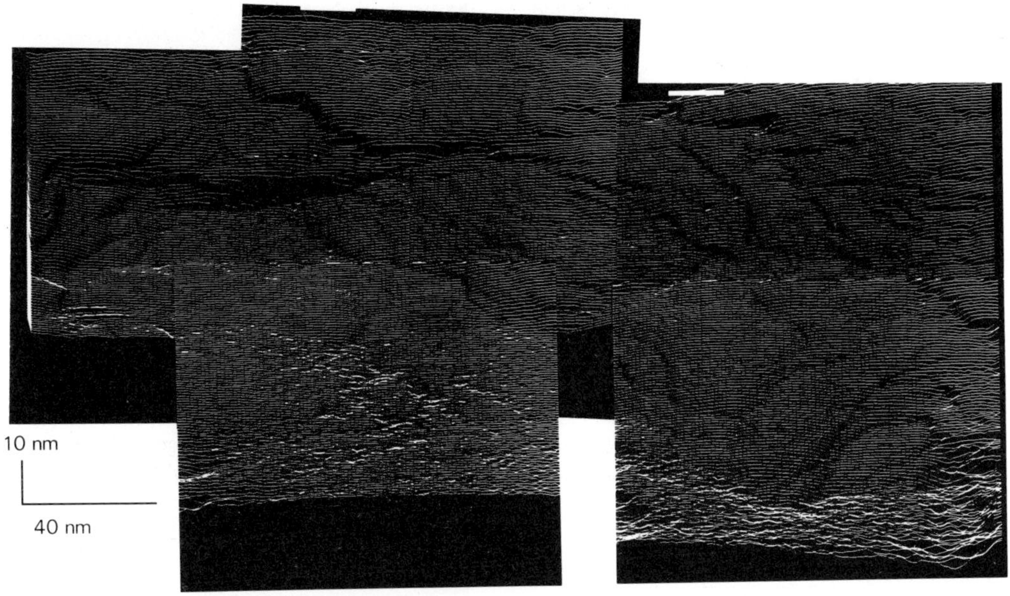

Fig. 6. Constant-current image of 0.20 μm × 0.25 μm area of $TiC_{0.95}(100)$ acquired in air with a PtIr tip. The heights of the facets observed on this surface range from 3 to 5 nm. The marker gives lateral (x, y) and vertical (z) scales.

Both $I-V$ characteristics (Fig. 7(a)) and conductance spectra (Fig. 7(b)) reveal the metallic component to bonding in the (100) plane. This is seen in the magnitude of the current at very low biases; it should be noted that the voltage scale is an order of magnitude smaller than that used for the SiC case. More convincing of the metallic contribution to bonding in this surface is the continuous, gradually decreasing density in both occupied and unoccupied states as the bias approaches zero. The bottoming out of this curve associated with a band gap is not observed in the case of TiC. The rectification observed in the SiC exposed to the atmosphere is not evident in the TiC exposed to air.

One region of the surface exhibited a surface feature which did not seem to be crystallographic in nature and which we suspect was an adsorbed contaminate. It appeared as a string-like feature with dimensions of 3 nm × 5 nm × 35 nm, as shown in Fig. 8(a). Interestingly, this feature changed during imaging. The initial changes occurred rapidly and the feature appeared as shown in Fig. 8(b) after scanning only four times. This initial sudden shape alteration can be attributed to tip contact with the adsorbed particle. The probe tip physically moved the particle, changing its shape and even separating pieces of the adsorbate as seen by the debris spread out on the surface. At this point the image was relatively stable; the general shape remained unchanged but its height gradually decreased until, after about $\frac{3}{4}$ h, it was gone. The gradual change in height with no accompanying change in particle geometry cannot be attributed to tip–sample contact. It remains that this latter alteration results from some reaction of the particle in the field between the sample and tip. Although these observations are preliminary in that the reaction was not fully characterized, they serve to demonstrate the capability of the STM monitor surface reactions *in situ* and its potential to cause reactions to occur on a spatial scale related to the size of the probe tip.

3.3. Al_2O_3–TiC phase boundary

A polycrystalline 50wt.%Al_2O_3–50wt.%TiC composite was produced by hot pressing compacted powders to full density. Surfaces were polished to a 0.5 μm finish, rinsed in acetone and then ethanol and analyzed in air with a tungsten tip. Images were acquired at bias voltages ranging from +0.5 to +1.5 V with the most stable images occurring at around +1.5 V. Two distinct

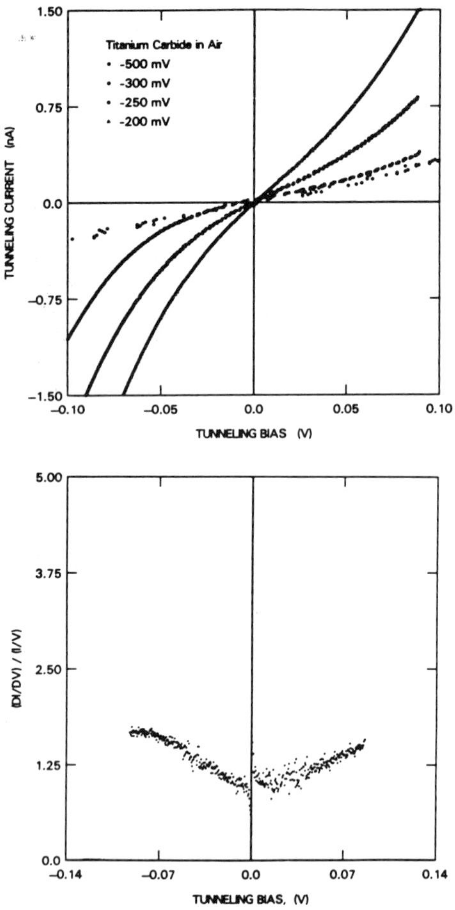

Fig. 7. STS of $TiC_{0.95}(100)$ expressed both (a) as $I-V$ characteristics and (b) in a logarithmic derivative spectrum. The metallic nature of the bonding in this plane is evident from the almost ohmic response at very low biases in the $I-V$ curve and in the continuous occupancy of states as the voltage approaches zero in the derivative spectrum.

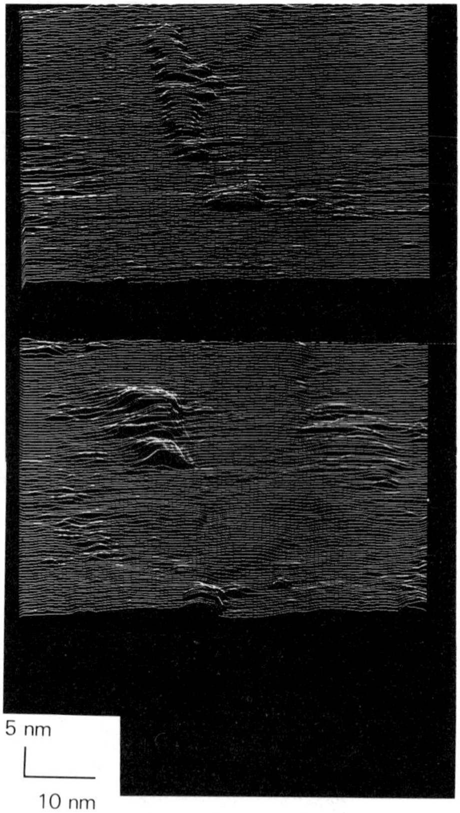

Fig. 8. Constant-current images of 50 nm × 50 nm area of $TiC_{0.95}$ acquired in air with a PtIr tip where a string-like adsorbate particle is observed on the surface. The surface structure as it appeared in the first two images is shown in (a) and as it appeared after three more scans in (b). The marker gives lateral (x, y) and vertical (z) scales.

tunneling responses were observed on the Al_2O_3–TiC, as shown in Fig. 9: a stable current, producing images similar to those on TiC (Figs. 6 and 8), and an unstable current, producing a noisy image. The transition between the two regions was sharply discontinuous and often accompanied by a spike in the current. The discontinuity arises from the sudden change in surface work function and state occupation on crossing a phase boundary. In fact, as there is predicted to be no density of states at the Fermi level in Al_2O_3, the question arises as to what the image actually represents. The effects of surface conductivity have been neglected in both theoretical and experimental considerations of tunneling but obviously make a large contribution to the current response in Al_2O_3. Unambiguous interpretation of images such as this awaits further work. The scanning tunneling spectra acquired in the more stable regions superimpose exactly on some of the TiC(100) curves in Fig. 7(a). Although a more detailed characterization of these phase boundaries is under way, it is clear that the images presented here represent the intersection of an Al_2O_3–TiC interface with the surface.

4. Discussion

An important conclusion from the results of this investigation is that new information is gained

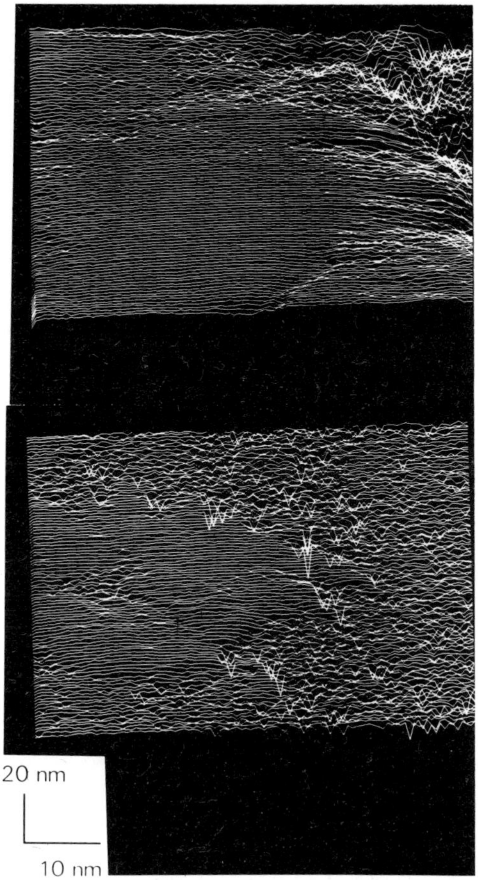

Fig. 9. Low resolution constant-current image of a 50 nm × 50 nm area of Al_2O_3–TiC acquired in air with a tungsten tip. Stable tunneling on the left-hand side of the area allows higher resolution topographical imaging than does the unstable tunneling on the right-hand side which produces a noisier and consequently lower resolution image. The areas are separated by sharp discontinuities expected when crossing a biphase boundary. Such differences in tunneling response and the interface discontinuity indicate differences in the material.

by the application of STM to hard materials such as carbides. These observations of the topographic and electronic structure of carbides have significance on two levels: one involving the characterization of the carbide surfaces presented here and the other involving the extension of STM to such general materials science investigations as the study of crystal growth, fracture, grain boundary structure, wear processes and machining damage.

Crystal growth mechanisms are often identified by characterization of the resulting surface structure. Optical microscopy is used to image single-crystal surfaces and transmission electron microscopy (TEM) has long been used to study crystal growth by imaging a plane perpendicular to the growth front. Very recently, new TEM contrast techniques [10] and high resolution reflected electron microscopy [11] have also been used to investigate steps on growth surfaces. However, the STM characterization of the SiC surface presented here provides the first direct real-space imaging of crystal growth surfaces at high resolution. The very fine faceted structure observed on this surface indicates that ledge growth mechanisms are operative on several scales. The macroscopic growth steps are themselves made of microscopic facets. While this is an intuitively satisfying observation, careful correlation of step sizes with growth process parameters is needed in order to study the atomistics of this process.

In situ cleavage is one method of preparing clean, atomically flat surfaces and therefore such surfaces are frequently analyzed by STM and STS [2, 12, 13]. Past observations of such surfaces, however, have concentrated on characterizing the reconstructions which occur on “perfect” regions of the cleaved surface rather than on the investigation of the fracture process itself. The analysis of the TiC is the first attempt to characterize a fracture surface at such high resolution, concentrating on how the features relate to the fracture process. The structure of the TiC(100) cleavage plane indicates that the fracture front (crack) is deflected in increments of the order of five to ten unit-cell dimensions. This deflection is expected to occur when the applied loading is not exactly normal to the fracture plane or the fracture origin is not parallel to the cleavage plane. The off-axis component of the local stress required to make the crack deviate five to ten unit cells should not be large, but a more detailed investigation of the structure of facets as a function on controlled loading is necessary to quantify these relations.

In addition to the characterization of carbide surfaces, the results presented here have strong implications in the use of STM in materials science. In the past, geometric constraints have limited the use of STM in structural studies of materials. The largest possible area of analysis was only about 100 nm × 100 nm, features with vertical magnitudes larger than about 5 nm exceeded the range of the instrumentation, and it was not possible to search around a surface for features of interest. Attempts to overcome these experimental limitations are presented in the mapping of a 0.3 μm × 0.3 μm area (TiC), in the imaging of features on the order of 50 nm in height (SiC in air), and in the location and imaging of a phase boundary (Al_2O_3–TiC). Further modifications of the instrument have increased the possible area of analysis by an additional four orders of magnitude and allow movement around the surface.

In extending STM to materials other than metals and conventional semiconductors, the question of limiting resolution arises. Imaging and spectroscopy can be performed at various levels of spatial resolution and the power of STM at atomic resolution has been amply demonstrated by some elegant experiments performed on semiconductors [1, 2]. In addition to instrumentation parameters, the resolution of a particular image will be influenced by the probe tip radius, magnitude of the applied bias, and sample surface condition. In the application of STM to a wider variety of materials, we may expect larger effects from these factors. In fact, analyses of ceramics have shown this to be true; however, there is no fundamental reason that these experimental difficulties cannot be overcome and higher resolution achieved in these materials as well [3].

Equally important, although perhaps not as well recognized, is the usefulness of topographic information gained in imaging below atomic resolution. Even at the lower resolution conditions presented here (of the order of 0.5 nm), STM provides previously inaccessible structural and spectroscopic information specific to the surface of interest, as is illustrated in the analysis of SiC, TiC and Al_2O_3–TiC. It is precisely this type of analysis which can provide direct measurement of surface properties important, for example, in the study of wear processes and machining damage.

The results presented here, although admittedly preliminary, have demonstrated how STM can be used to study such fundamental processes as single-crystal growth, cleavage fracture and surface reactions by characterization both before and after (in the case of the reaction of SiC with the atmosphere) and by *in situ* observation (in the case of particle desorption in the TiC). The potential of STS is seen in the identification of the metallic contribution to bonding in the TiC(100) plane, in the determination of midgap states on doped SiC and in its usefulness in differentiating the Al_2O_3 and TiC in the composite material. Perhaps the most exciting single result in this investigation is the imaging of an interface. With this observation, the application of the new STM techniques to such problems as the structure of interfaces and wear surfaces is proven.

4. Summary

The topographic and electronic surface structure of the SiC basal plane, $TiC_{0.95}$(100) plane and polycrystalline Al_2O_3–TiC composite have been characterized by STM and STS. The clean SiC surface exhibited facets of the order of 1.4–3.0 nm, which resulted from crystal growth processes. Three midgap states were identified, which allowed imaging at low biases relative to the size of the energy band gap of 6H SiC. A reaction of SiC with the atmosphere caused a large rectification in the tunneling I–V characteristics, presumably owing to the resulting asymmetric change in tunnel barrier. Imaging of a TiC(100) fracture surface over a wide spatial area indicated the presence of steps with heights varying from 3 to 5 nm. Such steps are probably the result of local loading conditions where a component of the stress is not normal to the cleavage plane during fracture. An *in situ* reaction was monitored by observing the geometric changes in an adsorbate under the field of the probe tip as a function of time. First images of phase boundaries in the Al_2O_3–TiC composite were presented. These observations have demonstrated that some of the previous limitations in the use of STM and STS to study materials structures have been overcome.

Acknowledgments

Valuable discussions with D. R. Clarke and R. M. Feenstra were greatly appreciated, as were critical comments on the manuscript by R. F. Cook and G. M. Pharr. SiC single crystals were provided by Frank Hungar, Standard Oil Engineered Materials Division, and TiC single crystals by Subhash Shinde, IBM, and Walter Precht, Martin Marietta.

References

1 R. J. Hamers, R. M. Tromp and J. E. Demuth, Electronic and geometric structure of Si(111) 7×7 and Si(001) surfaces, *Surf. Sci., 181* (1987) 346–355.

2 R. M. Feenstra, J. A. Stroscio and A. P. Fein, Tunneling spectroscopy of the Si(111) 2 × 1 surface, *Surf. Sci., 181* (1987) 295–306.
3 D. A. Bonnell and D. R. Clarke, Scanning tunneling microscopy and spectroscopy of ceramics: SiC and ZnO, *J. Am. Ceram. Soc., 71* (1988) 629–637.
4 G. Binnig and H. Rohrer, Scanning tunneling microscopy—from birth to adolescence, *Rev. Mod. Phys., 59* (1987) 615–625.
5 P. K. Hansma and J. Tersoff, Scanning tunneling microscopy, *J. Appl. Phys., 61* (1987) R1–R23.
6 J. Tersoff and D. R. Hamann, Theory of scanning tunneling microscopy, *Phys. Rev. B, 31* (1985) 805–813.
7 N. M. Miskovsky, S. J. Shepherd, P. H. Cutler, T. E. Sullivan and A. A. Lucas, Importance of geometry, field and temperature in tunneling and rectification behavior of point contact junctions of identical metals, *Appl. Phys. Lett., 35* (7) (1979) 560–562.
8 *Landolt–Bornstein, New Series*, Group III, Vol. 17a, Springer, Berlin, 1982, p. 133.
9 W. Precht and G. E. Hollox, A floating zone method for the growth of carbide single crystals, *J. Crys. Growth, 3* (4) (1968) 818–823.
10 J. M. Gibson, M. L. McDonald, J. L. Batstone and J. M. Phillips, High resolution electron microscopy of interfaces and surfaces, *Ultramicroscopy, 22* (1987) 35–46.
11 T. Hsu and L. M. Peng, Experimental studies of atomic step contrast in reflection electron microscopy, *Ultramicroscopy, 22* (1987) 217–224.
12 J. A. Stroscio, R. M. Feenstra and A. P. Fein, Imaging electronic surface states in real space on the Si(111) 2 × 1 surface, *J. Vac. Sci. Technol. A, 5* (4) (1987) 838–841.
13 R. M. Feenstra and J. A. Stroscio, Tunneling spectroscopy of the GaAs(110) surface, *J. Vac. Sci. Technol. B, 5* (4) (1987) 923–929.

Materials Science and Engineering, A105/106 (1988) 65–75

High Toughness Ceramics*

A. G. EVANS

Materials Department, College of Engineering, University of California, Santa Barbara, CA 93106 (U.S.A.)

(Received November 10, 1987)

Abstract

The principal microstructural sources of toughening are reviewed, with emphasis on comparisons between theory and experiment. Process zone mechanisms, such as transformation and microcracking, as well as bridging mechanisms induced by either ductile phases or fibers and whiskers are afforded primary consideration. Microstructural variables which allow toughness optimization are discussed.

1. Introduction

The last decade has witnessed major advances in the development of ceramics having enhanced toughness. All the mechanisms which provide appreciable toughening have the common feature that material elements at, or near, the crack surfaces exhibit non-linear behavior, with hysteresis, as schematically illustrated in Fig. 1. Indeed, the toughening can be explicitly related to the hysteresis, as will be elaborated for each of the important mechanisms. Furthermore, in most cases, the mode of toughening results in resistance curve characteristics (Fig. 2), wherein the fracture resistance systematically increases with crack extension. The individual mechanisms that have been established include displacive transformations, microcracking, ductile phase and brittle fiber or whisker reinforcement. The general philosophy adopted in the modelling of the toughening includes the classical concept of homogenizing the properties of the material around the crack and formulating a constitutive law which then characterizes the material response. This approach fully describes the toughening behavior when the length of the non-linear zone (Fig. 1) is appreciably larger than the spacing between the microstructural entities which dictate the non-linearity, as elaborated below. Otherwise, microstructural interactions must be treated discretely.

The known mechanisms can be conveniently considered to involve either a process zone or a bridging zone (Fig. 3). The former category exhibits a toughening fundamentally governed by a critical stress for the onset of non-linearity (σ_0 in Fig. 1) and by the permanent strain induced by the non-linear mechanism (ε_0 in Fig. 1). The hysteresis is dictated by the stress–strain behavior of composite elements within the process zone, such that integration over the zone gives

$$\Delta G_c \approx 2 f \sigma_0 E h \qquad (1)$$

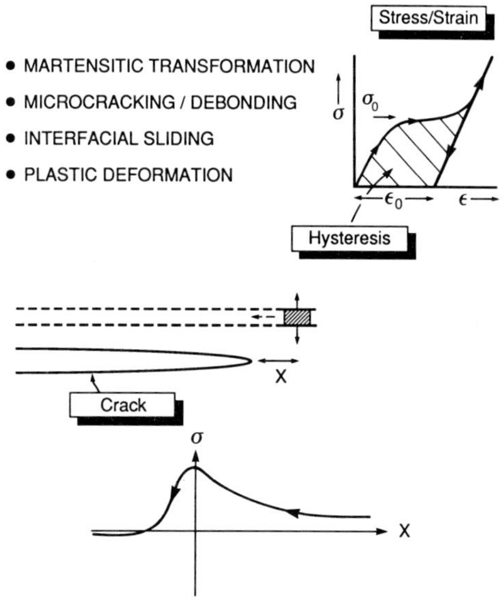

Fig. 1. A schematic diagram illustrating non-linear hysteretic elemental response and associations with enhanced toughness.

*Paper presented at the 3rd International Conference on the Science of Hard Materials, Nassau, The Bahamas, November 9–13, 1987.

0921-5093/88/$3.50

© Elsevier Sequoia/Printed in The Netherlands

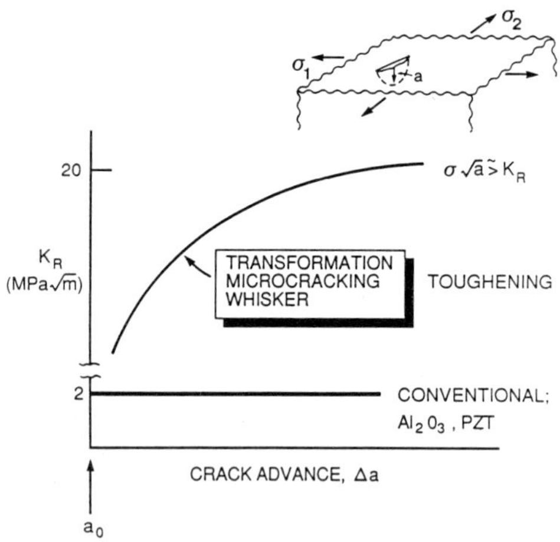

Fig. 2. Resistance curve behavior characteristically encountered in tough ceramics: K_R is the fracture resistance and Δa the crack advance.

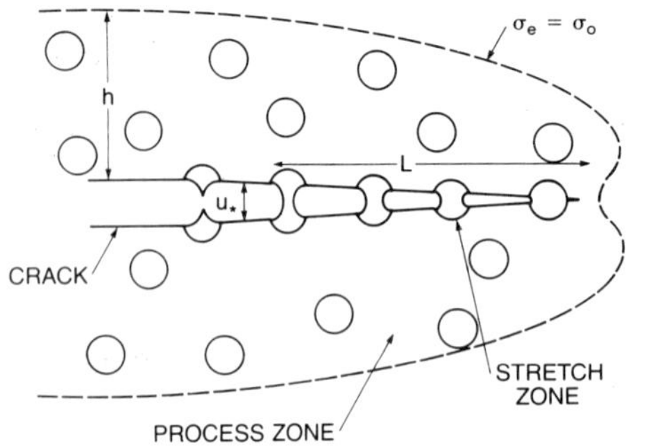

Fig. 3. A schematic diagram illustrating both process zone and bridging zone mechanisms of toughening.

where ΔG_c is the increase in critical strain energy release rate, f is the volume fraction of toughening agent and h is the width of the process zone (Fig. 3). Transformation and microcrack toughening are mechanisms of this type.

The latter category exhibits toughening governed by hysteresis along the crack surface:

$$\Delta G_c = f_A \int_0^{u_*} \sigma(u)\, du \tag{2}$$

where u is the crack opening, u_* the opening at the edge of the bridging zone, σ the tractions on the crack surfaces exerted by the intact toughening agent (Fig. 3) and f_A is the area fraction of reinforcements along the crack plane. The relationship between f_A and the volume fraction f depends on reinforcement morphology, on the residual stress and on the elastic mismatch (through crack attraction–repulsion effects). Ductile dispersions, as well as whiskers and fibers, toughen by means of bridging tractions.

In the toughest materials, a steady state cracking phenomenon has been identified, wherein the crack extension stress becomes independent of crack length. For such materials, the toughness is usually non-unique and not, therefore, a useful material parameter. It is also noted that some of the composite systems of interest have anisotropic toughness, causing the properties to depend on the loading axis and the mode of loading. However, the central problem that must first be resolved in order to have a successful tough composite concerns the mode I (tensile) crack propagation resistance on planes normal to the reinforcement axis. This mode of crack propagation is thus emphasized in the present paper.

2. Steady state cracking

In certain composite ceramics, matrix cracking occurs at a constant stress [1–3]. This phenomenon is referred to as steady state cracking. Such behavior is most widely known in fiber-reinforced ceramics but is also possible in ductile-phase-reinforced ceramics. Steady state cracking in fiber-reinforced ceramics involves mode I cracks which extend normal to the fiber axis, resulting in a crack bridged by strong reinforcements (Fig. 4). An equilibrium crack opening then develops, with the load fully supported by the intact reinforcements. For such cases, the stress intensity associated with matrix cracking becomes independent of crack length. The steady state stress depends on the response of the interface and matrix to the load on the reinforcements between the crack surfaces. For fibers which resist opening by friction, the cracking stress is [1–3]

$$\sigma_s = \left\{ \frac{6\tau K_0^{\,2} f^2 E_f E^2}{(1-f) E_m^{\,3} R} \right\}^{1/3} + \frac{\sigma_R E}{E_m} \tag{3}$$

where τ is the shear resistance of the interface, K_0 is the fracture resistance of the matrix, f is the volume fraction of fibers, E is Young's modulus, R is the fiber radius, σ_R is the residual stress in the matrix, and the subscripts f and m refer to the fiber and matrix respectively. This prediction

Fig. 4. Steady state cracking in a ceramic matrix fiber composite.

agrees well with various experimental results [2, 3] obtained on fiber-reinforced ceramics and glasses.

The equivalent result for partially debonded fibers without friction is [3]

$$\sigma_s = \frac{K_0 E}{E_m}\left\{\frac{6f^2 E_f}{E(1-f)^2 R(1+\nu_m)}\right\}^{1/4} \times \left\{1+\frac{4fl\theta}{R(1-f)}\right\}^{1/2}\left(1+\frac{\rho l}{R}\right)^{-1} + \frac{\sigma_R E}{E_m} \qquad (4)$$

where l is the debonded length, θ is the ratio of the debond to matrix toughness, ν is Poisson's ratio and $\rho = \{0.8/(1-f)\}\{6E/E_f(1+\nu_m)\}^{1/2}$.

Steady state crack growth conditions exist when the reinforcement strength $\bar{S}$ satisfies the inequality $\bar{S} > \sigma_s/f$. Such materials have very desirable properties, by virtue of the associated damage tolerance and significant non-linearity before ultimate failure [4, 5]. The transition from matrix cracking to toughness-controlled fracture can sometimes be conveniently represented as a map. For example, when frictional sliding dominates the composite fracture process, imposing $\bar{S} = \sigma_s/f$ onto eqn. (3) gives a transition condition.

3. Transformation toughening

The stress-induced transformations which can cause significant toughening include martensitic [6, 7] and ferroelastic [8] transformations, as well as twinning. The former involves both dilatational and shear components of the transformation strain, while the latter typically have only a shear component. Martensitic transformation toughening in ZrO_2 has been most extensively investigated and will be given primary emphasis.

At the simplest level, transformation toughening can be regarded as a process dominated by the dilatational component of the stress-free strain, e_{ii}^T as expressed by the associated supercritical* stress–strain curve. Then, on the basis of the path independence of the J integral, initial crack growth occurs without toughening (Fig. 2). However, on crack extension, J becomes path dependent, and toughening develops. For steady state conditions, the supercritical toughening is readily derived as [6, 9–11]

$$\Delta G_c = 2hf\int \sigma \, d\varepsilon \qquad (5)$$

where the integral represents the hysteresis area depicted in Fig. 1, such that (*cf.* eqn. (1))

$$\Delta G_c = 2hf\sigma_0 e_{ii}^T \qquad (6)$$

This steady state level is attained after crack extensions of about $5h$ [9]. A directly equivalent result for the increase ΔK_c in critical stress intensity factor can be derived by considering the residual stress field created by transformation, giving [9]

$$\Delta K_c = \frac{\lambda E e_{ii}^T f h^{1/2}}{1-\nu} \qquad (7)$$

where E is the composite modulus and $\lambda = 0.22$ is a coefficient.

Comparison with experimental data [6] (Fig. 5) has revealed that eqn. (6) consistently underestimates the toughness, because shear effects and zone widening have not been incorporated. One hypothesis regarding the shear strain which seemingly coincides with existing observations and measurements involves non-associated flow [6]. Specifically, it is presumed that the shear stress dominates the nucleation of the transformation, because of the large transformation shear strain

*Supercritical refers to the condition wherein all particles within the process zone fully transform.

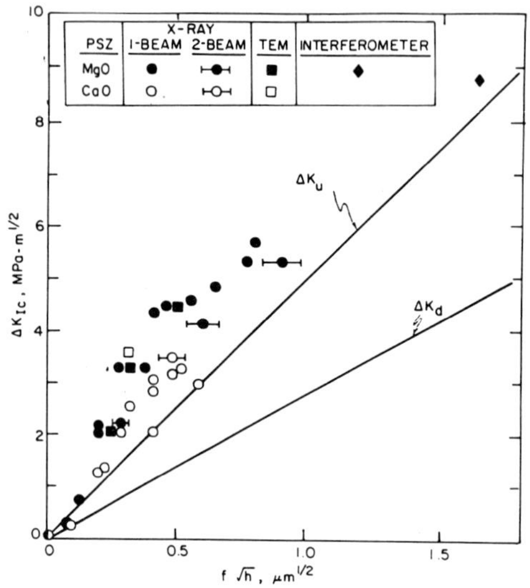

Fig. 5. A comparison between theory and experiment for various partially stabilized zirconia materials: ΔK_u refers to a zone shape dictated by the equivalent stress, whereas ΔK_d refers to a zone shape governed by the mean stress.

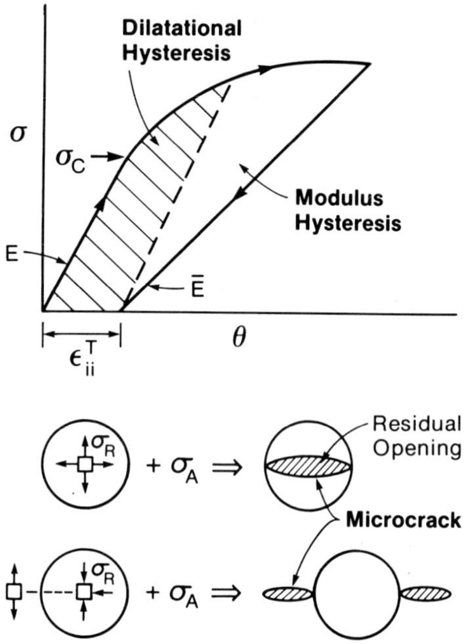

Fig. 6. The basic concepts of microcrack toughening.

associated with the nucleus, but that the residual particle strain is predominantly dilatational, because extensive twinning eliminates long-range shear strains [12]. This premise results in the zone profile, in plane strain, which, by virtue of the absence of the transformed material ahead of the crack, predicts a toughening that exceeds eqn. (6) and furthermore agrees well with experimental data (Fig. 5). Such zone profiles are also consistent with available observations [13, 14]. While the above notions offer a self-consistent explanation of the transformation toughening in some cases, it is emphasized that alternative models of the influence of shear strains must be used to address other transformation problems, especially twin and ferroelastic toughening.

The zone size h represents the major microstructural influence on toughness. Clearly, h is governed by a nucleation law [15]. However, a fully validated law for nucleation does not yet exist, because the nucleation sequence has not been established. Consequently, connections between h and the microstructure still cannot be specified. Nevertheless, certain trends are apparent, based on the free energy of the fully transformed product [16, 17]. Specifically, h invariably decreases with increase in temperature and decrease in particle size. A temperature- and particle-size-dependent toughness is thus inevitable for this mechanism.

Finally, it should be noted that h is not usually unique but varies with crack extension, especially during initial growth [9, 18, 19]. This behavior leads to resistance curves which increase over larger crack extension lengths than expected from the constant-h analysis [20]. The full simulation of this process has yet to be performed.

4. Microcrack toughening

The phenomenon of microcrack toughening has been postulated for over a decade [21–23] and, indeed, a range of materials exhibit trends in toughness with particle size, temperature, etc., qualitatively consistent with this mechanism. However, as yet there is only one fully validated example of this mechanism: Al_2O_3 toughened with monoclinic ZrO_2 [24]. The fundamental premise concerning the mechanism is depicted in Fig. 6. Microcracks occur within regions of local residual tension, caused by thermal expansion mismatch or by transformation [25, 26]. The microcracks locally relieve the residual tension and thus cause a dilatation governed by the volume displaced by the microcrack. Furthermore, the microcracks reduce the elastic modulus within the microcrack process zone. Consequently, the elemental stress–strain curve for a microcracking solid has the form depicted in Fig. 6. The hysteresis dictated by this curve, when the microcracks are activated by the passage of a macrocrack, contributes to the change in toughness, as elaborated below. However, this contribution is partially counteracted by a degradation of the material ahead of the microcrack. The full extent of the degradation is at present unknown.

The crack shielding can be conveniently separated into dilatational and modulus contributions. The former depends on the process zone size and shape, while the latter depends only on the zone shape. The dilatational contribution ΔK_d has essentially the same form as eqn. (7). Notably for steady state supercritical conditions [24, 26],

$$\Delta K_d = -\lambda E f \theta^T h^{1/2} \tag{8}$$

where θ^T is the volume strain, as governed by the microcrack shape and the prior residual tension. The steady state modulus contribution ΔK_m is [26]

$$\frac{(1-\nu)\,\Delta K_m}{K_c} = \left(k_1 - \tfrac{5}{8}\right)\left(\frac{\mu}{\bar{\mu}-1}\right) + \left(k_2 + \tfrac{3}{4}\right)\left(\frac{\bar{\nu}\mu}{\bar{\mu}-\nu}\right) \tag{9}$$

where $k_1 \approx 0.017$, $k_2 \approx -0.043$, μ is the shear modulus, the bar refers to microcracked material and K_c is the composite toughness. The contributions ΔK_d and ΔK_m are additive unless both are large compared with K_c, whereupon interaction effects occur and numerical procedures are then needed to determine the change in toughness [27].

Comparison between theory and experiment has been made for the Al_2O_3–ZrO_2 system [24], based on the microcrack density measurements presented in Fig. 7 (obtained using electron microscopy), and the θ^T results for annular microcracks with interface opening. For this case, the contributions to toughening from the dilatation and the modulus are evaluated as approximately 2.5 and 5 MPa $m^{1/2}$ respectively, compared with a measured composite toughness of about 6 MPa $m^{1/2}$. While this comparison is reasonable and validates microcracking as the prevalent toughening mechanism, present understanding of micro-

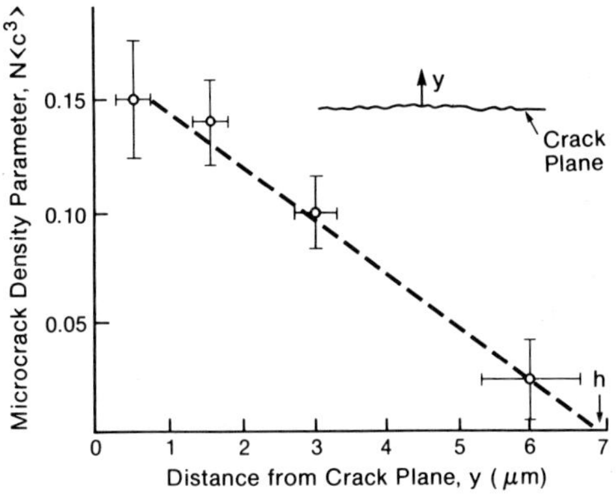

Fig. 7. Trends in microcrack density with distance from the crack surface for an Al_2O_3–ZrO_2 material with predominantly monoclinic ZrO_2.

crack toughening is nevertheless incomplete. In particular, interaction effects between the modulus and dilatational contribution, as well as near-tip degradation effects require further investigation.

Because of the experimental difficulty involved in detecting microcracks*, there are no other validated examples of microcrack toughening. However, it is noted that several particulate-reinforced systems exist wherein enhanced toughness only develops when the particulates exceed a critical size. Microcracking is a likely contribution to the toughening in several of these systems.

One potentially detrimental feature of microcrack toughening is the incidence of thermal microcracks at the largest particles in the distribution [24]. Such cracks can be strength limiting, resulting in material which is relatively tough but has only moderate strength [28]. Avoidance of such strength-limiting cracks requires stringent control of the size distribution of the reinforcing particles, just beneath the critical size for thermal microcracking.

Finally, it is noted that the dilatational contribution to microcrack toughening would usually be temperature dependent because of the reduction in residual stress with increase in temperature. However, the modulus contribution is temperature invariant, at least at temperatures below which the process zone size is relatively large compared with the microcrack spacing. Microcrack toughening is, typically, less potent than transformation toughening.

*Typical residual crack openings are less than 2 nm.

5. Ductile phase toughening

Practical ceramic systems toughened by ductile phases typically have interpenetrating networks of the brittle and ductile phases. For such systems, cracks in the brittle ceramic phase are required to intercept the ductile phase such that the area fraction of reinforcing phase on the crack plane is of the same order as the volume fraction. In these materials, two sources of toughening occur, as governed by plasticity in the ductile phase. Segments intercepted by the crack exhibit plastic stretching in the crack wake [29] (Fig. 3) and contribute to the toughness by inhibiting crack opening. Simultaneously, plastic straining of segments in a process zone causes crack shielding (Fig. 3). The former mechanism has the general characteristics that high toughness is encouraged by a large value of the product of the work of rupture and particle size [29]. By contrast, the crack shielding process tends to become larger for ductile sections having a small size and low yield strength. An estimate of the shielding has been derived by considering the residual strain in the crack wake caused by plastic distortion of ductile enclaves in an elastic matrix. This estimate indicates that, to achieve an appreciable shielding contribution to the toughness, the yield strength Y must be less than about 10 MPa and the ductile phase sections less than about 1 μm in diameter.

The contribution to the toughness from stretching between crack surfaces is explicitly governed by the stress–opening curve (eqn. (2)), as depicted in Fig. 8, such that [29, 30]

$$\Delta G_c = \chi f Y a \qquad (10)$$

where a is the radius of the ductile enclave and χ is a parameter which depends on the work-hardening rate n, the ductility of the ductile phase and the bonding between the ductile and brittle phases. Calculations conducted for fully bonded, fully ductile phases which neck to a point predict χ values of between 0.5 and 3 [29] (Fig. 9). However, larger values of χ are possible when limited debonding occurs [30]. The maximum χ observed by experiment is about 6. Based on this maximum value, experimental toughness data obtained for the Al_2O_3–Al system (Fig. 10) conform well with the values predicted by eqn. (10). However, the incidence of debonding in this system has not been established. Further understanding of this mode of toughening is thus predicated on a rigorous assessment of interface debonding.

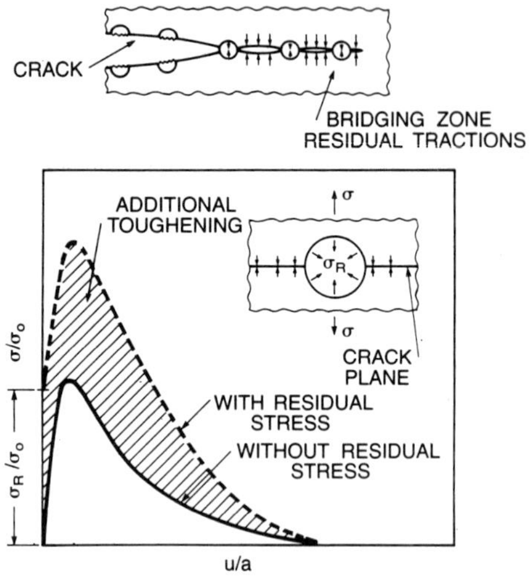

Fig. 8. Stress–crack opening relation for a crack bridged by a ductile particle.

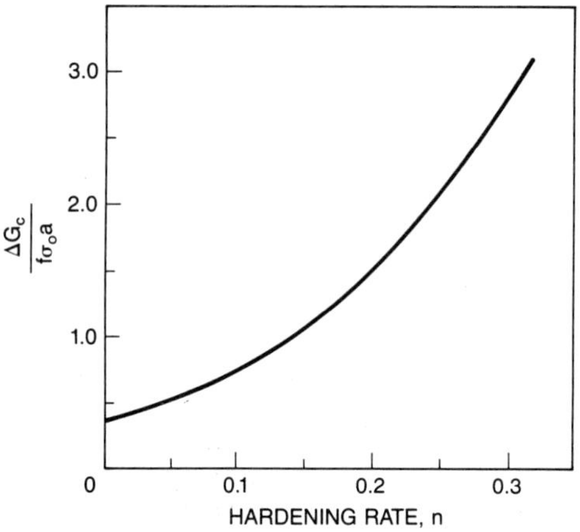

Fig. 9. Trends in toughness with hardening rate with $\Delta G_c/fYa$ being equivalent to χ in eqn. (10).

Finally, it is noted that ductile particle toughening should be temperature sensitive, because of the temperature dependence of the yield strength.

6. Fiber and whisker toughening

The toughening of ceramics by brittle fibers and/or whiskers occurs subject to debonding at the interface. In the absence of debonding, because the fiber and matrix typically have comparable toughnesses, the composite is brittle and

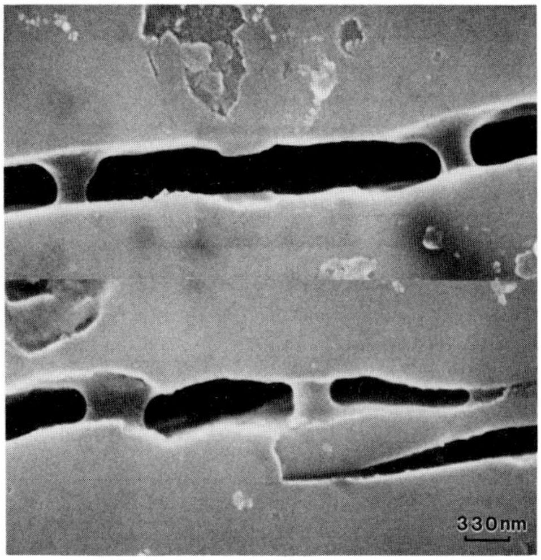

Fig. 10. Ductile rupture in Al_2O_3–Al revealing that the aluminum fails by necking.

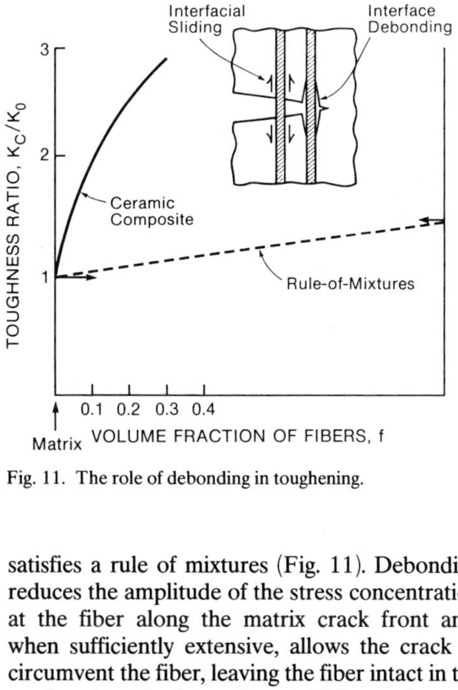

Fig. 11. The role of debonding in toughening.

satisfies a rule of mixtures (Fig. 11). Debonding reduces the amplitude of the stress concentration at the fiber along the matrix crack front and, when sufficiently extensive, allows the crack to circumvent the fiber, leaving the fiber intact in the crack wake. The intact fiber inhibits crack opening and allows a composite toughness exceeding that of either constituent (Fig. 11). The overwhelmingly important issue in fiber or whisker toughening thus concerns the extent of debonding, its dependence on interface properties, and its effect on crack opening and fiber failure.

The full debonding and fiber fracture problem has several different aspects which eventually require resolution. These include the characteristics of interface fracture properties and associated requirements for the initiation and extent of debonding at the matrix crack tip, as well as the influence of debond length on the crack-opening force and the relationship between debond length and the fiber fracture force. While some progress has been made toward an understanding of these problems, several important questions remain. The present status is briefly reviewed.

The central issue to be addressed in this article concerns the mode I propagation resistance for matrix cracks on a plane normal to the fiber axis. Once this problem has been confronted and resolved in terms of interface debonding etc., mixed-mode and delamination cracking effects require resolution, in order to describe the overall fracture behavior of the composite.

6.1. Initial debonding

Even though debonding at the matrix crack tip is a prerequisite to fiber toughening, there is minimal quantitative understanding of the phenomenon. Two mechanics solutions have relevance: one for forked or kinked cracks [31] and the other for debonding ahead of a crack tip [3]. The solution to the former clearly indicates that debonding is a mixed-mode problem (Fig. 12). Furthermore, debonding initiates when the interface fracture resistance is about quarter that of the fibers. However, further debonding becomes increasingly difficult. The extent of the difficulty is estimated from the latter solution, which reveals that, in order to achieve extensive debonding, the interface fracture resistance should be about an order of magnitude less than that of the matrix (Fig. 13). Further experimental and theoretical study of this important phenomenon would greatly facilitate the design of interfaces that allow optimal debonding.

6.2. Crack opening

Preliminary results have been obtained for the crack-opening forces associated with intact fibers normal to the crack plane. For the extreme condition of extensive debonding, with opening inhibited by friction, a shear lag analysis [2, 3] indicates that the initial segment is linear and represents behavior wherein the interfacial shear stress is

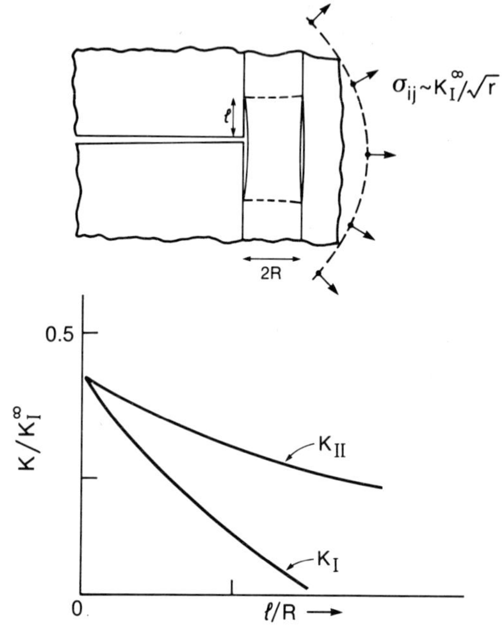

Fig. 12. A schematic diagram illustrating trends in local K with debond length.

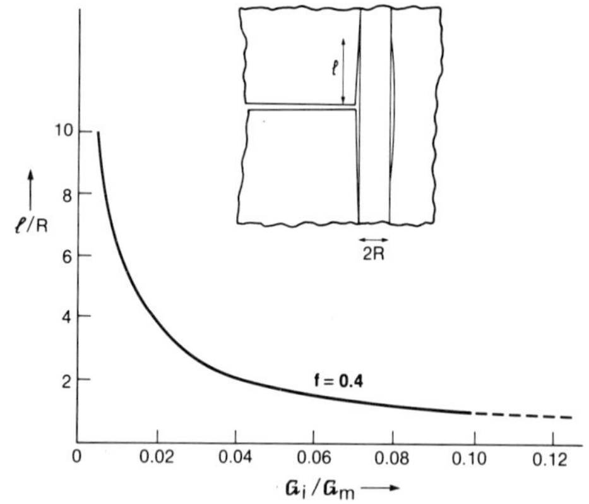

Fig. 13. Trends in debond length with relative interface fracture resistance.

less than the sliding stress τ such that the stress on the fiber end is

$$\sigma = \frac{lfE_f(1+\xi)}{R}u \tag{11}$$

where $\xi = E_f f/(1-f)E_m$. At larger stresses ($\sigma > 2\tau f E_f(1+\xi)/l$), the behavior is dominated by sliding and is given by

$$\sigma = 2f(1+\xi)\left(\frac{E_f\tau}{R}\right)^{1/2}\left(u - \frac{\tau R}{E_f}\right)^{1/2} \tag{12}$$

For debonds without friction, it is noted that a steady state condition exists when the debond length $l > 2R$, wherein the stress intensities and strain energy release rate for the debond become independent of l [32]. When the matrix has the same elastic properties, the steady state quantities are

$$\begin{aligned}\frac{K_I}{E\,\Delta\alpha\,\Delta TR^{1/2}} &= 0.28 + \frac{0.11t}{E\,\Delta\alpha\,\Delta T}\\ \frac{K_{II}}{E\,\Delta\alpha\,\Delta TR^{1/2}} &= -0.51 + \frac{0.45t}{E\,\Delta\alpha\,\Delta T}\\ \frac{G}{ER(\Delta\alpha\,\Delta T)^2} &= 0.31 - \frac{0.37t}{E\,\Delta\alpha\,\Delta T} + 0.20\left(\frac{t}{E\,\Delta\alpha\,\Delta T}\right)^2\end{aligned} \tag{13}$$

where t is the stress on the fiber between the crack surfaces, $\Delta\alpha$ is the thermal expansion mismatch (fiber minus matrix) and ΔT is the cooling range. Steady state debonding behavior dictates that the crack opening u varies linearly with debond length, such that [32, 33]

$$u \approx \frac{tl}{E}\beta(f, \Sigma) + u_o(\Delta\alpha\,\Delta T, f) \tag{14}$$

where β is a function of the volume fraction f of fibers and the modulus ratio Σ and u_o is an offset at zero stress.

When fibers fail, the above functions are clearly modified and the further opening of the crack often occurs subject to a declining stress, as elaborated below.

6.3. *Fiber failure*

The process of fiber failure is not yet understood in detail. Available analyses have focused on the problem of fracture behind the matrix crack front, neglecting fractures which might occur in the crack tip field. When fibers fail in the wake, large debond lengths encourage a statistical mode of fracture within the debond zone, as governed by the flaw size distribution in the fiber.

The statistical mode of fiber failure leads to a fiber failure probability which increases with increase in distance behind the crack tip [34], while the most probable fiber failure location displaces further from the crack surface. The associated stress–displacement curves have sliding and pull-out contributions which can be derived in terms of the fracture properties of the fibers and the shear resistance of the interface. For fibers normal to the crack having a flaw strength distribution,

$$\int g(S)\,dS = \left(\frac{S}{S_0}\right)^m \tag{15}$$

where S_0 is a scale parameter and m a shape parameter; weakest-link statistics give the crack-opening stresses, plotted in Fig. 14, in terms of the non-dimensional qualities σ/Σ and u/ν where, in this case, Σ and ν are defined by

$$\Sigma = \left\{\frac{S_0{}^m \tau(m+1)}{2\pi R^2}\right\}^{1/(m+1)}$$
$$\nu = \frac{\Sigma^2 R}{4E_f\tau(1+\xi)} \tag{16}$$

When the fibers or whiskers are inclined to the crack plane, additional effects occur. Clearly, the inclination angle affects the debond length and the stress at fiber fracture. However, following such fracture, crack-opening interference may persist, because of the trajectory of the associated fiber crack. An enhanced declining segment of the $\sigma(u)$ function may then obtain.

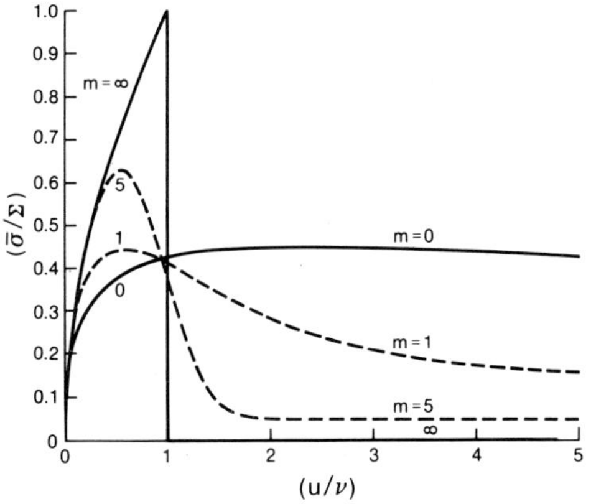

Fig. 14. Stress–crack-opening functions for statistical fiber failure with friction.

6.4. The toughening

Trends in toughening can be deduced from the $\sigma(u)$ function, using eqn. (2). However, resistance curve and steady state cracking effects must be considered as important related effects. Perhaps the most straightforward case involves the whisker-reinforced systems, Al_2O_3–SiC and Si_3N_4–SiC. For these materials, limited debonding along an amorphous interphase, without friction, seems to contribute to the toughness. Consequently, for aligned whiskers normal to the crack plane, eqns. (2) and (14) predict a toughening that varies linearly with debond length. The fiber fracture condition, as well as the initial crack-opening force, is also influenced by the residual field. Analysis of this problem has not yet been conducted and, hence, trends in toughness with residual stress are at present unknown. Furthermore, satisfactory solutions for toughness in the presence of non-aligned whiskers, which fail in the crack wake by bending plus tension, have yet to be developed.

Composites which exhibit extensive debonding and frictional sliding typically have pull-out contributions to the toughness [34, 35]. The associated $\sigma(u)$ functions (Fig. 14) may simply be integrated to predict the asymptotic toughness, using eqn. (2). Frequently, however, the mean pull-out length exceeds the crack mouth opening (except for very long cracks), indicating that asymptotic behavior does not occur and, instead, fracture is governed by a rising resistance curve. Simulation of trends in toughness with crack extension has yet to be conducted.

One important feature of fiber and whisker toughening is the absence of an obvious temperature-dependent parameter, except for possible temperature effects on the debond resistance. This toughening approach thus seems to have the greatest potential for toughness at elevated temperature.

7. Other mechanisms

A number of ceramic materials have a contribution to toughness from intact grains along the crack surface [36]. The phenomenon is evident in several large-grained polycrystals, most notably Al_2O_3, and in certain glass ceramics. The contribution of these intact grains to toughness is formally similar to whisker toughening. However, the corresponding $\sigma(u)$ law and the analogous criteria which govern grain intactness are not yet well understood. Nevertheless, by direct analogy to whisker toughening, grain boundaries weakened by additives have been demonstrated [37] to encourage bridging effects and thus to enhance toughness. The ability of certain grains to remain intact probably involves residual stresses, with associated grain size effects [38], perhaps coupled with statistical variability of the grain boundary fracture resistance.

8. Synergism

When bridging and process zone mechanisms occur in conjunction, synergistic toughening becomes possible [39, 40]. The basic concept is illustrated in Fig. 3. Specifically, the tractions exerted on the crack surface by the intact particles can locally widen the process zone. This zone widening, in turn, enhances the tip shielding above the level expected in the absence of bridging. The net effect of both processes on toughness is larger than the sum of the individual mechanisms. Indeed, cases exist wherein the toughening is multiplicative. This intriguing topic is expected to be the subject of appreciable future research.

9. Concluding remarks

A number of substantial issues remain to be addressed, particularly for fiber- and whisker-toughened materials. Present evidence suggests that coatings are needed on the fibers having a low debond resistance and a small subsequent sliding resistance along the debond. The coatings must also be thermodynamically and microstructurally stable. Coatings such as BN, SiO_2 and carbon have been widely used, but all have major limitations, especially at elevated temperatures. Other interphases are needed and primary attention is now being focused on the identification of appropriate materials and fiber coating approaches. Interesting approaches include refractory metal and intermetallic coatings as well as porous coatings.

References

1 J. Aveston, G. A. Cooper and A. Kelly, *National Physical Laboratory Conf. on the Properties of Fiber Composites*, IPC, Guildford, 1971, p. 15.
2 D. B. Marshall, B. N. Cox and A. G. Evans, *Acta Metall., 33* (1985) 2013.
3 B. Budiansky, J. W. Hutchinson and A. G. Evans, *J. Mech. Phys. Solids, 2* (1986) 167.
4 K. Prewo and J. J. Brennan, *J. Mater. Sci., 15* (1980) 463.
5 D. B. Marshall and A. G. Evans, *J. Am. Ceram. Soc., 68* (1985) 225.
6 A. G. Evans and R. M. Cannon, *Acta Metall., 34* (1986) 761.
7 N. Claussen, M. Rühle and A. H. Heuer (eds.), *Advances in Ceramics*, Vol. 12, American Ceramic Society, Columbus, OH, 1985.
8 A. Virkar and R. Matsumoto, *J. Am. Ceram. Soc., 69* (1986) C224.
9 R. McMeeking and A. G. Evans, *J. Am. Ceram. Soc., 65* (1982) 242.
10 D. B. Marshall, M. D. Drory and A. G. Evans, in R. C. Bradt, A. G. Evans, D. P. H. Hasselman and F. F. Lange (eds.), *Fracture Mechanics of Ceramics*, Vol. 5, Plenum, New York, 1983, p. 289.
11 B. Budiansky, J. W. Hutchinson and J. Lambropolous, *Int. J. Solid Struct., 19* (1983) 337.
12 M. Rühle and W. M. Kriven, in H. Aaronson, D. Laughlin, R. F. Sekerka and C. M. Wayman (eds.), *Solid–Solid Phase Transformations*, AIME, New York, 1982, p. 97.
13 M. V. Swain and L. R. F. Rose, *J. Am. Ceram. Soc., 69* (1986) 511.
14 M. Rühle, B. Kraus, A. Strecker and D. Waidelich, in N. Claussen, M. Rühle and A. H. Heuer (eds.), *Advances in Ceramics*, Vol. 12, Plenum, New York, 1985, p. 256.
15 M. Rühle and A. H. Heuer, in N. Claussen, M. Rühle and A. H. Heuer (eds.), *Advances in Ceramics*, Vol. 12, Plenum, New York, 1985, p. 14.
16 A. G. Evans, M. D. Drory, N. Burlingame and W. Kriven, *Acta Metall., 29* (1981) 447.
17 F. F. Lange, *J. Mater. Sci., 17* (1982) 225.
18 D. B. Marshall, *J. Am. Ceram. Soc., 69* (1986) 173.
19 A. H. Heuer, *J. Am. Ceram. Soc., 70* (1987) 689.
20 M. V. Swain and R. J. H. Hanninck, in N. Claussen, M. Rühle and A. H. Heuer (eds.), *Advances in Ceramics*, Vol. 12, Plenum, New York, 1985, p. 225.
21 R. G. Hoagland, J. D. Embury and D. J. Green, *Scr. Metall., 9* (1975) 907.
22 N. Claussen, J. Steeb and R. F. Pabst, *Am. Ceram. Soc., Bull., 56* (1977) 559.
23 R. G. Hoagland and J. D. Embury, *J. Am. Ceram. Soc., 63* (1980) 404.
24 M. Rühle, A. G. Evans, P. G. Charalambides, R. M. McMeeking and J. W. Hutchinson, *Acta Metall., 35* (1987) 2701.
25 A. G. Evans and K. T. Faber, *J. Am. Ceram. Soc., 67* (1984) 255.
26 J. W. Hutchinson, *Acta Metall.*, to be published.
27 P. G. Charalambides and R. M. McMeeking, *Mech. Mater., 6* (1987) 71.
28 M. Rühle, N. Claussen and A. H. Heuer, *J. Am. Ceram. Soc., 69* (1986) 195.
29 L. Sigl, P. Mataga, B. J. Dalgleish, R. M. McMeeking and A. G. Evans, *Acta Metall., 36* (1988) 945.
30 M. F. Ashby, to be published.
31 B. Cotterell and J. R. Rice, *Int. J. Fract., 16* (1980) 155.
32 P. G. Charalambides and A. G. Evans, *J. Am. Ceram. Soc.*, to be published.
33 M. Rühle, B. J. Dalgleish and A. G. Evans, *Scr. Metall., 21* (1987) 681.
34 M. D. Thouless and A. G. Evans, *Acta Metall., 36* (1988) 517.
35 D. C. Phillips, *J. Mater. Sci.*, 7 (1972) 1175.

36 P. L. Swanson, C. J. Fairbanks, B. R. Lawn, Y. W. Mai and B. J. Hockey, *J. Am. Ceram. Soc., 70* (1987) 289.
37 R. F. Cook, C. J. Fairbanks, B. R. Lawn and Y. M. Mai, *J. Mater. Res., 2* (1987) 345.
38 Y. Fu and A. G. Evans, *Acta Metall., 30* (1982) 1619.
39 B. Budiansky, in J. P. Lamb (ed.), *Micromechanics II, Proc. 10th U.S. Congr. of Applied Mechanics, Austin, TX, 1986*, p. 25.
40 A. G. Evans, P. W. R. Beaumont, Z. Ahmad and D. G. Gilbert, *Acta Metall., 34* (1986) 79.

Materials Science and Engineering, *A105/106* (1988) 77–82

Microcrack and Transformation Toughening of Zirconia-containing Alumina*

MANFRED RÜHLE

Materials Department, College of Engineering, University of California, Santa Barbara, CA 93106 (U.S.A.)

(Received January 13, 1988)

Abstract

Experimental studies on the microcrack and transformation toughening of zirconia(ZrO_2)-containing alumina are presented. The investigated ceramics contain about the same volume fractions of tetragonal and monoclinic ZrO_2 particles respectively. The microcrack parameters (microcrack length, microcrack density and microcrack opening) and the transformation behavior are determined by transmission electron microscopy studies. The microstructural results are used to explain the mechanical properties of the toughened material.

1. Introduction

For more than 10 years, scientists have assumed that the toughening of ceramics by a microcrack process zone should be possible [1–3]. Evans and coworkers [4–6] found considerable similarities between transformation toughening and microcrack toughening. Evans and coworkers showed that the change in toughness is caused by the residual dilatation of the material in the process zone and in the wake of a crack. The required dilatation arises in microcrack toughening because the microcracks relieve local residual tensile stresses which can be caused by thermal expansion mismatch, transformation etc. However, until now, no systematic experimental studies on microcrack toughening have been performed to allow a direct comparison between theory and experiment.

Experimental observations [7] on zirconia (ZrO_2)-toughened alumina (ZTA) suggest that this material is an ideal candidate for such a systematic investigation. Lindemann and Claussen [8] processed ZTA ceramics with a fixed total content of 15 vol.% ZrO_2. The processing parameters varied such that the volume fraction x_t of tetragonal ZrO_2 (t-ZrO_2) varied from 0.23 to 0.86. x_t was determined by X-ray diffraction analyses of polished surfaces of a bulk specimen. The flexural strength σ and the fracture toughness were measured [8] and the results are represented in Fig. 1. For $x_t = 0.46$, $\sigma = 950$ MPa and $K_{Ic} = 5.9$ MN m$^{-3/2}$. Surprisingly, the toughnesses of different materials with different x_t are nearly independent of x_t which explicitly implies that microcracking and transformation influence the toughness quite similarly. Until now, two materials have been studied, one primarily containing monoclinic ZrO_2 (m-ZrO_2) ($x_t = 0.23$) and the other containing primarily t-ZrO_2 ($x_t = 0.86$). Rühle *et al.* [7] showed that the toughness of the $x_t = 0.86$ material is increased by the t-ZrO_2→m-

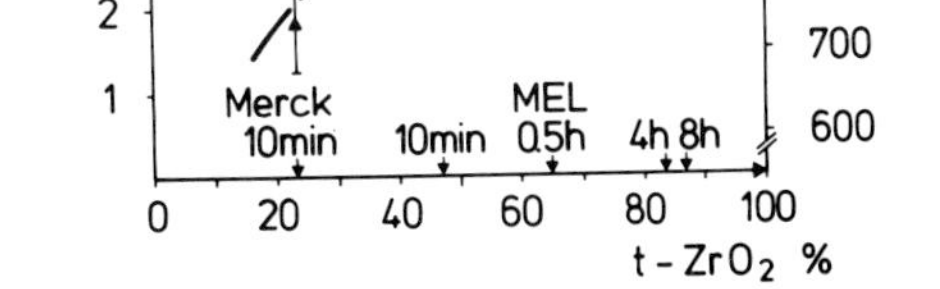

Fig. 1. Fracture toughness and flexural strength of ZTA with 15 vol.% ZrO_2 as a function of t-ZrO_2 volume fraction.

---

*Paper presented at the 3rd International Conference on the Science of Hard Materials, Nassau, The Bahamas, November 9–13, 1987.

0921-5093/88/$3.50

© Elsevier Sequoia/Printed in The Netherlands

ZrO_2 transformation which occurs in front of the crack tip. The transformation leads to a residual dilatation in the wake of the crack which is responsible for the increase in toughness. In the $x_t = 0.23$ material, mainly microcracking occurs. different microcrack parameters could be determined [9] and the change in toughness due to microcracking could be explained by a theory which has been developed recently by Hutchinson [10]. Microcrack toughening relates to quantities such as the number density of microcracks in the process zone, the microcrack length distribution and the residual openings of the microcracks [9, 10]. Microstructural quantities can be determined by transmission electron microscopy (TEM) [7, 11].

In this paper the results on a material which contains about equal volume fractions of t-ZrO_2 and m-ZrO_2 ($x_t = 0.46$) are presented. Both mechanisms (microcracking and transformation) should occur in this ZTA.

2. Microstructural analysis

TEM studies reveal that all ZrO_2 particles are present as intercrystalline facetted grains. The size distributions of Al_2O_3 grains and ZrO_2 grains were determined by a standard stereological analysis. The area A of each particle or grain was converted to an equivalent diameter $d = A^{1/2}$ appropriate to one of the 12 size groups defined in Table 1. The grain size distribution of Al_2O_3 (within the experimental error) is the same as for $x_t = 0.23$ and $x_t = 0.86$ ZTA [7, 11]. The volume fractions of ZrO_2 particles belonging to the different size groups are shown in Fig. 2. The partitioning of the volume fraction of the different size groups in Fig. 2 will be explained later in the text.

TABLE 1 Size range of the different size groups—logarithmic scale

Size group	*Lower bound* (μm)	*Upper bound* (μm)
1	0	0.10
2	0.10	0.13
3	0.13	0.18
4	0.18	0.24
5	0.24	0.32
6	0.32	0.44
7	0.44	0.59
8	0.59	0.79
9	0.79	1.06
10	1.06	1.42
11	1.42	1.91
12	1.91	∞

The stereological analysis allows the determination of the critical particle size d_c [11] for spontaneous t-ZrO_2→m-ZrO_2 transformation on cooling [11]. Figure 3 shows that all particles in size group 8 or a lower group had a tetragonal symmetry which results in $d_c = 0.71$ μm. The d_c value is larger than values obtained from earlier studies. TEM studies show that all grain boundaries are wetted by an amorphous grain boundary film (Fig. 4). Chemical analyses by energy-dispersive spectroscopy at triple junctions of grains demonstrate that the glass consists of SiO_2, Al_2O_3 and small amounts of ZrO_2 (less than 8%). The composition of the glass varies within the material.

In situ straining experiments were performed on the material. TEM specimens about 1 μm

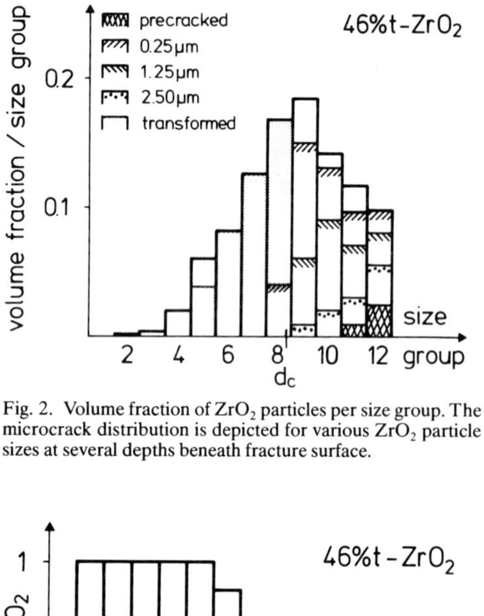

Fig. 2. Volume fraction of ZrO_2 particles per size group. The microcrack distribution is depicted for various ZrO_2 particle sizes at several depths beneath fracture surface.

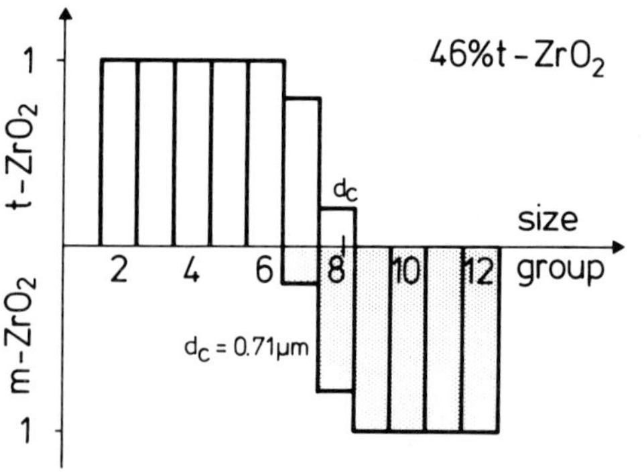

Fig. 3. Determination of critical particle size d_c. For each size group (see Table 1) the division into t-ZrO_2 and m-ZrO_2 particles is marked.

Fig. 4. (a) Bright field and (b) dark field TEM images. The dark field image is formed by diffuse scattering electrons. The faint lines in (b) prove the existence of an amorphous grain boundary film.

thick with a selected geometry [12] were deformed in a high voltage electron microscope and the propagation of stress-induced cracks was observed. For distances of up to 5 μm, the t-ZrO_2 grains transform in front of the crack tip (Fig. 5). Frequently, small t-ZrO_2 grains do not transform if they are close to a large m-ZrO_2 grain. A semi-quantitative analysis shows that the halfwidth of the transformation zone is about 3 μm, which is in accordance with earlier studies [7, 12]. However, similar to the $x_t = 0.23$ ZTA, hardly any microcracks at large m-ZrO_2 inclusions could be observed. Only crack branching could be seen in a few cases.

Rühle *et al.* [7] pointed out that *in situ* experiments are not suitable for the determination of microcracking since the high voltage electron microscopy foil is too thin to provide the mechanical constraints of a bulk specimen. The mechanical constraint is essential for microcrack nucleation. Therefore, rather tedious experiments had to be performed to assess the extent of microcracking and transformation associated with crack propagation.

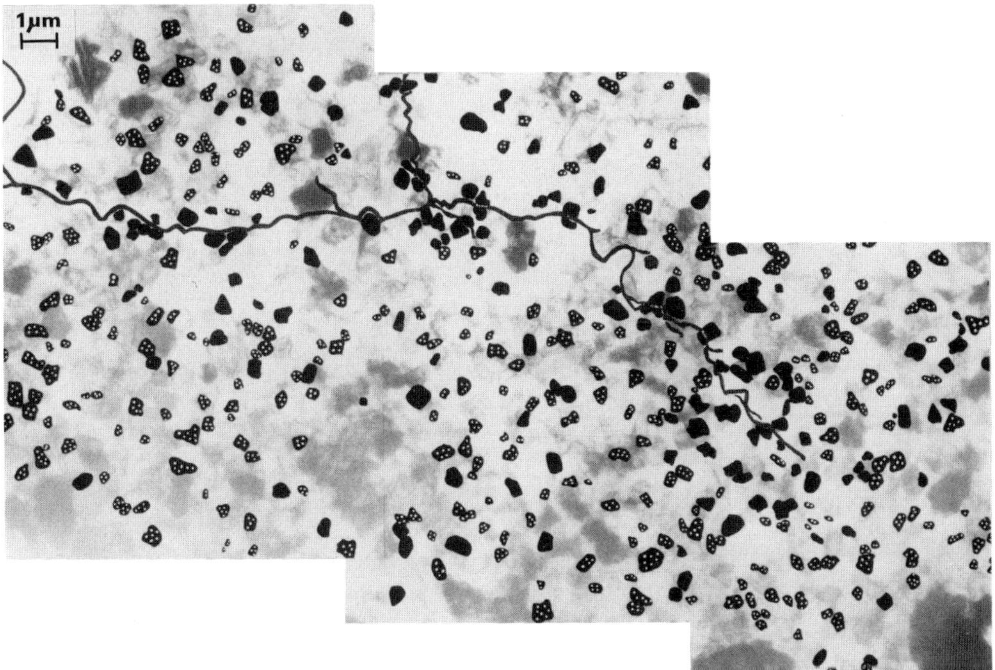

Fig. 5. *In situ* straining experiment for the $x_t = 0.46$ specimen: ▨, t-ZrO_2 transformed during crack propagation; ■, m-ZrO_2. The TEM micrograph shows a region close to a crack which was introduced within the transmission electron microscope. ZrO_2 particles with diameters smaller than 0.7 μm are marked.

For the preparation of the specimen, a layer of a selected thickness y is removed from the fracture surface by calibrated ion thinning or by grinding. The TEM specimen is prepared by the standardized back-thinning technique. Five values of y (0.25, 1.25, 2.5, 4 and about 20 μm) were selected. The accuracy of the selection of y is better than 10%.

Far from the fracture surface ($y = 20$ μm), no radial microcracks were analyzed. Only a very few cracks (about 20 μm) were observed in connection with very large clusters of m-ZrO_2. It has been shown [10] that the large cracks determine the flexural strength of the material. Microcracks and small mZrO_2 could be analyzed in foils with $y \leqslant 4$ μm; examples are shown in Fig. 6.

For $y = 0.25$, all ZrO_2 grains with diameters greater than 0.2 μm are of monoclinic symmetry. It can be concluded that the t-ZrO_2 with $0.2 < d < d_c$ transformed in the stress field of the propagating crack. With increasing distance y from the fracture surface, the fraction of m-ZrO_2 grains with $d < d_c = 0.71$ μm decreased. At a distance y of 2.5 μm, essentially no t-ZrO_2 grains transform for a diameter $d < 0.5$ μm. In principle, a transformation zone can be evaluated from the observations at different y. However, the results obtained so far are not sufficient for a quantitative evaluation similar to those in refs. 7 and 12.

However, a high density of microcracks could be analyzed in these specimens and the visibility of the microcracks depends very sensitively on their inclination with respect to the incoming electron beam. As noted in ref. 10, the visibility decreases strongly. Therefore, the interpretation of the TEM contrast becomes more difficult when the angle between the flanks of the crack and the direction of the incoming electron beam exceed the ratio of the microcrack opening δ to the height of the microcrack.

Quantitative evaluation allows the determination of the microcrack density parameter $\alpha = N\langle l^3 \rangle$, as described by Budiansky and O'Connell [13], where l is the length of the microcracks and N the number density per cubic centimeter. The results of the analysis are shown in Fig. 7. α is highest for small distances y from the bulk crack plane and decreases nearly linearly with increasing distance from the crack plane. Furthermore, the microcrack length l as well as the residual microcrack opening δ were determined for m-ZrO_2 particles of different radii. The results are shown in Figs. 8 and 9. δ and l depend only

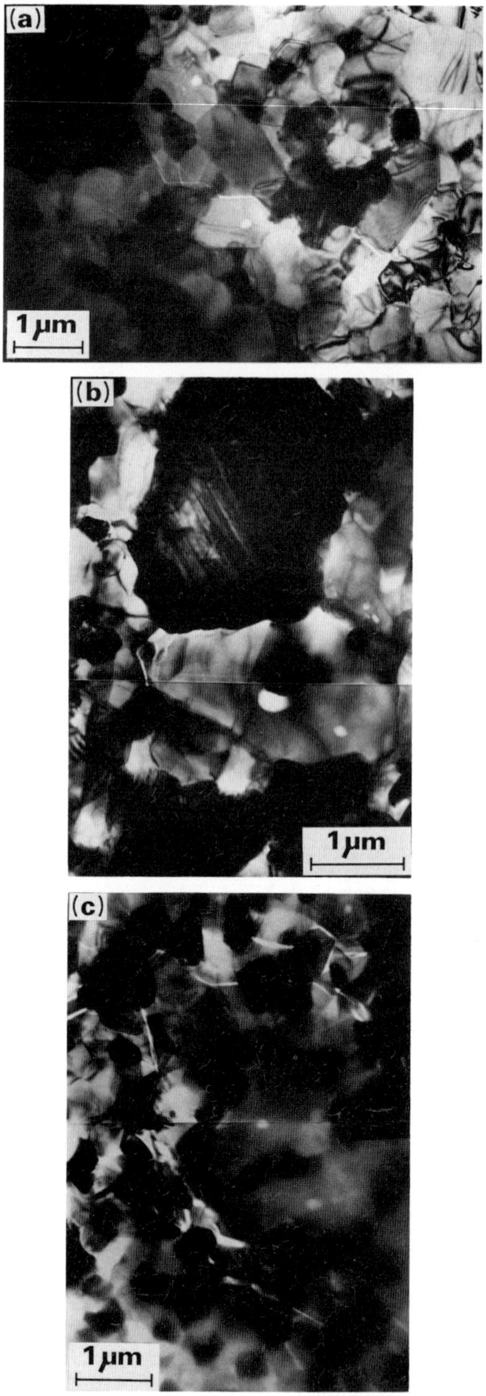

Fig. 6. (a)–(c) TEM micrographs of a specimen taken at a distance y of 0.25 μm from the bulk fracture surface. Microcracks can readily be identified.

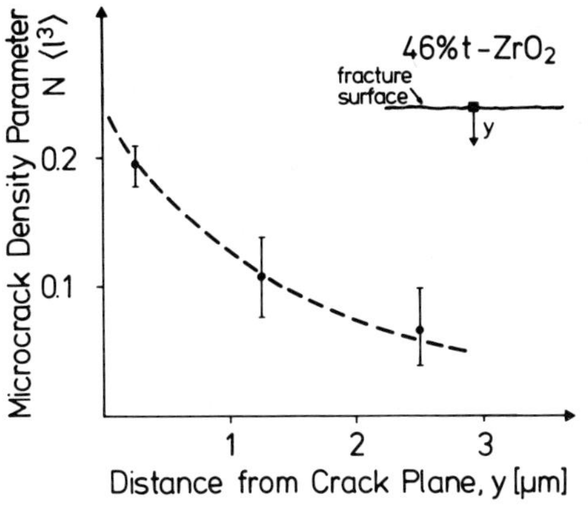

Fig. 7. Experimentally determined microcrack density parameter $\alpha = N\langle l^3 \rangle$ for various distances y from the crack plane.

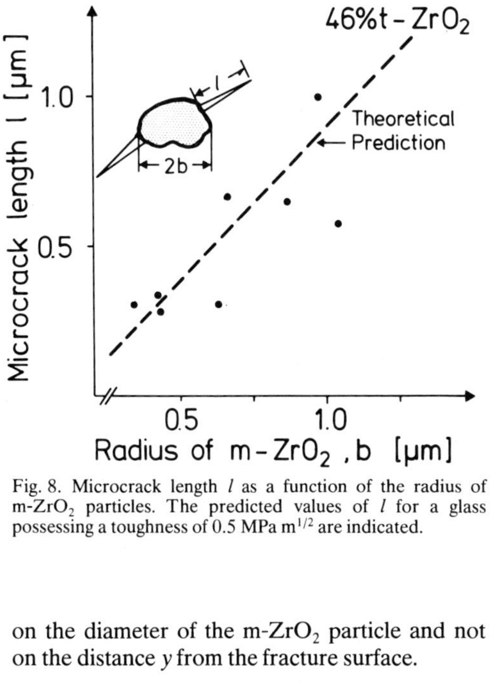

Fig. 8. Microcrack length l as a function of the radius of m-ZrO_2 particles. The predicted values of l for a glass possessing a toughness of 0.5 MPa $m^{1/2}$ are indicated.

on the diameter of the m-ZrO_2 particle and not on the distance y from the fracture surface.

3. Interpretation and discussion

The strength and toughness of the $x_t = 0.46$ ZTA ceramics are determined by t-$ZrO_2 \rightarrow$ m-ZrO_2 transformation and microcracking, both processes taking place during crack propagation. The experimental studies demonstrate that microcracking and transformation can be observed. Earlier studies [7, 10] suggest that radial

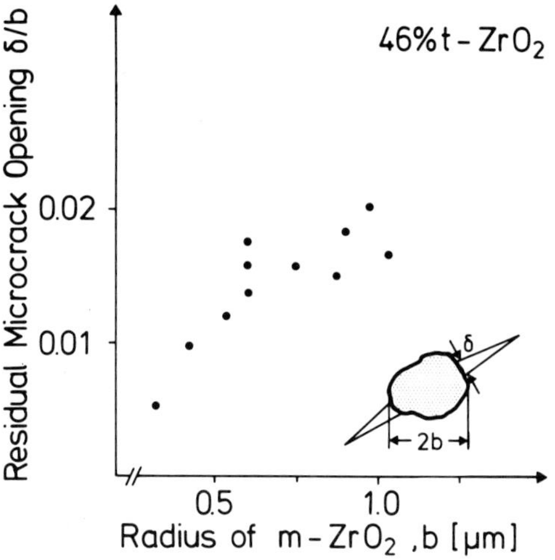

Fig. 9. Residual microcrack opening δ for m-ZrO_2 particles with different radii.

microcracking only occurs at those ZrO_2 grains which are present in monoclinic symmetry prior to the fracture experiment. There was no experimental evidence that radial microcracking occurred at those ZrO_2 grains which transformed during crack propagation. It was argued [7] that a critical stress is necessary to nucleate a microcrack and that this stress can no longer be effective after the t-$ZrO_2 \rightarrow$ m-ZrO_2 transformation has finished. This explanation is supported by the studies presented in this paper (Figs. 2 and 3).

For the interpretation, it will be assumed that the change ΔK in toughness (compared with pure Al_2O_3) is linearly composed of

$$\Delta K = \Delta K^{T} + \Delta K^{mc}$$

where ΔK^{T} is the change in toughness caused by the t-$ZrO_2 \rightarrow$ m-ZrO_2 transformation and ΔK^{mc} is the change in toughness caused by microcracking. Micromechanical models allow the determination of both quantities.

Evans and Cannon [6] argued that

$$\Delta K^{T} = \frac{\lambda}{1-\nu} E f \varepsilon_{ii}{}^{T} w^{1/2} \qquad (1)$$

where λ is a constant which depends on assumptions about the transformation mode, E is the modulus of elasticity, f is the volume fraction of transforming particles, $\varepsilon_{ii}{}^{T}$ is the unconstrained transformation strain and w is the width of the transformation zone. If f depends on w, then an integration over contributions of the different regions of the transformation zone results in the effective change of toughness. With $\lambda = 0.22$ [6], $f = 0.05$ and a gaussian-shaped transformation zone with a halfwidth of 3 μm,

$$\Delta K^{T} = 1.2 \text{ MN m}^{-3/2}$$

The evaluation of ΔK^{mc} is more tedious. Following the evaluation by Rühle *et al.* [12], we obtain

$$\Delta K^{mc} = \frac{2\pi\lambda}{3}\int_0^w \sigma_0 \frac{\xi^2(2+\xi)y^{1/2}(\partial\alpha/\partial y)}{1+\xi}\,dy - 1.1\left(1+\frac{3}{2}\xi\right)\alpha K \qquad (2)$$

where σ_0 is the initial stress in the particle, $\alpha = N\langle l^3\rangle$ is the microcrack density parameter, $\xi = l/b$ (with l the microcrack length and b the radius of the m-ZrO_2 particles) and K is the stress intensity of the glass present on the grain boundaries. An evaluation of the data presented in Figs. 5–9 results in an increase in toughness in the range

$$\Delta K^{mc} = 1.5\text{–}4 \text{ MN m}^{-3/2}$$

The large scattering of ΔK^{mc} is caused by experimental errors and by the uncertainty of different parameters in eqn. (2).

The calculated increase in toughness due to microcracking and transformation agrees with the measured quantity to within a factor of 2 (Fig. 1). A quantitative test of the micromechanical model requires a knowledge of the materials parameters, *e.g.* the stress σ_0 in the particle, the toughness of the glass covering the grain boundary and the mechanics of microcrack nucleations. Furthermore, the error bars of the experimental data are very large. It will be very difficult to test quantitatively the validity of the micromechanical model by experiments on ZTA materials.

Clarke [14] suggested an estimation of the toughness increase due to microcracking and/or simple transformation by using Eshelby's theoretical experiment. Clarke's results can also be used for an estimation of ΔK. Numerical results are obtained which agree with the measured values to within a factor of 2.

The experimental studies presented in this paper demonstrate that a process zone exists which contains a variable density of microcracks as well as stress-induced transformed t-ZrO_2 particles. Different theoretical concepts developed for these conditions can be applied for the evaluation. A test of different micromechanical models only makes sense if experimental data are available which possess smaller error bars.

Acknowledgments

The author acknowledges the assistance of Ms. C. Kohler, Ms. D. Waidelich and Mr. A. Strecker during the experimental part of the work. This work was supported by the U.S. Air Force Office of Scientific Research under Contract AFOSR-87-0291.

References

1 R. G. Hoagland, J. D. Embury and D. J. Green, *Scr. Metall., 9* (1975) 907.
2 F. E. Buresch, *Fract. Mech. Ceram., 4* (1975) 835.
3 A. G. Evans, *Scr. Metall., 10* (1976) 93.
4 A. G. Evans and K. T. Faber, *J. Am. Ceram. Soc., 67* (1984) 255.
5 A. G. Evans and Y. Fu, *Acta Metall., 33* (1985) 1525.
6 A. G. Evans and R. M. Cannon, *Acta Metall., 34* (1986) 761.
7 M. Rühle, N. Claussen and A. H. Heuer, *J. Am. Ceram. Soc., 69* (1986) 195.
8 G. Lindemann and N. Claussen, unpublished research, 1986.
9 M. Rühle, A. G. Evans, R. M. McMeeking, P. G. Charalambides and J. W. Hutchinson, *Acta Metall., 35* (1987) 2701.
10 J. W. Hutchinson, unpublished research, 1987.
11 A. H. Heuer, N. Claussen, W. M. Kriven and M. Rühle, *J. Am. Ceram. Soc., 65* (1982) 642.
12 M. Rühle, B. Kraus, A. Strecker and D. Waidelich, *Adv. Ceram., 12* (1984) 256.
13 B. Budiansky and R. J. O'Connell, *Int. J. Solids Struct., 12* (1976) 81.
14 D. R. Clarke, *J. Am. Ceram. Soc., 67* (1984) 15.

Resistance Curve Behavior of Supertough MgO-partially-stabilized ZrO_2*

A. H. HEUER, M. J. READEY† and R. STEINBRECH‡

Department of Materials Science and Engineering, Case Western Reserve University, Cleveland, OH 44106 (U.S.A.)

(Received March 1, 1988)

Abstract

The fracture mechanics of supertough MgO-partially-stabilized ZrO_2 (Mg-PSZ) exhibiting R curve behavior (the fracture resistance being a function of crack extension) is reviewed. In principle, fracture depends on the size of the initial flaw and the slope of the R curve but, because of wake shielding, the R curve depends on the geometry of the fracture mechanics test specimen, i.e. the R curve depends on crack history. The implications of this finding, and other aspects of the mechanical properties of supertough Mg-PSZ, are considered.

1. Introduction

MgO-partially-stabilized ZrO_2 (Mg-PSZ) ceramics are two-phase materials consisting of a solute-rich cubic ZrO_2 (c-ZrO_2) matrix that contains low solute tetragonal ZrO_2 (t-ZrO_2) precipitates (Fig. 1) [1]. They derive their high strength and high toughness from the stress-induced martensitic transformation of the t-ZrO_2 precipitates to monoclinic symmetry [2–5]. "Supertough" is used to describe this group of ceramics because their toughness—up to 20 MPa $m^{1/2}$—is the highest of any known monolithic ceramic. In detail, this high toughness derives from the "wake" of transformed particles adjacent to crack surfaces as the crack penetrates the matrix containing transformed particles [5], the transformation being induced by the high stresses ahead of the crack tip [4]. This wake of transformed particles, which is now known [6] to be diffuse and to extend to a distance of 1–2 mm in the toughest samples, exerts compressive transactions on the crack surface due to the dilational component of the transformation strain tensor ε_{ij}^{T} and thus "shields" the crack from the applied (far-field) stresses. This causes the toughness K to increase as the crack extends; $K_R = K_{TIP} + \Delta K(\Delta c)$, and the maximum or plateau toughness K_p can be expressed as $K_{TIP} + \Delta K_{max}$, where K_{TIP} is given by the Griffith equation for crack propagation through the matrix containing transformed particles [1]. All models [5] of transformation toughening show that ΔK_{max} is proportional to $E\varepsilon_{ij}^{T} V_f h^{1/2}/(1+\nu)$, where E and ν are the elastic modulus and Poisson's ratio respectively of the two-phase ceramic and V_f is the volume fraction of transforming phase within the transformation zone of extent h. The increase in toughness with crack extension causes an increase in crack resistance; this "resistance curve" ("R curve") behavior profoundly affects the strength of components made from such ceramics. In this paper, we review the fracture criteria for ceramics showing R curve behavior, report recent experiments determining R curves for several Mg-PSZ materials and discuss some troubling issues in our

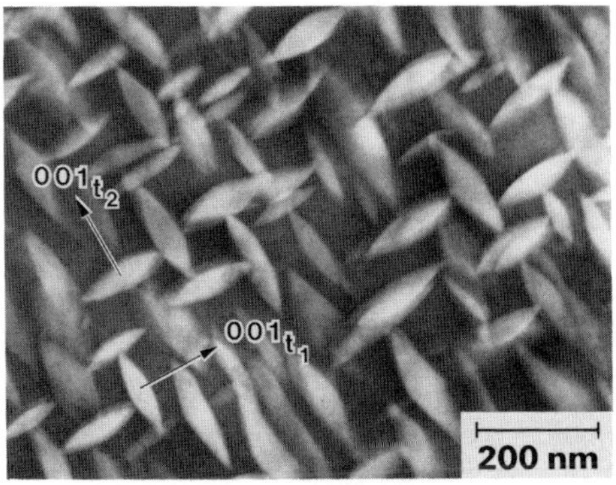

Fig. 1. Transmission electron micrograph showing t-ZrO_2 precipitates in a c-ZrO_2 matrix in Mg-PSZ. Only two of the three precipitate variables are visible in this micrograph [1].

*Paper presented at the 3rd International Conference on the Science of Hard Materials, Nassau, The Bahamas, November 9–13, 1987.

†Present address: Coors Ceramic Co., Golden, CO, U.S.A.

‡Present address: University of Dortmund, Dortmund, F.R.G.

0921-5093/88/$3.50

© Elsevier Sequoia/Printed in The Netherlands

present understanding of the fracture mechanics of such supertough ceramics.

2. Fracture of toughened ceramics showing *R* curve behavior

Figure 2 shows the salient mechanical behavior of a ceramic showing *R* curve behavior. The sample contains a flaw of size c_0 and at a stress intensity factor of K_{R_0}, given by $Y\sigma_0 c^{1/2}$, where σ_0 is the applied stress and Y a constant of order unity, whose exact value depends on the flaw geometry, the crack begins to grow. If *R* curve behavior were absent, K_{R_0} would be equal to K_{Ic} (the Griffith condition) and catastrophic fracture would ensue. Because of the *R* curve, however, the crack resistance increases as soon as crack propagation commences; an increase in *K* is needed for continued crack propagation. The crack thus ascends the *R* curve, and considerable stable crack growth precedes fracture.

From the definition given above of the stress intensity factor *K*, the curve of applied *K vs.* crack length at a constant applied stress is parabolic. The critical stress at which catastrophic crack propagation occurs is when the applied *K* curve becomes tangent to the K_R curve. Failure occurs when the crack has grown to c_f, at an applied K_R*, which is less than the maximum (plateau) *K* (Fig. 2). The strength thus depends on the slope of the K_R curve and the starting flaw size, and it is convenient to use the construction in Fig. 3, in which the initial flaw size increases to the left and the extent of stable crack propagation increases to the right. The failure condition now involves both a critical K_R (fail) and the tangency condition

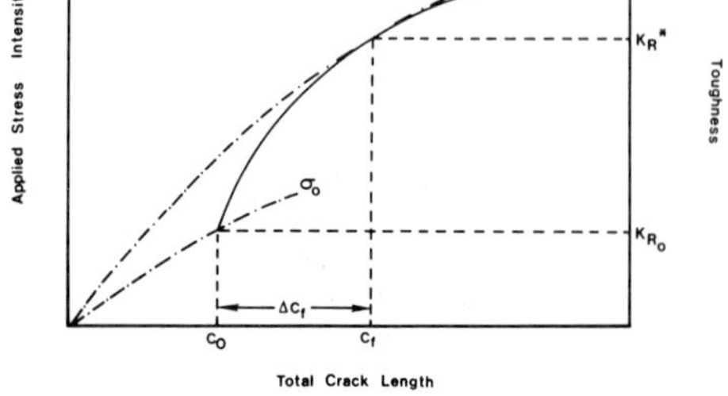

Fig. 2. Schematic diagram of K_R curve behavior. See text for discussion.

$$\frac{dK_{app}}{dc} = \frac{dK_R}{dc}$$

Use of *G*, the strain energy release rate rather than $K(G \equiv K^2/E)$, as in Fig. 4, is even more convenient, as the tangent condition of catastrophic failure is now represented by a straight line whose slope is determined only by the applied stress at failure.

$$K = \pi\sigma c^{1/2}$$

$$K^2 = \pi^2 \sigma^2 c$$

$$= GE$$

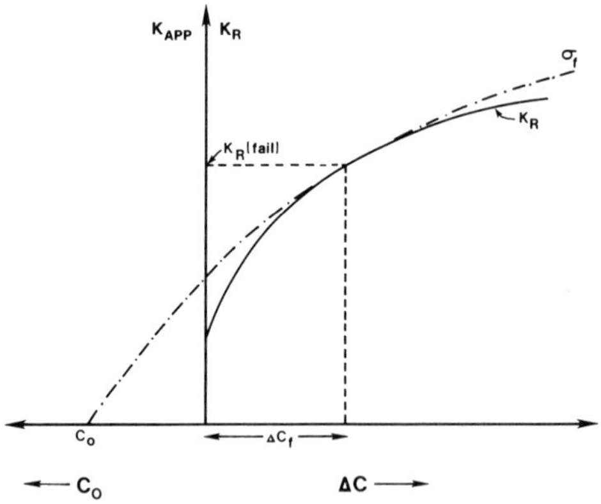

Fig. 3. Failure condition for ceramics exhibiting K_R curve behavior.

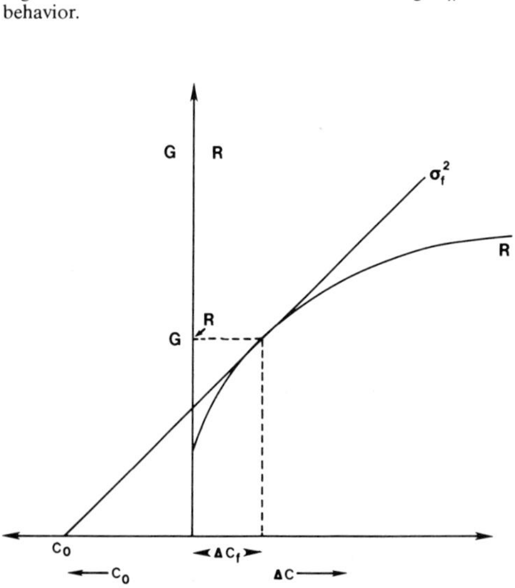

Fig. 4. *G* construction of *R* curve behavior. By analogy with Fig. 3, *G* is the energy equivalent of the crack driving force and *R* is the corresponding fracture resistance.

$$\frac{dG}{dc} = \frac{\pi^2 \sigma_2}{E}$$

where plane stress has been assumed. (As is now conventional, the crack resistance R is identified with G, the strain energy release rate, and the fracture toughness with the stress intensity factor K.)

As discussed elsewhere [7], counter-intuitive results can obtain if the flaw size is varied for two materials with different K_R curves; for a certain range of starting flaw sizes, the material with the lower plateau toughness can be stronger! R-curve-controlled fracture can also explain [8] the important result, first pointed out by Swain [9], that, for a given family of ZrO_2-toughened ceramics whose plateau toughness can be varied (by heat treatment that changes the particle transformability, for example), the maximum strength does not correspond to the maximum toughness but rather is observed at some intermediate toughness.

Having determined a K_R curve for a fracture mechanics sample containing an artificial flaw, and then measuring the fracture strength of the two halves of the broken specimen, a simple graphical procedure can be used to determine both the starting flaw size and the extent of stable crack growth. The square of the K_R curve is plotted and a straight line, whose slope is determined only by the fracture stress, is positioned tangent to the R curve; c_0 and Δc_f can be obtained immediately. An application of this approach will be shown below.

3. Determination of *R* curves in MgO-partially-stabilized ZrO_2

K_R curves of several commercial Mg-PSZ materials given different heat treatments [10] were determined using the double-cantilever-beam (DCB) geometry (2 mm × 25 mm × 75 mm). Load, load-point-opening displacement, and crack length were measured simultaneously (Fig. 5), allowing straightforward determination of K_R curves. From a given batch of commercial supertough Mg-PSZ, K_R curves were reproducible (see Fig. 10(a) (below)). These K_R curves are further noteworthy in two regards: firstly, the high value of the plateau toughness, 15 MPa $m^{1/2}$ or higher and, secondly, the shallowness of the curves—the fracture resistance is still increasing after a crack extension of several millimeters.

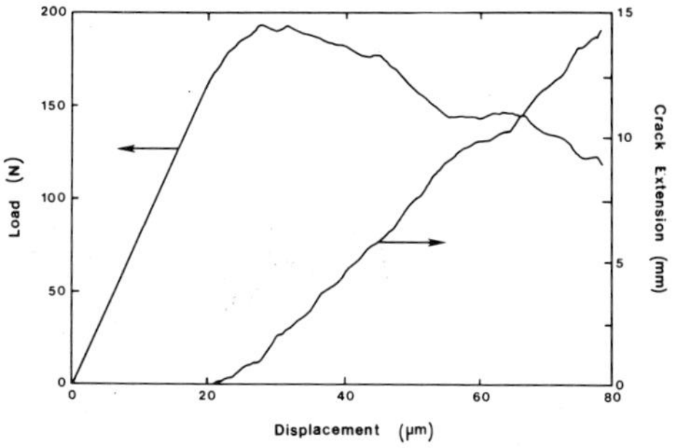

Fig. 5. Load and crack extension *vs.* load point displacement for a supertough Mg-PSZ specimen.

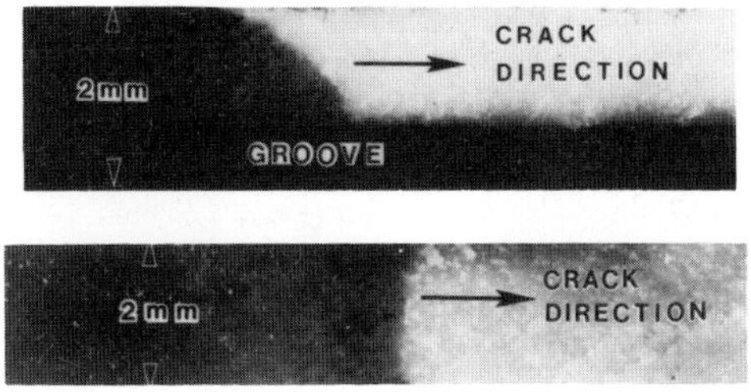

Fig. 6. Dye penetrant experiments to measure the crack profile: (upper photograph) the grooved "normal" DCB geometry; (lower photograph) the ungrooved "short" DCB geometry. See text for further explanation.

The testing procedure used initially involved specimens which had one face of the DCB beam grooved and one face polished; crack extension was measured on the polished face, and the groove ensured that the cracks would propagate along the midplane of the DCB specimen. To test whether the groove affected the course of crack propagation, a crack was propagated partly along the DCB specimen, the crack front was marked with dye penetrant while the specimen was under load, and the crack was then propagated completely through the specimen after the dye had dried. As shown in Fig. 6, the presence of the groove caused the crack front not to be accurately perpendicular to the polished face, which of course was unacceptable in a test designed to measure an *R* curve.

In most cases, ungrooved specimens could not be used, as the cracks would run out of the centre-line towards the top or bottom of the specimen (Fig. 7). By reducing the length of the specimen (making a "short" DCB specimen, which is actually similar to the "compact tension" geometry widely used for testing metals [11]) the

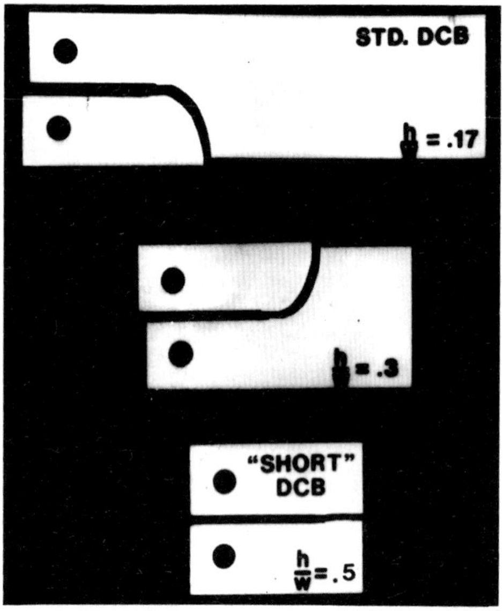

Fig. 7. Crack propagation in DCB beams with various geometries. See text for further explanation.

crack front was found to be accurately planar (Fig. 6); furthermore, extending the depth of the "starter notch" led to stable crack growth in ungrooved specimens (Fig. 7).

With this geometry, we have tested the graphical construction shown in Fig. 4 with two samples with significantly different R curves (Fig. 8(a)). After the R curves had been determined, the two DCB arms were fractured in four-point bending. The stronger material, with a σ of 700 MPa, derives from a material with a 10 MPa m$^{1/2}$ plateau toughness. It failed from a 72 μm flaw, with almost no stable flaw growth, and failure occurred at a K_R* of 6.5 MPa m$^{1/2}$ (Fig. 8(b)). The second material, with a plateau toughness of 11 MPa m$^{1/2}$, had a strength of 475 MPa, and failed at a K_R* of 8.5 MPa m$^{1/2}$ from a starting flaw of 185 μm after flaw growth of 65 μm (Fig. 8(c)). The microstructures of the two materials are shown in Fig. 9; the lower strength, higher toughness material has a much more heterogeneous microstructure, consistent with a larger starting flaw size. However, the high toughness provided by transformation toughening gives a remarkable and useful flaw tolerance; materials with such poor microstructures as that of Fig. 9(b) yield ceramics with quite useful and respectable mechanical properties.

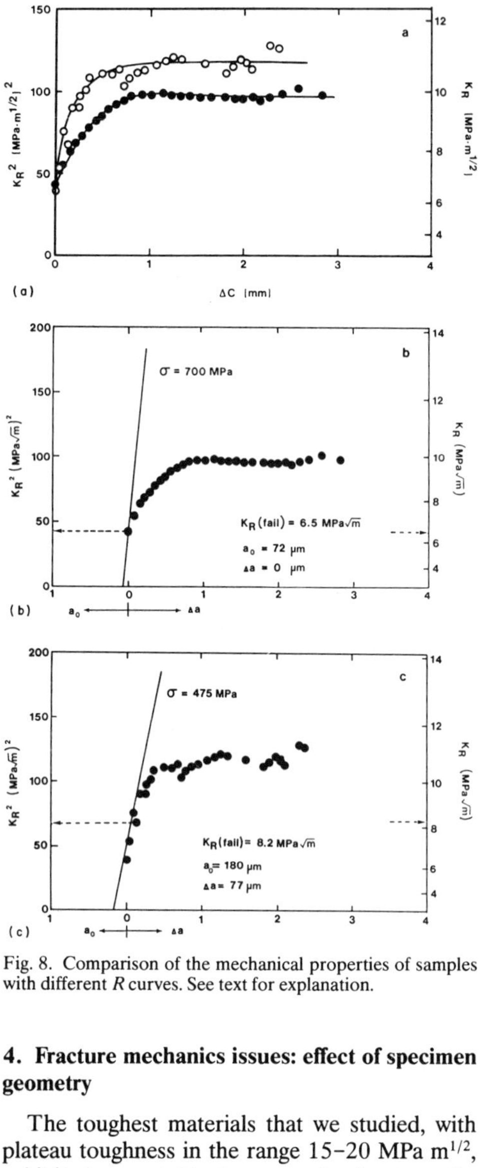

Fig. 8. Comparison of the mechanical properties of samples with different R curves. See text for explanation.

4. Fracture mechanics issues: effect of specimen geometry

The toughest materials that we studied, with plateau toughness in the range 15–20 MPa m$^{1/2}$, exhibited very stable transgranular fracture, the zone size being much larger than the grain size. In fact, crack growth was so stable that we were able to determine R curves in a few unnotched specimens with the "normal" DCB geometry (2 mm × 25 mm × 75 mm) and are thus able to compare R curves of nearly identical materials tested using the normal and short (2 mm × 25 mm × 30 mm) DCB geometries (Fig. 10). The results are exceedingly troubling—the slope of the R curve

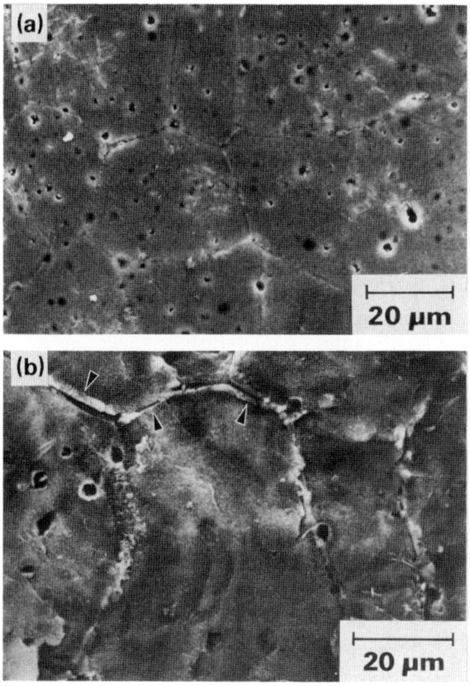

Fig. 9. Microstructure of samples in Fig. 8: (a) the lower toughness, higher strength material. (The arrowed features in (b) show a grain boundary phase formed by eutectoid decomposition of the cubic matrix [12].)

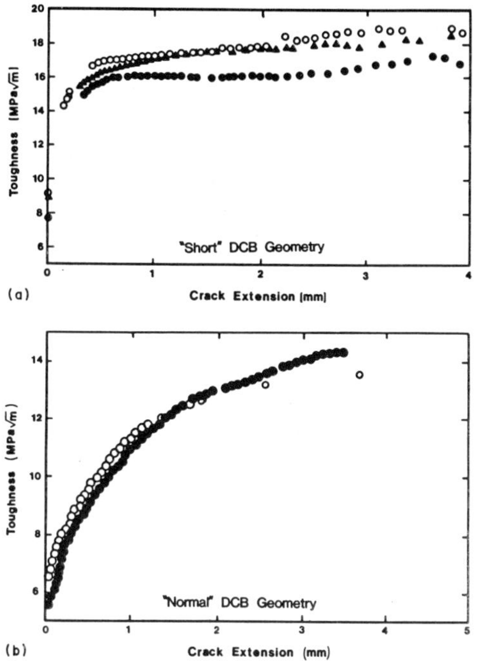

Fig. 10. Comparison of K_R curves for same material tested in the (a) normal and (b) short DCB geometries.

depends on sample geometry, but the strength of a component depends on the slope of the *R* curve! This is not believed to be a plane strain–plane stress issue, as the thicknesses of both samples were the same, and the slopes of *R* curves were only marginally different for short DCB specimen thicknesses varying between 1 and 4 mm. (In fact, the plateau toughness systematically increased with increasing specimen thickness, in one set of experiments from 13.5 to 15.5 to 16.5 MPa $m^{1/2}$, as the thickness increased from 1 to 2 to 4 mm. It should be noted that this trend is opposite to that found in tough metals, where the toughness decreases as the specimen thickness is increased from that giving plane stress to that giving plane strain conditions.)

The geometry dependence of the shape of the *R* curves derives from the zone shielding itself and is probably an essential feature in supertough Mg-PSZ and in other ceramics toughened by wake effects, *i.e.* materials exhibiting microcrack toughening, whisker or fiber reinforcement etc. In fact, such effects should be ubiquitous whenever wake toughening occurs, as a crack history effect will always be present whenever toughening is achieved by such an extrinsic microstructural feature. Devising a satisfactory testing geometry for such toughened ceramics, which possess a crack history dependence on fracture resistance, is essential if such toughened ceramics are to find wide acceptance by engineers in advanced structural applications.

The toughest materials are extraordinarily flaw tolerant and actually do not fail from processing or machining flaws. Marshall [8] has shown that grain-sized (about 50 μm) microcracks pop in at a critical stress (below the fracture stress) owing to transformation shear bands in favorably oriented grains impinging on a grain boundary with an unfavorably oriented grain. At this stress, the load–deflection curve will deviate from linear elasticity [8]; the stress-induced transformation provides real plasticity, albeit one which differs significantly from dislocation plasticity in tough metals. Many such microcracks nucleate and begin to grow stably as the applied stress is increased, and failure occurs either by the linking up of separately nucleated cracks, or by the dominance of a single growing microcrack with a particularly shallow R curve. Fracture mechanics testing of materials showing R curve behavior, and where flaw nucleation and the slope of the R curve both depend on transformation plasticity, may require other constructions than that shown in Fig. 4.

5. Fatigue and aging effects in supertough MgO-partially-stabilized ZrO_2

Transformation plasticity, being basically irreversible, might be expected to lead to severe fatigue effects and this has been found to be the case. Although detailed cyclic fatigue studies have not yet been conducted, copious evidence pointing to strong fatigue, even during single-cycle loading, has been found. Figure 11(a) is a typical example and shows load–displacement curves taken in four-point bending (these data have not been converted into stress–strain curves because the transformation plasticity prevents the use of beam formulae derived from linear elasticity, and strain gauges were not used in these tests). One sample was loaded directly to failure; the onset of generalized transformation plasticity is indicated by the change in slope of the load–displacement curve. The second sample was held at a constant displacement (past yield) for some time and then was reloaded; it failed at a lower load than the first sample did, indicating that the failure stress depends on load history.

Similar fatigue effects were found in DCB tests. The load under a load–displacement curve in such a test is a straightforward measure of the work of fracture (WOF), which in the example shown in Fig. 11(b) is 417 J m^{-2}. In a second test, an identical sample was periodically unloaded (this is a common procedure employed to measure specimen compliance if load, crack-opening displacement and crack length are not simultaneously measured), and the WOF decreased to 326 J m^{-2}.

These supertough ceramics were also sensitive to room temperature aging. This was discovered when a specimen annealing (at 1000 °C for 20 min) prior to testing (an annealing intended to reverse any tetragonal-to-monoclinic transformation that had occurred during specimen machining) was intentionally omitted; the plateau toughness changed to 9 MPa m$^{1/2}$ from the value of 18 MPa m$^{1/2}$ for the annealed sample (Fig. 12). In spite of extensive characterization efforts, no obvious difference between the tetragonal-to-monoclinic ratios for as-polished or as-ground samples and those for samples annealed at 1000 °C was found. However, a sample polished after annealing, and showing high toughness, developed surface rumpling after storage at room temperature for 8 months (Fig. 13); such rumpl-

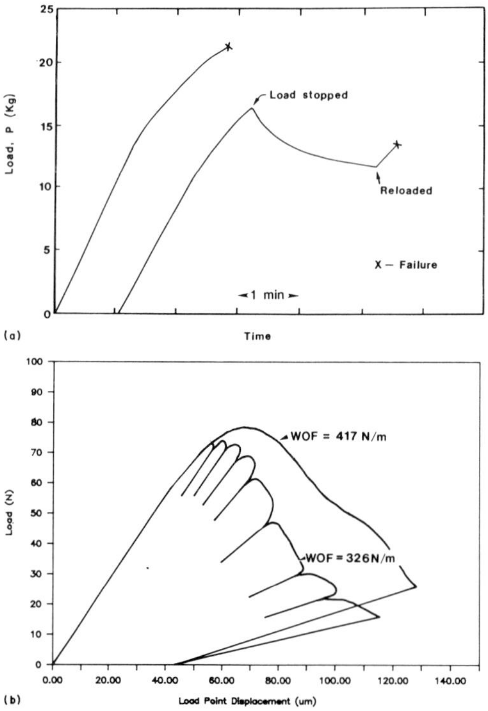

Fig. 11. Load–displacement curves for (a) four-point loading and (b) DCB geometry for supertough Mg-PSZ showing fatigue. See text for explanation.

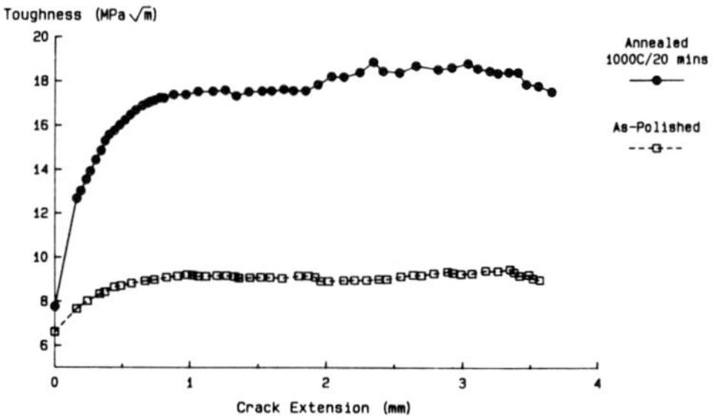

Fig. 12. K_R curves for supertough Mg-PSZ before and after annealing for 20 min at 1000 °C.

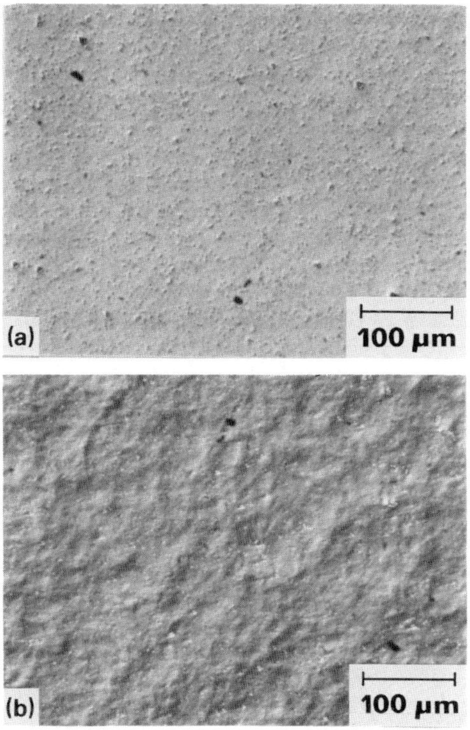

Fig. 13. Surface rumpling occurring during room temperature aging; the polished sample is shown (a) before and (b) after storage for 8 months (Normaski interference contrast).

ing is very good evidence that some transformation, albeit too modest to be detected by routine X-ray diffraction characterization, had occurred during the room temperature aging and dramatically decreased the plateau toughness. The annealing is now viewed as reversing the bulk transformation occurring during storage rather than reversing the machining-induced transformation which is confined to near-surface regions.

These fatigue and aging effects are unfortunate because they are deleterious, but they are a concomitant accompaniment to the ease of transformation inherent in producing supertough Mg-PSZ. Possibly an intermediate toughness, in the range 8–12 MPa $m^{1/2}$, may obviate both deleterious effects and may turn out to be the most useful for engineering applications.

Acknowledgments

The research was supported by the National Science Foundation under Grants DMR-82-14128 and INT 86-10359.

M. J. Readey was the Timken Fellow at Case Western Reserve University during this work.

References

1 A. H. Heuer, Transformation toughening in ZrO_2-containing ceramics, *J. Am. Ceram. Soc., 70* (10) (1987) 689–698.
2 R. C. Garvie, R. H. Hannink and R. T. Pascoe, Ceramic steel?, *Nature (London), 258* (5537) (1975) 703–704.
3 D. L. Porter and A. H. Heuer, Mechanisms of toughening partially stabilized zirconia (PSZ), *J. Am. Ceram. Soc., 60* (1977) 183–184.
4 D. L. Porter, A. G. Evans and A. H. Heuer, Transformation toughening in partially stabilized zirconia (PSZ), *Acta Metall., 27* (2) (1979) 1649–1654.
5 A. G. Evans and R. M. Cannon, Toughening of brittle solids by martensitic transformations, *Acta Metall., 34* (5) (1986) 761–800.
6 D. B. Marshall, personal communication, December 1987.
7 R. W. Steinbrech and A. H. Heuer, *R*-curve behavior and the mechanical properties of transformation-toughened ZrO_2-containing ceramics, in Y. Chen, W. P. Kengery and R. J. Stokes (eds.), *Defect Properties and Processing of High-technology Nonmetallic Materials, Materials Research Society Symp. Proc.,* Vol. 60, Materials Research Society, Pittsburgh, PA, 1986, pp. 469–481.
8 D. B. Marshall, Strength characteristics of transformation-toughened zirconia, *J. Am. Ceram. Soc., 69* (3) (1986) 173–180.
9 M. V. Swain, Inelastic deformation of Mg-PSZ and its significance in strength–toughness relationships of zirconia toughened ceramics, *Acta Metall., 33* (1985) 2083–2088.
10 M. J. Readey, *Ph.D Thesis,* Case Western Reserve University, Cleveland, OH, 1988.
11 Standard method of test for plane strain fracture toughness of metallic materials, *ASTM Stand., 399-381,* 1981.
12 S. C. Farmer, A. H. Heuer and R. H. J. Hannink, Eutectoid decomposition of MgO-partially-stabilized ZrO_2, *J. Am. Ceram. Soc., 70* (6) (1987) 431–440.

Materials Science and Engineering, *A105/106* (1988) 91–95

Physicochemical Aspects of the Fracture Rupture of Ceramics under Intensive Mechanical and Thermal Loading*

V. Ya. SHEVCHENKO and V. A. FJODOROV

Kurnakov Institute of General and Inorganic Chemistry, U.S.S.R. Academy of Sciences, Moscow (U.S.S.R.)

(Received November 12, 1987)

Abstract

A model analysis is presented for the processes of mechanical and thermal loading of ceramics. The analysis is based on a new approach to the wave processes which develop in elastic (brittle) materials. An analytical form is found for the relation between the physicochemical constants and the characteristics of these processes. This approach provides a more meaningful analysis of the fracture character of ceramics in a number of applied problems.

1. Introduction

In recent years an increasing amount of attention has been given to analysing the strength of ceramic materials. This is because, until recently, there were comparatively few theoretical studies of the failure of "brittle" bodies, such as ceramics, and also because the field of applications of ceramics has significantly expanded. Of special interest is the fact that ceramics are now used under extreme loading conditions, *i.e.* at very large energy densities and dynamic loading, *e.g.* as in a gas turbine or in a special tool.

2. Failure of ceramics under high rates of mechanical loading

Shevchenko *et al.* [1] were the first to call attention to some specific details of the process of the failure of ceramics under high rates of mechanical loading and to propose a model of the so-called "dissociative" failure. This model satisfactorily describes the behaviour of a ceramic at a loading speed of the same order as the speed of sound in a ceramic, as well as its behaviour in a region of "weak" loading where the failure parameters are functions of the ratio of the kinetic loading energy to elastic impedance [2].

Formulae were obtained for calculations of the ultimate strength σ_{max}, Young's modulus E and ultimate strain ε_{max}

$$\sigma_{max} = 0.25\, v_e \rho \left(\frac{4D}{n_b M}\right)^{1/2}$$

$$E = \rho v_e^2$$

$$\varepsilon_{max} = \frac{b}{v_e}\left(\frac{4D}{n_b M}\right)^{1/2}$$

where ρ is the density, v_e is the velocity of sound, D is the dissociation energy, M is the molar mass, and n_b and b are coefficients. By use of these expressions a number of theoretical constants were calculated for diamond (Table 1). For other ceramics the calculated values of E and σ_{max} are in good agreement with experimental results and values calculated by other methods (Tables 2 and 3). In these calculations it was assumed that $n_b = 4$.

It can be seen that the maximum theoretical strength is higher than the experimental value, because of the calculation model and non-uniformity of the experimental conditions. Never-

TABLE 1 Theoretical constants calculated for diamond

Calculation method	*Reference*	σ_{max} (GPa)
$\sigma_{max} = E/10$	[3]	121
$\sigma_{max} = E/20$	[4]	60
E	[5]	205
J	[6]	270
A	[3]	110
This work		125

*Paper presented at the 3rd International Conference on the Science of Hard Materials, Nassau, The Bahamas, November 9–13, 1987.

0921-5093/88/$3.50

© Elsevier Sequoia/Printed in The Netherlands

TABLE 2 Experimental and calculated values of Young's modulus E for a number of ceramics

Material	*Young's modulus E* (GPa)			
	Experiment			*Calculation*
	[4]	*[3]*	*[7]*	
Diamond	1200	1210	1030	1214
SiC	485	—	487	401
SiO_2	—	73	122	121
Al_2O_3	510	460	447	467
Si	159	188	177	204
α-Si_3N_4	—	—	382	185

TABLE 3 Experimental and calculated values of σ_{max} for a number of materials

Calculation method	*Reference*	σ_{max} (GPa)				
		SiC	SiO_2	Al_2O_3	*Si*	α-Si_3N_4
$E/20$	[4]	11.3	—	29	6.5	—
$E/10$	[4]	22.6	—	58	13	—
E	[5]	—	16	46	32	—
This work		49.6	24.8	58.9	21.7	31.4
Experiment	[3]	20.6	13.7	28.9	6.9	13.7

theless, it should be stressed that the equation [1]

$$l_K = b^{3/2} l_c \left(\frac{2\rho_c v^2}{nb\rho v_c^2} \right)^{1/3}$$

gives an excellent qualitative description of the collision process of two moving brittle bodies for a wide range of velocities. It is in fact the ultimate case of total brittle destruction, which is characteristic of many ceramics and a wide variety of loading conditions. (In this relation, $b = \ln^2 - 2$ where l_c is the length of an undeformed rod with a density ρ_c and a velocity v_c which interacts with a target of density ρ.)

3. Failure of ceramics under large thermal loads

Analysis of the failure of ceramics under great thermal loads (thermal shocks) is more complicated. When the temperature near one of the boundaries of a solid changes rapidly, thermal stresses develop in its bulk, which lead, when a limit σ_1 is exceeded to catastrophic failure of the body. Numerous estimates of a thermal endurance criterion are based on these ideas; it is commonly understood as the maximum difference (ΔT) between the mean temperature of a body and the temperature on its boundary which causes its failure. Hasselman [8] has given an analytical form of the criterion:

$$\Delta T = \frac{\sigma_e(1-\mu)}{\alpha E} \qquad (1)$$

where μ is Poisson's ratio, α is the linear expansion coefficient and E is Young's modulus. Hasselman also gave some modifications of the equation for a number of particular cases. However, comparisons of the experimental values of the thermal endurance criterion and those calculated according to eqn. (1) show systematic differences, the experimental values being, as a rule, greater [2]. As well as some difficulties in obtaining these estimates (data for the same specimens are not always available and, in addition, thermal loading conditions differ in different papers), the differences discovered show that, in obtaining criterion (1), some aspects of energy dissipation under thermal loading are not taken into account, *e.g.* the development of wave processes.

Let us consider a simple problem for a homogeneous rod of length l at temperature T_0, one of whose ends comes into contact with a medium whose temperature is $T_1 > T_0$. The heat current through the rod's boundary is proportional to the difference in temperature:

$$-\kappa \left.\frac{\partial T}{\partial x}\right|_{x=0} = h(T_1 - T|_{x=0})$$

where κ is the coefficient of heat conductivity, h is the heat transfer coefficient and $x = 0$ is the origin of the coordinates on the boundary of the rod; the initial conditions are

$$T|_{t=+0} = T_0$$

and

$$\left.\frac{\partial T}{\partial t}\right|_{t=+0} = 0$$

The solution of the heat equation for this problem is

$$T = T_0 + \gamma(T_1 - T_0)\exp(\pi\gamma^2 t)$$

$$\times \int_0^t \frac{\exp(-\pi\gamma^2 z - x^2/4\chi z)\,\mathrm{erfc}[\gamma\{\pi(t-z)\}^{1/2}]}{z^{1/2}}\,dz \qquad (2)$$

where $\chi=\kappa/c_p\rho$ is the heat conductivity coefficient, c_p is the specific heat, ρ is the density and $\gamma=h/(\pi\kappa c_p\rho)^{1/2}$.

We assume that such thermal loading will be accompanied by propagation of elastic compression waves along the rod. When deducing the wave equation, we assume, for simplicity, that $\mu=0$ and we make use of the system of coordinates Ox' which is related to the old Ox coordinates by the relation $x'=l/2-x$, *i.e.* we transfer the system of coordinates to the mass centre of the rod. We denote by ξ the coordinate of any arbitrary cross-section of the rod at $t=0$ and by $x'(\xi, t)$ its coordinate at the moment t. The thermodynamically equilibrated displacement

$$u_{\mathrm{eq}}(\xi, t)=x'_{\mathrm{eq}}(\xi, t)-\xi$$

then is given by

$$u_{\mathrm{eq}}(\xi, t)=\int_0^{\xi}\alpha\{T(\xi, t)-T_0\}\,\mathrm{d}\xi-\Delta\xi_{\mathrm{mc}}(t) \tag{3}$$

where α is the thermal expansion coefficient:

$$\Delta\xi_{\mathrm{mc}}(t)=\xi_{\mathrm{mc}}(t)-\xi_{\mathrm{mc}}(t=0)$$

The actual displacement of the cross-section at any moment t is given by

$$u_{\mathrm{act}}(\xi, t)=\int_0^t\frac{\partial x'(\xi, t)}{\partial t}\,\mathrm{d}t \tag{4}$$

while Hooke's law assumes the form

$$\sigma(\xi, t)=E\frac{\partial}{\partial\xi}(u_{\mathrm{act}}-u_{\mathrm{eq}}) \tag{5}$$

where the non-vanishing component of stress tensor is denoted by $\sigma(\xi, t)$.

Substituting eqns. (4) and (3) into eqn. (5), we obtain

$$\frac{\sigma(\xi, t)}{E}=\int_0^t\frac{\partial^2x'(\xi, t)}{\partial\xi\,\partial t}-\alpha\{T(\xi, t)-T_0\}$$

Using the equation of motion

$$\rho\frac{\partial^2x'(\xi, t)}{\partial t^2}=\frac{\partial\sigma(\xi, t)}{\partial\xi}$$

we have

$$\frac{\partial^2\sigma(\xi, t)}{\partial t^2}-v^2\frac{\partial^2\sigma(\xi, t)}{\partial\xi^2}=-\alpha E\frac{\partial^2T(\xi, t)}{\partial t^2} \tag{6}$$

where $v=(E/\rho)^{1/2}$ is the speed of sound in the rod.

Since only the second derivative with respect to ξ occurs in eqn. (6), reverting to the old coordinate system $x=l/2-\xi$ does not change the form of the equation

$$\frac{\partial^2\sigma(x, t)}{\partial t^2}-v^2\frac{\partial^2\sigma(x, t)}{\partial x^2}=-E\alpha\frac{\partial^2T(x, t)}{\partial t^2} \tag{7}$$

Now, let us investigate the development of the wave process during the time interval $t\lesssim l/v$, *i.e.* without accounting for the wave reflected from the cold end of the rod. Thus, it is sufficient to take only the first-order term of $\gamma t^{1/2}$ in eqn. (2). (Indeed, $\gamma(l/v)^{1/2}\approx hl^{1/2}/(\kappa c_p\rho v)^{1/2}\ll 1$ for all practically important values of the coefficients in eqn. (2).) In this approximation,

$$\begin{aligned}T(x, t)&=T_0+\gamma(T_1-T_0)\int_0^t\frac{\exp(-x^2/4\chi z)}{z^{1/2}}\,\mathrm{d}z\\&=T_0+2\gamma t^{1/2}\left[\exp\left(-\frac{x^2}{4\chi t}\right)-\frac{\pi^{1/2}x}{2(\chi t)^{1/2}}\right.\\&\qquad\left.\times\,\mathrm{erfc}\left\{\frac{x}{2(\chi t)^{1/2}}\right\}\right]\end{aligned}$$

$$\frac{\partial^2T(x, t)}{\partial t^2}=\gamma(T_1-T_0)\exp\left(-\frac{x^2}{4\chi t}\right)\frac{1}{2t^{3/2}}\left(\frac{x^2}{2\chi t}-1\right) \tag{8}$$

Next, we make use of d'Alembert's formula [9]. Taking into account the fact that, for $t<x/v$, the wave reflected from the heated end of the rod will not influence stresses at x, we have

$$\begin{aligned}\sigma(x, t)=&-\frac{E\gamma\alpha(T_1-T_0)}{2v}\int_0^t\mathrm{d}\tau\int_{x-v(t-\tau)}^{x+v(t-\tau)}\mathrm{d}\xi\\&\times\exp\left(-\frac{\xi^2}{4\chi\tau}\right)\frac{1}{2\tau^{3/2}}\left(\frac{\xi^2}{2\chi\tau}-1\right)\end{aligned}$$

or

$$\begin{aligned}\sigma(x, t)=&-E\alpha(T_1-T_0)\frac{\gamma(\pi\chi)^{1/2}}{2v}\\&\times\left[\exp\left(-\frac{vx+v^2t}{\chi}\right)\mathrm{erfc}\left\{\frac{x-2vt}{2(\chi t)1/2}\right\}\right.\\&\left.-\exp\left(\frac{vx+v^2t}{\chi}\right)\mathrm{erfc}\left\{\frac{x-2vt}{2(\chi t)1/2}\right\}\right]\end{aligned} \tag{9}$$

For $t > x/v$, it is necessary to take into account the reflected wave. Then

$$\sigma(x,t) = -\frac{E\gamma\alpha(T_1 - T_0)}{2v}\left\{\int_{t-x/v}^{t} \mathrm{d}\tau \int_{x-v(t-\tau)}^{\infty} \mathrm{d}\xi \times \exp\left(-\frac{\xi^2}{4\chi\tau}\right)\frac{1}{2\tau^{3/2}}\left(\frac{\xi^2}{2\chi\tau} - 1\right) - \int_0^{t-x/v} \mathrm{d}\tau \times \int_0^{x+v(t-\tau)} \mathrm{d}\xi \exp\left(-\frac{\xi^2}{4\chi\tau}\right) \times \frac{1}{2\tau^{3/2}}\left(\frac{\xi^2}{2\chi\tau} - 1\right)\right\} \quad (10)$$

The second term in this expression is the contribution of the wave reflected from the heated end. The limit ∞ in the integral over ξ means neglecting the contribution of the wave coming from the cold region of the rod. This is justifiable for $x^2 \gg \chi t$. Integrating eqn. (10), we obtain

$$\sigma(x,t) = -E\alpha(T_1 - T_0)\frac{\gamma(\pi\chi)^{1/2}}{2v}\exp\left\{\frac{v(vt-x)}{\chi}\right\} \times \left[1 + \operatorname{erfc}\left\{\frac{2vt - x}{2(\chi t)^{1/2}}\right\} - 2\operatorname{erfc}\left\{\frac{v(vt-x)}{\chi}\right\}^{1/2}\right] \quad (11)$$

Since $x^2 \gg \chi t$ and $2vt - x > x$, the expression in square brackets can be written with good accuracy as

$$2\operatorname{erfc}\left[\frac{v(vt-x)}{\chi}\right\}^{1/2}$$

while, for $vt \gtrsim 2x$, $v(vt-x) \gg \chi$ and using $\operatorname{erfc} z = \exp(-z^2)/z\pi^{1/2}$ for large z, we have

$$\sigma(x,t) = -E\alpha(T_1 - T_0)\frac{\gamma\chi}{v^{3/2}(vt-x)^{1/2}}$$

$$T_1 - T_0 = \Delta T \quad (12)$$

Rewriting eqn. (12) in a convenient form, we have

$$\Delta T_{\mathrm{cr}}(x,t) = \frac{\sigma_{\mathrm{e}} v^{3/2}(vt-x)^{1/2}}{E\alpha\gamma\chi} \quad (13)$$

Taking into account the relations

$$\gamma = \frac{h'}{(\kappa c_p \rho)^{1/2}}$$

$$\chi = \frac{\kappa}{c_p \rho}$$

$$E = \rho v^2$$

$$h' = \frac{h}{\pi^{1/2}}$$

we have

$$\Delta T_{\mathrm{cr}}(x,t) = \frac{\sigma_{\mathrm{e}} c_p}{h'\alpha}\left(\frac{vt-x}{\chi v}\right)^{1/2} \quad (14)$$

Since eqn. (11) is obtained under the assumption that $vt > x$, eqn. (12) can be written

$$\Delta T_{\mathrm{cr}}(x,t) = \frac{\sigma_{\mathrm{e}} c_p t^{1/2}}{h'\alpha\chi^{1/2}}$$

or, as $v_{\mathrm{T}} = (\chi/t)^{1/2}$, where v_{T} is the speed of the heat wave, we have

$$\Delta T_{\mathrm{cr}} = \frac{\sigma_{\mathrm{e}} c_p}{h'\alpha v_{\mathrm{T}}} \quad (15)$$

It is interesting to note that the results obtained transform in the limit $v \to 0$ into expressions similar to eqn. (1). For example, eqn. (9) tending to the limit $v \to 0$ yields

$$\lim_{v\to 0}\{\sigma(x,t)\} = -E\alpha(T_1 - T_0)2\gamma t^{1/2}\left[\exp\left(-\frac{x^2}{4\chi t}\right) - \frac{\pi^{1/2}x}{2(\chi t)^{1/2}}\operatorname{erfc}\left\{\frac{x}{2(\chi t)^{1/2}}\right\}\right]$$

or, comparing this with eqn. (8),

$$\lim_{v\to 0}\{\sigma(x,t)\} = -E\alpha\{T(x,t) - T_0\}$$

the latter corresponding to eqn. (1) at $\mu = 0$.

In spite of some simplifications assumed in the calculations, eqn. (13) or eqn. (15) obtained within the framework of our model allows us to estimate the thermal endurance value and its variation with time and with coordinate for other materials as well as ceramics under thermal loading and also other thermal and mechanical properties.

4. Conclusions

The analysis presented here is based on a new approach to the wave processes which develop in elastic (brittle) materials. This approach provides a more meaningful analysis of the fracture character of ceramics in a number of applied problems.

References

1 V. Ya. Shevchenko, A. D. Izotov, V. B. Lazarev and N. M. Zhavoronkov, *Izv. Akad. Nauk S.S.S.R., Neorg. Mater., 20* (6) (1984) 1047–1052 (in Russian).

2 A. G. Evans and T. G. Langton, *Construction Ceramics*, Metallurgiya, Moscow, 1980 (in Russian).

3 A. Kelly, *Strong Solids*, Oxford University Press, London, 1966.

4 E. Moelvin-Hughes, *Physical Chemistry*, Pergamon, Oxford, 1962.

5 E. Orowan, Fracture and strength of solids, *Rep. Prog. Phys., 12* (1949) 185.

6 J. Gilman, *Physics and Chemistry of Ceramics*, Gordon and Breach, New York, 1963, p. 240.

7. L. Liebowitz (ed.), *Fracture and Advanced Treatise*, Vol. 7, Academic Press, New York, 1972.

8 D. P. H. Hasselman, *Am. Ceram. Soc. Bull., 49* (12) (1970) 1033.

9 V. S. Vladimirov, *Equations of Mathematical Physics*, Nauka, Moscow, 1971.

Materials Science and Engineering, A105/106 (1988) 97–104

Strengthening and Toughening of Boride and Carbide Hard Material Composites

R. TELLE and G. PETZOW

Max-Planck-Institut für Metallforschung, Institut für Werkstoffwissenschaften, Pulvermetallurgisches Laboratorium, Heisenbergstrasse 5, D-7000 Stuttgart 80 (F.R.G.)

(Received January 13, 1988)

Abstract

Many of the well-established strengthening and toughening mechanisms for oxide ceramics are also applicable to hard materials based on borides and carbides. Chemical complications, however, limit a similar successful combination of phases, yielding metal matrix cemented borides and transformation toughening by ZrO_2 additives. Hence the most encouraging methods for improvement of the mechanical properties are grain size refinement by growth-retarding additives and crack-propagation-influencing mechanisms such as crack deflection, microcrack interaction and crack impediment.

Ternary cemented borides, liquid-phase-sintered TiB_2, B_4C- and SiC-based composites as well as ZrO_2-toughened TiB_2 are discussed as successful examples.

1. Introduction

In spite of their extreme hardness and excellent wear resistance, the applicability of borides and carbides as structural ceramics is still restricted to some very specific problems in which their strongest disadvantage, their tremendous brittleness, is not the limiting factor. Moreover, single-phase boride and carbide ceramics usually suffer from a poor sinterability. Both high temperatures close to the melting point and high pressures have to be applied for sufficient densification. Prolonged high temperature treatments result in an exaggerated coarsening, however. Grain sizes of 50–200 μm, which are common for commercial boron carbide (B_4C), titanium diboride (TiB_2) and pressureless-sintered silicon carbide (SiC), create strong internal stresses owing to anisotropy effects and hence severe cracking on cooling. Thus, in order to extend the field of applicability, strengthening and toughening strategies have to be utilized which are rather well established for advanced oxide and nitride ceramics. Consequently, boride and carbide composite materials become more and more important because the unsatisfactory properties of a single-phase ceramic may be overcome by a suitable choice of phases as additives enhancing the densification, preventing coarsening, generating crack deflection or microcracking, and (under certain circumstances) enabling transformation toughening.

The topic of this work is to discuss which reinforcing mechanisms might be applicable for borides and carbides. After a more general introduction of possible combinations of materials, the most encouraging examples of successful improvement are presented.

2. Materials

B_4C is one of the most important hard materials which urgently requires drastically improved strength and toughness in order to make appropriate use of its superior high temperature hardness. Moreover, its poor sinterability and low oxidation resistance limit its widespread application. A considerable increase in strength was obtained by grain size refinement as a result of hot isostatic pressing [1, 2]. A bending strength of 500 MPa and a K_{Ic} of 3–4 MPa m$^{1/2}$ seem to be the ultimate upper limit, however, even if sophisticated processing methods are used.

SiC exhibits excellent oxidation resistance and wear resistance combined with a high creep resistance if glass phases can be avoided. One of its disadvantages is its brittleness and tendency to

*Paper presented at the 3rd International Conference on the Science of Hard Materials, Nassau, The Bahamas, November 9–13, 1987.

0921-5093/88/$3.50

© Elsevier Sequoia/Printed in The Netherlands

undergo strong coarsening during pressureless sintering at temperatures between 2000 and 2200 °C. Although the strength can be significantly improved by grain size reduction due to low temperature sintering of submicron powders, the toughness remains a subject for further optimization.

Titanium carbide (TiC) and tungsten carbide (WC) possess metallic character and, thus, high thermal and electrical conductivity combined with excellent hardness and relatively high toughness. Their most important advantage is the chemical compatibility with metals, enabling easy production. Although the superior toughness of the cermets does not require significant further improvement, the presence of a metal limits the applicability at high temperatures owing to softening. Thus, TiC and WC remain interesting materials for ceramic composites, especially as sintering additives.

TiB_2 is a subject for intensive efforts for improving the sinterability, strength and toughness. Because of its physical, chemical and structural relationship to zirconium diboride (ZrB_2), tantalum diboride (TaB_2) and W_2B_5, it will be discussed as an example for all transition metal diborides. Although TiB_2 is considered to exhibit suitable mechanical properties as a major constituent of highly wear-resistant and temperature-resistant structural ceramics, it may also be used as an additive for the improvement of the hard materials mentioned above because of its chemical compatibility with B_4C, SiC, TiC and WC [3]. One of the important properties of the diborides is the strong anisotropy of the thermal expansion coefficients of about 42% up to 1000 °C [4, 5], which may be detrimental to coarse single-phase ceramics and beneficial for generating crack deflection in composite materials.

3. Applicability of strengthening and toughening strategies

In addition to the material-specific first-order approaches to optimize mechanical properties such as sintering to high density, avoiding flaws and microstructural inhomogeneities, reducing grain growth by applying low sintering temperatures and devitrification of amorphous phases, this study focuses on basic strengthening and toughening mechanisms which are effective in composites, namely metal matrix reinforcement, grain size refinement by sintering additives, transformation toughening, crack impediment and crack deflection or branching by misfit effects. All these strategies have already been described extensively theoretically and experimentally (see for example refs. 6–8). Contrary to oxide ceramics, however, their applicability to borides and carbides is much more limited because of chemical complications.

3.1. Metal matrix composites: cemented borides

Contrary to the hard metals, a similar simple association of binary borides with metallic matrices fails because of chemical reactions. For instance, B_4C decomposes to form $M_xB_y + C$ if it is combined with metals M other than with tin, zinc, copper and silver. Only if the reaction velocity is slow enough, *e.g.* at low temperatures, can cermets be prepared. As an example, the infiltration of porous B_4C compacts with liquid aluminium yields composites of high toughness and wear resistance [9]. TiB_2 reacts with transition metals such as iron, nickel, or cobalt to form the metallic borides Fe_3B, Ni_3B, Ni_2B, Co_3B etc. which may even dissolve titanium [10, 11]. These reaction products melt typically between 900 and 1200 °C and exhibit an extremely low fracture toughness (*e.g.* K_{Ic} for Ni_3B is 1.4–1.9 MPa $m^{1/2}$) combined with a relatively high Vickers hardness (H_V = 10.7 GPa) [12].

A more encouraging type of cemented boride involves the ternary ϕ and ω phases as the major constituents. The stoichiometries are $M^IM^{II}B$ (ϕ) or $M^I_2M^{II}B_2$ (ω) where $M^I \equiv$ Ta, W, Cr or Mo and $M^{II} \equiv$ Fe, Ni or Co and solid solutions thereof, respectively. Since these phases coexist with melts of M^{II} saturated in both M^I and B, these materials can be pressureless liquid phase sintered typically in the temperature range between 900 and 1200 °C. Another type of ternary boride, namely the τ phase $M^I_{21}M^{II}_2B_6$ with $M^I \equiv$ Fe, Ni or Co and $M^{II} \equiv$ Ti, Nb or W is applied as plasma-sprayed wear-resistant coatings with the M^I matrix on large steel parts [13]. In Fig. 1 an isothermal section of the B–Co–W system [14] is shown in which both the ϕ and the τ phases are linked with metallic cobalt. If cobalt is substituted by nickel, the ϕ and τ phases do not form, but the ω phase will be in equilibrium with liquid nickel. The composites consist of two phases in analogy to hard metals. Depending on the volume fraction of the liquid phase, the ternary compound forms faceted crystals embedded in a metal matrix or a rigid skeleton with a small amount of binder in

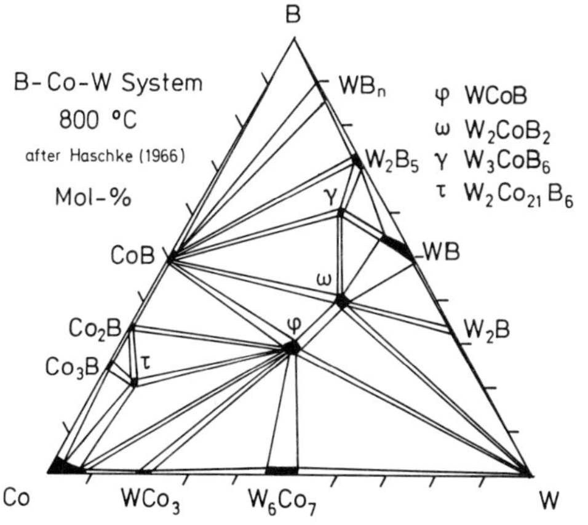

Fig. 1. Isothermal section of the B–Co–W phase system after ref. 14. Examples of cemented borides are the τ and ϕ phases which are in equilibrium with metallic cobalt and hence can be liquid phase sintered.

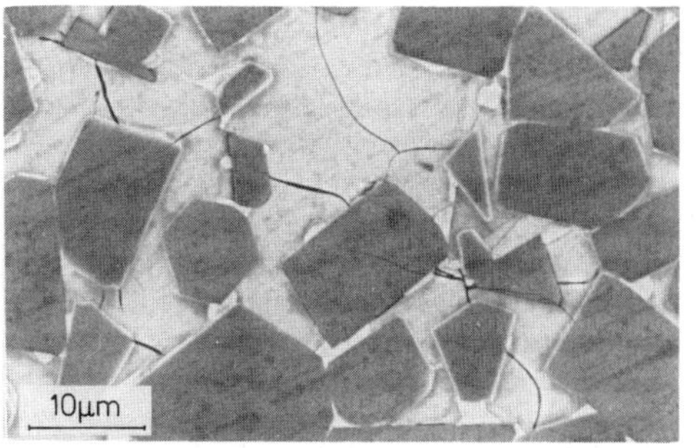

Fig. 2. Microstructure of TiB_2 sintered with liquid nickel. Severe cracking indicates misfit stresses between the TiB_2 particles and the resulting Ni_3B and Ni_2B matrix phase.

between. In spite of the extreme brittleness of the boride phase, the composite exhibits excellent toughness, strength (1.5–2.5 GPa) and solderability. Interesting wear-resistant materials have been produced in the Mo_2FeB_2–Fe system with nickel and chromium additives [15]. They can be applied as bearings, valve seats, drawing dies, spinning nozzles, and parts for injection moulding machines.

3.2. Grain growth retardation by sintering additives

Because of their low melting points and the suitable wetting behaviour, the borides of iron, nickel and cobalt can be used as additives for a liquid phase sintering of TiB_2-type borides. The densification mechanism observed in these systems is dissolution and reprecipitation which generates an accelerated grain growth of faceted crystals, however. A typical microstructure of a hot-pressed TiB_2–Ni_3B–Ni composite is shown in Fig. 2. The severe cracking is due to shear stresses at the phase boundaries caused by the misfit of the thermal expansion and enhanced by the anisotropic behaviour of TiB_2.

Grain size refinement below the critical diameter for spontaneous microcracking is achieved by the addition of transition metal borides such as TaB_2 and W_2B_5 which form $(Ti, M)B_2$ solid solution layers on the TiB_2 particles. An even more efficient retardation of the rapid grain growth is obtained by transition metal carbide additives such as TiC or WC which partially dissolve in the liquid phase. In the first stage of reprecipitation, a $(Ti, W)B_2$ solid solution forms, whereas in a second stage ϕ- or ω-type ternary phases (WCoB or W_2NiB_2) grow from the melt. The residual liquid phase solidifies as Ni_3B with η phase or (Ti, W)C solid solution [16]. These composites are pressureless sinterable at temperatures of 1600–1700 °C, yielding grain sizes of 0.8–1.5 μm. In Fig. 3 a typical microstructure is compared with hot-pressed TiB_2 without any additives and with 16 wt.% Ni. The significant decrease in average grain size due to growth-inhibiting additives and its dependence on sintering temperature is shown in Fig. 4. The relationship between grain size and flexural strength (Fig. 5) reveals a critical particle diameter of 5 μm, which is considered to correspond to the critical flaw size. In a perfect microstructure, the production of composites with a rupture strength greater than 1000 MPa and $K_{Ic} > 6.5$ MPa $m^{1/2}$ seems to be feasible [17].

It has to be noted, however, that sintering of borides with liquid metals is much more complex than for cemented carbides and requires a detailed knowledge of the phase diagrams. Furthermore, extremely brittle phases are involved, and the rapid grain growth of strongly anisotropic compounds implies problems concerning thermal shock, fatigue properties and strength whereas internal stresses and microcracking may even be beneficial to the fracture toughness.

Another example of improved strength by grain size refinement is B_4C with silicon or aluminium additives [18]. Coarse microstructures with 50–200 μm particles which are common in commercial B_4C ceramics limits the strength to 200–300 MPa whereas the addition of

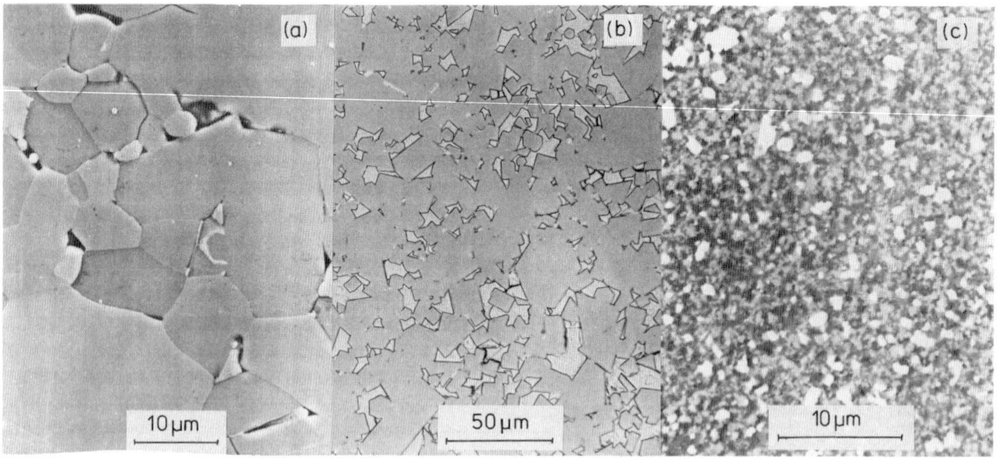

Fig. 3. Microstructures: (a) TiB_2 without additives, hot pressed, 1900 °C, 1 h, 40 MPa; (b) TiB_2 + 16 wt.% Ni, hot pressed, 1600 °C, 1 h, 40 MPa; (c) TiB_2–WC–Ni–Co mixture, pressureless sintered, 1600 °C, 2 h.

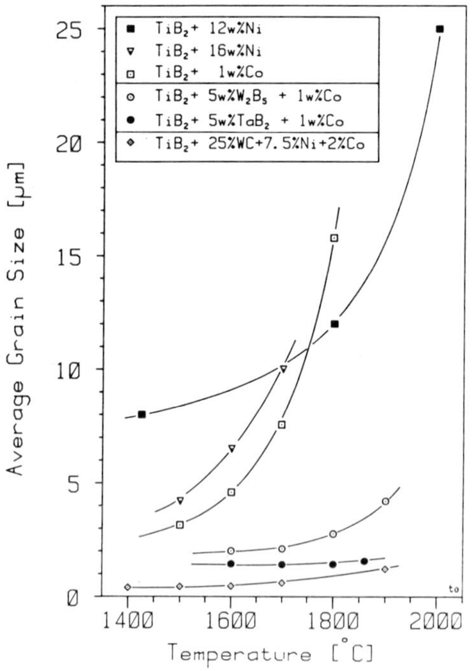

Fig. 4. Sintering temperature dependence of the average grain size in TiB_2-based ceramics.

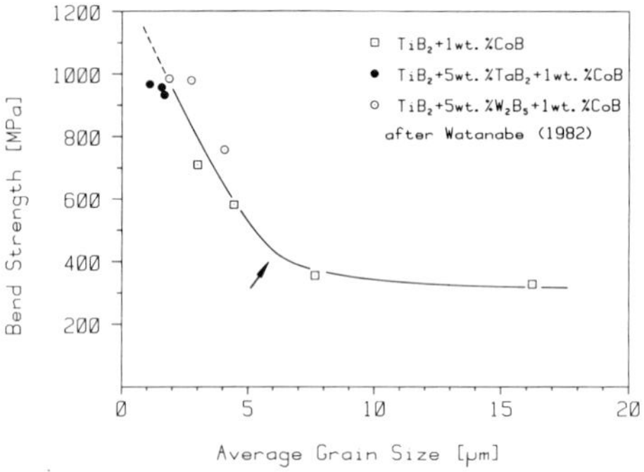

Fig. 5. Dependence of strength on the average grain size. (Data after ref. 17.) The arrow indicates the critical grain size, above which spontaneous cracking is considered to occur.

1–5 at.% Si or Al results in a decrease in the grain size of more than one order of magnitude to 5–7 μm and thus in an increase in strength by a factor of 2. The grain-growth-retarding reaction in the B_4C–Si system is the formation of a silicon-containing B_4C solid solution and the simultaneous precipitation of a 100–200 nm SiC layer around the particles which reduces the grain boundary mobility [19] (Fig. 6). Similar grain boundary pinning is also obtained in B_4C composites with incorporated SiC particles [20].

3.3. Crack deflection, microcracking and crack impediment

The improvement of strength by the incorporation of particles of another phase is usually accompanied by an increase in fracture toughness due to crack deflection, microcracking and crack impediment. To influence the crack propagation,

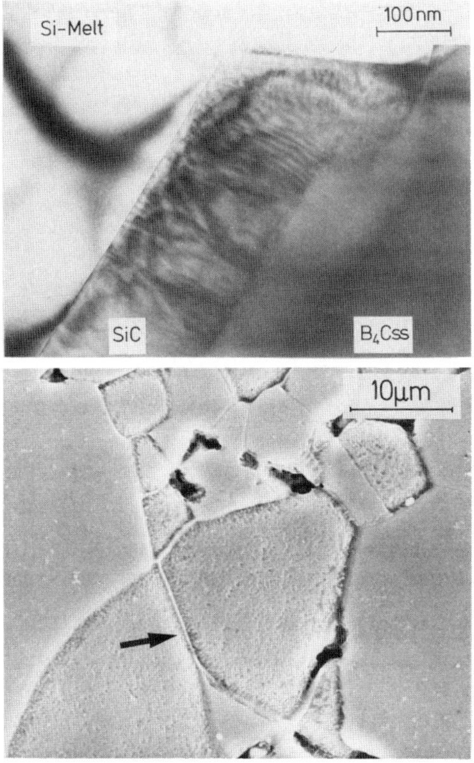

Fig. 6. Grain size refinement of B_4C by introducing SiC as a grain boundary phase (liquid silicon reacts with B_4C to form SiC as a layer on $B_{12}(B, C, Si)_3$ solid solution): (top) transmission electron micrograph of a sample with silicon residue; (bottom) scanning electron micrograph of hot-pressed B_4C with 2 at.% Si.

internal misfit stresses due to the mismatch of the thermal expansion coefficients or Young's moduli of the matrix phase and dispersed particles are required. The desired radial tensile or compressive stresses around particles can thus be generated by a suitable choice of phases. The most usual phase combination involves a low expansion matrix such as B_4C and SiC and a high expansion additive such as TiC, TiB_2 or other transition metal borides. Examples are B_4C–TiB_2–MB_2 composites ($\sigma_b = 800$–1000 MPa; $K_{Ic} = 4$–6 MPa $m^{1/2}$) [21, 22] with a high wear resistance, and also SiC-based materials such as SiC–TiC ($\sigma_b = 550$–750 MPa; $K_{Ic} = 5$–6 MPa $m^{1/2}$) [8, 23, 24] and SiC–TiB_2 composites ($\sigma_b - 450$–700 MPa; $K_{Ic} = 6$–8 MPa $m^{1/2}$) [4–6]. In all cases, crack deflection and crack branching are observed as the toughening mechanisms.

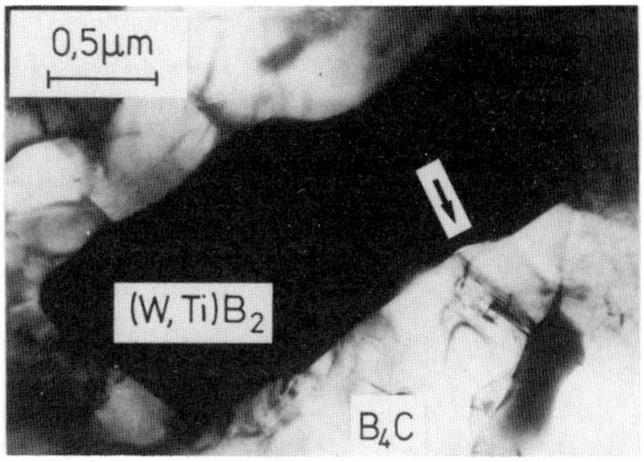

Fig. 7. Transmission electron micrograph of a B_4C–TiB_2–W_2B_5 composite. The formation of a microcrack (arrow) in the stress field between two $(Ti, W)B_2$ particles and the distortion of the B_4C matrix should be noted.

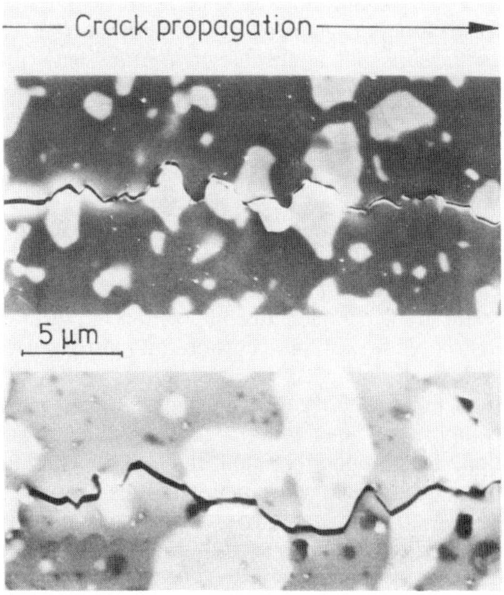

Fig. 8. Crack deflection in SiC matrix composites: (top) Si–TiC; (bottom) SiC–TiB_2.

Figure 7 shows the typical features of crack deflection around TiC and TiB_2 particles in a SiC matrix, resulting in a strong increase in fracture surface.

If we consider "additive" contents exceeding 50 vol.%, crack impediment can be employed as the opposite effect since the tensile stresses then change to compressive stresses. Moreover, the increase in toughness can be enhanced by using coarse incorporated particles which have to have a diameter below the critical flaw size, of course (Fig. 8) [24].

Spontaneous microcracking is considered to occur in B_4C–TiB_2–W_2B_5 and TiB_2–ZrO_2 composites which gives rise to fracture energy dissipation by microcrack opening and crack branching [22, 27]. In Fig. 9, a B_4C particle is distorted by the tensile stresses between two (Ti, W)B_2 plates and reacts by the formation of stacking faults and twins and by debonding. In spite of the considerably improved toughness, spontaneous microcracking is generally not desired, however, for cutting tools because dynamic loading may result in microfracture at the cutting edge [22].

3.4. Transformation toughening of carbides and borides

The most important toughening strategy, transformation toughening by ZrO_2 or HfO_2 dispersed particles, is not applicable to borides and carbides with as spectacular success as for oxide ceramics. The restricting factors are the limited chemical compatibility and the chemical, geometrical and mechanical destabilization of tetragonal ZrO_2.

Owing to the low oxidation resistance of most of the borides and carbides, coupled reduction and oxidation reactions involving ZrO_2 may result in the complete decomposition of both the matrix phase and the incorporated particles; this is further enhanced by the requirement of a reducing or inert gas atmosphere to sinter the composites. Depending on the overall composition, the temperature and the oxygen vapour pressure, the following reactions were observed.

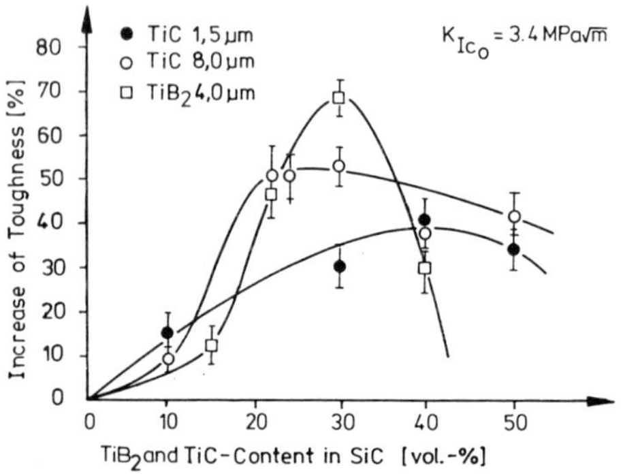

Fig. 9. Increment of toughness in SiC–TiC and SiC–TiB_2 composites. The increase in toughness due to the larger grain sizes of incorporated particles should be noted.

(a) Bulk reduction is followed by combined carbide–oxide or boride–oxide decomposition, *e.g.*

$$3SiC + 5ZrO_2 + 2C \longrightarrow Zr_5Si_3 + 5CO_2 \qquad [28]$$

$$13SiC + 5ZrO_2 \longrightarrow Zr_5Si_3 + 10CO + 10Si \qquad [28]$$

$$B_4C + 2ZrO_2 \longrightarrow 2ZrB_2 + B_2O_3 + CO \qquad [21]$$

(b) A phase boundary redox reaction occurs with subsequent formation of grain boundary phase, *e.g.*

$$2TiB_2 + 2ZrO_2 \longrightarrow (Ti, Zr)B_2 + TiZrO_4 \qquad [27]$$

$$TiB_2 + ZrO_2 \longrightarrow (Ti, Zr)B_2 + (Zr, Ti)O_2 \qquad [27]$$

An example of both oxidation effects and mechanical destabilization is the TiB_2–ZrO_2 system [27]. Despite these rather inconvenient conditions, a significant increase in strength ($\sigma_b = 800$ MPa) and fracture toughness ($K_{Ic} = 8.4$ MPa $m^{1/2}$) was obtained [29].

The high K_{Ic} was attributed to the stress-induced tetragonal-to-monoclinic transformation of ZrO_2 which was partially stabilized by the diffusion of titanium from the TiB_2. Moreover, microcrack mechanisms were observed and also related to an increase in toughness of compositions with less than 30 wt.% ZrO_2. For higher amounts of dispersed ZrO_2, microcrack interactions result in a decrease in both fracture toughness and strength [29]. Similar experiments by the present authors and coworkers [27] verified the strong increase in bend strength up to 809 ± 16 MPa between 30 and 40 vol.% ZrO_2, whereas the K_{Ic} values (indentation method) ranged between 6.5 and 9.5 MPa $m^{1/2}$ and did not show any significant correlation with the volume fraction of ZrO_2 (Fig. 10). This behaviour is attributed to radial tensile stresses perpendicular to the TiB_2–ZrO_2 interfaces due to the mismatch of the thermal expansion. These tensile stresses enhance the spontaneous tetragonal-to-monoclinic transformation but contribute to an increasing toughness by generating crack deflection mechanisms.

Summarizing the properties of TiB_2–ZrO_2 composites, grain size refinement and high sintering densities due to dispersed ZrO_2 particles markedly contribute to the increase in flexural strength. The high fracture toughness is suggested to be attributed more to crack deflection than to stress-induced transformation. More investigations are required for a better understanding of

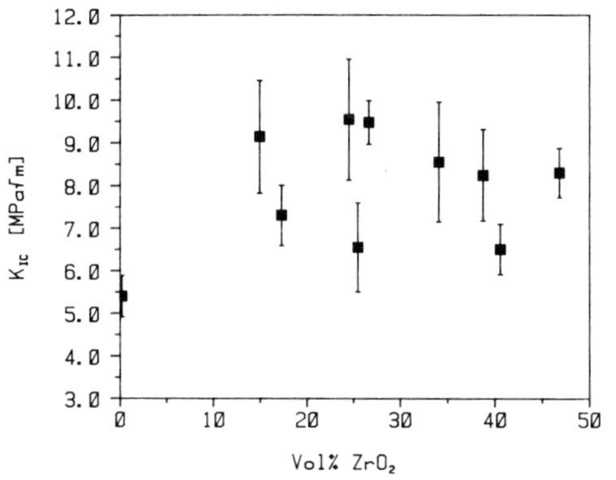

Fig. 10. Volume dependence of K_{Ic} in hot-pressed TiB_2–ZrO_2 composites.

the sintering kinetics, and the use of finer TiB_2 powders, suitable sintering aids or other preparation techniques may result in even better materials with finer grain size, and probably active transformation toughening.

4. Conclusions

Most of the important strengthening and toughening strategies which are successfully utilized for oxide ceramics are not applicable to borides and carbides in a similar easy way. This is in particular the case for metal matrix composites and transformation toughening because of chemical complications. Crack deflection, crack impediment and microcrack interactions are the most effective mechanisms to date. Residual stresses imply fatigue problems, however, during thermal and mechanical cycling.

Many opportunities for a further improvement in the sintering behaviour and the mechanical properties arise from knowledge of the constitution of the particular systems and new processing techniques. Multicomponent systems which may also involve high pressure modifications such as diamond and cubic boron nitride may yield a large number of new composite materials, possibly even with a metallic matrix. New processing techniques imply both improved material preparation, *e.g.* fabrication of ultrapure and ultrafine powders by pyrolysis of metallic–organic precursors, and conditioning of materials by physicochemical methods such as physical vapour deposition, chemical vapour deposition, plasma spraying and ion implantation. These methods may even bring about the combination of materials which are normally incompatible. Improved processing in the field of coatings of ceramic powders may create unique composites by introducing diffusion barriers and enhancing the wetting behaviour. Hence, there is still a high potential for new developments of high performance boride and carbide materials.

References

1 K. A. Schwetz, W. Grellner and A. Lipp, in E. A. Almond, C. A. Brookes and R. Warren (eds.), *Proc. 2nd Int. Conf. on the Science of Hard Materials, Rhodes, September 23–28, 1984*, in *Inst. Phys. Conf. Ser., 75* (1986) 413.
2 H. T. Larker, L. Hermansson and J. Adlerborn, *Ind. Ceram., 8*(1) (1988) 17.
3 H. Holleck, in D. Rowcliff, R. K. Viswanadham and J. Gurland (eds.), *Proc. 1st Int. Conf. on Science of Hard Materials, Jackson, WY, 1981*, Plenum, New York, 1983, p. 849.
4 Y. S. Touloukian, R. K. Kirby, R. E. Taylor and T. Y. R. Lee (eds.), *Thermophysical Properties of Matter*, Vol. 13, *Thermal Expansion, Nonmetallic Solids*, Plenum, New York, 1977.
5 B. Lönnberg and T. Lundström, in H. Werheit (ed.), *Proc. 9th Int. Symp. on Boron, Borides and Related Compounds, Duisburg, 1987*, University of Duisburg, Duisburg, 1988, p. 330.
6 R. W. Rice, in F. D. Gac (ed.), *Proc. 9th Annu. Conf. on Composites and Advanced Ceramic Materials, Cocoa Beach, FL, 1985*, American Ceramics Society, Columbus, OH, 1985, p. 589.
7 K. T. Faber and A. G. Evans, *Acta Metall., 31* (4) (1983) 565.
8 G. C. Wei and P. F. Becher, *J. Am. Ceram. Soc., 67* (8) (1984) 571.
9 D. C. Halverson, J. W. Bridge, A. J. Vidlak, R. S. Glass and V. D. Frechette, *Proc. 89th Annu. Meet., Pittsburgh, PA, 1987*, American Ceramic Society, Columbus, OH, 1987, p. 200.
10 V. J. Tennery, C. B. Finch, C. S. Yust and G. W. Clark, in D. Rowcliff, R. K. Viswanadham and J. Gurland (eds.), *Proc. 1st Int. Conf. on Science of Hard Materials, Jackson, WY, 1981*, Plenum, New York, 1983, p. 891.
11 P. S. Sklad and C. S. Yust, in D. Rowcliff, R. K. Viswanadham and J. Gurland (eds.), *Proc. 1st Int. Conf. on Science of Hard Materials, Jackson, WY, 1981*, Plenum, New York, 1983, p. 911.
12 C. B. Finch, O. B. Cavib and P. F. Becher, *J. Cryst. Growth, 67* (1966) 556.
13 E. Lugscheider and H. Eschnauer, in H. Werheit (ed.), *Proc. 9th Int. Symp. on Boron, Borides and Related Compounds, Duisburg, 1987*, University of Duisburg, Duisburg, 1988, p. 202.
14 H. Haschke, H. Nowotny and F. Benesovsky, *Monatsh. Chem., 97* (1966) 1459.
15 K. Tagaki, S. Ohira, T. Ide, T. Watanabe and Y. Kondo, *Proc. Int. Powder Met. Conf., Toronto, 1984, Modern*

Developments in Powder Metallurgy, p. 153.
16 G. Petzow and R. Telle, in S. Somiya (ed.), *Advances in Ceramics, Proc. Lectures on Advanced Ceramics, Yokohama, 1983*, Terra Scientific Publishers, Tokyo, 1987, p. 131.
17 T. Watanabe and S. Kouno, *Ceram. Bull., 61* (9) (1982) 970.
18 L. B. Ekbom and C. O. Amundin, *Sci. Ceram., 10* (1980) 183.
19 R. Telle and G. Petzow, in P. Vincenzini (ed.), *High Tech Ceramics, Proc. 6th Int. Conf. on High Technology Ceramics, Milan, 1986*, Elsevier, Amsterdam, 1987, p. 961.
20 K. A. Schwetz, K. Reinmuth and A. Lipp, *Sprechsaal, 116* (12) (1983) 1063.
21 D. Lange and H. Holleck, *Proc. 11th Int. Plansee Semin., 1985*, Metallwerk Plansee, Reutte, 1985, p. 747.
22 R. Telle, G. Petzow, J. Alderborn and K. Weiss, in H. Werheit (ed.), *Proc. 9th Int. Symp. on Boron, Borides and Related Compounds, Duisburg, 1987*, University of Duisburg, Duisburg, 1988, p. 453.
23 M. A. Janney, *Am. Ceram. Soc., Bull., 65* (2) (1986) 357.
24 R. Telle and D. Ly Ngoc, to be published.
25 M. A. Janney, *Am. Ceram. Soc., Bull., 66* (2) (1987) 322.
26 C. H. McMurtry *et al., Am. Ceram. Soc., Bull., 66* (2) (1987) 325.
27 R. Telle, S. Meyer, G. Petzow and E.-D. Franz, *Mater. Sci. Eng., A105/106* (1988) 125–129.
28 G. Petzow and J. Lorenz, *BMFT Rep. 01 Zc 038 A-Za/NT/NTS 1016,* 1980 (Bundesministerium für Forschung und Technologie).
29 K. Shoubu, T. Watanabe, J. Drennan, R. H. J. Hannik and M. V. Swain, *Proc. 4th Int. Conf. on the Science and Technology of Zirconia, Tokyo, 1986.*

Materials Science and Engineering, A105/106 (1988) 105–116

Mechanical Properties and Wear Resistance of a Whisker-reinforced Zirconia-toughened Alumina*

M. BOHMER† and E. A. ALMOND

National Physical Laboratory, Teddington, Middx. (U.K.)

(Received November 9, 1987)

Abstract

The influence of SiC whisker reinforcement on the mechanical properties of a dense zirconia-toughened alumina has been investigated. Results of four-point bend tests showed enhanced fracture toughness and flexural strength at room temperature. High temperature properties were measured by hot indentation tests at temperatures up to 900 °C and by creep tests up to 1400 °C. Friction and wear were measured at room temperature in abrasion and dry sliding tests using a pin-on-disc geometry. Abrasion resistance was enhanced by whisker reinforcement, and sliding wear was reduced.

1. Introduction

Increasing attention has been paid recently to the possibilities of reinforcing ceramics with whiskers in order to tailor their physical, chemical and mechanical properties to the requirements of specific applications. Reinforcement of ceramic matrices with suitable whiskers can provide improvements in mechanical properties such as fracture toughness, hardness and sometimes strength, while decreasing variability in properties and performance [1–5].

Preliminary results on such materials indicated that the addition of SiC whiskers to a conventional zirconia-toughened alumina-based ceramic could lead to the development of a very strong and tough ceramic [6, 7] as a result of combining the two toughening mechanisms. However, so far only limited data have been published on properties generally, and particularly on high temperature mechanical properties. For example, it might be expected that these materials would be vulnerable to degradation processes in oxidizing atmospheres at elevated temperatures.

The purpose of the present paper is to indicate some of the beneficial effects of SiC whisker reinforcement of a zirconia-toughened alumina by describing the results from mechanical property tests as well as high temperature performance and friction and wear behaviour.

2. Experimental procedures

For most experiments in the present work, samples of a fine-grained zirconia-toughened alumina containing approximately 17 wt.% of commercial SiC whiskers, and samples of the unreinforced ceramic were provided by the manufacturer, Krupp Widia G.m.b.H., Essen, F.R.G. The whisker-reinforced ceramics were manufactured by axial hot pressing at 1600 °C under a pressure of 30 MPa [7]. The as-received SiC whiskers, consisting of approximately 5 mass% α-SiC and approximately 95 mass% β-SiC, were first treated to disperse agglomerates. They were then mixed with an alumina powder 3 μm in diameter containing 13 wt.% zirconia powder with a particle size of 1 μm. The processing technique for similar materials has been described recently by a number of researchers [5, 8].

The microstructural examinations were performed by optical microscopy, acoustic microscopy, transmission electron microscopy (TEM) and scanning electron microscopy (SEM).

The mechanical tests were performed in a hard beam testing machine using a four-point bend test

*Paper presented at the 3rd International Conference on the Science of Hard Materials, Nassau, The Bahamas, November 9–13, 1987.

†Present address: RWTH Aachen, Institut für Werkstoffkunde, Lehrstuhl B, Templergraben 55, D-5100 Aachen, F.R.G.

0921-5093/88/$3.50

© Elsevier Sequoia/Printed in The Netherlands

fixture manufactured from hard metal. The rollers were set at an outer span of 40 mm and an inner span of 13 mm. Fracture toughness was measured in four-point bend tests on chevron-notched bend specimens 3 mm thick, 8 mm high and 50 mm long. The specimens were oriented with the hot-pressing direction in the plane of the V-notch, from the vertex to the midbase of the triangle. Fracture toughness was calculated from the maximum load observed, using the Bluhm approximation. The tensile strength of the ceramics was measured in four-point bend on specimens with the same nominal dimensions as the fracture toughness specimens. The reinforced ceramics were oriented with both the longitudinal axis of the specimen and the loading axis perpendicular to the hot-pressing axis. Displacement was measured with a linear variable transducer resting on the beam's longitudinal axis at the midpoint between the central rollers. Young's modulus was determined from measurements of the slope of the elastic portion of the load–displacement curve. The shear modulus and Poisson's ratio were calculated from measurements of longitudinal and shear wave velocities using ultrasonic pulses.

A measurement of the materials' edge toughness was obtained by indenting rectangular blocks with a Rockwell indenter at increasing loads at different distances from an edge until a flake broke off and measuring the dependence of load on distance from an edge [9].

For measurement of high temperatures properties, computer-controlled hot indentation tests were performed using a Vickers pyramid diamond indenter at temperatures up to 900 °C in vacuum. The creep rate was determined by maintaining a load on the indenter and measuring the size of the indent for increasing times [10].

To study the effect of SiC whisker reinforcement on creep at temperatures above 900 °C, specimens were deformed in three-point bend in air. The specimens were 22 mm long and 1 mm thick and had widths of 2.5 or 5 mm. The span of the lower support pins was 18.4 mm and the tensile surfaces were perpendicular to the hot-pressing direction. The specimens were heated to the test temperature and allowed to stabilize for about 1 h, before starting the tests. The nominal stress–displacement curves were derived from load–displacement data. The displacements were measured with a linear-displacement transducer, the core of which formed part of the load train.

The oxidation kinetics were studied in thermogravimetric analysis equipment in the atmosphere at a test temperature of 1200 °C. In addition, chemical analyses were performed in order to determine the sensitivity to oxidation at elevated temperatures. The amount of carbon released during oxidation was determined by coulometry. The furnace was set at temperatures of 900 and 1200 °C. No flux was used. Released oxygen was determined with a nitrogen and oxygen analyser at temperatures of approximately 1300 and 2300 °C.

To examine the wear resistance, which is an important property in many industrial tooling applications, the specimens were tested in an abrasive wear test machine similar to a conventional pin-on-disc test but modified by using aqueous slurries of various compositions of abrasives. The wear resistance in various orientations was measured as a function of the hot-pressing axis. The abrasive slurry was contained in a Petri dish mounted on a polishing wheel which rotated at a speed of 100 rev min^{-1}. The 5 mm × 5 mm face of a vertically mounted specimen was pressed into the slurry under a load of 10 N and the specimen holder was moved in an arc across the wheel for the duration of a 10 min test. The calculated relative speed between the specimen and disc had a mean value of 0.35 m s^{-1}. The grits used were alumina, SiC and boron carbide with a nominal particle size of 29 μm. For comparison, tests were also performed on other commercial grade ceramics including pure alumina, and aluminas reinforced with 5–25 wt.% SiC.

Specimens were also tested in unlubricated sliding wear at ambient temperatures over a wide range of test loads in a pin-on-disc machine. The specimens were flat discs with a polished surface of an average roughness of 0.3 μm centre-line average and pins 6 mm in diameter with chamfered edges 0.5 mm by 45° and the same surface roughness as the disc. The test surfaces were perpendicular to the hot-pressing direction, and in each test the pin and discs were made from the same material. The test equipment consisted of a pivoted loading arm, to which the pin holder was attached. The pin pressed against the disc specimen which rotated in a horizontal plane and caused the loading arm to push against a load cell which was mounted on a vertical pillar and measured the frictional force. Normal loads of 10–200 N were used and the sliding velocity was 0.3 m s^{-1}.

3. Results and discussion

3.1. Microstructure

A wide variety of specimen shapes were required for the different mechanical tests and, whilst this introduced no problems for unreinforced samples which were manufactured by conventional sintering treatments, a problem with microstructural reproducibility arose with the reinforced material. Thus the hot-pressing route that was employed to consolidate the whisker-reinforced ceramic produced a directionality in whisker orientation which varied in extent depending on the percentage reduction and final thickness of the specimen blanks. The reinforced and unreinforced ceramics had densities of approximately 3.95 Mg m^{-3} and 4.16 Mg m^{-3} respectively. The densities were obtained by liquid displacement. Differences in densities for the various specimen shapes of several thicknesses were less than 1%.

The microstructures of the reinforced ceramics consisted of very fine SiC whiskers in a ceramic matrix. Figure 1 shows the microstructure of a polished section perpendicular to the hot-pressing axis. Some whiskers (the bright phase) of high aspect ratios are randomly oriented in a grey matrix of alumina and zirconia. The distribution of the whiskers in the matrix was uniform and there was some evidence of a preferred orientation of the whiskers on planes perpendicular to the hot-pressing axis in polished sections parallel to the hot-pressing axis.

3.2. Mechanical properties

The room temperature fracture toughness and flexural strength of SiC-whisker-reinforced and unreinforced alumina–zirconia matrices are given in Fig. 2. The fracture toughness values as well as the flexural strength of the 17 wt.% SiC whisker ceramic were significantly greater than those of the matrix material. The enhancement of both toughness and strength demonstrates the advantage of combining the two toughening mechanisms dispersion toughening by zirconia particles together with reinforcement by SiC whiskers. It has been shown that the addition of zirconia to alumina is more effective in increasing the strength than in increasing the toughness whereas whisker reinforcement can be very effective in enhancing toughness [11]. However, in some systems, whisker reinforcement leads to an increase in fracture toughness but decreases the strength

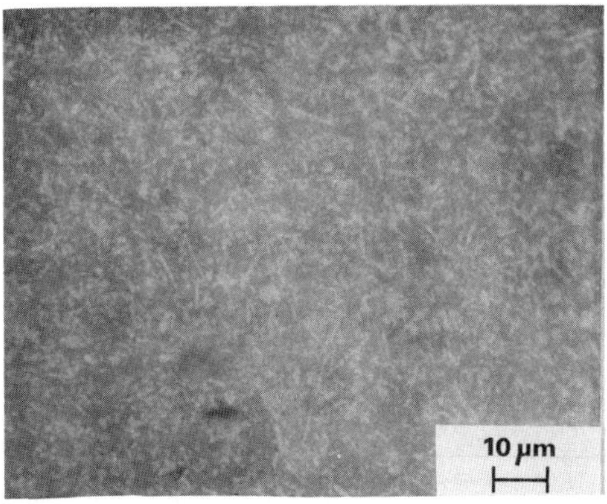

Fig. 1. Photomicrograph of polished section of SiC whiskers in a matrix of alumina and zirconia.

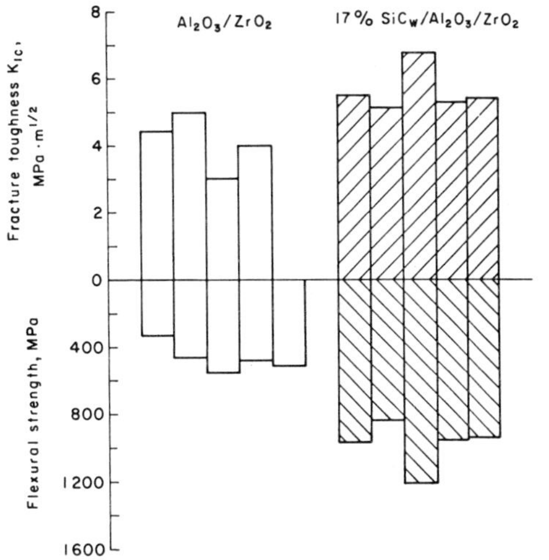

Fig. 2. Room temperature fracture toughness and flexural strength of alumina–zirconia ceramics with and without SiC whisker reinforcement.

[2, 12]. The reduction in strength has been attributed either to flaws introduced with the whiskers or to poor whisker–powder mixing [12].

Observations of fracture surfaces did not reveal any inhomogeneities or clumps of whiskers which could be interpreted as fracture origins. Many whiskers were observed to lie in the plane of the fracture surface, indicating that cracks were locally deflected around the whiskers. Apart from the whisker debonding which accompanied crack deflection, interface debonding resulting from whisker pull-out was observed but not very frequently (Figs. 3(b), 3(c) and 3(d)).

Figure 3(c) shows hexagonal cavities which were probably the result of whisker pull-out. However, the relatively high incidence of crack deflection compared with whisker pull-out indicates that the main mechanisms of toughening were probably crack deflection by SiC whiskers [13] and matrix microcracking around zirconia particles in the reinforced ceramics.

Observations of fracture surfaces from flexural strength tests provided macroscopic evidence of crack deflection along the whiskers. Thus, whereas the fracture plane of the alumina–zirconia ceramic was always normal to the bending axis, the whisker-reinforced ceramic broke always at an angle of about 45° and had a rougher fracture surface, indicating the deviation of the crack through the whisker interfaces. Here also some whisker pull-out was observed; this was associated with hexagonal cavities and occurred more frequently near the tensile surface of the specimens (Figs. 4(a) and 4(b)).

Comparison of the fracture surfaces of the reinforced ceramic with those of the unreinforced ceramic provided evidence of a difference in grain sizes. This supports the suggestion that the incorporation of SiC whiskers inhibits grain growth during the production process (Figs. 4(c) and 4(d)). The practical advantage of a finer grain size is that it enables sharper cutting edges to be maintained for longer service lives than a coarse grain size does.

Another advantage of whisker reinforcement is that it can increase stiffness. Young's moduli

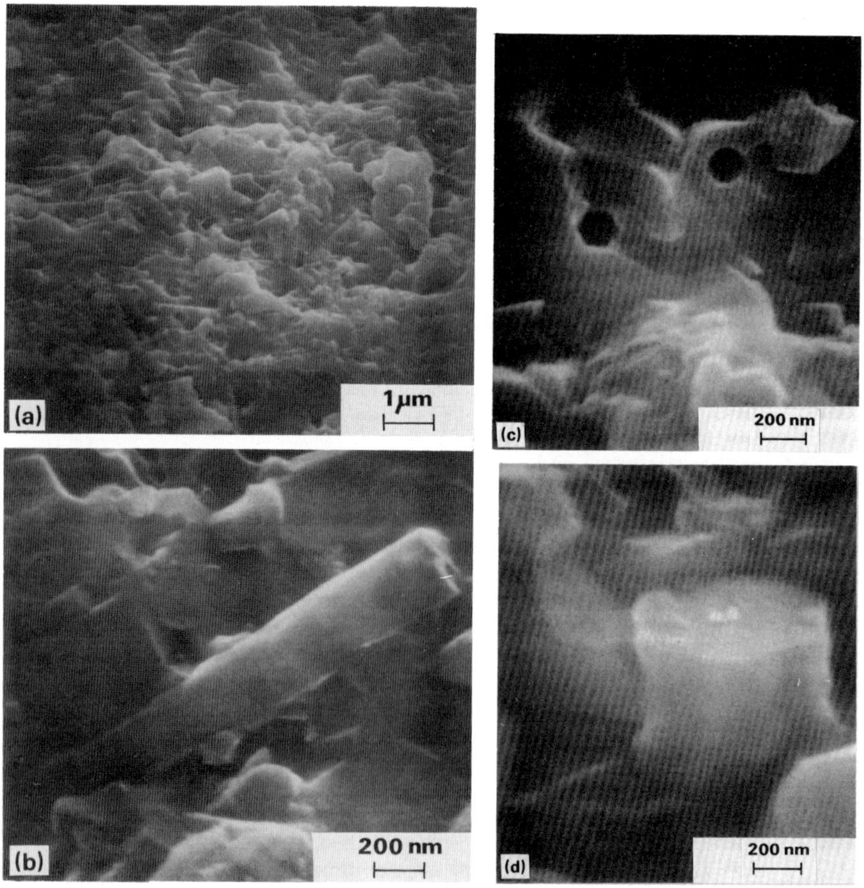

Fig. 3. SEM micrographs (fracture surface) of alumina–zirconia ceramic with SiC whiskers.

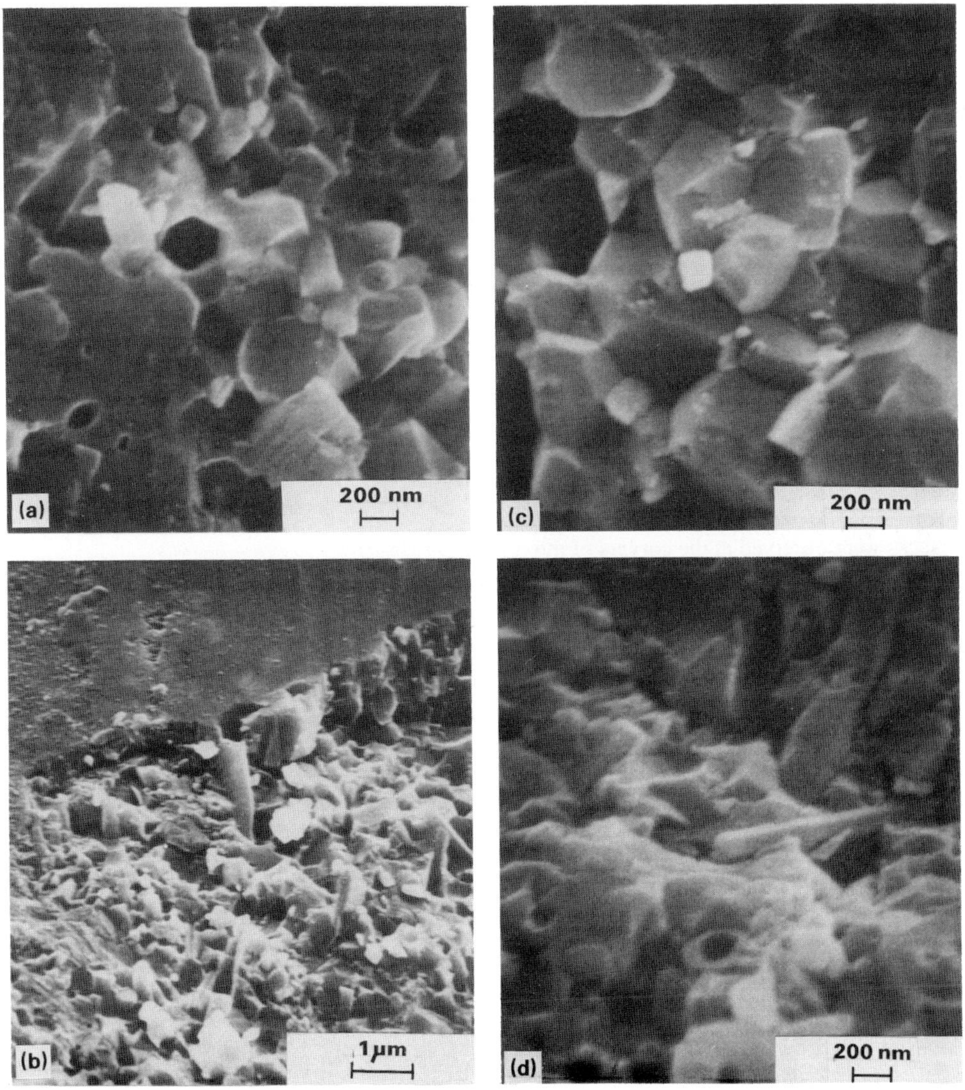

Fig. 4. SEM micrographs of fracture surfaces: (a), (b), (d) alumina–zirconia ceramic with SiC whiskers; (c) alumina–zirconia ceramic.

determined from the linear portion of the load–displacement curves in the strength tests had average values of 342 GPa for the SiC-whisker-reinforced alumina–zirconia ceramic and 325 GPa for the alumina–zirconia ceramic. The shear modulus and Poisson's ratio, determined by the ultrasonic technique, were 128 GPa and 0.28 respectively for the whisker-reinforced ceramic. Thus, whisker reinforcement resulted in a higher stiffness but by a smaller amount than would be expected from the rule of mixtures since the whiskers have a relatively small aspect ratio.

The results of the edge toughness tests showed the same trends as the results from the fracture toughness measurements. As can be seen in Fig. 5, a higher load was necessary to break off a flake from the edge of a specimen of the SiC-whisker-reinforced ceramic than was needed for the unreinforced material at the same distance from the edge. The results indicate that tools made from the SiC-whisker-reinforced ceramic should have better edge retention than unreinforced ceramics do in applications where the cutting edge is subjected to impact loading.

The Vickers hardnesses (load, 30 kgf) of the materials were found to be 1700 HV 30 and 1905 HV 30 respectively for the unreinforced and the whisker-reinforced alumina–zirconia ceramics in tests performed with a Vickers pyramid indenter. The resistance to indentation at elevated temperatures was studied in hot indentation tests at 500, 700 and 900 °C. It was found that, compared with the zirconia-toughened alumina, the whisker-reinforced material was more creep resistant at all test temperatures. Figure 6 shows the increase in indent diameter with increasing time at 900 °C; similar results were obtained at 500 and 700 °C.

No cracks were observed at the corners of the indents at elevated temperatures or room temperature for loads of 3 N in the hot-indentation tests. In order to obtain indentation cracking, it was necessary to raise the indentation load to 25 N. The crack paths exhibited numerous deviations. Figure 7, obtained by scanning acoustic microscopy, shows that the indent was outlined by several fringes which are caused by ridging of the surrounding material; the phenomenon is similar to the plastically deformed zone observed around indents in metals and is indicative of the material's capacity for undergoing large permanent strains without fracturing. (The circular features are either small pores beneath the specimen surface or an artefact caused by the presence of zirconia.)

The high temperature creep behaviour of the SiC-whisker-reinforced ceramic and of the unreinforced alumina–zirconia ceramic were compared at temperatures up to 1400 °C (Figs. 8

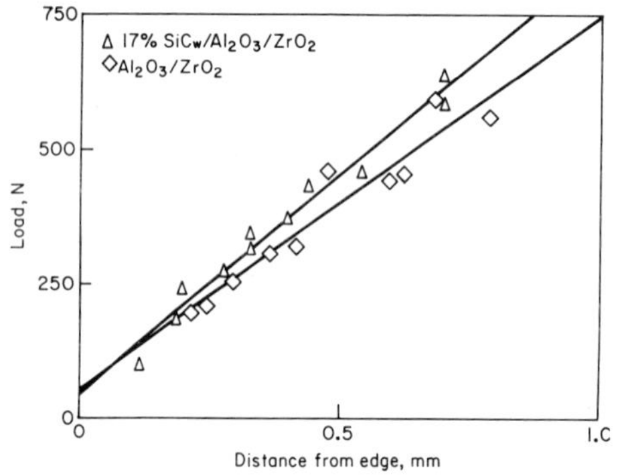

Fig. 5. Load *vs.* distance from edge for alumina–zirconia ceramics with and without SiC whisker reinforcement.

Fig. 7. Acoustic microscope image of a hardness indentation in an alumina–zirconia ceramic with SiC whisker reinforcement.

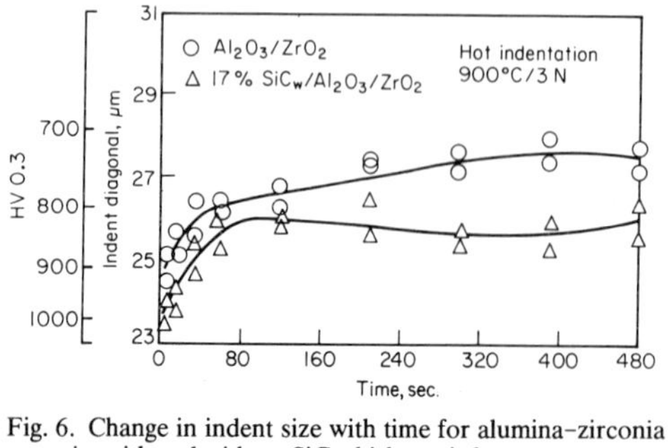

Fig. 6. Change in indent size with time for alumina–zirconia ceramics with and without SiC whisker reinforcement.

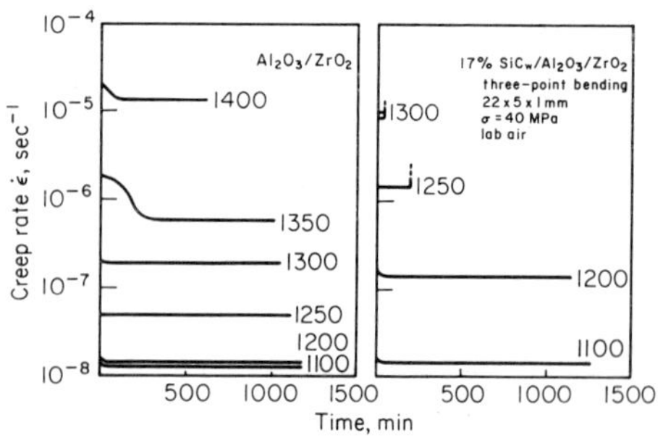

Fig. 8. Creep rate *vs.* time curves for creep deformation of alumina–zirconia ceramics with and without SiC whisker reinforcement, at temperatures up to 1400 °C.

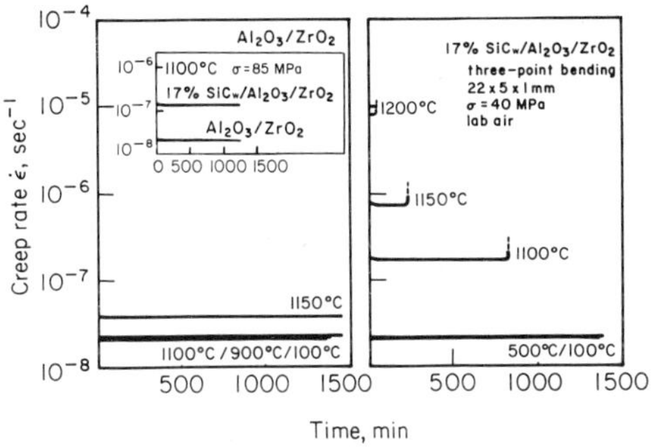

Fig. 9. Creep rate *vs.* time curves for creep deformation of alumina–zirconia ceramics with and without SiC whisker reinforcement, at temperatures up to 1200 °C.

and 9). It can be seen that a short primary creep stage occurs in all the creep curves. Similar creep rates for both materials were obtained for temperatures up to 1100 °C (Fig. 8). Above 1100 °C, the alumina–zirconia ceramic showed better creep resistance, and there were significant differences at 1350 and 1400 °C. Thus, while the alumina–zirconia ceramic showed maximum displacement without rupture at 1250 °C, the SiC-whisker-reinforced ceramics entered the tertiary creep and creep rupture stages (Fig. 8).

Tests under higher nominal stresses emphasized the creep resistance of both materials up to temperatures of 1000 °C (Fig. 9); higher temperatures led to creep rupture of the whisker-rein-

Fig. 10. SEM micrographs: (a) surface structure of an alumina–zirconia ceramic after a creep test at 1300 °C; (b) surface structure of an alumina–zirconia ceramic with SiC whisker reinforcement after a creep test at 1300 °C; (c), (d) fracture section of an alumina–zirconia ceramic with SiC whisker reinforcement after a creep test at 1200 °C.

forced material, while the alumina–zirconia ceramic showed better temperature resistance. The creep ruptures of the reinforced ceramics were strongly stress dependent as the small graph in Fig. 9 demonstrates.

Work elsewhere gave similar results, suggesting that the creep resistance of the whisker-reinforced materials would be retained to approximately 1000 °C and would decrease at 1500 °C for nominal stresses under 100 MPa [1, 14]. Some authors report that Al_2O_3–SiC whisker reinforced composites are very resistant to creep even at high temperatures up to 1500 °C. The failure of the composites that we have examined at lower temperatures is probably a result of the high temperature instability of the whiskers. The present observations suggested that a reaction between the whiskers and the ambient atmo-

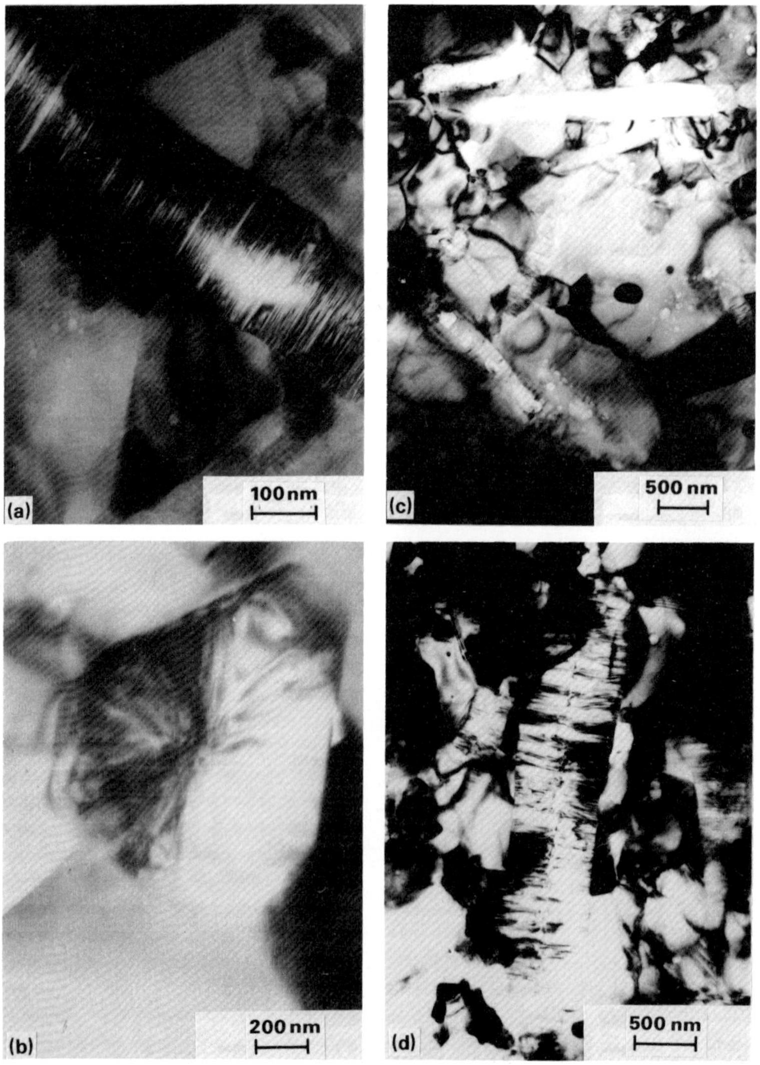

Fig. 11. TEM photomicrographs of SiC whiskers in alumina–zirconia ceramics: (a) longitudinal image of an SiC whisker; (b) cross-sectional image of an SiC whisker; (c) microstructure; (d) microstructure of the ceramic after heat treatment.

sphere might be occurring above 1000 °C. This was supported by SEM examinations of the surfaces of the creep test specimens at 1300 °C, which showed a marked change in the surface appearance of the SiC-whisker-reinforced ceramic (Fig. 10(b)) compared with the alumina–zirconia ceramic exposed to the same temperature (Fig. 10(a)).

An additional feature was a significant increase in the incidence of whisker pull-out (Fig. 10(c)), which seems to indicate that a bond-weakening mechanism was probably occurring through a reaction between the whiskers and the matrix, or between the whiskers and the atmosphere at the higher temperatures. Another characteristic of the fracture surfaces of the SiC-whisker-reinforced creep test specimens was the presence of surface inhomogeneities and craters of different shapes.

Surprisingly, chemical analyses indicated that the ceramics had not undergone bulk oxidation even under the aggressive conditions of heating to 2300 °C in air. A possible reason for this could be that there was a reaction between the SiC whiskers and alumina on the specimens' surface to give a mullite which formed a surface layer and protected the interior of the specimen against further oxidation [14]. Some support for this suggestion was found in the results of the thermogravimetric analyses in atmosphere at 1200 °C, which showed there was only a 0.1% increase in the weight of the SiC-whisker-reinforced specimens, while the alumina–zirconia ceramic was completely stable.

To see whether the high temperature treatments had affected the microstructure, TEM specimens were examined. Figures 11(a) and 11(b) show cross-sectional and longitudinal images respectively of SiC-whiskers. The structure appeared to be fully dense, but there were occasional voids (Fig. 11(a)), which would have reduced the strength of the whiskers and could have been nucleation sites for fracture. From the SEM and TEM observations, it was estimated that the bulk of the whiskers had a diameter of 200–600 nm and their aspect ratio was 5–10.

No significant structural changes resulting from the heat treatment at 1200 °C for 24 h in air could be detected (Fig. 11(d)). However, there appeared to be a thin interfacial layer and small pockets of what was possibly a glassy phase associated with the whisker–matrix interface; these features could have lowered the shear strength of the interfacial bond and facilitated whisker pull-out in the high temperature creep tests; similar explanations have been proposed previously for enhanced creep in whisker-reinforced ceramics [8, 9, 15].

Other features revealed by the TEM examination (Fig. 11) were the random orientation of the whiskers and the apparent absence of agglomeration.

3.3. Friction and wear

Using a high purity alumina as the basis for comparison, improved abrasive wear resistance was obtained by reinforcement with whiskers or zirconia particles. However, the best wear resistance was obtained by combining the two reinforcement processes for both whisker orientations. The rating of the materials abraded with SiC (Fig. 12) was the same for abrasion with alumina and boron carbide. The superior wear resistance of whisker-reinforced alumina–zirconia ceramic was probably due to its greater hardness and toughness since a combination of both properties is needed for abrasion resistance according to proposed wear mechanisms based

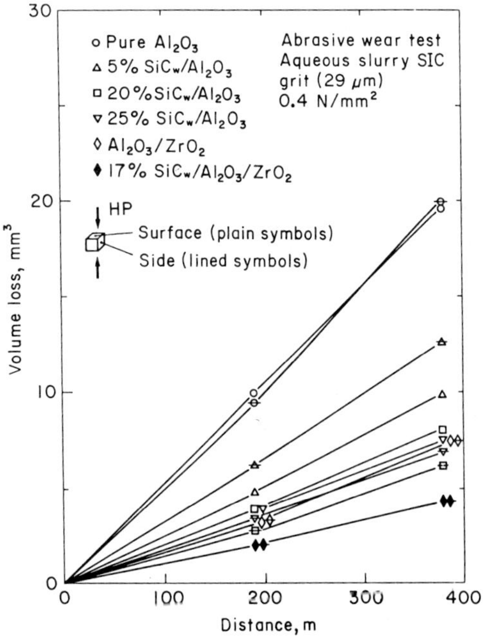

Fig. 12. Variation in abrasive wear with wear distance for different types of wear-resistant ceramics.

Fig. 13. SEM micrographs of wear surfaces from alumina–zirconia ceramics with SiC whisker reinforcement: (a) abraded with boron carbide grit; (b) abraded with SiC grit; (c) abraded with alumina grit.

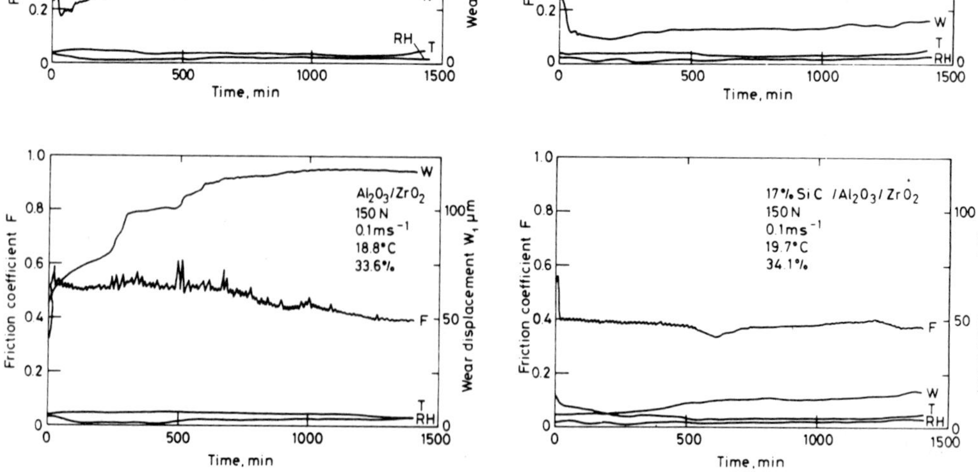

Fig. 14. Variation in friction coefficients and wear with sliding distance for alumina–zirconia ceramics with and without SiC whisker reinforcement: T, test temperature; RH, relative humidity.

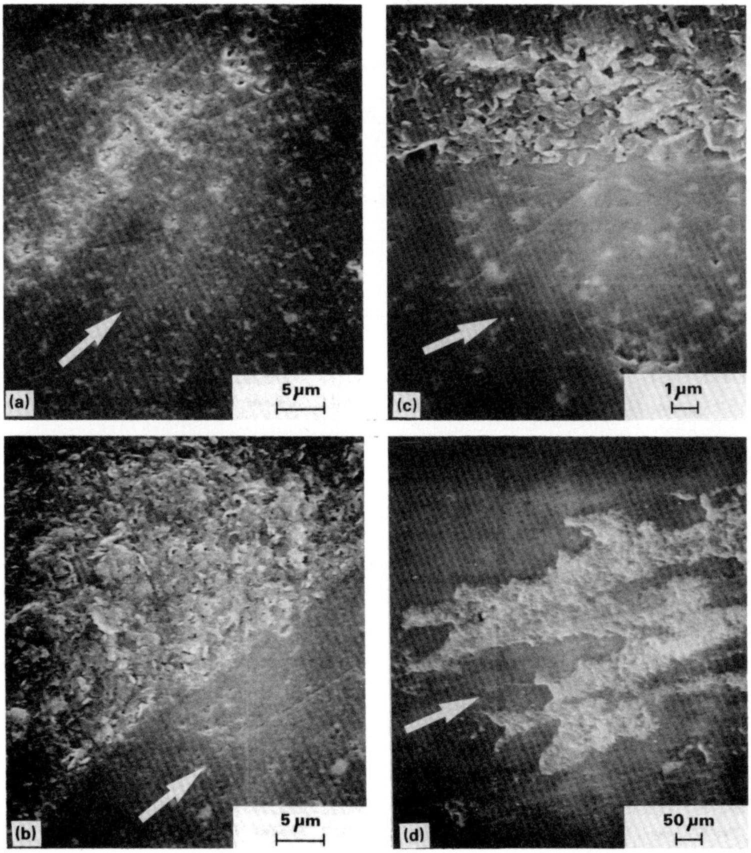

Fig. 15. SEM micrographs of wear surfaces: (a) wear track on an alumina–zirconia ceramic with SiC whisker reinforcement (load, 40 N); (b) wear track on an alumina–zirconia ceramic (load, 40 N); (c) wear track on an alumina–zirconia ceramic with SiC whisker reinforcement (load, 150 N); (d) wear track on an alumina–zirconia ceramic (load, 150 N). The arrows denote the direction of relative motion between pin and disc.

on indentation cracking as the main material removal process (Fig. 13) [16].

In dry sliding wear, there was no significant difference between the frictional resistance in tests on wear couples of the alumina–zirconia ceramic with and without whisker reinforcement and a 95% alumina (Fig. 14). Also, the alumina and the alumina–zirconia ceramic had similar wear resistances. However, the whisker-reinforced alumina–zirconia ceramic had the best wear resistance and gave a more reproducible performance than was obtained from the unreinforced ceramic. For the latter, the high wear rate for a load of 150 N (Fig. 14) was due to the break-away of a chip from the pin and subsequent fragmentation of the chip to form an abrasive wear debris. This resulted in a striated wear surface compared with the relatively smooth surfaces obtained in other tests on the reinforced and unreinforced alumina–zirconia ceramic (Fig. 15).

4. Conclusions

Fracture toughness values of 5.6 MPa $m^{1/2}$ and flexural strengths of 980 MPa were obtained by combining SiC whisker reinforcement with zirconia toughening in a fine-grained alumina. Measurement of creep rates in indentation and bend tests indicated creep resistance up to about 1000 °C in the whisker-reinforced ceramic. Creep resistance decreased at higher temperatures but there was no significant degradation of structure at temperatures up to 2300 °C. Examination of fracture surfaces indicated that the major contributions to toughening appeared to be crack deflection at room temperature, and whisker pull-out occurred at temperatures above 1000 °C.

Stiffness and hardness were also raised by whisker reinforcement, and the resistance to abrasion was greater than that of aluminas reinforced by whiskers alone. In dry sliding conditions the wear losses were decreased and the friction was unaltered.

Acknowledgments

Support by the German Science Foundation (Deutsche Forschungsgemeinschaft) is gratefully acknowledged by M.B.

The authors are indebted to Mr. Kolaska and Dr. Dreyer of Krupp Widia G.m.b.H., Essen, F.R.G., for supplying specimen materials. Thanks are due to Mr. Clinton, Dr. Gee, Mr. Lay, Mr. McCormick, Dr. Morrell and Mr. Paddon of the National Physical Laboratory for experimental assistance and to Professor Knotek, Rheinisch-Westfälische Technische Hochschule, Aachen, F.R.G., for helpful discussions.

References

1 R. W. Davidge, *Composites, 18* (1987) 92.
2 R. Warren, *Proc. 17th CIMAC Conf., Warsaw, 1987*, p. 1–29.
3 T. Mah, M. G. Mendiratta, A. P. Katz, R. Ruh and K. S. Mazdiyasni, *J. Am. Ceram. Soc., 68* (1985) C27.
4 P. F. Becher and G. C. Wei, *J. Am. Ceram. Soc., 67* (1984) C267.
5 G. C. Wei and P. F. Becher, *Am. Ceram. Soc. Bull., 64* (1985) 298.
6 T. N. Tiegs and P. F. Becher, in W. Bunk and H. Hausner (eds.), *Proc. 2nd Int. Symp. on Ceramic Materials and Composites for Engines*, Deutsche Keramische Gesellschaft, Bad Honnef, 1986, p. 193.
7 H. Grewe, K. Dreyer and J. Kolaska, *CFI Ber. DKG8/9-87*, 1987, p. 303.
8 N. Claussen, K. L. Weisskopf and M. Ruehle, *J. Am. Ceram. Soc., 69* (1986) 288.
9 E. A. Almond, *Powder Metall., 30* (1987) 23.
10 B. Roebuck and E. A. Almond, *J. Mater. Sci. Lett., 1* (1982) 519.
11 N. Claussen and G. Petzow, in W. Bunk and H. Hausner (eds.), *Proc. 2nd Int. Symp. on Ceramic Materials and Composites for Engines*, Deutsche Keramische Gesellschaft, Bad Honnef, 1986, p. 649.
12 R. Lundberg, B. Nyberg, K. Williander, M. Persson and R. Carlsson, *Composites, 18* (1987) 125.
13 K. T. Faber and A. G. Evans, *Acta Metall., 31* (1983) 565.
14 A. H. Chokshi and J. R. Porter, *J. Am. Ceram. Soc., 68* (1985) C144.
15 V. K. Sarin and M. Ruehle, *Composites, 18* (1987) 129.
16 E. A. Almond, *Mater. Sci. Technol., 2* (1986) 641.

Influence of Mechanical and Tribochemical Properties on the Wear Behaviour of B_4C–TiB_2–W_2B_5 Composites*

R. LÖRCHER, R. TELLE and G. PETZOW

Max-Planck-Institut für Metallforschung, Institut für Werkstoffwissenschaften, Pulvermetallurgisches Laboratorium, Heisenbergstraße 5, D-7000 Stuttgart 80 (F.R.G.)

(Received December 18, 1987)

Abstract

Pin-on-disc tests with B_4C composites yield an adhesion wear process when cutting AlSi17 discs. In contact with C15 steel discs, abrasive and adhesive wear dominates while abrasion and surface fatigue are the operating wear mechanisms if sheets of clinker are cut. Additional tests with ZrO_2 and SiC–TiC ceramics revealed that the pin materials must have a high fracture toughness to exhibit good wear resistance. The tribochemical influence is proposed to be a weakening of the grain boundaries strength due to an oxidation of the grain boundaries.

1. Introduction

Cutting processes require superior mechanical properties of the tool materials. Promising candidates for these purposes are hard materials such as B_4C or SiC ceramics. In particular, B_4C with its excellent high temperature hardness may be well suited, if it is possible to improve its oxidation resistance at temperatures exceeding 500 °C and to reduce its tendency to brittle fracture. The dispersion of TiB_2 and W_2B_5 in a B_4C matrix densified via hot isostatic pressing resulted in an improved toughness up to 5.5 MPa $m^{1/2}$. The bending strength was found to be 1030 MPa while the hardness is in the range 30–38 GPa [1, 2].

Pin-on-disc tests were conducted with these B_4C composites against different disc materials to identify the operating wear mechanisms. In addition, to determine the influence of mechanical properties on the wear resistance of ceramic materials, ZrO_2- and SiC-based ceramics were tested under the same conditions. The influence of tribochemistry on the wear behaviour of these materials was investigated by Auger electron spectroscopy (AES) and fracture analysis of the pin specimens made by scanning electron microscopy (SEM).

2. Experimental apparatus

A specially designed pin-on-disc apparatus (Fig. 1) was used to test the ceramic samples under

Fig. 1.Pin-on-disc apparatus.

*Paper presented at the 3rd International Conference on the Science of Hard Materials, Nassau, The Bahamas, November 9–13, 1987.

severe conditions. The load is applied by means of a spring, and a micrometer screw controlled the distance between the relaxed and compressed positions of the spring. The sample holder is mounted on a dynamometer, which measures the frictional force. The disc held by a four-jaw chuck is moved by an electric motor which keeps the speed constant even if a load is applied. The speed range is 100–800 rev min^{-1}. An additional feature of the apparatus is a cross-feed drive so that the pin can be moved in a spiral track over the disc. To study the influence of different atmospheres, facilities are available to perform experiments under argon, N_2, and wet and dry air.

3. Materials and sample preparation

Ground bars (3 mm × 3 mm × 11 mm) of B_4C–TiB_2–W_2B_5, SiC–TiC and ZrO_2 ceramics with one lapped and polished face and a cutting angle of 5° were used as the pin materials. The mechanical properties and fabrication processes are listed in Table 1.

The disc materials were as-ground Al–Si alloy with a silicon content of 17 wt.% (AlSi17), steel alloy with 0.15 wt.% C, 0.25 wt.% Si and 0.40 wt.% Mn (C15 steel) 150 mm in diameter and 10–15 mm thick and commercially available sheets of tiles with burned glassy surfaces (clinker) and dimensions 200 mm × 200 mm × 8 mm. Hardness values of 0.9 GPa for the AlSi17 discs, 2 GPa for the C15 steel discs and 6 GPa for the clinker discs were measured by the Vickers indentation method under a load of 50 N.

4. Experimental procedure

Tests with a duration of 300 s were performed under loads of 20, 30 and 50 N in an ambient dry air atmosphere with a sliding velocity of 0.5 m s^{-1} and without any lubrication. The run was interrupted after 20, 90, 160 and 300 s, the weight loss of the pin samples was determined and a check of the specimens made by optical microscopy to estimate the damage of the pins. Before the experiment was started again, the wear track was cleaned of wear debris by air blasting. Subsequently, investigations of pin fracture surfaces by SEM and of the wear tracks on the AlSi17 and C15 steel discs by AES were conducted.

5. Results

5.1. Wear mechanisms

5.1.1. B_4C pin on an AlSi17 disc

Figure 2 is an SEM micrograph of a B_4C pin after cutting an AlSi17 disc. A strong material transfer, resulting in a built-up edge, from the AlSi17 disc can be recognized. The built-up edge has a layered structure. This type of material transfer is considered to be typical of an adhesion-type wear process. Most of the material was transferred within the first 20 s while the main damage to the ceramic pin was caused by chipping of large fragments behind the cutting edge and occurred during the first 90 s. The appearance of fracture was conchoidal. The very rough wear tracks on the AlSi17 discs had depths between 5 and 9 mm.

5.1.2. B_4C pin on a C15 steel disc

In Fig. 3 the SEM micrograph of a B_4C pin after contacting a C15 steel disc is shown, reveal-

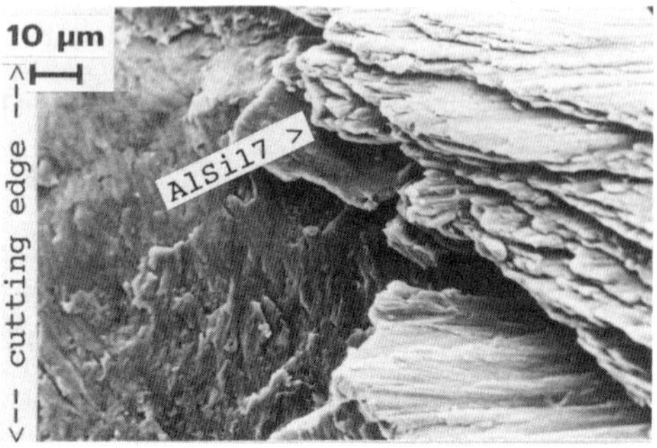

Fig. 2. SEM micrograph of a B_4C–TiB_2–W_2B_5 pin after cutting an AlSi17 disc. Strong material transfer was observed. The operating wear mechanism was adhesion.

TABLE 1 Mechanical properties of the ceramics used as pin materials

Material	*H* (GPa)	K_{Ic} (MPa m$^{1/2}$)	G_B (MPa)	*E* (GPa)
Hot isostatically pressed B_4C–TiB_2–W_2B_5	29–25	3.6–5.2	880–960	400–436
Hot-pressed SiC–TiC	24	5.1–5.8	620–730	441–462
Slip-cast ZrO_2	11	7	400	218

Fig. 3. SEM micrograph of a $B_4C-TiB_2-W_2B_5$ pin which was in contact with a C15 steel disc. Less material transfer to the disc occurred. The wear mechanisms were adhesion and abrasion.

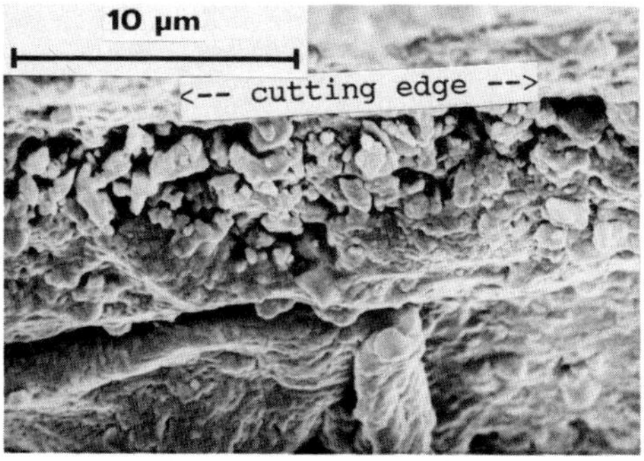

Fig. 4. SEM micrograph of a $B_4C-TiB_2-W_2B_5$ pin which cut a sheet of clinker. The pin was subjected to abrasive wear and surface fatigue.

ing adhesion and abrasion as the operating wear mechanisms.

The amount of material transferred from the steel to the ceramic was less than from the AlSi17 discs. The fracture behaviour was the same as for a B_4C pin on an AlSi17 disc. However, in comparison with the AlSi17 the wear tracks were not as deep (2–4 mm).

5.1.3. B_4C pin on a clinker disc

In this couple the wear mechanisms were identified as abrasion and surface fatigue, as can be seen in Fig. 4. In contrast with Fig. 2 and Fig. 3, no material transfer to the pin was observed. The damage to the pins was restricted to grain pull-out from the cutting edge. No chipping of large fragments behind the cutting edge occurred. The damage of the ceramics decreased in the order AlSi17, C15 steel and clinker as disc materials.

5.2. Influence of the mechanical properties

In Fig. 5 the relationship between the wear volume (which is determined by the measured weight loss of the pin specimens after the experiments divided by the density of the considered materials) and the fracture toughness is plotted for different loads and disc materials. Generally the wear decreases for high values of fracture toughness and low loads if no adhesive wear is

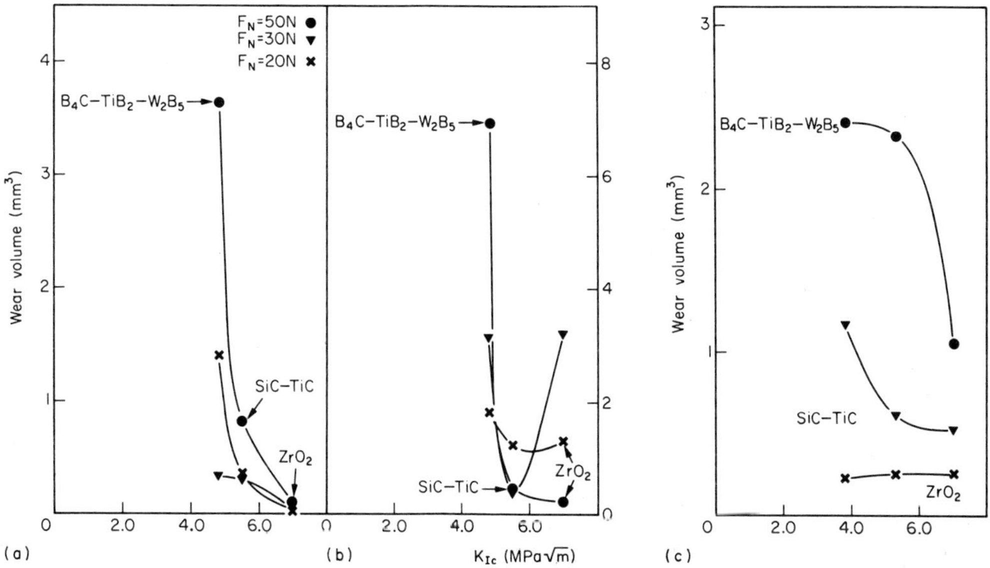

Fig. 5. Relationship between the volume wear and fracture toughness for different ceramic pins in contact with (a) C15 steel, (b) AlSi17 and (c) clinker as disc materials.

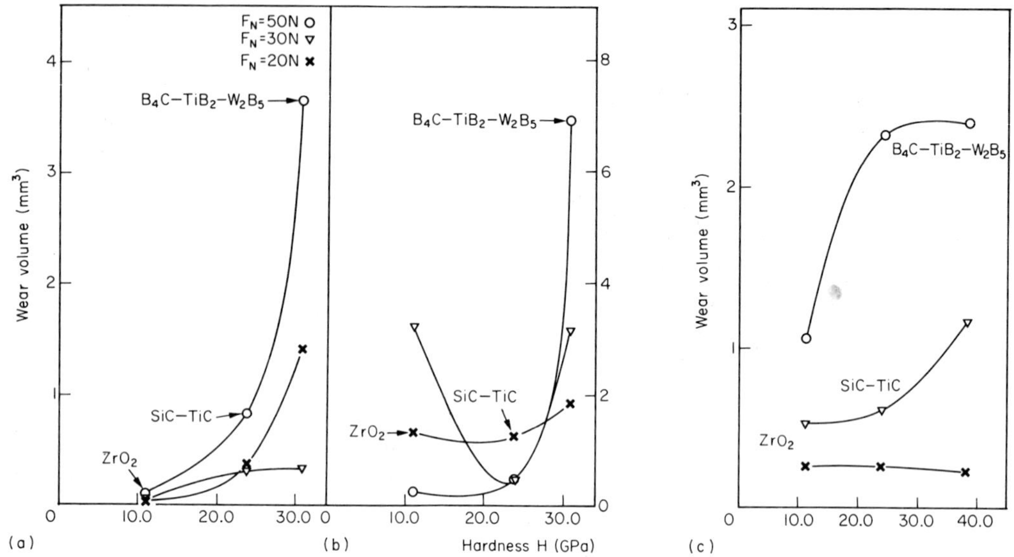

Fig. 6. Volume wear plotted against hardness of different pin materials cutting (a) C15 steel, (b) AlSi17 and (c) clinker.

observed. When adhesion is the wear mechanism, no clear load dependence exists. Figure 6 shows the relationship between the wear volume and the hardness for AlSi17, C15 steel and clinker as the disc materials and for different loads. If AlSi17 is cut by the ceramics, then no strong load dependence could be seen in the case of SiC–TiC and ZrO_2. It is obvious only for B_4C that, with increasing load, the wear increases too. For C15 steel as the disc material and B_4C as the pin material, at a load of 20 N the wear is higher than at a load of 30 N. There is almost no difference between the wear at loads of 30 and 20 N for SiC–TiC and ZrO_2. Using clinker as the disc material the wear of the ceramics increased for loads of 30 and 50 N with increasing pin hardness but remained nearly constant for a load of 20 N. A common feature of all ceramic materials was that, independent of the disc material, a higher wear volume of the ceramic pin resulted with increasing pin hardness.

5.3. *Influence of tribochemistry*

AES spectra were recorded from parts of the wear tracks cut by a B_4C pin on the AlSi17 and C15 steel discs. In Fig. 7 the spectra of the unsputtered wear track surfaces are shown. Both spectra reveal a high oxygen content and a low concentration of boron at the surfaces. After the

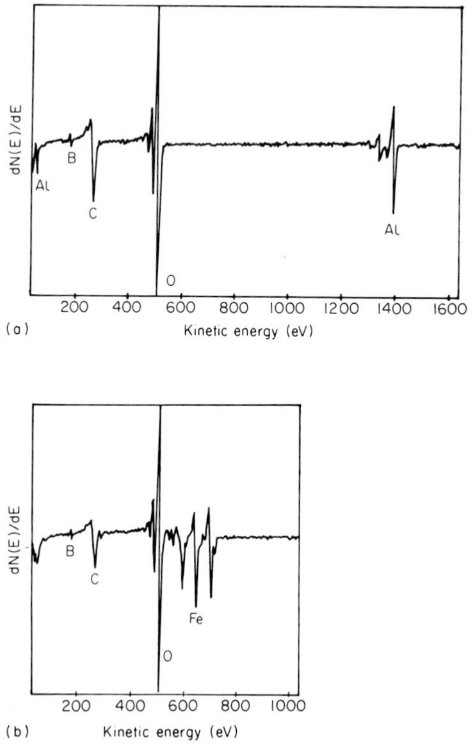

Fig. 7. AES spectra from unsputtered (a) AlSi17 and (b) C15 steel wear surfaces caused by $B_4C–TiB_2–W_2B_5$ pins.

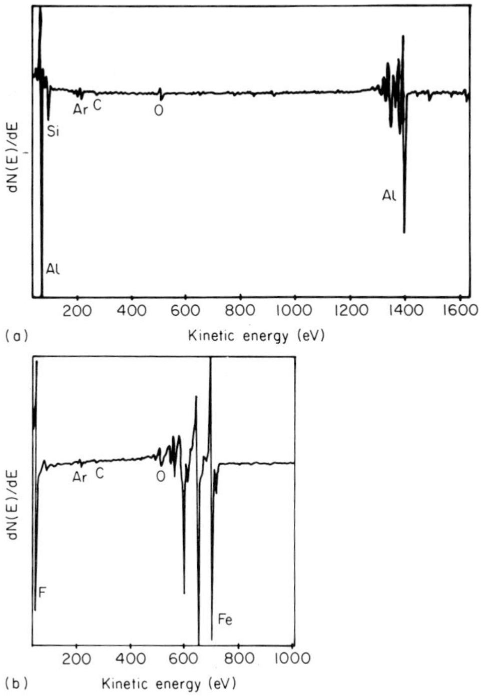

Fig. 8. AES spectra from (a) AlSi17 and (b) C15 steel wear surfaces. The surfaces were sputtered for 10 min with Ar^+ at a sputter rate of 3 nm min^{-1}. The discs were cut by B_4C–TiB_2–W_2B_5 pins.

surfaces had been sputtered for 10 min with Ar^+ at a sputter rate of 3 nm min^{-1}, the oxygen decreased considerably while boron could not be detected (Fig. 8). Small amounts of boron, titanium and tungsten were detected on the wear tracks of AlSi17 and C15 steel discs by performing additional wavelength-dispersive X-ray investigations.

6. Discussion

6.1. Wear mechanisms

The cutting of AlSi17 discs is accompanied by a strong material transfer from the discs to the pins. Obviously material not only from the AlSi17 surface but also from the bulk volume is transferred to the pins. Such a material is regarded as having a negative gradient in the mechanical properties which is described, according to Kragelski [3] by

$$\frac{d\tau_{xc}}{dz} < 0$$

where τ_{xc} is the critical shear stress parallel to direction of the contact face and z is the direction perpendicular to the contact face. This means that the strength of the bonding between the B_4C and the AlSi17 is stronger than the strength of the layers in the AlSi17. Consequently the damage occurs in the weaker regions to a greater depth in AlSi17. Therefore the deformation is not restricted to the surface but a larger volume is deformed. This is confirmed by the very rough wear tracks of the AlSi17. The gradient of the mechanical properties depends on the temperature in the contact zone. Surfaces always have asperities and the contact between two bodies occurs in these asperities, which undergo plastic deformation. Bowden and Ridler [4] have demonstrated that, if the adhesive bonds in the asperities were destroyed by sliding processes, the available temperatures can exceed the melting point of many metal alloys and therefore softening of the surface regions occurs. This can lead to a change in surface strength and hardness. Furthermore, existing oxide layers could be destroyed and the formation of stronger chemical bondings between the two bodies is possible because surfaces without any oxide layer are very reactive. In addition, the diffusion could be higher if the temperature increases.

The chipping of large fragments behind the cutting edge of the B_4C pin can be explained by the following arguments. Different thermal expansion coefficients exist between the B_4C matrix and the TiB_2 or W_2B_5 particles, and so internal thermal mismatch stresses are induced. These stresses produce a crack deflection process [5] which is responsible for the higher fracture toughness of 5.5 MPa $m^{1/2}$. If shear stresses from the sliding process are superposed, they may cause chipping of the large fragments. The internal shear stresses at the grain boundaries were estimated from the following equation [6]:

$$\sigma_{mr} = \frac{(\alpha_p - \alpha_m)\Delta T}{\{(1+\nu_m)/2E_m\} + \{(1-\nu_p)/E_p\}}$$

where σ_{mr} is the radial stress in the matrix, α_m the thermal expansion coefficient of the matrix, α_p the thermal expansion coefficient of the particles, E Young's modulus and ν Poisson's ratio (subscript m represents the matrix and subscript p the particles).

For B_4C–TiB_2–W_2B_5 the residual stresses σ_{mr} are calculated to be $1.69\Delta T$ MPa. On the assumption that a sample is cooled down from 1200 °C which is proposed as the minimum temperature for thermal stress relaxation, the residual internal thermal mismatch stresses σ_{mr} at room temperature are ($\Delta T = 1200$ °C) 2.0 GPa which is about a factor of 2 higher than the bending strength of the material. In addition, the wear track on AlSi17 is very deep (greater than 5 mm), and bending moments, tensile and compressive stresses have to be superposed; hence, they may amplify the external shear stresses. Overall a very complicated stress state results which is considered to be responsible for the chipping. The same mechanisms are proposed for the cutting of C15 steel discs.

When sheets of clinker were used as the disc material, then a change in the wear mechanisms to abrasion and surface fatigue occurred. The wear changed from severe (AlSi17 and C15 steel) to mild wear. One reason may be that the hardness difference between the clinker and the B_4C is much less than for the other two disc materials. Hence the depth of penetration by the B_4C is not so great and the external shear stresses are concentrated in a smaller volume so that the wear is restricted to the cutting edge.

6.2. Mechanical properties

As demonstrated by Figs. 5 and 6, generally the wear decreased with decreasing values of the hardness and with increasing values of the fracture toughness. However, the hardness of the pin material must be greater than a certain critical value depending on the disc material; otherwise, no material removal from the discs is possible. As an example, C15 steel and clinker discs could not be cut by ZrO_2. If the surface loading exceeds a critical value, then microchipping dominates and the fracture toughness of the wearing material determines the abrasive weight loss. Moore [7] showed for abrasive wear that the material removal caused by fracture mechanisms is several orders higher than for deformation processes. Adewoye and Page [8, 9] confirmed the superposition of a plastic and a brittle wear process in ceramics when performing sliding friction tests with conical diamond indenters on (0001) planes of SiC single crystals, on polycrystalline SiC and Si_3N_4. Using low loads the damage of the stressed surfaces was caused by grooving and plastic deformation with little or no cracking. The applied hydrostatic stress field had tensile components which were substantially lower than the stresses necessary for cleavage. With increasing applied load the available shear stress for plastic deformation and the tensile component of the applied stress also increased. When the tensile components exceeded the cleavage stress, microfracture and only a little plastic deformation occur. In the case of ceramics with an exceptional hardness, almost no plastic deformation is possible [10]. Consequently the wear of B_4C composites is caused predominantly by fracture mechanisms. Fracture mechanisms dominate for low values of the ratio of fracture toughness to hardness [7].

6.3. Tribochemistry

From AES investigations, there is evidence that the unsputtered wear surfaces on the AlSi17 and C15 steel discs reveal a small amount of boron and a high oxygen content. If coupled with B_4C ceramic, it may be concluded that the reaction

$$B_4C + 4O_2 \rightarrow 2B_2O_3 + CO_2$$

is initiated during the sliding process. Adsorbed oxygen at the surfaces of the discs has unsaturated bonds and is very reactive so that a thin layer of B_2O_3 is formed. During cutting the discs the B_2O_3 layer is removed. Oxygen from the air can be adsorbed on the fresh and very reactive wear track again. Since the pin is moving, a new layer of B_2O_3 is formed if the pin is in contact with the new adsorbed oxygen. A further conclusion is that local temperatures are expected to exceed 500 °C which is the temperature of the oxidation of B_4C.

An additional assumption is that B_2O_3 is also formed at the grain boundaries in the B_4C composites. Therefore the grain boundary strength decreases, and break-out of the grains from the ceramic can occur owing to impacts excited by vibrations from the apparatus. As mentioned before, small amounts of boron, titanium and tungsten were detected on the AlSi17 and C15 steel wear surfaces. This may prove that, during the sliding process, TiB_2 and W_2B_5 grains break out and are deposited in the wear track, *i.e.* they become abrasive particles. Dynamic impact studies [11] have confirmed the separation of TiB_2 and W_2B_5 grains from the B_4C matrix due to a crack deflection mechanism which favours break-out of grains (Fig. 9).

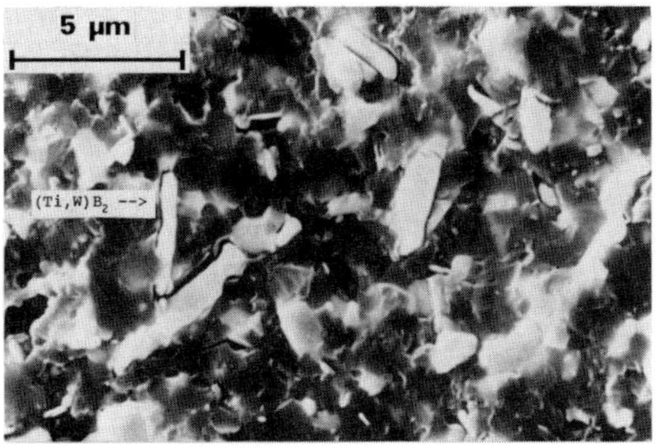

Fig. 9. SEM micrograph of a B_4C–TiB_2–W_2B_5 sample which was subjected to dynamic impact.

7. Conclusions

(1) The operating wear mechanisms are obviously determined by the ratio of the hardness of the pin to the hardness of the disc. With decreasing values of this ratio, no adhesive wear but abrasion and surface fatigue occur.

(2) Wear caused by adhesion is greater than wear caused by abrasion.

(3) Fracture mechanisms (*i.e.* chipping) are predominantly responsible for the damage of the B_4C–TiB_2–W_2B_5 ceramics. So high fracture toughness values are required for this material to prevent chipping and to exhibit good wear resistance.

(4) It is suggested that tribochemistry influences the wear behaviour by the oxidation of the grain boundaries in the B_4C composites, decreasing the strength of the grain boundaries and resulting in break-out of grains, which can work as abrasive particles.

References

1 H. Hofmann and G. Petzow, *J. Less-Common Met., 117* (1986) 121–127.
2 R. Telle, G. Petzow, J. Adlerborn and K. Weiss, in H. Werheit (ed.), *Proc. 9th Int. Symp. on Boron, Borides and Related Compounds, Duisburg, 1987*, p. 453.
3 I. W. Kragelski, *Reibung und Verschleiß*, Carl Hanser, Munich, 1971.
4 F. P. Bowden and K. E. Ridler, The surface temperature of sliding metals, *Proc. R. Soc. London, 154* (883) (1936) 640.
5 G. C. Wei and P. F. Becher, *J. Am. Ceram. Soc., 67* (8) (1984) 571–574.
6 J. Selsing, *J. Am. Ceram. Soc., 44* (8) (1961) 419–422.
7 M. A. Moore and F. S. King, *Wear, 60* (1980) 123–140.
8 O. O. Adewoye and T. F. Page, *Wear, 70* (1981) 37–51.
9 O. O. Adewoye and T. F. Page, *Wear, 73* (1981) 247–260.
10 K. H. Habig, *Verschleiß und Härte von Werkstoffen*, Carl Hanser, Munich, 1980.
11 B. Ilschner and R. Zohner, *Ann. Rep.*, 1986 (European Concerted Action COST 503).

Sintering Behaviour and Phase Reactions of TiB_2 with ZrO_2 Additives*

R. TELLE

Max-Planck-Institut für Metallforschung, Institut für Werkstoffwissenschaften, Pulvermetallurgisches Laboratorium, Heisenbergstrasse 5, D-7000 Stuttgart 80 (F.R.G.)

S. MEYER

Institut für Mineralogie und Kristallchemie, Universität Stuttgart, Pfaffenwaldring 55, D-7000 Stuttgart 80 (F.R.G.)

G. PETZOW

Max-Planck-Institut für Metallforschung, Institut für Werkstoffwissenschaften, Pulvermetallurgisches Laboratorium, Heisenbergstrasse 5, D-7000 Stuttgart 80 (F.R.G.)

E. D. FRANZ

Institut für Mineralogie und Kristallchemie, Universität Stuttgart, Pfaffenwaldring 55, D-7000 Stuttgart 80 (F.R.G.)

(Received November 12, 1987)

Abstract

The sintering behaviour of TiB_2–ZrO_2 composites with ranging from 10 to 50 vol.% of unstabilized ZrO_2 was investigated by pressureless sintering up to 2100°C in a dilatometer as well as by hot pressing up to 1700°C, resulting in a maximum density of 98% of the theoretical density.

Phase determinations by X-ray diffraction and transmission electron microscopy revealed the formation of (Ti, Zr)B_2 and (Zr, Ti)O_2 solid solutions in layers close to the grain boundaries. Moreover, there is evidence of the formation of the zirconium titanate (Zr, Ti)$_2O_4$ as a grain boundary phase. The phase relations are discussed in respect to their contribution to expected transformation toughening.

1. Introduction

Titanium diboride (TiB_2) is one of the few borides which are expected to be chemically compatible with zirconia (ZrO_2) at high temperatures. Thus, a composite of both of these materials is considered to combine the excellent hardness and thermal conductivity of TiB_2 with a high fracture toughness due to the tetragonal-to-monoclinic transformation of dispersed ZrO_2 particles.

Recent investigations in the TiB_2–ZrO_2 system [1, 2] revealed a significant increase in both strength (800 MPa) and K_{Ic} (8–9 MPa $m^{1/2}$) which was claimed to be due to a grain size refinement and transformation toughening. The partial stabilization of ZrO_2 was found to result from a titanium diffusion from TiB_2 into ZrO_2 which is accompanied by the simultaneous formation of a (Ti, Zr)B_2 solid solution.

In this study, the sintering behaviour of TiB_2–ZrO_2 composites containing up to 50 vol.% of unstabilized ZrO_2 was investigated by sintering dilatometry and hot pressing as a function of various milling treatments. Temperature-dependent phase reactions are discussed and related to the suggested toughening mechanisms.

2. Experimental details

Powder batches of 100 g were prepared by attrition milling of TiB_2 (H. C. Starck, 2467 grade A; average grain size, 1.75 μm; Brunauer–Emmett–Teller (BET) specific area, 0.4 $m^2\ g^{-1}$; impurities less than 1 mass% O, C, N and Fe, 1.16 mass% W†) and 10, 20 and 35 vol.% unstabilized ZrO_2 (Magnesium Electron Ltd., SC 20; average grain size, 0.2–0.6 μm; agglomerate size, 6–8 μm; BET specific area, 3.61 $m^2\ g^{-1}$; no impurities given). Attrition milling was carried out in a 500 ml stainless steel vessel using a hard

*Paper presented at the 3rd International Conference on the Science of Hard Materials, Nassau, The Bahamas, November 9–13, 1987.

†Tungsten was not listed in the specification provided by H. C. Starck.

0921-5093/88/$3.50 © Elsevier Sequoia/Printed in The Netherlands

metal rotor and 1700 g of 3 mol.% Y_2O_3 tetragonal stabilized ZrO_2 milling balls of 3 and 5 mm diameter. The milling time in isopropanol was 30, 60 or 120 min, with a rotor velocity of 1000 rev min^{-1}. Pick-up from the vessel and the rotor was negligible whereas the eroded material from the milling balls was found to contribute significantly to the overall composition of the powder mixture (Fig. 1).

After drying in a rotation evaporator, the powder was cold isostatically pressed in silicon rubber moulds applying 630 MPa for 1 min. The average green density was 60%–70% of the theoretical density. Pressureless sintering was performed with a heating rate of 50 K min^{-1} up to 2150 °C in an argon atmosphere in a graphite tube furnace with a differential dilatometer attached. In separate experiments, sintering times of 1 h at 1700 and 2100 °C were used. Axial hot pressing was conducted at 1700 °C at 24 MPa for 1 h in an argon atmosphere using graphite dies and plungers coated with hexagonal boron nitride.

The microstructure and phase composition of ground and polished sections was analysed by scanning electron microscopy (SEM), transmission electron microscopy (TEM), scanning transmission electron microscopy (STEM) and wavelength-dispersive X-ray analysis (WDXA). The theoretical density of the particular specimens was calculated by the amount of residual porosity measured from SEM micrographs.

3. Results

Dilatometric measurements of TiB_2–ZrO_2 mixtures revealed an onset of shrinkage between 1200 and 1300 °C regardless of the ZrO_2 content and the milling procedure. The maximum densification rate was observed between 1500 and 1600 °C and increased with increasing amount of ZrO_2. For mixtures milled for 60 or 120 min the shrinkage was almost complete at 1750 °C and was followed by a severe bloating above 2100 °C whereas the powder milled for 30 min exhibited a continuous shrinkage up to 2100 °C. For isothermal sintering, a maximum temperature of 1700 °C was chosen in order to avoid the strong swelling. The final densities after 1 h at 1700 °C ranged between 93% for the lowest ZrO_2 content and 98% for the highest ZrO_2 content (Fig. 2).

With prolonged milling, the microstructure of pressureless-sintered materials became increasingly homogeneous. Cluster formation of ZrO_2 particles—a characteristic feature of powders milled for a short time which is due to both undestroyed ZrO_2 agglomerates and coarse TiB_2 particles—was reduced. Furthermore, the grain growth of TiB_2 and ZrO_2 was retarded because of the enhanced distribution of the particles. In Fig. 3, the microstructure of a TiB_2–20vol.%ZrO_2 composite sintered at 1700 °C is shown. The characteristic diffusion layer of (Ti, Zr)B_2 solid solutions is visible because of the atomic number contrast.

Irrespective of the ZrO_2 content or the milling time, samples sintered slightly above 2100 °C exhibited a large amount of round 1–15 μm pores because of vapour phase reactions which gave rise to the pronounced bloating. As the mass loss was always very small (typically 1.5–2.5 mass%), it is concluded that, owing to the completed densification at lower temperatures, the

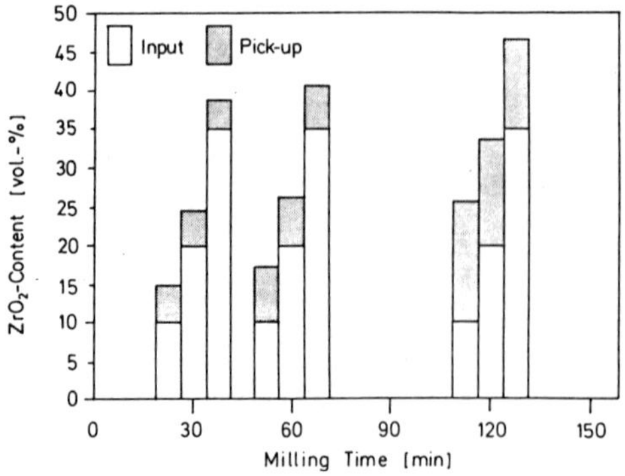

Fig. 1. ZrO_2 pick-up from milling balls during attrition milling.

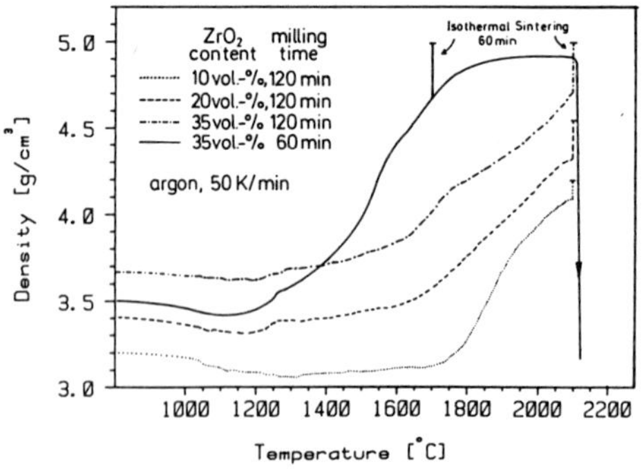

Fig. 2. Sintering behaviour of various TiB_2–ZrO_2 mixtures. The final densities after isothermal treatment corresponds to 93%–98% of the theoretical density.

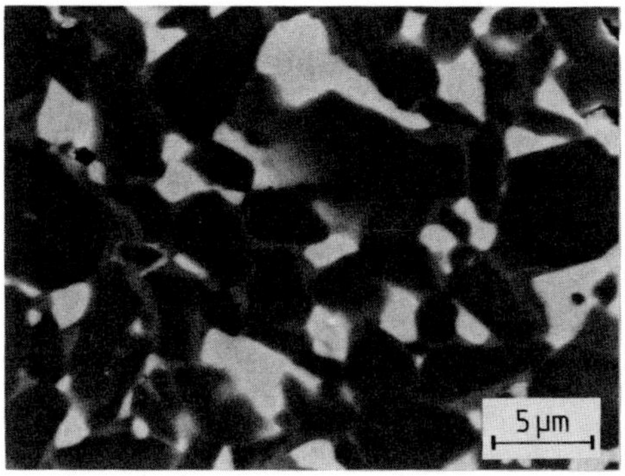

Fig. 3. Microstructure of a 10 vol.% ZrO_2 material milled for 60 min with (Ti, Zr)B_2 diffusion layers around TiB_2 particles visible by the atomic number contrast: light phase, ZrO_2; dark phase, TiB_2; intermediate phase, (Ti, Zr)B_2.

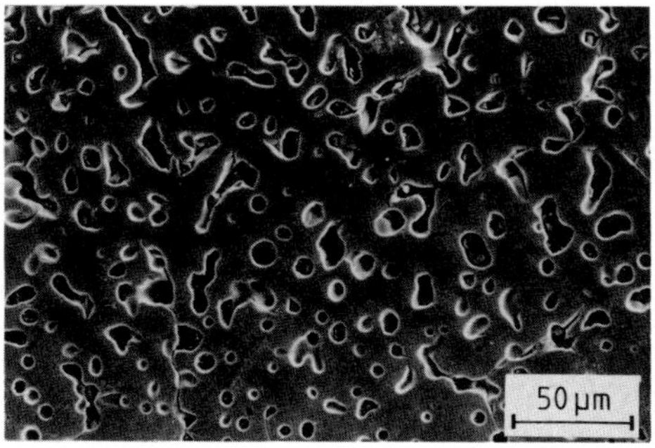

Fig. 4. Entrapped B_2O_3-gas pores due to internal oxidation above 2100 °C.

gas was entrapped rather than evaporated (Fig. 4). The condensed reaction products TiO_2 (rutile), B_2O_3 and $TiBO_3$ were identified in the pores after cooling.

Since axial hot pressing was carried out at the same temperature as pressureless sintering, the results of both treatments thus are comparable. Shrinkage started at approximately 1100 °C and yielded 98% of the theoretical density. No influence of ZrO_2 content or milling procedure on the densification behaviour could be detected. The microstructures show similar characteristics to those of pressureless-sintered material, namely a decreasing grain size and an improved homogeneity with prolonged milling time (Fig. 5). Furthermore, the formation of a (Ti, Zr)B_2 solid solution around the TiB_2 particles was also observed as in the pressureless-sintered material. On increasing the volume fraction of ZrO_2, the material cracked severely.

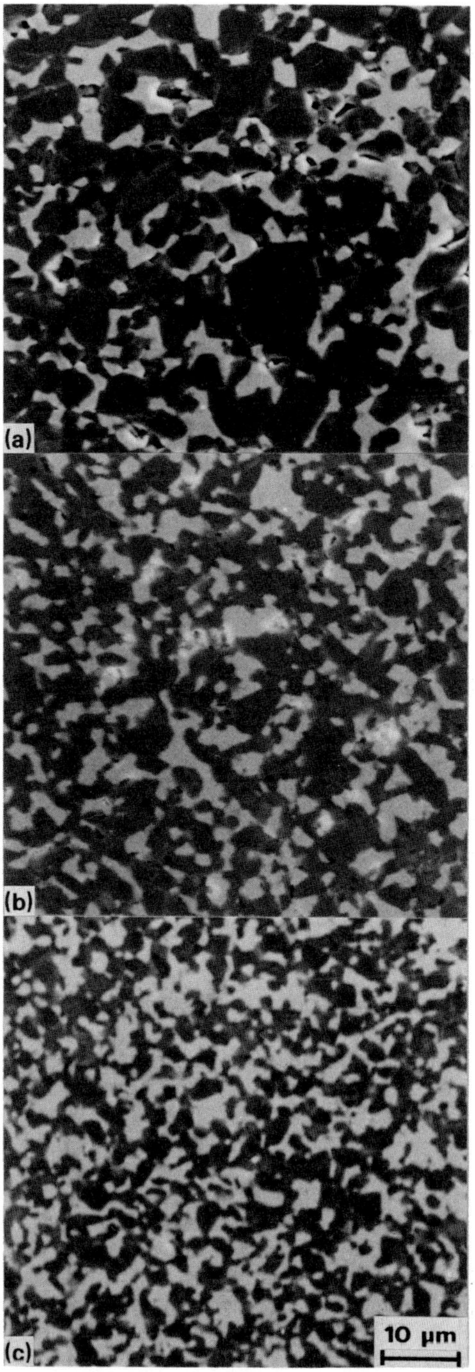

Fig. 5. Microstructural characteristics of hot-pressed 20 vol.% ZrO_2 composites as a function of attrition milling time: (a) 30 min; (b) 60 min; (c) 120 min.

Phase determinations by X-ray diffraction and TEM revealed the simultaneous formation of (Ti, Zr)B_2 and (Zr, Ti)O_2 solid solutions in areas close to the phase boundaries independent of the method of densification. STEM and WDXA clearly established TiO_2 concentrations of up to 9–10 mass% (14–16 mol.%) in the outer layers of the ZrO_2 particles. The maximum ZrB_2 content in the (Ti, Zr)B_2 reaction layers was 5.3 mass% (3.2 mol.%). Towards the cores of the particles, the ZrB_2 concentration decreased discontinuously. Moreover, there is evidence for the existence of an intergranular phase at (Ti, Zr)B_2–(Zr, Ti)O_2 junctions which is suggested to be zirconium titanate (Zr, Ti)$_2O_4$ because TEM diffraction patterns (Fig. 6) are similar to those of tetragonal ZrO_2 (t-ZrO_2) but exhibit a lower symmetry which may be attributed to the space group P*bcn*. Although a considerable amount of t-ZrO_2 was detectable especially for ZrO_2-poor compositions (Fig. 7), repeated heating and cooling of hot-pressed material revealed the well-known hysteresis due to the phase transformation of unstabilized ZrO_2. Hence, untransformable (Zr, Ti)$_2O_4$, titanium-stabilized t-ZrO_2, transformable t-ZrO_2 and monoclinic ZrO_2 are present.

4. Discussion

The addition of ZrO_2 to TiB_2 is obviously beneficial to the sintering behaviour by enhancing the densification and retarding the grain growth of TiB_2. An increasing amount of ZrO_2 results, however, in severe cracking. Prolonged attrition milling influences the microstructure much more than the densification process does.

One of the most important results of the sintering experiments is the evidence for the thermodynamic instability of TiB_2 together with ZrO_2. Approaching equilibrium conditions, mutual diffusion of both titanium and zirconium takes place which may be enhanced by a preceding loss of oxygen. Thus, (Ti, Zr)B_2 and (Zr, Ti)O_2 solid solutions form with a pronounced concentration gradient between the outer regions and the cores of the particles (Fig. 8). This may cause the TiO_2

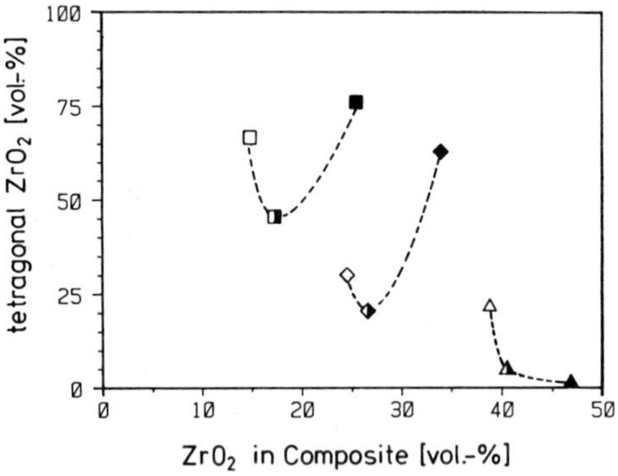

Fig. 7. Volume fraction of t-ZrO_2 as a function of the overall initial ZrO_2 content (□, 10 vol.%; ◇, 20 vol.%; △, 35 vol.%) and milling time (□, 30 min; ◨, 60 min; ■, 120 min).

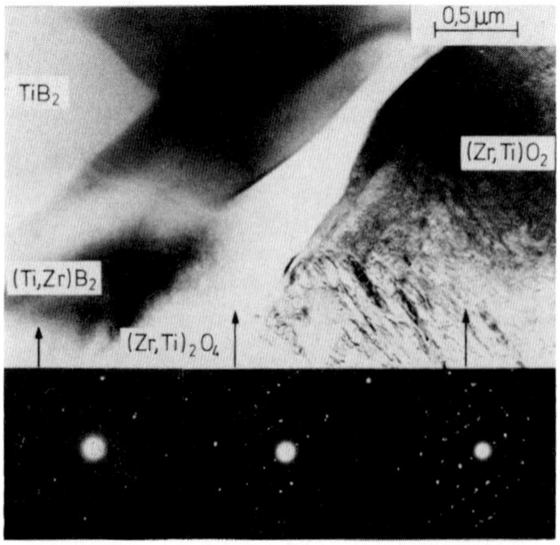

Fig. 6. TEM micrograph of a TiB_2–ZrO_2 junction. The diffraction patterns and STEM analysis yield evidence of a (Zr, Ti)$_2O_4$ grain boundary phase.

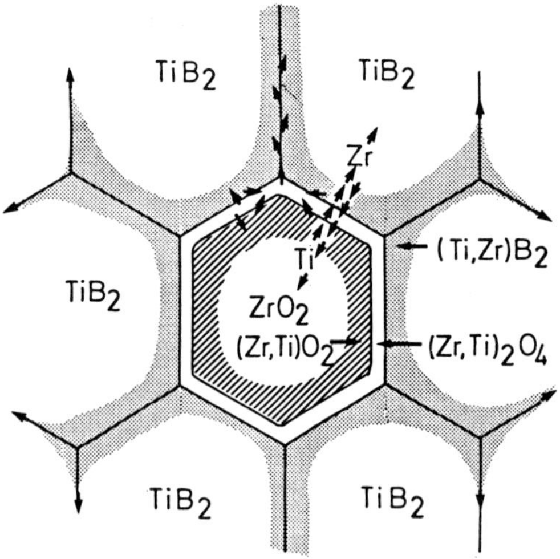

Fig. 8. Schematic view of diffusion paths and phase interactions between TiB_2 and ZrO_2 below 2100 °C.

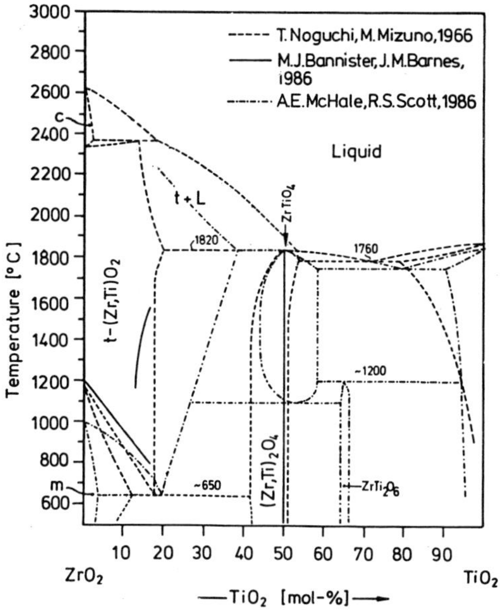

Fig. 9. Superposition of ZrO_2–TiO_2 phase diagrams [3–5].

content of the ZrO_2 grains to exceed 16 mol.% and thus may lead to a full stabilization of the tetragonal modification in the surface layers of the particles. Hence, a heterogeneous nucleation of the tetragonal-to-monoclinic transformation is suppressed. Moreover, the TiO_2 content may locally even increase to above 18–20 mol.% at the grain boundaries. The composition then is situated in the two-phase field t-ZrO_2+(Zr, Ti)$_2O_4$ of the quasi-binary TiO_2–ZrO_2 phase diagram [3–5] (Fig. 9) which results in the formation of intergranular zirconium titanate solid solution.

In addition to these reactions, substantial decomposition of both TiB_2 and ZrO_2 is considered to occur above 2100 °C by internal oxidation. The formation of titanium oxides, titanium borates and boron oxides as residual reaction products at room temperature has been established. It is questionable, however, whether this may be attributed to a solid-state-diffusion-induced release of boron oxides or to a liquid phase reaction between (Ti, Zr)B_2 and a TiO_2–ZrO_2-containing melt created by the peritectic decomposition of (Zr, Ti)$_2O_4$ (Fig. 9). Generally, the influence of boron on the phase equilibria of the quasi-binary TiO_2–ZrO_2 system is not known.

From the phase reactions discussed before, some considerations concerning transformation toughening can be derived. Although there is a certain amount of transformable t-ZrO_2 proved by *in situ* TEM observation, its contribution to the significant increase in fracture toughness from 5.4 MPa $m^{1/2}$ for single-phase TiB_2 to 9.5 MPa $m^{1/2}$ for TiB_2–25vol.%ZrO_2 is believed to be relatively small. Apart from the fact that most of the ZrO_2 is either tetragonally stabilized by titanium diffusion from the neighbouring TiB_2 or unstabilized by being associated in clusters, there is another reason to assume that transformation toughening is not operational, namely mechanical destabilization. Since the thermal expansion of TiB_2 is markedly anisotropic, the larger value being similar to and the smaller value being less than that of ZrO_2, radial tensile stresses develop on dispersed ZrO_2 particles on cooling. Thus the t-ZrO_2 is not constrained but transforms spontaneously, also giving rise to severe cracking in the sample. This behaviour is even enhanced by the relatively coarse grain size of ZrO_2 compared with other transformation-toughened ceramics. However, the tensile stresses generate crack deflection and microcracking which are proposed to be the predominant toughening mechanisms rather than the tetragonal-to-monoclinic transformation whereas the high strength is attributed to the enhanced densification and the grain size refinement of TiB_2.

Acknowledgments

The STEM work was carried out by M. Ruhle (formerly of Max-Planck-Institut, Stuttgart, and now at Santa Barbara, CA) and coworkers; the WDXA was performed by Cambridge Instruments, Dortmund.

References

1 T. Watanabe and K. Shoubu, *J. Am. Ceram. Soc., 68* (2) (1985) C-34.
2 K. Shoubu, T. Watanabe, J. Drennan, R. H. J. Hannink and M. V. Swain, *4th Int. Conf. on Science and Technology of Zirconia, Advanced Ceramics, Tokyo, 1986.*
3 T. Noguchi and M. Mizuno, *Bull. Chem. Soc. Jpn., 41* (12) (1968) 2898, cited in M. K. Reser (ed.), *Phase Diagrams for Ceramists,* 1975 supplement, American Ceramics Society, Columbus, OH, 1975, p. 169, Fig. 4452.
4 M. J. Bannister and J. M. Barnes, *J. Am. Ceram. Soc., 69* (11) (1986) C-269.
5 A. E. McHale and R. S. Scott, *J. Am. Ceram. Soc., 69* (11) (1986) 827.

Hardness and Fracture Toughness of Semiconducting Materials Studied by Indentation and Erosion Techniques*

FREDRIC ERICSON, STEFAN JOHANSSON and JAN-ÅKE SCHWEITZ

Materials Science Division, Uppsala University, Box 534, S-751 21 Uppsala (Sweden)

(Received November 9, 1987)

Abstract

In recent years, the growing field of semiconductor micromechanics has created an increasing demand for strength data on semiconductors and for adequate tests and evaluations of their mechanical properties. In a recently published paper, the present authors have demonstrated that the solid particle erosion rate can be taken as a simple and highly reproducible statistical measure of the susceptibility of silicon and GaAs to contact damage in the micron range. In the present work the scope is broadened to include several new crystal orientations (and one new doping level), as well as three other materials: germanium, InP and InAs, for which the hardness and fracture toughness K_{Ic} values are determined by means of the indentation technique. K_{Ic} values are also derived from erosion data by means of a recently reported brittle fracture model, based on non-lateral spalling in single-crystal semiconductors. These values are compared with results obtained by the indentation technique and conventional test methods reported in the literature. Fracture surface energies are deduced from the experimental K_{Ic} results.

The materials tested are ranked with respect to elastic properties, microhardnesses, fracture toughnesses, and sensitivities to contact damages in general. The influence of crystallographic orientation on room temperature microfracture properties is clearly established, but the corresponding influence on microhardness is found to be rather limited. The influence of doping on the room temperature mechanical properties is noticeable but small.

*Paper presented at the 3rd International Conference on the Science of Hard Materials, Nassau, The Bahamas, November 9–13, 1987.

1. Introduction

In the present study, two different techniques for investigating mechanical properties of surfaces have been applied to a number of semiconducting materials. The growing field of semiconductor micromechanics [1, 2] has created an increasing demand for strength data on semiconductors and for adequate tests and evaluations of their mechanical properties. In a recently published paper [3], the present authors have demonstrated that the solid particle erosion rate can be taken as a simple and highly reproducible statistical measure of the susceptibility of silicon and GaAs to contact damage in the micron range. In the present work the scope is broadened to include some new crystal orientations and one new doping level, as well as three other materials: germanium, InP and InAs, for which hardness and fracture toughness K_{Ic} values are determined by means of the indentation radial crack technique [4–8]. K_{Ic} values are also derived from erosion data by means of a brittle fracture model, based on a non-lateral spalling mechanism [3], which in the present experiments is considered to dominate over the lateral mechanism used in previous erosion models based on materials parameters [9–13]. The values obtained by the two techniques are compared with results obtained by the indentation technique and conventional test methods and reported in the literature. The Griffith–Irwin criterion is utilized to deduce fracture surface energies from fracture toughness data. The materials tested are ranked with respect to elastic properties, microhardnesses, fracture toughnesses, and sensitivities to contact damages in general.

2. Materials and experimental details

Details of the materials investigated are given in Table 1. All specimens were investigated by the

0921-5093/88/$3.50 © Elsevier Sequoia/Printed in The Netherlands

following means: microhardness measurements by Vickers indentation; fracture toughnesses determined by the Vickers indentation technique; solid particle erosion rate measurements; fracture toughnesses determined by the erosion technique. All experiments and measurements were performed at room temperature. Before all experiments, the specimens were polished to optical grade, washed in acetone and rinsed in nitrogen gas. In the erosion experiments, the specimens were rinsed in nitrogen gas before each mass loss measurement.

3. Microhardness measurements

The microhardness measurements were performed at room temperature in a Leitz tester, and the diagonals of the Vickers diamond were oriented along crystallographic directions according to Table 2, in which the hardness results are presented.

In our measurements, neither the surface orientation nor the doping had any significant influence on the room temperature Vickers hardness values. The fact that H_V is weakly influenced by the surface orientation is in agreement with previous observations, indicating that the plastic zone below the impression is not strongly affected by the anisotropy of semiconductor single crystals [14]. The hardness ranking of the five materials investigated is (from hardest to softest) silicon, germanium, GaAs, InP, InAs, where the Vickers hardness of silicon is more than three times that of InAs.

4. Fracture toughness determined by the indentation technique

K_{Ic} values were determined from crack length measurements near Vickers indentations, using the expressions [6]

$$K_{Ic} = \gamma \frac{F_i}{c^{3/2}} \tag{1}$$

$$\gamma \approx \kappa \left(\frac{E}{H_V} \right)^{1/2} \tag{2}$$

TABLE 1 Materials investigated in the experiments

Material	*Surface orientation*	*Type*	*Dopant*	*Crystal growth*	*Dopant concentration* (atoms cm^{-3})	*Density* ($g\ cm^{-3}$)
Si	(111)	n	P	FZ	$(2.8–8.5)\times 10^{13}$	2.33
Si	(100)	n	P	CZ	$(0.65–1.2)\times 10^{14}$	2.33
Si	(110)	n	P	CZ	$(0.9–1.5)\times 10^{15}$	2.33
Si	Polycrystalline					2.33
Ge	(111)					5.32
GaAs	(100)		Undoped			5.32
GaAs	(100)	n			7×10^{17}	5.32
GaAs	(111)		Undoped			5.32
InP	(100)		Undoped			4.79
InAs	Polycrystalline		Undoped			5.66

CZ, Czochralski grown; FZ, floating zone grown.

TABLE 2 Results of microhardness measurements by Vickers indentation (load, 50 gf) (polarity effects neglected)

Specimen	H_V (GPa)	*Indentation orientation* 1	2
Si(111)	10.8 ± 1.3	$[\bar{1}10]$	$[11\bar{2}]$
Si(100)	11.2 ± 1.0	$[011]$	$[0\bar{1}1]$
Si(110)	11.3 ± 1.1	$[001]$	$[1\bar{1}0]$
Ge(111)	9.2 ± 0.5	Undefined	
GaAs(100)	6.9 ± 0.3	$[011]$	$[0\bar{1}1]$
GaAs(100) (doped)	6.9 ± 0.1	Undefined	
GaAs(111)	7.0 ± 0.2	$[\bar{1}10]$	$[11\bar{2}]$
InP(100)	4.3 ± 0.2	$[011]$	$[0\bar{1}1]$
InAs(polycrystalline)	3.5 ± 0.1	Undefined	

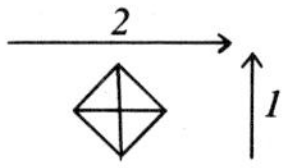

where F_i is the load, c is the length of a radial crack measured in the surface, E is Young's modulus, H_V is the Vickers hardness and κ is a calibration constant. Equations (1) and (2) are intended for pure radial–median cracking in isotropic materials and can therefore be somewhat risky to apply to our highly anisotropic single crystals, where the low energy crack planes might not be perpendicularly oriented to the surface, and where the plastic zone (generating residual stresses which cause cracking) might not be of semispherical symmetry. However, transmission electron microscopy (TEM) studies of vertical sections through Vickers indentations have revealed that more or less pure radial–median cracking actually occurs in single-crystal silicon, irrespective of crystal orientation, even for fairly low loads (Fig. 1). This observation is in accordance with previous investigations [7]. For that reason we shall use eqns. (1) and (2) to estimate K_{Ic} values, since these are scarce in the literature, especially for III–V semiconductors.

Anstis *et al.* [6] evaluated the calibration constant κ in eqn. (2) by adjusting the results of the indentation method to fit the K_{Ic} results obtained by "conventional" techniques, for seven different glasses and polycrystalline ceramics. They found the value $\kappa = 0.016 \pm 0.004$. However, this value is the result of a compromise between amorphous and crystalline materials. Adjustment to fit the glass data alone would yield a lower κ value, whereas adjustment to fit the crystalline materials alone would result in a higher κ value.

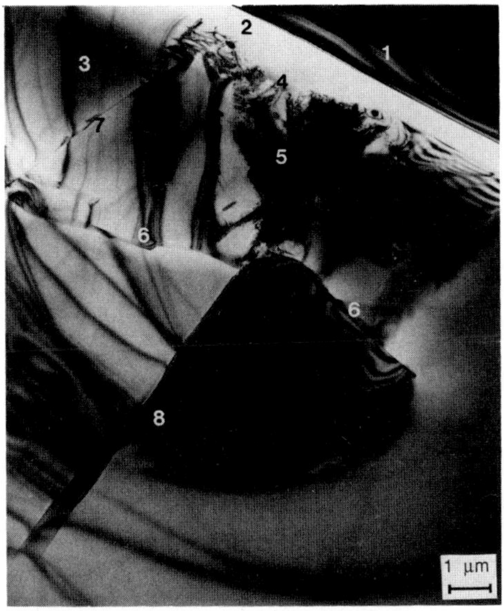

Fig. 1. TEM image of cross-section through a Vickers indentation in an Si(111) surface, showing the following main features: 1, scrap silicon; 2, adhesive (removed in the ion-milling process); 3, investigated silicon specimen; 4, Vickers impression; 5, plastically deformed zone; 6, lateral crack; 7, hertzian ring or cone crack; 8, radial–median crack. The contrasts not annotated in the figure are mainly buckling fringes.

In order to determine an appropriate κ value for monocrystalline semiconductors, we utilized results for the K_{Ic} of silicon given in the literature (Table 3). The most reliable of these range from 0.82 to 0.95 MPa m$^{1/2}$. Adjusting our own indentation results for silicon to fit this K_{Ic} range, we obtain a new value for the calibration constant: $\kappa = 0.0185$. This value lies within the κ range previously estimated by Anstis *et al.* and is close to the value that they would have obtained, had they used crystalline reference materials exclusively. Hence all K_{Ic} results (determined by indentation) presented in Table 4 are based on the value $\kappa = 0.0185$.

In Section 3, we noted that H_V is not orientation sensitive in the surface; so we do not expect problems with anisotropy effects for that parameter. Young's modulus E, in contrast, is known to be strongly anisotropic in single crystals, and we must decide which E value to use in eqn. (2). In the derivation of eqn. (2), E is introduced via Hooke's law, which is applied to the elastic strain field transmitting residual stresses from the plastic zone to the median crack. The direction of interest in this context is the stress direction normal to the plane of cracking. Hence we should use an E value determined for a direction normal to that plane, *i.e.* for the direction defined by the following vectorial product:

$$\hat{\sigma} = [ijk] \times [lmn] \tag{3}$$

where (lmn) is the plane of the indented surface and $[ijk]$ is the direction in which the crack propagates in that surface.

In calculating E values for various directions, we have utilized the elastic stiffness constants c_{11}, c_{12} and c_{44} compiled in refs. 24 and 25 (for InP the stiffness constants were estimated from the average reduced constants). The values for polycrystals are taken as the mean values of the Hashin and Shtrikman bounds [26].

The low energy crack planes of silicon and GaAs are well known ({111} and {110} respectively). Hence pure radial–median cracking of low

TABLE 3 Fracture toughness K_{Ic} and fracture surface energy Γ_{Ic} values from previous measurements (K_{Ic} and Γ_{Ic} are related by eqn. (11))

	K_{Ic} (MPa m$^{1/2}$)	Γ_{Ic} (J m^{-2})	*Reference*	*Method*
Si	0.83	1.2	[15]; re-evaluated [16]	Double cantilever beam
Si	0.94	2.7	[16]	Double cantilever beam (crack initiation data)
Si	0.86	2.3	[16]	Double cantilever beam (crack arrest data)
Si	0.82–0.95	1.9–2.6	[17]	Indentation pre-cracking
Si	0.95	2.6	[18]	Indentation pre-cracking
Si	0.67–0.82	1.1–1.9	[19]	From surface energy data
Si	0.91–0.95	2.4–2.6	[20]	Indentation pre-cracking
Si	0.70–0.78	1.2–2.1	[21]	From surface energy data
Si	0.79	1.81	[22]	Double torsion specimen
Ge	0.58–0.64	1.1–1.8	[21]	
GaAs	0.50	0.9	[19]	
GaAs	0.31–0.45	0.4–0.8	[23]	Indentation pre-cracking

TABLE 4 Experimental values of microhardness H_V, radial crack length c, erosion rate e and fracture toughness values K_{Ic} deduced from indentation and erosion results (Young's moduli E were calculated from elastic stiffness constants)

Specimen	H_V (GPa)	E (GPa)	c (μm)	e (mg g^{-1})	K_{Ic}^{indent} (MPa m$^{1/2}$)	K_{Ic}^{eros} (MPa m$^{1/2}$)
Si(111)	11.0	181	20.1	1.08	0.83	0.95
Si(100)	11.0	151	17.8	0.99	0.91	0.91
Si(110)	11.0	164	18.0	1.00	0.94	0.94
Si (polycrystalline)	11.0	165	—	1.01	—	0.94
Ge(111)	9.2	146	24.5	5.21	0.61	0.60
GaAs(100)	7.0	103	30.4	7.72	0.43	0.44
GaAs(100) (doped)	7.0	103	26.7	7.21	0.51	0.45
GaAs(111)	7.0	131	31.7	7.44	0.45	0.50
InP(100)	4.3	84	29.5–33.7	9.76	0.42–0.51	0.36
InAs (polycrystalline)	3.5	76	37.1	13.60	0.38	0.32

The E values given are the mean value of some low index directions in the (lmn) surface.

energy types should be possible to obtain for suitable orientations of the Vickers diamond diagonal in Si(110) surfaces and GaAs(001) surfaces. We performed these experiments and, in a subsequent attempt to evaluate low energy K_{Ic} values from eqns. (1) and (2), we utilized E values for the directions defined by eqn. (3). The results were disappointing; in fact, we obtained slightly higher K_{Ic} values in these experiments than were obtained on the average for randomly oriented Vickers indentations. We concluded that, even if low energy cracking actually did occur in our experiments, disturbances from secondary effects (additional cracking in other planes etc.) were too strong to make evaluations of well-defined low energy K_{Ic} values possible with this method.

For that reason, we confined ourselves to the evaluation of an "average" K_{Ic} value for each surface orientation; we used the mean of the radial crack lengths for a series of indentations where the Vickers diamond was turned in 30° intervals. The corresponding E value was taken to be equal to the mean of the E values of a few low index directions in the indented surface. Each one of the resulting K_{Ic} values is associated with an impaired surface of a particular crystallographic orientation but not with some particular fracture plane. These "average" K_{Ic} values should not be confused with polycrystalline values; sometimes the latter are somewhat lower owing to intercrystalline cracking.

Only one indentation load was used (1 N) since the K_{Ic} values for silicon determined by Anstis *et al.* [6] proved to be rather insensitive to variations in the load. The crack lengths c were measured on photographs taken in an optical microscope. The scatter was quite large, of the order of 20% or more. In many cases the radial cracking was very asymmetric or otherwise badly defined, and a certain arbitrariness in the measurements was unavoidable.

The K_{Ic} results for indentation are displayed in Table 4 and can be compared with previous measurements in Table 3. Naturally, as silicon is our reference material, the results for silicon agree well with previous measurements, but also the K_{Ic} values for germanium and GaAs are in good agreement with the few results available in the literature. Somewhat surprisingly, our result for doped GaAs is significantly higher than the undoped GaAs results.

The K_{Ic} ranking (from toughest to brittlest) of the five materials investigated is nearly equal to that previously found for hardness, with a possible exception for the InP values, which were of the same magnitude as for GaAs. The K_{Ic} results of InP and InAs, however, should be regarded with scepticism; in some of the InP and InAs samples, it was observed that the median cracks did not seem to develop properly into halfpenny shapes. Hence the model of eqn. (1) would tend to yield too high K_{Ic} values in these cases.

5. Erosion rate measurements

The erosion experiments were performed with alumina sand in a slinger-type erosion tester [3] with normal incidence of the sand particles onto the target surfaces. The particle size (106 ± 18 μm) and the velocity of incidence (43.7 m s^{-1}) were chosen in ranges where the maximum impact force of each particle was of the same order of magnitude as the loads used in the Vickers indentation measurements.

The specimens were mounted in specimen holders with circular openings, in order to avoid edge erosion effects. The sand doses were determined by collecting and weighing the sand passing through the opening of an empty holder. All experiments were performed in air, and the possible influence of air resistance on the impact velocities for different particle sizes was investigated. A theoretical analysis yields (see Appendix A)

$$v = v_0 \exp\left(-\frac{\delta x}{D}\right) \tag{4}$$

where v_0 is the initial velocity, D is the particle diameter, x is the flight distance and δ is an "air resistance factor" which is inversely proportional

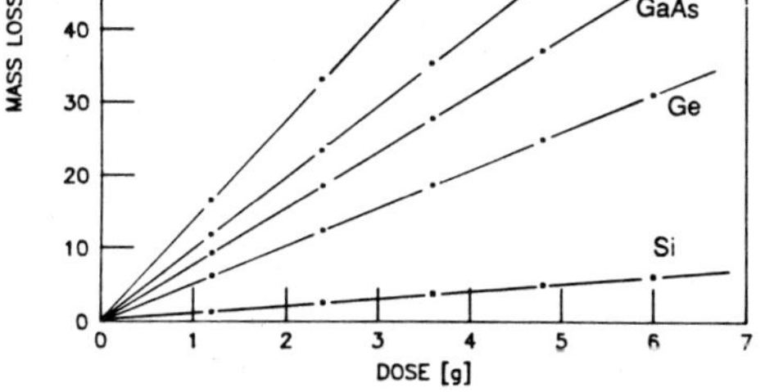

Fig. 2. Mass loss as a function of erodent dose for semiconducting materials (scatter of less than 1%).

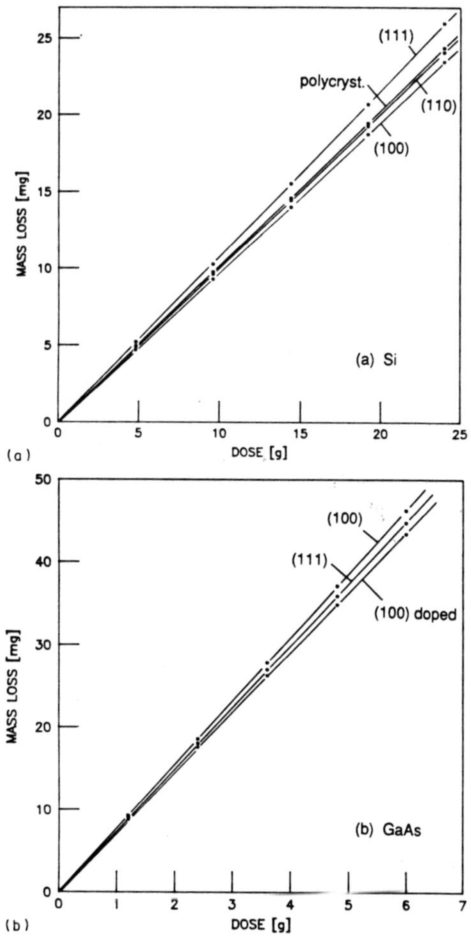

Fig. 3. Mass loss as a function of erodent dose for surfaces of varying orientations in (a) silicon and (b) GaAs.

to the mass density of the particle and also depends on aerodynamic properties and air pressure. In vacuum, δ equals zero. v_0 was determined from the periphery velocity of the slinger disc and from the exit angle of the particles. It was found that all particles used in our experiments had equal exit velocities: $v_0=43.7$ m s^{-1}. The possible influence of the air resistance was checked by comparing erosion rates for targets placed at two different distances from the slinger disc. When the proper geometric factors were taken into account, no significant change in the erosion rate was found, and we concluded that δ was so small that air resistance effects could be neglected in our experiments. This conclusion was further supported by an experiment discussed in Section 6 (see Fig. 4).

The main results of the erosion experiments are displayed in Figs. 2 and 3 and in Table 4. Figure 2 illustrates the mass loss as a function of erodent dose for the five different target materials. It is seen that the curves are almost exactly linear. The experimental scatter is small, normally about 1%. The slope of each curve is a dimensionless measure of the erosion rate *e* (mass loss per weight unit erodent). It is seen that the general ranking order for erosion resistance (from low *e* to high *e*) is the same as those obtained for hardness and indentation fracture toughness. The erosion rate of silicon is one order of magnitude lower than that of InAs.

In Fig. 3 the influence of surface orientation on the erosion rate is demonstrated for silicon. The differences between different orientations are clearly resolvable, and the order of ranking (from low *e* to high *e*) is

Si:(100), (110), polycrystalline, (111)

The (111) orientation yields an approximately 10% higher erosion rate than the (100) orientation does. For GaAs, two different surface orientations were studied. The order of ranking is

GaAs: (111), (100)

The difference between *e* for the two GaAs orientations is less than 4%. Table 4 summarizes the results of our erosion rate measurements on different target materials and surface orientations.

The influence of doping on the room temperature erosion rate has previously been investigated [3] and was found to be barely resolvable for silicon. For p-type silicon no influence on *e* was detected, but for very high n doping levels (10^{19} atoms cm^{-3}) a very small but noticeable increase in the room temperature erosion rate could be detected. For comparison, we tested an n-doped GaAs(100) surface (10^{17} atoms cm^{-3}) and detected a slight but noticeable decrease in *e*, which is in qualitative agreement with the increased K_{Ic} value obtained with the indentation technique.

6. Fracture toughness determined by the erosion technique

In a recently published paper [3], the present authors have presented an analytical model expressing the erosion rate in terms of materials and testing parameters:

$$e=(1-f)\,Y\frac{\rho_t\rho_p{}^{2/9}E^{2/3}v^{22/9}D^{2/3}}{H_V{}^{5/9}K_{Ic}{}^{4/3}} + fZ\frac{\rho_t\rho_p{}^{1/3}Ev^{8/3}D}{H_V{}^{1/3}K_{Ic}{}^{2}} \quad (5)$$

where *e* is the dimensionless erosion rate (mass loss per unit weight of erodent), ρ_t and ρ_p are the mass densities of the target and of the erodent particles respectively, *D* is the average diameter of the particles, *v* is the velocity of normal incidence, and *E*, H_V and K_{Ic} are Young's modulus, the Vickers hardness and the fracture toughness respectively of the target material.

The first term in eqn. (5) has been derived previously by others [9–13] and accounts for a mechanism of mass loss by lateral spalling. The second term in eqn. (5) is based on the observation that spallings of non-lateral types frequently occur in silicon and GaAs. The factors $1-f$ in the "lateral term" and f in the "non-lateral term" determine the balance between these two main types of spalling. The factors *Y* and *Z* can also be expressed as $Y=\kappa^{4/3}Y'$ and $Z=\kappa^2Z'$ where κ is the calibration constant in eqn. (2), and Y' and Z' contain a complex variety of unmonitored statistical information related to the average geometrical shapes (not sizes!) of the sand particles and the spalls. For convenience, κ is here included in *Y* and *Z* and will therefore be implicitly redetermined as *Y* and *Z* are evaluated. The essential influence of the materials properties and of the testing conditions should be accounted for by the seven materials and testing parameters explicitly appearing in eqn. (5). Therefore, Y and Z are only weakly dependent on such factors, and we may consider them as constants, at least as long as the group of target materials is not too heterogeneous, or the testing conditions are not too drastically varied.

In order to determine a value of the f factor, we utilize the fact that the two terms in eqn. (5) depend on the particle size D with two different exponents: $\frac{2}{3}$ and 1 respectively. By varying D while keeping all other parameters constant, we obtain erosion curves of the type illustrated in Fig. 4 for Si(100). We note that in a D interval ranging from about 60 to 70 μm the *e vs. D* curve is linear with negligible scatter. (We also note that this linear behaviour extrapolates through the origin of the diagram, which is possible only when the "air resistance factor" δ in eqn. (4) is close to zero). Hence we conclude that, in this D range, erosion is controlled by the second term of eqn. (5) alone, *i.e.* $f=1$. This conclusion does not necessarily imply that lateral cracking does not occur; it simply means that other types of cracking are dominant in the mass removal process, at least in the cases investigated here. To some extent this conclusion is supported by observations made by the present authors in vertical TEM sections at single-particle impact sites in silicon. For large enough particles, some lateral cracking can be observed, but the frequently occurring multiple non-lateral cracking penetrates more deeply and certainly appears to be the most serious cause of spalling. The scanning electron microscopy (SEM) image in Fig. 5 shows a typical steady state topography of an eroded GaAs surface, with ample signs of non-lateral cracking. Equation (5) now is reduced to

$$e = Z\frac{\rho_t {\rho_p}^{1/3} E v^{8/3} D}{{H_V}^{1/3} {K_{Ic}}^2} \tag{6}$$

for D values larger than 70 μm.

The K_{Ic} exponent in eqn. (6) is equal to -2, which differs from the exponent predicted for lateral spalling ($-\frac{4}{3}$) but agrees better with the K_{Ic} exponent evaluated by Wiederhorn and Hockey [12] from experimental data by means of linear regression analysis; they found a value of -1.9.

In the D range below about 60 μm the *e vs. D* curve in Fig. 4 deviates from the linearity displayed in the D region above this transition point. We have interpreted this phenomenon as a transition from pronounced elastic–plastic behaviour for large particles ($D>70$ μm) to nearly elastic fracture behaviour for smaller particles ($D<60$ μm). This interpretation is supported by TEM observations of small-particle impact sites, where the plastic regions are too small to cause lateral cracking (Fig. 6).

In order to evaluate the constant Z in eqn. (6), we need a reference value of K_{Ic}. Analogously, in the indentation technique, Anstis *et al.* utilized reference materials with known K_{Ic} values to determine a value of the calibration constant in eqn. (2). We choose silicon as our reference material, and we select the Si(110) specimen as our reference surface. The value determined in this manner is

$$Z = 12.9\times10^{-5}\ m^2$$

which is considered to be constant for all mater-

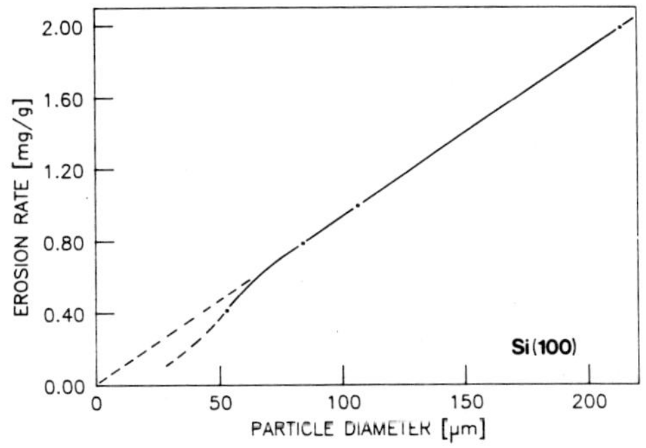

Fig. 4. Erosion rate e as a function of particle diameter D for Si(100) ($v=43.7$ m s^{-1}).

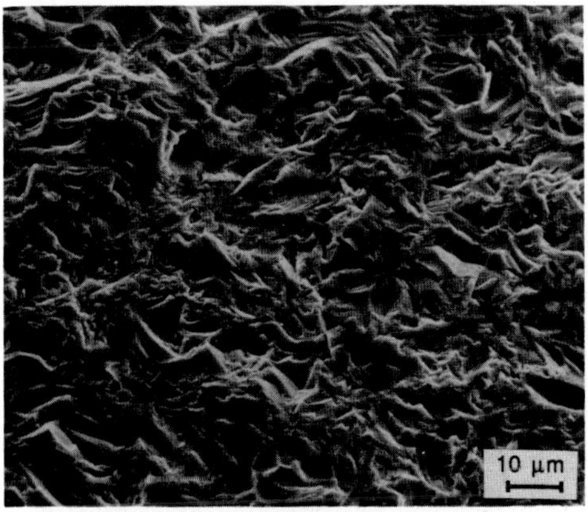

Fig. 5. SEM image of steady state topography of an eroded GaAs(001) surface ($D=94$ μm).

ials and surface orientations investigated in the present study. Equation (6) can now be used to calculate K_{Ic} values for all other surface orientations and target materials.

The fracture toughness values determined in this manner are listed in Table 4 and agree well with previous measurements (Table 3). Comparing the erosion values with the indentation values in Table 4, we find general agreement with a few notable exceptions. The erosion result for Si(111) is close to the results for all other silicon orientations, whereas the Si(111) result obtained by indentation is significantly lower. This difference cannot be explained by the use of different E modules for different surface orientations; the two methods simply yield slightly different results in this case. Another anomaly is found for the GaAs case; for erosion the K_{Ic} results of the undoped and doped cases respectively hardly differ at all, but for the indentation method a significant difference is detected. Apart from this anomaly the GaAs results of the two techniques agree well, and so do the germanium results. The InP and InAs results, however, obtained by the erosion technique, are somewhat lower than the corresponding indentation results. The erosion results seem to agree well with the observed fragile behaviour of InP and InAs, whereas the indentation method appears to overestimate their fracture toughnesses, as previously discussed.

Fig. 6. TEM image of a cross-section through the impact site of a particle (35–38 μm) impingement in a Si(111) surface.

The general ranking orders (from high strength to low strength) of E modulus, hardness, erosion rate and fracture toughness (determined by both methods) are all identical: silicon, germanium, GaAs, InP and InAs.

In eqn. (6), we have assumed that the erosion process is controlled by non-lateral spalling mechanisms ($f=1$). However, it should be noted that, even if we assume the opposite extreme case (pure lateral spalling, $f=0$), we obtain K_{Ic}^{eros} values just a few per cent lower than those in Table 4. Hence the spalling mechanism is not of critical importance in this context.

7. Fracture surface energies

In Irwin's [27] notation of the Griffith [28] crack propagation condition, we have

$$K_{Ic} = \left(\frac{G_{Ic}E}{1-\nu^2}\right)^{1/2} \quad \text{(plane strain)} \tag{7}$$

$$K_{Ic} = (G_{Ic}E)^{1/2} \quad \text{(plane stress)} \tag{8}$$

These expressions relate the fracture toughness K_{Ic} to the strain energy release rate ("critical crack extension force") G_{Ic}. The latter is by definition related to the fracture surface energy Γ_{Ic} according to

$$G_{Ic} = 2\Gamma_{Ic} \tag{9}$$

For ideally brittle fracture, Γ_{Ic} is equal to the reversible surface energy γ but, if plastic deformation takes place at the crack tip during cracking, we have $\Gamma_{Ic} > \gamma$. From TEM studies of room temperature cracking in silicon and germanium, it has been concluded that the crack tips are atomically sharp; no sign of plasticity was observed [29]. Hence, in this case it seems reasonable to assume that $G_{Ic} = 2\gamma$. However, in an eroded surface layer, cracks will propagate through regions which have been plastically deformed by previous particle impacts. We cannot disregard the possibility that a high stress concentration at the tip of a propagating crack might cause some local plastic flow in regions of high dislocation density. In our case it seems appropriate to assume that $G_{Ic} \gtrsim 2\gamma$, or

$$\gamma \lesssim G_{Ic}/2 = \Gamma_{Ic} \tag{10}$$

In our evaluation of fracture surface energies from fracture toughness data, we used the plane strain relationship (eqn. (7)) and, combining it with eqn. (9), we obtain

$$\Gamma_{Ic} = \frac{K_{Ic}^2(1-\nu^2)}{2E} \qquad (11)$$

Using selected K_{Ic} values from Table 4 and polycrystalline values of E and ν, we obtain the fracture surface energies listed in Table 5.

Experimental data on surface energies of a few semiconductors are available in the literature [15, 16, 21], although the results scatter rather badly. Using eqn. (11) to transform the previous experimental K_{Ic} values into fracture surface energies, the results displayed in Table 3 are obtained. Our silicon value (Table 5) agrees fairly well with most other measurements (Table 3), and so do our germanium and GaAs results. No values are available in the literature for InAs and InP.

TABLE 5 Fracture surface energies Γ_{Ic} calculated from K_{Ic} values deduced from erosion results

Specimen	E (GPa)	ν	K_{Ic} (MPa m$^{1/2}$)	Γ_{Ic} (J m^{-2})
Si	165	0.22	0.94	2.55
Ge	133	0.21	0.60	1.29
GaAs	117	0.24	0.44	0.78
InP	95	0.25	0.36	0.63
InAs	76	0.28	0.32	0.61

The E and ν values are for polycrystalline specimens.

8. Summary and conclusions

Vickers microhardness values have been determined for various surface orientations in five different semiconductor materials (Table 2). The Vickers hardness values, which range from about 1100 HV in silicon to about 350 HV in InAs, are only weakly orientation dependent for any given material, and the effect of doping on the hardness was found to be small. These room temperature observations agree with previous findings.

Fracture toughness values were determined by means of the indentation technique for the different materials and surface orientations (Table 4), taking into account the anisotropy of the elastic modulus. The K_{Ic} values obtained range from 0.94 MPa m$^{1/2}$ for silicon down to 0.38 MPa m$^{1/2}$ for InAs and display small but significant differences for the various surface orientations, owing to different angles between the surface and the low energy fracture planes. We should be careful not to associate the K_{Ic} values obtained in this manner with some particular well-defined fracture plane; these K_{Ic} values should simply be regarded as values characteristic for indentations in the respective surfaces. They agree well with K_{Ic} values for silicon, germanium and GaAs obtained by other techniques. For InP and InAs, no values are available in the literature. In two cases, we attempted to determine K_{Ic} values of well-defined low energy fracture planes by means of the radial crack method. These attempts were not successful.

The erosion experiments resulted in linear relationships between target mass loss and erodent dose. The erosion rate determinations were highly reproducible and of low scatter character; hence the erosion rates can be considered to be good statistical measures of the susceptibilities of these materials to contact damage in the micron range. The ranking of the materials with respect to erosion resistance is the same as for the E modulus, hardness, and fracture toughness, *i.e.* silicon, germanium, GaAs, InP and InAs, where the erosion rate of InAs is some 14 times higher than that of silicon (Table 4). The erosion rate is sensitive to changes in crystallographic surface orientation. At room temperature, the influence of doping on the erosion rate is very small but noticeable.

Fracture toughness values were calculated from the erosion results by means of an analytical relationship between erosion rate and K_{Ic}. In these calculations, isotropic H_V values (Table 2) and anisotropic E values were utilized, resulting in a set of anisotropic K_{Ic} values (Table 4). These results agree in general with the K_{Ic} results obtained by the indentation technique, which is not surprising since both ways of determining K_{Ic} are based on the same relationship (eqn. (1)) between crack length and fracture toughness. The main advantage of the erosion technique is its high degree of statistical confidence; K_{Ic} values are calculated from experimental results (erosion rates) which are highly reproducible and of low scatter nature. In the indentation technique, however, K_{Ic} is calculated from the results of direct measurements of radial cracks, which often are difficult to identify and measure, and which display a considerable scatter in length. For crystals of low strength, the indentation method appears to yield overestimated K_{Ic} values.

Both the indentation and the erosion techniques, when applied to single crystals, suffer from the disadvantage that it is not possible to associate the resulting K_{Ic} values with well-defined crack planes in an unambiguous manner. In contrast "average" K_{Ic} values, associated with specific crystallographic orientations of the damaged surface, might be just as useful in many applications.

Finally, the Griffith–Irwin criterion was utilized to deduce fracture surface energies from K_{Ic} data (Table 5). The results show a general agreement with data deduced from available results in the literature for K_{Ic}.

Acknowledgments

The authors wish to express their gratitude to the Swedish National Board for Technical Development for financial support and to ASEA AB (Sweden), Semitronics AB (Sweden) and Wacker–Chemitronic G.m.b.H. (F.R.G.) for providing materials for the experiments.

References

1 K. E. Petersen, *Proc. IEEE, 70* (1982) 420–457.
2 J. B. Angell, S. C. Terry and P. W. Barth, *Sci. Am., 248* (4) (1983) 36–47.
3 S. Johansson, F. Ericson and J.-Å. Schweitz, *Wear, 115* (1987) 107–120.
4 D. B. Marshall, B. R. Lawn and P. Chantikul, *J. Mater. Sci., 14* (1979) 2225.
5 B. R. Lawn, A. G. Evans and D. B. Marshall, *J. Am. Ceram. Soc., 63* (1980) 574.
6 G. R. Anstis, P. Chantikul, B. R. Lawn and D. B. Marshall, *J. Am. Ceram. Soc., 64* (1981) 533.
7 B. R. Lawn, D. B. Marshall and P. Chantikul, *J. Mater. Sci., 16* (1981) 1769.
8 D. B. Marshall, B. R. Lawn, and A. G. Evans, *J. Am. Ceram. Soc., 65* (1982) 561.
9 A. G. Evans and T. R. Wilshaw, *J. Mater. Sci., 12* (1977) 97.
10 A. G. Evans, M. E. Gulden and M. Rosenblatt, *Proc. R. Soc. London, Ser. A, 361* (1978) 343.
11 W. F. Adler, *Rep. CR 79-680*, 1979 (U.S. Army Research Office).
12 S. M. Wiederhorn and B. J. Hockey, *J. Mater. Sci., 18* (1983) 766.
13 D. B. Marshall, in F. L. Riley (ed.), *Progress in Nitrogen Ceramics*, Martinus Nijhoff, The Hague, 1983, pp. 635–656.
14 S. G. Roberts, P. D. Warren and P. B. Hirsch, *J. Mater. Res., 1* (1986) 162.
15 J. J. Gilman, *J. Appl. Phys., 31* (1960) 2208.
16 C. St. John, *Philos. Mag., 32* (1975) 1193.
17 C. P. Chen and M. H. Leipold, *Am. Ceram. Soc., Bull., 59* (1980) 469.
18 P. Chantikul, G. R. Anstis, B. R. Lawn and D. B. Marshall, *J. Am. Ceram. Soc., 64* (1981) 539.
19 C. Messmer and J. C. Bilello, *J. Appl. Phys., 52* (1981) 4623.
20 K. Yasutake, M. Iwata, K. Yoshii, M. Umeno and H. Kawabe, *J. Mater. Sci., 21* (1986) 2185.
21 R. J. Jaccodine, *J. Electrochem. Soc., 110* (1963) 524.
22 S. B. Bhaduri and F. F. Y. Wang, *J. Mater. Sci., 21* (1986) 2489.
23 C. P. Chen, *Publ. 84–81*, 1984 (Jet Propulsion Laboratory, Pasadena, CA).
24 J. R. Drabble, in R. K. Willardson and A. C. Bee (eds.), *Semiconductors and Semimetals*, Vol. 2, Academic Press, New York, 1966, p. 110.
25 G. Simmons, *J. Grad. Res. Cent., 34* (1), (2) (1965) (Southern Methodist University Press, Dallas, TX).
26 G. Simmons, *J. Grad. Res. Cent., 36* (1) (1967) (Southern Methodist University Press, Dallas, TX).
27 G. R. Irwin, *Weld. J., 31* (1952) 450; *Handb. Phy., 6* (1958) 551.
28 A. A. Griffith, *Philos. Trans. A, 221* (1920) 163.
29 B. R. Lawn, B. J. Hockey and S. M. Wiederhorn, *J. Mater. Sci., 15* (1980) 1207.

Appendix A

A.1. The influence of air resistance on particle velocity

According to elementary aerodynamics, the air resistance force acting on a flying particle is given by a relationship of the following type:

$$F_r = K_1 v^2 D^2 \qquad \text{(A1)}$$

where v is the velocity of the particle, D is its cross-sectional diameter and K_1 is proportional to the density of the air and to a geometric shape factor. The opposing force of inertia is given by

$$\begin{aligned} F_i &= ma \\ &= K_2 D^3 a \qquad \text{(A2)} \end{aligned}$$

where a is the acceleration and K_2 is proportional to the mass density of the particle. In an inertial frame of reference, we have

$$F_r + F_i = 0 \qquad \text{(A3)}$$

which yields a negative value of the acceleration (*i.e.* a retardation):

$$a = -\frac{K_1 v^2}{K_2 D} \qquad \text{(A4)}$$

If v_0 is the initial velocity of the particle, its instantaneous velocity is given by

$$v^2 = v_0^2 + 2\int_0^x a \, dx \qquad \text{(A5)}$$

where x is the momentaneous flight distance. Inserting the acceleration of eqn. (A4) into eqn.

(A5) and differentiating with respect to x, we obtain the following differential equation:

$$\frac{d(v^2)}{dx} = -2\frac{K_1 v^2}{K_2 D} \tag{A6}$$

which has the solution

$$v^2 = v_0^2 \exp\left(-\frac{2K_1 x}{K_2 D}\right) \tag{A7}$$

or

$$v = v_0 \exp\left(-\frac{\delta x}{D}\right) \tag{A8}$$

The "air resistance factor" $\delta = K_1/K_2$ is directly proportional to the air density and inversely proportional to the mass density of the particle. Of course, δ also is strongly dependent on the aerodynamic properties of the particle.

Some Fundamental Aspects of the Mechanical Wear of Hard Ceramic Crystals due to Sliding*

C. A. BROOKES and A. R. PARRY

Department of Engineering Design and Manufacture, University of Hull, Hull HU6 7RX (U.K.)

(Received November 9, 1987)

Abstract

Here the fundamental criteria for a purely mechanical process of the wear of hard ceramic crystals by softer lubricated sliders are outlined. Essentially, it is shown that microplasticity is produced when the pressure due to point loading is insufficient to cause spontaneous cracking but is high enough to develop a resolved shear stress in excess of the critical level to move dislocations in the crystal. Since the applied pressure is directly related to the hardness of the slider, then it can simply and effectively be controlled through the use of different softer materials. Repeated traversals over the same surface lead to an increase in the density of dislocations in the crystal, i.e. work hardening, and may ultimately lead to fracture and fragmentation through a fatigue mechanism. This type of wear is exemplified through studies on the behaviour of magnesium oxide and titanium carbide, and the application of static point loading of softer cones on harder crystals to determine the critical resolved shear stress of titanium carbide and diamond is demonstrated. Finally, a predictive methodology for this type of cumulative deformation and wear, in ceramic crystals in general, is outlined.

1. Introduction

The wear of hard engineering ceramic surfaces, under conditions of sliding, can be divided into two main categories. One category includes those wear processes which involve chemical reactions, *e.g.* between cutting tool and workpiece, and where the difference in the hardness of the sliding surfaces may be largely irrelevant. The other category includes purely mechanical wear, but this category can be subdivided into two divisions. The first division is where penetration and abrasion of the softer surface are immediately observed when traversed by another solid which is not less than 20% harder than that surface. When both solids are ceramics deformed at room temperature, or low homologous temperatures, then the resultant mode of wear is often dominated by brittle fracture and much research has been published on this subject. However, whilst helpful to an understanding of the performance of abrasives in grinding, the catastrophic rates of wear would be such as to exclude serious consideration of this combination for engineering components in tribological applications. The second division is where the applied stress is insufficient to nucleate cracks spontaneously but where the critical resolved shear stress in the bulk of the crystal is exceeded. Under these conditions, localized dislocation movement and multiplication is possible, *i.e.* microplasticity. Given repeated traversals, the density of dislocations in the deformed crystal may increase, leading to significant work hardening and, ultimately, fracture and fragmentation. It will be shown here that the most important criterion which governs this mechanism of wear is the ratio of the flow stress of the softer solid, which in metals would be about one-third of the indentation hardness, to the critical resolved shear stress of the harder solids which is never more than one-sixth of its indentation hardness value. If the potential of engineering ceramics is to be fully realized, there can be little doubt that a better understanding of this particular mechanism of wear is required.

2. Microplasticity during quasi-static loading

The resolution of shear stresses on the active slip systems in crystals subjected to point loading

---

*Paper presented at the 3rd International Conference on the Science of Hard Materials, Nassau, The Bahamas, November 9–13, 1987.

0921-5093/88/$3.50 © Elsevier Sequoia/Printed in The Netherlands

has been modelled for a number of indenters (Knoop [1–3] and Vickers [4, 5]) and sliders [6–8]. Generally, the magnitude of the resolved shear stress is calculated in terms of the orientation of the active slip systems with respect to the shape of a particular indenter, or in terms of the coefficient of friction with respect to the crystallographic direction of sliding. In such models, and for cubic crystals having six or more slip systems, the maximum resolved shear stress is in the range of 0.3–0.5 times the mean pressure P_m for all surfaces corresponding to low index crystallographic planes. Most often the indenter or slider is harder than the crystal and a visible indentation or groove is produced, the mean pressure being considerably in excess of that required to initiate dislocation movement. Previously the most appropriate model for the purposes of this work is that applied to the deformation of a LiF(001) plane loaded with a nylon sphere [9]. There the peak resolved shear stress at a distance below the surface corresponding to the radius of the contact area was shown to be about $0.4P_m$. In that work, whilst no visible indentation was formed on the surface of the lithium fluoride, etching techniques were used to demonstrate the initiation and movement of dislocations in the bulk of the crystal below the contact area.

Since we wished to investigate the lowest level of stress capable of initiating dislocation movement, thereby enabling repeated stressing to lead to localized work hardening and wear, we have subjected ceramic crystals to point loading using a significantly softer cone. The cone was made from either a metal or another ceramic, but in this latter case the experimental temperature had to be high enough to allow full plastic deformation of the cone. Thus the cone macroscopically plastically deformed and flattened so that the contact area developed was sufficiently large to support the applied load elastically. The nominal mean pressure then applied to the ceramic was roughly constant, *i.e.* it was essentially independent of the applied load, since it was determined by the flow stress of the material used for the cone. With polycrystalline metal cones after repeated sliding traversals, the nominal mean pressure was approximately one-third of the indentation hardness of that metal [10]. However, because a similar relationship has not yet been established for quasi-static indentation, then it is advisable to determine the mean nominal pressure P_m by dividing the applied normal load by the measured area of the flattened tip, as we have done in this work. The application of a very recent model [11] enables the nominal mean pressure, for circular contact loading on a particular crystallographic plane, to be resolved onto the active slip systems. Thus, for a (001) plane and $\{110\}\langle 1\bar{1}0\rangle$ slip systems, the maximum resolved shear stress is about $0.35P_m$ and is concentrated at a point around 0.33 of the contact diameter beneath the surface. For the same plane but $\{111\}\langle 1\bar{1}0\rangle$ slip systems, the relevant stress is about $0.32P_m$ and this lies on a circular line, concentric with the centre-line of the contact area, which is approximately 0.29 of the contact diameter beneath the surface. In Table 1, we have collected a limited amount of data from our own work in which we show the measured P_m for a given cone used to load specific planes of titanium carbide, magnesium oxide and diamond crystals. These data, together with the predictions of the model [11], have been used to calculate the resolved shear stress which proved sufficient to produce visible plastic deformation in the ceramic. In the following, we have epitomized some of the more important observations made with a range of ceramic crystals by reference to the softest (magnesium oxide) and the hardest (diamond).

2.1. Magnesium oxide crystals

Much of the basic work on deformation under point loading has been carried out on magnesium oxide and the availability of a reliable dislocation etchant, together with the simple geometry of its $\{110\}\langle 1\bar{1}0\rangle$ slip systems, enables ready recognition of the initiation of dislocation movement.

TABLE 1 Conditions of nominal mean pressure P_m and temperature, observed to initiate plastic flow in some ceramic crystals

Softer cone material	P_m (GPa)	*Temperature* (°C)	*Ceramic crystal*	τ_c (GPa)	H_K (GPa)
Cu	0.7	20	MgO(100)	0.06	3.5–7.3
Phosphor bronze	1.4	20	MgO(100)	0.06	3.5–7.3
Stainless steel	1.4	750	MgO(100)	0.02	1.4–1.8
Ir	0.84	850	TiC(100)	0.23	3.7
Cubic BN	7.8	1000	Diamond(111)	(2.2)	31

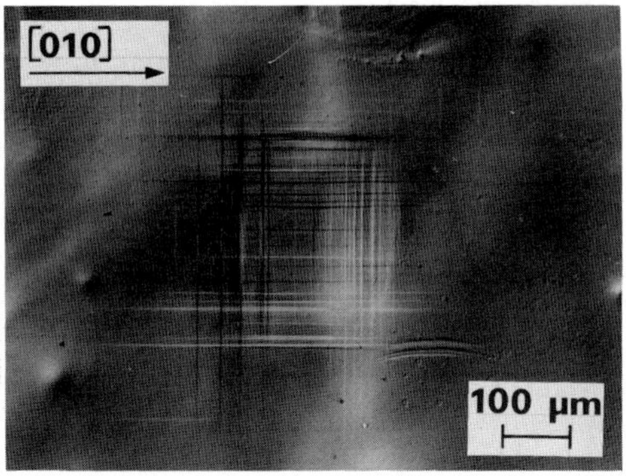

Fig. 1. Slip lines formed on an MgO(001) surface, at room temperature, due to a 115 N load applied by a flattened phosphor bronze cone ($P_m = 1.4$ GPa).

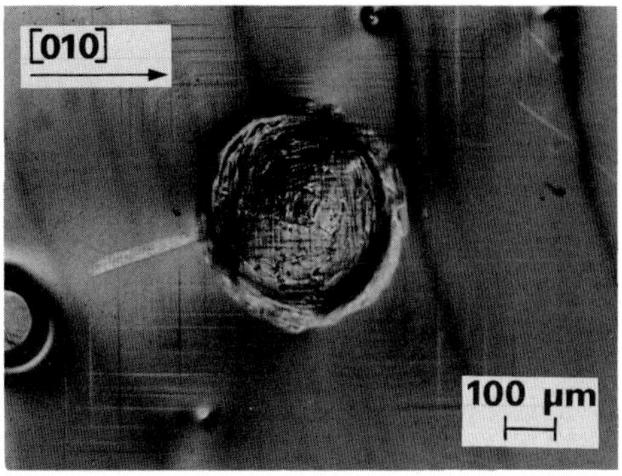

Fig. 2. Deformation on an MgO(001) surface due to a load applied by a stainless steel cone at 750 °C ($P_m = 1.4$ GPa).

Fuller details will be published elsewhere, but the principal observations of experiments using a range of metal cones on MgO(001) surfaces at room temperature can be summarized.

(a) With metals which developed a nominal pressure just sufficient to exceed the bulk critical resolved shear stress (*e.g.* cadmium and zinc), there was no evidence of an indentation, or slip lines, on the surface but dislocations were revealed on etching.

(b) As harder metals were used for the cones, slip lines became apparent within the contact area (*e.g.* copper, $P_m = 0.7$ GPa), then slip lines extended outside the contact area as shown in Fig. 1 (phosphor bronze, $P_m = 1.4$ GPa), and eventually (austenitic stainless steel, $P_m = 4$ GPa) slip lines were apparent inside and outside the contact area with the type of crack formation on those {110} planes which are normal to the (001) specimen surface first described by Keh *et al.* [12].

(c) The extent and depth of dislocation movement in the bulk of the crystal, beneath all these cones, were essentially dependent on the magnitude of the load applied and independent of the nominal contact pressure. However, in earlier work [10], the density of dislocations within the dislocated zone (as reflected by a measured increase in hardness) was shown to be directly related to the nominal contact pressure.

On the basis of these observations, we shall therefore conclude that the presence of slip lines inside or outside the contact area, or etched dislocations where appropriate, confirms that the nominal pressure applied by the softer cone is sufficient to exceed the critical resolved shear stress in the bulk of the harder crystal.

The value of this simple type of experiment, in studying the deformation of ceramics, will be enhanced by its application at higher temperatures. Preliminary results indicate that, provided that adhesion between the two surfaces can be avoided, basic information can readily be obtained. For example, Fig. 2 shows a MgO (001) surface after it has been subjected to a nominal pressure of 1.4 GPa developed by a flattened stainless steel cone at 750 °C. The slip lines produced correspond to the emergence, on that surface, of edge dislocations from the 45° {110} slip planes and lie in ⟨100⟩-type directions on the (001) plane. Screw dislocations emerging on this surface from the 90° {110} slip planes are naturally not apparent as slip steps but the associated edge dislocations running parallel to the surface were revealed by dislocation etchants. It can therefore be concluded that, under these particular conditions, dislocations were interacting on all six of the intersecting slip systems. Clearly these interactions would result in work hardening and this implies that the actual resolved shear stress was significantly in excess of the critical value for dislocation movement. This is consistent with the fact that the critical resolved shear stress of magnesium oxide is 0.02 GPa at 750 °C [13] with a resolved shear stress of about 0.49 GPa due to the nominal pressure of 1.4 GPa developed by the stainless steel cone (see Table 1). It is also worth noting that cracks of the type described by Keh *et al.* formed at room temperature, with the same combination of cone and crystal, were not now formed although a greater amount of strain has been generated in the contact area.

2.2. Titanium carbide crystals

Whilst the primary slip systems of titanium carbide at temperatures up to about 600 °C are variously reported as $\{110\}\langle 1\bar{1}0\rangle$ [14] or $\{111\}\langle 1\bar{1}0\rangle$ [15], there is little doubt that $\{111\}\langle 1\bar{1}0\rangle$ slip prevails at higher temperatures. From the work of Williams [16], it is known that the critical resolved shear stress on these systems, at 850 °C, is around 0.23 GPa. At such temperatures, there are few metals which are sufficiently hard to apply a mean pressure in excess of the critical level. The most likely candidates are the transition metals which are inclined to react chemically with the carbide to form strongly adherent junctions. However, an f.c.c. iridium cone was used to apply a mean pressure of 0.84 GPa, for 1 h at 850 °C, to a cleaved TiC(001) surface. This pressure corresponds to a resolved shear stress of 0.27 GPa on $\{111\}\langle 1\bar{1}0\rangle$, slightly in excess of the critical value, but there was no evidence of an indentation or slip lines. However, after etching in boiling sulphuric acid for 1 min, dislocation etch pits were produced around the area of contact, as shown in Fig. 3. These dislocations are in rows along $\langle 110\rangle$ directions on the cube plane and correspond to dislocations lying on $\{111\}$, intersecting that surface.

2.3. Diamond crystals

The only measurement of the critical resolved shear stress for type I diamond crystals, using conventional compressive data, was made at an experimental temperature of 1800 °C and was reported as 1.2 GPa [17].

There are no satisfactory dislocation etchants for diamond and our particular experiments were based simply on the observation of slip lines. Cones made from either a single crystal of titanium carbide or a polycrystalline aggregate of cubic boron nitride were loaded on the (111) surface of a natural type I diamond macle at experimental temperatures of 1000 or 1200 °C. The titanium carbide cone did not produce slip steps on the diamond surface but it is worth observing that, as a direct result of macroscopic plastic flow, its flattened tip formed an exact replica of the diamond macle surface that it had been loaded against. For example, Fig. 4 shows a part of the flattened tip of a titanium carbide single-crystal cone which has been loaded for 30 min against the diamond macle at 1000 °C. The triangular features correspond to trigons, natural truncated pyramidal pits bounded by $\{111\}$ planes, commonly observed on as-grown $\{111\}$ diamond surfaces. The larger trigons are approximately 1 μm deep and the precision of replication is self-evident. At 1200 °C, after 30 min, a firmly adherent junction was formed when a cone made from a cubic boron nitride aggregate cutting tool material was loaded against another diamond macle. Subsequent fracture of that junction, at room temperature, left particles of cubic boron nitride over the contact area. However, slip lines were observed, adjacent to the contact area, corresponding to the intersection of $\{111\}$ slip planes on the (111) surface. A similar experiment but carried out at a temperature of 1000 °C for about 1 h and where adhesion was not developed again produced evidence of slip steps outside the contact area (Fig. 5). Once more, trigons on the

Fig. 3. Dislocation array formed on a TiC(001) surface due to an iridium cone loaded at 850 °C for 1 h (P_m = 0.84 GPa).

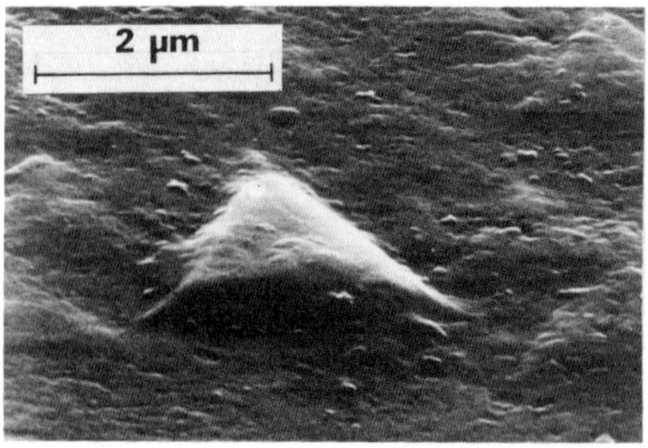

Fig. 4. Replication of trigons, from the original diamond(111) surface, on a flattened titanium carbide cone (P_m = 0.67 GPa). The load was applied for 30 min at 1000 °C.

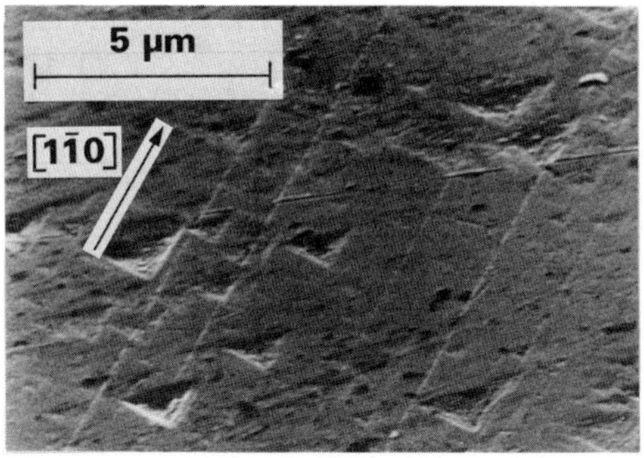

Fig. 5. Slip steps adjacent to contact area on diamond(111) macle caused by a load of 115 N applied for about 1 h through a cubic boron nitride cone at 1000 °C (P_m = 7.8 GPa).

diamond surface were replicated on the flattened tip of the cubic boron nitride cone. The intimacy of contact apparent in these micrographs verifies that there can be little difference between the apparent and real areas of contact under these conditions. Also, the behaviour observed here clearly implies that a degree of conformability between ceramic materials can be anticipated and, conceivably, utilized in their tribological applications at elevated temperatures.

The geometric factor relating an applied force normal to a (111) surface to the resolved shear stress on $\{111\}\langle 1\bar{1}0\rangle$ slip systems is about 0.27 [11]. On the basis of the experimental data summarized in Table 1 for cubic boron nitride cones on diamond, and given the assumptions made earlier, the maximum value of the critical resolved shear stress for diamond at 1000 °C is estimated to be 2.1 GPa. It appears that cubic boron nitride is the only material hard enough to develop microplastic deformation in diamond.

3. Microplasticity and work hardening during sliding

When the softer cones are caused to slide on the ceramic crystal under conditions as described above which favour limited dislocation movement rather than fracture, the depth of the dislocated zone does not increase with increasing number of traversals [10]. However, each traversal may cause other dislocations to move and multiply such that the dislocation density within the dislocated zone increases. Then, the harder crystal will become work hardened to a level determined by the nominal mean pressure developed in the contact area, which is proportional to the indentation hardness of the cone. The maximum degree of work hardening, for a given slider, may be reached after relatively few traversals, typically about 20. Subsequent traversals may, through the production of work-hardening debris and the consequent development of stress concentrations, cause the initiation of cracks which immediately provoke extensive fracture and fragmentation. The critical number N_c of traversals required to cause this surface fragmentation and accelerated wear is inversely related to P_m, is independent of the normal load but is dependent on the crystallographic plane and the direction of sliding. A plot of P_m *vs.* N_c resembles a conventional fatigue curve and indicates that the surface will not fail when sliders below a certain limiting hardness are used. This type of behaviour has been observed, to a greater or lesser degree, in a range of ceramic materials, *e.g.* lithium fluoride, magnesium oxide, nickel oxide, titanium carbide, strontium titanate and yttrium aluminium garnet [10, 18, 19]. Here we shall be primarily concerned with recent observations on the work hardening and surface deformation of magnesium oxide and titanium carbide under these conditions.

3.1. *Magnesium oxide crystals*

Figure 6(a) provides clear evidence that microplasticity, due to softer lubricated metal sliders at room temperature, can be readily observed on MgO(001) crystal surfaces through the emergence of edge dislocations along ⟨100⟩ directions forming visible slip steps, similar to those shown in Fig. 1. These slip lines appeared to increase in height as the number of traversals carried out with a particular metal slider increased, and in number when the indentation hardness of the metal used for slider material was increased. Also, cleaved and etched cross-sections readily demonstrated the distribution and extent of dislocation movement beneath the contact area during sliding (Fig. 6(b)). Indeed, the dislocated zone immediately beneath the contact area has become so heavily populated with dislocations that individual etch pits can no longer be resolved. Knoop indentations placed within this dislocated zone, using a normal load of 1.0 N for comparative hardness measurements in this zone and in the bulk of the crystal remote from it, established that the increased dislocation density was reflected in an increased hardness of the

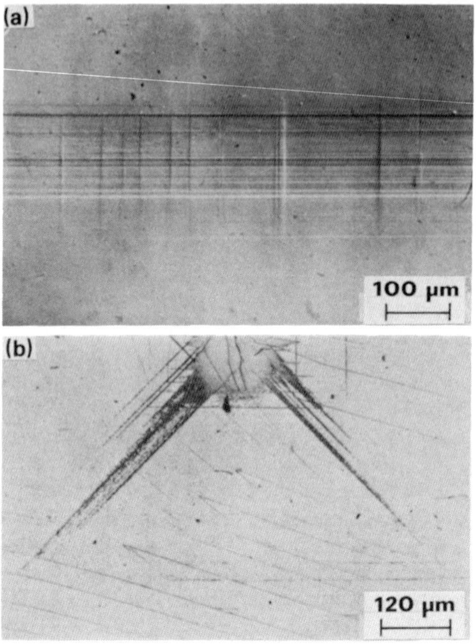

Fig. 6. (a) Slip steps produced on an MgO(001) surface due to repeated traversals of a lubricated 70–30 brass slider, in the [100] direction at room temperature; (b) cross-section of (a), on a (010) cleavage plane, to reveal the dislocation distribution. (Reproduced from *Proc. R. Soc. London, Ser. A, 409*, by kind permission of the Royal Society.)

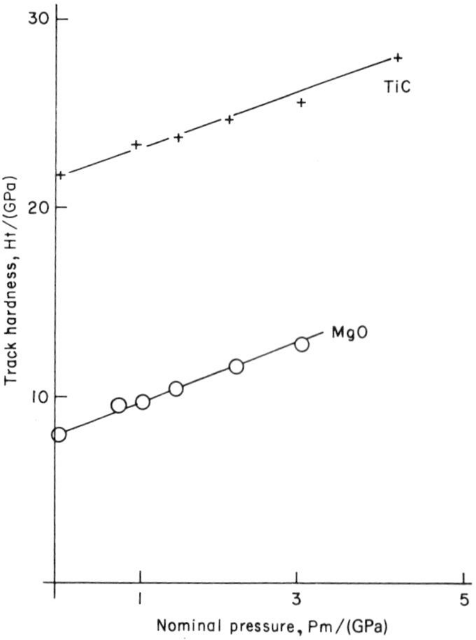

Fig. 7. Work hardening due to repeated traversals of lubricated softer metal sliders, at room temperature, on TiC(001) and MgO(001) surfaces.

crystal. The increased hardness of the cumulatively deformed magnesium oxide was directly related to the nominal mean pressure applied by each of the metal sliders, as shown in Fig. 7. Hence, we have assumed that, in the absence of a suitable dislocation etchant or visible slip lines (as below for titanium carbide), a measured increase in hardness is also indicative of the operation of this mechanism.

3.2. Titanium carbide crystals

Slip steps on the surfaces of titanium carbide deformed at room temperature in this way are not readily apparent and the dislocation etchant used above is not as effective as those available for magnesium oxide crystals. Nevertheless, the application of Knoop microindentation hardness measurements have demonstrated that work hardening due to repeated traversals with softer metal sliders can be developed in this material (Fig. 7). Further evidence to support the microplasticity mechanism has been obtained using direct phase-detecting microscopic interferometry [20]. For example, Fig. 8 shows a TiC(001) surface which has been subjected to 10 000 traversals with a hardened steel lubricated slider at room temperature, *i.e.* a nominal pressure of about 3 GPa has been applied. A shallow groove has clearly been developed by the softer metal. The consequent work-hardening effect is responsible for a measured increase in Knoop hardness, again using a normal load of 1.0 N in the track and in the bulk, from 23 to 28 GPa. (The "blistered" appearance of the Knoop indentations at both sites should be noted. This effect, presumably due to elastic recovery after removal of the normal load, was not observed using the direct-phase multiple-beam interferometry tech-

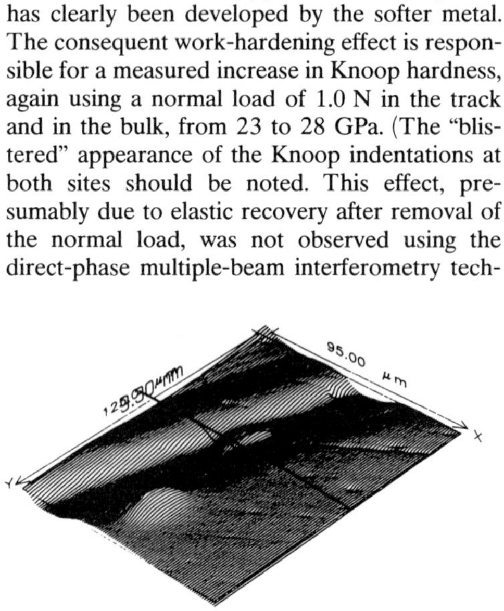

Fig. 8. Direct phase-detecting microscopic interferometry map of a TiC(001) crystal after repeated traversals of a lubricated steel slider, in a [110] direction, at room temperature. The depth of the groove is about 75 nm.

nique on other hard ceramic crystals but seems worthy of further investigation.)

From the reasoning outlined earlier, it follows that the upper bound of the critical resolved shear stress, for a given crystal, may be estimated from the lowest nominal pressure which produces measurable work hardening under these conditions. For example, this work has shown that measurable work hardening is developed on the cube face of titanium carbide by using a steel slider which develops a nominal pressure of about 1 GPa at room temperature. On the assumption of $\{110\}\langle 1\bar{1}0\rangle$ slip systems at room temperature [14], this corresponds to a critical resolved shear stress of not more than 0.5 GPa. This compares with a measured value, at 800 °C, of 0.26 GPa [16], for slip on the $\{111\}\langle 1\bar{1}0\rangle$ systems. Therefore, it may be concluded that the room temperature critical resolved shear stress of titanium carbide is between 0.26 and 0.5 GPa.

3.3. *A predictive methodology*

Rates of wear in engineering ceramics will essentially depend on the nature of bulk deformation. Most often, once cracks are initiated, brittle behaviour will be dominant and wear becomes catastrophic at room temperature. At elevated temperatures, the fracture process could become one of ductile shear and wear rates might then be significantly reduced. In any event, the microplasticity described above may precede fracture and can be anticipated when the following conditions apply: (a) where the softer of the two surfaces transmits a mean pressure which exceeds the critical resolved shear stress of the harder crystal but not enough to cause it to crack; (b) when the pressure is repeatedly applied at a level to give an increase in dislocation density and work hardening; (c) at temperatures above those required to enable the softer ceramic to deform macroscopically plastically but below those which enable annealing processes to occur in the harder crystal and/or diffusion and welding of the two surfaces.

Some further guidance on the above criteria can be obtained from the knowledge that thermal recovery and/or diffusion and welding are probable at temperatures in excess of half the melting point. If we allow for the fact that many ceramics do not have a melting point as such, this upper temperature is likely to be about 1500 °C for all materials. Also, the data for the indentation hardness and critical resolved shear stress given in Table 2 can be used to indicate the probability of microplastic deformation for a given pair of materials, although it must be emphasized that too much reliance should not be placed on the hardness data alone since the relationship between that parameter and the critical resolved shear stress is neither universal nor simple for these particular materials. It should be noted that the Knoop hardness measurements shown in this table are based on indentations in the load-independent regime and that the range shown for a particular crystal is due to anisotropic effects [25].

4. Conclusions

The following principal conclusions may be reached, concerning the microplasticity of ceramic crystals, on the basis of this work.

(1) Incremental dislocation movement and interaction leading to work hardening and, ultimately, fracture is possible whenever the applied nominal mean pressure develops a resolved shear stress above the critical value for a given crystal but is insufficient to cause immediate cracking.

(2) An estimate of the upper bound critical resolved shear stress can be made, for a given crystal, by determining either the lowest mean pressure required to develop slip steps in quasi-static loading or to produce work hardening after

TABLE 2 The ratio of indentation hardness (Knoop indenter, 1 kgf) to critical resolved shear stress for some ceramic crystals

Crystal	*Temperature* (°C)	*Hardness* (GPa)	τ_c (GPa)	H/τ_c	*References*
MgO	20	3.5–7.3	0.06	58.3–121.7	[13, 21]
MgO	750	1.4–1.8	0.02	70.0–90.0	[13, 21]
TiC	20	27.8	(0.5)	55.6	[10, 22]
TiC	850	3.7	0.23	16.1	[16, 23]
TiC	1000	2.4	0.13	18.5	[16, 23]
Diamond	1000	31	(2.1)	14.1	[24], this work
Diamond	1800	—	1.2	—	[17]

repeated reciprocating traversals under lubricated conditions.

(3) Precise replication of the surface of a ceramic by another softer ceramic can be developed at sufficiently high temperatures, implying the possibility of a high degree of conformability between contacting surfaces.

Acknowledgments

The authors would like to express their gratitude to De Beers Industrial Diamond Distributors Ltd. for a grant to the Department and thank Mr. G. Robinson for technical help.

References

1 F. W. Daniels and C. G. Dunn, *Trans. Am. Soc. Met., 41* (1949) 419–442.
2 C. A. Brookes, J. B. O'Neill and B. A. W. Redfern, *Proc. R. Soc. London, Ser. A, 322* (1971) 73–88.
3 S. G. Roberts, *Philos. Mag. A, 58* (1988) 347–364.
4 C. A. Brookes, E. L. Morris and B. Moxley, *Proc. Iron and Steel Conf.*, Iron and Steel Institute, London, 1971, pp. 29–33.
5 B. Moxley, *Ph.D. Thesis*, University of Exeter, 1974.
6 F. P. Bowden and C. A. Brookes, *Proc. R. Soc. London, Ser. A, 295* (1966) 244–258.
7 P. Green, *PhD Thesis*, University of Exeter, 1974.
8 C. A. Brookes and P. Green, *Proc. R. Soc. London, Ser. A, 368* (1979) 37–57.
9 M. V. Swain, *Wear, 48* (1978) 173–180.
10 C. A. Brookes, M. P. Shaw and P. E. Tanner, *Proc. R. Soc. London, Ser. A, 409* (1987) 141–159.
11 S. G. Roberts, personal communication, 1988, cited in C. A. Brookes, V. R. Howes and A. R. Parry, *Nature (London), 332* (1988) 139–141.
12 A. S. Keh, J. C. M. Li and Y. T. Chou, *Acta Metall., 7* (1959) 694–699.
13 C. O. Hulse and J. A. Pask, *J. Am. Ceram. Soc., 43* (7) (1960) 373–378.
14 R. J. Hannink, E. Kohlstedt and M. J. Murray, *Proc. R. Soc. London, Ser. A, 326* (1972) 409–420.
15 S. L. Shinde, R. Sinclair and D. Rowcliffe, *Mater. Sci. Eng., A105/106* (1988) 299.
16 W. S. Williams, *J. Appl. Phys., 35* (1964) 1329–1338.
17 T. Evans and R. K. Wild, *Philos. Mag., 12* (1) (1965) 479.
18 M. P. Shaw, *Ph.D. Thesis*, University of Exeter, 1977.
19 P. E. Tanner, *M.Sc. Thesis*, University of Exeter, 1983.
20 D. M. Perry, P. J. Moran and G. M. Robinson, *J. Inst. Electr. Radio Eng., 55* (4) (1985) 145–150.
21 P. H. Harrison, *Ph.D. Thesis*, University of Exeter, 1973.
22 G. V. Samsonov, M. S. Kovalchenko, V. V. Dzemelinskii and P. Upadyaya, *Phys. Status Solidi A, 1* (1970) 327.
23 D. L. Kohlstedt, *J. Mater. Sci., 8* (1973) 777–786.
24 C. A. Brookes, in E. A. Almond, C. A. Brookes and R. J. Warren (eds.), *Proc. 2nd Int. Conf. on the Science of Hard Materials, Rhodes, September 23–28, 1984*, in *Inst. Phys. Conf. Ser., 75* (1986) 207–219.
25 C. A. Brookes, in R. K. Viswanadham, D. Rowcliffe and J. Gurland (eds.), *Science of Hard Materials*, Plenum, New York, 1983, pp. 181–197.

On the α-to-β Phase Transformation in Silicon Nitride*

V. K. SARIN

GTE Laboratories Incorporated, 40 Sylvan Road, Waltham, MA 02254 (U.S.A.)

(Received November 20, 1987)

Abstract

The importance of the α-to-β conversion during consolidation to the strength of Si_3N_4 based ceramics is well established. However, the exact relationship between α and β, the kinetics of the transformation and its effect on microstructural development are still not fully resolved. This study focuses on the relationship between these two phases, and the influence of the α-to-β conversion during liquid phase consolidation on the microstructure of the Si_3N_4-based ceramic. Additionally, structural analyses are utilized to investigate the effects of consolidation time and temperature on the morphology of the β phase in the densified body. The influence of the presence of Al_2O_3 on this transformation in the Si_3N_4–Y_2O_3 system is also examined. Furthermore, a model illustrating microstructural development and its relationship to the α-to-β transformation in Si_3N_4-based ceramics is put forward.

1. Introduction

Silicon nitride (Si_3N_4) has two polymorphic forms designated as α and β. Both of these forms are hexagonal, with the *c* axis of the α unit cell approximately twice that of the β unit cell. Furthermore, they both have similar densities, and first-order bonding is the same in each phase. The monotropic transformation temperature from α to β has been reported to be in the range 1573-1723 K [1]. Although it has been postulated that α- and β-Si_3N_4 are the low and high temperature forms respectively, only transformation from α to β has been observed, but never the reverse. Madelung energy calculations [2] have indicated that β is more stable than α above 293 K, and therefore it has been suggested that transformation from α to β is possible, but not the reverse.

Structural considerations suggest that the α-to-β transformation can be classified as a reconstructive transformation of secondary coordination [3, 4]. Although the first coordination is the same in each structure, Si–N bonds have to be broken for such a transformation to occur. Typically, such transformations are sluggish and only occur in the presence of solvents, suggesting that the mechanism of the α-to-β transformation consists of dissolution of the less stable, more soluble α form in the presence of a liquid phase, followed by precipitation of the less soluble, more stable β form. While results confirming this mechanism have been reported [5, 6], evidence suggesting that the α-to-β transformation can occur in high purity powder in the absence of a liquid phase [7] has led to some uncertainty as to the actual mechanism of the α-to-β transformation.

Ever since the existence of the two forms of Si_3N_4 was established [8], the relationship between them has been the subject of much controversy. Apart from disagreement as to the kinetics of the transformation, it is still unresolved as to whether α- and β-Si_3N_4 are true polymorphs. Evidence is against the proposed oxynitride formula of α-Si_3N_4 [9, 10] and, although under certain circumstances oxygen may substitute for nitrogen in α-Si_3N_4, there seems to be no structural requirement for its presence. Whether or not one of these two forms of Si_3N_4 is metastable, or perhaps stabilized by impurities, is still unclear. The importance of α-to-β conversion in the densification step to the strength of Si_3N_4-based ceramics is well established. However, since the exact relationship between α and β, the kinetics of the transformation and its effect on microstructural development during liquid-phase consolidation are still not fully understood, therefore it is diffi-

*Paper presented at the 3rd International Conference on the Science of Hard Materials, Nassau, The Bahamas, November 9-13, 1987.

0921-5093/88/$3.50

© Elsevier Sequoia/Printed in The Netherlands

cult to try to control properties systematically via microstructural tailoring. In this study, structural analyses are utilized to investigate the effects of densification time on the morphology of the β phase and to establish a correlation between structure and toughness. Additionally, the influence of the presence of Al_2O_3 in the binder phase on this transformation is also examined. From these results, a model illustrating microstructural development and its relationship to the α-to-β transformation in Si_3N_4-based ceramics is developed.

2. Experimental procedure

Structural evaluations were performed on three compositions, as shown in Table 1.

Sintered AY-6 billets were fabricated from a premixed-and-blended powder batch under a N_2 overpressure of 1.4 MPa at 1880 °C. Identical processing parameters were used for the (β-)AY-6 material. The 100% β-Si_3N_4 powder was made from SN-502 powders in experimental batches for these sintering studies. The PY-6 samples were made from similarly processed powder blends without Al_2O_3 and sintered at 1880 and 1950 °C in a 1.4 MPa N_2 overpressure. Since the starting Si_3N_4 powders are identical for all three compositions, the nominal impurity level can be assumed to be similar.

X-ray diffraction was utilized to determine the polymorphic form of the Si_3N_4, for both the powders and the consolidated samples. However, it should be noted that this technique cannot accurately detect the presence of a phase when it is below 5 wt.% of the total composition. Structural characterization was made with the aid of electron microscopy using both scanning electron microscopy (SEM) and transmission electron microscopy (TEM). Owing to the acicular nature of the grains in the consolidated samples, only grains where both length and diameter could be measured on the polished surface were considered. Grain size measurements were made on SEM photomicrographs of etched samples using a digital image analyser. Aspect ratio calculations from these measurements were used to characterize the coarsening behavior of the β-Si_3N_4 grains.

The fracture toughness was determined using an indentation technique on precracked annealed samples in a four-point loading fixture placed in a tungsten mesh furnace. A cross-head speed of 0.05 cm min^{-1} was used. The fracture toughness K_{Ic} was calculated using the relationship [11]

$$K_{Ic} = \sigma M \left(\frac{\pi a}{Q}\right)^{1/2}$$

where σ is the maximum outer fiber tensile stress, M is a numerical factor related to flaw and specimen geometry (assumed to be 1.03), a is the flaw depth and $Q = \phi^2 - 0.202(\sigma/\sigma_y)^2$ (σ_y is the tensile yield stress and ϕ is the elliptical integral).

TABLE 1 **Nominal composition of the materials investigated**

Designation	*Composition* (wt.%)
AY-6[a]	Si_3N_4–$6Y_2O_3$–$2Al_2O_3$
PY-6[a]	Si_3N_4–$6Y_2O_3$
(β-)AY-6[b]	Si_3N_4–$6Y_2O_3$–$2Al_2O_3$

[a]Made with SN-502 Si_3N_4 powder (α + about 10 wt.% β).
[b]Made with 100% β-Si_3N_4 powder.

3. Results and discussion

Depending on the method and process parameters used in the synthesis of Si_3N_4 powders [12], they can contain varied ratios of α or β, including the two extremes, 100% α or β. A detailed study on the α-to-β conversion in these powders [13] has indicated that conversion of pure chemically vapor-deposited α-Si_3N_4 powder (stubbly hexagonal discs) could only be achieved with the aid of a flux and resulted in the formation of elongated whisker-like acicular grains of β. Although powder morphology is dependent on the polymorphic form of the powder, milling of the Si_3N_4 powder with the binder before consolidation generally equalizes differences in morphology (Fig. 1). In spite of this, however, the polymorphic form of the powder still has a major influence on the microstructural development during the early stages of consolidation, as the results discussed below clearly demonstrate.

The densification behavior and resulting microstructure, especially α-to-β transformation rates, of hot-pressed Si_3N_4 with sintering aids have been the subject of several investigations [14–17]. It has been proposed that both densification and α-to-β conversion follow first-order kinetics, which are controlled by a solution–diffusion–reprecipitation mechanism. Furthermore, it has been postulated that a major part of the

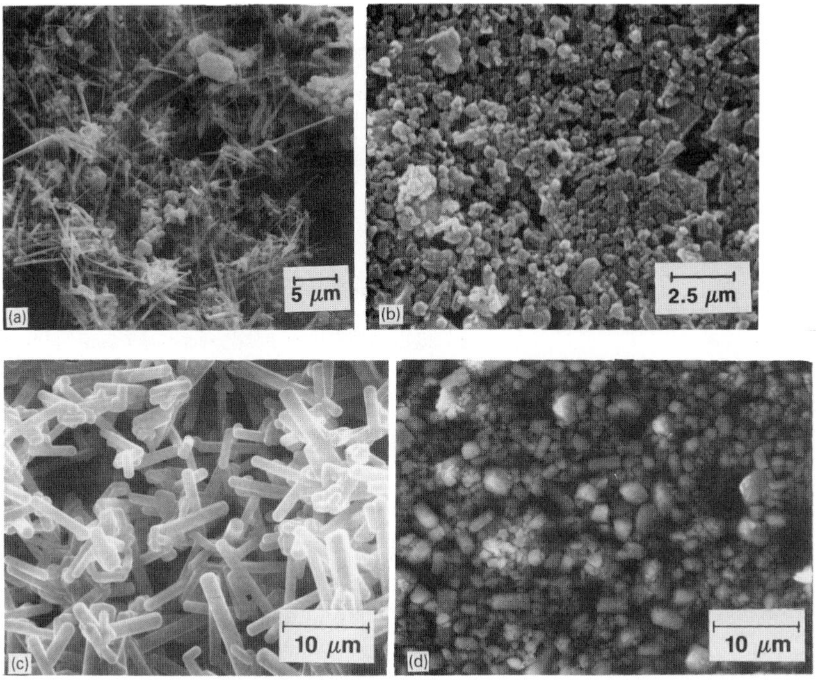

Fig. 1. SEM photomicrograph showing the morphology of the powders (a), (b) $\alpha + 10\%\ \beta$ in the as-received-and-milled condition and (c), (d) $100\%\ \beta$ in the as-transformed-and-milled condition.

α-to-β conversion takes place after densification is complete. However, results from this study suggest that transformation of α to β during consolidation and the densification process are not directly related. The presence of the liquid phase, which promotes densification via three successive stages, namely particle rearrangement, solution–diffusion–precipitation and coalescence, also acts as the solvent necessary for the α-to-β conversion. Both processes occur simultaneously during the early stages of liquid phase sintering, with the α-to-β transformation aiding in particle rearrangement.

Since the presence of a liquid flux with some specific components is necessary for the α-to-β transformation to occur, its availability is essential for this conversion to continue. Therefore, even though a liquid phase is present, aiding in the completion of densification, localized depletion of the appropriate flux can result in incomplete conversion of α to β. It should be noted that, as in the case of densification, the contribution of different additives to the rate of α to-β conversion will vary. Liquid phase sintering studies on Si_3N_4–6wt.%Y_2O_3–2wt.%Al_2O_3 clearly illustrate

TABLE 2 Effect of consolidation time on α-to-β transformation

	Density (% of the theoretical density)	*X-ray analysis result*
Sintered at 1880 °C for 1 h	97.6	All β
Sintered at 1880 °C for 3 h	99.2	All β
Sintered at 1880 °C for 15 h	99.6	All β
Hot pressed at 1725 °C for 1.5 h	100	All β

this phenomenon. Within the detection limits of this technique, X-ray diffraction analysis on samples sintered at 1880 °C (Table 2) shows complete conversion of α to β even after 1 h.

The densification curve at 1880 °C and the resulting microstructures are shown in Fig. 2. Hot pressing also resulted in complete densification and full conversion of α to β at 1725 °C after 1.5 h. The PY-6 material (no Al_2O_3) also yielded complete α-to-β conversion when sintered at 1880 and 1950 °C for 4 h, with resulting densities of 96.3% and 99.8% respectively of the theoretical density. As in the case of AY-6 (Fig.

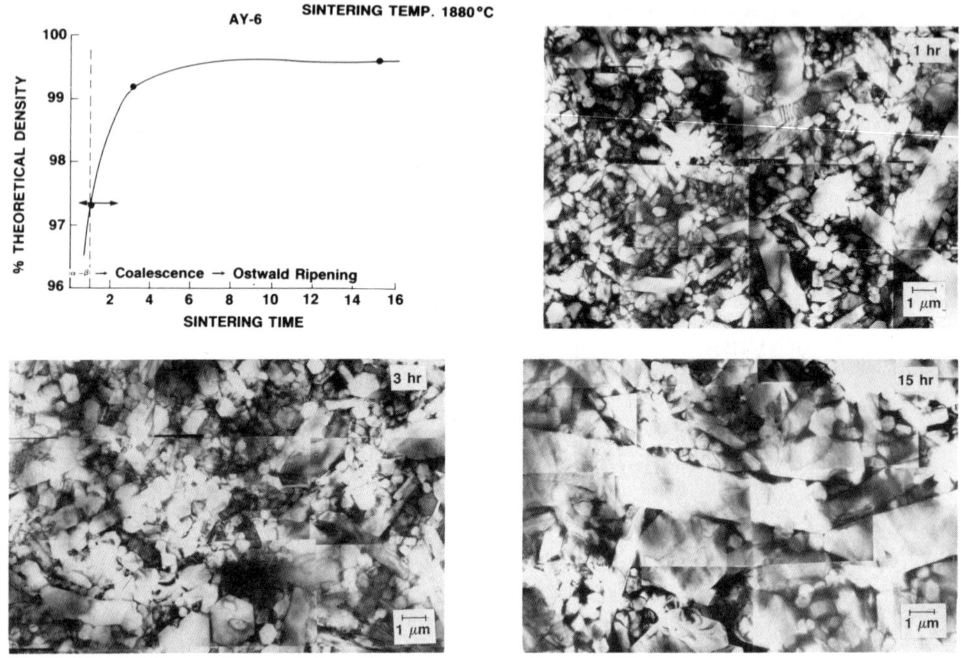

Fig. 2. Sintering time *vs.* attained density and resulting microstructure for the AY-6 composition.

2), similar microstructural development, with large acicular β grains, was also obtained for the PY-6 material (Fig. 3). In the case of (β-)AY-6, a 4 h sintering time resulted in a density of 98.2% of the theoretical density, whereas 100% of the theoretical density was obtained in 12 h. It is note worthy that, in spite of no α-to-β conversion in this case, extended sintering time resulted in the coalescence and growth of the β grains in an acicular morphology (Fig. 4). Microstructural development of (β-)AY-6 is then similar to that of AY-6 and PY-6. These results are in direct conflict with the suggestion [16] that, once complete α-to-β conversion has occurred, the structure becomes relatively stable, resulting in restricted β grain growth. Furthermore, the present results disprove the concept that, if only β-Si_3N_4 particles are present, they would then grow under nearly equilibrium conditions and tend toward becoming spherical particles to minimize surface energy [12, 17]. The present results indicate that β-Si_3N_4 also seems to go through a dissolution–precipitation reaction in the presence of a liquid flux, similar to that previously reported to be exclusive to α-Si_3N_4. Furthermore, the microstructural development in the liquid phase consolidation of Si_3N_4 (specifically the growth of large acicular β grains) does not seem to be dependent on the presence and amount of α-Si_3N_4 in the starting powder, as has been suggested [16].

Structural analysis via selective measurement of the length and diameter of the acicular β-Si_3N_4 grains was used to quantify the coalescence and growth behavior during liquid phase sintering. The effect of both consolidation time and temperature on the large acicular β-Si_3N_4 grain size for the three compositions investigated is shown in Table 3.

Although twofold to threefold increases in length are observed, modest increases in diameter and therefore aspect ratio (length divided by width) also occur. This suggests that the growth kinetics of the acicular β-Si_3N_4 grains are planar and controlled by the deposition rate on both the basal and the prism faces. For this to occur, the acicular β-Si_3N_4 grains must be completely encapsulated by a liquid flux throughout their growth cycle. Therefore, it becomes obvious that the growth rate of these grains is controlled by the amount, viscosity and supersaturation limits of the liquid phase and should be highly dependent

Fig. 3. Photomicrographs ((a) optical microscopy and (b) TEM) showing the effect of sintering temperature on the structure of the PY-6 composition: upper, 1880 °C; lower, 1950 °C.

on both consolidation time and temperature. The present results (Table 3) confirm this since a variation in the size of the β-Si_3N_4 grains was observed with different consolidation times and temperatures. Additionally, the presence of Al_2O_3 in the binder phase seems to accelerate the growth rate of the β grains, since the Al_2O_3-containing composition (AY-6) has a grain size after sintering for 1 h similar to that of PY-6 after 4 h at 1880 °C. The indentation fracture toughness of AY-6 at room temperature (Fig. 5(a)) was found to be linearly proportional to the length of the β-Si_3N_4 grains in the consolidated structure. However, since eventual growth of the β grains is restricted by impingement, a maximum toughness value for a particular composition at a given temperature is probably attained. A similar magnitude of increase in K_{Ic} was found to be maintained up to 1200 °C when measured on pre-cracked samples in a four-point bend test (Fig. 5(b)).

The polymorphic form of the starting powder (α or β) seems to have a definite influence on the morphology of the β grains in the consolidated structure. When the starting powder contains mainly α phase material, the milled particles (Fig. 1(b)) first convert into long whisker-like acicular grains of β, which then grow via the dissolution–precipitation mechanism. However, the milled β phase particles (Fig. 1(d)) merely coalesce and grow. Figure 6 schematically illustrates these two cycles. The nucleation-and-growth step related to the transformation of α to β greatly impacts the consolidated structure since the aspect ratio of the β grain when the starting powder is predominantly α, as in AY-6, is much larger than when it was β, as in (β-)AY-6 (Fig. 7). Although the growth rate dl/dt in the AY-6 material (1.29 μm h^{-1}) is faster than that for (β-)AY-6 (0.86 μm h^{-1}), the change in aspect ratio $d(l/d)/dt$ in this material (0.11 μm h^{-1}) is about the same as that in (β-)AY-6 (0.13 μm h^{-1}).

In other words, the diameter of the β grains in the AY-6 material, which starts off smaller owing to the initial α-to-β transformation, during consolidation continues to maintain this difference in aspect ratio even after complete α-to-β conversion has been obtained and only β

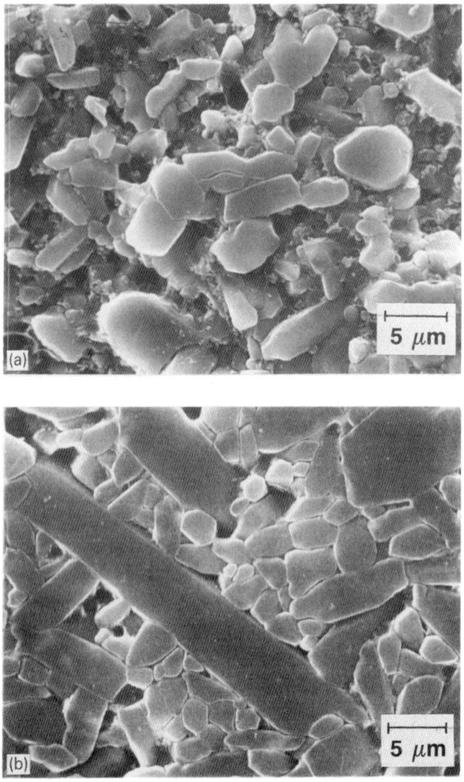

Fig. 4. SEM micrographs showing the structure of (β-)AY-6 after sintering for (a) 4 h and (b) 12 h at 1880 °C.

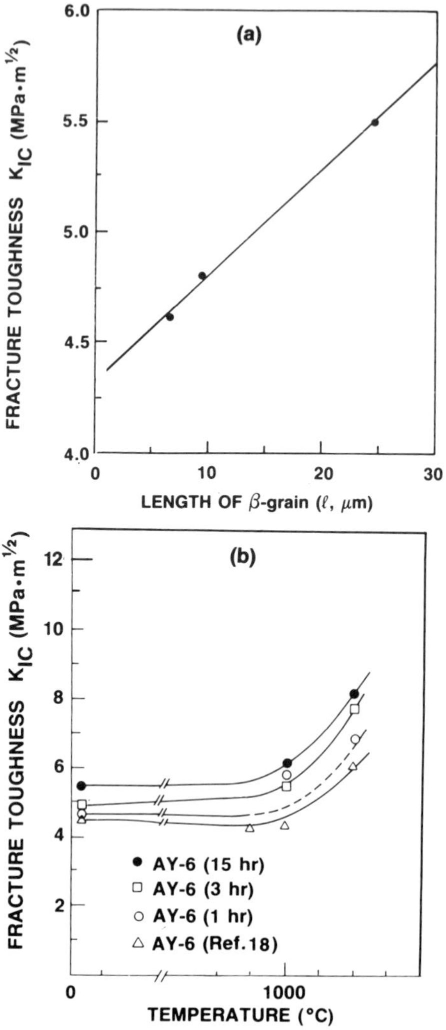

Fig. 5. (a) Effect of the length of β-Si_3N_4 grains on K_{Ic} at room temperature; (b) variation in K_{Ic} with temperature.

TABLE 3 Effect of consolidation time and temperature on the size and shape of β-Si_3N_4 grains

Sample	*Consolidation temperature* (°C)	*Consolidation time* (h)	*β grain size*	
			Average length l (μm)	*Aspect ratio l/d*
AY-6	1880	1	6.6 ± 1.1	6.2 ± 1.3
AY-6	1880	3	9.5 ± 1.4	6.1 ± 1.2
AY-6	1880	15	24.8 ± 6.5	7.7 ± 1.9
AY-6	1725	1.5	2.82 ± 0.5	5.35 ± 1.4
(β-)AY-6	1880	4	8.0 ± 1.6	3.3 ± 0.1
(β-)AY-6	1880	12	15.2 ± 4.6	4.3 ± 1.4
PY-6	1880	4	6.4 ± 1.7	5.5 ± 1.0
PY-6	1950	4	18.2 ± 6.2	7.2 ± 1.0

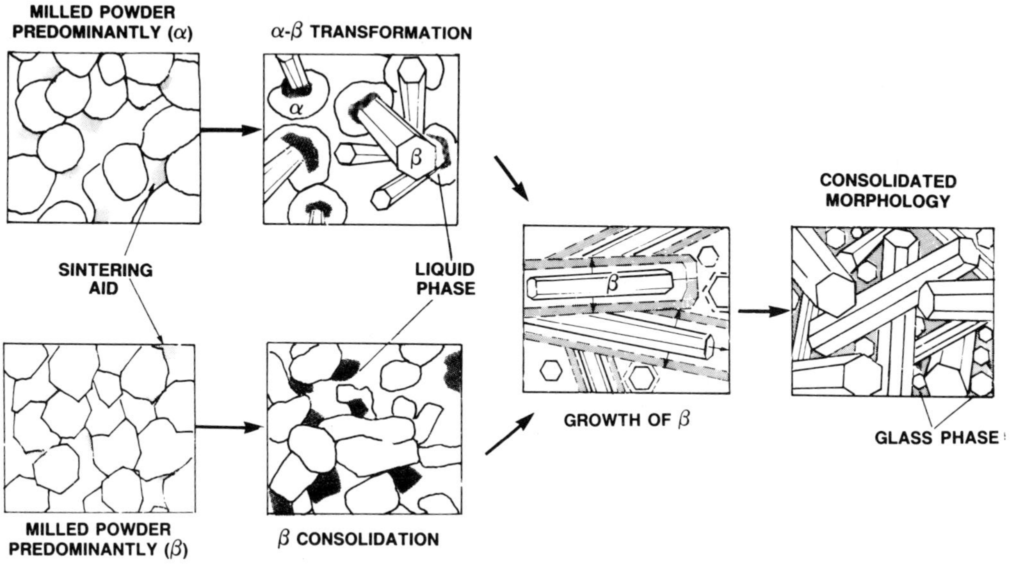

Fig. 6. Schematic diagram illustrating the morphological development of β-Si_3N_4 grains during consolidation.

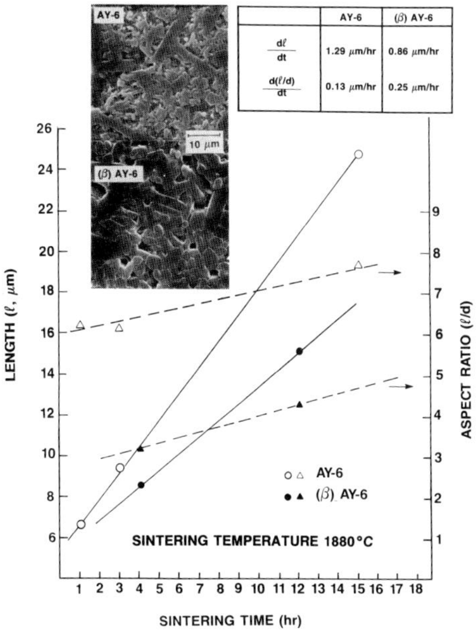

Fig. 7. The effect of sintering time on the length l and aspect ratio l/d of the β-grains of AY-6 and (β-)AY-6.

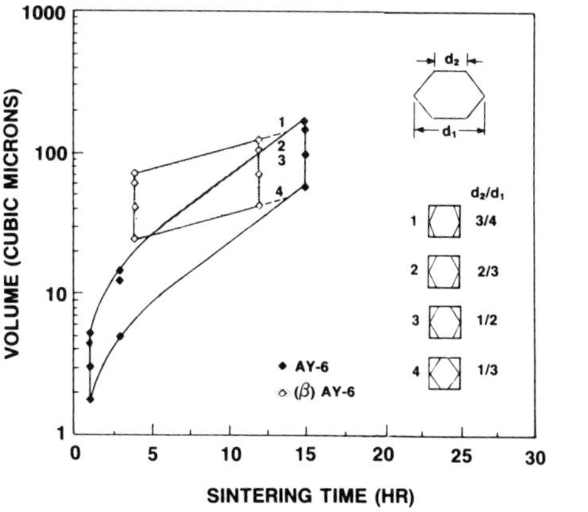

Fig. 8. Calculated increase in the volume of β grains with increasing sintering time.

grains are growing. However, in both AY-6 and (β-)AY-6, the acicular morphology of the β grains is maintained.

Using grain growth data (Table 3), and on the assumption of idealized hexagonal cross-sections with a range of shapes (d_2/d_1), the average volumetric increase of the β grains with sintering time (Fig. 8) was calculated. After an initial rapid rate of volumetric increase of the β grains in AY-6, both AY-6 and (β-)AY-6 compositions tend toward similar rates of volume increase of the β grains. Therefore, independent of the cross-sectional size and shape of the β grains, at a given temperature, an equilibrium growth rate is

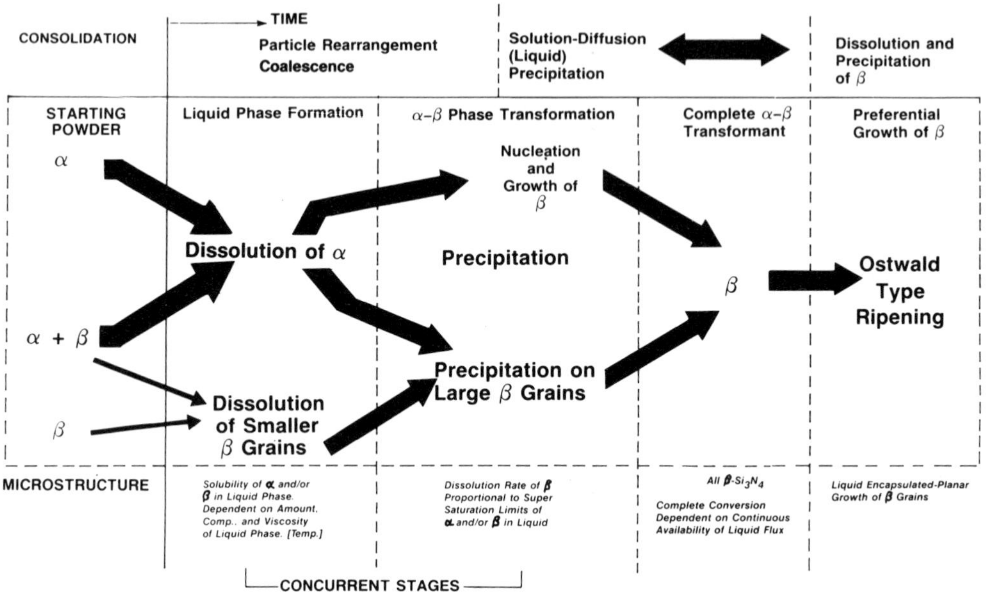

Fig. 9. Microstructural development during liquid phase consolidation of various types of Si_3N_4 powder.

TABLE 4 Transmission electron microscopy compositional analysis of AY-6 and PY-6 [20]

Material	*Large β grains* (wt.%)		*Average glass phase* (wt.%)		
	Al	*Y*	*Y*	*Si*	*Al*
AY-6 (Si_3N_4–Y_2O_3–Al_2O_3)	0.6–1.6	0.5–0.8	62.6	32.4	4.8
PY-6 (Si_3N_4–Y_2O_3), 1880 °C	—	0.5–1.1	70.5	29.3	
PY-6 (Si_3N_4–Y_2O_3), 1950 °C[a]	—		83.4	16.6	

[a]Crystallized.

attained. However, even under equilibrium growth conditions, β-Si_3N_4 maintains its acicular morphology.

From the results of this investigation, the consolidation, phase transformation and microstructural development of Si_3N_4-based materials in the presence of a liquid phase can be modified [18] into several regimes as shown in Fig. 9. At the onset of liquid phase formation, α-to-β transformation commences via a solution (α-Si_3N_4)–reprecipitation (β-Si_3N_4) mechanism. Therefore, the higher the solubility in the liquid phase (*i.e.* higher consolidation temperature or different phase composition), the faster should be the rate of conversion; this, in fact, has been generally observed. The growth of β grains overlaps the α-to-β transformation step. In fact, precipitation of β from the saturated solid solution on existing β-Si_3N_4 grains can occur concurrently with the nucleation and growth of new β-Si_3N_4 grains form a α-Si_3N_4 particles.

Unlike the α-Si_3N_4 structure, β-Si_3N_4 [19] is relatively unstrained and contains large channels. It should therefore be easier to accommodate foreign atoms in the available voids in the β structure. This, indeed, was found to be the case as indicated by the results of compositional analysis inside β grains using both electron energy loss spectroscopy and energy-dispersed X-rays (Table 4). The constituent ions of the liquid phase were found to be contained within all the β-Si_3N_4 grains examined.

Since both the α-to-β transformation and the growth of β-Si_3N_4 are via reprecipitation from a liquid solution containing other atoms, aluminum and yttrium in this case, their possible incorporation into the structure makes the growth of β-Si_3N_4 both kinetically and energetically favorable.

Furthermore, once such foreign atoms have been accommodated into the β-Si_3N_4 structure, its transformation back to α is not possible. It is suggested that this, rather than the Madelung energy difference between α and β [2], is a more probable reason why the reverse β-to-α transformation has never been observed because the magnitude of the latter is only 1.3% of the absolute value.

4. Conclusions

The results of this study clearly indicate that the α-to-β transformation, under the consolidation conditions investigated, occurs in the early stages of the liquid phase formation and is not dependent on, but rather aids, the densification process. However, this conversion drastically influences the structural development owing to the formation of whisker-like (large aspect ratio) acicular β-Si_3N_4 grains. Both these transformed β-Si_3N_4 grains, and those present in the initial powder were found to grow via coalescence and a dissolution–precipitation reaction similar to that previously reported to be exclusive to the conversion of α to β. Additionally, it was observed that extended consolidation times after complete conversion of α to β has occurred do not result in the spheroidization of β-Si_3N_4 grains. A model illustrating microstructural development during liquid phase consolidation as influenced by the polymorphic form of the starting powder and thus the α-to-β transformation was developed.

Acknowledgments

The author is indebted to Dr. S. Natansohn, without whose help this study would not have been possible. Thanks are also due to Professor M. Ruhle, Dr. H. Rebenne, Mr. J. Neil and Mr. J. Sanchez for their assistance with various aspects of this investigation.

References

1 W. D. Forgeng and B. F. Decker, *Trans. AIME, 212* (1958) 343–348.
2 R. Grun, *Acta Crystallogr. B, 35*(1979) 800–804.
3 M. J. Buerger, in R. Smoluchowski *et al.* (eds.), *Phase Transformations in Solids,* Wiley, New York, 1951, p. 183.
4 D. R. Messier, F. L. Riley and R. J. Brook, *J. Mater. Sci., 13*(1978) 1199.
5 H. F. Priest, F. C. Burns, G. L. Priest and E. Skaar, *J. Am. Ceram. Soc., 56*(1973) 395.
6 D. R. Messier and R. L. Riley, in F. L. Riley (ed.), *Nitrogen Ceramics,* Noordhoff, Leyden, 1977, p. 141.
7 C. Greskovich and S. Prochazka, *J. Am. Ceram. Soc., 60* (1977) 60.
8 E. T. Turkdogan, P. M. Bills and V. A. Tippet, *J. Appl. Chem., 8*(1958) 296–302.
9 P. Grieveson, K. H. Jack and S. Wild, *Spec. Ceram., 4* (1968) 237–238.
10 H. Feld, P. Ettmayer and I. Petzenhauser, *Ber. Dtsch. Keram. Ges., 51*(1974) 127–131.
11 J. J. Petrovic, *J. Am. Ceram. Soc., 58*(3–4) (1976) 113.
12 G. Wötting and G. Ziegler, *Powder Metall. Int., 18* (1986) 25–32.
13 S. Natansohn and V. K. Sarin, to be published.
14 D. R. Messier and W. J. Croft, *Rep. AMMRC TR 82-42,* 1982 (Army Materials and Mechanics Research Center, Watertown, MA).
15 L. J. Bowen, R. J. Weston, T. G. Carruthers and R. J. Brook, *J. Mater. Sci., 13*(1978) 341–350.
16 J. L. Iskoe and F. F. Lange, in R. M. Fuhalti and J. A. Pask (eds.), *Ceramic Microstructures,* Westview Press, Boulder, CO, 1977.
17 H. Knoch, K. A. Schwetz and A. Lipp, in F. L. Riley (ed.), *Nitrogen Ceramics,* Noordhoff, Leyden, 1977, p. 381.
18 H. Knoch and G. E. Gazza, *Ceramurgia Int., 6*(1980) 51.
19 D. R. Messier, *Rep. AMMRC TR 77-15,* 1977 (Army Materials and Mechanics Research Center, Watertown, MA).
20 V. K. Sarin and M. Ruhle, to be published.

Materials Science and Engineering, A105/106 (1988) 161–168

Mixed α- and β-(Si–Al–O–N) Materials with Yttria and Neodymia Additions*

T. EKSTRÖM

AB Sandvik Hard Materials, P.O. Box 42056, S-126 12 Stockholm and Department of Inorganic Chemistry, Arrhenius Laboratory, University of Stockholm, S-106 91 Stockholm (Sweden)

P.-O. KÄLL, M. NYGREN AND P.-O. OLSSON

Department of Inorganic Chemistry, Arrhenius Laboratory, University of Stockholm, S-106 91 Stockholm (Sweden)

(Received November 9, 1987)

Abstract

A number of ceramic materials have been fabricated on a semipilot plant scale at different overall compositions in the (Y, Nd)–Si–Al–O–N system at 1775 °C. Constant molar amounts of oxide mixtures of Y_2O_3:Nd_2O_3 in the ratios 100:0, 75:25, 50:50, 25:75 or 0:100 have been added. Dense materials were obtained for all compositions except those corresponding to mixed α- and β-(Si–Al–O–N) with higher α-(Si–Al–O–N) contents and high Nd_2O_3 contents. At the preparation temperature used in this study, the formation of an α-(Nd–Si–Al–O–N) seems prohibited and, thus, with increasing Nd_2O_3 content the amount of α-(Si–Al–O–N) decreased. The Nd_2O_3 added mainly formed crystalline intergranular phases such as the N-melilite phase, which increased in amount with increasing Nd_2O_3 in the starting mix. Hardness and indentation fracture toughness measurements were made and are discussed in relation to the phase composition and the microstructure. Some of the high Nd_2O_3 content Si–Al–O–N materials have as high fracture toughness values as the pure Y_2O_3 Si–Al–O–N materials do.

1. Introduction

Silicon nitride (Si_3N_4) is one of the engineering ceramics which has attracted much attention in the past because of the expected outstanding properties, such as high wear resistance and high strength [1]. However, these properties cannot be reached unless the ceramic material is fabricated with a near-theoretical density and this cannot be done by pressureless sintering without the use of additives. The most successful sintering aids used for commercial materials are yttria (Y_2O_3) or Y_2O_3 plus alumina (Al_2O_3)–AlN. If aluminium is also present in sufficient amounts, then α- and/or β-(Si–Al–O–N) crystals are formed. The composition of the β-(Si–Al–O–N) crystals can vary according to the formula $Si_{6-z}Al_zO_zN_{8-z}$ with $0<z\lesssim4.2$. The extent of this solid solution is indicated in Fig. 1(a). In the Y–Si–Al–O–N system the α-(Si–Al–O–N) crystals can also have a variable composition, $Y_x(Si, Al)_{12}(O, N)_{16}$, as indicated.

Mixed α- and β-(Si–Al–O–N) materials have been found to yield benefits compared with pure β-(Si–Al–O–N) ceramics in engineering applications as these materials combine a good fracture toughness, generated by the elongated β-(Si–Al–O–N) grains, with a significantly higher hardness from the α-(Si–Al–O–N) component. One example is the use as metal-cutting tools where a considerable increase in the lifetime can be seen for the mixed α- and β-(Si–Al–O–N) tools compared with β-(Si–Al–O–N) cutting tools [2, 3]. The overall compositions which allow pressureless sintering of fully dense ceramic materials consisting of mixed α- and β-(Si–Al–O–N) phases (with Y_2O_3 as sintering aid) are fairly well mapped out [4]. With normal large-scale fabrication routines, materials with an α-Si–Al–O–N contents up to about 70% can be made. With the use of carefully selected raw materials, compositions and processing parameters, pure yttrium α-(Si–Al–O–N) materials with densities exceeding 98% of the theoretical density can be achieved [5]. We

*Paper presented at the 3rd International Conference on the Science of Hard Materials, Nassau, The Bahamas, November 9–13, 1987.

0921-5093/88/$3.50

© Elsevier Sequoia/Printed in The Netherlands

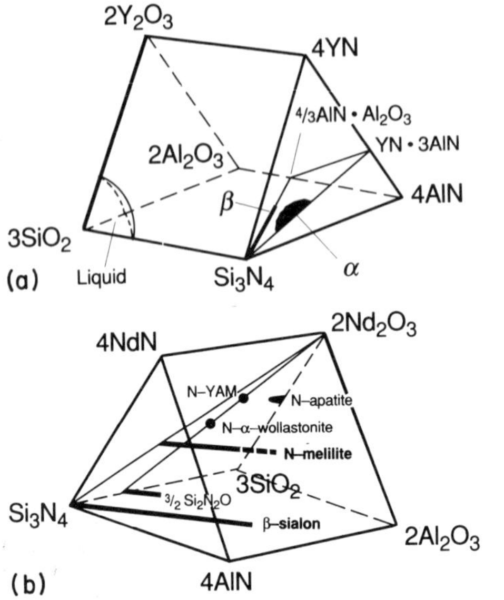

Fig. 1. (a) A schematic representation of the Y-Si-Al-O-N system to illustrate the linear β-(Si-Al-O-N) solid solution range $Si_{6-z}Al_zO_zN_{8-z}$ and the two-dimensional α-(Si-Al-O-N) solid solution domain $Y_x(Si, Al)_{12}(O, N)_{16}$ found in the plane Si_3N_4–YN.3AlN–$\frac{4}{3}$AlN.Al_2O_3. (b) The Nd-Si-Al-O-N system at high temperatures [7]. The N-melilite structure ($Nd_2O_3.Si_3N_4$) has a solid solution range extended far into the Nd-Si-Al-O-N system as a result of simultaneous substitution of aluminium for silicon and oxygen for nitrogen.

do not know of any commercial materials of this type.

Y_2O_3 is a rather expensive sintering aid, as it is normally used in quantities of 5–15 wt.%, and cheaper sintering aids have been searched for to replace this oxide without too significant a loss of performance. Lanthanum oxide (La_2O_3), cerium oxide (CeO_2) and neodymium oxide (Nd_2O_3) have all been mentioned as possible candidates, with prices about one-tenth of the price of Y_2O_3. However, the element neodymium is the only one of these three elements which has been reported to substitute into the α-Si_3N_4 structure to stabilize an α-(Si-Al-O-N) like yttrium does and thereby allow mixed α- and β-(Si-Al-O-N) materials to be prepared [6].

Phase relationships in the Nd-Al-O-N and the Nd-Si-Al-O-N systems have been carefully studied by Slasor *et al.* [7]. A neodymium-containing N-melilite phase $Nd_2O_3.Si_3N_4$, isostructural with the yttrium form of this quaternary silicon oxynitride, was observed to have an extended homogeneity range along the tie line $Nd_2Si_3O_3N_4$–

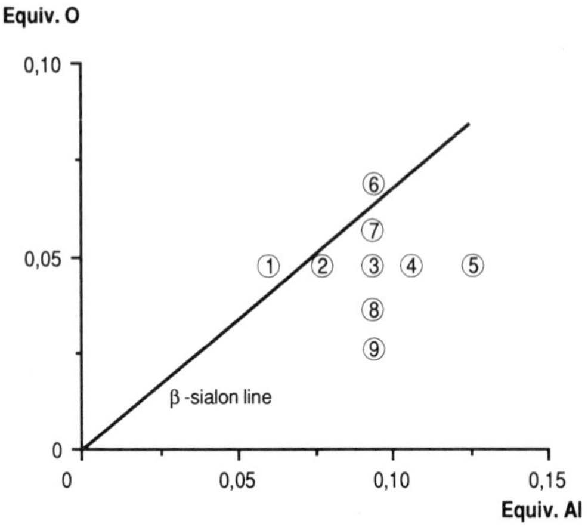

Fig. 2. The overall compositions of the samples made in this study represented as added equivalents of oxygen and aluminium to Si_3N_4; see text.

$Nd_2Si_2AlO_4N_3$, and it is therefore expected to become dominant in the Nd-Si-Al-O-N system (Fig. 1(b)).

To examine the possibilities of replacing Y_2O_3 with Nd_2O_3 as a sintering aid in normal fabrication routines, a series of Si-Al-O-N materials was prepared on a semipilot plant scale. The results of these preparations with evaluation of phase composition, microstructure and some physical properties are reported below.

2. Materials and experimental procedure

The overall compositions of all the samples made in this study are close to the Si_3N_4 corner of the M-Si-Al-O-N phase diagram, and compositions were selected as illustrated in Fig. 2 as a function of equivalents of oxygen and aluminium. The different samples were prepared from carefully weighed-in additions of Al_2O_3–AlN, Y_2O_3 and Nd_2O_3 to Si_3N_4. The small amount of oxygen present as oxides or oxynitrides in the Si_3N_4 and AlN raw materials is also included in the calculations. It should be noted, however, that the oxygen from the added Y_2O_3 and/or Nd_2O_3 is not included in the equivalents shown in Fig. 2, which are defined as

$$\text{equivalent Al} = \frac{3 \times \text{at.\% Al}}{3 \times \text{at.\% Al} + 4 \times \text{at.\% Si}}$$

$$\text{equivalent O} = \frac{2 \times \text{at.\% O}}{2 \times \text{at.\% O} + 3 \times \text{at.\% N}}$$

The same constant molar amount of Y_2O_3 and Nd_2O_3 sintering aid was added to all samples, corresponding for the pure Y_2O_3 specimen and the pure Nd_2O_3 specimen to 6.0 wt.% and 8.9 wt.% respectively. For each composition represented in Fig. 2, samples were prepared with mixtures of Y_2O_3:Nd_2O_3 in the ratios 100:0, 75:25, 50:50, 25:75 or 0:100.

The powder mixtures were made with Si_3N_4 (LC1; H. C. Starck, Berlin), Al_2O_3 (A16SG; Alcoa), AlN (H. C. Starck, Berlin), Y_2O_3 (purity, 99.99%; H. C. Starck, Berlin) and Nd_2O_3 (purity, greater than 99.5%; Molycorporation Inc.). The analysed oxygen content of Si_3N_4 corresponds to 2.9 wt.% SiO_2 and the oxygen content of the AlN to 1.9 wt.% Al_2O_3. The Nd_2O_3 was found to contain substantial amounts of neodymium hydroxide, and this material was therefore calcined at 1000 °C for 2 h before use.

The starting materials were mixed in water-free propanol, and milled in a vibratory mill for 16 h with Si–Al–O–N milling media. The sample batch size used was 500 g dry weight of the starting materials, *i.e.* semipilot plant scale. The slurry was then dried, and powder compacts of size 16 mm × 16 mm × 6 mm were pressed at a load of 125 MPa. The specimens were embedded in micron-sized BN powder in graphite crucibles and sintered at 1775 °C for 2 h in nitrogen gas at atmospheric pressure. A selected number of compositions were also hot isostatically pressed at 1775 °C for 1 h using a glass encapsulation technique and argon gas at 200 MPa pressure.

After sintering, specimens were prepared for physical characterization by grinding and polishing with standard techniques. The Vickers hardness (HV 10) and indentation fracture toughness K_{Ic} at room temperature were obtained with a Vickers diamond indenter using a 98 N (10 kgf) load, and the fracture toughness was evaluated using the method of Anstis *et al.* [8] assuming a value of 300 GPa for Young's modulus. The accuracy in the density measurement, made in toluene by Archimedes' principle, was ± 0.002 g cm^{-3}.

X-ray diffraction (XRD) analyses were performed both by direct analysis of polished surfaces of as-sintered materials with a diffractometer (Rigaku rotating copper anode, 10 kW) and by a Guinier–Hägg camera techniquc using a slice of sintered material crushed to a powder. The two XRD methods gave identical results. From the lattice parameters of the hexagonal β-(Si–Al–O–N) phase $Si_{6-z}Al_zO_zN_{8-z}$ the z values were obtained from the equations $a = 0.7603 + z \times 0.00297$ nm and $c = 0.2907 + z \times 0.00255$ nm [9]. Polished surfaces of as-sintered materials were also studied by scanning electron microscopy (SEM) (using a JEOL JSM 820 equipped with a Link AN 10000 energy-dispersive spectroscopy (EDS) analyser) after the application of a carbon coating. For transmission electron microscopy (TEM) studies (using a JEOL 200 CX) the as-sintered specimens were cut into thin (less than 0.2 mm) slices and the final thinning was done by ion milling.

3. Results

3.1. Densification

To establish the theoretical density, some of the samples were densified by glass encapsulation and hot isostatic pressing (HIP), as this sintering procedure gives dense materials. Samples 1, 3, 5 and 9, the compositions of which are defined in Fig. 2, with varying Y_2O_3–Nd_2O_3 additions were hot isostatically pressed at 1775 °C and 200 MPa for 1 h. The sintered samples were controlled in an optical microscope on carefully polished sections of the specimens and the observed porosity is clearly well below 0.5% of the total sample volume. Therefore the hot isostatically pressed materials can be considered as fully dense Si–Al–O–N ceramics.

The observed densities of the hot isostatically pressed ceramics varied with the Y_2O_3:Nd_2O_3 ratio used, but no significant variation between samples with different overall compositions but fixed Y_2O_3:Nd_2O_3 ratios was observed. Thus the densities of the pure Y–Si–Al–O–N materials were found to be 3.24 g cm^{-3} and to increase linearly with increasing Nd_2O_3 content to 3.30 g cm^{-3} for pure Nd–Si–Al–O–N materials.

All the sample compositions were made by pressureless sintering at 1775 °C for 2 h. It was found that all pure Y–Si–Al–O–N materials sinter to near-theoretical density, but with increasing neodymium content those samples intended to have higher α-(Si–Al–O–N) contents, *i.e.* samples 3, 4, 5, 8 and 9 have a reduced relative density, as seen in Fig. 3(a). Samples 2 and 7, which were intended to have a small content of α-(Si–Al–O–N) material, are found to be almost 100% dense even for pure Nd–Si–Al–O–N materials. The observed densities for samples 6 and 1, which were both intended to be α-(Si–Al–O–N)-

free materials, differ as seen in Fig. 3(b). Thus, pressureless sintering of materials with the overall composition of sample 6 yields fully dense materials independent of the Y_2O_3:Nd_2O_3 ratio used, while for sample 1 the relative density decreases with increasing neodymium content.

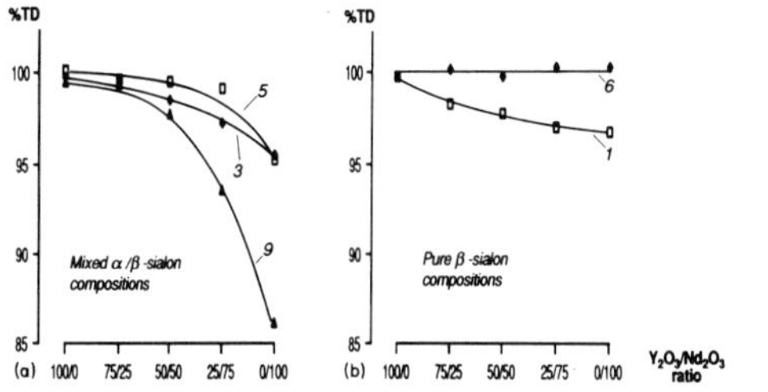

Fig. 3. The variation in the observed density as a percentage of the theoretical density after sintering at 1775 °C for 2 h with different additions of the sintering aids Y_2O_3–Nd_2O_3: (a) three mixed α- and β-(Si–Al–O–N) materials; (b) two β-(Si–Al–O–N) materials.

3.2. X-ray diffraction phase analysis

The XRD results showed that samples 1 and 6 are β-(Si–Al–O–N) materials, whereas the remaining samples are mixed α- and β-(Si–Al–O–N). The relative amount of α-(Si–Al–O–N), represented as the α-(Si–Al–O–N):α- and β-(Si–Al–O–N) ratio, increases as the aluminium content increases in the sample series 2, 3, 4 and 5 and as the nitrogen content increases in the sample series 7, 3, 8 and 9, as seen in Table 1.

For the Si–Al–O–N materials containing only Y_2O_3, the intergranular phase consists generally of a glassy phase. As Nd_2O_3 is added, however, crystalline melilite is found, the amount of which increases for the same overall composition with increasing Nd_2O_3 content. In the mixed α- and β-(Si–Al–O–N) materials the amount of α-(Si–Al–O–N) clearly decreases with increasing neodymium content, as seen from the shift in the $\alpha:\alpha+\beta$ ratio in Table 1. The lattice parameters of the hexagonal α-(Si–Al–O–N) phase are not significantly influenced by the overall composi-

TABLE 1 X-ray diffraction results obtained by the Guinier–Hägg camera technique

Sample	*Yd_2O_3:Nd_2O_3 ratio*	*Phases obtained by XRD*	*$\alpha:\alpha+\beta$ ratio*	*z, β-(Si–Al–O–N)*	*Lattice parameters, (α-Si–Al–O–N) (nm)*	
					a	*c*
1	100:0	β-(Si–Al–O–N)	—	0.50	—	—
1	50:50	β-(Si–Al–O–N), very weak melilite	—	0.47	—	—
1	0:100	β-(Si–Al–O–N), very weak melilite, very weak U phase	—	0.45	—	—
2	100:0	α- and β-(Si–Al–O–N)	0.15	0.53	0.7800	0.5681
2	0:100	β-(Si–Al–O–N), weak melilite	—	0.47	—	—
3	100:0	α- and β-(Si–Al–O–N)	0.30	0.60	0.7802	0.5684
3	50:50	α- and β-(Si–Al–O–N), weak mililite	0.25	0.57	0.7801	0.5685
3	0:100	α- and β-(Si–Al–O–N), medium melilite	0.06	0.50	—	—
4	100:0	α- and β-(Si–Al–O–N), very weak b phase	0.36	0.62	0.7801	0.5683
4	0:100	α- and β-(Si–Al–O–N), medium melilite	0.07	0.51	—	—
5	100:0	α- and β-(Si–Al–O–N), very very weak 12H	0.47	0.66	0.7804	0.5687
5	50:50	α- and β-(Si–Al–O–N), weak melilite	0.39	0.54	0.7803	0.5682
5	0:100	α- and β-(Si–Al–O–N), medium melilite	0.13	0.49	0.7802	0.5690
6	100:0	β-(Si–Al–O–N)	—	0.67	—	—
6	0:100	β-(Si–Al–O–N), weak melilite	—	0.54	—	—
7	100:0	α- and β-(Si–Al–O–N)	0.11	0.55	0.7798	0.5679
7	0:100	α- and β-(Si–Al–O–N), weak melilite	0.02	0.50	—	—
8	100:0	α- and β-(Si–Al–O–N)	0.43	0.58	—	—
8	0:100	α- and β-(Si–Al–O–N), medium melilite	0.02	0.53	—	—
9	100:0	α- and β-(Si–Al–O–N)	0.58	0.41	0.7800	0.5680
9	50:50	α- and β-(Si–Al–O–N), very weak melilite	0.48	0.47	0.7804	0.5680
9	0:100	α- and β-(Si–Al–O–N), medium melilite	<0.01	0.41	—	—

The relative amounts of α- and β-(Si–Al–O–N) phases were estimated from the peak areas of the 102 plus 210 and of the 101 and 210 reflections respectively. The z values of the β-(Si–Al–O–N) $Si_{6-z}Al_zO_zN_{8-z}$ are obtained from refined lattice parameters. The error in the calculated lattice parameters for the hexagonal α-(Si–Al–O–N) phase is ± 0.0003 nm.

tion or by the neodymia content. Thus the observed α-(Si–Al–O–N) can be said to have the unit-cell dimensions $a=0.7802$ nm and $c=0.5682$ nm independent of the overall composition of the starting material used in this study. Finally, it can be seen that the z value of the β-(Si–Al–O–N) decreases slightly with increasing neodymium content from the Y–Si–Al–O–N to the Nd–Si–Al–O–N material.

3.3. Microstructure analysis

The overall structural features were most easily viewed by scanning electron microscopy on polished surfaces. Both secondary electron and backscattered electron detectors were used. The latter, however, gives a more pronounced difference between the α- and β-(Si–Al–O–N) crystals. Typical microstructures of a mixed α- and β-(Si–Al–O–N) ceramic are demonstrated in Fig. 4, as the added sintering aid ranges from pure Y–Si–Al–O–N to a pure Nd–Si–Al–O–N. The micrographs are recorded with high contrast in order to emphasize the white contrast originating from the intergranular phase. It is seen that the intergranular phase increases with the increasing neodymium content and that the Si–Al–O–N grains become smaller.

Using a simple counting technique with a grid put on the micrographs, results show that the intergranular phase volume of the mixed α- and β-(Si–Al–O–N) sample 3 in Fig. 4 shifts from about 6% to about 15%, moving from a Y–Si–Al–O–N to an Nd–Si–Al–O–N. The two β-Si–Al–O–N ceramics, samples 1 and 6, show a similar trend of increasing amount of intergranular phase and decreasing grain size with increasing neodymium content. The volume of the intergranular phase in the β-(Si–Al–O–N) sample 1 changes from about 10% to about 20%.

A qualitative EDS analysis of the pure yttrium-containing sample 3 shown in Fig. 4(a) shows that no yttrium is found in the dark β-(Si–Al–O–N) grains but that some yttrium is found in the medium-grey α-(Si–Al–O–N) grains and a 5–10 times stronger yttrium signal is recorded from the intergranular phase of white contrast in the micrograph. As neodymium is added, the specimen with a Y_2O_3:Nd_2O_3 ratio of 50:50 and of the same overall composition shows that only a small amount of the added neodymium is found in the α-(Si–Al–O–N) together with yttrium and that most of the neodymium is found in the intergranular phase. The decreased grain size and the

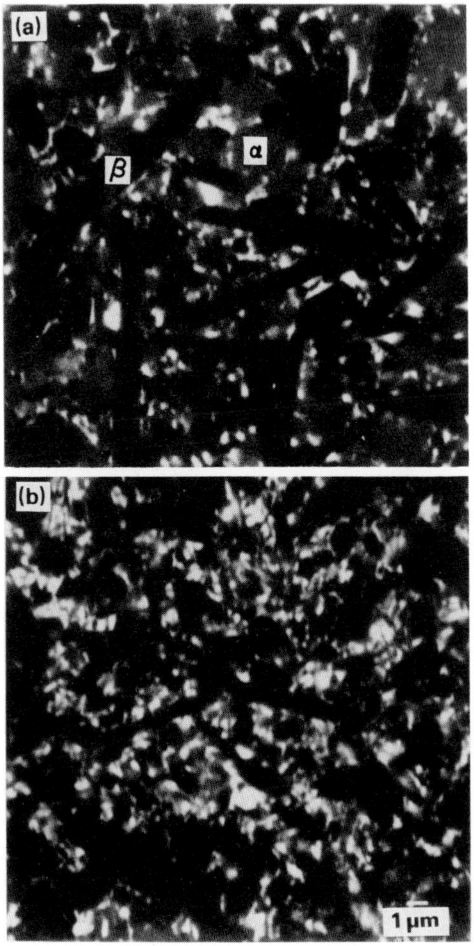

Fig. 4. Hard-contrast SEM micrographs in order to illustrate the variations in amount and distribution of the intergranular phase (white contrast): (a) mixed α- and β-(Si–Al–O–N) sample 3 with pure Y_2O_3 added as a sintering aid; (b) mixed α- and β-(Si–Al–O–N) sample 3 with pure Nd_2O_3 added as a sintering aid.

fine distribution intergranular phase areas generally less than 1 μm in size introduce some resolution problems for the EDS analysis of bulk specimens. This is especially so for the pure neodymium-containing specimen, where it can only be said that the major part of the neodymium is found in the intergranular phase.

Some preliminary TEM studies were made on only neodymium-containing mixed α- and β-(Si–Al–O–N) materials with a high melilite content in order to establish the extent to which the

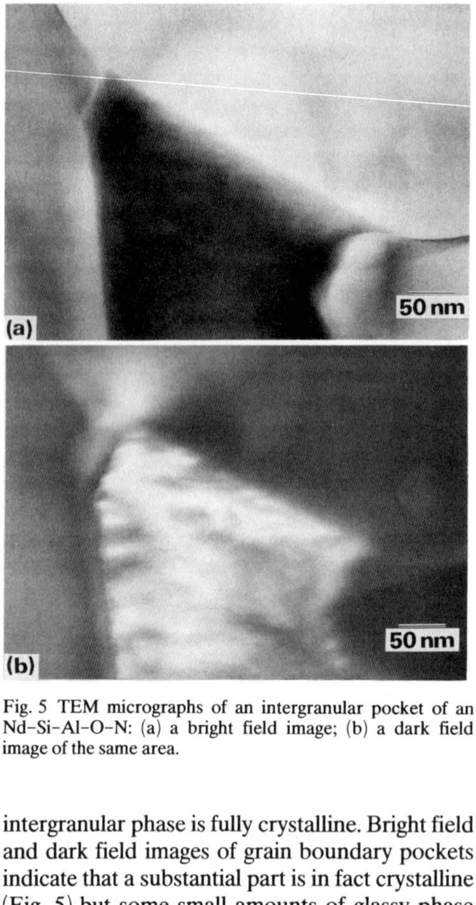

Fig. 5 TEM micrographs of an intergranular pocket of an Nd–Si–Al–O–N: (a) a bright field image; (b) a dark field image of the same area.

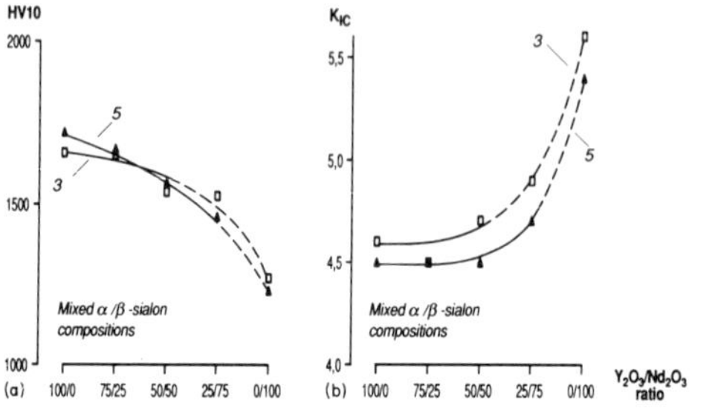

Fig. 6. The variation in measured Vickers hardness (load, 98 N) (HV 10 scale) and indentation fracture toughness K_{Ic} with various Y_2O_3:Nd_2O_3 ratios for mixed α- and β-(Si–Al–O–N) materials.

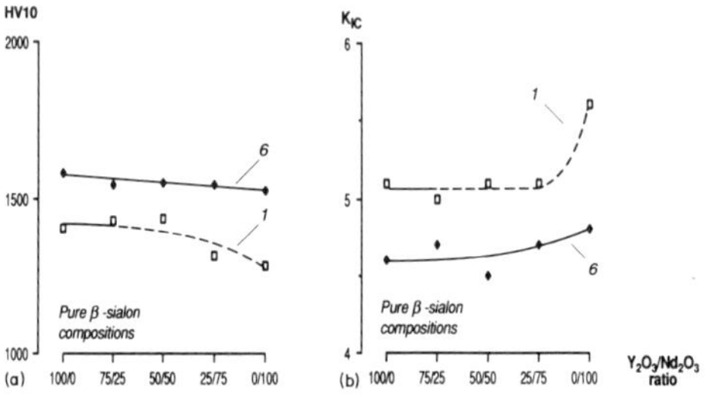

Fig. 7. The variation in measured Vickers hardness (load, 98 N) (HV 10 scale) and indentation fracture toughness K_{Ic} with various Y_2O_3:Nd_2O_3 ratios for β-(Si–Al–O–N) materials.

intergranular phase is fully crystalline. Bright field and dark field images of grain boundary pockets indicate that a substantial part is in fact crystalline (Fig. 5) but some small amounts of glassy phase were also detected. The crystalline phase was identified as melilite from its electron diffraction pattern.

3.4. *Physical properties*

It is known that hardness (HV 10) and K_{Ic} measurements on materials with high porosity levels yield less reliable values. For comparison, however, we performed such measurements on some samples which were quite porous, as revealed by examination in an optical microscope. Roughly it can be said that the values given for samples with a relative density of less than 98% of the theoretical density are somewhat doubtful.

In Fig. 6 the behaviour of two typical mixed α- and β-(Si–Al–O–N) materials is shown, and in Fig. 7 the corresponding results for β-(Si–Al–O–N) materials. The figures show that the hardness of mixed α- and β-(Si–Al–O–N) decreases with increasing neodymium content, but not that of β-(Si–Al–O–N). In addition, the fracture toughness is roughly constant or increases slightly with increasing neodymium content for both types of Si–Al–O–N material. However, the sharp increase in K_{Ic} and decrease in Vickers hardness noted for pure Nd–Si–Al–O–N might be attributed to the porosity of these samples. Measurements on fully dense materials (by HIP) of the overall compositions of samples 5 and 9 showed that pure Nd–Si–Al–O–N have hardness and K_{Ic} values of 1410 and 4.8 and of 1690 and 4.6 respectively.

4. Discussion

From the unit-cell dimensions and the composition the theoretical density of a β-(Si–Al–O–N), $Si_{6-z}Al_zO_zN_{8-z}$, with $z = 0.5$ can be calculated to be 3.18 g cm^{-3} [9] and similarly an α-(Si–Al–O–N) of the composition $Y_{0.3}Si_{10.2}Al_{1.8}O_{1.2}N_{14.8}$ should have a density of 3.28 g cm^{-3} [10]. The density of a Y–Si–Al–O–N glass can vary between 3.82 and 4.00 g cm^{-3} when the nitrogen content varies between 6 and 15 at.% [11]. Nd–Si–Al–O–N glass can dissolve as much nitrogen as the yttrium-containing glass [12], and the density can correspondingly vary between 4.1 and 4.5 g cm^{-3} [13]. Thus, both yttrium and neodymium glass phases have substantially higher density values than those of the two Si–Al–O–N phases given above. This implies that the major cause of the observed linear increase in density from 3.24 to 3.30 g cm^{-3} when Y_2O_3 is gradually replaced by Nd_2O_3 is a combined effect of increasing density of the intergranular phase and a change in the volume of intergranular phase. Estimations from SEM micrographs indicate that the volume is roughly doubled by totally replacing yttrium by neodymium. This observation is in agreement with the findings by Slasor *et al.* [7], who found the liquid region in the Nd–Si–Al–O–N system at high temperatures to be quite extensive.

One aim of this study was to examine the possibilities of replacing Y_2O_3 by Nd_2O_3 as a sintering aid. Mixtures of Y_2O_3 and Nd_2O_3 or Nd_2O_3 alone can be used as sintering aids to obtain Si–Al–O–N materials of near-theoretical density. In particular, the β-(Si–Al–O–N) sample 6 and the mixed α- and β-(Si–Al–O–N) samples 2 and 7, which are closest to the liquid-forming region at the sintering temperature (Figs. 1(a) and 2) and thus contain a high fraction of the liquid phase, sintered well for 2 h at 1775 °C. However, for the materials intended to form mixed α- and β-(Si–Al–O–N) with higher α-(Si–Al–O–N) contents, and having less liquid phase present, the densities showed a clear trend to decrease with increasing Nd_2O_3 content. This is because the formation of crystalline neodymium melilite will further reduce the amount of "available" liquid phase. One possible solution of this problem might be to increase the sintering time or the sintering temperature. In the work by Slasor and coworkers [5, 7] it was found that melilite dissolves in the liquid present at 1800 °C. This indicates a path to more complete α-(Nd–Si–Al–O–N) formation.

The phase analysis of the ceramics sintered with Y_2O_3 showed that the synthesis results were quite similar to those of previous studies [4, 14], but the Y_2O_3–Nd_2O_3 additions produced materials which were substantially different from the above materials. Thus, with pure Y_2O_3, a glassy intergranular phase is normally formed. Depending on the cooling rate, minor amounts of a crystalline α-wollastonite type such as the b phase Y_2SiAlO_5N may be formed [15], but crystallization of the intergranular phase gives in general a material showing a more brittle behaviour. However, in materials with larger Nd_2O_3 contents, crystalline intergranular phases were noted without any notable drop in toughness. The phase occurring most frequently and in some specimens in substantial amounts were the N-melilite $Nd_2O_3.Si_3N_4$. However, it is well known that small additions of Al_2O_3 to the starting mix remove the low temperature oxidation problem [16]. Oxidation experiments will therefore be carried out to see whether this effect is of importance in the present materials.

5. Conclusions

(1) Nd_2O_3 can replace Y_2O_3 partly or completely as a sintering aid to form dense Si–Al–O–N materials.

(2) The neodymium-containing Si–Al–O–N materials have as high fracture toughness values as the Y–Si–Al–O–N materials do.

Acknowledgments

AB Sandvik Hard Materials is thanked for permission to publish this article. This work has in part been financed by the Swedish Board for Technical Development and in part by AB Sandvik Hard Materials.

References

1 K. H. Jack, *Met. Technol.*, *9* (1982) 297.
2 J. Aucote and S. R. Foster, *Met. Sci. Technol.*, *2* (1986) 700.
3 N. Ingelström and T. Ekström, *J. Phys. (Paris)*, *47* (1986) 347.
4 T. Ekström and N. Ingelström, *Proc. Int. Conf. on Hot Isostatic Pressing, Luleå, June 15–16, 1987*, p. 367.
5 S. Slaser and D. P. Thompson, in S. Hampshire (ed.), *Non-Oxide and Technical and Engineering Ceramics*, Elsevier, London, 1986, p. 223.
6 H. K. Park, D. P. Thompson and K. H. Jack, *Sci. Ceram.*, *10* (1980) 251.
7 S. Slasor, K. Liddell and D. P. Thompson, *Br. Ceram. Proc.*, *37* (1986) 51.
8 G. R. Anstis, P. Chantikul, B. R. Lawn and D. P. Marshall, *J. Am. Ceram. Soc.*, *64* (1981) 533.
9 T. Ekström, P.-O. Käll, M. Nygren and P.-O. Olsson,

J. Mater. Sci., in print.
10 C. Chatfield, T. Ekström and M. Mikus, *J. Mater. Sci., 21* (1986) 2297.
11 D. R. Messier, *Int. J. High Technol. Ceram., 3* (1987) 33.
12 S. Hampshire, R. A. L. Drew and K. H. Jack, *Phys. Chem. Glasses, 26* (1985) 182.
13 K. H. Jack, in S. Hampshire (ed.), *Non-Oxide and Technical and Engineering Ceramics*, Elsevier London, 1986, p. 1.
14 T. Ekström and N. Ingelström, in S. Hampshire (ed.), *Non-Oxide and Technical and Engineering Ceramics*, Elsevier, London, 1986, p. 231.
15 D. P. Thompson, in R. E. Tressler, G. L. Messing, C. G. Pantano and R. E. Newham (eds.), *Tailoring Multiphase and Composite Ceramics*, Plenum, New York, 1986, p. 79.
16 J. K. Patel and D. P. Thompson, *Institute of Ceramics Annu. Conv. Meeting, Canterbury, April 1987.*

The Effect of Temperature and Yttria Content on the Properties of Hot Isostatically Pressed Si_3N_4*

LARS PEJRYD†

SECO TOOLS AB, S-773 01 Fagersta (Sweden)

(Received December 19, 1987)

Abstract

Additive-free Si_3N_4 and Si_3N_4 with Y_2O_3 additions up to 7 wt.% has been sintered in the temperature range 1550–1750 °C at a total pressure of 160 MPa using a glass-encapsulated hot isostatic pressing technique. Densities are more than 99% of the theoretical value for samples containing 1 wt.% Y_2O_3 or more sintered at 1600 °C or above. The Vickers hardness of fully densified samples decreases with increasing Y_2O_3 content and sintering temperature. For hot hardness up to 1000 °C, the effects of Y_2O_3 additions and sintering temperature are the same as for room temperature hardness. Fracture toughness increases with increasing Y_2O_3 content. Temperature differences above 1650 °C have no marked effect on the fracture toughness. The transformation of α-Si_3N_4 to β-Si_3N_4 is enhanced by the temperature while the Y_2O_3 content has no great influence.

1. Introduction

Silicon-nitride-based materials are among the most important materials developed for high temperature structural and wear resistant applications. It has, for example, proved to be excellent as cutting tool material especially for cutting cast irons and nickel-based alloys.

As a result of its strong covalent bond nature, silicon nitride is very difficult to sinter. An additional problem is that Si_3N_4 is thermodynamically unstable above 1800 °C under atmospheric pressures of nitrogen. Silicon nitride ceramics are therefore usually fabricated either by sintering or by hot pressing of Si_3N_4 powder mixed with oxides such as Al_2O_3, CaO, MgO or Y_2O_3. However, the presence of oxides reduces the mechanical strength of the material especially at temperatures above 1000 °C. Therefore, it is desirable to reduce the amount of oxide additions as much as possible.

The difficulty of pressureless sintering of low oxide-containing silicon nitride has, in most cases, put major restrictions on the geometries of the ceramic parts produced. A method by which to circumvent these problems is to use a technique with encapsulated hot isostatic pressing (HIP) which makes it possible to sinter Si_3N_4 to fully dense bodies in complicated geometries even with small amounts of sintering aid.

Of the above-mentioned oxides used as sintering aids, Y_2O_3 and Al_2O_3 give the most refractory silicate grain boundary phases. The effects of the oxide content (mainly Y_2O_3) and the process parameters have, to a varying extent, been studied by different authors such as Yamada *et al.* [1], Quackenbush *et al.* [2], Heinrich *et al.* [3] and Yeheskel *et al.* [4]. Of these, the work of Yamada *et al.* and of Heinrich *et al.* concerns glass-encapsulated HIP while that of Yeheskel *et al.* deals with HIP without encapsulation, and that of Quackenbush *et al.* with hot pressing. The pressure effect was the major concern of Heinrich *et al.*, while Yamada *et al.* concentrated on additive-free Si_3N_4.

This investigation was conducted to establish the limits of the glass-encapsulated HIP method and to clarify the effects of sintering temperature and Y_2O_3 content on the material properties of fully dense Si_3N_4 materials.

2. Experimental details

Si_3N_4 (Denka, SN-9FW) and Y_2O_3 (Rhone Poulenc, Ceramic Grade) were mixed and milled

*Paper presented at the 3rd International Conference on the Science of Hard Materials, Nassau, The Bahamas, November 9–13, 1987.

†Present address: Volvo Flygmotor AB, S-461 81 Trollhättan, Sweden.

0921-5093/88/$3.50
© Elsevier Sequoia/Printed in The Netherlands

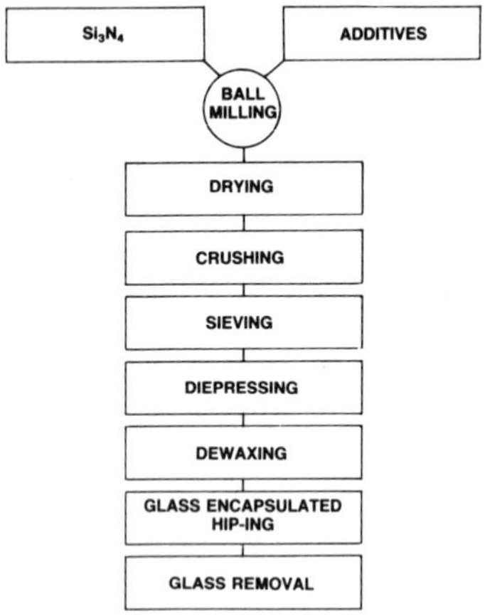

Fig. 1. Sample preparation method.

TABLE 1 Chemical composition of Si_3N_4 starting powder

Element	Si	N	Fe	Al	Ca	Mg	C	O
Amount (wt.%)	59.7	38.8	0.04	0.2	0.1	0.005	0.25	0.9

(by wet milling in methylenechloride) in a plastic jar with Si_3N_4 milling balls. Polyethyleneglycol was used as a binder. The treatment of the samples is described in Fig. 1. The HIP was performed with the glass encapsulation technique described by Larker [5]. With this method, very low diffusion rates between the sample and the encapsulation material are obtained. The HIP was carried out at a pressure *P* of 160 MPa, with a holding time at sintering temperature of 1.5 h, and at sintering temperatures *T* of 1550, 1600, 1650, 1700 and 1750 °C.

The sample compositions were of Si_3N_4 without additions and with Y_2O_3 additions of 1, 3, 5 and 7 wt.%. The composition of the Si_3N_4 powder is given, as the makers' values, in Table 1. The oxygen content was measured to 1.0 wt.% and during milling the oxygen content is increased by 0.6 wt.%.

This study focuses on bulk properties of the materials and does not concern differences between bulk and surface conditions resulting from interface reactions between the sample and the capsule material. However, the results are representative of cutting tool inserts made by this technique. Analysis of the materials was therefore performed on the central part of a polished surface of a cutting tool insert (ISO 120400) cut at half height.

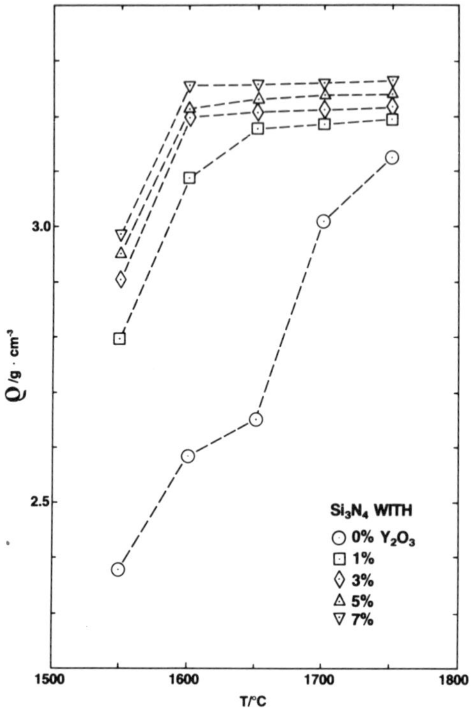

Fig. 2. Density *vs.* sintering temperature for Si_3N_4 with different Y_2O_3 contents. The theoretical density (TD) for pure Si_3N_4 is 3.18 g cm^{-3}. For samples containing 1 wt.%, 3 wt.%, 5 wt.% and 7 wt.% Y_2O_3, TD is 3.192 g cm^{-3}, 3.215 g cm^{-3}, 3.239 g cm^{-3} and 3.264 g cm^{-3} respectively. (For definition of TD see text.)

3. Results and discussion

3.1. Density

The density was measured using Archimedes' principle, weighing the samples dry and in H_2O. The densities of the samples are shown in Fig. 2. For samples containing 1 wt.% Y_2O_3 sintered at 1650 °C or higher, and samples containing 3–7 wt.% Y_2O_3 sintered at 1600 °C or higher, densities of 99% or more of the TD are obtained. In this case, TD is defined as

$$\mathrm{TD} = \frac{{}^{m}\mathrm{Si_3N_4} + {}^{m}\mathrm{Y_2O_3}}{{}^{m}\mathrm{Si_3N_4}/{}^{\rho}\mathrm{Si_3N_4} + {}^{m}\mathrm{Y_2O_3}/{}^{\rho}\mathrm{Y_2O_3}}$$

Samples without Y_2O_3 additions or those sintered at low temperatures did not produce fully dense bodies. These results differ from those of Yamada *et al.* [1] in that they needed 3.2 eq.% Y_2O_3 (about 2.5 wt.%) and 1700 °C to reach 97% density. In contrast, Heinrich *et al.* [3] seem to have reached high densities at 1750 °C and pressures higher than 50 MPa with samples containing 0.7–7.0 wt.% Y_2O_3. For HIP without glass encapsulation, Yeheskel *et al.* [4] did not obtain high densities until 10 wt.% Y_2O_3 was added and $T=1800$ °C was reached.

3.2. *Microstructure and crystallography*

Crystallographic properties were determined using X-ray powder diffractometry and the α phase and β phase contents were evaluated using the X-ray technique described by Ekström and Aronsson [6]. It should be noted that the phases present in amounts less than a few weight per cent are not detected with X-ray techniques. The microstructure was studied on polished and on etched carbon-coated surfaces using scanning electron microscopy (SEM). Phases with different Y_2O_3 contents can be separated by the difference in their atomic-number contrast giving the Y_2O_3-containing "glass phase" a bright appearance while both α-Si_3N_4 and β-Si_3N_4 appear dark and hard to separate from each other. For samples with very low Y_2O_3 contents, even the different grains are hard to separate owing to a very low content of "glass phase". Therefore, etching of the samples was needed to reveal the differences in their grain morphologies. Examples of the microstructural differences are shown in Fig. 3, while the crystallographic properties are shown in Fig. 4 and Table 2.

Samples sintered at 1600 °C or lower (Fig. 3(a)) mostly contain equiaxed particles identified using the X-ray technique as being mainly α-Si_3N_4 and small amounts of a Y_2O_3-containing intergranular phase which, in most cases, is a glass (*cf.* Table 2). This "glass phase" cannot be seen in the etched samples. For additive-free Si_3N_4, the microstructure is similar even at higher sintering temperature although the Si_3N_4 is fully converted to β-Si_3N_4 at 1750 °C. The reason that acicular β phase grains are not formed is probably that the amount of liquid formed in this system is too low to allow preferential directional growth. The liquid formed is due to the presence of aluminium and calcium in the starting powder and of oxygen both from the starting powder and from milling.

Figures 3(b) and 3(c) are good representations of Y_2O_3-containing samples sintered at 1650 °C or above. Here the β-Si_3N_4 content is high and the "glass phase" content higher. Thus, there is a larger volume of liquid at the sintering tempera-

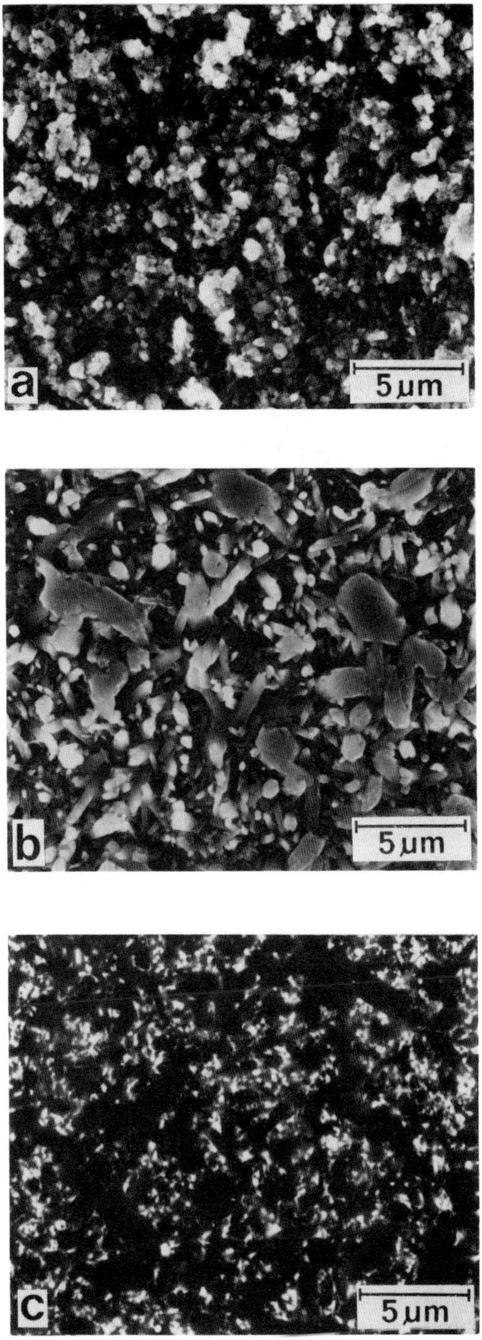

Fig. 3. Microstructure of Si_3N_4 (a) containing 3 wt.% Y_2O_3 sintered at 1600 °C and (b), (c) containing 7 wt.% Y_2O_3 sintered at 1700 °C. (a) and (b) are polished and etched while (c) is only polished.

ture T, which allows the grains to grow more freely giving rise to the typical fibrous microstructure of Si_3N_4 ceramics. The effect of increasing the temperature above 1650 °C is an increase in α–β conversion and some grain growth. The effects of Y_2O_3 on the microstructure and crystallographic properties are mainly increases in the grain size of Si_3N_4 and in the volume fraction of "glass phase" with an increase in Y_2O_3.

The major factor governing the α–β conversion is temperature. Almost no conversion occurs below 1600 °C. The effect of oxide content is small. This abrupt change in β phase content is in agreement with investigations of CaO and MgO additions to Si_3N_4 [7]. However, in that case, this conversion takes place at lower temperatures. The temperature at which 50% conversion is obtained is related to the solidus temperature of the system in question; a low solidus temperature giving a low "conversion" temperature, the diffusion rates within the melt being the major governing factor. However, a high degree of α–β conversion is not a prerequisite for complete consolidation during the conditions used in this study. These results are in agreement with those of Sarin [8] who also found that α–β conversion and densification were not directly related during gas-pressure sintering of Y_2O_3–Si_3N_7 and Y_2O_3–Al_2O_3–Si_3N_7 samples.

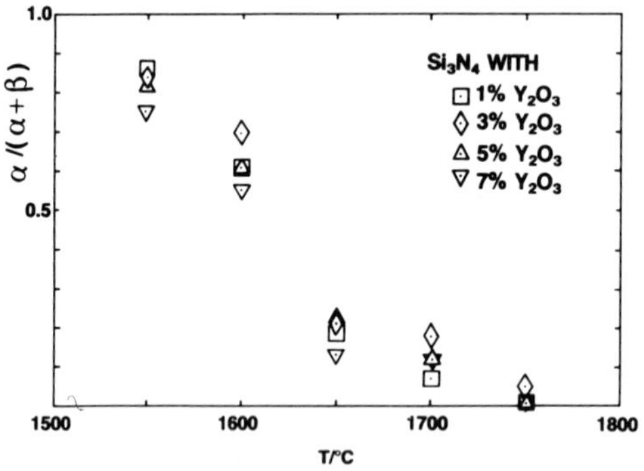

Fig. 4. α-Si_3N_4 content of hot isostatically pressed Si_3N_4 containing Y_2O_3.

3.3. *Hardness and fracture toughness*

Hardness was determined as Vickers hardness with a load of 3 kgf and the fracture toughness was determined with the indentation technique according to Evans and Charles [9]. Hardness and fracture toughness values for fully dense bodies TD > 99% are shown in Fig. 5. The standard deviations of hardness and of fracture toughness are 25 kgf mm^{-2} and 0.20 Pa $m^{1/2}$ respectively. The hardness decreases with increasing temperature and with increasing Y_2O_3 content. This may seem contradictory to

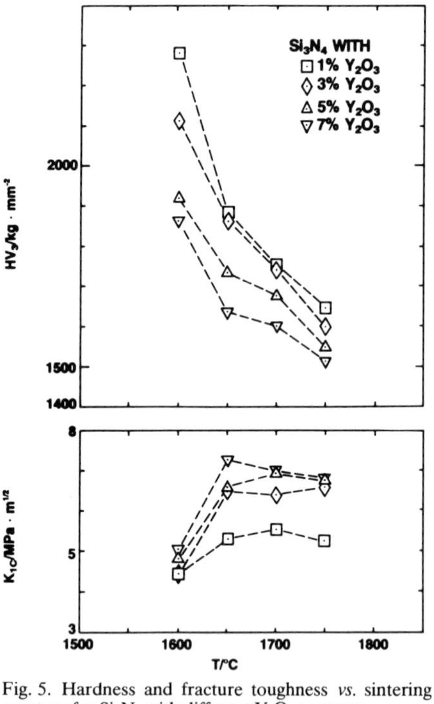

Fig. 5. Hardness and fracture toughness *vs.* sintering temperature for Si_3N_4 with different Y_2O_3 contents.

TABLE 2 Crystallographic properties

Y_2O_3 content (wt.%)	*Phases at 1550 °C*	*Phases at 1600 °C*	*Phases at 1650 °C*	*Phases at 1700 °C*	*Phases at 1750 °C*
1	α, β	α, β, Si_2N_2O	α, β, Si_2N_2O	α, β, Si_2N_2O	α, β, Si_2N_2O
3	α, β	α, β	α, β, Si_2N_2O	α, β, Si_2N_2O	α, β
5	α, β	α, β	α, β, Si_2N_2O	α, β	α, β
7	α, β, Ap	α, β, Ap	α, β	α, β	α, β

Ap ≡ Y-apatite.

Yeheskel *et al.* [4] who reported hardness as increasing with increasing Y_2O_3 content. However, the densities in their samples varied between 65% and 98% of TD and a figure with hardness *vs.* density might have been more appropriate in their case. In the present study, the hardness decreased rather smoothly with increasing sintering temperature T and Y_2O_3 content. However, a tendency of a more marked degree between 1600 and 1650 °C can be seen, this coinciding with a drastic decrease in α-Si_3N_4 content. According to Chakraborty and Mukerji [10] single-crystal investigations have shown that the hardness of α-Si_3N_4 is greater than that of β-Si_3N_4 which could also be expected for polycrystalline materials. The factors influencing hardness seem to be the grain size and volume fraction of the "glass phase" and, to some extent, the $\alpha/(\alpha+\beta)$ ratio.

The fracture toughness increases with temperature up to 1650 °C and it also increases with Y_2O_3 content. The major factor influencing fracture toughness is the grain morphology and, in particular, the grain shape. For samples sintered at 1650 °C or higher, the fracture toughness is high and there is a large amount of fibrous β-Si_3N_4 grains. For samples sintered at temperatures lower than 1650 °C, the fracture toughness is low and the samples contain mainly equiaxed α-Si_3N_4 particles. The effect of Y_2O_3 content is similar. The samples containing 1 wt.% Y_2O_3 have a low fracture toughness and small mostly equiaxed grains while the samples containing more Y_2O_3 have a high fracture toughness and larger more fibrous grains (higher aspect ratio). The major factor controlling grain size is Y_2O_3 content, the temperature having less effect. The effect on fracture toughness of grain size is in agreement with the results of Tani *et al.* [11] who found an increase in K_{IC} with grain size for gas-pressure-sintered Si_3N_4.

3.4. *Hot hardness*

Hot hardness was measured up to 1000 °C with a load of 0.5 kgf. For all samples the hardness decreases quite linearly with temperature up to 1000 °C. This phenomenon of a hardness decrease starting at low temperatures is significant for Si_3N_4-based ceramics. In addition to this system, it has been observed in CaO- and MgO-doped Si_3N_4 [7], in Al_2O_3- and Y_2O_3-doped Si_3N_4 [12], and for commercial Si_3N_4 cutting tools. The reason for this behaviour is not fully understood although a discussion of possible contributions has been presented by Ekström

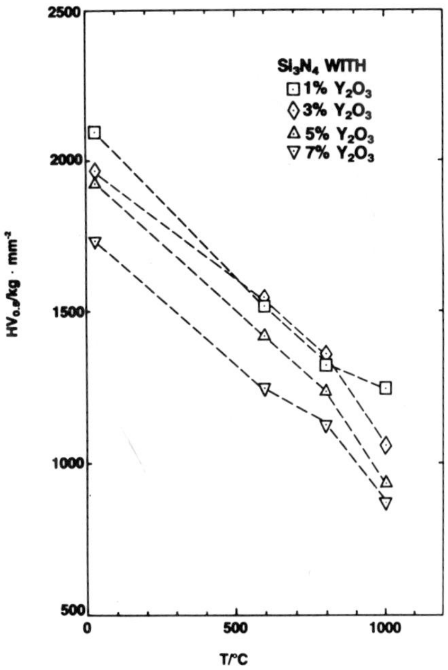

Fig. 6. Hot hardness of Si_3N_4 with different Y_2O_3 contents sintered at 1650 °C.

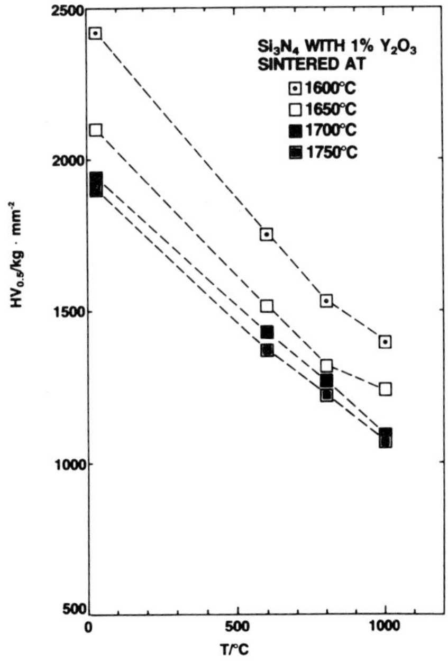

Fig. 7. Hot hardness for Si_3N_4 with 1 wt.% Y_2O_3 sintered at different temperatures.

and Ingelström [12]. The effects of Y_2O_3 content and sintering temperature on hot hardness (Figs. 6 and 7) are the same as for room temperature hardness. In summary, it can be said for temperatures up to 1000 °C that the higher the room temperature hardness, the higher is the hot hardness. It could be expected that high temperature hardness of samples with the same amount of Y_2O_3 prepared at different temperature should be similar [7]. The reason for this not being the case is probably that softening of the "glass phase" does not become substantial until the melting temperature is approached. In this system, this temperature is between 1300 and 1400 °C.

4. Conclusions

(i) Fully dense Si_3N_4 ceramic bodies can be obtained with Y_2O_3 additions of 1 wt.% or more, sintered at 1600–1650 °C or higher, using an HIP technique with glass encapsulation.

(ii) The effects of increasing sintering temperature are decreasing hardness, decreasing hot hardness, increasing toughness (up to 1650 °C), increasing β phase content, increasing grain size and increasing volume per cent of "glass phase".

(iii) The effects of increasing Y_2O_3 content are decreasing hardness, decreasing hot hardness, increasing toughness (up to 3 wt.% Y_2O_3), small effect on β phase content, increasing grain size and increasing volume per cent of "glass phase".

Acknowledgments

I thank SECO TOOLS AB for the permission to publish these results and Ms E. Andersen and Ms C. Pejryd for valuable technical assistance.

References

1 T. Yamada, M. Shimada and M. Koizumi, Fabrication of Si_3N_4 by hot isostatic pressing, *Am. Ceram. Soc. Bull., 60* (1) (1981) 1225–1228.

2 C. L. Quackenbush, J. T. Smith, J. T. Neil and K. W. French, Sintering, properties and fabrication of $Si_3N_4 + Y_2O_3$-based ceramics, in F. L. Riley (ed.), *Progress in Nitrogen Ceramics*, Martinus Nijhoff, The Hague, 1983, pp. 669–682.

3 J. Heinrich, N. Henn and M. Böhmer, The pressure dependence of microstructure and mechanical properties of hot isostatically pressed Si_3N_4, *Mater. Sci. Eng., 71* (1985) 131–136.

4 O. Yeheskel, Y. Gefan and M. Talianker, Properties and microstructure of hot isostatically pressed Si_3N_4, *Mater. Sci. Eng., 78* (1985) 209–216.

5 H. Larker, Dense ceramic parts hot pressed to shape by HIP, *Mater. Sci. Res., 17* (1984) 571–582.

6 T. Ekström and B. Aronsson, X-ray powder diffraction techniques in the studies of high performance ceramics, *Chem. Scr. A, 25* (1986) 33–35.

7 L. Pejryd, Microstructure and mechanical properties of CaO/MgO doped Si_3N_4 sintered by HIP, *Adv. Ceram. Mater., 3* (4) (1988) 403–405.

8 V. K. Sarin, On the α to β phase transformation in silicon nitride, *Proc. 3rd Int. Conf. on the Science of Hard Materials, Nassau, November 9–13, 1987*, in *Mater. Sci. Eng., A105–A106* (1988).

9 A. G. Evans and E. A. Charles, Fracture toughness determinations by indentation, *J. Am. Ceram. Soc., 59* (7–8) (1976) 371, 372.

10 D. Chakraborty and I. Mukerji, Characterization of silicon single crystals and polycrystalline reaction-sintered silicon nitride by micro-hardness measurements, *J. Mater. Sci., 45* (1980) 3051–3056.

11 E. Tani, S. Umebayashi, K. Kishi, K. Kobayashi and M. Nishijima, Effect of size of grains with fibre-like structure of Si_3N_4 on fracture toughness, *J. Mater. Sci. Lett., 4* (1985) 1454–1456.

12 T. Ekström and N. Ingelström, The relation between phase composition and hardness of different yttrium sialons, in T. Garvare (ed.), *Proc. Int. Conf. on HIP, Luleå, 1987*, Centek Publishers, Luleå, 1988.

Materials Science and Engineering, *A105/106* (1988) 175–182

Room Temperature Strength and Microstructure of Si_3N_4–Y_2O_3–ZrO_2–Al_2O_3 Ceramics*

K. HAYASHI

Institute of Industrial Science, The University of Tokyo, Tokyo (Japan)

A. YAMAKAWA

Sumitomo Electric Industries Ltd., Itami (Japan)

(Received November 8, 1987)

Abstract

Si_3N_4-(0-24)mol.%Y_2O_3-(0-27)mol.%ZrO_2-(4.0-4.6)mol.%Al_2O_3 compacts were sintered in nitrogen gas at 0.5 MPa and 2073 K for 7.2 ks. The room temperature mechanical properties, and in particular the transverse-rupture strength (TRS), were investigated in relation to the Y_2O_3 and ZrO_2 contents. It was found that the compact showed a high strength of about 0.7 MPa in the composition range of low Y_2O_3, low to high ZrO_2 or low to medium Y_2O_3 + ZrO_2 contents, and it showed a low strength of about 0.45–0.55 GPa in the other composition ranges. The dependences of both fracture toughness and hardness on the composition were nearly the same as that of the TRS. The low values of TRS in the above composition range were discussed in relation to the microstructure.

1. Introduction

Si_3N_4 ceramics have received considerable attention because of their applicability as wear-resistant materials as well as high temperature structural materials. For example, these ceramics are used as a guide-roll for the hot drawing of steel wire, cutting tools of cast iron, bearings and mechanical seals and vanes [1–3]. In order to prepare dense Si_3N_4 sintered compacts, oxides such as MgO [4–7], Al_2O_3, Y_2O_3 [8, 9] and/or ZrO_2 [10] are usually added as sintering aids. During sintering, these oxides form a liquid phase which flows to wet Si_3N_4 powder particles and then form a so-called grain boundary phase in the sintered compact. The thickness of the grain boundary phase is well known to be less than 0.2 μm. The size of the region where the composition and/or the crystallographic orientation of the grain boundary phase are/is uniform is also known to be very large (up to 1–2 mm) [11–13].

The optimum amount of the sintering aid such as MgO, MgO + Al_2O_3 and Al_2O_3 + Y_2O_3 to give the maximum room temperature strength of Si_3N_4 ceramics and the cause for the existence of such optimum amount have been studied by the present authors, taking into consideration various parameters such as the relative density of the sintered compact, the crystal type or grain shape of the Si_3N_4 phase, the type and microstructure of the grain boundary phase and the fracture source [11–14]. In this study, the strength of the Si_3N_4–Y_2O_3–ZrO_2–Al_2O_3 system was investigated from the same viewpoint.

2. Specimens and experimental procedures

Commercially available α-Si_3N_4 (Starck Co. Ltd.; average grain size, 1.2 μm), Y_2O_3 (0.5 μm) and ZrO_2 (0.95 μm) were used as starting powders. These powders were mixed by ball milling in an Al_2O_3 pot so that the composition of the mixed powder was Si_3N_4-(0-24)mol.%-Y_2O_3-(0-27)mol.%ZrO_2-(4.0-4.6)mol.%Al_2O_3, as shown in Fig. 1. The composition of the mixed powder was varied in the following three ways at a fixed Al_2O_3 content of about 4.3 mol.%.

(1) The Y_2O_3 content is varied at a fixed ZrO_2 content of about 7 mol.% (Fig. 1, specimens A–E).

(2) The ZrO_2 content is varied at a fixed Y_2O_3 content of 3.5 mol.% (specimens F–J).

*Paper presented at the 3rd International Conference on the Science of Hard Materials, Nassau, The Bahamas, November 9–13, 1987.

0921-5093/88/$3.50

© Elsevier Sequoia/Printed in The Netherlands

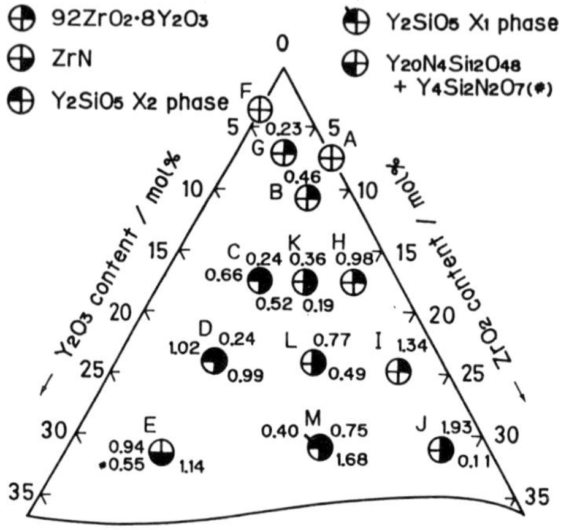

Fig. 1. Compositions of the specimens. The specimens always contain about 4.3 mol.% Al_2O_3 as well as Y_2O_3 and ZrO_2. The type of grain boundary and its relative X-ray intensity to β-Si_3N_4 (210) are also shown. The index or spacing of diffractive lattice plane is (311) for $92ZrO_2.8Y_2O_3$, (311) for ZrN, 0.1731 nm for Y_2SiO_5 (X_2 type), 0.2753 nm for Y_2SiO_5 (X_1 type), 0.1933 nm for $Y_{20}N_4Si_{12}O_{48}$ and 0.2828 nm for $Y_4Si_2N_2O_7$. (Co Kα X-rays were used.)

(3) The $Y_2O_3+ZrO_2$ content is varied at a nearly fixed ZrO-to-Y_2O_3 molar ratio of 1–2 (specimens G–M).

The mixed powder was compacted by die pressing. Then the compacts were sintered in nitrogen gas at 0.5 MPa and 2073 K for 7.2 ks. The room temperature mechanical properties (the transverse-rupture strength (TRS), the fracture toughness K_{Ic} and the Vickers hardness) and also the fracture surface and the optical or scanning electron microscopy (SEM) microstructures of the sintered compact were investigated as a function of Y_2O_3, ZrO_2 or $Y_2O_3+ZrO_2$ content. The surface of the sintered compact was ground with a diamond wheel of grade 250 for the TRS specimen. The specimens for K_{Ic}, hardness or microstructure studies were polished with a diamond wheel of grade 2000 after the grinding. The size of the specimen for TRS test was 3 mm × 4 mm × 35 mm (span, 30 mm; three-point method). K_{Ic} was measured using the indentation method [15].

3. Experimental results and discussion

Figure 2 shows the effect of the Y_2O_3, ZrO_2 or $Y_2O_3+ZrO_2$ content on TRS of Si_3N_4 ceramics. When the Y_2O_3 content increased at a fixed ZrO_2 content of 7 mol.%, the TRS increased and had a

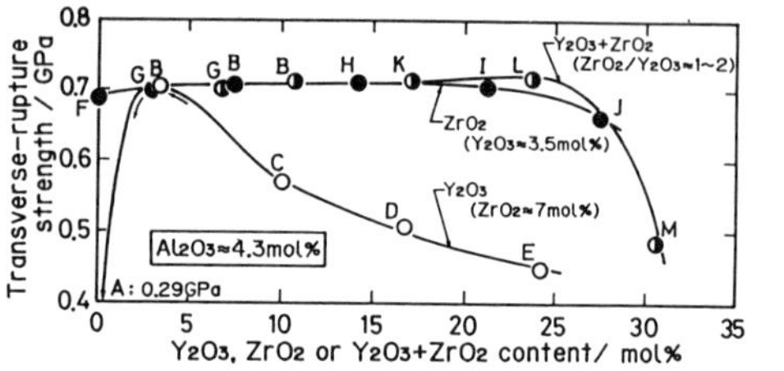

Fig. 2. TRS as a function of the type and amount of sintering aids.

maximum of about 0.7 GPa at about 4 mol.%; it then considerably decreased. In contrast, the TRS hardly varied with increasing ZrO_2 content up to 27 mol.% at a fixed Y_2O_3 content of 3.5 mol.%. When the $Y_2O_3+ZrO_2$ content increased at a fixed Y_2O_3-to-ZrO_2 molar ratio of 1–2, the TRS also scarcely varied up to about 25 mol.% but sharply decreased at about 30 mol.%.

The TRS is mainly affected by two factors, *i.e.* the size of the fracture source and K_{Ic}, as suggested by the formula for K_{Ic} [15]: $K_{Ic}=Y\sigma_m C^{1/2}$ where Y is the shape factor of the crack or fracture source, σ_m the fracture strength and C half the crack length of the crack or fracture source. The fracture source of the specimens was a region containing a few 2–4 μm micropores in the composition range where a high TRS was obtained, as shown in Figs. 3(a) and 3(a′) (Fig. 3(a′) is an enlarged micrograph of Fig. 3(a)), or an aggregate of Si_3N_4 particles. The sizes of both fracture sources were as small as about 15 μm. However, the fracture source in the composition range where a low TRS was obtained was a skeleton of Si_3N_4 particles among which the grain boundary phase disappears or continuous pores exist, as shown in Figs. 3(b) and 3(b′). The size of such skeleton regions was as large as about 50–100 μm. The value of K_{Ic} is shown in Fig. 4. The K_{Ic} dependence on the Y_2O_3, ZrO_2 or $Y_2O_3+ZrO_2$ content is quite similar to that of the TRS. Therefore, it can be said that the low TRS in the above composition range can be substantially attributed to a low K_{Ic}, although it is partly attributed to the large fracture source.

With respect to the relation between K_{Ic} or TRS for Si_3N_4 ceramics and the relative density or the shape of Si_3N_4 particles, K_{Ic} or TRS is generally said to increase with increasing relative density D_s [16] and/or with increasing amount of the needle-shaped Si_3N_4 particles produced by the α-to-β transformation [17]. Figure 5 shows D_s as a func-

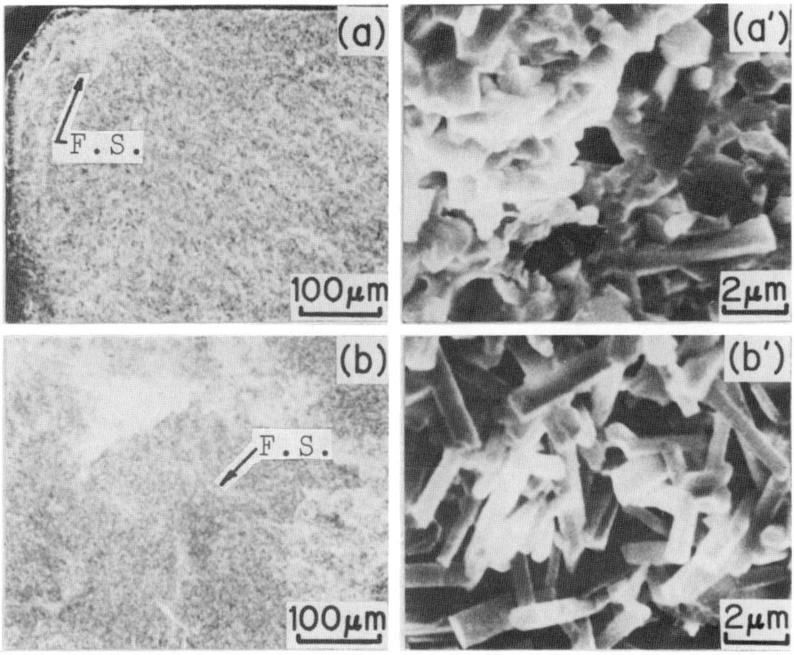

Fig. 3. SEM micrographs of the fracture surfaces of the compacts, showing a typical fracture source (FS): (a), (a') region containing micropores; (b), (b') Si_3N_4 particle skeleton having no grain boundary phase.

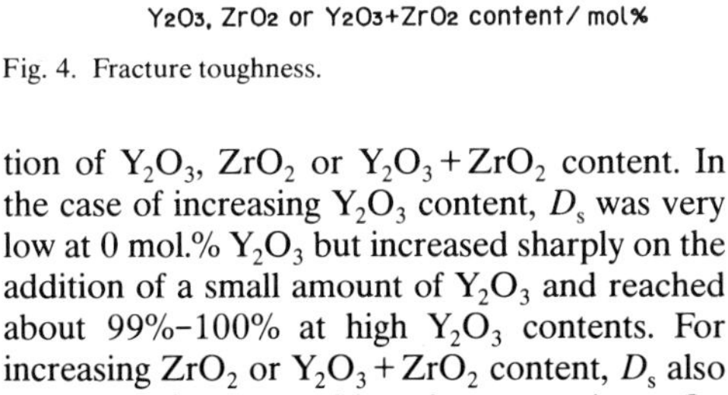

Fig. 4. Fracture toughness.

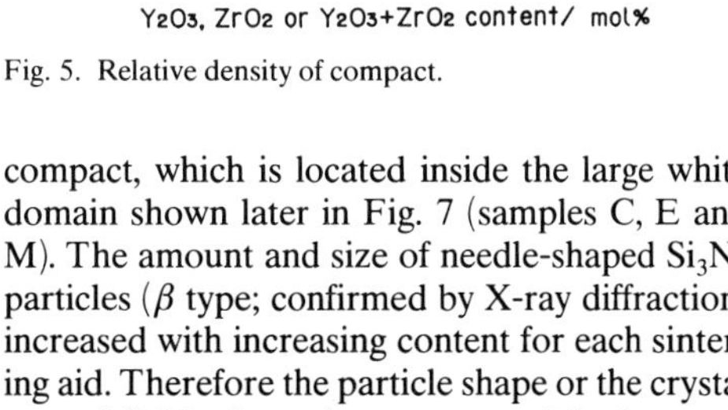

Fig. 5. Relative density of compact.

tion of Y_2O_3, ZrO_2 or $Y_2O_3+ZrO_2$ content. In the case of increasing Y_2O_3 content, D_s was very low at 0 mol.% Y_2O_3 but increased sharply on the addition of a small amount of Y_2O_3 and reached about 99%–100% at high Y_2O_3 contents. For increasing ZrO_2 or $Y_2O_3+ZrO_2$ content, D_s also tended to increase with a few exceptions. On comparison of Fig. 4 with Fig. 5, it is clear that K_{Ic} does not necessarily increase with increasing D_s. Therefore the dependence of K_{Ic} on the Y_2O_3, ZrO_2 or $Y_2O_3+ZrO_2$ content as mentioned above cannot be explained only in terms of D_s.

Figure 6 contains examples of SEM micrographs of the polished surface of the sintered compact, which is located inside the large white domain shown later in Fig. 7 (samples C, E and M). The amount and size of needle-shaped Si_3N_4 particles (β type; confirmed by X-ray diffraction) increased with increasing content for each sintering aid. Therefore the particle shape or the crystal type of Si_3N_4 phase also cannot explain the result for K_{Ic}.

The result for the type and relative amount of the crystalline grain boundary phase which were investigated by the X-ray diffraction method is shown in the previous composition diagram (Fig. 1). The crystalline grain boundary phase detected was $92ZrO_2.8Y_2O_3$ (cubic type) at small Y_2O_3

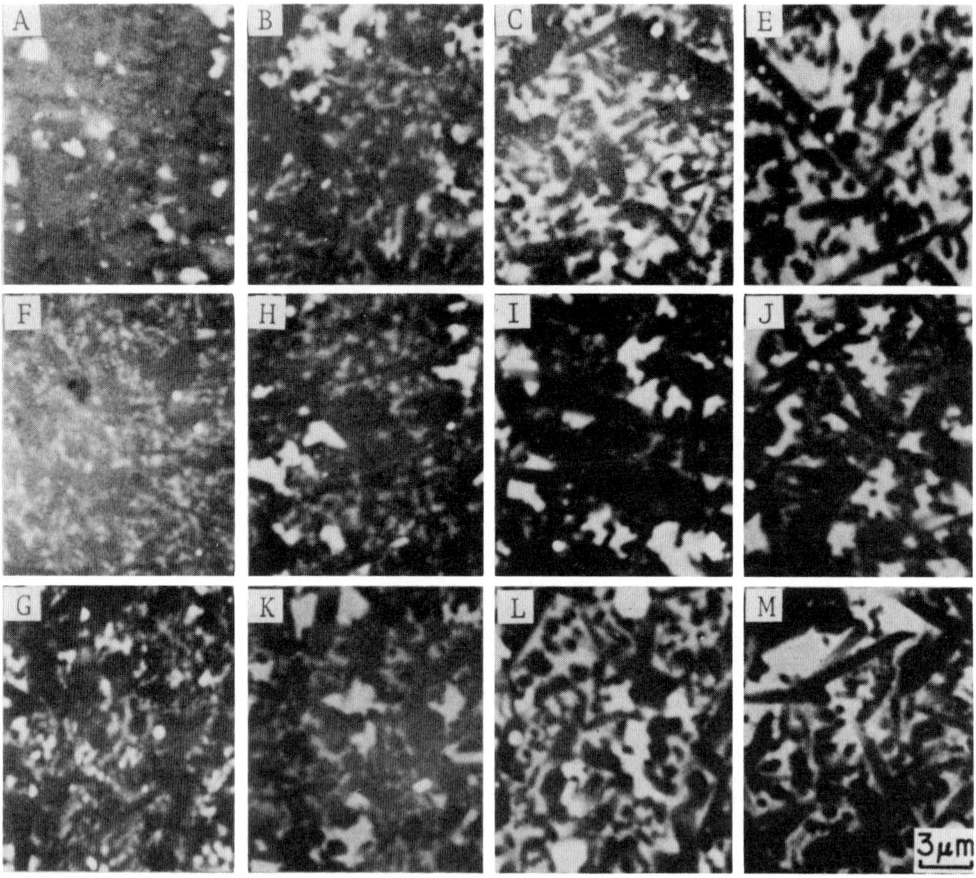

Fig. 6. SEM micrographs of the polished surfaces of compacts. The composition of each specimen is shown in Fig. 1. (Magnifications, 2500 ×.)

contents of about 3.5 mol.%. Y_2SiO_5 (X_2 type), $Y_{20}N_4Si_{12}O_{48}$, $Y_4Si_2N_2O_7$ and/or ZrN were detected as well as $92ZrO_2.8Y_2O_3$ at medium or high Y_2O_3 contents where the TRS sharply decreased. In contrast, the grain boundary phase was in general $92ZrO_2.8Y_2O_3$ at low to high ZrO_2 contents where the TRS was high, and its amount only increased with increasing ZrO_2 content. In the case of increasing $Y_2O_3 + ZrO_2$ content, the amount of both $92ZrO_2.8Y_2O_3$ and ZrN only increased up to medium contents but Y_2SiO_5 (X_1 type) also appeared at high contents where the TRS decreased.

Figure 7 consists of optical micrographs of the polished surface of the sintered compacts. The microstructures are generally very heterogeneous, owing to the appearance of many grain boundary phases. In the upper micrographs which show the effect of Y_2O_3 content on the microstructure at a fixed ZrO_2 content (specimens A–E), the white heterogeneous region is not observed in the specimen without Y_2O_3 (specimen A); however, a small white heterogeneous region of size 5–10 μm start to appear at small Y_2O_3 contents (sample B), and large white or black regions of 50–900 μm size are observed at medium to high Y_2O_3 contents (specimens C and E). The heterogeneous region was suggested to be the region where the chemical composition or the crystallographic orientation of the grain boundary phase is uniform or the region where the porosity level or the degree of crystallinity is different, taking into consideration our previous results for Si_3N_4–MgO, Si_3N_4–MgO–Al_2O_3 and

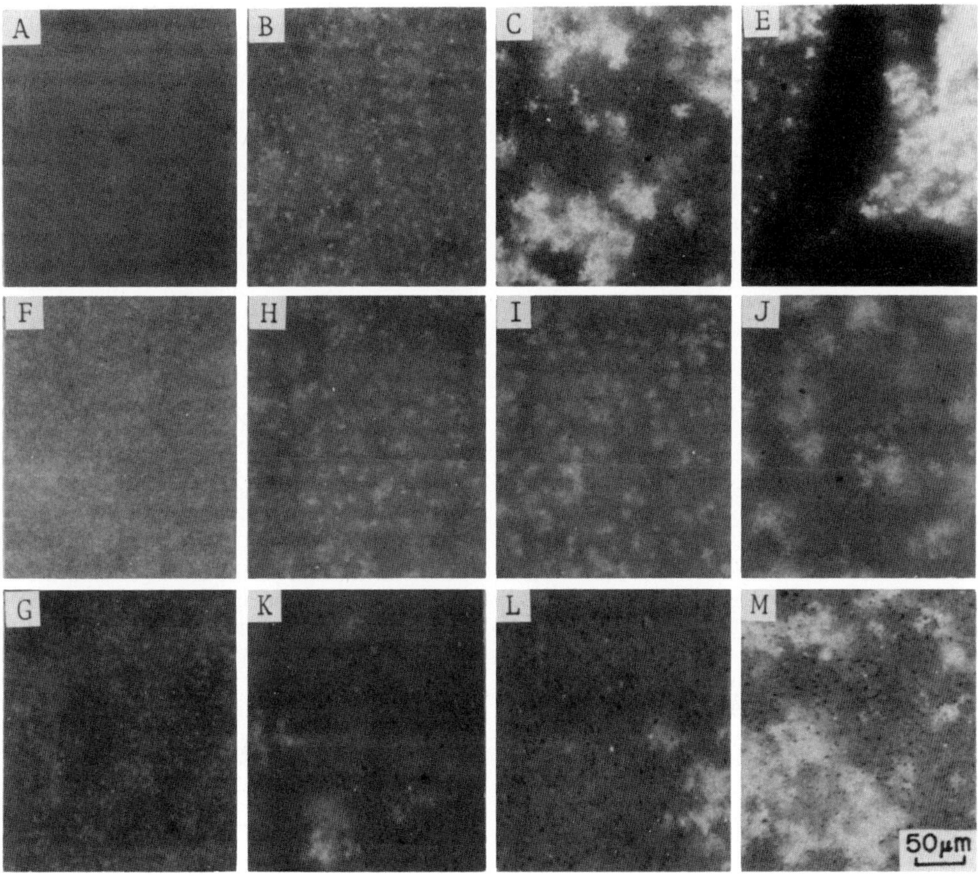

Fig. 7. Optical micrographs of the polished surfaces of compacts. (Magnifications, 150 ×.)

Si_3N_4–Al_2O_3–Y_2O_3 systems [12–14, 18] as well as those for WC–Co or TiC–Mo_2C–Ni cemented carbides [19, 20]. Hereafter, the heterogeneous region is called the domain in this paper. Also, small black particles (the actual colour is golden; the size is 3–5 μm) are observed at medium to high Y_2O_3 contents.

The large white or black domain observed in the optical micrograph could not be clearly observed in the SEM micrograph (Fig. 6). The reason is considered to be that the image of the optical micrograph is produced by light penetrating into the inside of the sintered compact and reflecting at the interface of the neighbouring domains owing to the difference in the indexes of refraction [12]. By comparing the area of each domain in Fig. 7 with the X-ray intensity of each grain boundary phase in Fig. 1, the types of grain boundary phase in each domain were suggested to be the following: the grain boundary phases of the white domain and the grey matrix at low Y_2O_3 contents (specimens A and B) are a glass phase and $92ZrO_2.8Y_2O_3$ respectively; the phase of the large white domain at medium Y_2O_3 contents (specimens C and D) is Y_2SiO_5 (X_2 type) and the phases of the white and black domains at high Y_2O_3 contents (specimen E) are $Y_{20}N_4Si_{12}O_{48}$ and $Y_4Si_2N_2O_7$ respectively. The black grains were considered to be ZrN.

For increasing ZrO_2 content, which is shown in the middle micrographs in Fig. 7, the white domain can be observed even at small contents and its size increases with increasing ZrO_2 content. However, it is noted that the size of the domain even at high ZrO_2 contents is much smaller than those at medium to high Y_2O_3 contents as mentioned above. When the fact that the melting point of $92ZrO_2.8Y_2O_3$ is as high as 2853 K is taken into account [21], the grain boundary phases of the white domain and the grey matrix were regarded to be glass and $92ZrO_2.8Y_2O_3$ respectively. In the lower micrographs (specimens G–M) showing the effect of $Y_2O_3+ZrO_2$ content, it is clear that large white domain appears at high $Y_2O_3+ZrO_2$ contents. By comparing the micrograph with Fig. 1, the grain boundary phase in the white domain was regarded as Y_2SiO_5 (X_1 type).

By correlating the results of the types of the grain boundary phase and the domain sizes with K_{Ic}, the following relations were established. K_{Ic} is always low irrespective of the type of the grain boundary phase for the specimens in which the large domains appear, and it is high for the specimens in which such large domains do not appear. Therefore, it could be said that K_{Ic} or TRS of the Si_3N_4–Y_2O_3–ZrO_2–Al_2O_3 ceramics is low when the size of the domain is large. We shall discuss this in the following.

Figure 8 shows the macrohardness (load, 98 N) of the sintered compact, and Table 1 shows the microhardness (load, 9.8 N) of the large domain. It can be seen that the macrohardness shows nearly the same composition dependence as those of TRS and K_{Ic} and that the microhardness of the large white domains, *i.e.* the microhardness of the grain boundary phase such as X_1-type and X_2-type Y_2SiO_5 (specimens C, D and M) and $Y_{20}N_4Si_{12}O_{48}$ (specimen E) is low. It is also known that the macrohardness of the specimens is low when the microhardness of the domain is low. The low macrohardness of the specimen A with a low Y_2O_3 or low ZrO_2 content, in which such a large domain of low hardness did not appear, can be attributed to the low relative density of the sintered compact.

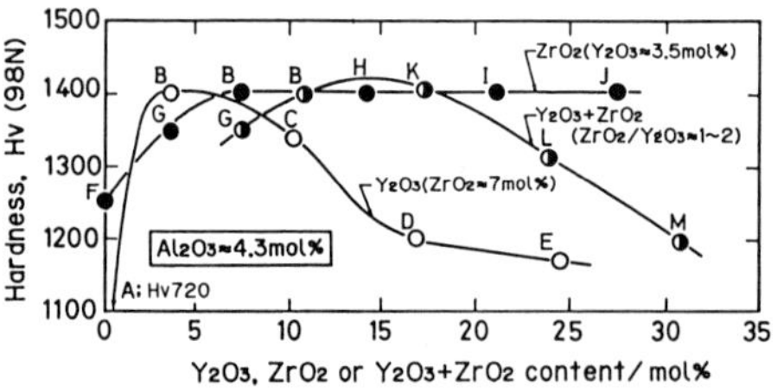

Fig. 8. Macrohardness of a compact (load, 98 N).

4. Discussion

As described above, TRS correlates with K_{Ic}, and K_{Ic} is strongly affected by the type of the grain boundary phase or the domain size. These results can be explained as follows. As mentioned above, K_{Ic} in this study was measured by using the H_V indentation method. As shown in Fig. 9, the length of cracks formed around the indentation increases with decreasing K_{Ic}. Figure 10 shows an example of the cracks around the indentation. Two or three cracks are formed in the large white domain; this differs from the other region in which a single crack is formed. The length of each crack in the large white domain is relatively short; however, the total length of the cracks was greater than that of the single crack

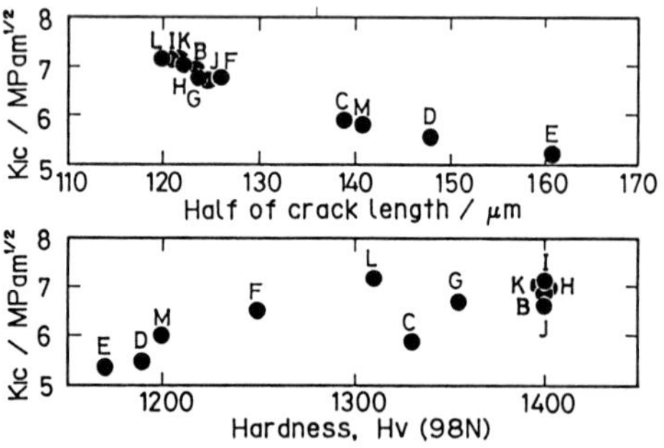

Fig. 9. Relation between K_{Ic} and half the crack length or macrohardness.

TABLE 1 Microhardness of the domain of the grain boundary phase (load, 9.8 N)

	Vickers hardness (HV 9.8) *for the following specimens*						
	B	*C*	*D*	*E*	*G*	*J*	*M*
Grey matrix	1580	1560	1510	1500	1540	1560	1670
White domain	1580	990	860	680	1540	1560	920
Black domain	—	—	—	1460	—	—	—

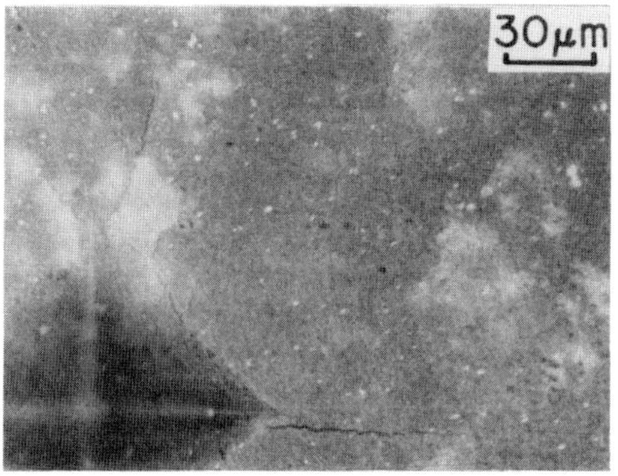

Fig. 10. Cracks formed near the Vickers hardness indentation.

formed in the other region. Thus, it is clear that the large white domain or the region containing the grain boundary phase of X_1-type Y_2SiO_5, X_2-type Y_2SiO_5 and/or $Y_{20}N_4Si_{12}O_{48}$ is brittle or allows cracks to form or propagate easily. Therefore, K_{Ic} or TRS for the specimens having such large domains was considered to be low. Here, the fact that cracks could easily form or propagate in the large domains is of course caused by the brittleness of the grain boundary phase, but it is also considered to be affected by the largeness of the domain. Also, it is already known that, for Si_3N_4–MgO–Al_2O_3 ceramics, the TRS decreases with increasing domain size of the same grain boundary phase because of decreasing cooling rate after sintering [18], and the thermal stress caused by the difference between the thermal coefficient of the different domains or the stress concentration caused by the difference in elastic modulus naturally becomes larger with increasing domain size.

With respect to the optimum amount of Y_2O_3 and ZrO_2 to give a maximum or high room temperature strength, it can be said the content should be such that the relative density of the sintered compact is as high as possible above 96%–98% and also that large domains do not appear. The same is true for the other sintering aids such as MgO, MgO + Al_2O_3, Al_2O_3 + Y_2O_3 and ZrO_2 + CaO + Al_2O_3 [12–14].

The types of grain boundary phase formed in the specimens can be explained as follows. The specimens are actually a five-component system of Si_3N_4–Y_2O_3–ZrO_2–Al_2O_3–SiO_2, because the Si_3N_4 powder contains SiO_2. The phase diagram of this five-component system has not been reported. However, Y_2SiO_5 (Y_2O_3.SiO_2), $Y_{20}N_4$-$Si_{12}O_{48}$ (10Y_2O_3.9SiO_2.Si_3N_4) and $Y_4Si_2N_2O_7$ (4Y_2O_3.SiO_2.Si_3N_4), which were detected at medium to high Y_2O_3 and low ZrO_2 contents in the specimens, appear in the Si_3N_4–Y_2O_3–SiO_2 three-component system [22], and 92ZrO_2.8Y_2O_3, which was detected at low Y_2O_3 and low to high ZrO_2 contents in the specimen, appears at low Y_2O_3 contents in the Y_2O_3–ZrO_2 two-component system [21]. The correlation between the appearance of these phases and the composition of the specimens, however, is not well understood. The appearance of various grain boundary phases as well as the formation mechanism of the domain or the heterogeneous region and the cause for the pore formation are still being investigated.

5. Summary

The effect of the composition on the room temperature mechanical properties of Si_3N_4–(0–24)mol.%Y_2O_3–(0–27)mol.%ZrO_2–(4.0–4.6)-mol.%Al_2O_3 ceramics was investigated in relation to the microstructure, and the following results were obtained.

(1) The effect of the Y_2O_3 content on the fracture strength (TRS) was quite different from the effects of the ZrO_2 and Y_2O_3 + ZrO_2 contents; the strength when the Y_2O_3 content is increased at a fixed ZrO_2 content was as low as about 0.45–0.55 GPa at medium to high Y_2O_3 contents, but it was as high as 0.7 GPa at low to high ZrO_2 or low to medium Y_2O_3 + ZrO_2 contents for increasing ZrO_2 or Y_2O_3 + ZrO_2 contents.

(2) The size of the fracture source which affects the fracture strength, together with the fracture toughness, was small at the composition at which the high strength was obtained, and it was large at the composition at which the low strength was obtained.

(3) The fracture toughness showed quite similar composition dependence to that of TRS. Therefore the low fracture strength was attributed substantially to the low fracture toughness and partly to the largeness of the fracture source.

(4) The relative density of the sintered compact, the crystal type or grain shape of the Si_3N_4 phase and the type of the crystalline grain boundary phase (92ZrO_2.8Y_2O_3, ZrN, $Y_4Si_2N_2O_7$, Y_2Si_5O or $Y_{20}N_4Si_{12}O_{48}$) could not necessarily explain the results on the strength or fracture toughness.

(5) The size of the heterogeneous region, in which the chemical composition or crystallographic orientation of the grain boundary phase was suggested to be uniform or where the porosity level or the degree of crystallinity was suggested to be different, was as small as 3–10 μm for the specimens of high fracture toughness and as large as 50–900 μm for the specimens of low fracture toughness. The microhardness of the large heterogeneous region was generally low compared with that of the matrix.

(6) To judge from the correlation of the fracture strength or toughness with the various parameters discussed above, the low value of the strength in the above composition was suggested to be related to the appearance of the large heterogeneous region. This region, however, should be further investigated in detail.

References

1 G. G. Deeley, J. M. Herbert and N. C. Moore, *Powder Metall., 8* (1961) 145–151.
2 K. H. Jack, *J. Mater. Sci., 11* (1976) 1135–1158.
3 R. N. Katz, in F. L. Riley (ed.), *Progress in Nitrogen Ceramics*, Martinus Nijhoff, Boston, MA, 1983, pp. 3-20.
4 G. E. Gazza, *J. Am. Ceram. Soc., 56* (1973) 662.
5 G. R. Terwilliger and F. F. Lange, *J. Am. Ceram. Soc., 57* (1974) 25–29.
6 D. R. Clark and G. Thomas, *J. Am. Ceram. Soc., 60* (1977) 491–495.
7 O. I. Krivanek, T. M. Shaw and G. Thomas, *J. Am. Ceram. Soc., 62* (1979) 585–590.
8 F. F. Lange, S. C. Singhal and R. C. Kuznicki, *J. Am. Ceram. Soc., 60* (1977) 249–252.
9 A. Tuge, H. Kudo and K. Komeya, *Ceram. Bull., 57* (1978) 424–431.
10 S. Dutta and B. Buzek, *J. Am. Ceram. Soc., 67* (1984) 89–92.
11 H. Suzuki, K. Hayashi, K. Tokumoto and Y. Onomura, *J. Jpn. Soc. Powder Powder Metall., 31* (1984) 14–19.
12 K. Hayashi and Y. Onomura, *J. Jpn. Soc. Powder Powder Metall., 31* (1984) 272–278.
13 K. Hayashi and M. Kosakai, *J. Jpn. Soc. Powder Powder Metall., 33* (1986) 210–215.
14 K. Hayashi and J. Ogiwara, *J. Jpn. Soc. Powder Powder Metall., 34* (1987) 175–180.
15 K. Nihara, *Ceramics (Jpn.), 20* (1985) 12–17.
16 G. Ziegler, in K. Somiya (ed.), *Ceramic Science and Technology at the Present and in the Future*, Uchida-Rokakuho-Tokyo, 1982, pp. 53–89.
17 L. J. Bowen and T. G. Carruthers, *J. Mater. Sci., 13* (1978) 684–671.
18 K. Hayashi and Y. Onomura, *J. Jpn. Soc. Powder Powder Metall., 32* (1985) 224–228.
19 H. Suzuki, K. Hayashi and Y. Fuke, *J. Jpn. Soc. Powder Powder Metall., 19* (1973) 346–353.
20 H. Suzuki, K. Hayashi and O. Terada, *J. Jpn. Soc. Powder Powder Metall., 21* (1974) 51–55.
21 P. S. Duwez, F. H. Brown, Jr. and F. Odell, *J. Electrochem. Soc., 98* (1951) 360–367.
22 R. R. Wills, S. Holmqist and J. M. Wimmer, *J. Mater. Sci., 11* (1976) 1307–1315.

Materials Science and Engineering, A105/106 (1988) 183–192

Pressureless-sintered Al_2O_3–TiC Composites*

RAYMOND A. CUTLER and ANDREW C. HURFORD

Ceramatec Inc., 2425 South 900 West, Salt Lake City, UT 84119 (U.S.A.)

ANIL V. VIRKAR

Department of Materials Science, University of Utah, Salt Lake City, UT 84112 (U.S.A.)

(Received November 9, 1987)

Abstract

Titanium carbide (TiC) (typically 30 wt.% TiC is used in cutting tools) dispersed in an alumina (Al_2O_3) matrix improves the hardness over that of an Al_2O_3 cutting tool, principally by limiting Al_2O_3 grain growth. Al_2O_3–TiC composites have been traditionally processed by intimately mixing the oxide and carbide powders and subsequently hot pressing at temperatures near 1700 °C and pressures of 35 MPa. Pressureless sintering of Al_2O_3–30wt.%TiC was accomplished by adding titanium hydride (TiH_2) and heating at moderately high rates (40–50 °C min^{-1}) in a graphite furnace under inert gas. Both free titanium and a eutectic liquid, formed by the carbothermal reduction of Al_2O_3, aided sintering. Conventional heating rates permit free titanium to combine with TiC at temperatures below the melting point of titanium and therefore do not take advantage of the liquid phase. Short sintering times above 1850 °C are necessary owing to the high vapor pressure of the phases produced by the carbothermal reduction of Al_2O_3.

A density greater than 99% of the theoretical density, a hardness in excess of 22 GPa and a four-point flexural strength higher than 600 MPa were measured on pressureless-sintered-and-hot-isostatically-pressed bars. These properties were comparable with those measured on composites made by hot pressing. There were no observable differences (microstructural or mechanical) between the hot-pressed and the pressureless-sintered parts. Machining tests showed that the performance of the pressureless-sintered composites was comparable with that of conventional tools.

It was also shown that Al_2O_3–30wt.%TiC powders synthesized by the aluminothermic reduction of titania in the presence of carbon could be pressureless sintered without TiH_2 additions. The mechanical properties of these composites were comparable with those of hot-pressed Al_2O_3–TiC composites.

1. Introduction

Ceramic cutting tools composed primarily of 2–5 μm alumina (Al_2O_3) have been used for over 30 years in finishing operations owing to their ability to run at higher speeds than cemented carbides [1]. A major improvement in ceramic cutting tools was a hot-pressed Al_2O_3-based composite containing 1–5 μm titanium carbide (TiC) uniformly dispersed throughout the matrix. Commercially available cutting tools are nominally Al_2O_3–30wt.%TiC and are electrically conducting owing to a continuous carbide phase which forms at about 25 vol.% TiC. The thermal conductivity of Al_2O_3–30wt.%TiC, however, is lower than that of either Al_2O_3 or TiC. The major advantage of adding TiC to Al_2O_3 is that the carbide limits the grain growth of the Al_2O_3 matrix and results in a higher strength and hardness. In addition, a small improvement in fracture toughness with increasing TiC content has been reported [2].

Recently, Borom and Lee [3] and Lee and Borom [4] demonstrated that heating rates in excess of 100 °C min^{-1}, between 1550 and approximately 1950 °C, allowed densification of Al_2O_3–30wt.%TiC composites to greater than 95% of the theoretical density. They were not successful, however, in obtaining high densities in graphite furnaces because of the reactions between Al_2O_3 and graphite, which inhibit densi-

*Paper presented at the 3rd International Conference on the Science of Hard Materials, Nassau, The Bahamas, November 9–13, 1987.

0921-5093/88/$3.50

© Elsevier Sequoia/Printed in The Netherlands

fication [3, 4]. Rapid heating permitted densification by kinetically limiting volatilization.

It is recognized that, as Al_2O_3 is carbothermically reduced to Al_4C_3, the oxycarbides Al_4O_4C and Al_2OC form [5, 6] and gaseous species such as Al_2O [6, 7] and CO [8], are produced. Huang *et al.* [9] showed that the eutectic liquid formed between Al_2O_3 and Al_4O_4C can be used to densify SiC–Al_2OC–AlN composites if volatilization is limited by containment. Recent work has shown that the Al_2O_3–Al_4O_4C eutectic liquid is transient and that clean grain boundaries and good mechanical properties at elevated temperatures can be obtained for SiC sintered using Al_2O_3 and Al_4C_3 [10].

In the present work, titanium hydride (TiH_2) was added to Al_2O_3 and TiC powders in order to form titanium as a transient liquid phase when sintering rates were moderately high (40–50 °C min^{-1}). Since densification to closed porosity did not occur until temperatures above 1800 °C, there is strong evidence that the Al_2O_3–Al_4O_4C eutectic also aids in sintering Al_2O_3–TiC composites. The objective of this paper is to communicate a method for pressureless sintering Al_2O_3–TiC composites which permits heating in graphite furnaces at rates on the order of 40–50 °C min^{-1}, thereby allowing continuous batch feeding through conventional equipment.

2. Experimental procedures

2.1. Titanium as a sintering aid

Al_2O_3 (A-16 Superground, Alcoa Aluminum Co., Pittsburgh, PA), TiC (Grade CA, Hermann, C. Starck Inc., New York, NY) and TiH_2 (TI 353, Atlantic Equipment Engineers, Bergenfield, NJ) were dispersed and milled in a stainless steel mill 150 mm in diameter (at a speed of 75 rev min^{-1}) with hexane for 24 h at a ratio of the ball (WC–6wt.%Co) to charge of 10 to 1. Larger batches were attrition milled in deionized water and a binder was added prior to spray drying. All powders were wet screened −325 mesh and dry screened −80 mesh prior to uniaxial pressing. Specimens were sintered on graphite plates in a graphite-resistance-heated laboratory batch furnace (model 1000-3560-FP20, Astro Industries, Santa Barbara, CA) or using continuous feeding through a graphite tube 100 mm in diameter in a horizontal furnace. Sintering was typically performed in flowing (1 l min^{-1}) argon gas (nitrogen (N_2) and helium (He) were alternatively used) at temperatures between 1860 and 1875 °C (at a partial pressure of oxygen of approximately 5×10^{-14} atm, fixed by the graphite in the furnace) for 10 min. The heating rate above 1000 °C was 40–50 °C min^{-1}. Hot isostatic pressing (HIP) was performed by stacking the parts on a graphite setter plate and heating for approximately 1 h while increasing the argon pressure, and holding at temperature for 15–30 min.

The density was determined by water displacement. The bend strength was measured on specimens (five or six for each mean value) approximately 4 mm × 5 mm × 50 mm (30 μm surface finish) using a support span of 40 mm and a loading span of 20 mm at a constant cross-head speed (0.5 mm min^{-1}) on a universal testing machine (model 1123, Instron Corporation, Canton, MA). The hardness and fracture toughness [11] were measured by indenting (five or six indents for each mean value) polished specimens with a Vickers diamond indenter at loads between 50 and 75 N.

2.2. Powders synthesized by exothermic reactions

Al_2O_3–TiC powders were synthesized by dispersing TiO_2 (P-25, Degussa Corporation, Teterboro, NJ), aluminum (123, Alcoa Aluminum Co., Pittsburgh, PA), carbon (Shawinigan Acetylene Black, Chevron Chemical Co., Houston, TX) and Al_2O_3. The mixture was then milled in hexane, dried and screened as described above. Powders were wrapped in graphite foil and individually reacted in a quartz tube by heating to approximately 660 °C, when the powders spontaneously reacted to form TiC and Al_2O_3, as identified by X-ray diffraction [12]. The reacted powders (with a surface area of about 5 m^2 g^{-1}) were ball milled as above, except without TiH_2 additions. The reacted powders were hot pressed at 1650 °C for 10 min in argon with a 35 MPa applied uniaxial pressure. The powders were alternatively pressureless sintered by heating to temperatures between 1875 and 1950 °C for 10 min, with HIP and characterization as described above.

3. Results and discussion

3.1. Titanium as a sintering aid

The Ti–C phase diagram [13] (Fig. 1) shows that TiC exists over a wide stoichiometry range, being always substoichiometric in carbon at low temperatures. Titanium and carbon both diffuse quickly through TiC, which has the NaCl structure. Using diffusion couples, Quinn and Kohlstedt [14] showed that the reaction between titanium and TiC, given by

$$TiC_x + nTi \rightarrow (1+n)TiC_y \qquad (1)$$

where $y = x/(1+n)$ and $0.47 < y < 0.97$, requires the diffusion of carbon to the Ti–TiC interface. The reaction interface advances in the direction of the titanium metal. They found that the diffusivity for carbon in TiC was 1.4×10^{-9} cm^2 s^{-1} at 1400 °C, which was an order of magnitude greater than lattice diffusivity [14]. Quinn and Kohlstedt observed the formation of Ti_2C platelets between the TiC and titanium, allowing short-circuit diffusion paths.

Regardless of the diffusion path taken, it is clear that titanium and TiC will interdiffuse, allowing no liquid phase above the eutectic temperature (approximately 1650 °C) unless heating rates are rapid. On the assumption of a diffusion distance x of 5 μm and a diffusion coefficient D of approximately 1×10^{-8} cm^2 s^{-1} at 1600 °C [14], the diffusion time t using the relationship

$$x \approx (Dt)^{1/2} \qquad (2)$$

is only 25 s. It is not surprising, therefore, that Quinn and Kohlstedt [15] observed the formation of a TiC skeleton, which prevented significant particle rearrangement, when they attempted to pressureless sinter TiC using titanium as a liquid phase. They concluded, however, that some particle rearrangement does take place at heating rates of 50 °C min^{-1}, although it was not enough to densify high TiC compositions [15]. They were able to hot press TiC–Ti mixtures containing between 10 and 30 vol.% metal to greater than 95% of the theoretical density at temperatures below 1600 °C, whereas TiC densified without the active metal had densities below 85% of the threoretical density [16]. The pressure applied during hot pressing enhanced densification, and hold times of 30 min were adequate to form substoichiometric TiC.

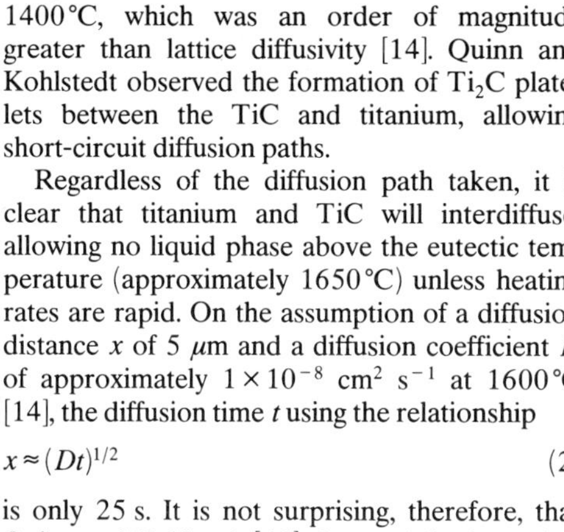

Fig. 1. The titanium-rich portion of the Ti–C phase diagram [13].

The potential disadvantage of using titanium as a transient liquid phase for aiding the pressureless sintering of Al_2O_3–TiC composites is the possible degradation of mechanical properties. Quinn and Kohlstedt [16] observed titanium at triple points and Ti_2C platelets within TiC grains. Williams [17] showed that the hardness of TiC single crystals decreased significantly with carbon substoichiometry. The critical resolved shear stress of TiC single crystals also decreased with decreasing carbon content [18].

In order to take a preliminary look at the properties of Al_2O_3–30wt.%TiC as a function of carbon stoichiometry, the four compositions listed in Table 1 were prepared using TiH_2 as the titanium source, since the brittle hydride can be intimately mixed with TiC and Al_2O_3 powders during milling. All four compositions were hot pressed (uniaxial pressure, 35 MPa) at 1700 °C for 10 min in N_2, resulting in similar microstructures. X-ray diffraction showed only TiC and Al_2O_3 peaks, although small amounts of TiN and TiO could be in solid solution with TiC. The density, hardness, strength and fracture toughness were all independent of carbon stoichiometry (see Table 1). Assuming a hardness of 27 GPa for

TABLE 1 Effect of carbon stoichiometry on hot-pressed Al_2O_3–30wt.%TiC

Carbon in TiC[a] (wt.%)	*Density* (g cm^{-3})	*Strength* (MPa)	*Hardness* (GPa)	*Fracture toughness* (MPa m$^{1/2}$)
20.0[b]	4.26	488 ± 51	19.9 ± 0.7	4.01 ± 0.46
19.0	4.27	518 ± 75	20.1 ± 0.2	4.25 ± 0.09
18.0	4.25	469 ± 104	20.0 ± 1.4	4.29 ± 0.20
17.0	4.26	595 ± 69	19.9 ± 0.9	4.26 ± 0.42

[a]All compositions were made neglecting the substoichiometry of starting TiC powder.
[b]Starting TiC contained 19.42% total carbon and 0.1% free carbon.

near-stoichiometric TiC and 22 GPa for $TiC_{0.85}$, based on single-crystal data [17], and taking the hardness of polycrystalline Al_2O_3 as 18 GPa, we would expect a decrease in hardness from 20.3 to 19.0 GPa with increasing carbon substoichiometry. The scatter in the hardness measurements preclude firm conclusions, but no trend in hardness with carbon content was observed, possibly owing to the formation of TiN in the N_2 environment.

None of the compositions was pressureless sintered to closed porosity below 1840 °C. In flowing N_2 at 1840 °C, the control composition (without titanium) sintered to 88% of theoretical density, while specimens with 3.7 wt.% TiH_2 (17.5 wt.% C (*i.e.* $TiC_{0.875}$)) sintered to 95% of the theoretical density and closed prorosity. Specimens sintered in N_2 showed a golden coating, indicating TiN formation, although no shift in lattice parameter was noted. Pressureless sintering to closed porosity in argon gas occurred at slightly higher temperatures (between 1860 and 1875 °C) with argon sintering generally giving slightly higher densities. It was not possible to sinter specimens without TiH_2 additions to closed porosity at any temperature using heating rates of 40–50 °C min^{-1} above approximately 1000 °C.

It is believed that the need to reach temperatures of approximately 1850 °C to ensure densification is due to a sintering boost given by the Al_2O_3–Al_4O_4C eutectic liquid (Fig. 2) since a small amount of free carbon exists in TiC powders. Sintering at temperatures in excess of approximately 1925 °C lowered the density of the

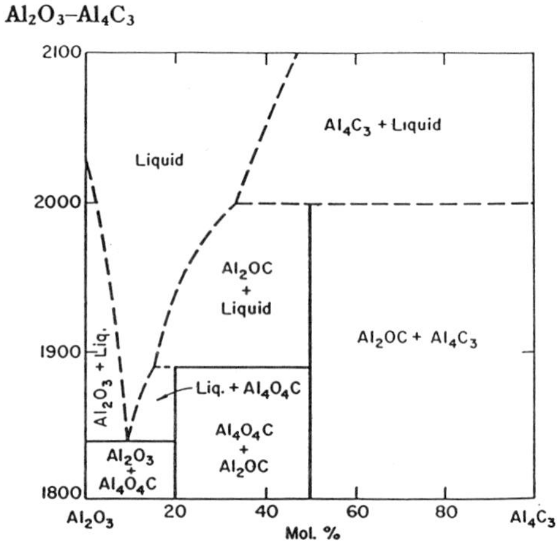

Fig. 2. The Al_2O_3–Al_4C_3 pseudobinary phase diagram [5].

TABLE 2 Effect of pressing pressure on densification of Al_2O_3–30wt.%$TiC_{0.875}$

Pressing pressure (MPa)	*Green density* (% of the theoretical density)	*Shrinkage* (%)	*Density* (% of the theoretical density)	
			Sintered[a]	*Hot isostatically pressed*[b]
35	54.0	16.9	92.3	95.3
70	56.2	16.2	93.9	96.5
105	57.7	15.8	95.6	99.3
135	58.5	15.7	96.5	99.8
205	59.5	15.3	97.3	99.8
275	60.2	14.9	97.0	99.8
345	60.3	14.5	97.0	99.8

[a]Pressureless sintered at 1875 °C for 10 min in argon.
[b]Hot isostatically pressed at 1600 °C and 200 MPa in argon.

TABLE 3 Comparison of properties of Al_2O_3–26.3wt.%TiC–3.7wt.%TiH_2 sintered in N_2 at 1840 °C and subsequently hot isostatically pressed in argon

HIP conditions	*Density* ($g\ cm^{-3}$)	*Density* (% of the theoretical density)	*Strength* (MPa)	*Hardness* (GPa)	*Toughness* ($MPa\ m^{1/2}$)
As sintered	4.22	97.5	415 ± 55	17.3 ± 0.2	4.3 ± 0.4
1300 °C, 100 MPa	4.24	97.9	555 ± 58	17.8 ± 0.4	4.2 ± 0.4
1600 °C, 100 MPa	4.33	100.0	681 ± 28	19.4 ± 0.5	4.1 ± 0.2
1700 °C, 200 MPa	4.31	99.5	717 ± 10	19.7 ± 0.2	4.2 ± 0.3

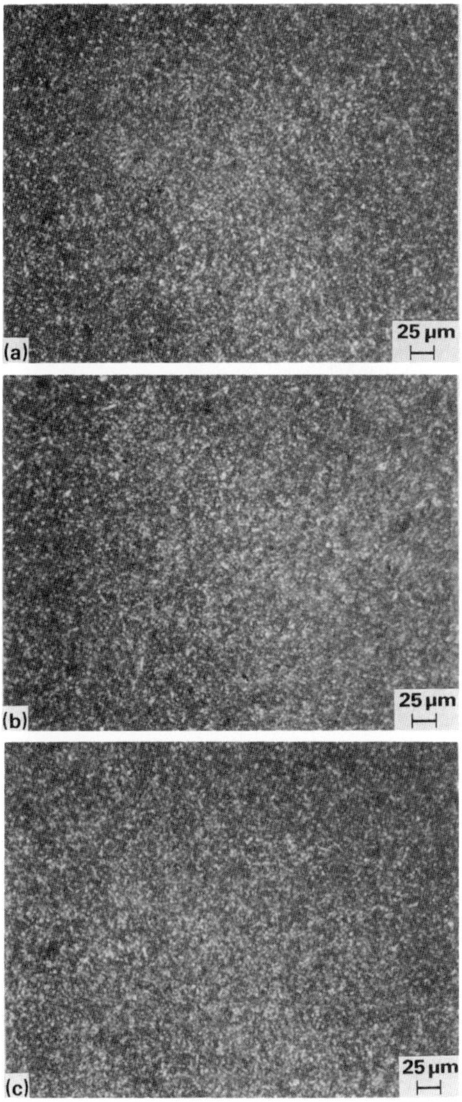

Fig. 3. Optical micrographs of Al_2O_3–26.3wt.%TiC–3.7wt.%TiH_2 sintered at 1840 °C in N_2: (a) as sintered; (b) sintered and subsequently hot isostatically pressed at 1300 °C and 100 MPa in argon; (c) sintered and subsequently hot isostatically pressed at 1600 °C and 100 MPa in argon.

composite owing to the reaction between Al_2O_3 and C. As in the work of Borom and Lee [3] and Lee and Borom [4], rapid heating limits the carbothermal reduction of Al_2O_3, which is significant above 1550 °C [19]. This is consistent with results in the SiC–Al_2OC system where a short time above the Al_2O_3–Al_4O_4C eutectic temperature aids sintering, but isothermal hold times are kept to a minimum to limit volatilization [9, 10]. The advantage of adding a titanium-forming species, such as TiH_2, is that it allows sintering in graphite furnaces. For specimens sintered in argon, Al_2O_3–30wt.%TiC powders without titanium additions sintered to a maximum density of 93% of the theoretical density, resulting in hardness values of 14 GPa, while titanium-doped powders (designed to give $TiC_{0.875}$) had typical densities of 97% of the theoretical density and hardness values of 17–19 GPa. More importantly, pressureless-sintered composites made with TiH_2 additions could be cladless hot isostatically pressed to densities greater than 99% of the theoretical density, while conventional powders typically showed only small increases in density (a maximum of 95% of the theoretical density) owing to open porosity in the sintered parts.

Powder pressing pressures in excess of 100 MPa were typically required to achieve closed porosity (Table 2) in pressureless-sintered parts made using titanium-doped powders. A cursory investigation of HIP conditions showed that little change occurred at 1300 °C but that 1600 °C was sufficient to ensure densification (Table 3). A Rockwell A hardness of hot isostatically pressed parts was typically higher than 95. As expected, HIP increased the density, resulting in improved hardness, and decreased the critical flaw size, leading to improved strength, but had no effect on the fracture toughness. Microstructures of composites after sintering (or post-sintering HIP at 1300 °C) still had closed porosity, while HIP at 1600 °C was sufficient to remove the porosity (Fig. 3).

Al_2O_3–26.3wt.%TiC–3.7wt.%TiH_2 powders prepared in aqueous slurries, spray dried, sintered in argon by heating to temperatures between 1860 and 1890 °C in a continuous-feed process and hot isostatically pressed at 1600 °C in argon had properties identical with those of the hot-pressed material (Table 4 and Fig. 4). It was not possible to distinguish between the pressureless-sintered-and-hot-isotatically-pressed compositions and the conventionally hot-pressed Al_2O_3–30wt.%TiC on the basis of X-ray diffraction or optical micrographs (Fig. 5).

The pressureless-sintered-and-hot-isostatically-pressed tools containing substoichiometric TiC made using the TiH_2 additive were tested by turning grey cast iron (Brinell hardness of 170

TABLE 4 Properties of Al_2O_3–26.3wt.%TiC–3.7wt.%TiH_2 sintered in argon

Description	*Density* (% of the theoretical density)	*Strength* (MPa)	*Hardness* (GPa)	*Toughness* ($MPa\ m^{1/2}$)
Sintered	94.9	480 ± 94	18.7 ± 0.6	4.35 ± 0.24
Sintered and hot isostatically pressed	99.5	638 ± 127	22.0 ± 1.1	4.42 ± 0.52
Hot pressed	98.6	583 ± 73	23.1 ± 0.7	4.53 ± 0.24

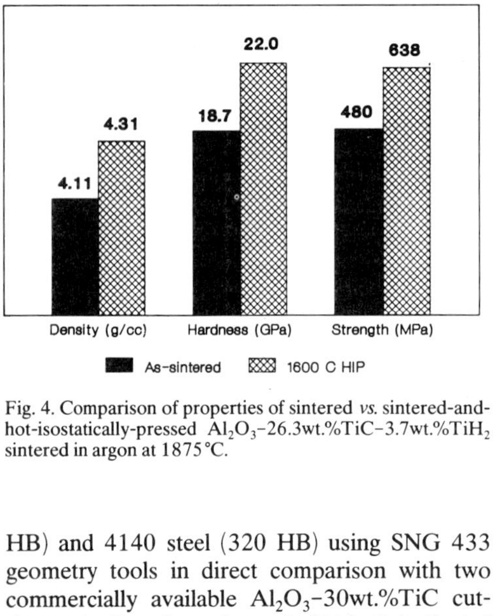

Fig. 4. Comparison of properties of sintered *vs.* sintered-and-hot-isostatically-pressed Al_2O_3–26.3wt.%TiC–3.7wt.%TiH_2 sintered in argon at 1875 °C.

HB) and 4140 steel (320 HB) using SNG 433 geometry tools in direct comparison with two commercially available Al_2O_3–30wt.%TiC cutting tools. Test data, shown in Table 5, indicated that the pressureless-sintered tools performed comparably with the hot-pressed conventional tools.

3.2. Powders synthesized by exothermic reactions

It is possible to make oxide–carbide composites using inexpensive starting materials and equipment via thermite-type reactions [12]. The reactions are strongly exothermic and the short reaction times result in powders with ultimate particle sizes near 0.3 μm with agglomerate sizes of 1–10 μm which are easily broken down in milling. Pressureless sintering of Al_2O_3–TiC powders prepared by thermite-type reactions was investigated since these powders probably contained small amounts of free carbon and titanium uniformly dispersed throughout the reacted powders due to short reaction times [12]. Three powders were prepared using the reaction

$$3TiO_2 + 4Al + 3C + xAl_2O_3 \rightarrow 3TiC + (2 + x)Al_2O_3 \qquad (3)$$

with $x = 0.00$ (no Al_2O_3 filler, 41.4 vol.% TiC and 46.9 wt.% TiC), $x = 0.65$ (35.0 vol.% TiC and 40.0 wt.% TiC) and $x = 2.11$ (25.7 vol.% TiC and 30.0 wt.% TiC). Higher dilutions were attempted in order to make 25 and 20 wt.% TiC powders, but no visible reactions were observed. The reaction initiation temperatures were at the melting point of aluminum (660 °C). The calculated reaction temperatures are at the melting point of Al_2O_3 (2055 °C) for $x = 0.00$ and $x = 0.65$ and at 1590 °C for $x = 2.11$. Free titanium and carbon were below the detectability limits of X-ray diffraction where only TiC and Al_2O_3 phases were observed.

The reacted powders were hot pressed, resulting in densities ranging from 4.25 g cm^{-3} for 30 wt.% TiC to 4.54 g cm^{-3} for 47 wt.% TiC. The strengths (600–750 MPa), hardnesses (19–23 GPa) and indentation fracture toughnesses (3.5–5.5 MPa $m^{1/2}$) were equal to or exceeding those of samples made from conventionally prepared powder (TiC and Al_2O_3 ball milled prior to hot pressing). The strength and fracture toughness of exothermically reacted composites increased with increasing TiC content, whereas the hardness decreased with increasing TiC content. The excellent hardness and strength of the samples made from exothermically reacted powders was due to their fine grain size, typically 1–3 μm in diameter.

The major advantage of the exothermically reacted powders is their sinterability. The sintering range of these powders is between 1875 and 1950 °C. The sinterability is believed to be due to three factors: (1) the fine particle size of the powders after the reaction (typically 0.1–0.5 μm) and the relative ease with which agglomerates are milled, making the powder reactive; (2) free titanium (below detection by X-ray diffraction)

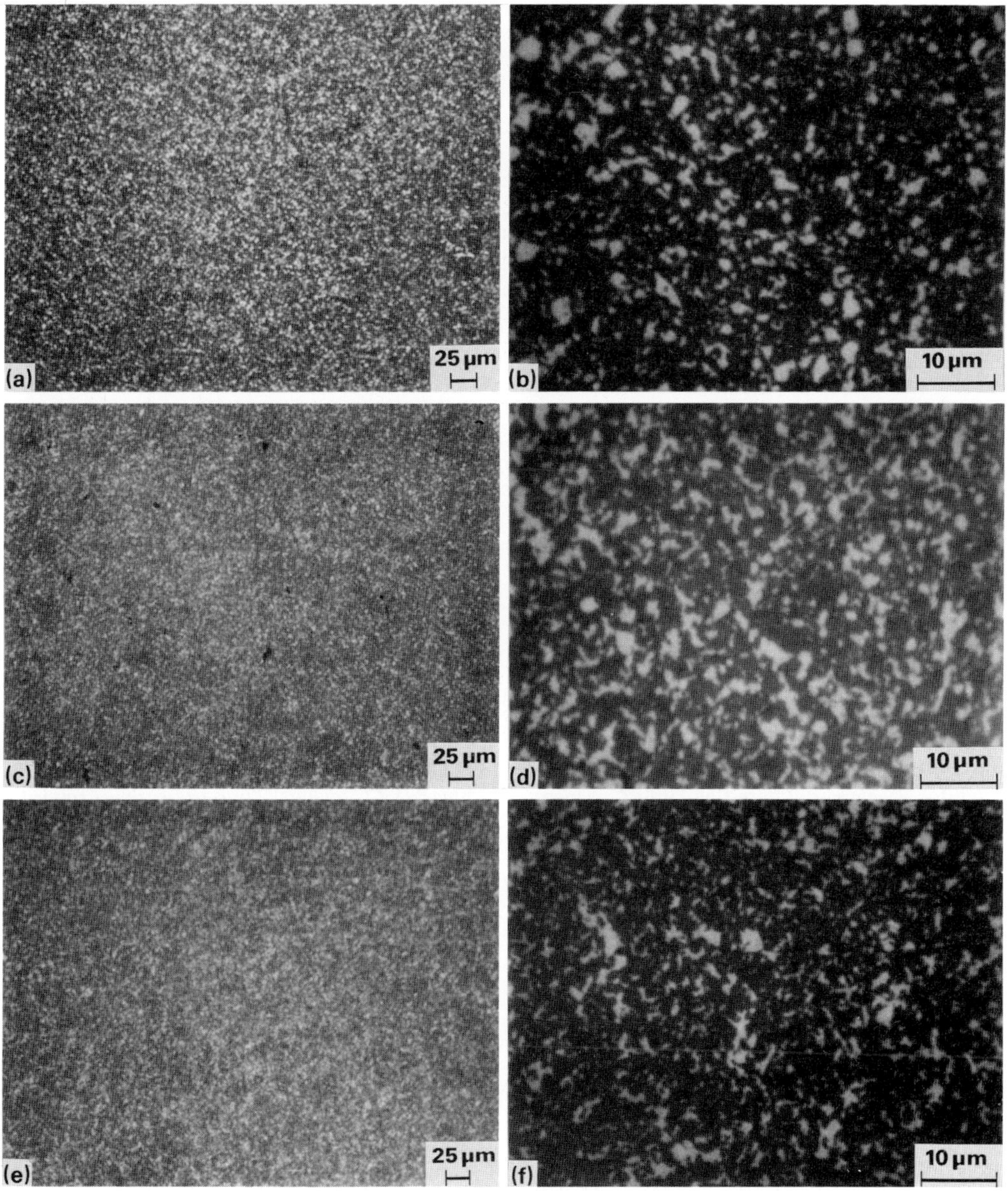

Fig. 5. Optical micrographs of Al_2O_3–30wt.%TiC composites: (a), (b) conventionally milled and hot-pressed Al_2O_3–30wt.%$TiC_{0.97}$; (c), (d) pressureless-sintered Al_2O_3–30wt.%$TiC_{0.875}$; (e), (f) pressureless-sintered-and-hot-isostatically-pressed Al_2O_3–30wt.%$TiC_{0.875}$.

which acts as a transient liquid phase above 1650 °C; (3) free carbon which reduces some Al_2O_3 to form a transient eutectic liquid above 1850 °C. Relatively rapid (above 30 °C min^{-1} above 1000 °C) heating is important for the reasons discussed above.

As expected, the shrinkage and sintered density are dependent on particle packing. The green

TABLE 5 Machining data on cast iron and steel

Brinell hardness (HB)	*Ceramic composition* (wt.%)	*Test material*	*Speed*[a] (m min^{-1})	*Flank wear*	
				Depth (mm)	*Time at which measurement taken* (min)
170	Al_2O_3–26.3TiC–3.7TiH_2[b]	Cast iron	305	0.305	12
170	Al_2O_3–30TiC (commercial material A)	Cast iron	305	0.279	12
170	Al_2O_3–30TiC (commercial material B)	Cast iron	305	0.318	12
320	Al_2O_3–26.3TiC–3.7TiH_2[b]	4140 steel	490	0.457	6
320	Al_2O_3–30TiC (commercial material A)	4140 steel	490	—[c]	—[c]
320	Al_2O_3–30TiC (commercial material B)	4140 steel	490	0.457	6

[a]All tests performed at a feed rate of 0.20 mm rev^{-1} and depth of cut of 0.635 mm.
[b]Composition prior to sintering in argon.
[c]Chipped out after 5 min (similar flank wear to other two materials up to 5 min).

TABLE 6 Properties of pressureless-sintered Al_2O_3–30wt.%TiC composites made by reaction (3) and sintered in argon

Sintering conditions	*Density* (g cm^{-3})	*Strength* (MPa)	*Hardness* (GPa)	*Fracture toughness* (MPa m$^{1/2}$)
Sintered at 1875 °C	4.23	387 ± 80	22.0 ± 1.3	4.4 ± 0.3
Sintered and hot isostatically pressed	4.25	548 ± 113	22.8 ± 0.9	4.9 ± 0.8

density increased from 54.7% to 60.0% of the theoretical density as the uniaxial pressing pressure increased from 35 MPa to 375 MPa respectively for specimens sintered at 1900 °C. Linear shrinkage decreased from 17.8% to 15.8%, as the green density increased from 54.7% to 60.0% of the theoretical density, and the sintered density increased from 93.3% to 98.7% of the theoretical density over the same range. Pressing pressures of 135 MPa were required to reach closed porosity at temperatures below 1925 °C. Table 6 gives the properties of sintered-and-hot-isostatically-pressed Al_2O_3–30wt.%TiC, showing that strength, hardness and toughness (bulk fracture toughness measurements are required before it can be concluded that sintered-and-hot-isostatically-pressed specimens have a higher toughness than do the hot-pressed composites) are equivalent between hot-pressed and sintered-and-hot-isostatically-pressed composites. Figure 6 shows micrographs of pressureless-sintered Al_2O_3–30wt.%TiC composites, as well as similar cutting tools which were subsequently containerless hot isostatically pressed. Cutting tools with properties as shown in Table 6 have been sent to commercial cutting tool manufacturers for evaluation.

4. Conclusions

The addition of 3–4 wt.% TiH_2 to Al_2O_3–30wt.%TiC powders allows pressureless sintering to closed porosity at moderate heating rates (40–50 °C min^{-1}) in graphite furnaces. The titanium permits the attainment of closed porosity at densities greater than about 94% of the theoretical density, thereby permitting cladless HIP. Commercial powders sintered without the titanium additive had densities ranging from 88% to 93% of the theoretical density with significant open porosity. It is believed that, during rapid heating, the titanium permits a small amount of liquid phase to aid densification. The liquid phase reacts with TiC to form substoichiometric TiC. Temperatures in excess of 1850 °C were typically required to densify the oxide–carbide composites. A eutectic between Al_2O_3 and Al_4O_4C corresponds to this densification temperature. Sintering above 1925 °C resulted in volatilization of Al_2O_3 due to a reaction with carbon. Physical and mechanical property data for pressureless-sintered-and-hot-isostatically-pressed composites were comparable with commercially available hot-pressed Al_2O_3–30wt.%TiC composites. The machinability of cast iron and steel

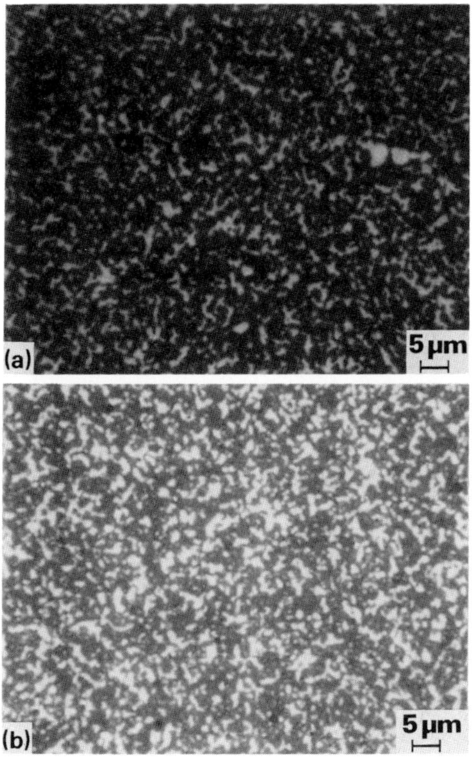

Fig. 6. Optical micrographs of Al_2O_3–30wt.%TiC made via reaction (3): (a) as sintered; (b) sintered and hot isostatically pressed.

using pressureless-sintered composites made with TiH_2 additions was comparable with that of commercially available tools.

Al_2O_3–30wt.%TiC powders were made by aluminothermically reducing titania in the presence of carbon and Al_2O_3 filler at 660 °C. The resulting powder was wet milled and pressureless sintered to closed porosity without TiH_2 additions. The physical properties were similar to those obtained using the titanium addition.

It is believed that small amounts of free titanium and carbon in the exothermically reacted powder preclude the need to add TiH_2 in order to achieve closed porosity. The high heating rates (greater than 100 °C min^{-1}) used by Borom and Lee [3] and Lee and Borom [4] may have taken advantage of small amounts of free titanium and carbon in commercial materials, as well as limiting volatilization. TiH_2 additions to commercially available Al_2O_3 and TiC powders or the synthesis of Al_2O_3–TiC powder by exothermic reactions result in powders which can be sintered to closed porosity within graphite furnaces at moderately high heating rates (40–50 °C min^{-1}).

Acknowledgments

The work on exothermic reactions was supported by Lawrence Livermore National Laboratory under Subcontract 5839605.

Kevin Rigtrup of Ceramatec helped with powder synthesis and characterization. Dr. Ron Peters of Fansteel performed machining tests.

References

1 A. G. King, Ceramics for cutting metals, *Bull. Am. Ceram. Soc., 43* (5) (1965) 395–401.
2 R. P. Wahi and B. Ilschner, Fracture behavior of composites based on Al_2O_3–TiC, *J. Mater. Sci., 15* (1980) 875–885.
3 M. P. Borom and M. Lee, Effect of heating rate on densification of alumina–titanium carbide compositions, *Adv. Ceram. Mater., 1* (4) (1986) 335–340.
4 M. Lee and M. P. Borom, Rapid rate sintering of Al_2O_3–TiC composites for cutting-tool applications, *Adv. Ceram. Mater., 3* (1) (1988) 38–44.
5 L. M. Foster, G. Long and M. S. Hunter, Reactions between aluminum oxide and carbon, *J. Am. Ceram. Soc., 39* (1) (1956) 1–11.
6 W. L. Worrell, Carbothermal reduction of alumina, *Can. Metall. Q., 4* (1) (1965) 870–895.
7 H. Yanagida and F. A. Kroger, The system Al–O, *J. Am. Ceram. Soc., 51* (12) (1968) 700–706.
8 J. H. Cox and L. M. Pidgeon, X-ray diffraction patterns of Al_4C_3 and Al_4O_4C, *Can. J. Chem., 41* (6) (1963) 1414–1416.
9 J. L. Huang, A. C. Hurford, R. A. Cutler and A. V. Virkar, Sintering behavior and properties of SiCAlON ceramics, *J. Mater. Sci., 21* (1986) 1448–1456.
10 T. B. Jackson, A. C. Hurford, S. L. Bruner and R. A. Cutler, SiC-based ceramics with improved strength, in J. W. Cawley (ed.), *Silicon Carbide,* American Ceramics Society, Columbus, OH, in the press.
11 G. R. Anstis, P. Chantikul, B. R. Lawn and D. B. Marshall, A critical evaluation of indentation techniques for measuring fracture toughness, *J. Am. Ceram. Soc., 64* (9) (1981) 533–543.
12 R. A. Cutler, A. V. Virkar and J. B. Holt, Synthesis and densification of oxide–carbide composites, *Ceram. Eng. Sci. Proc., 6* (7–8) (1985) 715–728.
13 E. K. Storms, *Refractory Materials,* Vol. 2, *The Refractory Carbides,* Academic Press, New York, 1967, p. 3.
14 C. J. Quinn and D. L. Kohlstedt, Solid state reaction between titanium carbide and titanium metal, *J. Am. Ceram. Soc., 67* (5) (1984) 305–310.
15 C. J. Quinn and D. L. Kohlstedt, Reactive processing of titanium carbide with titanium, Part 1: Liquid phase sintering, *J. Mater. Sci., 19* (1984) 1229–1241.

16 C. J. Quinn and D. L. Kohlstedt, Reactive processing of titanium carbide with titanium, Part 2: Solid state hot pressing, *J. Mater. Sci., 19* (1984) 1242–1250.
17 W. S. Williams, *Brittle–Ductile Behavior of the Transition Metal Carbides, Propriétés Thermodynamiques, Physiques et Structurales des Dérivés Semi-métalliques*, Centre National de la Recherche Scientifique, Paris, 1967, pp. 181–189.
18 W. S. Williams, Influence of temperature, strain rate, surface condition and composition on the plasticity of transition-metal carbide crystals, *J. Appl. Phys., 35* (1964) 1329.
19 F. J. Klug, W. D. Pasco and M. P. Borom, Microstructural development of aluminum oxide: graphite mixture during carbothermic reduction, *J. Am. Ceram. Soc., 65* (12) (1982) 619–624.

Fracture Behaviour of Selected Mixed-ceramic Tool Materials up to 1200 °C*

GUNNAR BRANDT

AB Sandvik Coromant, S-126 12 Stockholm (Sweden)

BENGT JOHANNESSON and RICHARD WARREN

Chalmers University of Technology, S-412 96 Göteborg (Sweden)

(Received November 8, 1987)

Abstract

The fracture behaviour of selected mixed-ceramic tool materials up to 1200 °C has been studied. The materials examined were Al_2O_3–4wt.%ZrO_2, Al_2O_3–30wt.%Ti(N, C), β-(Si–Al–O–N). Fracture studies included measurement of load–bending displacement behaviour in four-point bend of rectangular bars pre-cracked using bridge indentation. The resulting fracture surfaces were characterized fractographically by scanning electron microscopy. The transformation state of the fracture surface of the ZrO_2-containing ceramic was examined by X-ray diffraction. At displacement rates of 60 and 600 μm min^{-1}, all materials exhibited linear elastic behaviour up to 1000 °C. The room temperature K_{Ic} values were 4.2 MN m$^{-3/2}$, 3.5 MN m$^{-3/2}$, 4.5 MN m$^{-3/2}$ and 4.3 MN m$^{-3/2}$ for the four materials respectively and these did not change significantly up to 1000 °C. At 1200 °C, subcritical crack growth was observed. The β'-(Si–Al–O–N) exhibited an increase in toughness, whereas a slight decrease was observed for the other materials tested.

1. Introduction

In the past decade, considerable progress has been made in the development of advanced engineering ceramics through the use of the principles of alloying, *i.e.* the judicious combination of various refractory compounds to enhance mechanical properties. A number of such ceramic "alloys" have found application as tool materials and, in particular, for metal-cutting operations. The multiplicity of tool ceramics that have been developed largely reflects the complexity of metal cutting, each grade having been optimized for a relatively restricted area of application. The tool life is usually determined by one dominating wear mechanism operating among many simultaneously. The controlling wear process can be of a mechanical, thermomechanical, chemical or physical nature depending on the conditions of cutting (*e.g.* fast or slow, intermittent or continuous) and the compositions of the tool and work-piece materials. It can be readily appreciated that the mechanisms of wear and the tool wear resistance will normally be dependent on a complex combination of properties such as hardness, thermal conductivity, thermal expansion, chemical inertness as well as mechanical strength. Similarly, metal-cutting performance can rarely be predicted on the basis of an isolated material parameter such as fracture toughness. Thus, although fracture toughness and high temperature fracture behaviour (the subjects of this work) are not expected alone to be decisive for tool life, they can be expected to play a part in combination with other properties. It is not within the scope of this work to give a detailed account of specific examples of tool failure; however, two relevant examples will be mentioned below (see Section 4).

In this work, the fracture behaviour and fracture toughness of four commercially developed materials have been investigated between room temperature and 1200 °C. These were as follows: Al_2O_3 with 4 wt.% ZrO_2 added to give some toughening; Al_2O_3 combined with 30 wt.% Ti(N, C); a material consisting of β-(Si–Al–O–N) grains

*Paper presented at the 3rd International Conference on the Science of Hard Materials, Nassau, The Bahamas, November 9–13, 1987.

0921-5093/88/$3.50 © Elsevier Sequoia/Printed in The Netherlands

in a glassy phase intended to aid sintering; a mixed (α–β)-(Si–Al–O–N) also with additions forming a glassy phase. Apart from establishing the fracture toughness of these materials, the purpose of this work was to study the influence of the various phases on the modes of fracture. Since the number of samples available was limited, this survey of the behaviour should be considered as preliminary.

2. Experimental details

2.1. *Materials*

The properties of the four materials are given in Table 1 and their microstructures as determined by scanning electron microscopy (SEM) are shown in Figs. 1–4. Briefly they are as follows (these materials were produced by Sandvik AB, Stockholm, and have the following company designations respectively: CC620; CC650; 980; CC680).

2.1.1. Al_2O_3–4wt.%ZrO_2

Since it has a very fine grain size, most of the ZrO_2 in this material is in the metastable tetragonal form. The main purpose of the ZrO_2 is to provide some toughening through transformation to the monoclinic form.

2.1.2. Al_2O_3–30wt.%Ti(N, C)

The Ti(N, C) content was formed by the addition of 23 wt.% TiN and 7 wt.% TiC. It forms a second phase and is intended to improve hardness and thermal conductivity relative to pure Al_2O_3.

2.1.3. β-Si–Al–O–N

This material was formed from Si_3N_4 with additions of alumina and yttria as sintering aids. The microstructure that develops consists of β-(Si–Al–O–N) grains interspersed with about 15 vol.% Y-Si–Al–O–N glass. The β-(Si–Al–O–N) has an Si_3N_4 structure; its composition was determined via lattice parameter measurements as

$$Si_{5.74}Al_{0.26}O_{0.26}N_{7.74}$$

The ratio of the metals Si:Al:Y in the glass was determined in a similar material to be 46:32:21.

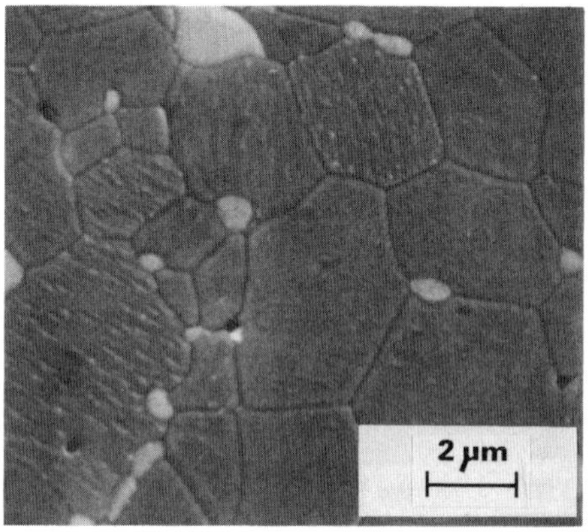

Fig. 1. SEM micrograph of Al_2O_3–4wt.%ZrO_2; the ZrO_2 is the light phase (thermally etched in H_2).

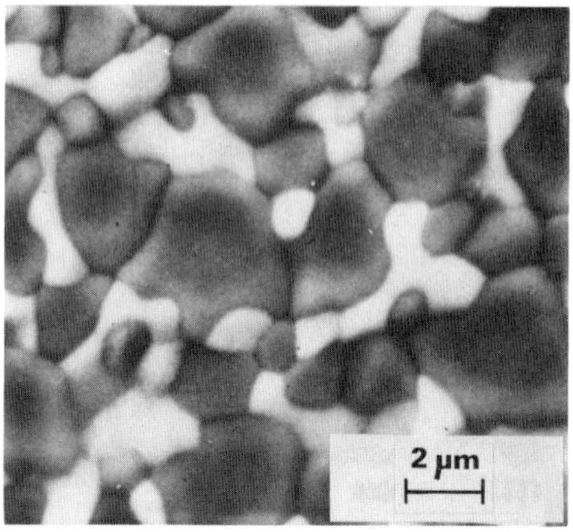

Fig. 2. SEM micrograph of Al_2O_3–30wt.%Ti(N, C); the Ti(N, C) is the light phase (thermally etched in H_2).

TABLE 1 Properties of the ceramics studied

Material	*Density* (g cm^{-3})	*Coefficient of thermal expansion* ($\times 10^6$ K^{-1})	*Thermal conductivity* (W mK^{-1})	*Young's modulus*[a] (GPa)	*Hardness (load, 10 kgf)* (HV 10)	K_{Ic}[a] (MN m$^{-3/2}$)
Al_2O_3–4wt.%ZrO_2	4	8.7	0.22	380	1650	4.2
Al_2O_3–30wt.%Ti(N, C)	4.3	8.6	0.24	400	1750	3.5
β-(Si–Al–O–N)	3.2	3.6	0.17	300	1450	4.5
α-β-(Si–Al–O–N)	3.2	3.5	0.10	300	1750	4.3

[a] Approximate value at room temperature.

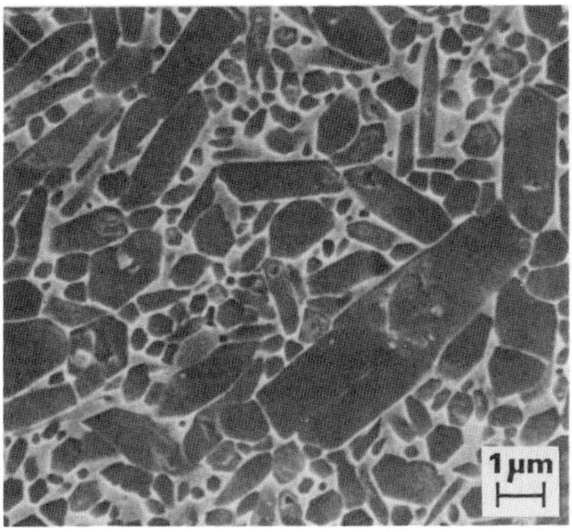

Fig. 3. SEM micrograph of the β-(Si–Al–O–N) alloy; the light phase is the glass phase (plasma etched).

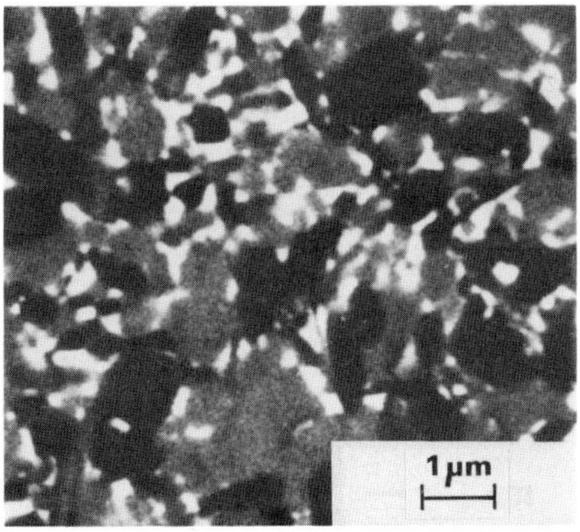

Fig. 4. SEM micrograph of the α–β-(Si–Al–O–N); the dark phase is β and the grey phase α. The white phase is glass and b phase (as polished).

2.1.4. α–β-(Si–Al–O–N)

This material was prepared from Si_3N_4 with additions of Al_2O_3, Y_2O_3 and AlN [1]. The microstructure consists of both α- and β-(Si–Al–O–N) grains, again interspersed with a glass phase. The α-(Si–Al–O–N) has the α-Si_3N_4 structure with a proportion of the silicon and nitrogen substituted by aluminium and oxygen and with yttrium occupying interstitial positions to maintain electronic neutrality. Its formula is

$$Y_x(Si, Al)_{12}(N, O)_{16}$$

where $x<2$. The ratio of the metals Si:Al:Y in the glassy phase was 37:25:38. The microstructure also contains small amounts of both b phase (a distorted yttrium aluminium garnet with the formula Y_2SiAlO_5N) and a 12H Si–Al–O–N polytype. The relative amounts of the crystalline phases are 35% α-(Si–Al–O–N), 58% β-(Si–Al–O–N), 5% b phase and 2% 12H.

2.2. Preparation of materials

All materials were prepared by vibratory milling of the components and isostatic cold pressing of the dried powders into required shape. They were all pressureless sintered in various atmospheres. The Al_2O_3–4wt.%ZrO_2 was sintered in air at 1600 °C, the Al_2O_3–30wt.%Ti(N,C) was sintered in nitrogen at 1750 °C and the Si–Al–O–N materials were sintered in nitrogen at 1725 °C. During sintering, the two Si–Al–O–N materials were packed in boron nitride powder to reduce loss by evaporation. The values of density and hardness after sintering can be found in Table 1.

2.3. Fracture testing

Fracture tests were carried out at room temperature, 800 °C, 1000 °C and 1200 °C in vacuum using a four-point single-notch bend test on samples with dimensions 3 mm × 6 mm × 45 mm. The bend test roll spacings were 19 and 38 mm. The apparatus used for the tests up to 1000 °C has been described in detail by Johannesson [2]. The 1200 °C tests were performed in similar equipment designed for higher temperatures.

The samples were pre-cracked with sharp straight-through notches using bridge indentation [3]. The sample surfaces were first polished to permit measurement of the pre-crack length on the sides of the sample prior to testing. Crack lengths a of between 1.5 and 3 mm ($a/w=$ 0.25–0.5) were used. Examination of the fracture surface sometimes indicated that the pre-crack was larger than observed on the side of the sample. Except for tests at 1200 °C when significant subcritical crack growth occurred, this discrepancy generally led to only small uncertainties in K_{Ic} values. When subcritical crack growth occurred, leading to a non-linear load–deflection behaviour, effective K_{Ic} values K_c were calculated on the basis of the maximum load and the initial crack length.

Room temperature tests were performed with a cross-head speed of 60 μm min^{-1}, while at 800 and 1000 °C the speed was raised to 600 μm min^{-1}. At 1200 °C, both speeds were used. Only two samples of each material were tested at each temperature. However, earlier tests on WC hard metals indicated that the test methods used have intrinsically good reproducibility. Some variation with pre-crack length was observed in some cases, possibly owing to the *R* curve behaviour of the pre-crack [2].

2.4. Fractography

The fracture surfaces of each sample were examined using SEM. The fracture surfaces of the Al_2O_3–ZrO_2 samples were analysed by X-ray diffraction to assess the relative proportions of the tetragonal and monoclinic forms of ZrO_2.

3. Results

The fracture toughness of the four materials are shown as a function of temperature in Fig. 5. The fractographs of their fracture surfaces are shown in Figs. 6–9. Their fracture behaviour will be described in turn.

3.1. Al_2O_3–4wt.%ZrO_2

The room temperature toughness of this material was 4.3 MN $m^{-3/2}$ and this fell only slightly up to 1000 °C. At room temperature the fracture exhibited roughly similar proportions of grain boundary fracture and transgranular cleavage. With increasing temperature the proportion of intergranular failure increased (Fig. 6). Between 1000 and 1200 °C, there was a significant drop in K_c accompanied by the onset of significant subcritical crack growth; in the subcritical growth region the fracture remained intergranular and was not affected by the testing rate between 60 and 600 μm min^{-1}. The deviation from linearity in the load–deflection curve began at a stress intensity as low as 2 MN $m^{-3/2}$. It has not been possible to ascertain to what extent this was due to plastic deformation or crack growth. However, it is clear that crack resistance is reduced markedly at 1200 °C. A possible explanation for this behaviour is that this material is intrinsically weak and that the toughening created by the ZrO_2 is lost above 1000 °C [4]. The X-ray diffraction measurements were inconclusive in that they indicated that about 20–30 wt.% of the monoclinic ZrO_2 did not change significantly with temperature. Stable intergranular crack growth has been observed earlier in fine-grained Al_2O_3 [5, 6].

3.2. Al_2O_3–30wt.%Ti(N, C)

Scatter was observed in the K_{Ic} measurements for this material. This seemed to be associated

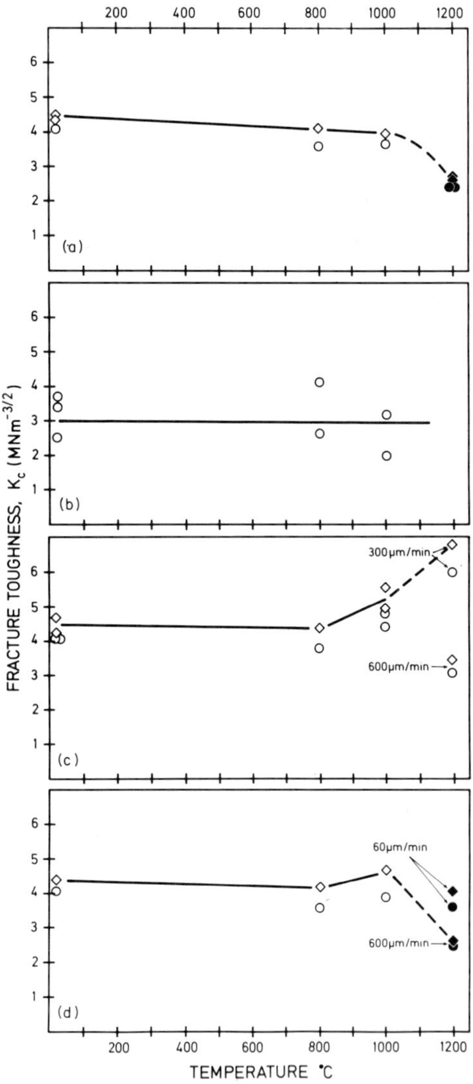

Fig. 5. Fracture toughnesses of (a) Al_2O_3–4wt.%ZrO_2, (b) Al_2O_3–30wt.%Ti(N, C), (c) β-(Si–Al–O–N) and (d) α-β-(Si–Al–O–N) as functions of temperature: ○, K_c based on initial crack length measured on specimen sides and maximum load; ◇, K_c based on stable crack length observed on fracture surface and maximum load; ●, ◆, significant non-linearity in load–deflection curve.

with difficulty in identifying the true pre-crack length. If we allow for this uncertainty, the fracture toughness is not affected significantly by temperature up to 1000 °C; the fracture remained linear elastic. As in the Al_2O_3–ZrO_2 ceramic, the

Fig. 6. Fracture surfaces of Al_2O_3–4wt.%ZrO_2 at (a) 20 °C, (b) 800 °C, (c) 1000 °C and (d) 1200 °C. (Magnifications, 3500 ×.)

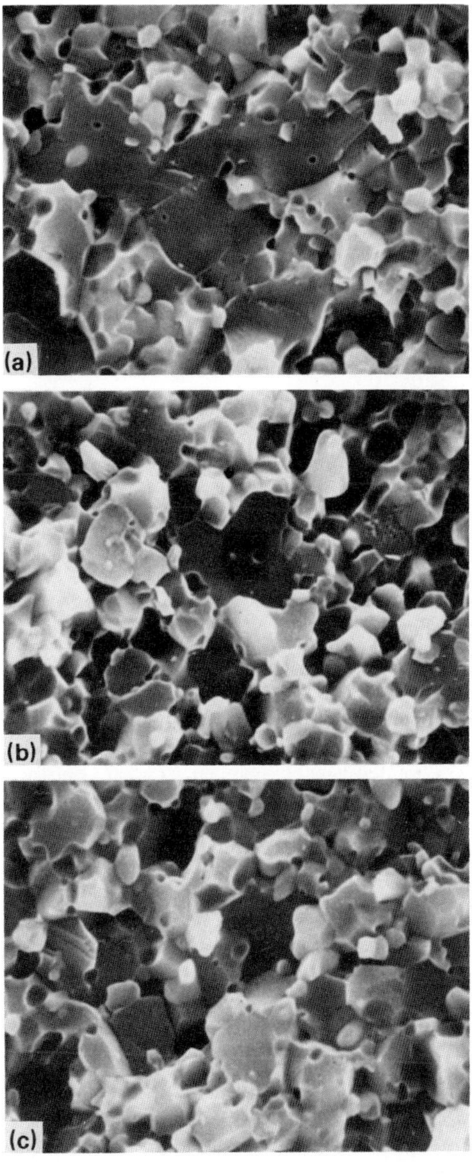

Fig. 7. Fracture surfaces of Al_2O_3–30wt.%Ti(C, N) at (a) 20 °C, (b) 800 °C and (c) 1200 °C. (Magnifications, 5000 ×.)

fracture at room temperature consisted of roughly equal proportions of transgranular and intergranular fracture and the proportion of intergranular fracture increased with increasing temperature (Fig. 7).

3.3. β-(Si–Al–O–N)

The fracture toughness of this material was approximately 4.3 MN $m^{-3/2}$ up to 800 °C, increasing slightly at 1000 °C; the fracture exhibited both transgranular and intergranular modes, the proportions not changing significantly up to 1000 °C (Fig. 8). Although the fracture was linear elastic up to 1200 °C, the fracture surface appearance at 1200 °C was very different from that at lower temperatures. The majority of the surface had a smoothed-out appearance, suggesting passage of the crack through a soft glass film or through a glass layer which smoothed out directly after passage of the crack (Fig. 8(d)). A narrow intermediate fracture zone between the pre-crack and the main fracture surface had a more irregular appearance with evidence of the formation of voids between 0.1 and 1 μm in diameter (Fig. 8(e)). This is interpreted as a process zone which developed at the crack tip through void formation in the glass phase. In certain circumstances, such a process could lead to rising crack resistance and slow crack growth [7–10]. In the present case the zone growth transferred at an early stage to fast fracture through the glass. The significantly higher apparent K_{Ic} for the slower test speed could reflect that the formation of the process zone causes tip blunting.

3.4. α-β-(Si–Al–O–N)

The fracture behaviour of this material was very similar to that of the β-(Si–Al–O–N) (Fig.9). Thus, up to 1000 °C, the fracture was linear elastic and exhibited both transgranular and intergranular modes. At 1200 °C, a significant proportion of the fracture surface again had a smoothed-out appearance (Fig. 9(d)) while a more irregular intermediate zone was observed close to the pre-crack (Fig. 9(e)). Important dif-

Fig. 8. Fracture surfaces of the β-(Si–Al–O–N) alloy at (a) 20 °C, (b) 800 °C, (c) 1000 °C and (d) 1200 °C. (e) The appearance of the intermediate zone between the pre-crack and main fracture surface at 1200 °C. All were tested at 600 μm min^{-1} except the 20 °C sample which was tested at 60 μm min^{-1}. (Magnifications, 3500 ×.)

ferences were that the fracture occurred by slow crack growth and the intermediate zone was more extensive. The differences can be explained by the facts that there is a smaller fraction of glass in this material and that the higher yttrium content of the glass gives it a higher viscosity. This would cause a delay in the transition from fracture by cavity growth to fast fracture through the glass phase.

4. Discussion

One aspect of metal cutting in which the moderate temperature fracture toughness of ceramic tool materials can be of importance is the reliability of ceramic inserts in automated cutting operations. A common method of assessing reliability with respect to resistance to bulk brittle failure of the insert is to carry out a severe interrupted cutting test [11]. In such a test (Fig. 10), a cemented carbide exhibits a very reliable behaviour, *i.e.* low frequency of failure up to a threshold number of cuts beyond which all the inserts fail within a very short time. This is indicative of a stable and reproducible crack growth up to a critical size, possibly the result of a thermal shock process. The mixed ceramic Al_2O_3–30wt.%Ti(N, C) exhibits a similar behaviour but at a very low number of cuts; this is consistent with the very low fracture toughness of this material which would tolerate very limited stable crack growth. Both the Al_2O_3–ZrO_2 and the β-(Si–Al–O–N) material show a finite frequency of failure during the first few cuts but thereafter the cumulative number of failures increases slowly with the number of cuts. This may indicate a sensitivity to the existing flaw distribution in the material in association with an erratic crack growth response to thermal shock. The number

Fig. 9. Fracture surfaces of the α–β–(Si–Al–O–N) alloy at (a) 20 °C, (b) 800 °C, (c) 1000 °C and (d) 1200 °C. (e) Intermediate zone between the pre-crack and main fracture at 1200 °C. All were tested at 600 μm min^{-1} except at the 20 °C sample which was tested at 60 μm min^{-1}. (Magnifications, 3500 ×.)

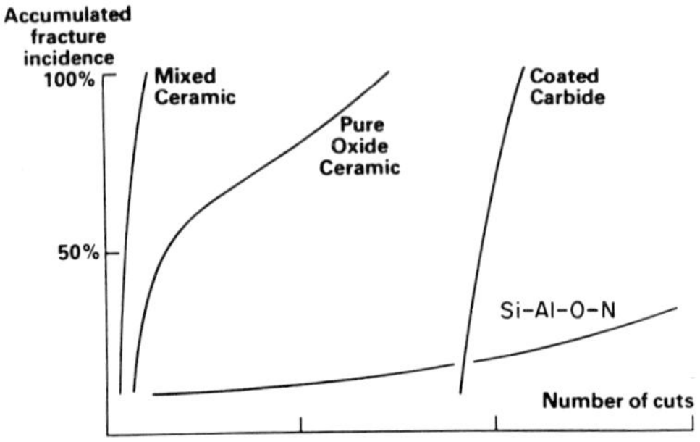

Fig. 10. Relative bulk fracture behaviour of various insert materials in a severe intermittent cutting test (cutting speed, 500 m min^{-1}; feed, 0.5 mm rev^{-1}; depth of cut, 3.0 mm). Pure oxide refers to the Al_2O_3–4wt.%ZrO_2 alloy.

of failures of Si–Al–O–N inserts increases much more slowly than that of the oxide ceramic which is consistent with the higher thermal shock resistance of Si–Al–O–N (lower coefficient of thermal expansion).

A tool failure mechanism which is related to high temperature fracture has been identified by Aucote and Foster [12] in Si–Al–O–N inserts when cutting nickel-based superalloys at high speeds. This involves the breaking off of thin flakes from the rake face of the insert. Aucote and Foster proposed that this occurs by fracture initiation on the flank face just below the cutting edge, followed by crack growth parallel to the rake face driven by an oscillating compression load from the chip which is serrated. The loads on the flank face would also encourage this crack growth. It was found that resistance to this type of failure increased with the addition of α-(Si–Al–O–N) to the ceramic. This is consistent with the results of this study which demonstrate an increased resistance to fast fracture at high temperature in the α–β-(Si–Al–O–N).

Acknowledgment

The 1200 °C fracture tests were performed by Dr. G. Grathwohl, Nuclear Research Centre, Karlsruhe.

References

1 C. Chatfield, T. Ekström and M. Mikus, Microstructural investigation of α–β yttria SiAlON materials, *J. Mater. Sci., 21* (1986) 2297.
2 B. Johannesson, The fracture behaviour of hardmetals, *Ph.D. Thesis,* Chalmers University of Technology, Göteborg, 1987.
3 R. Warren and B. Johannesson, Creation of stable cracks in hardmetals using “bridge” indentation, *Powder Metall., 27* (1984) 25–29.
4 F. F. Lange, Transformation toughening, *J. Mater. Sci., 17* (1986) 255–262.
5 H. Wieninger, K. Kromp and R. F. Pabst, Crack resistance curves of alumina and zirconia at room temperature, *J. Mater. Sci., 21* (1986) 411–418.
6 W. Blumenthal and A. G. Evans, High temperature failure of polycrystalline alumina: II creep crack growth and blunting, *J. Am. Ceram. Soc., 67* (1984) 751–759.
7 R. L. Tsai and R. Raj, Creep fracture in ceramics containing small amounts of a liquid phase, *Acta Metall., 30* (1982) 1043–1058.
8 A. G. Evans and W. Blumenthal, High temperature failure mechanisms in ceramic polycrystals, in R. E. Tressler and R. C. Bradt (eds.), *Deformation of Ceramic Materials II,* Plenum, New York, 1984, p. 487.
9 S. H. Knickerbocker, Z. Zangvil and S. D. Brown, Displacement rate and temperature effects in fracture of a hot-pressed silicon nitride at 1100 °C to 1325 °C, *J. Am. Ceram. Soc., 68* (1984) 365–368.
10 S. H. Knickerbocker, A. Zangvil and S. D. Brown, High temperature mechanical properties and microstructures for hot-pressed silicon nitrides with amorphous and crystalline intergranular phases, *J. Am. Ceram. Soc., 68* (4) (1985) 99–101.
11 D. H. Jack, Ceramic cutting tool materials, *Mater. Des., 7* (1986) 267–273.
12 J. Aucote and S. R. Foster, Performance of SiAlON cutting tools when machining nickel-base aerospace alloys, *Mater. Sci. Technol., 2* (1986) 700–708.

Joining of Silicon Nitride Ceramics to Metals: The Phase Diagram Base*

J. C. SCHUSTER

Istituto G. Donegani, Gruppo Montedison, I-28100 Novara (Italy)

F. WEITZER

Institut für Physikalische Chemie, Universität Wien, A-1090 Vienna (Austria)

J. BAUER

Laboratoire de Métallurgie et Physicochimie des Materiaux, Institut National des Sciences Appliquées Rennes, F-35043 Rennes Cédex (France)

H. NOWOTNY

Institute of Materials Science, University of Connecticut, Storrs, CT 06268 (U.S.A.)

(Received November 9, 1987)

Abstract

Chemical compatibility of silicon nitride–metal (M) joins at elevated temperatures are assessed by the evaluation of phase diagrams for M–Si–N systems. Joining strategies are discussed by considering the coexisting phases and their thermal expansion behavior.

1. Introduction

Silicon nitride (Si_3N_4) is one of the most important structural ceramics for high temperature applications. It has good hardness, wear resistance and high strength up to high temperatures. In many applications, Si_3N_4 is used together with structural metals and needs to be bonded to these metals. Thus the development of joining techniques which allows us to overcome the problem of chemical incompatibility as well as thermal mismatch is crucial for these applications.

Any adherence of a metal to a sintered ceramic is a secondary result of a chemical interaction at the interface. However, such intermediate adherence, promoting reaction layers, strongly affects in turn the mechanical properties of the join. On thermal cycling, the expansion mismatch causes an internal mechanical stress across the interface which must be considered as additive to any applied stress. If the sum of the internal and applied stresses exceeds the fracture strength of the ceramic or the yield strength of the metal, the join fails. Thus the design for strong metal–ceramic joins needs coexisting material combinations with a minimum thermal mismatch and a high failure strength of each component.

On the direct bonding of Si_3N_4 to various metals M or the metallization of Si_3N_4 at elevated temperatures, the occurrence of new phases at the interface have frequently been observed. These reaction products between Si_3N_4 and the metals were in most cases binary metal silicides [1–4], but also more complex structures such as Si_3N_4/TiN/Ti_5Si_3/Ti [2, 5–7] or Si_3N_4/Nb_2N/Nb_5Si_3/Nb [2] have been reported. Typically the bonding strengths obtained did not exceed 50 MPa [7]. The failure of the join occurred in the Si_3N_4 ceramic [1, 2], but delamination at the ceramic–metal interface ("peel-off") is reported as well [2, 5].

In order to identify the coexisting phases and to predict interfacial reactions at elevated temperatures in Si_3N_4–M joins, a survey of phase equilibria in ternary M–Si–N systems is pre-

*Paper presented at the 3rd International Conference on the Science of Hard Materials, Nassau, The Bahamas, November 9–13, 1987.

0921-5093/88/$3.50 © Elsevier Sequoia/Printed in The Netherlands

sented. The thermal mismatches for these coexisting material combinations are calculated from thermal expansion data [1, 8–11] in order to discuss the mechanical behavior observed in Si_3N_4–M joins and strategies for their improvement.

2. Experimental procedures

Binary M–Si alloys were prepared from powders or ingots of the constituting elements by arc melting under argon (99.999% pure). Ternary compositions were prepared by mixing powders of these master alloys and binary nitride powders. These ternary mixtures were cold pressed and sintered under pure argon in an r.f. furnace at temperatures between 1150 and 1400 °C. Heat treatment at temperatures up to 1000 °C was performed in evacuated quartz tubes lined with molybdenum or tantalum foil.

X-ray diffraction (Cu Kα_1 radiation or Cr Kα radiation) was performed on all alloys using Guinier or Debye–Scherrer cameras. These data (*e.g.* those presented in Table 1 for the system Hf–Si–N) were used to establish isothermal sections of the ternary M–Si–N phase diagrams.

In several ternary systems, Si_3N_4 coexists with the transition metal at 1000 °C. In order to determine the temperature at which reaction takes place a large pellet (5–10 g) of the cold-compacted powder mixture (Si_3N_4–30at.%M) was placed in an r.f. furnace under vacuum and heated slowly. Monitoring the increase in temperature with an optical pyrometer *vs.* the gas pressure (on a logarithmic scale with arbitrary units) enables us to observe directly the onset of gas evolution resulting from the reaction $Si_3N_4 + M \rightarrow MSi_x + N_2$. Furthermore the nitrogen gas acts as transporting agent for metallic vapors which are deposited at the quartz window, thereby virtually bringing to an end the optical temperature measurements at the onset of the reaction. A typical curve (for the system Si_3N_4–W) is shown in Fig. 1.

3. Results and discussion

For virtually all transition metals TM, isothermal sections in TM–Si–N systems were established: scandium, cerium, holmium, titanium, zirconium, hafnium, vanadium, niobium, tantalum, chromium, molybdenum, tungsten, manganese, rhenium, iron, ruthenium, cobalt and nickel.

TABLE 1 Solid state reaction products observed in the Hf–Si–N system on annealing at 1573 K for 72 h (r.f. furnace; argon at 10^5 Pa)

Si_3N_4 + 20 at.% Hf	→ Si_3N_4 + HfN
Si_3N_4 + 40 at.% Hf	→ $HfSi_2$ + HfN
Si_3N_4 + 50 at.% Hf	→ HfSi + HfN
Si_3N_4 + 52 at.% Hf	→ HfSi + Hf_5Si_4 + HfN
Si_3N_4 + 60 at.% Hf	→ Hf_2Si + HfN + Hf_3N_2
Si_3N_4 + 80 at.% Hf	→ Hf_2Si + α-Hf(solid solution N)
HfN + $Hf_{0.50}Si_{0.50}$ (15 at.% N)	→ HfN + HfSi
HfN + $Hf_{0.56}Si_{0.44}$ (15 at.% N)	→ HfN + Hf_5Si_4 + HfSi
HfN + $Hf_{0.60}Si_{0.40}$ (15 at.% N)	→ HfN + Hf_3Si_2
HfN + $Hf_{0.67}Si_{0.33}$ (15 at.% N)	→ HfN + Hf_2Si

In several samples, traces of HfO_2 and $D8_8$–$Hf_5Si_3O_x$ were observed.

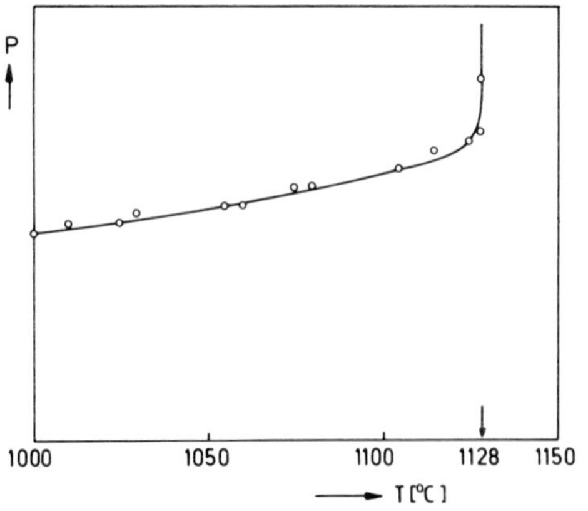

Fig. 1. Pressure (logarithmic scale) *vs.* temperature data for an Si_3N_4–30at.%W alloy (heated *in vacuo* in an r.f. furnace).

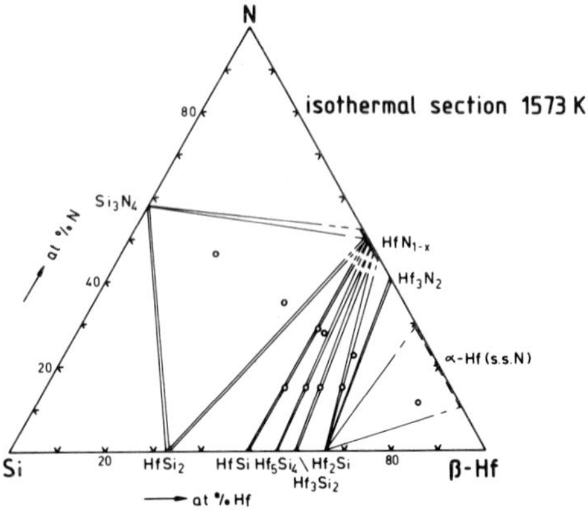

Fig. 2. Isothermal section of the Hf–Si–N system at 1573 K.

TABLE 2 Solid phases found in equilibrium with Si_3N_4 (under argon)

Ternary system	*Temperature* (K)	*Equilibrium phases*	*Reference*
Sc–Si–N	1273	ScN, $ScSi_{2-x}$	[16]
Ce–Si–N	1273	CeN, $CeSi_{2-x}$, $CeSi_2$	[16]
Ho–Si–N	1273	HoN, $HoSi_{2-x}$, $HoSi_2$	[16]
Ti–Si–N	1273	TiN, $TiSi_2$	[12]
Zr–Si–N	1273	ZrN, ZrSi, $ZrSi_2$	[12]
Zr–Si–N	1573	ZrN, $ZrSi_2$	[12]
Hf–Si–N	1273, 1573	HfN, $HfSi_2$	This work
V–Si–N	1273	VN, V_9N_4, $V_5Si_3N_{1-x}$, VSi_2	[12]
Nb–Si–N	1273	NbN, Nb_4N_3, Nb_2N, Nb_5Si_3, $NbSi_2$	[12]
Ta–Si–N	1273	TaN, Ta_2N, Ta_5Si_3, $TaSi_2$	[12]
Cr–Si–N	1273	Cr_3Si, Cr_5Si_3, CrSi, $CrSi_2$	[14]
Cr–Si–N	1673	Cr_5Si_3, CrSi, $CrSi_2$	[14]
Mo–Si–N	1273, 1673	Mo_5Si_3, $MoSi_2$	[14]
W–Si–N	1273	W, W_5Si_3, WSi_2	[14]
W–Si–N	1673	W_5Si_3, WSi_2	[14]
Mn–Si–N	1273	$MnSiN_2$, Mn_3Si, Mn_5Si_3, MnSi, $MnSi_{2-x}$	[13]
Re–Si–N	1273	Re, $Re_{17}Si_9$, $ReSi_2$	[14]
Re–Si–N	1673	$Re_{17}Si_9$, $ReSi_2$	[14]
Fe–Si–N	1173	Fe, Fe_3Si, Fe_5Si_3, FeSi, $FeSi_2$	[13]
Fe–Si–N	1423	Fe (solid solution), Fe_3Si, Fe_2Si, FeSi, Fe_2Si_5	[13]
Ru–Si–N	1273, 1473	Ru, Ru_4Si_3, RuSi, Ru_2Si_3	[15]
Ru–Si–N	1520	RuSi, Ru_2Si_3	[15]
Co–Si–N	1273	Co, Co_2Si, CoSi, $CoSi_2$	[13]
Ni–Si–N	1173	Ni, Ni_3Si, $Ni_{31}Si_{12}$, Ni_2Si, Ni_3Si_2, NiSi, $NiSi_2$	[13]

TABLE 3 Couples consisting of Si_3N_4 plus metal couples chemically inert at 1273 K (under argon)

Couple	*Limiting temperature T* (K)	*Reaction*
$Si_3N_4 + W$	1398	Reacts to form W_5Si_3, N_2
$Si_3N_4 + Re$	$1273 < T < 1473$	Reacts to form $Re_{17}Si_9$, N_2
$Si_3N_4 + Fe$	1393	Reacts to form Fe (solid solution), Fe_3Si, N_2
$Si_3N_4 + Ru$	1483	Reacts to form Ru_4Si_3
$Si_3N_4 + Co$	1423	Reacts to form Co (solid solution), Co_2Si, N_2
$Si_3N_4 + Ni$	1443	Reacts to form Ni (solid solution), Ni_3Si, N_2
$Si_3N_4 + Cu$	1358	Melts

Table 2, extracted from these phase diagrams, is a compilation of all phases found to coexist with Si_3N_4 at subsolidus temperatures. With the exception of the system Hf–Si–N (Fig. 2), all phase diagrams have been published in detail elsewhere [12–16].

For structural applications of Si_3N_4 the temperature range of 1273 K and above is of primary interest. Si_3N_4 reacts with most metals at $T < 1273$ K under decomposition and formation of new phases. The pairs of Si_3N_4 plus metal, for which the onset of this reaction is found above 1273 K, are listed in Table 3. Among these metals, tungsten has the smallest thermal mismatch ($\Delta\alpha_{M\text{-}Si_3N_4} = 2.0 \times 10^{-6}$ K^{-1}) followed by rhenium ($\Delta\alpha_{M\text{-}Si_3N_4} = 3.5610^{-6}$ K^{-1}). The maximum temperature of coexistence is observed for Si_3N_4 plus ruthenium (1483 K). Above this temperature the growth of adherence, promoting reaction layers, takes place at the Si_3N_4–M interface in all systems investigated in the present work. The reaction proceeds via silicon diffusion from Si_3N_4 to the metal to form metal silicide followed in many systems by the reaction of the metal with the free nitrogen to binary metal nitrides on the ceramic side of the join. (In the few cases where ternary phases occur, these are formed by nitridation of primary metal silicides.) These nitride phases act as diffusion barriers for further silicon diffusion, resulting in metastable Si_3N_4/metal nitride/metal silicide/metal layered structures for such ceramic–metal junctions (Table 4). This confirms and explains the earlier observations by Orent and Wagner [5] for the

TABLE 4 Sequence of phases in the microstructure of the reaction zone for Si_3N_4–M joints where binary metal nitrides are formed

Couple	*Temperature* (K)	*Phase sequence*
Si_3N_4 + Sc	1273	Si_3N_4–ScN–Sc_5Si_3–Sc
Si_3N_4 + Ce	1273	Si_3N_4–CeN–Ce_5Si_3–Ce
Si_3N_4 + Ho	1273	Si_3N_4–HoN–$Ho_5Si_3N_{1-x}$–Ho
Si_3N_4 + Ti	1273	Si_3N_4–TiN–Ti_5Si_3–Ti
Si_3N_4 + Zr	1273, 1573	Si_3N_4–ZrN–Zr_2Si($Zr_5Si_3N_{1-x}$)–Zr
Si_3N_4 + Hf	1273, 1573	Si_3N_4–HfN–Hf_2Si–Hf
Si_3N_4 + V	1273	Si_3N_4–V_9N_4–V_3Si($V_5Si_3N_{1-x}$)–V
Si_3N_4 + Nb	1273	Si_3N_4–Nb_2N–Nb_5Si_3–Nb
Si_3N_4 + Ta	1273	Si_3N_4–Ta_2N–Ta_3Si(Ta_2Si, Ta_5Si_3)–Ta
Si_3N_4 + Al	873	Si_3N_4–AlN–Al

couple Si_3N_4 plus titanium and by Suganuma *et al.* [2] for Si_3N_4 plus niobium. In these structures, adjacent phases are in equilibrium with each other but not with the phases beyond. If no stable metal nitride exists at or above the onset temperature of the reaction of Si_3N_4 with the metal, interface structures of the type Si_3N_4/metal silicide/metal will occur as reported for the Si_3N_4–Cr [3], Si_3N_4–Mo [4], Si_3N_4–Fe [2, 3] and Si_3N_4–Ni [2, 3] systems.

Analysis of data in the literature for coefficients of thermal expansion shows that the mismatch $\Delta\alpha_{\text{silicide-}Si_3N_4}$ or $\Delta\alpha_{\text{nitride-}Si_3N_4}$ is generally of the same order of magnitude as $\Delta\alpha_{M\text{-}Si_3N_4}$ (Table 5). Thus the reaction products of Si_3N_4 with the low α metals tungsten or rhenium, W_5Si_3 and $Re_{17}Si_9$ respectively, can be expected to exhibit small thermal mismatches, although no experimental data for their coefficient of thermal expansion could be found in the literature. Thus, from a thermal mismatch point of view, the strength of the Si_3N_4–M join is not very sensitive to the formation of a reaction zone at the interface. However, the brittleness of the binary or ternary phases coexisting with Si_3N_4 is considerably greater than the brittleness of the pure metals. The internal stress caused by thermal mismatch cannot be accommodated by plastic deformation of the intermetallic phases. Thus the strength of the joins in systems where Si_3N_4 reacts with the metal under formation of binary or ternary intermediate phases is severely limited. A thermal mismatch $\Delta\alpha$ of 8×10^{-6} K^{-1} causes failure of the join by fracture within the Si_3N_4 at an external stress of 50 MPa [2, 3] regardless of the metal involved (*e.g.* titanium or iron). Reducing the thermal mismatch $\Delta\alpha$ to $(2\text{–}3) \times 10^{-6}$ K^{-1} (*e.g.* $\Delta\alpha_{W\text{-}Si_3N_4}$ or $\Delta\alpha_{Mo\text{-}Si_3N_4}$) results in doubling the strength of the join to 100 MPa [1]. It should be

TABLE 5 Thermal mismatch $\Delta\alpha$ between Si_3N_4 and metals, metal nitrides and metal silicides calculated using $\alpha_{Si_3N_4} = 3.2 \times 10^{-6}$ K^{-1} [1] (range, ambient temperature to 1200 K)

M	$\Delta\alpha \times 10^6$ (K^{-1})			
	M–Si_3N_4	*M_3Si–Si_3N_4*	*M_5Si_3(N)–Si_3N_4*	*Nitride–Si_3N_4*
Ti	8.3		7.8	6.2
V	5.1[a]	7.0	6.8	4.9
Cr	8.6	7.3	7.9	6.2
Mo	3.0	3.8	3.5	3.0
W	2.0	—		—
Re	3.5	—		—
Fe	8.5[a]	7.8[b]		
Co	13.1	13.4[b]		
Ni	10.1[a]	11.6[b]		

[a]Room temperature value.
[b]High temperature value.

noted that, to obtain such a high strength, composite (cermet) interlayers were applied. The gradual change in composition from pure ceramic to pure metal (silicide) in a series of composite interlayers equilibrates the rates of thermal dilatation and contraction of the components of the join. Failure still occurred within the Si_3N_4. Other concepts for reducing the residual stress in the ceramic–metal interface, such as the use of interlayers having α values identical with or smaller than Si_3N_4, are not applicable because there are no such low α metals [17, 18]. The introduction of a further soft metal interlayer adjacent to the ceramic [19] again yields a bending strength of only 50 MPa for a Si_3N_4/Fe/W/steel system [2]. The thermal expansion mismatch is thus a limiting factor for the strength of Si_3N_4–M joins.

As well as identifying the thermal expansion matching phases coexisting in Si_3N_4–M joins at elevated temperatures, the TM–Si–N phase diagrams investigated provide information about the temperature stability limits of these phases. Phase transitions cause intergranular weakening and rupture of bonding in the join. Thus a suitable pair of materials should coexist without transformations or reactions with Si_3N_4 within the thermal cycle of the anticipated application. This requirement eliminates most metals as possible candidates for applications between 1273 and about 1673 K. In the systems with titanium, zirconium, hafnium, vanadium, niobium and tantalum, the nitrogen-rich phase boundaries of the metals (solid solutions), nitrides and silicides ($M_5Si_3N_x$) are extremely temperature sensitive and shift above 1273 K with increasing temperature considerably towards increasing nitrogen content [12]. In the Cr–Si–N system the reaction between Si_3N_4 and Cr_3Si to form Cr_5Si_3 and N_2 takes place [14], while in the Mn–Si–N system the ternary phase $MnSiN_2$ decomposes at around 1573 K [20]. In the systems with tungsten, rhenium, iron, ruthenium, cobalt and nickel, the onset of the Si_3N_4–M reaction occurs (Table 3). Only within the Mo–Si–N system does no phase change take place at 1273 K $< T <$ 1673 K. Thus, once Si_3N_4 has reacted with molybdenum to Mo_5Si_3 (plus N_2), thermal cycling will not cause degradation of the bonding strength.

In commercial Si_3N_4 grades the onset temperature for the reaction with metal is shifted upwards by the presence of densification, promoting oxide additions owing to the formation of oxynitride solid solutions phases [21]. However, high temperature enrichment of the interface with glassy impurities [22] reduces the fracture toughness of the join and thus affects the mechanical behavior of the join in the same manner as silicide formation at the interface does.

4. Conclusions

The ideal strong Si_3N_4–M join for structural applications at temperatures above 1273 K requires the coexistence over the entire temperature range 1273 K $< T <$ 1673 K of a ductile phase, which diffusion bonds to Si_3N_4 and matches the thermal expansion coefficient of Si_3N_4 closely. It has been shown, that no metal fulfils all these requirements simultaneously.

(1) Ruthenium is a ductile metallic phase coexisting with Si_3N_4 up to the maximum temperature of 1483 K, but it is difficult to bond since no mutual solid solution of ruthenium and Si_3N_4 takes place. The thermal mismatch $\Delta\alpha_{Ru\text{-}Si_3N_4}$ is rather large (5.9×10^{-6} K^{-1} at ambient temperature).

(2) Tungsten is less ductile and coexists with Si_3N_4 only up to 1398 K. It is difficult to bond to Si_3N_4 below this temperature but has the lowest thermal mismatch ($\Delta\alpha_{W\text{-}Si_3N_4} = 2.0 \times 10^{-6}$ K^{-1}) to Si_3N_4 of all metals.

(3) Rhenium behaves very similarly to tungsten. The thermal mismatch $\Delta\alpha_{Re\text{-}Si_3N_4}$ is 3.5×10^{-6} K^{-1}.

(4) Molybdenum reacts with Si_3N_4 even at $T <$ 1273 K with the formation of Mo_5Si_3 and N_2, but Mo_5Si_3 coexists with Si_3N_4 over the entire temperature range 1273 K $< T <$ 1673 K without undergoing any phase transformation. Furthermore, Mo_5Si_3 has a relatively low thermal mismatch $\Delta\alpha_{Mo\text{-}Si_3N_4}$ of 3.5×10^{-6} K^{-1} but is brittle, however.

Thus, although the goal of bond properties equivalent to those of the bulk material, at both ambient and elevated temperatures, might never be achieved in Si_3N_4–M joins, acceptable compromises will most probably be found by the use of these metals.

Acknowledgment

This work was supported in part by the U.S.A. European Research Office through Contract DAJA45-84-C-0038 of the University of Vienna.

References

1 A. Müller-Zell and H. W. Hennicke, *Forschungsber. BMFT FB T79-124*, 1979 (Fachinformationszentrum G.m.b.H., Eggenstein-Leopoldshafen 2).
2 K. Suganuma, T. Okamoto, Y. Miyamoto, M. Shimada and M. Koizumi, *Mater. Sci. Technol., 2* (1986) 1156.
3 A. Kohno, S. Hioki, Y. Yamada and N. Nakae, in S. P. Howlett and D. Taylor (eds.), *Special Ceramics 8*, Institute of Ceramics, Stoke-on-Trent, 1986, 125.
4 K. Suganuma, T. Okamoto, M. Shimada and M. Koizumi, *J. Am. Ceram. Soc., 66* (1983) C117.
5 T. W. Orent and R. A. Wagner, *J. Vac. Sci. Technol. B, 1* (1983) 844.
6 Y. Kagawa, *J. Mater. Sci. Lett., 4* (1985) 1062.
7 M. Naka, M. Kubo and I. Okamoto, *J. Mater. Sci. Lett., 5* (1986) 855.
8 *Handbook of Metals*, Vol. 1, 8th edn., American Society for Metals, Metals Park, OH, 1961, p. 54.
9 G. V. Samsonov, *Plenum Press Handbook of High Temperature Materials*, Vol. 2, Plenum, New York, 1964, p. 128.
10 G. V. Samsonov, W. S. Fomenko, I. A. Podtschernjaewa and L. N. Ochremtschuk, *Rev. Int. Hautes Temp. Réfract., 11* (1974) 269.

11 S. Sivaram, P. J. Ficalora and K. C. Cadien, *J. Appl. Phys.*, *58* (1985) 1314.
12 J. C. Schuster and H. Nowotny, *Proc. 11th Int. Plansee Semin.*, Vol. 1, Metallwerk Plansee, Reutte, 1985, p. 899.
13 F. Weitzer and J. C. Schuster, *J. Solid State Chem.*, *70* (1987) 178–184.
14 J. C. Schuster, *J. Mater. Sci.*, *23* (1988).
15 F. Weitzer, P. Rogl and J. C. Schuster, *Z. Metallkd.*, *79* (1988) 154.
16 F. Weitzer, J. Bauer and J. C. Schuster, to be published.
17 K. Suganuma, T. Okamoto, M. Koizumi and M. Shimada, *J. Am. Ceram. Soc.*, *67* (1984) C256.
18 K. Suganuma, T. Okamoto, M. Koizumi and M. Shimada, *J. Nucl. Mater.*, *133–134* (1985) 773.
19 K. Suganuma, T. Okamoto, M. Koizumi and M. Shimada, *J. Mater. Sci. Lett.*, *4* (1985) 648.
20 R. Marchand, M. Maunaye and L. Lang, *C. R. Acad. Sci. Paris, Sér. C*, *272* (1971) 1654.
21 J. C. Fletcher and W. M. Phillips, *U.S. Patent 4.072.532*, 1978.
22 G. Elssner, W. Diem and J. C. Wallace in J. Pask and A. Evans (eds.), *Surfaces and Interfaces in Ceramic and Ceramic–Metal Systems*, Plenum, New York, 1981, p. 629.

Indentation Fracture Toughness and Dynamic Young's Modulus of Ceramic Biomaterials*

D. W. JONES, A. S. RIZKALLA and E. J. SUTOW

Dalhousie University, Halifax, Nova Scotia (Canada)

H. W. KING

Technical University of Nova Scotia, Halifax, Nova Scotia (Canada)

(Received December 18, 1987)

Abstract

The strain-energy-absorbing quality of ceramic biomaterials can be evaluated from plain strain fracture toughness measurements. This method was used to compare the mechanical properties of several commercial ceramic dental materials. Cylindrical specimens were produced from as-received commercial feldspathic glass powders and tetrasilicic mica glass materials. The plain strain fracture toughness K_{Ic} *was evaluated by the Vickers indentation flaw method using a Tukon hardness tester. The dynamic Young's modulus parameter which was used to characterize the heating schedule of these materials was determined by means of a sonic method.*

The Student–Newman–Keuls rank test analysis for the mean fracture toughness values was able to separate these materials into several significant groups at the 0.99 level of confidence. The findings were compared with published data on similar feldspathic materials.

1. Objective

Ceramic or porcelain materials are being increasingly used as dental biomaterials in the form of thin veneers bonded to the surface enamel of anterior (front) teeth and as inlays to replace areas of the enamel in molar (back) teeth. The objective of the present investigation was to compare a range of different ceramic biomaterials in terms of the plain strain fracture toughness using an indentation method.

2. Introduction

Ceramics, silver–amalgam alloys, glass–resin composites, acrylic and glass polyalkenoate materials are used as dental biomaterials to replace natural tooth. During function, such materials may be subjected to high stress levels and abrasion. Ceramics have the advantage that they are harder, have a greater resistance to abrasion and can simulate the colour and translucency of natural tooth. The Vickers hardness values for natural tooth enamel and dentine and five dental porcelain materials have been reported previously [1]. No direct relationship was found between the hardness, the modulus of rupture and the modulus of elasticity. All five of the dental porcelain materials were found to be significantly harder than samples of natural tooth enamel. However, the limited elastic strain energy which can be absorbed by such biomaterials prior to fracture can result in catastrophic failure.

Fracture toughness has become established as an effective method of evaluating the strain-energy-absorbing quality of ceramic materials by measuring the critical stress intensity factor K_{Ic}. There are a number of different fracture toughness test methods which use different test specimen configurations. It is generally accepted that different test configurations may result in slightly different K_{Ic} values. One advantage of the indentation plain strain fracture toughness method is that it is independent of the specimen geometry and only small specimens are required.

Typical fracture toughness values (plain strain fracture toughness K_{Ic}) in the literature for various biomaterials range from $K_{Ic} = 2.0$ MPa m$^{1/2}$ for Al_2O_3 (sapphire) [2] and poly(methyl methacrylate) [3] to 1.98–1.26 for silver amalgam

*Paper presented at the 3rd International Conference on the Science of Hard Materials, Nassau, The Bahamas, November 9–13, 1987.

0921-5093/88/$3.50 © Elsevier Sequoia/Printed in The Netherlands

alloys [4], 1.02–1.84 for glass–resin composite systems and down to as low as 0.50 for the glass polyalkenoate materials [5]. This compares with reported natural tooth enamel values which are in the range 0.7–1.27 [6] and natural tooth dentine with a value of 3.08 [7]. Dental porcelain materials have been reported as having K_{Ic} values ranging from 0.90 to 1.56 MPa $m^{1/2}$ [8]. In contrast, dental casting alloys have fracture toughness values some 30–100 times higher than that of dental porcelain [2].

The indentation fracture toughness method was first suggested by the work of Parmquist [9] who used it to study the fracture of hard metals. Wiederhorn [10] later suggested that the method had application to fracture toughness of glass materials. The method was further developed by Evans and Charles [11] and by Evans [12]. The technique has been further evaluated and a theoretical basis for the procedure established [13–16]. The method involves indentation with a Vickers indenter into a smooth polished surface of the glass or ceramic. A load is selected which will cause readily observable fractures at the four corners of the diamond pyramid indenter. The indentation method has considerable advantage for dental porcelain materials in that only small specimens are required to produce an acceptable number of data for statistical analysis.

3. Experimental procedure

3.1. Materials

Six commercial feldspathic porcelain powders and cylindrical specimens of a tetrasilicic–mica glass in both the as-cast glass and the "cerammed" glass–ceramic condition were supplied by the manufacturers.

3.2. Specimen fabrication

Cylindrical specimens of six commercial dental porcelains were made by mixing the as-received powder with distilled water to form a slurry. A two-part metal mould 12 mm in diameter lined with polytetrafluorethylene tape to prevent oxidation and adhesion of the sample to the wall was used to form the specimen. The slurry was vibrated into the mould and excess water was removed by means of a dry cellulose paper. The specimen was ejected from the die and air dried for 2 h. The green specimens were heated under a vacuum of 1×10^{-3} Torr at a temperature starting from 500 °C to a temperature of 40 °C lower than that recommended by the manufacturers at a rate of 40 °C min^{-1}; then the vacuum was released and the temperature was raised to reach the required value and held for a specific time. A series of specimens were produced using different holding times ranging from 3.5 to 10 min at 1 min intervals. The holding time at the specified temperature for enamel A materials ranged from 0 to 3.5 min at 0.5 min intervals. At the end of the heating schedule, the specimens were removed from the furnace and cooled under a glass beaker so as to minimize the thermal stresses generated from cooling. No heat treatment was performed on the as-cast glass and glass–ceramic specimens.

Cylindrical specimens (about 10 mm in diameter and 5 mm in height) were used for evaluation. The end surface of the specimens was ground flat and parallel using SiC under water. The grit size was varied from 400 to 600 μm and polished with a 6 μm diamond paste. The density of the glass specimens was determined by means of the water displacement technique.

3.3. Dynamic Young's modulus

The dynamic Young's modulus was evaluated by an ultrasonic method [17] using lithium niobate crystals for the transmitting and receiving signals which were generated at 10 MHz resonant frequency. The travelling time of the longitudinal compression waves across the sample was determined using the following formula:

$$t = T - \Delta t \qquad (1)$$

where T (μs) is the total travelling time of the compression waves across the sample and the receiving crystal and Δt is the travelling time across the receiving crystal which equals about 0.05 μs for 10 MHz lithium niobate crystals. The value of Δt (0.05 μs) was calculated from the velocity of sound in the crystal (4000 m s^{-1}) and its thickness (0.2 mm).

The experimental value of Δt was also determined by using two cylindrical specimens of similar materials but different heights. The time T measured from the oscilloscope was recorded for each sample separately as $T_1 = 0.92$ μs and $T_2 = 1.12$ μs and then the two specimens were stuck together by means of silicon grease and the time $T_3 = 1.99$ μs was measured.

Therefore the travelling time of the compression waves across each specimen can be calculated using the following equations:

$$t_1 + \Delta t = 0.92 \qquad (2)$$

$$t_2 + \Delta t = 1.12 \tag{3}$$

$$(t_1 + t_2) + \Delta t = 1.99 \tag{4}$$

Solving the above three equations simultaneously gave $t_1 = 0.87$ μs, $t_2 = 1.07$ μs and $\Delta t = 0.05$ μs. This shows excellent agreement between the measured Δt and calculated Δt values.

The dynamic Young's modulus of the cylindrical specimens was calculated using the relationship:

$$C_L = \frac{\text{thickness of specimen}}{t} \tag{5}$$

$$C_L = \left\{ \frac{E(1-\nu)}{\rho(1+\nu)(1-2\nu)} \right\}^{1/2} \tag{6}$$

where C_L (m s^{-1}) is the velocity of the longitudinal wave in the material, ρ (kg m^{-3}) the density of the material, E (GPa) the dynamic Young's modulus and ν (≈ 0.25) Poisson's ratio.

3.4. Fracture toughness

Fracture toughness values were only determined on those specimens fired to produce the optimum modulus of elasticity (*i.e.* the plateau region). The cylindrical specimens were sectioned in half longitudinally and mounted in a resin, ground and polished using the procedure described above. Indentation fracture toughness tests were performed on the two sections of each specimen using a Vickers diamond indenter. For six materials, indentation data were obtained using a load of 500 gf (4.9 N). For both opaque materials, a load of 1000 gf (9.8 N) was required in order to measure the cracks at a magnification of 50 ×. A diagram illustrating the radial cracks generated at the corners of the diamond indentation is shown in Fig. 1. These cracks are semicircular perpendicular to the surface of the indentation [14]. A total of 30 indentations was measured for each specimen.

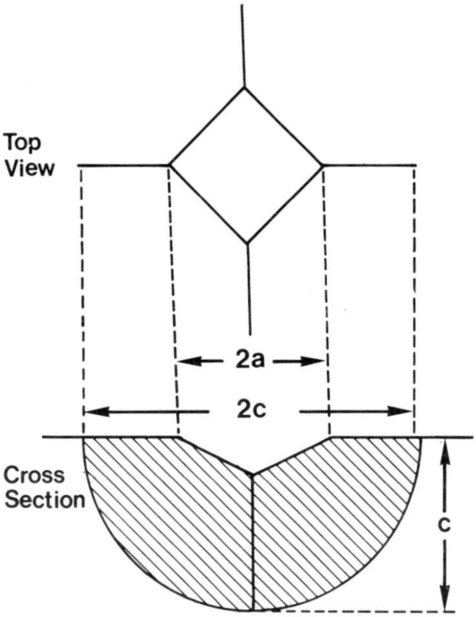

Fig. 1. Schematic diagram of the Vickers indentation.

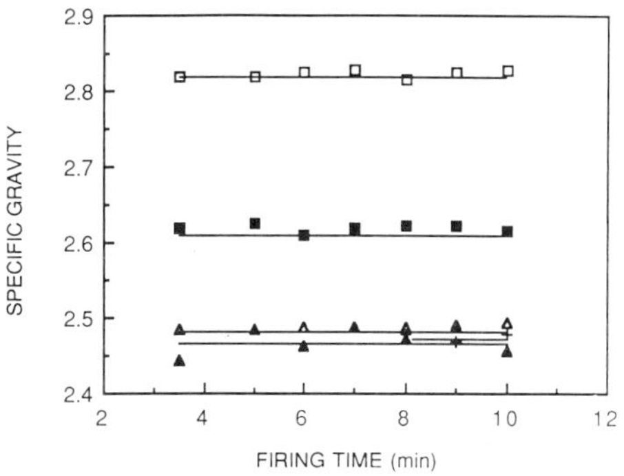

Fig. 2. Effect of firing time at the recommended temperatures (given in parentheses) on the specific gravity of commercial dental porcelain materials: □, opaque A material (982 °C); ■, opaque B material (982 °C); △, enamel B (982 °C); +, enamel C (960 °C); ▲, body material (960 °C).

4. Results and discussion

The density of each of the commercial feldspathic dental materials did not vary significantly with the different firing times at the specified firing temperature. Figure 2 illustrates the consistency of the specific gravity values for the various firing temperatures. The Student–Newman–Keuls (SNK) rank order test for the mean relative density for all the firing times at the specified temperature for each of the eight different materials is shown in Table 1. This test separated the eight different materials into six significant groups at the 0.99 level of confidence.

Although the specific gravity values did not show any significant change with firing time, it was interesting to find that the velocity of sound was very sensitive to the firing time. The velocity of sound increased with longer firing times, until

TABLE 1 The Student–Newman–Keuls rank order test comparing the relative density of eight commercial dental ceramic materials

Materials	*Firing temperature* (°C)	*Specific gravity*
Opaque A	982	2.823 ± 0.004
Glass-ceramic		2.686 ± 0.003
Opaque B	982	2.619 ± 0.006
As-cast glass		2.563 ± 0.001
Enamel B	982	2.488 ± 0.004 }
Enamel C	960	2.475 ± 0.006 }
Body	960	2.470 ± 0.017 }
Enamel A	871	2.419 ± 0.010

Level of confidence, 0.99.
Opaque material is a feldspathic glass which contains opacifiers such as TiO_2 or ZrO_2 particles dispersed in a glassy matrix. Enamel is a feldspathic glass which contains a leucite phase dispersed in a glassy matrix. Body material is feldspathic glass which contains a combination of colour frits and enamel. Cerammed glass is a tetrasilicic mica glass that contains a mica phase dispersed in a glassy matrix.

in most cases a plateau was reached. Since the modulus of elasticity is directly proportional to the velocity of sound, optimum dynamic Young's modulus values could be obtained for all but the opaque materials. For the opaque materials the modulus of elasticity at the longest firing time was used to calculate the fracture toughness values K_{Ic}.

Density (specific gravity) measurements have a limitation in sensitivity when using the standard immersion technique. The glass frit (dental porcelain) when sintered at the shorter times may exhibit interfacial faults between the frit particles. Increasing the holding time at the same temperature will allow for increased fusion and bridging between frit particles, thus improving the mechanical continuity across the specimen. These small interfacial defects would have only a very limited effect on the density measurements but would readily influence the velocity of sound measurements. Incomplete sintering between individual frit particles would be seen as defects by the ultrasound. As the firing time is increased, it progressively improves the interfacial fusion attainable at that particular temperature. Once the optimum fusion between particles has taken place, a plateau would be obtained on the firing time *vs.* Young's modulus graph.

The reason for plotting the dynamic Young's modulus *vs.* firing time was to determine the optimum firing schedule which would produce the maximum modulus of elasticity values. Indentation fracture toughness tests were then only per-

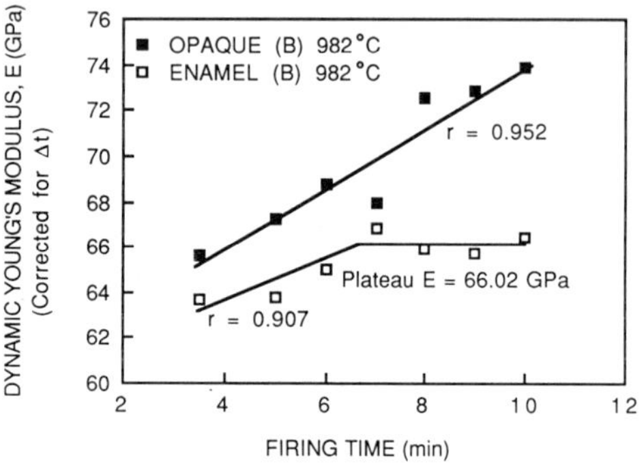

Fig. 3 Effect of firing time at the recommended temperatures on the dynamic Young's modulus for opaque B material and enamel B.

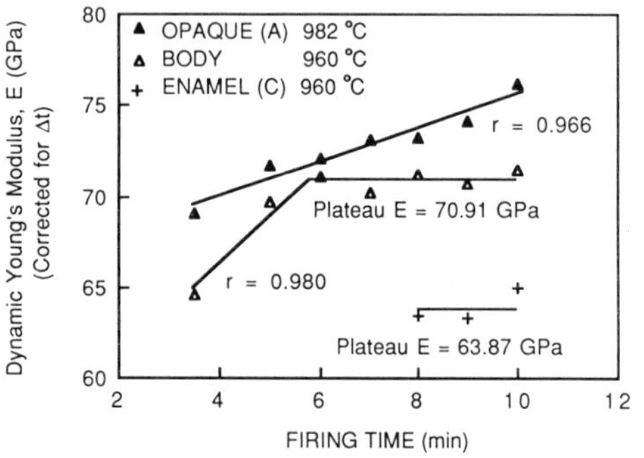

Fig. 4. Effect of firing time at the recommended temperatures on the dynamic Young's modulus for opaque material A, body material and enamel C.

formed on specimens which exhibited the maximum dynamic Young's modulus values. The dynamic Young's modulus (or the velocity of sound) values when plotted against a range of holding times at a specific temperature provide an excellent method for optimizing the firing schedule of the various frit materials.

Relationships between the dynamic Young's modulus and firing time for the six commercial feldspathic materials are illustrated in Figs. 3–5. It can be seen that enamel B and body materials exhibited a linear relationship between Young's modulus and firing time followed by a plateau region. The point of intersection between the regression line and the plateau indicates the minimum time required to firing the material at the specified temperature. Furthermore, opaque A and opaque B materials did not exhibit a plateau up to 10 min of firing which indicates that a

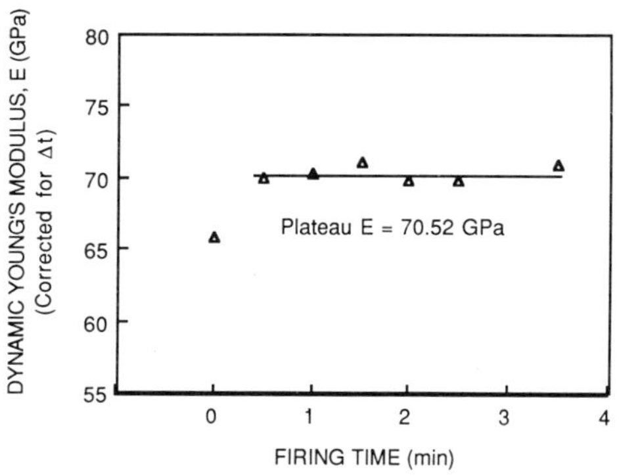

Fig. 5. Effect of firing time at the recommended temperature on the dynamic Young's modulus for enamel A.

longer firing time is required at these specified temperatures or a shorter time at higher temperatures.

It can also be seen from Fig. 4 that enamel C was fired in the plateau region. No data are shown at the shorter firing times below the plateau in this graph because an insufficient amount of material was available for evaluation.

It should be emphasized that, if the travelling time Δt of the compression waves is not taken into account, Young's modulus values of the eight commercial materials will exhibit a lower E value ranging from 10.2% to 12.5%, as shown in Table 2.

A total of 30 indentations were made on each specimen. It is shown in Table 3 that the mean Vickers hardness of the eight commercial materials ranged from as low as 3.85 GPa for a glass–ceramic (in the cerammed condition) to as high as 6.19 GPa for a feldspathic porcelain material; the coefficient of variation for the hardness ranged from 3 to 9. Furthermore, the variation in crack length for the various materials with 30 cracks per material is surprisingly low for a test of this type. The mean crack lengths varied from as low as 35.4 ± 3.18 μm up to 48.32 ± 2.79 μm. The coefficient of variation for the crack length ranged from 3 to 9.

The plain strain fracture toughness was calculated using the formula developed by Blendell [18]:

$$K_{\mathrm{Ic}} = \frac{Ha^{1/2}}{\phi}\left\{0.55 \log\left(8.4\frac{a}{c}\right)\left(\frac{H}{E\phi}\right) - 0.4\right\} \quad (7)$$

where H (GPa) is the Vickers hardness, a (μm) the half-diagonal of indentation, ϕ ($\simeq 3$) the constraint factor, C (μm) the crack length and E (GPa) Young's modulus. The fracture toughness K_{Ic} is given in megapascales per (metre)$^{1/2}$.

The Student–Newman–Keuls (SNK) rank test for the fracture toughness of the various materials is shown in Table 4. This test separated the eight different materials into five groups at the 0.99

TABLE 2 Mean dynamic Young's modulus E of eight commercial ceramic dental materials (correction for Δt)

Material	*Firing temperature* (°C)	*Mean dynamic Young's modulus E* (GPa)		*Difference* (%)
		Corrected	*Uncorrected*	
Opaque A	982	75.67	68.06	11.2
Opaque B	982	73.96	65.13	13.6
Body	960	70.91	63.04	12.5
Enamel A	871	70.52	62.73	12.4
Enamel B	982	66.02	59.83	10.4
Enamel C	960	63.87	57.89	10.3
Cerammed glass		64.66	57.49	12.5
As-cast glass		61.03	55.38	10.2

TABLE 3 The Vickers hardness and crack lengths of eight commercial ceramic dental materials

Materials	*Firing temperature* (°C)	*Hardness* (GPa)	*Crack lengths* (μm)
Opaque A	982	5.43 ± 0.47	46.33 ± 1.24
Opaque B	982	5.03 ± 0.24	48.32 ± 2.79
Body	960	5.99 ± 0.37	35.41 ± 3.12
Cerammed glass		3.85 ± 0.12	34.89 ± 0.69
Enamel A	871	5.18 ± 0.13	38.23 ± 1.64
Enamel B	982	6.19 ± 0.27	40.38 ± 3.30
Enamel C	960	5.63 ± 0.18	37.07 ± 2.40
As-cast glass		5.31 ± 0.22	38.89 ± 2.44

TABLE 4 The Student–Newman–Keuls rank test comparing the fracture toughness of eight commercial ceramic dental materials

Material	*Firing temperature* (°C)	*Fracture toughness* K_{Ic} (MPa m$^{1/2}$)
Opaque A	982	1.70 ± 0.05
Opaque B	982	1.64 ± 0.06
Body	960	1.34 ± 0.08
Cerammed glass		1.27 ± 0.01 }
Enamel A	871	1.27 ± 0.04 }
Enamel C	960	1.25 ± 0.05 }
Enamel B	982	1.19 ± 0.07 }
As-cast glass		1.18 ± 0.05 }

Level of confidence, 0.99.

TABLE 5 Mean fracture toughness K_{Ic} of eight commercial ceramic dental materials (correction for E)

Material	*Firing temperature* (°C)	*Mean fracture toughness* K_{Ic} (MPa $m^{1/2}$)		*Difference* (%)
		Corrected	*Uncorrected*	
Opaque A	982	1.70	1.63	4.3
Opaque B	982	1.64	1.56	5.1
Body	960	1.34	1.28	4.7
Cerammed glass		1.27	1.22	4.1
Enamel A	871	1.27	1.21	5.0
Enamel C	960	1.25	1.20	4.2
Enamel B	982	1.19	1.14	4.4
As-cast glass		1.18	1.13	4.4

level of confidence. Enamel A, enamel C and cerammed glass were not significantly different. Enamel B had a K_{Ic} value comparable with that of the as-cast glass. The highest fracture toughness K_{Ic} value was 1.70 ± 0.05 MPa $m^{1/2}$ and the lowest K_{Ic} value was 1.18 ± 0.05 MPa $m^{1/2}$. The coefficient of variation for the fracture toughness values for the various materials ranged between 2.7 and 5.6. It should be further emphasized that, if the corrected value of Young's modulus is not taken into account, the resultant plain strain fracture toughness values will be lower by 4.2%–5.1% for the various materials, as shown in Table 5.

No correlation was found between the Vickers hardness, the dynamic Young's modulus and the plain strain fracture toughness. The glass–ceramic following heat treatment exhibited an increase in plain strain fracture toughness of only 6% as a result of the ceramming process. The fracture toughness of the glass–ceramic in the cerammed state was not significantly higher than those of enamel A and enamel C.

The values of K_{Ic} obtained in the present work are considerably higher than the values reported by Morena *et al.* [8] for two of the same commercial ceramic dental materials. The values differ for body material by 43% and for opaque B material by 66%. Such differences may possibly be explained on the basis of temperature and time of firing of the specimens as well as the frequency used for obtaining the sonic measurements of Young's modulus. Morena *et al.* [8], in fact, used a sonic value (66.7 GPa) from one material only in calculating the fracture toughness values. In the present work, individual sonic modulus measurements were made on all samples and these values were then used to calculate the K_{Ic} values. In addition, Morena *et al.* [8] used a dilute HF etch to accentuate cracks prior to measuring. This etching could lead to crack propagation owing to the residual stresses generated during loading at the crack tip. No such etch was used in the present investigation. The data presented by Morena *et al.* were only able to separate fracture toughness values of the feldspathic from the aluminous ceramic materials statistically. In the present study the method proved to be more discriminating in being able to separate six commercial feldspathic-type porcelains, one glass–ceramic and one as-cast glass into five groups at the 0.99 level of confidence using a SNK multiple-comparison test.

Concern has been expressed by others [19] about the validity of the indentation fracture toughness technique. The major area of difficulty has been said to be due to the residual stresses generated at the area surrounding the indentation which would cause the crack to propagate with time. In the present investigation the crack lengths were measured within 1 min of indentation. However, it was also observed that no crack propagation took place up to 24 h for any of the eight materials studied. The low coefficient of variation of the crack length which ranged from 3 to 9 and the fracture toughness which ranged between 3 and 6 for the data observed in this investigation suggest that the technique is not sensitive to residual stresses generated during the Vickers indentation when using loads of 4.9 N and 9.8 N for the translucent and opaque feldspathic materials used in the present study.

5. Conclusions

(1) The dynamic Young's modulus determined by the sonic method has been found to be a useful technique for optimizing the firing schedule of the materials.

(2) The indentation fracture toughness technique has been found to be a valuable tool in the characterization of the mechanical behaviour of dental bioceramic materials. The main advantage of the indentation fracture toughness technique is that it requires only a small volume of test material in order to produce sufficient data for adequate statistical analysis. A range of eight materials were separated into five significantly different levels at the 0.99 level of confidence using the SNK rank order test.

(3) The fracture toughnesses of commercial dental porcelain materials in the present investigation are higher than values reported for natural tooth enamel and are in the same general range of values reported for the dental glass–resin composite systems.

References

1 D. W. Jones, Ceramics and glass as restorative materials, in D. C. Smith and D. F. Williams (eds.), *Biocompatibility of Dental Materials,* Vol. IV, CRC Press, Boca Raton, FL, Chapter 5, 1983, pp. 79–122.

2 J. C. Bittence (ed.), *Guide to Engineered Materials, Advanced Materials and Processes and Metal Progress,* Vol. 1, No. 1, American Society for Metals, Metals Park, OH, 1986.

3 G. D. Stafford, R. Huggett and B. E. Causton, Fracture toughness of denture base acrylics, *J. Biomed. Mater. Res., 14* (1980) 359–371.

4 D. W. Cruckshank-Boyd and N. Roswati, A comparison of fracture toughness testing methods for dental amalgam, *J. Dent. Res., 60* (1981) 1148, Abstract 34.

5 M. Goldman, Fracture properties of composite and glass ionomer dental restorative materials, *J. Biomed. Mater. Res., 19* (1985) 771–783.

6 R. Hassan, A. A. Caputo and R. F. Bunshah, Fracture toughness of human enamel, *J. Dent. Res., 60* (1981) 820–827.

7 O. M. El Mowafy and D. C. Watts, Fracture toughness of human dentine, *J. Dent. Res., 65* (1986) 677–681.

8 R. Morena, P. E. Lockwood and C. W. Fairhurst, Fracture toughness of commercial dental porcelains, *Dent. Mater., 2* (1986) 58–62.

9 S. Parmquist, Occurrence of crack formation during Vickers indentation as a measure of the toughness of hard metals, *Arch. Eisenhüttenwes., 33* (1962) 629–633.

10 S. M. Wiederhorn, Prevention of failure in glass by proof-testing, *J. Am. Ceram. Soc., 56* (1973) 227–287.

11 A. G. Evans and E. A. Charles, Fracture toughness determination by indentation, *J. Am. Ceram. Soc., 59* (1976) 371.

12 A. G. Evans, Fracture toughness: the role of indentation techniques, in S. W. Freiman (ed.), *Fracture Mechanics Applied to Brittle Materials, ASTM Spec. Tech. Publ., 678,* 1979, pp. 112–135.

13 G. R. Anstis, P. Chantikul, B. R. Lawn and D. B. Marshall, A critical evaluation of indentation techniques for measuring fracture toughness: I, Direct crack measurements; II, Strength method, *J. Am. Ceram. Soc., 64* (1981) 533, 539.

14 B. R. Lawn and M. U. Swain, Microfracture beneath point indentation in brittle solids, *J. Mater. Sci., 10* (1975) 113–122.

15 D. B. Marshall, Controlled flaws in ceramics; a comparison of Knoop and Vickers indentation, *J. Am. Ceram. Soc., 66* (1983) 127.

16 B. R. Lawn, The indentation crack as a model surface flaw, in R. C. Bradt, A. G. Evans, D. P. H. Hasselman and F. F. Lange (eds.), *Fracture Mechanics of Ceramics,* Vol. 5, Plenum, New York, 1983, pp. 1–25.

17 O. I. Babikov, *Ultrasonics and its Industrial Applications,* Russian Publishers, Consultants Bureau, New York, Chapter V, p. 123.

18 J. E. Blendell, The origins of internal stress in polycrystalline Al_2O_3 and their effects on mechanical properties, *D.Sc. Thesis,* Massachusetts Institute of Technology, 1979.

19 P. S. Nicholson, personal communication, 1987.

Kinetics of Liquid Phase Sintering*

R. M. GERMAN, S. FAROOQ and C. M. KIPPHUT

Materials Engineering Department, Rensselaer Polytechnic Institute, Troy, NY 12180-3590 (U.S.A.)

(Received November 9, 1987)

Abstract

Densification by liquid phase sintering requires that several criteria be satisfied. These criteria include various particle characteristics and system thermodynamic attributes. In this paper a review is given of the state of classic liquid phase sintering theory and this theory is updated to include the effects of heating rate, solubility and initial homogeneity. As a consequence of experiments and modeling, the effects of several material and process variables are included in a qualitative update to guide future practice and research.

1. Introduction

Liquid phase sintering (LPS) is a widely used process for the fabrication of both ceramic and metallic materials [1, 2]. It is a primary fabrication technique for hard materials intended for high performance applications [3]. The product is a composite consisting of a hard compound interwoven in a ductile matrix phase, giving a synergistic combination of properties.

One advantage of LPS is the fast sintering rate. A wetting liquid exhibits sufficient capillary force that external forces are not required to attain full density. The liquid also reduces the interparticle friction and dissolves sharp particle edges, thereby aiding rearrangement of the solid particles. The liquid phase provides for faster atomic diffusion than concurrent solid state processes do.

There are some disadvantages with LPS. A common problem is shape distortion which occurs when too much liquid is formed during sintering. Additionally, the parameters which control the sintered microstructure also control the final properties. Separation of these effects is sometimes difficult. Furthermore, in the initial compact there are at least three phases; vapor, liquid and solid. Accordingly, several interfaces and energies are associated with such a structure. The complex solubility, viscosity, surface energy and diffusivity effects have hindered a full analytic treatment of LPS.

The sintering kinetics are controlled by the interaction between the constituents as dictated by basic wetting, solubility, segregation and diffusivity considerations. Phase diagram information provides a good first basis for estimating the relative suitability of a given system for processing in the presence of a liquid phase [4, 5]. Furthermore, the sensitivity to various processing factors depends on the material characteristics. This presentation brings these factors together to give an update on the kinetics of LPS.

2. Classic model of liquid phase sintering

The classic example of LPS is attributed to Price *et al.* [6] in their report on tungsten heavy alloys. An example W–Ni–Fe sintered microstructure is shown in Fig. 1. The large tungsten grains were solid at the sintering temperature, while the matrix persisted as a liquid throughout the high temperature portion of the sintering cycle. Considerable densification and grain growth occurred during this LPS treatment. In this case the tungsten grains are 25 times larger than the original particle size and have become flattened along neighboring faces. Also, the tungsten grains are interconnected, providing the solid skeleton important to shape retention during sintering.

The classic theory of LPS assumes densification occurs in three overlapping stages [1, 2, 7–9], with no densification prior to liquid formation. Figures 2 and 3 illustrate the classic sequence of steps. Initially, the mixed powders are heated to a temperature at which a liquid forms. With liquid formation, rapid densification occurs owing to

*Paper presented at the 3rd International Conference on the Science of Hard Materials, Nassau, The Bahamas, November 9–13, 1987.

0921-5093/88/$3.50 © Elsevier Sequoia/Printed in The Netherlands

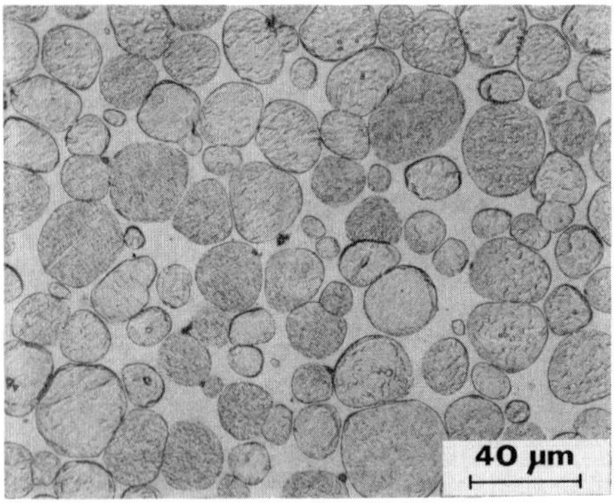

Fig. 1. An optical micrograph showing the microstructure existing after LPS a W–Ni–Fe heavy alloy.

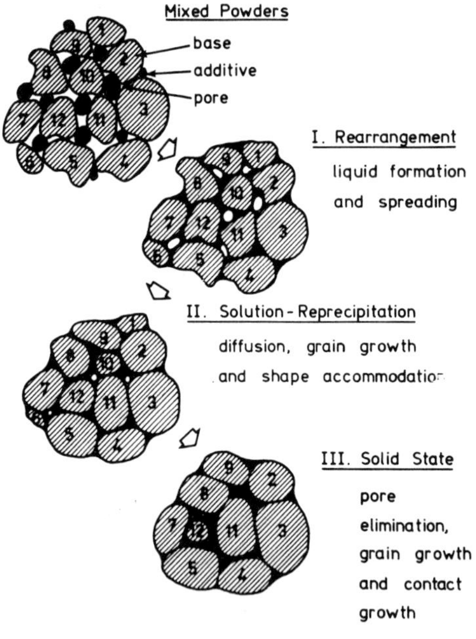

Fig. 2. The classic stages of LPS involving mixed powders which form a liquid on heating.

the capillary force exerted by the wetting liquid on the solid particles. The elimination of porosity occurs as the system minimizes its surface energy. During rearrangement, the compact responds as a viscous mass to the capillary action. The elimination of porosity increases the compact viscosity. As a consequence, the densification rate continuously decreases [10, 11].

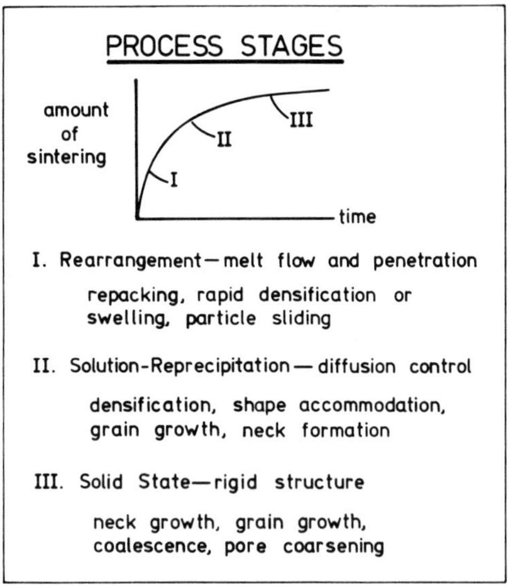

Fig. 3. The process stages associated with classic LPS, giving the main microstructural changes.

The amount of densification attained by rearrangement is dependent on the amount of liquid, the particle size and the solubility of the solid in the liquid. Usually, finer particles give better rearrangement. Full density is possible by rearrangement if enough liquid is formed; however, rearrangement can be inhibited by a high green density or irregular particle shape. The particle contacts resulting from compaction form solid state bonds during heating, thereby preventing rearrangement. Concurrent with the rearrangement stage are various other events, but the kinetics of rearrangement initially are dominant.

As densification by rearrangement slows, solubility and diffusivity effects become dominant. This second stage is termed solution–reprecipitation. The solubility of a grain in its surrounding liquid varies inversely with the grain size; small grains have a higher solubility than coarse grains do. The difference in solubilities establishes a concentration gradient in the liquid. Material is transported from the small grains to the large grains by diffusion. The net result is a progressive growth of the larger grains, giving fewer grains with a wider spacing. In the models, solution–reprecipitation contributes not only to grain coarsening but also to densification [12–14].

The grain shape is altered by diffusion to allow tighter packing of the grains, thereby leading to pore elimination. The amount of liquid affects solution–reprecipitation in terms of the diffusion distance and amount of grain shape accommodation. Both solubility of the solid in the liquid and diffusive transport are necessary criteria. The actual grain shape is determined by the relative solid–liquid and solid–solid interfacial energies, the amount of liquid and any anisotropy in surface energy of the solid [1]. Although the solid–liquid interface area is enlarged by a deviation from a spherical shape, grain shape accommodation results in elimination of porosity and the higher energy interfaces associated with pores.

The last stage of classic LPS is referred to as solid-state-controlled sintering. Densification is slow in this stage because of the solid skeleton. Processes dominant in the final stage are also active throughout the entire LPS cycle; however, because of the slow nature, solid state sintering is not dominant until late in the sintering cycle. The rigidity of the solid skeleton inhibits further rearrangement, although microstructural coarsening continues by diffusion. The residual pores will enlarge if they contain entrapped gas, giving compact swelling. In general, the properties of LPS materials are degraded by prolonged final stage sintering. Hence, short sintering times are preferred in practice. These three stages of classic (persistent) LPS are summarized in Fig. 3 in terms of the key microstructural changes.

There are many possible variants of these stages and events [1, 15]. Such variations affect the rate of sintering and the microstructure evolution. The cross-interaction of the material and processing factors contributes to the difficulty in studying LPS. This interaction also makes for a highly flexible manufacturing technique with widespread application.

3. Problems with classic theory

In spite of its importance, the theory of LPS has remained idealistic over the past 30 years. The classic theory assumes monosized spheres with minimal densification and microstructural coarsening prior to liquid formation. With liquid formation, rapid densification takes place, followed by a continual decrease in densification rate. An inherent assumption is that the contact angle at the solid–liquid–vapor interface is low (for capillarity [16]) and that the dihedral angle is zero (so that particle sliding can occur [9]).

In reality, several problems exist with this view of LPS. First, the initial microstructure is not ideal since most processing is based on mixed powders with a range of particle sizes, leading to varying levels of homogeneity [17]. In practice, heating is slow because of the constraints of binder removal and thermal inertia of furnaces. During slow heating, significant shrinkage is observed prior to liquid formation [18]. In many systems, considerable solid state intersolubility exists; thus, activated sintering occurs prior to liquid formation [19]. Figure 4 illustrates this behavior for tungsten treated with 0.1 wt.% Ni. The shrinkage is shown *vs.* sintering temperature together with the binary W–Ni phase diagram. For pure tungsten, only slight shrinkage occurs with temperatures up to 1500 °C. However, the addition of nickel gives considerable shrinkage at temperatures below the occurrence of the first melt. This implies that the microstructure at the liquid formation temperature is considerably densified and coarsened beyond that assumed in current models.

The reconciliation of theory with experiment and practical experience has proven difficult. Heating, solubility and homogeneity problems are assumed to be non-existent in the previous theories; in practice, these are not small problems. For example, it is possible to form liquid locally in the microstructure and not to attain uniform densification. The local liquid pools give high density regions surrounded by poorly sintered regions [20]. Another factor is the intersolubility which provides a large chemical driving

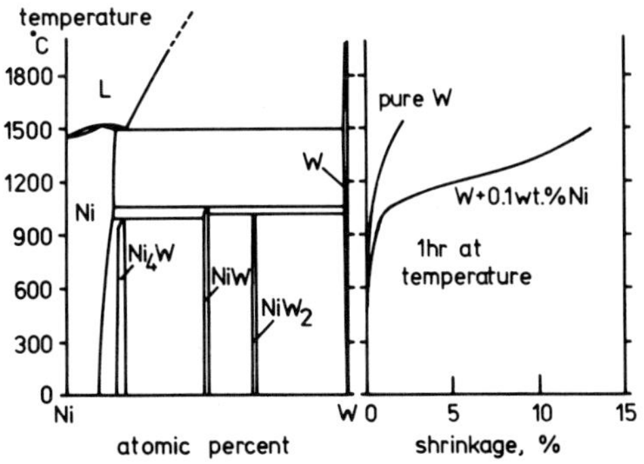

Fig. 4. The W–Ni binary phase diagram and the corresponding shrinkages measured *vs.* temperature for pure tungsten and W–0.1wt.%Ni; considerable shrinkage occurs in the binary system prior to the formation of a liquid phase.

force for mass flow during heating. Depending on the solubility ratio, either swelling or shrinkage can be observed [1]. In a system with no solid state intersolubility, there is little densification prior to liquid formation. Alternatively, in a system with a unipolar solubility of base in additive, there is considerable solubility-driven densification prior to liquid formation.

Densification and microstructural coarsening during heating depend on diffusion rates, solubilities and heating rate. In hard material systems, the LPS process is actually composed of four stages. Initially, during heating, solid state densification occurs as driven by chemical potential gradients. When the first liquid forms, additional densification occurs since the volume of liquid increases rapidly. This liquid is not in equilibrium with the solid. Consequently, it spreads rapidly, penetrating along interparticle contacts where it lubricates the sliding of particles past one another. This allows pore filling by both solid and liquid flow, termed rearrangement. Subsequently, solution–reprecipitation occurs. This is the predominant cause of grain growth after liquid formation; however, grain growth may be very advanced prior to liquid formation. Finally, during solution–reprecipitation the solid–solid contacts form as the equilibrium dihedral angle is established. The attainment of a rigid solid structure interferes with further densification, giving a low final stage densification rate.

4. Model phase diagrams

One goal of recent research has been prediction of systems with favorable characteristics for LPS [4, 5]. In Section 3, several desirable characteristics which have been related to phase diagram features and the underlying thermodynamics [21, 22] were introduced. Since favorable interactions are evident in the phase diagram, computer calculations have been used to isolate the underlying interaction characteristics [4, 5, 23]. Although LPS is not an equilibrium process, phase diagrams do provide useful information with respect to the behavior of mixed components during sintering. Obviously, the melting temperature guides the selection of a sintering temperature.

Figure 5 shows an ideal system for LPS [5]. The desired characteristics are related to the diffusivity, solubility and segregation tendencies as noted in the phase diagram. The system is a binary combination of a base B (the major phase) and an additive A (the minor phase). Only the case of persistent LPS will be treated; however, similar but modified arguments can be made for several related processing techniques [24]. Transport of B through the liquid rich in A requires a high solubility for the base material in the liquid. A high solubility aids wetting of the base particles [1]. Additionally, the base should have a low solubility for the additive. This will lower the amount of additive necessary as well as promote segregation of liquid to the interparticle boundaries to aid densification. If the solubility of A in B is large, then swelling may result during sintering. Decreasing liquidus and solidus lines indicate segregation of the liquid to the interparticle regions between particles. Segregation is also aided by choosing A so that its atom size is smaller than B. A large melting difference between the liquid and the base is essential for enhanced diffusive transport of the base. Thus, optimum phase diagrams for LPS will have deep eutectics, and no high melting intermediate compounds. It should be noted that these characteristics are identical with those desired for activated sintering, except for a lower process temperature.

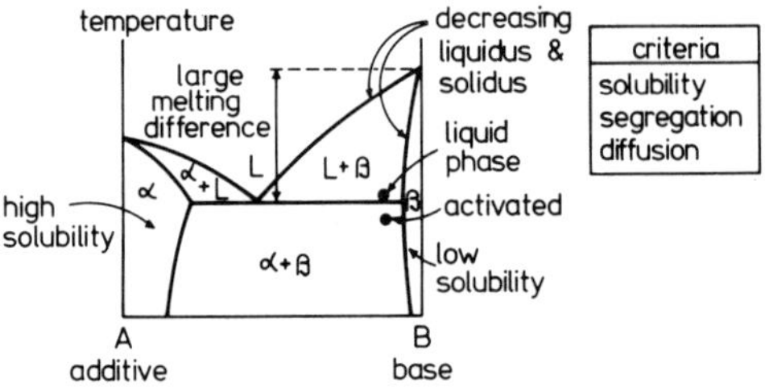

Fig. 5. The ideal binary phase diagram for LPS showing the critical elements of solubility, diffusivity and segregation. LPS occurs in the two-phase field, at a temperature slightly above that for activated sintering.

Phase diagrams constructed using computer simulations [25] provide a link between the thermodynamics and sintering behavior. A key factor in these calculations is the interaction parameter *C*. This term in the regular solution thermodynamic model influences the stability range, melting range and slopes of the liquidus and solidus lines; thus the interaction parameter is an indication of phase stability. In turn, the interaction parameters are related to the electron structure of the binary combination [26].

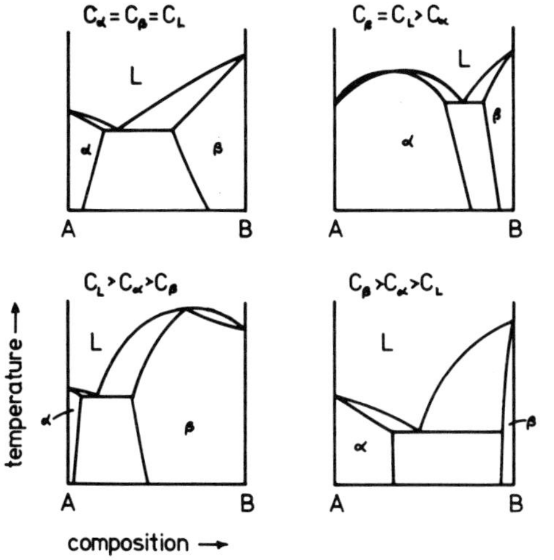

Fig. 6. Four binary phase diagrams from computer simulations showing the effects of variations in the interaction parameters.

Figure 6 shows four phase diagrams based on variations in the interaction parameters for the α, β and liquid phases. Many simulations were conducted using systematic variations in the interactions parameters. On the basis of these calculations, the interrelations that provide the desired phase diagram features of Fig. 5 were isolated. The resulting rank order of interaction parameters was as follows [27]:

$$C_\beta > C_\alpha > C_{\text{liquid}}$$

where the β phase is the stable crystal structure for the base B, and the α phase is the stable crystal structure for the additive A. A low solubility of additive in the base and high solubility of the base in the additive requires the α phase to be more stable than the β phase (stability is implied by a negative interaction parameter since lower free energies are more stable). The desired decreasing liquidus and solidus for the base as it is alloyed with additive is attained by making the liquid more stable than the β phase. Finally, the desired low melting temperature liquid implies stability for the liquid with respect to both phases. Consequently, the rank order of interaction parameters indicates that the liquid is the most favorable phase and the β phase is the least favorable. Obviously, this ranking biases mass flow from the base into the α phase and liquid.

The identification of candidate LPS systems is aided by inspection of binary phase diagrams, where favorable features have been isolated. A favorable system will have the greatest thermodynamic stability (as measured by a relatively low interaction parameter) associated with the liquid, and the least stability for the base. This sequence of interaction parameters suggests a pattern for selection of LPS systems based on electron structure.

5. Updated theory

Beyond basic theoretical advances, experiments were conducted to isolate the effect of three major variables previously not treated by theory — solubility, homogeneity in the green microstructure and heating rate [28]. Previous experiments showed these factors were important to LPS [29–31]. To conduct these studies, the W–Ni–Cu system was selected because of the systematic change in tungsten solubility with the copper-to-nickel ratio. The tungsten powder had a mean size of 2.5 μm. Variations in the solubility were obtained by forming alloys with 90 or 95 wt.% W, and the balance composed of 100% Cu (100 Cu), 70% Cu (70 Cu), 30% Cu (30 Cu) or 0% Cu (0 Cu). The respective liquid formation temperatures for these systems were 1075 °C, 1229 °C, 1293 °C and 1483 °C. The lack of solid state solubility in the Cu–W binary, in contrast with the high solubility of the Ni–W binary, allows solubility adjustment through the composition (nickel and copper have total intersolubility). The best homogeneity was obtained by coating layers onto tungsten powder through nitrate salt decomposition [19]. Lower values of homogeneity were obtained by comixing nickel of 1 or 36 μm size with the tungsten powder. Generally, compaction was at 150–200 MPa giving a green density of 50% of the theoretical density.

Dilatometry provides insight into the densification process. Figure 7 shows a plot of the shrinkage rate and cumulative shrinkage *vs.* homologous temperature for the 30 Cu alloy at three different heating rates. Solid state shrinkage increases with a slower heating rate while rearrangement at the liquid formation temperature increases with increasing heating rate.

The solubility effect on densification is demonstrated in Fig. 8 for a heating rate of 6 K min^{-1}. The greatest shrinkage during heating occurs for the pure nickel binder (0 Cu) and the least

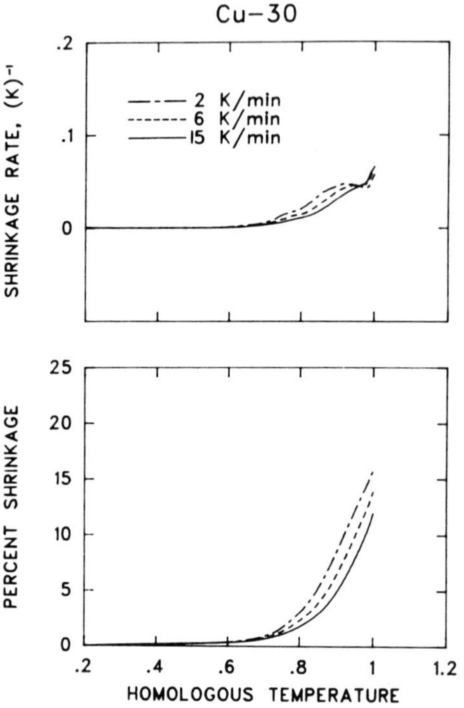

Fig. 7. Dilatometer data for the 30 Cu alloy heated with three rates. The upper plots show the instantaneous shrinkage rates and the lower plots give the cumulative shrinkages.

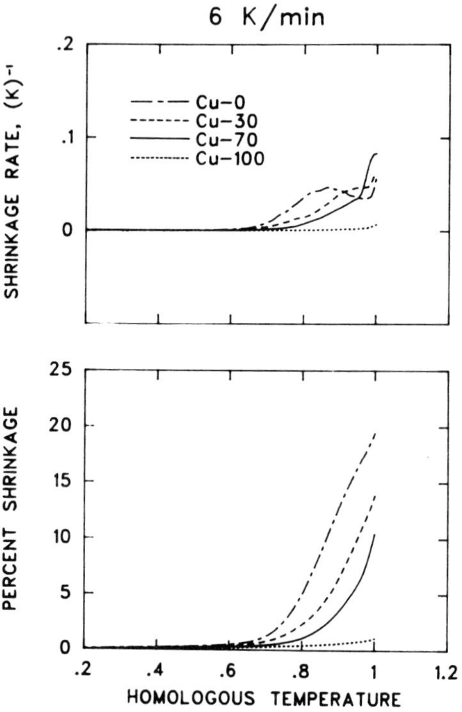

Fig. 8. Shrinkage and shrinkage rate data for four W–Ni–Cu alloys heated at 6 K min^{-1} to the liquid phase formation temperatures.

shrinkage occurs for the pure copper binder (100 Cu). Such results at all three heating rates indicate that solubility has a major effect on densification prior to liquid formation. After the liquid forms, the shrinkage rate is highest for the intermediate solubility compositions. This indicates that the amount of densification by rearrangement varies with both the porosity and the solubility increase at liquid formation; rearrangement occurs by particle dissolution and liquid penetration of the solid state sinter bonds. The largest net shrinkage was for the 0 Cu composition at all heating rates. The shrinkage rate after liquid formation was highest for the 70 Cu composition, but the net shrinkage never matched that of the 0 Cu or 30 Cu compositions. This shows the dominant effect of solid-state-activated sintering at all heating rates investigated. Experiments were also conducted with variations in the liquid content. No significant effect was noted in contrast with the large solubility role. Thus the differences shown in Fig. 8 are dominated by solid state solubility.

The third parameter investigated was the initial green compact homogeneity. Figure 9 shows the variation in shrinkage behavior for 0 Cu heated at 6 K min^{-1}, formed from coated and comixed powders. Up to the liquid formation temperature the comixed powders are inferior to the coated powder. As expected, a greater homogeneity in the green compact promotes densification by solid-state-activated sintering. After liquid formation, the shrinkage rate was higher for the lower density compact formed from the coarse nickel powder. Figure 10 presents optical micrographs of the 0 Cu samples formed from coated and comixed powders. The densification with the coated powder is substantially greater than with the comixed powder, and the coarse nickel powder creates a very inhomogeneous sintered structure. With the same content of nickel and a small particle size, rapid densification is attained. This points to a major role of initial homogeneity with respect to densification. Such a role has not been previously incorporated in LPS theory, although

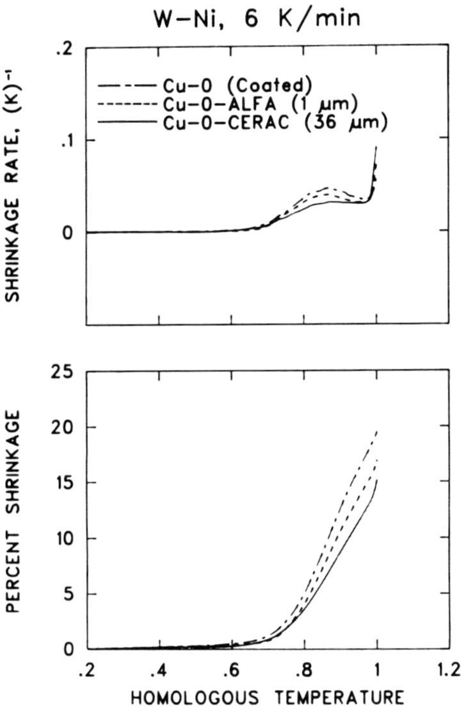

Fig. 9. The effect of homogeneity on the shrinkage behavior of 95wt.%W–5wt.%Ni alloys composed of coated tungsten powder, and comixed powders with nickel sizes of 1 or 36 μm.

it has been previously recognized [17]. Best sintering occurs with coated tungsten powders which represent a high initial homogeneity.

Measurements of total porosity and open porosity were coupled with pore size distribution measurements. Open pore structures were present up to approximately 90% density. However, closed porosity began to form at slightly over 50% density, and equal populations of closed and open pores were present at 85% density. As the nickel-to-copper ratio was increased, the porosity and pore size decreased at all temperatures prior to liquid formation. Also, for densities below approximately 70% of the theoretical density, the pore size distribution shifts to finer pore sizes with decreasing porosity. Further sintering was accomplished by an increase in pore size due to coarsening. Thus the microstructural evolution initially showed densification with pore shrinkage, followed by grain growth and pore coarsening in the latter stages of densification. All these changes took place prior to the formation of

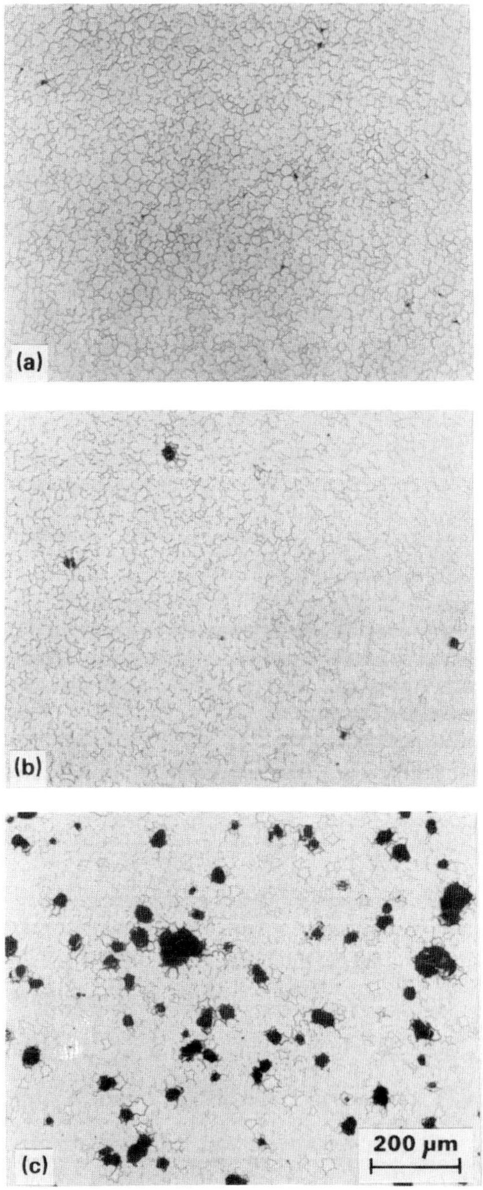

Fig. 10. Optical micrographs of the sintered structures for three 95wt.%W–5wt.%Ni compacts having varying levels of initial homogeneity (see Fig. 9): (a) coated; (b) comixed 1 μm nickel; (c) comixed 36 μm nickel.

the first liquid, contradicting the fundamental assumptions of classic LPS theory.

For rearrangement to take place in LPS, liquid films have to attack the solid state bonds, thereby converting the compact into a viscous mass. To establish that this behavior occurs under conditions of slow heating, a sample was loaded in tension and sintered in a vertical furnace. Liquid penetration of the solid–solid contacts resulted in rupture. Prior calculations and observations suggest that such transient liquid wetting of grain boundaries does occur because of mass transport when the solid dissolves into the liquid [29]. The structure at liquid formation was captured by quenching the failed sample. Figure 11 is a scanning electron micrograph of this fracture. It shows smooth round tungsten grains with no evidence of solid–solid bonding. This establishes that rearrangement can occur under conditions of dissolution at liquid formation, in spite of prior solid state sintering.

This research provides a basis for formulating a more realistic theory of LPS. In this spirit, a computer model of LPS is being formulated which includes heating rate, solid state sintering, particle size, solubility and homogeneity factors. The main features of the updated theory are incorporation of solid state sintering via activated sintering during heating and homogeneity effects on densification and grain growth prior to liquid formation. The model uses a solid-state-activated sintering model [32] as the basis for creating the computer simulation. Early results indicate that this approach will provide a more realistic representation of LPS.

6. Microstructural models

Microstructural evolution is a very important concern in LPS. The final microstructure depends largely on the dihedral angle and liquid content [1]. The contiguity and connectivity are dictated by the combination of these two features. Figures 12 and 13 illustrate this behavior as applied to LPS. Figure 12 plots the contiguity of various carbides *vs.* volume fraction of solid phase. The contiguity is the amount of solid–solid contact between grains after sintering. The calculated behavior from a recent model [33] is included for dihedral angles of 15°, 30°, 45°, 60° and 75°. The experimental variations in contiguity are well matched with the theoretical curves. In Fig. 13 the two-dimensional contacts per grain are shown *vs.* the volume fraction of solid phase for two liquid-phase-sintered systems. Again the model calculations [34] represented by the full lines fit the experimental pattern well. Both connectivity and contiguity, as well as the

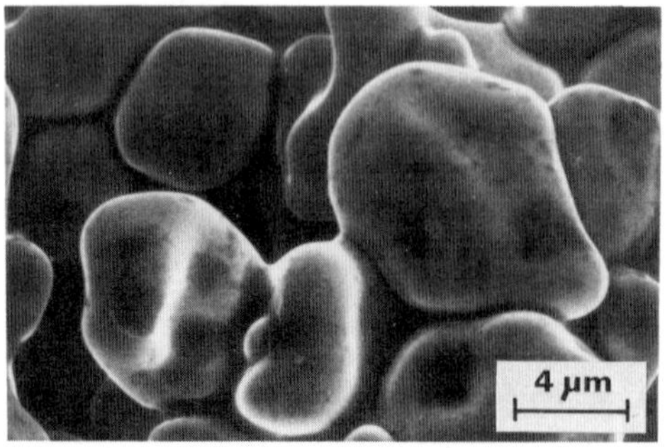

Fig. 11. The fracture surface of a 95wt.%W–5wt.%Ni sample which failed on liquid formation, showing no solid–solid bonding.

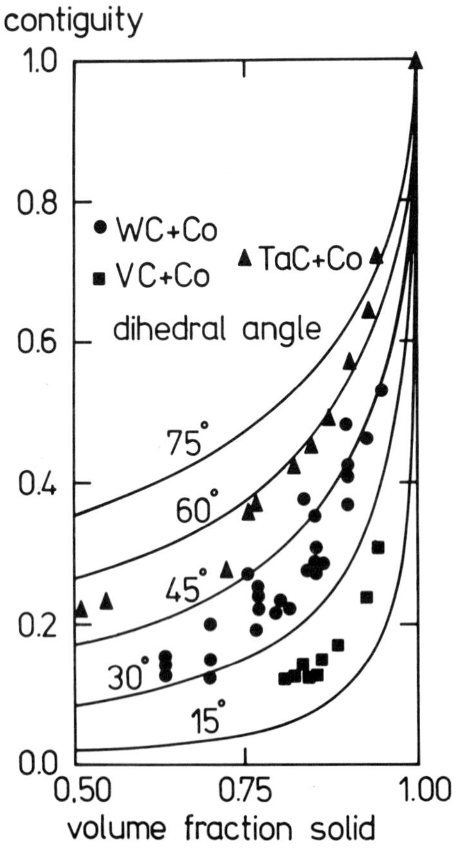

Fig. 12. A plot of the contiguity for three carbides *vs.* the volume fraction of solid phase. Curves predicted by a recent model [33] are shown for dihedral angles ranging from 15° to 75°.

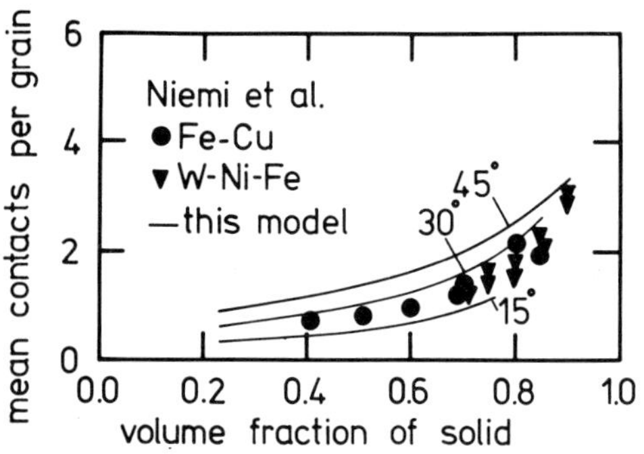

Fig. 13. The two-dimensional connectivity of liquid-phase-sintered materials shown *vs.* the volume fraction of solid phase. Curves predicted by recent computer simulations [34] are shown for dihedral angles of 15°, 30° and 45°.

grain size, are important to the properties of LPS materials.

Material properties depend on the microstructure after sintering [35–39]. Thus, variations in the processing path provide means for altering the sintered properties. Because of the improved understanding, additional microstructural control factors have become accessible through processing or chemistry adjustments.

7. Conclusions

The kinetics of LPS are better understood in terms of system characteristics, densification stages and practical processing concerns, including solubility, heating rate and compact homogeneity effects. A favorable combination of characteristics is visible in the binary phase diagram. Solid state sintering (activated sintering) must be incorporated to predict accurately densification and microstructure evolution. The refinements discussed here reflect actual LPS systems better. A critical observation is that, even with high sintered densities, the newly formed liquid penetrates solid–solid bonds. The penetration depends on an increased solid solubility in the liquid at the melting point. Thus, rearrangement is possible even after slow heating. These experiments establish a basis for bringing theory closer to reality in LPS.

Acknowledgments

The concepts presented in this paper represent inputs from various research projects; the comments of Fritz Lenel, Animesh Bose, Barry Rabin and Deepak Madan were most helpful.

This study was performed under funding from the U.S. Army Research Office (Contract DAAG29-84-K-0026) and the National Aeronautics and Space Administration (Grant NAG 3-744).

References

1 R. M. German, *Liquid Phase Sintering*, Plenum, New York, 1985.
2 V. N. Eremenko, Y. V. Naidich and I. A. Lavrinenko, *Liquid Phase Sintering*, Consultants Bureau, New York, 1970.
3 P. Schwartzkopf and R. Kieffer, *Cemented Carbides*, Macmillan, New York, 1960.
4 R. M. German, in E. N. Aqua and C. I. Whitman (eds.), *Modern Developments in Powder Metallurgy,* Vol. 15, Metal Powder Industries Federation, Princeton, NJ, 1985, p. 253.
5 R. M. German and B. H. Rabin, *Powder Metall., 28* (1985) 7.
6 G. H. S. Price, C. J. Smithells and S. V. Williams, *J. Inst. Met., 62* (1938) 239.
7 F. V. Lenel, *Trans. AIME, 175* (1948) 878.
8 H. S. Cannon and F. V. Lenel, in F. Benesovsky (ed.), *Proc. Plansee Semin.*, Metallwerk Plansee, Reutte, 1953, p. 106.
9 W. D. Kingery, *J. Appl. Phys., 30* (1959) 301.
10 L. Froschauer and R. M. Fulrath, *J. Mater. Sci., 11* (1976) 142.
11 L. Froschauer and R. M. Fulrath, *J. Mater. Sci., 10* (1975) 2146.
12 J. White, in G. C. Kuczynski, N. Hooton and C. Gibbon (eds.), *Sintering and Related Phenomena*, Gordon and Breach, New York, 1967, p. 81.
13 T. J. Whalen and M. Humenik, in G. C. Kuczynski, N. Hooton and C. Gibbon (eds.), *Sintering and Related Phenomena*, Gordon and Breach, New York, NY, 1967, p. 15.
14 G. H. Gessinger, H. F. Fischmeister and H. L. Lukas, *Powder Metall., 16* (1973) 119.
15 M. Jeandin, J. L. Koutny and Y. Bienvenu, *Powder Metall., 17* (1983) 26.
16 K. S. Hwang, R. M. German and F. V. Lenel, *Metall. Trans. A, 18* (1987) 11.
17 W. J. Huppmann and W. Bauer, *Powder Metall., 18* (1975) 249.
18 C. J. Li and R. M. German, *Int. J. Powder Metall. Powder Technol., 20* (1984) 149.
19 C. J. Li and R. M. German, *Metall. Trans. A, 14* (1983) 2031.
20 O. J. Kwon and D. N. Yoon, in G. C. Kuczynski (ed.), *Sintering Processes*, Plenum, New York, 1980, p. 203.
21 G. V. Samsonov and V. I. Yakovlev, *Z. Metallkd., 62* (1971) 621.
22 R. M. German and Z. A. Munir, *Rev. Powder Metall. Phys. Ceram., 2* (1982) 9.
23 R. M. German, in W. A. Kaysser and W. J. Huppmann (eds.), *Horizons of Powder Metallurgy*, Part II, Schmid,

Freiburg, 1986, p. 1239.
24 R. M. German, *J. Met.*, *38* (1986) 26.
25 L. Kaufman and H. Bernstein, *Computer Calculation of Phase Diagrams*, Academic Press, New York, 1970.
26 L. Brewer, in V. F. Zackay (ed.), *High Strength Materials*, Wiley, New York, 1965, p. 12.
27 C. M. Kipphut and R. M. German, *Sci. Sintering, 20* (1988) 31.
28 S. Farooq, A. Bose and R. M. German, *Prog. Powder Metall., 43* (1987) 65.
29 R. M. German, *Int. J. Powder Metall., 22* (1986) 31.
30 P. Ramakrishnan and B. K. Agrawal, *Int. J. Powder Metall., 5* (1969) 79.
31 F. J. Puckert, W. A. Kaysser and G. Petzow, *Z. Metallkd., 74* (1983) 737.
32 R. M. German, *Sci. Sintering, 15* (1983) 27.
33 R. M. German, *Metall. Trans. A, 16* (1985) 1247.
34 R. M. German, *Metall. Trans. A, 18* (1987) 909.
35 G. Grathwohl and R. Warren, *Mater. Sci. Eng., 14* (1974) 55.
36 J. Gurland, *Mater. Sci. Eng., 40* (1979) 59.
37 K. S. Churn and R. M. German, *Metall. Trans. A, 15* (1984) 331.
38 G. J. Witter and W. R. Warke, *IEEE Trans. Parts Hybrids Packag., 11* (1975) 21.
39 J. Larsen-Basse, *J. Met., 35* (1983) 35.

Materials Science and Engineering, *A105/106* (1988) 225–231

Kinetics of Compositional Modification of (W, Ti)C–WC–Co Alloy Surfaces*

M. SCHWARZKOPF, H. E. EXNER and H. F. FISCHMEISTER

Max-Planck-Institut für Metallforschung, Institut für Werkstoffwissenschaften, Seestraße 92, D-7000 Stuttgart (F.R.G.)

W. SCHINTLMEISTER

Metallwerk Plansee G.m.b.H. A-6600 Reutte (Austria)

(Received December 14, 1987)

Abstract

A tough surface zone can be formed on cemented carbide tool bits by means of a special treatment involving outgassing of nitrogen. In this paper, the influences of heat treatment conditions and alloy composition on the thickness of this zone are investigated and compared with theoretical predictions. It is shown that the kinetics of zone formation conform to a model in which the inward diffusion of metal atoms dissolved in the binder phase is considered to be the rate-controlling step.

1. Introduction

The lifetimes of cemented-carbide tool bits in metal working can be considerably increased by coating the wear faces with a thin layer of a hard material such as TiC, TiN or Al_2O_3 or with multilayers of such hard phases. However, catastrophic failure can occur more easily than with uncoated bits since cracks are easily formed in the brittle coatings which can proceed into the bulk material. It has become industrial practice to form tough surface zones on the carbide bits before coating using chemical vapour deposition or physical vapour deposition. These zones are depleted of cubic carbides and enriched with the metal binder. Because WC–Co has a higher crack resistance than WC–cubic-carbide–Co alloys, unavoidable cracks in the brittle coatings are stopped. Knowledge of the kinetics of layer formation is important for controlling the sintering process in order to achieve a zone of optimum thickness at low additional sintering cost.

Techniques for forming such zones have been described by various workers [1–3]. Nitrogen plays a decisive role and several mechanisms have been proposed to explain the observations [4–6]. Of these, the model by Suzuki *et al.* [6] tries to establish a quantitative basis for describing the kinetics of formation of cubic-carbide-free zones on WC–TiC–TiN–Co and WC–(W,Ti)(N,C)–Co alloys. While the experimental results of Suzuki *et al.* [6] seem to confirm this model, some qualitative and quantitative evidence will be given in the following that the kinetics are more complex than anticipated by earlier researchers.

2. Experimental details

Mixtures of WC, cobalt, cubic carbides ((W,Ti)C and (Ta,Nb)C), carbon black and nitrogen-containing phases (TiN, Ti(C,N), AlN and TaN) were prepared by the usual route of cemented-carbide technology. To ensure good mixing, extended ball or attritor milling (120 h and 10 h respectively) was applied. Sintering and annealing were carried out in a laboratory induction furnace under a controlled atmosphere (argon at 27 Pa). Pressed samples (245 MPa) were heated in a carefully controlled cycle including a holding time of 0.5 h at 1260 °C and heating for 0.25 h to a final sintering temperature between 1390 and 1540 °C. Holding times at maximum temperature ranged from immediate cooling to 16 h.

Optical microscopy of polished cross-sections was used to measure the thickness of the surface zone at approximately ten positions along the well-preserved edge at a magnification of 1000 ×. The standard deviation of these measurements was approximately ± 1 μm. Reactively sputtered

*Paper presented at the 3rd International Conference on the Science of Hard Materials, Nassau, The Bahamas, November 9–13, 1987.

0921-5093/88/$3.50

© Elsevier Sequoia/Printed in The Netherlands

interference layers were used for contrast enhancement which, compared with chemical etching, gives a much clearer differentiation of the phases. Scanning electron micrographs were used to determine the area fractions of the carbide, carbonitride and binder phases. The distribution of the cobalt, titanium and tungsten contents perpendicular to the surface was measured by energy-dispersive X-ray spectroscopy. Results for 10–20 fields (20 μm wide) arranged parallel to the surface were averaged and normalized with respect to the composition of the alloy at a large depth below the surface. The nitrogen contents of the nitrogen-containing additions, of the pressed compacts and of the sintered alloys were analysed by hot-extrusion analysis. A more detailed description of the experimental procedures and the preparation of the alloys [7] is available from the present authors.

3. Models for the formation of tough surface zones

Earlier workers [4–6] consider the stability of the nitrogen-containing phase as the criterion for the dissolution of the cubic mixed crystals when nitrogen is expelled from the alloy during sintering in vacuum. Suzuki *et al.* [6] tried to describe the kinetics quantitatively by assuming that the decrease in the nitrogen content in the liquid binder phase determines the decay of the cubic phase and that the outward diffusion of nitrogen through the binder phase is the rate-controlling step for the formation of the cubic-carbide-free zone. This mechanism results in the typical parabolic dependence of zone thickness X and sintering time t:

$$X^2 = 2k_1 V_v \frac{D_N}{C_N}(c_B - c_S)t \qquad (1)$$

where V_V is the volume fraction of the binder phase, k_1 is a numerical factor (approximately 1.5–2 depending on the alloy composition) which considers the enrichment of cobalt in the region close to the surface, D_N is the diffusivity of nitrogen in the liquid binder phase, C_N is the nitrogen content in the bulk alloy, and c_B and c_S are the nitrogen concentrations in the liquid binder at the boundary of the cubic-carbide-free zone and at the surface respectively.

There are two qualitative arguments which demonstrate that this simple model does not hold for the general case of a cubic-carbide-free zone.

(i) There is no reason why the cubic mixed carbides (W, Ti, Ta, Nb)C should dissolve in the binder when nitrogen outgases during liquid phase sintering. The denitrification process just determines the decay of nitrogen-containing phases (nitrides and carbonitrides).

(ii) At the edges of the tool bit, the cubic carbides extend further out to the surface, *i.e.* the thickness of the zone free of cubic crystals is smaller at the edges than at the plane faces (Fig. 1). In contrast, the thickness of the zone free from nitrides shows the expected increase towards the edges (Fig. 2).

As shown schematically in Fig. 3, this observation (among others) strongly suggests that the decay of nitrogen-containing phases present in the bulk of the alloy (nitrides and carbonitrides

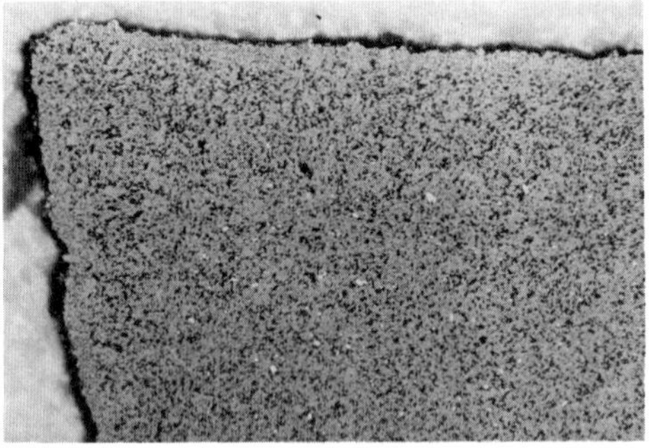

Fig. 1. Shape of the front of the cubic-carbide-free zone at the surface of a technical alloy (6wt.%Co–3.5wt.%(W,Ti)C–5.5wt.%(Ta,Nb)C–83.5wt.%WC–1.5wt.%TiN) after it had been annealed for 1 h at 1540 °C in vacuum. (The larger thickness away from the edge should be noted.)

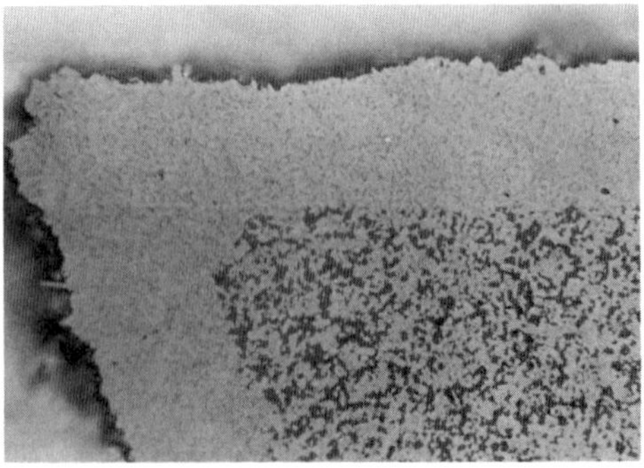

Fig. 2. Shape of the front of the TiN-free zone in an experimental alloy (9wt.%Co–5.5wt.%TiN–85.5wt.%WC) after it had been annealed for 1 h at 1540 °C in vacuum. (The larger thickness at the edge should be noted.)

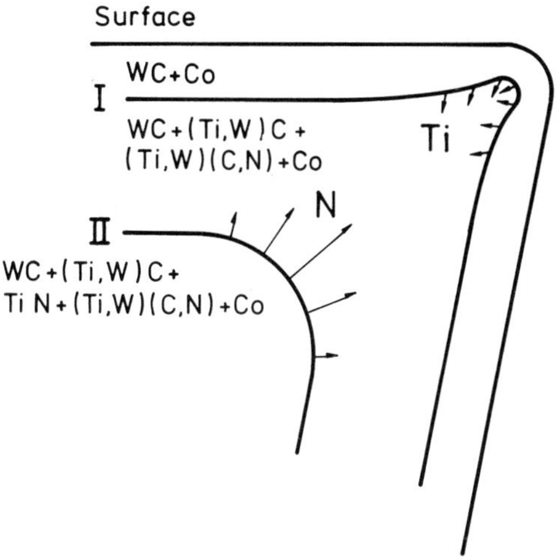

Fig. 3. Schematic plot of the fronts where the cubic carbide crystals (I) and where the nitrides and the carbonitrides present in the bulk alloy (II) become unstable. (The arrows indicate directions of nitrogen and titanium fluxes.)

formed from the non-equilibrium phases during densification) is controlled by the outward diffusion of nitrogen while the rate of depletion in cubic phases at the surface (cubic carbides and carbonitrides formed close to the surface) is determined by the inward diffusion of titanium (and of tantalum, niobium or the other transition metals contained in the cubic carbides of the alloy). In our model, we suggest that the flux of the metal atoms is driven by the activity gradient set up in the liquid binder phase by the concentration gradient of nitrogen. The cubic carbonitrides (in equilibrium in the bulk) decay when the solubility product $L=[W][Ti][N][C]$ in the binder is reached at a certain distance from the surface owing to the decrease in the nitrogen concentration [N]. As a result of the presence of free carbon, the activity of carbon is unity throughout the binder and [W][C] is constant owing to the presence of WC. Thus, $[Ti][N]=$ constant. Taking the derivative with respect to the distance x from the surface, we find that the titanium gradient is proportional and opposite to the nitrogen gradient, *i.e.* $dc_{Ti}/dx=-(c_{Ti}/c_N)dc_N/dx$.

For alloy compositions of the type WC–TiC–TiN–Co, all those processes which are important for the formation of the cubic-carbide-free zone can be quantitatively assessed by a model the details of which have been given elsewhere [7, 8]. Here, we give a qualitative description of our view together with the result of the analytical derivation.

When nitrogen outgases at the surface during vacuum annealing of the dense alloy, the nitrogen-containing phase which is the least stable dissolves in the binder. This will usually be the shells of a nitrogen-rich cubic carbonitride formed by reaction of the carbide and nitride powders during sintering. The formation of such shells (which are found in most technical alloys containing cubic carbides) has been studied in detail for nitrogen-free alloys by May [9]. Occasionally, original nitride particles may be protected from dissolution by more stable shells. The metallic elements form a carbon-rich carbonitride if the alloy is carbon rich, or a substoichiometric cubic phase (and eventually η phase) if the alloy is carbon deficient. We consider here only the case when there is free carbon, yielding stoichiometric phases. When the original nitride and all carbonitrides are fully dissolved at the surface, a nitrogen gradient develops in the liquid binder phase. If other nitrogen-containing phases (*e.g.* TiN) are present, these will decay when the nitrogen concentration of the liquid phase at the surface decreases further. Eventually, a zone free of nitrogen-containing phases forms and a nitrogen profile, which decreases from a high nitrogen concentration in the bulk alloy (determined by the solubility product of the least stable nitrogen-containing phase) to a low concentration at the surface (determined by the partial pressure of nitrogen in the furnace atmosphere according to Sieverts' law), develops in the binder phase.

In our model, we suggest that a flux of those metal atoms which have a high affinity for nitrogen, *e.g.* titanium, tantalum or niobium, is set up in the binder phase. Owing to the depletion of these elements in the binder phase, the solubility product of the cubic carbide will be reached at the surface and the cubic carbide will start to dissolve in the binder. When all cubic crystals have been dissolved, the cubic-carbide-free zone is formed. The rate at which the carbonitride, nitride and cubic-carbide-free zones grow are different, but they are coupled by the nitrogen gradient, which controls the outgassing as well as the diffusion of titanium and those of other transition elements forming cubic crystals, *e.g.* tantalum or niobium, if present in the alloy. Farther away from the surface, these metallic elements react with carbon, nitrogen and tungsten to form carbonitrides, the composition of which corresponds to the equilibrium with the nitrogen composition of the liquid cobalt binder in the multicomponent phase diagram. According to earlier studies [5], the composition of $(Ti_{1-x}, W_x)(N_{1-y}, C_y)$ will vary with decreasing distance from the surface such that a lower nitrogen concentration in the liquid phase corresponds to a lower nitrogen content (*i.e.* a large y) and a lower tungsten content (*i.e.* a small x) (or a higher TiC content) of the carbonitride.

In alloys which have a sufficiently high content of hard phases, a skeleton exists throughout the sintering and vacuum annealing process. Since this rigid skeleton fixes the total volume, the precipitating carbonitride pushes away the liquid cobalt which then flows back to the zone in which the loss of cubic carbide by diffusion causes a volume deficit. This means that cobalt is enriched at the surface. Figure 4 shows the composition profiles obtained by X-ray spectroscopy (it should be noted that the content of the elements and the distance from the surface are normalized with respect to the composition of the bulk and the width of the cubic-carbide-free zone respectively). The compositional changes due to the processes described above are clearly revealed. The decrease in cobalt concentration close to the surface and the corresponding increase in the tungsten content are due to cobalt evaporation during vacuum annealing.

The rate of this complex reaction is thus controlled by both the nitrogen and the titanium diffusion. This results in the following dependence of the thickness of the cubic-carbide-free zone on the process parameters:

$$X^2 = 2k_2 A_A^* \frac{D_{Ti}^2}{D_N} C_N(c_A - c_S)\left(\frac{c_{Ti}^*}{c_N^*}\right)^2 t \qquad (2)$$

where k_2 is a constant related to the composition of the alloy, A_A^* is the minimum cobalt concentration, *i.e.* at the location where the volume of precipitated carbonitrides reaches its maximum. c_{Ti}^* and c_N^* are the concentrations of titanium and nitrogen respectively at the same location, *i.e.* where at a given time t no precipitation or dissolution takes place and, therefore, Fick's first law holds for the diffusion of both nitrogen and titanium. D_{Ti} is the diffusivity of titanium in the liquid binder, c_A is the nitrogen concentration of the binder in the bulk, and D_N, C_N, c_S and t have the same meanings as in eqn. (1). For a given alloy composition, A_A^*, c_{Ti}^* and c_N^* are constant

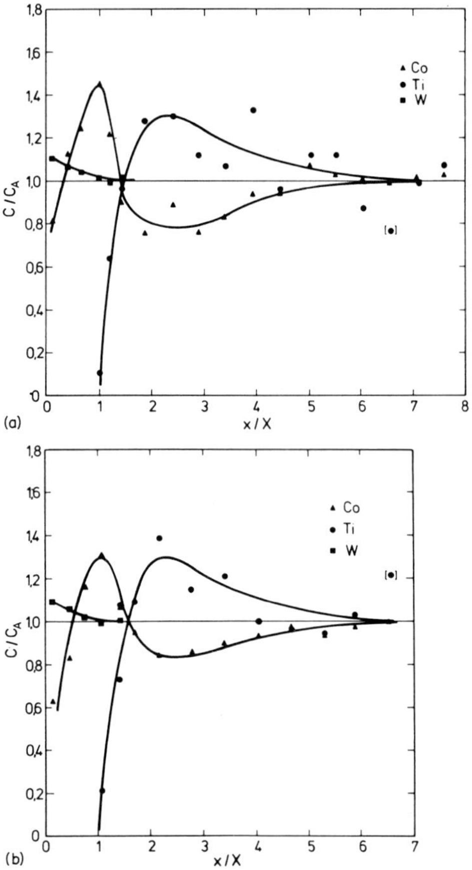

Fig. 4. Concentration profiles of elements in the near-surface region of nitrogen-containing alloys annealed for 1 h at 1540 °C in vacuum. The distances x are normalized with respect to the thickness X of the carbide-free layer and the concentrations C with respect to the content C_A of the bulk alloy. (The composition of the compact is (a) WC–12.9wt.%-(70WC–30TiC)–1.5wt.%TiN–9wt.%Co–0.4wt.%C and (b) WC – 3.5wt.%(70WC – 30TiC) – 5.5wt.%(90TaC – 10NbC) – 9wt.%Co–0.4wt.%C.)

throughout the isothermal part of the sintering cycle and eqn. (2) simplifies to

$$X^2 = k \frac{D_{Ti}^2}{D_N} C_N(c_A - c_S)t \qquad (3)$$

where k is a constant depending on the alloy composition as specified previously [7, 8]. In section 4, some predictions of this rate equation are analysed and compared with the experimental results.

4. Discussion

Although most physical and thermochemical data needed for a fully quantitative comparison of the rate of cubic-carbide-free zone are not available, the influence of the individual processing parameters can be predicted from eqn. (3) in a semiquantitative way. Here we discuss the effect of time and temperature of vacuum annealing and of some aspects of alloy composition.

Examples for the time dependence on the thickness of the cubic-carbide-free zone are shown in Fig. 5 where the square X^2 of the thickness is plotted *vs.* the time t after the annealing temperature had been reached. The results for three annealing temperatures—the highest and lowest temperatures applied and an intermediate temperature typical of the technical range of sintering temperatures—are shown. Slight deviations from zone thickness at zero time are observed. At 1390 °C, a short incubation time is ascribed to the fact that the zone formation starts only when full density is reached and nitrogen does not escape through open pores. At higher temperatures, a thin layer forms during heating to annealing temperatures after the alloy has reached full density at a lower temperature. Apart from these circumstances, the parabolic relationship predicted by our model as well as by Suzuki *et al.* [6] is closely observed.

The temperature dependence of the growth rate is determined by the exponential terms of the diffusivities and solubilities of titanium and nitrogen in the liquid binder, while the influence

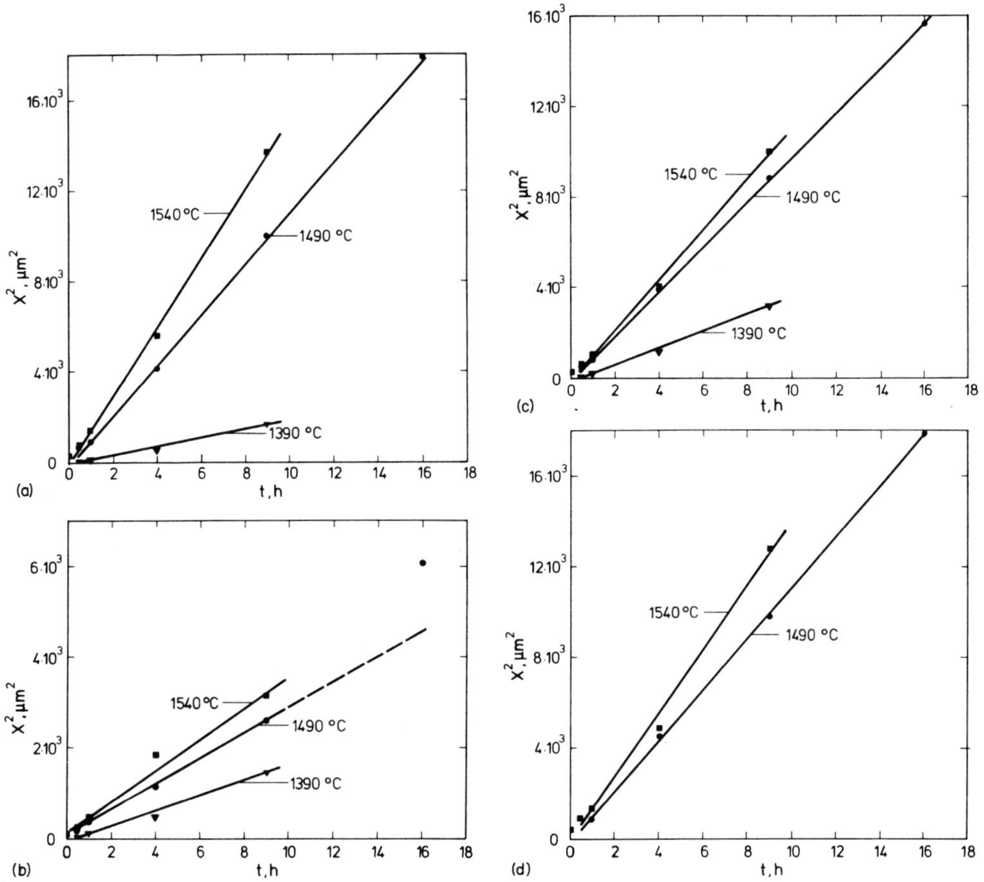

Fig. 5. Time dependence of the thickness of the cubic-carbide-free zones at three annealing temperatures. (The composition of the compact is (a) as in Fig. 4 (a), (b) WC–3.5wt.%(70WC–30TiC)–5.5wt.%(90TaC–10NbC)–0.75wt.%TiN–9wt.%Co–0.4wt.%C, (c) as in Fig. 4 (b) and (d) WC–0.6wt.%(70WC–30TiC)–5.5wt.%(90TaC–10NbC)–2.5wt.%Ti(C,N)–9wt.%Co–0.8wt.%C.)

of temperature on all other terms can be neglected. The term $c_A - c_S$ thus is represented by the difference of two exponential functions and, as a consequence, cannot be expressed as a single exponential. Therefore the temperature dependence of the growth rate can be described by an Arrhenius-type relation only if the concentration c_S of nitrogen in the liquid binder at the surface is very small compared with the concentration c_A in the bulk of the alloy. This is the case if the partial pressure of nitrogen in the furnace atmosphere is very low, or if the temperature is high, since the heat of solution of gaseous nitrogen (approximately 50 kJ mol^{-1}) is smaller than that of carbonitride (approximately 200 kJ mol^{-1}). The activation energies for the diffusivities of metallic elements in liquid cobalt are of the order of 30 kJ mol^{-1}; that of nitrogen is of the order of 50 kJ mol^{-1}. Therefore the major effect of temperature is that on the stability of the nitrogen-containing phase in the bulk, *i.e.* of the carbonitride shells formed during sintering. The heat of fusion of $(W_x–Ti_{1-x})(C_y–N_{1-y})$ is estimated to be of the order of 100–300 kJ mol^{-1} which compares favourably with the value obtained from an Arrhenius plot in the high temperature range ($T > 1700$ K) for the alloy made from a mixture of 12.9 wt.% WC–TiC, 1.5 wt.% TiN, 9% wt.% Co and 0.4 wt.% C, with the balance WC (193 kJ mol^{-1}). At lower temperatures ($T < 1700$ K), a very high effective activation energy is obtained for this alloy (higher than 500 kJ mol^{-1}); this has no physical meaning but is an effect of the rapidly decreasing difference $c_A - c_S$. The rate decreases to zero when the nitrogen partial pressure in the atmosphere is in equilibrium with the nitrogen activity in the binder corresponding to that of the carbonitride.

The compositional effects on the thickness of the cubic-carbide-free zone are a result of the complicated interplay between the stability of the nitrogen-containing phase and the amount of nitrogen in the alloy. Together, they determine the nitrogen gradient, the composition and the amount of cubic carbides which influence the amount of material diffusing away from the surface, and of the binder phase content which essentially controls the cross-section in which the diffusion takes place. It is interesting to note that no cubic-carbide-free layer is formed if the cobalt content is below approximately 4 wt.% since all binder phase channels are closed by the precipitated carbonitride. Also, no zone is formed if the nitride-containing phase is very unstable and decays before full density is reached (*e.g.* TaN), or if the amount of nitrogen becomes so small that the front of nitrogen-containing phases present in the bulk alloy moves away from the surface very fast.

Here we shall focus on the amount of the rate-controlling species in the alloys prepared from a mixture of WC, (W, Ti)C, (Ta, Nb)C, cobalt and TiN or TiCN, *i.e.* nitrogen, titanium and tantalum or niobium. Comparing the rates for alloys with a constant content of titanium and various amounts of nitrogen (Fig. 6) demonstrates that our model describes the experimental findings, correctly predicting an increasing growth rate of the cubic-carbide-free zone with increasing C_N. The functional relationship is not simply linear as suggested by eqn. (3), because the relative distances from the surface of the fronts of the cubic carbides and the nitrogen-containing phases vary with the ratio of titanium and nitrogen in a complex way which reduces the constant k_2 in eqn. (2), counteracting the effect of a higher nitrogen content. Obviously, Suzuki's eqn. (1) does not consider these effects in a realistic way. This failure of the model proposed by Suzuki *et al.* [6] to describe the formation of cubic-carbide-free zones is also demonstrated in our results with alloys in which the amount of nitrogen-containing phase was kept constant (1.5 wt.% TiN) and the amount of TiC is varied or different amounts of TaC–NbC are added (Fig. 7). The hyperbolic relationship between the growth rate X^2/t (or the square of the zone thickness at constant t (see Fig.

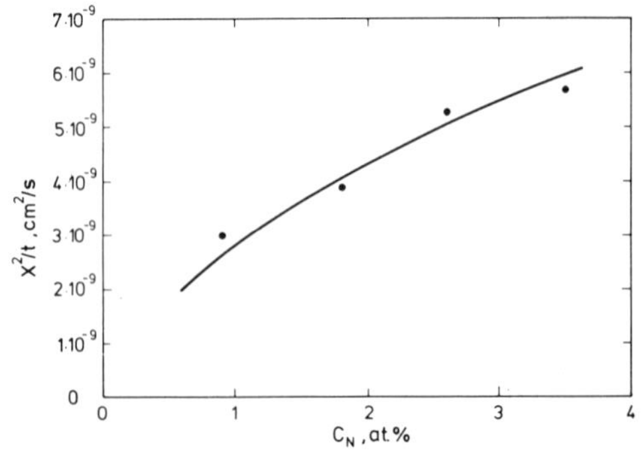

Fig. 6. Growth rate X^2/t of the cubic-carbide-free zone at 1540 °C as a function of nitrogen content at a constant titanium content of the alloy. (The composition of the alloys was varied by exchanging (W,Ti)C by TiN at a constant atomic content of 7.4 at.% Ti.)

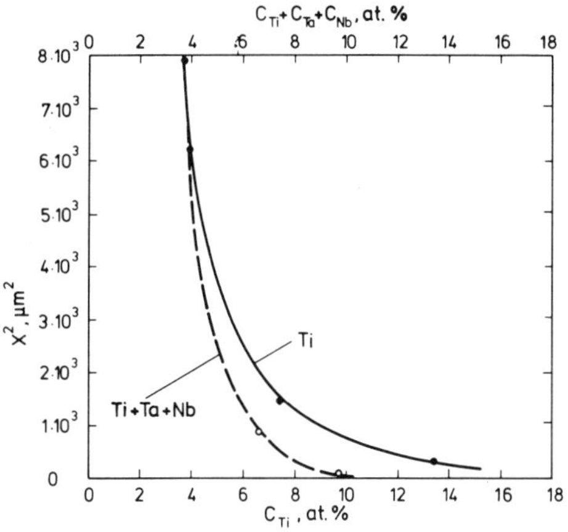

Fig. 7. Thickness of the cubic-carbide-free zone as a function of titanium and of combined titanium, tantalum and niobium contents (in atomic per cent) at a constant nitrogen content after the sample had been annealed for 1 h at 1540 °C in vacuum.

7) and the amount C_{Ti} of titanium is predicted by our model which takes into account both the increasing amount of titanium which has to diffuse for a given increase in zone thickness and the reduction in the cross-section of diffusion due to the larger amount of cubic phase precipitation. The experiments of Suzuki *et al.* [6] failed to expose this influence since the amounts of titanium and nitrogen were varied simultaneously, with the result that the rate decreases with increasing nitrogen content similarly to a hyperbolic function predicted by eqn. (1). However, this is clearly a misinterpretation because the nitrogen diffusion is assumed to be rate controlling.

Figure 7 also shows the thickness of the cubic-carbide-free layer for alloys containing tantalum and niobium (weight ratio, 90 to 10). At a constant total atomic percentage of elements forming cubic carbides, *i.e.* tantalum, niobium and titanium, the growth rate is definitely lower for alloys containing tantalum and niobium. The model of Suzuki *et al.* [6] seems to indicate that the addition of mixed cubic phases should have no influence other than that on c_B in eqn. (1). Since the elements of group V reduce the stability of nitrides and carbonitrides, their concentration in the binder phase is higher than that of titanium is; the diffusion of titanium, therefore, will remain the rate-controlling step. In addition, however, the solubility of titanium is decreased by the presence of tantalum and niobium in the binder. Thus the ratio c_{Ti}^*/c_N^* decreases, which overcompensates the effect of the nitrogen concentration c_A in the binder of the bulk alloy (the latter increases owing to the lower stability of (Ta, Nb, Ti, W)(C, N) compared with (Ti, W)-(C, N)). This explains the somewhat smaller thickness of the cubic-nitride-free zone in the alloys containing tantalum and niobium instead of titanium.

5. Conclusions

The theoretical and experimental results on the formation of surface zones free of cubic carbides show the following.

(i) Such zones are formed by annealing the alloys in vacuum only if the amount and stability of the nitride-containing phase are high enough that nitrogen is still present in the alloy when the compact has become dense and a nitrogen gradient develops during further annealing.

(ii) The rate-controlling mechanism is the inward diffusion of titanium.

(iii) A complex coupling of the nitride and titanium gradients must be considered in order to describe the growth kinetics of the cubic-carbide-free zone.

(iv) The phenomenological model yields a parabolic rate equation which describes the effects of all important variables (annealing time, temperature and alloy composition) semiquantitatively in full agreement with the experimental results while earlier models fail in some important aspects.

References

1 B. J. Nemeth and G. P. Grab, *Germ. Patent Appl. DE 3211047 A1*, 1982; *U.S. Patent 4,610,931*, 1987.
2 W. C. Yohe, *U.S. Patent 4,548,786*, 1985.
3 T. E. Hale, *U.S. Patent 4,497,874*, 1985.
4 M. Tobioka, M. Kodama, T. Asai and T. Yamamoto, *U.S. Patent 4,277,283*, 1981.
5 A. Doi, T. Nomura, M. Tobioka, K. Takahashi and A. Hara, *Proc. 11th Plansee Semin., Reutte, 1985*, Metallwerk Plansee, Reutte, 1985, p. 825.
6 H. Suzuki, K. Hayashi and Y. Taniguchi, *J. Jpn. Inst. Met., 22* (1981) 758.
7 M. Schwarzkopf, *Ph.D. Thesis*, Max-Planck-Institut für Metallforschung, Stuttgart, and Montanuniversität, Leoben, 1987.
8 H. E. Exner and M. Schwarzkopf, to be published.
9 W. May, *Tech. Mitt. Krupp. Forschungsber., 30* (1972) 15.

Materials Science and Engineering, *A105/106* (1988) 233–236

Anomalous Growth of the Complex Carbide Phase in Carbon-deficient WC–Co Hard Metal during the Infiltration of Eutectic Liquid*

JONG-KU PARK and KWANG YONG EUN

Division of Materials Science and Engineering, Korea Advanced Institute of Science and Technology, P.O. Box 131, Cheongryang, Seoul (South Korea)

D. N. YOON

Department of Materials Science and Engineering, Korea Advanced Institute of Science and Technology, P.O. Box 131, Cheongryang, Seoul (South Korea)

(Received November 11, 1987)

Abstract

During infiltration of the carbon-deficient 97WC–3Co bodies with 25WC–75Co infiltrants (where the compositions are in weight per cent), some large grains (about a few hundreds of micrometres in size) of η phase (Co_3W_3C) have been formed in the infiltrated region. A distinct microstructural difference existed between infiltrated and uninfiltrated regions. The large η grains in the infiltrated region contained many sound WC grains and they had a very angular shape. In the uninfiltrated region, the fine η and CoW_3C phases were distributed homogeneously. Some elongated η grains grew at the infiltration boundary. The anomalous growth of η phase is attributed to the composition change in the specimen from the WC–η–κ region to the WC–κ–liquid region during infiltration.

1. Introduction

In the sintering of tungsten carbide (WC) materials, control of the carbon content is very important in order to obtain strong sintered bodies. When a carbon content is too low, the carbon-deficient phases such as Co_3W_3C (η), Co_2W_4C (θ) and CoW_3C (κ) phases are precipitated. Contrarily, when the carbon content is too high, graphite is precipitated. Both η phase and graphite are very harmful to the mechanical properties of sintered WC materials.

Generally the η phase is finely distributed in the carbon-deficient cemented carbides but, in some cases, a few large η grains containing many sound WC grains inside are formed [1–3]. However, it is still not known how these large η grains can grow to such a large size in the three-phase region consisting of WC, η and liquid.

The same phenomenon was observed during infiltration of a carbon-deficient WC–Co skeleton with eutectic liquid. The purpose of this study is to determine the growth process of large η grains during the infiltration process.

2. Experimental details

The WC powder used had a high purity and the size was about 4 μm. It contained combined carbon (6.13 wt.%) and free carbon (0.03 wt.%). The cobalt powder was also of high purity having a 2 μm size. The carbon powder used was of lamp black grade. All powder mixtures were wet milled for 75 h. The skeletons and infiltrants were pre-sintered at 1100 °C for 1 h in vacuum. Infiltration was done by superimposing an infiltrant on a skeleton in vacuum above the liquid formation temperature (1440 °C). The skeleton compositions were 97WC–3Co and 96.9WC–3Co–0.1C, and the infiltrant was 25WC–75Co where all compositions are in weight per cent. The final composition of infiltrated bodies was fixed at 85WC–15Co. The microstructures on a vertical section of the infiltrated specimen were observed. Phase analysis was done by X-ray diffraction.

*Paper presented at the 3rd International Conference on the Science of Hard Materials, Nassau, The Bahamas, November 9–13, 1987.

0921-5093/88/$3.50

© Elsevier Sequoia/Printed in The Netherlands

3. Results and discussion

The 97WC–3Co skeletons were slowly infiltrated by 25WC–75Co infiltrants at 1440 °C. The microstructures on the vertical section are shown in Fig. 1. The upper part is the infiltrated region (Fig. 1(a)) and the lower part is the uninfiltrated part (Fig. 1(c)). In the upper part, there are a few large grains which have a very angular shape (Fig. 1(a)). In the lower part, fine black grains are distributed homogeneously in mainly WC matrix (Fig. 1(c)). Some large elongated grains were grown near the infiltration boundary (Figs. 1(b) and 2(b)).

The magnified microstructures of Fig. 1 are shown in Fig. 2. Figure 2(a) shows that the large η grains contain many WC grains. In Fig. 2(b), the elongated η grains seemed to grow at the infiltration boundary towards the infiltrated region. Near the infiltration boundary, some larger black grains grew among the fine black grains.

It was identified from the X-ray diffraction patterns that the types of phase in each part are clearly different, as shown in Fig. 3. The diffraction pattern on the top surface (the infiltrated region) shown at the top in Fig. 3 indicates that an η phase exists with WC and cobalt phases. The cobalt phase was a liquid phase at the infiltration temperature. In contrast, a κ phase exists with WC and η phases without cobalt phase in the uninfiltrated region (at the bottom in Fig. 3, which is the diffraction pattern on the bottom surface). The composition of the uninfiltrated part is within a three-phase region of WC, η and κ, while the infiltrated part is within a three-phase region of WC, η and liquid.

Generally the infiltration rate of cobalt liquid in WC–Co system is very fast. The 97WC–3Co skeleton was slowly infiltrated, while the 96.9WC–3Co–0.1C skeleton was rapidly infiltrated with the same 25WC–75Co infiltrants. The

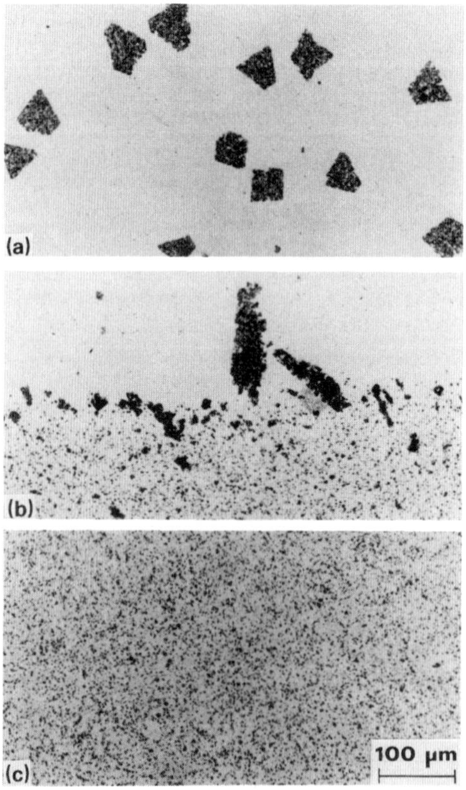

Fig. 1. Microstructures of the infiltrated specimen with 25WC–75Co infiltrant at 1440 °C in vacuum (original skeleton was 97WC–3Co): (a) upper part; (b) central part; (c) lower part. (The infiltration direction is from upper to lower parts.)

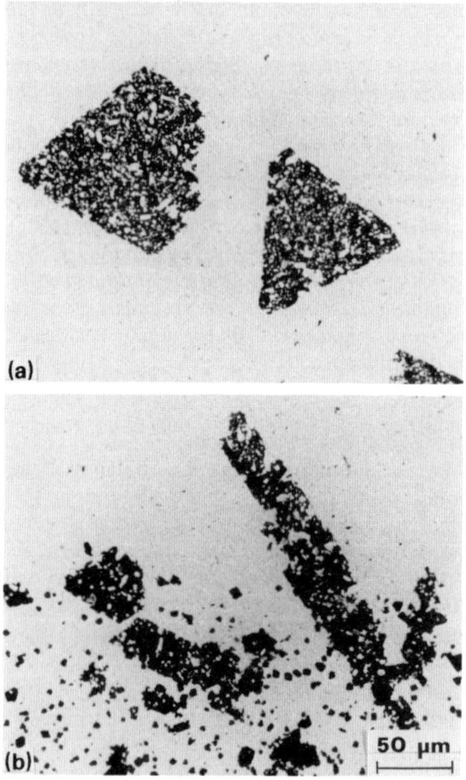

Fig. 2. Magnified microstructures of Fig. 1: (a) upper boundary region; (b) infiltration boundary region.

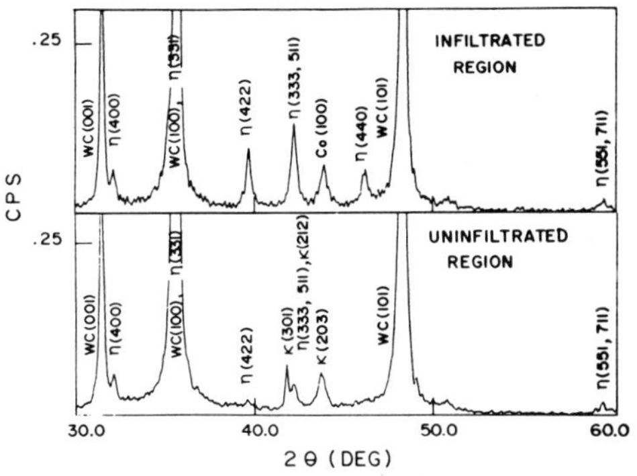

Fig. 3. X-ray diffraction patterns of the infiltrated region (upper part) and uninfiltrated region (lower part).

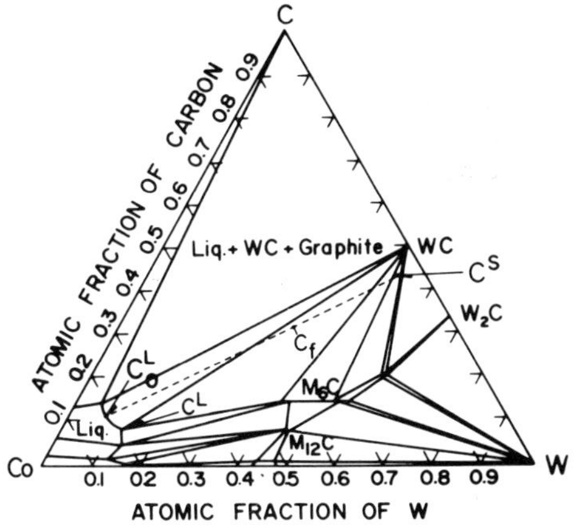

Fig. 4. Isothermal section of the W–Co–C phase diagram at 1425 °C [4].

97WC–3Co skeletons sintered at 1440 °C contained much carbon-deficient phase. This corresponds to the uninfiltrated part in Fig. 1(c). On the assumption that the isothermal section of the WC–Co phase diagram at 1440 °C is similar to that at 1425 °C (Fig. 4) [4] and from the results of X-ray diffraction, the initial concentration C^s of the 97WC–3Co skeleton is in the three-phase region of WC, CoW_3C and M_6C. The 96.9WC–3Co–0.1C skeleton is in the three-phase region of WC, liquid and M_6C. A liquid channel existed before infiltration in the case of 96.9WC–3Co–0.1C. So the 96.9WC–3Co–0.1C skeleton could be rapidly infiltrated through the pre-existing liquid channel. However, the 97WC–3Co skeleton could not be infiltrated so rapidly because it was totally in the solid state and had no liquid channel.

Infiltration rate changes continuously since liquid composition is not constant. The initial liquid composition is on the liquidus line because liquid is formed from a 25WC–75Co infiltrant which is near the eutectic composition and some excess WC grains remained on the top of the skeleton after infiltration. From Fig. 4, the concentration changes of liquid and skeleton can be explained conceptually. C^L_0 and C^s are the concentrations of initial liquid and solid skeleton respectively. When the liquid of concentration C^L_0 contacts with a skeleton of concentration C^s, the liquid dissolves the κ phase and penetrates into a solid skeleton. During liquid penetration, the liquid concentration changes from C^L_0 to C^L along the liquidus line, which is this equilibrium concentration of liquid in the three-phase region of WC, η and liquid. Since the initial skeleton was in the region of WC, η and κ, the average composition of the specimen deviates from the two-phase region of WC and liquid. When the concentration of the infiltrated region moves to C_f, the liquid concentration can change rapidly from C^L_0 to C^L because of fast diffusion in the liquid and the κ phase must be dissolved by the supply of liquid cobalt. Therefore the infiltration rate is fast at the beginning and then becomes slow with saturation of liquid. A two-phase region of WC and liquid was observed at the top of the specimen, since the initial liquid dissolved both η and κ phases.

Sarin [1] also observed the large η grains and called this η phase a glass-type defect-free crystal. He explained that η grains could grow to a large size in a decarburizing condition because of the difficult nucleation of η phase. Åkesson [3] also observed the same phenomenon in which the temperature was only a critical variable for growth of large η grains. He explained that the liquid phase could act as a fast diffusion path for the growth of η grains. However, it still has not been explained why only a few large η grains which incorporate the small η grains could survive.

During infiltration, the skeleton concentration varies continuously from the WC–η–κ region to the WC–η–liquid region. The unstable κ phase must disappear during infiltration as follows:

$$CoW_3C + 2Co - Co_3W_3C$$

This reaction needs cobalt atoms. It can occur only at the infiltration boundary because the liquid cobalt is supplied to this boundary. The unstable κ phase is dissolved by the supply of liquid cobalt and precipitates as η phase on the η grains since the original skeleton contained fine η grains.

The large angular η grains are thought to be formed at the beginning of the infiltration. Figure 5 shows the initial microstructure of a specimen infiltrated at 1440 °C for 5 min after fast heating. Large angular η grains begin to grow at the infiltration boundary instead of elongated η grains such as Fig. 2(b). The morphological difference between the two types of η grain at the infiltration boundary may be due to the concentration difference of penetrating liquid at each moment. As the liquid penetrates deeper and deeper, it becomes more saturated. On comparison of Figs. 2(b) and 5, it is thought that η grains are formed when the liquid is highly saturated, because they are formed only in the deeper region.

It is frequently observed that small η grains are finely dispersed in carbon-deficient sintered bodies. There is no general reason why only particular η grains can grow to such a large size, incorporating the smaller η grains. However, when the unstable phase is dissolved among the η grains during infiltration, certain η grains can grow to a relatively larger size than other η grains, as in Fig. 2(b). If large η grains, which depart from the normal grain size distribution can be formed through dissolution of the κ phase, these large η grains continue to grow, incorporating the surrounding fine η grains even after complete dissolution of κ grains. Additionally the energy of the WC–η interface seems to be low enough not to prevent the growth of η grains from entrapping the WC grains [1].

In some concentration ranges the growth of large η grains can occur even under the normal sintering conditions, because the specimen passes through various types of multiphase field, and it can pass through an unstable carbide region just below the liquid formation temperature during heating to the sintering temperature. When a specimen is heated to above the liquid formation temperature through an unstable carbide region, the unstable carbide must be dissolved and can trigger the occurrence of large grains. This experiment critically shows the occurrence of large η grains when a specimen passes through an unstable carbide region.

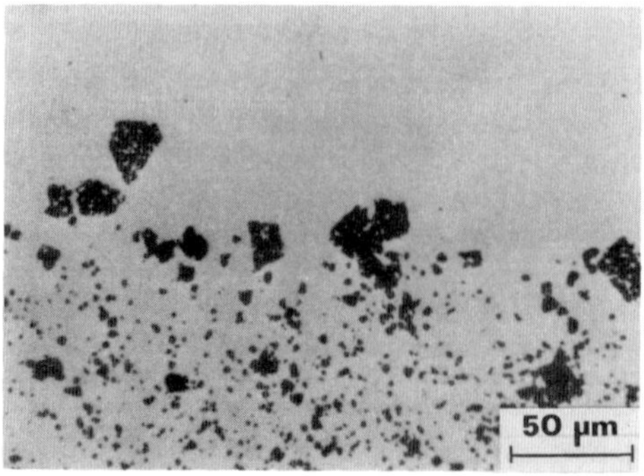

Fig. 5. Microstructure of the specimen infiltrated at 1440 °C for 5 min.

Previously [5], it was shown that some WC grains could grow abnormally in the presence of large η grains under carburizing conditions (which means that the η phase is not stable and must be dissolved) and also in the presence of large seed WC grains which do not have a normal grain size distribution. Therefore there is a possibility that both growth of large η grains and abnormal growth of WC grains can be treated as the same phenomenon, because the existence of unstable carbide and the appearance of large grains are common in both cases.

4. Conclusion

The growth of large η grains during infiltration is attributed to a concentration change in the infiltrated region. The unstable κ phase must be dissolved and precipitates as an η phase by the supply of liquid cobalt because the concentration of the infiltrated body changes from the WC–η–κ region to the WC–η–liquid region during infiltration. A few larger η grains formed by dissolution of the κ phase seem to grow and survive as large η grains, incorporating the fine η grains. The same phenomenon [1–3] can occur under normal sintering conditions through dissolution of unstable carbide during heating to a temperature above the liquid formation temperature. This observation might be a clue to understanding the reason for the abnormal grain growth of WC grains in the WC–Co system.

References

1 V. K. Sarin, *Mod. Dev. Powder Metall., 10* (1976) 553.
2 J. Gurland, *J. Met.,* (1954) 285.
3 L. Åkesson, *Thermochim. Acta, 29* (1979) 327.
4 L. Åkesson, *Proc. Int. Conf. on the Science of Hard Materials,* 1981, p. 71.
5 K. Y. Eun, *Ph.D. Thesis,* Korea Advanced Institute of Science and Technology, 1983.

Materials Science and Engineering, A105/106 (1988) 237–248

Identification of Optimum Binder Phase Compositions for Improved WC Hard Metals*

E. A. ALMOND and B. ROEBUCK

National Physical Laboratory, Teddington, Middx. (U.K.)

(Received November 9, 1987)

Abstract

The possibility has been investigated of using measurements of the mechanical properties of alloys with binder phase compositions to identify binder phases that will provide high hardness and toughness in WC hard metals based on cobalt, nickel, Co–Ni, Ni–Al, Ni–Cr–Mo and Ni–Cr–Mo–Al. The results showed that, in WC–Co and WC–(Co–Ni) hard metals with constant grain size and binder phase content, an optimum combination of fracture toughness and compressive strain could be obtained at particular tungsten and carbon contents which could be determined from measurements of magnetic saturation and lattice parameters. As predicted from tests on Ni–W–C alloys, it was found that WC–Ni hard metals had lower toughness and strength than WC–Co hard metals do. Additions of aluminium to nickel binder phases to form γ' precipitates raised the strength and conferred creep resistance but decreased the fracture toughness. Solid solution strengthening by chromium and molybdenum raised the hardness without reducing toughness as measured by Palmqvist tests and gave properties which matched those of WC–Co hard metals. By combining γ' hardening and solid solution hardening by chromium and molybdenum in nickel-based binder phases, better values of hardness and toughness were obtained than those of WC–Co hard metals. The new compositions also offered the possibilities of enhanced resistance to creep and corrosion, and properties that could be varied by heat treatments to meet specific requirements.

*Paper presented at the 3rd International Conference on the Science of Hard Materials, Nassau, The Bahamas, November 9–13, 1987.

1. Introduction

The binder phase in WC hard metals is an alloy containing small amounts of dissolved tungsten and carbon and may also contain elements to meet the requirements of a specific application. The composition can vary significantly between the boundaries for graphite and η phase precipitation and is important in determining the mechanical properties of the hard metal. Thus, industry obtains reproducibility in properties of hard metals based on cobalt, nickel and iron binder phases by controlling the total carbon content and by keeping the magnetic saturation constant; these measures ensure that the carbon content of the binder phase is repeatable and the tungsten (or non-magnetic) content does not alter from batch to batch.

The earliest published evidence that the mechanical properties of hard metals depend on binder phase composition was the demonstration that the transverse rupture strength exhibited a maximum value at a particular total carbon content in WC–Co hard metals [1–3]. Later work established that there was a significant dependence of fracture toughness on binder phase composition [4, 5]. The possibility that the compositional dependence of the properties of WC–Co hard metals is related to the properties of Co–W–C alloys with binder phase compositions was first examined by Rudiger *et al.* [3]. Later, for the same purpose, Roebuck and Almond [6, 7] investigated the properties of Co–W–C and Ni–W–C alloys with binder phase compositions. They identified alloy compositions which gave optimum combinations of strength and ductility and suggested that binder phases with these compositions would give the best properties in a WC hard metal. Data on the effect

0921-5093/88/$3.50

© Elsevier Sequoia/Printed in The Netherlands

of variation in tungsten and carbon content on the lattice parameter and magnetic saturation of Co–W–C alloys were also given. The present paper is an extension of the latter work and describes results on hard metals based on binder phase with varying compositions in the WC–Co, WC–Ni and WC–(Co–Ni) systems, and on binder phase alloys. The possibility of replacing the binder phase in WC–Co hard metals by nickel alloys based on superalloy compositions has also been explored. At an early stage of the work it was realised that some form of hardening mechanism would need to be adopted before nickel-based hard metals could be considered as competitors to cobalt-based hard metals in mechanical applications. Rather than devise entirely new nickel alloy compositions it was decided to concentrate on compositions which had already proved successful in other applications. These were the superalloys strengthened by γ' precipitates and the creep-resistant Hastelloy Ni–Cr–Mo alloys. A similar approach by other workers has met with mixed success [8, 9].

2. Experimental details

2.1. Materials

The materials used for manufacturing the alloys were high purity (99.7%) nickel shot, electrolytic flake cobalt, spectroscopically pure graphite, and tungsten and WC powders manufactured from ammonium paratungstate. The alloys were made by adding appropriate amounts of each constituent to an alumina crucible in a vacuum induction furnace to give a 1 kg charge which was melted under hydrogen at a partial pressure of 66 mbar, cast into a copper mould 38 mm in diameter under vacuum and cooled in hydrogen. The billets were heated in hydrogen at 1200 °C for 1 h and then reduced to 16 mm in diameter by direct extrusion with a molten glass lubricant.

The materials used for manufacturing the hard metals were a medium-grained WC powder, an extra-fine-grained cobalt powder and a fine-grained nickel powder. For three hard metals the binder phases were required to contain aluminium, titanium and tin. These were added in the form of Al_3Ni, Ni_2TiAl and Ni_3Sn_4. The powders were mixed in the required proportions, milled, mixed with paraffin wax, pressed into bars, pre-sintered at 900 °C and given a final sinter under vacuum at 1380 °C for the WC–Co hard metals, at 1420 °C for the WC–Ni and WC–(Co–Ni) hard metals and at 1425 °C for the hard metals with nickel alloy binder phases.

2.2. Testing

The tensile properties of the alloys were measured on waisted specimens with a minimum diameter of 6 mm and a waist radius of 100 mm. The load was applied at 50 N s^{-1} and the strain was measured with a calibrated clip gauge attached to the waist.

Fracture toughness measurements were made on the hard metals in four-point bend on single-edge-notched beam specimens, 40 mm long by 3 mm thick by 12 mm deep, which had been pre-cracked by a wedge indentation technique [10]. The toughness of some hard metals was also assessed from measurements of the Palmqvist radial cracks at the corners of Vickers hardness indentations. Before testing, the specimens were annealed at 800 °C for 1 h to eliminate surface stresses. The compression properties of the hard metals were measured on cylindrical specimens, 8.74 mm high by 4.84 mm in diameter, in a hard beam rig at a strain rate of 10^{-5} s^{-1}. The specimens were strain gauged and the ductility was measured by measuring the strain at which a sudden load drop and loud cracking occurred.

The lattice parameters of the alloys and of the binder phase of the hard metals were measured by X-ray diffraction using Cu Kα radiation. The alloys were electropolished to obtain a stress-free surface, and the surface grains of WC of the hard metal were removed with hot Murakami's reagent; WC powder was used as the reference lattice parameter. Magnetic saturation measurements were made with a Sucksmiths balance and an electromagnet with a field strength of 800 kA m^{-1}. A chemical analysis of the alloys was made using X-ray fluorescence to measure the tungsten content and coulometric analysis for the carbon content. The binder phases of the hard metals were analysed with a non-dispersive X-ray spectrometer attached to a scanning electron microscope.

3. Results

3.1. Cobalt and nickel binder phases

In previous work the properties of 16 Co–W–C alloys and nine Ni–W–C alloys were described and it was concluded that the optimum combinations of tensile properties were found in

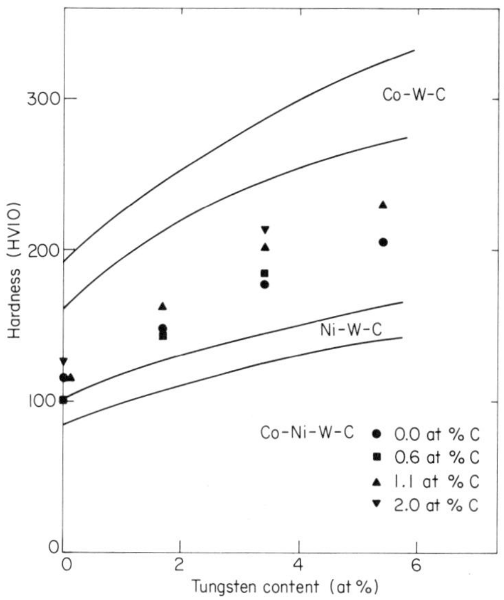

Fig. 1. Hardness values for the Co–Ni–W–C alloys with different tungsten and carbon contents compared with range of hardness values for Co–W–C and Ni–W–C alloys.

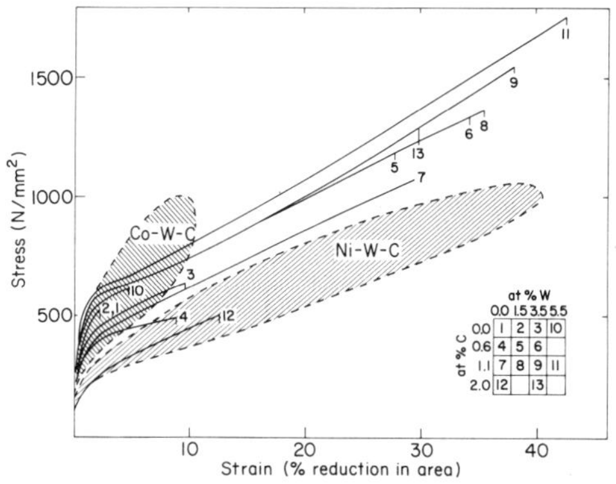

Fig. 2. Tensile stress–strain curves for Co–Ni–W–C alloys compared with envelopes of tensile stress–strain curves of Co–W–C and Ni–W–C alloys.

alloys containing approximately 0.5 at.% C and 3–6 at.% W in both systems [6, 7]. Subsequently, 13 70wt.%Co–30wt.%Ni alloys containing various amounts of tungsten and carbon were prepared and tested. The results of hardness tests and some of the tensile tests on the Co–Ni–W–C alloys are shown in Figs. 1 and 2, where it can be seen that the hardnesses and flow stresses are intermediate between those of the Co–W–C and Ni–W–C alloys, and the ductilities were higher but the work-hardening rates were lower than

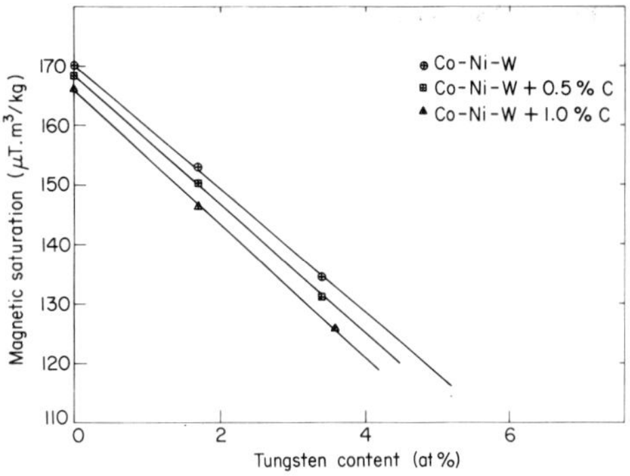

Fig. 3. Magnetic saturation values for the Co–Ni–W–C alloys with different tungsten and carbon contents.

those of the Co–W–C alloys. The optimum tensile properties in the Co–Ni–W–C alloys were obtained at 1.0 at.% C and 3.5–5.5 at.% W.

The results of chemical analyses indicated that the carbon and tungsten contents were within ±0.1 at.% of the nominal values shown in the inset in Fig. 2. It was found previously in the Co–W–C system that magnetic saturation decreased linearly with increase in tungsten content independently of carbon content, but present results indicated that in the Ni–W–C and Co–Ni–W–C alloys both tungsten and carbon decreased the magnetic saturation, although carbon had a much smaller effect (Fig. 3). From complementary X-ray measurements, it was found that the lattice parameters of the alloys varied linearly with the amounts of dissolved tungsten and carbon. Consequently, it proved possible to construct sets of equations of the form

$$S_m = 4\pi W_f\{s_0 - Am_W - m_C(B + Cm_W)\} \qquad (1)$$

and

$$a_b = a_0 + Dm_W + Em_C \qquad (2)$$

where S_m is the magnetic saturation per unit mass of sample, $4\pi s_0$ the magnetic saturation of pure cobalt, nickel or the 70wt.%Co–30wt.%Ni alloy, W_f, the weight fraction of binder phase in the hard metal, m_W (at.%) the amount of tungsten, m_C (at.%) the amount of carbon, a_0 the lattice parameter of pure cobalt, nickel or 70wt.%Co–30wt.%Ni alloy and a_b the lattice parameter of alloy or binder phase. A, B, C, D and E are constants and are listed in Table 1. These equations enabled the carbon and tungsten contents of the alloys to be expressed in terms of the measurements of the magnetic saturation and

TABLE 1 Values for the constants in eqns. (1) and (2)

System	$4\pi s_0$ (μT m^3 kg^{-1})	*A*	*B* (μT m^3 kg^{-1})	*C*	a_0 (nm)	*D*	*E*
Co-W-C	202.3	0.8	0	0	0.3548	3.6	12.0
Co-Ni-W-C	170.0	0.83	0.24	0.04	0.3541	4.2	4.9
Ni-W-C	69.7	0.8	0.25	0.18	0.3525	4.0	12.0

TABLE 2 Properties of hard metals with cobalt, nickel and Co-Ni binder phases

Hard metal	*Binder content* (wt.%)	*Density range* (Mg m^{-3})	*Hardness range* (HV 30)	*Magnetic saturation range* (μT m^3 kg^{-1})	*Mean grain size range* (μm)
B6F	5.7Co		1620-1710	8.7-11.6	0.66-0.81
B10C	10.0Co	14.55-14.65	1275-1390	15.4-20.2	1.67-2.13
B12C	11.6Co		1205-1300	16.6-22.8	1.71-2.04
B10CN	6.8Co-3.2Ni	14.43-14.69	1195-1435	9-17	1.41-1.71
B10N	9.6Ni-0.4Co	14.34-14.69	1075-1285	0.1-8.1	1.64-1.91

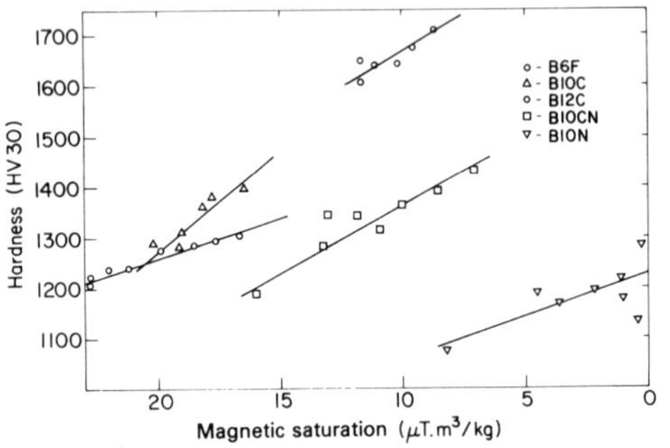

Fig. 4. Variation in hardness with magnetic saturation for the WC-Co, WC-(Co-Ni) and WC-Ni hard metals.

the lattice parameter and provided a method for calculating the tungsten and carbon contents of the hard metals.

Five hard metals were made at six carbon levels spanning the two-phase region between the formation of graphite and η phase. The physical properties are shown in Table 2 where it can be seen that grain sizes of the 10 wt.% and 12 wt.% binder phase hard metals were similar; their grain size distributions were also similar (see Fig. 9 below).

The hardness values are given in Fig. 4 and show a typical increase with increasing tungsten content of the binder phase (decrease in magnetic saturation). The envelope of the compressive stress-strain curves lay close to that for the WC-Co hard metals (Fig. 5) and the WC-(Co-Ni) hard metals exhibited greater ductility. The compressive failure strains and fracture toughness values are plotted against magnetic

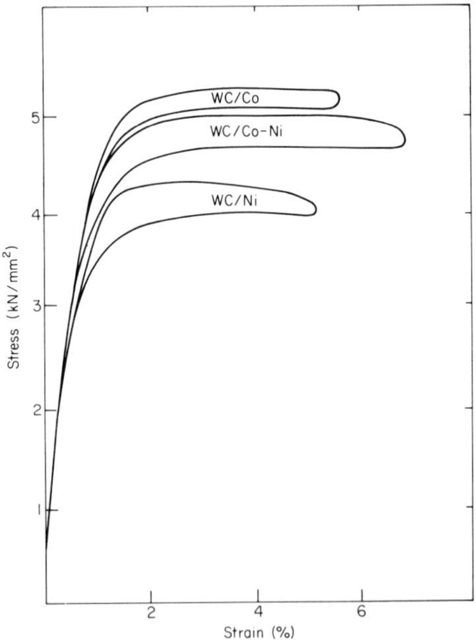

Fig. 5. Envelopes of compressive stress-strain curves for the WC-Co, WC-(Co-Ni) and WC-Ni hard metals.

saturation in Figs. 6 and 7 where each data point represents the average of five to six tests on the WC-Co hard metals and two to four tests on the WC-Ni and WC-(Co-Ni) hard metals. The results show a maximum value within the two-phase region in the WC-Co hard metals and indications of possible maxima in the other two sets of hard metals.

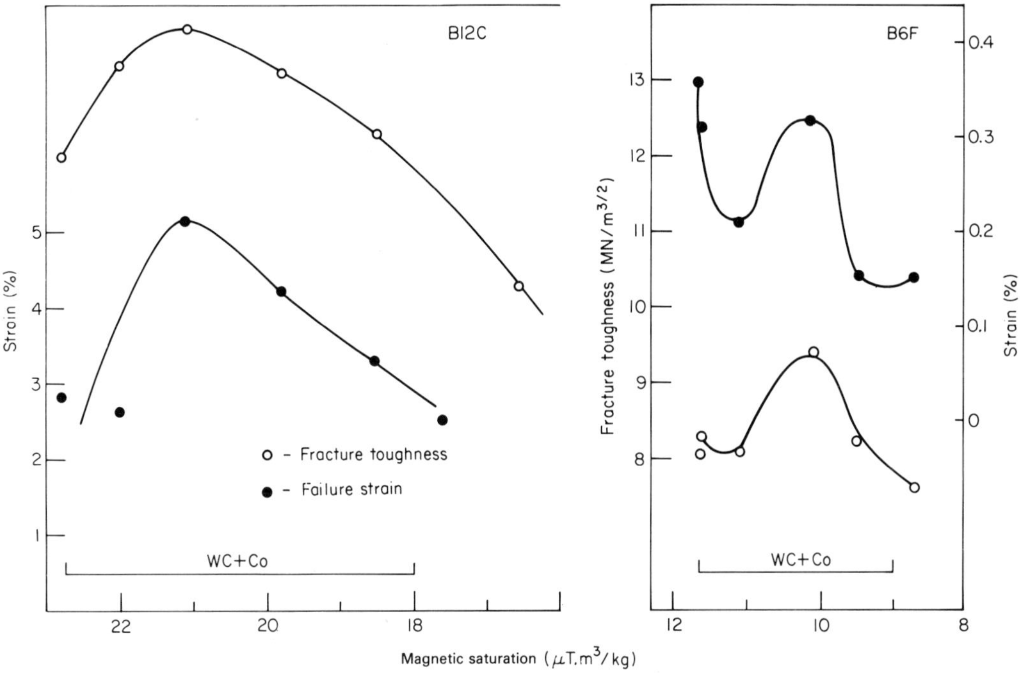

Fig. 6. Variation in fracture toughness and compressive failure strain with magnetic saturation for the WC–Co hard metals B12C and B6F.

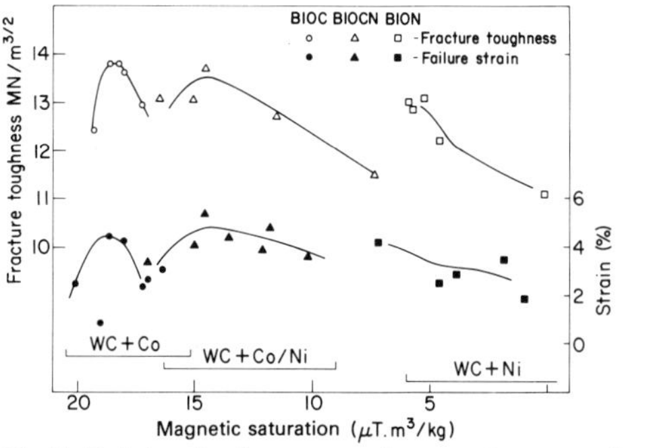

Fig. 7. Variation in fracture toughness and compressive failure strain with magnetic saturation for the WC–Co hard metal B10C, the WC–(Co–Ni) hard metal B10CN and the WC–Ni hard metal B10N.

3.2. *Nickel alloy binder phases*

Initially the possibilities were studied of strengthening Ni–W–C alloys using coherent precipitates of γ', Ni_3M, where M ≡ Al, Al(Ti), Sn or Si. From the results on small 300 g melts, it was established that an addition of 7–8 at.% Al, Al + Ti or Sn was sufficient to produce Vickers hardnesses (load, 10 kgf) after age hardening at 660 °C of 250–400 HV10 which were comparable with the values of 200–300 HV10 for the Co–W–C alloys. The Ni–W–C–Si alloys contained a precipitate but showed no significant age hardening. To enable a more detailed examination of properties to be made, 1 kg melts were prepared using a base composition of nickel plus 3 at.% W and 1 at.% C. To this was added 7.9 at.% Al to make alloy NWCAl, 4.0 at.% Al plus 4.0 at.% Ti for alloy NWCAlTi, and 7.9 at.% Al plus 8.5 at.% Mn for alloy NWCAlMn. The manganese was added to restore the magnetic properties of the nickel. Tensile tests were performed in the extruded condition and after aging at 700 °C for about 64 h to obtain the peak hardness. The results in Fig. 8 show that the flow stresses in the extruded and age-hardened conditions were similar to those of the Ni–Co–W–C alloys. The addition of manganese produced further increases in flow stress but there was a reduction in ductility. Additions of cobalt would be equally effective as manganese in improving the magnetic properties. Another beneficial effect of adding cobalt was that it was found it could be used to replace up to 50% of

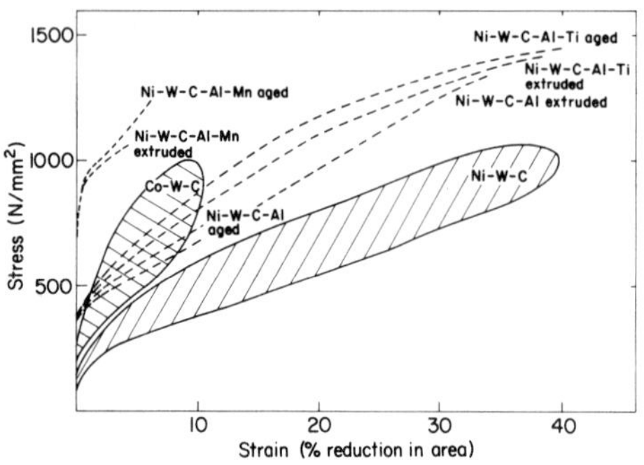

Fig. 8. Tensile stress–strain curves for the Ni–Al-based alloys in the extruded and aged conditions compared with envelopes of tensile stress–strain curves of cobalt- and nickel-based alloys.

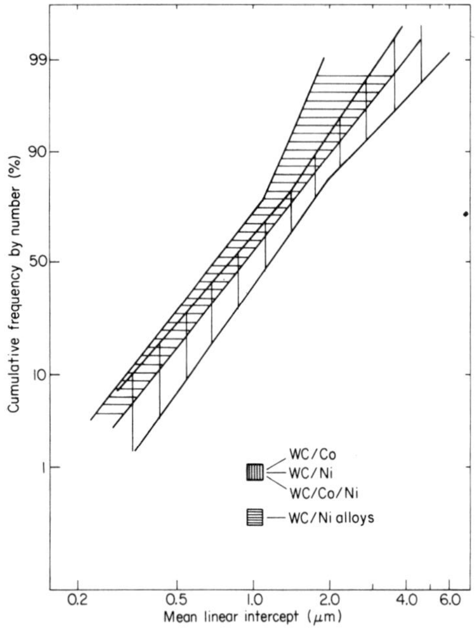

Fig. 9. Cumulative frequency distribution of carbide intercept lengths in the WC–Co, WC–(Co–Ni), WC–Ni and WC–Ni alloy hard metals.

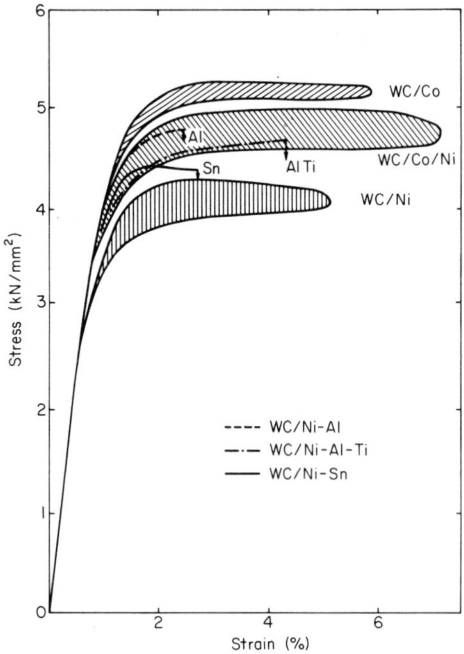

Fig. 10. Compressive stress–strain curves for WC–Ni alloy hard metals compared with envelopes of compressive stress–strain curves for the WC–Co, WC–(Co–Ni) and WC–Ni hard metals.

TABLE 3 Fracture toughness of nickel-based hard metals containing γ'

Hard metal	*Fracture toughness* (MN m$^{-3/2}$)
NiAla[a]	8.7, 9.5
NiAld	9.7, 9.4
NiAld[a]	8.5
NiAlTid	10.5
NiAlTid[a]	10.0

a and d refer to small differences in composition resulting from different sintering treatments.
[a]Resintered.

the nickel without affecting γ' precipitation. (A Ni–W–C–Sn alloy was also prepared but it broke up during hot extrusion.)

Using the results from the tests on the alloys to plan the compositions, hard metals were manufactured with a 10wt.%Ni–W–C binder phase containing nominal amounts of 4.3 wt.% Al and 12 wt.% Sn at three different carbon levels, and 3.4 wt.% Ti plus 1.9 wt.% Al at a medium carbon level. They were made with a similar grain size distribution to the WC–Co, WC–Ni and WC–(Co–Ni) hard metals, to enable a direct comparision of properties to be made independent of grain size effects (Fig. 9). The results of compressive tests are shown in Fig. 10 where it can be seen that the hard metals with strengthened binder phases had similar flow stresses to those of the WC–(Co–W–C) hard metals but the ductilities were lower. The WC–(Ni–Al) hard metals had the highest flow stresses in the as-received condition. The low ductilities of the hard metals were reflected in the poor values for fracture toughness (Table 3) which were between 9.5

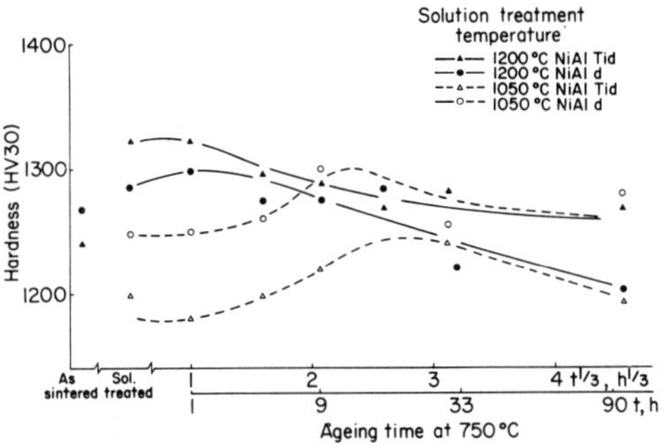

Fig. 11. Variation in hardness with aging time at 750 °C for WC–(Ni–Al) hard metals solution treated at 1050 and 1200 °C.

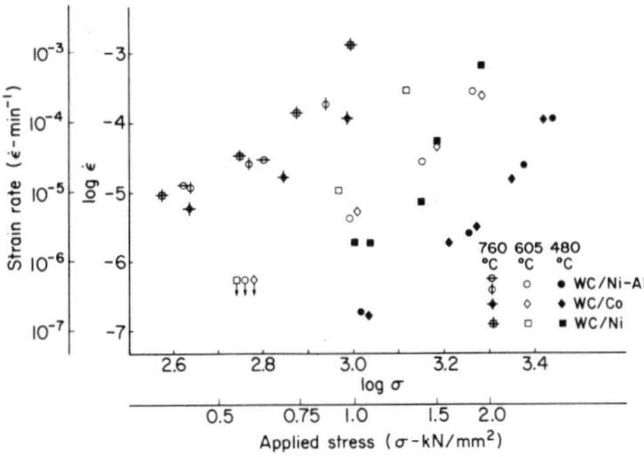

Fig. 12. Compressive creep rate *vs.* applied stress for WC–Co, WC–Ni and WC–(Ni–Al) hard metals.

and 10.5 MN $m^{-3/2}$ compared with about 13 MN $m^{-3/2}$ for the WC–Co, WC–(Co–Ni) and WC–Ni hard metals. Aging did not produce a significant change in compressive properties but it proved possible to alter the hardness by aging at 750 °C after solution treatments at 1050 and 1200 °C (Fig. 11). To see whether γ' precipitates provided creep resistance as they do in superalloys, tests were performed on a WC–(Ni–Al) hard metal, the WC–Co hard metal B10C and the WC–Ni hard metal B10N. The results showed that the WC–(Ni–Al) hard metal was equal in compressive creep resistance to the WC–Co hard metal and was better than the WC–Ni hard metal between 480 and 760 °C (Fig. 12). In indentation creep tests the WC–(Ni–Al) hard metal had bettter resistance than the other two hard metals at 500–700 °C.

The alternative method of strengthening nickel binder phases by solid solution hardening with molybdenum and chromium was explored initially by manufacturing alloys with and without

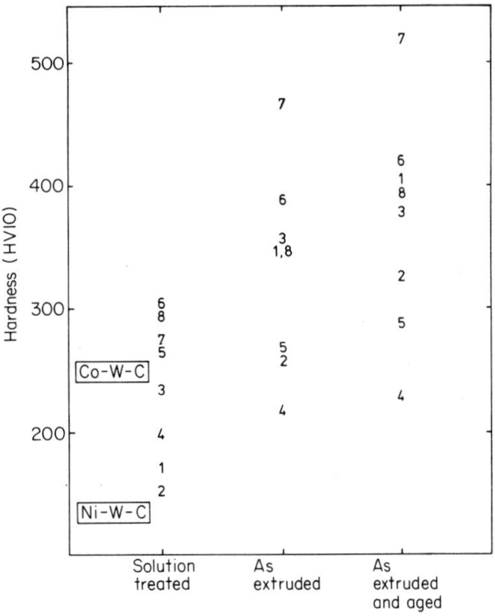

Fig. 13. Hardness values of Ni–Cr–Mo-based alloys in the solution-treated, extruded and extruded-and-aged conditions, compared with hardnesses of Co–W–C and Ni–W–C alloys (see Table 4 for key to alloy numbers).

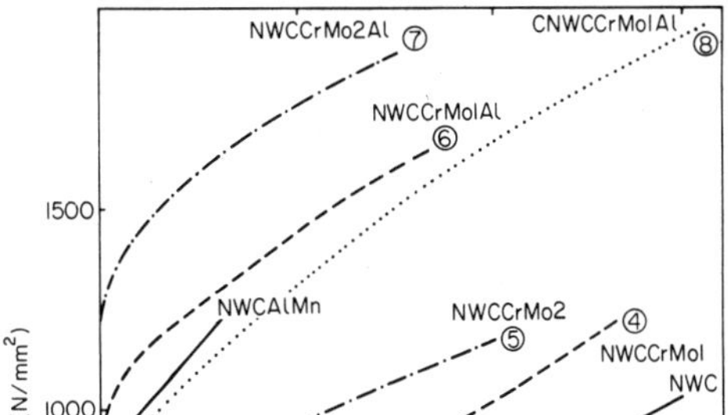

Fig. 14. Tensile stress–strain curves for Ni–Cr–Mo alloys compared with those for Co–W–C, Ni–W–C and Ni–W–C–Al–Mn alloys.

aluminium and cobalt. The results of hardness and tensile tests are shown in Figs. 13 and 14 for the compositions listed in Table 4. The stress–strain curves show that the Ni–

TABLE 4 Nominal compositions of nickel-based alloys

Number	*Alloy*	*Composition* (wt.%)						
		Ni	*Co*	*Cr*	*Mo*	*W*	*C*	*Al*
1	CNWC Al1	45	45	—	—	5	0.2	4
2	CNWC Al2	72	18	—	—	5	0.2	4
3	CNWC AlCr	65	15	10	—	5	0.2	4
4	NWC CrMo1	75	—	15	5	5	0.2	—
5	NWC CrMo2	65	—	15	15	5	0.2	—
6	NWC CrMo1Al	70	—	15	5	5	0.2	4
7	NWC CrMo2Al	60	—	15	15	5	0.2	4
8	CNWC CrMo1Al	35	35	15	5	5	0.2	4

Cr–Mo–W–C alloys had comparable properties with the Co–W–C alloys and that Ni–Cr–Mo–Al–W–C alloys with and without cobalt had the best properties. Clearly, these results and those obtained on the WC-(Ni–Al) hard metals indicated that there were numerous possibilities for combining chromium, molybdenum, aluminium and cobalt in different amounts in nickel-based binder phases to produce WC hard metals with equal or better properties than those of WC–Co hard metals. To circumvent the problems attached to developing individual processing routes for producing acceptable microstructures in such a wide range of experimental compositions, an infiltration method was devised for making small samples [11]. This consisted of manufacturing an alloy of suitable composition, placing a particular amount of the alloy on top of a hard metal in a furnace and heating the combination at a sufficiently high temperature for the alloy to diffuse into the hard metal. The resultant microstructures were uniform and the samples were sufficiently large to enable hardness and Palmqvist cracking to be measured.

The infiltration technique was used to prepare nine WC-(Ni–Cr–Mo) hard metals, 12 WC-(Ni–Co–Cr–Mo) hard metals and eight WC-(Ni–Cr–Mo–Al) hard metals with different compositions and free from graphite and η phase. The basic hard metals were a 6wt.%Co–WC hard metal with a grain size of 2.75 μm and a 6wt.%Ni–WC hard metal with a grain size of 1.75 μm. Samples of the latter were subjected to the same heat treatment as the infiltrated hard metals and it was found that they experienced grain coarsening whereas grain growth was inhibited in the hard metals with high molybdenum contents and to a lesser extent in those containing high chromium or low carbon contents. Analysis of the binder phase of the WC-(Ni–Cr–Mo) hard metals and examination of the microstructures established that the solubility limit for chromium was 8 at.% at high carbon contents and 10–12 at.% at low carbon contents. The molybdenum appeared to be partitioned between the binder phase and the carbide in the ratio 2:3. Palmqvist tests on the WC-(Ni–Cr–Mo) hard metals indicated that with additions of 8 wt.% Mo they were equally as tough as WC–Co hard metals of the same hardness and were tougher on addition of 12 wt.% Mo. The partial replacement of nickel by cobalt in the WC-(Ni–Cr–Mo) hard metals did not affect the solubility of chromium or molybdenum in the binder phase but the hard metals with high molybdenum contents contained a Cr–Mo-rich phase and they had poorer Palmqvist toughness than do WC–Co hard metals of the same hardness. The compositions of the WC-(Ni–Cr–Mo) hard metals which gave better properties than WC–Co and WC–Ni hard metals in Palmqvist tests are given in Table 5, and their properties are shown in Fig. 15 together with data illustrating the wide range of properties that were obtained for the WC-(Ni–Cr–Mo) and WC-(Ni–Co–Cr–Mo) hard metals.

The additional presence of aluminium in Ni–Cr–Mo binder phases (Table 6) enabled a

TABLE 5 Compositions of Ni–Cr–Mo binder phases of hard metals with good Palmqvist toughness

Alloy	*Composition* (at.%)		
	Ni	*Cr*	*Mo*
h2	89.5	7.1	3.4
h3	90.0	6.8	3.2
j2	87.8	10.4	1.8
j3	91.0	6.8	2.2

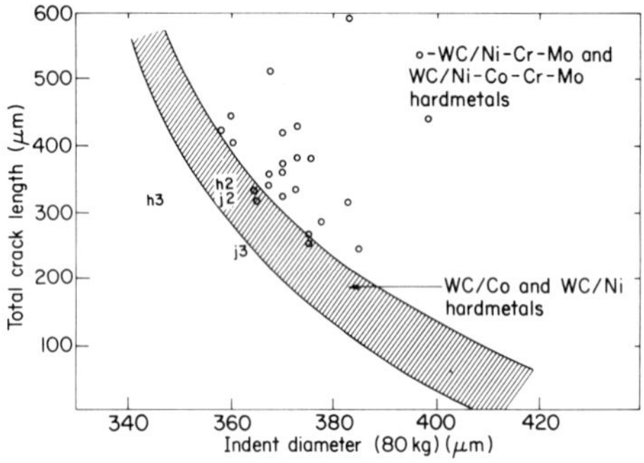

Fig. 15. Palmqvist crack length *vs.* indent diameter for infiltrated WC-(Ni–Cr–Mo) hard metals compared with the range of typical results for WC–Co and WC–Ni hard metals.

TABLE 6 Nominal compositions of Ni–Cr–Mo–Al binder phases

Alloy	*Binder phase composition* (wt.%)		
	Cr	*Mo*	*Al*
A	3.0	6.7	2.9
B	6.2	6.3	3.1
C	2.1	17.6	2.2
D	4.3	18.4	2.4
E	3.2	6.6	4.4
F	6.1	6.4	4.6
G	2.0	17.8	3.3
H	4.3	18.1	3.5

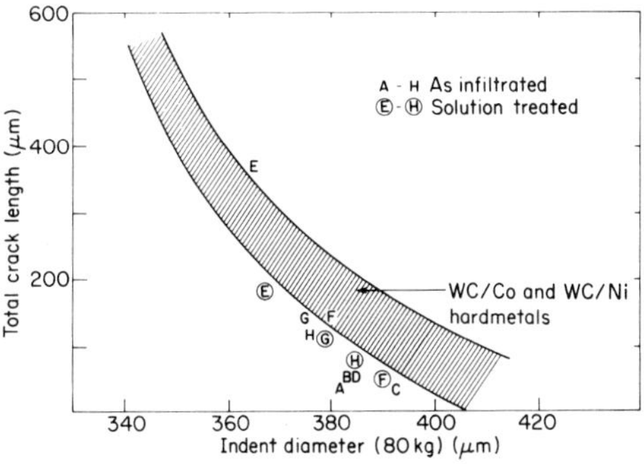

Fig. 16. Indent diameter and Palmqvist crack lengths for infiltrated WC–(Ni–Cr–Mo–Al) hard metals in the sintered, solution-treated and aged conditions.

range of properties to be obtained from one composition by solution treating at 1150 °C followed by age hardening at 750 °C for various times to alter the size and characteristics of the γ' precipitates (Figs. 16 and 17). It can be seen in Fig. 16 that the majority of WC–(Ni–Cr–Mo–Al) hard metals within the composition range studied had a better Palmqvist toughness than had WC–Co hard metals of the same composition. Examples of the shapes and sizes of γ' precipitates are shown in Fig. 18.

4. Discussion

4.1. Cobalt and nickel binder phases

The results in Figs. 4, 6 and 7 indicate some of the dangers of using several sources in the literature to generate empirical relations between hardness and fracture toughness of hard metals. Thus, it can be seen that close to the composition limit for graphite formation there is a range of binder phase compositions, where both the fracture toughness and the hardness can be increased by increasing the magnetic saturation value (decreasing the tungsten content). This is contrary to the common belief that increases in hardness inevitably result in a decrease in toughness. The results demonstrate that fracture toughness can

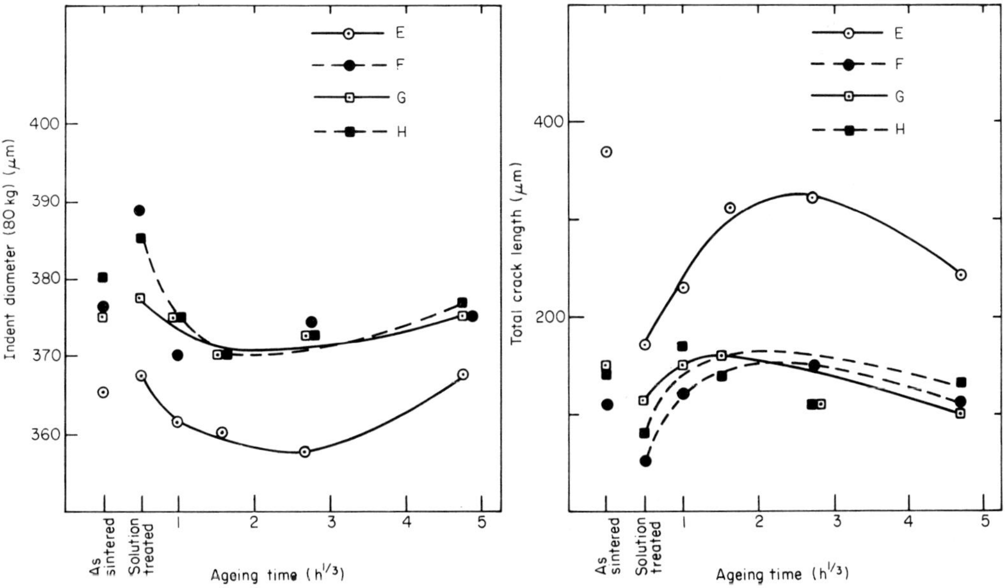

Fig. 17. Variation in hardness with aging time at 750 °C for infiltrated WC–(Ni–Cr–Mo–Al) hard metals solution treated at 1150 °C.

vary by 30% within the two-phase region. This also illustrates the danger of using data in the literature from several sources to test a fracture toughness model since the effects of binder phase composition can often predominate over those of WC grain size and binder phase content, but values for binder phase composition are seldom included as a variable in models for the strength or fracture toughness of hard metals.

The data on the variation in mechanical properties of the hard metals with magnetic saturation can be converted to a dependence on binder phase composition by inserting the data on the magnetic saturation and lattice parameters of the hard metals into eqns. (1) and (2). The results (Fig. 19 and Table 7) show that the tungsten content increased from close to zero to about 4–6 at.% in the coarse-grained hard metals, whilst the carbon content changed little within the two-phase range. This is contrary to the expected reciprocal relationship between dissolved carbon and tungsten, governed by an equilibrium constant [12]. A possible reason for the disagreement is that the cooling rate of the hard metals from the sintering temperature was probably too fast for the equilibrium compositions to be reached. Data obtained from using eqns. (1) and (2) also provide a means of calculating the levels of tungsten and carbon which gave the maximum toughness and ductilities in the hard metals. These compositions are listed in Table 7 together with the alloy compositions which gave the best properties in the binder phase alloys and it can be seen that there is good correspondence between the two sets of values. An interesting feature of the results shown here and in previous work on binder phase alloys is that they suggest that a minimum carbon con-

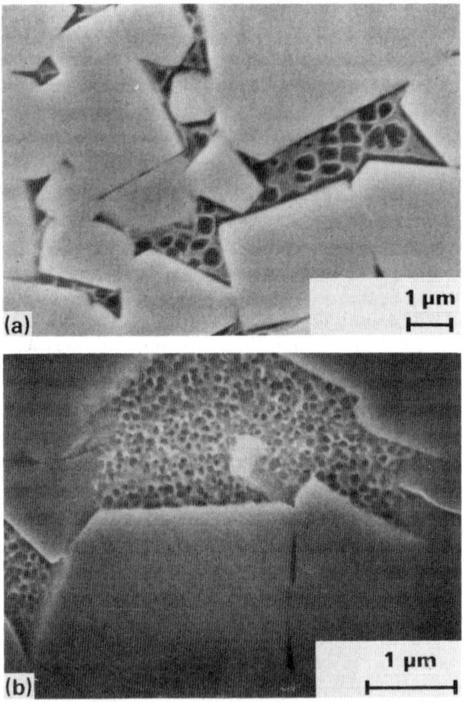

Fig. 18. Scanning electron micrographs showing γ' precipitates in hard metals electrolytically etched in dilute HF solution: (a) hard metal G as sintered; (b) solution treated at 1150 °C and rapidly cooled.

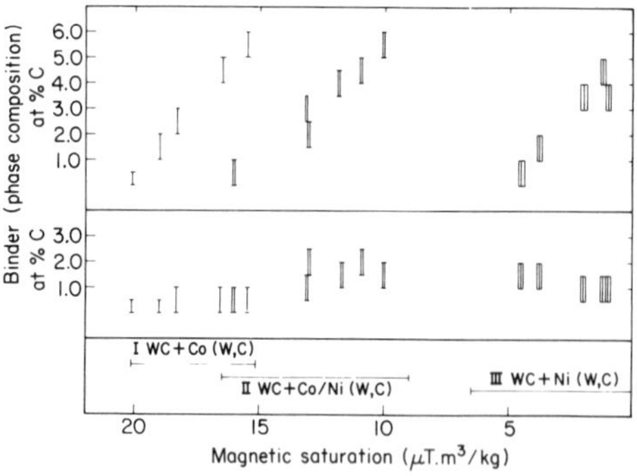

Fig. 19. Calculated tungsten and carbon contents in binder phases of WC–Co hard metal B10C and WC–(Co–Ni) and WC–Ni hard metals *vs.* magnetic saturation.

TABLE 7 Binder phase compositions for optimum properties

Hard metal	*Binder phase composition* (at.%)			
	Optimum alloy properties	*Maximum fracture toughness*	*Maximum failure strain*	*Maximum W content in binder*
B6F	(0.3–0.7)C–(1.4–2.7)W	2.3W	2.3W	3.8
B10C	(0.3–0.7)C–(1.4–2.7)W	1.8W–0.7C	1.8W–0.7C	4.3
B12C	(0.3–0.7)C–(1.4–2.7)W	2.0W–0.3C	2.0W–0.3C	5.5
B10CN	3.5W–0.5C	2.0W–1.0C	2.0W–1.0C	5.5
B10N	5.5W–0.5C	—	—	5.4

The uncertainty in the values is approximately ±0.5 at.%.

tent of about 0.5 at.% needs to be present to ensure good properties. A carbon level of 0.5 at.% was also the minimum amount found in the binder phases of the hard metals and may in some way be important in explaining why cobalt and nickel-based WC hard metals have good properties.

The above analyses indicate that it is possible to use alloys with approximate binder phase compositions to identify hard metal compositions which give good properties. However, there are some limitations to this procedure. For example, the alloys should have good wettability with WC since this will favour a fracture path which passes through the binder phase rather than along the WC–binder phase interface. Thus, fracture should be controlled by the strength and ductility of the binder phase rather than by the cohesive strength of the interface. Another difficulty arises when comparing the properties of different systems rather than examining one system with varying tungsten and carbon contents. For example, it is clear from Fig. 5 that the compressive properties of the WC–Co hard metals are superior to those of the other hard metals although the WC–(Co–Ni) hard metals are only slightly inferior. However, the stress–strain curves of the binder phase alloys indicate that the Co–Ni–W–C alloys have the best properties if a high failure strain and strength are used as a guide. A possible explanation why this criterion is insufficient is that in a hard metal, provided that the binder phase has a certain ductility, it may be more important to have a high hardness and a high work-hardening coefficient under the prevailing triaxial stress conditions. A high work-hardening coefficient is usually associated with a resistance to necking, and this has been observed for binder phase ligaments in the fracture path in a WC–Co hard metal [13].

4.2. Nickel alloy binder phases

The results on the WC hard metals with nickel alloy binder phases indicate that particular compositions should be capable of providing mechanical properties better than or equal to those of WC hard metals with cobalt-based binder phases. Basic strengthening with γ' precipitates alone did not provide a satisfactory combination of properties since the strength increments were obtained at the expense of ductility and fracture toughness. However, the solid solution strengthening of nickel by some combinations of molybdenum and chromium produced both good hardness and toughness as measured by Palmqvist tests. The improvement was probably due mainly to molybdenum since, compared with chromium, it produces a higher strengthening increment per weight per cent addition to nickel [14]. Also the partition of molybdenum between binder phase and carbide appears to have maintained a high toughness, possibly by forming a tough rim around WC grains as in TiC-based hard metals [15]. This result has been discussed in detail elsewhere [16]. The benefits of adding chromium were not examined comprehensively but they would include enhanced resistance to grain coarsening during sintering and improved corrosion resistance.

The best combinations of properties, however, were consistently better than those of WC–Co hard metals and were obtained by combining solid solution hardening by molybdenum and chromium with precipitation hardening by γ'. An attraction of using such compositions is that the properties can be altered by heat treatment or further alloying to meet a specific requirement. For example, the small precipitates in the microstructure in Fig. 18(b) can be aged to give a dispersion of particles with a suitable size, spacing and composition to give the optimum hardness as shown in Fig. 17. Similarly, both microstructures in Figs. 18(a) and 18(b) would be suitable for applications where creep resistance is important and the coarseness of the precipitate is not a major consideration. Also, various elements can be added to alter the characteristics of the γ' precipitate. Thus, suitable additions can improve strength by increasing the antiphase boundary energy, the solid solution strengthening and the lattice misfit [17, 18]. Enhanced creep properties could be obtained by lowering the lattice misfit and adding elements with high melting points to stabilise the γ' phase at high temperatures. Some of these changes could probably be brought about by simply varying the carbon and tungsten contents of the binder phase, but this has not been investigated in detail yet.

The achievement of good properties in the WC hard metals with a Ni–Cr–Mo–Al binder phase is further evidence that new binder phase compositions can be predicted from studying the properties of the corresponding alloys. Thus the Ni–Cr–Mo–Al–W–C alloys had high strength and good ductility. They also had a high work-hardening rate which was proposed above as a favourable characteristic for Co–W–C binder phases but has not been reported as a characteristic of superalloys.

5. Conclusions

It has been shown that the mechanical properties of WC hard metals can be related to the properties of alloys with compositions corresponding to those of binder phases of the hard metals. By varying the tungsten and carbon contents of the binder phase in WC–Co and WC–(Co–Ni) hard metals, the fracture toughness and compressive failure strength were varied by up to 30%, and it was found that these properties reached maximum values at compositions intermediate between graphite and η phase formation. From measurements of the effect of varying the tungsten and carbon contents on the magnetic saturation and lattice parameters of binder phase alloys and WC hard metals, expressions were derived which enabled the binder phase compositions of the hard metals to be calculated. The results showed that the tungsten content could be as high as 5.9 at.% in the binder phase of a 12 wt.% Co hard metal whilst the carbon content ranged from 0.5 to 2.0 at.% with little dependence on tungsten content in the binder phases of WC–Co, WC–(Co–Ni) and WC–Ni hard metals.

Compared with WC–Co hard metals, the hardness and compressive strengths were inferior in WC–Ni hard metals but were only slightly poorer in WC hard metals with a 70wt.%Co–30wt.%Ni binder phase. It was found that WC–Ni hard metals could be strengthened by adding aluminium, aluminium plus titanium, or tin to form γ' precipitates but, although creep resistance was also improved, the fracture toughness fell from 13 to 10 MN $m^{-3/2}$. The results from hardness and tensile tests on a range of Ni–W–C alloys containing various combinations of additions of cobalt, aluminium, chromium and molybdenum enabled binder phase compositions with promising mechanical properties to be identified. Hard metals with corresponding binder phase compositions were made by an infiltration technique, and it was found that some WC–(Ni–Cr–Mo) hard metals had comparable hardness and Palmqvist toughness properties to those of WC–Co hard metals. By combining γ' strengthening with solid solution hardening by chromium and molybdenum in the binder phase of a range of WC–Ni hard metals, better combinations of hardness and Palmqvist toughness were obtained than with WC–Co hard metals. The presence of chromium in the alloy binder phases should confer corrosion resistance while the size and form of the γ' can be altered by heat treatments to give enhanced strength or creep resistance.

Acknowledgments

The authors wish to thank Mr. E. G. Bennett for assistance with the experimental work.

Part sponsorship of the research by a group of U.K. hard metal manufacturers is gratefully acknowledged.

References

1 J. Gurland, *J. Met., 6* (1954) 285.
2 H. Suzuki and H. Kubota, *Planseeber. Pulvermetall., 14* (1966) 96.
3 O. Rudiger, D. Hirschfeld, A. Hoffman, J. Kolaska, G. Ostermann and U. Willbrand, *Int. J. Powder Metall., 7* (1971) 29.
4 J. R. Tingle, C. A. Shumaker, Jr., D. P. Jones and R. A. Cutler, in J. H. Underwood, S. W. Freiman and F. J. Baratta (eds.), *Chevron-notched Specimens: Testing and Stress Analysis, ASTM Spec. Tech. Publ. 855,* 1984, p. 281.
5 D. Bolton and R. J. Keely, *Int. J. Refract. Hard Met., 1* (1982) 103.
6 B. Roebuck, E. A. Almond and A. M. Cottenden, *Mater. Sci. Eng., 66* (1984) 179.
7 B. Roebuck and E. A. Almond, in H. M. Ortner (ed.), *Proc. 10th Plansee Semin., June 1981,* Vol. 1, Metallwerk Plansee, Reutte, 1981, p. 493.
8 K. Nishigaki, H. Doi, N. Yukawa, M. Morinaga and T. Koie, *Proc. 11th Plansee Semin., 1985,* Vol. 2, Metallwerk Plansee, Reutte, 1985, p. 487.
9 R. K. Viswanadham, P. G. Lindquist and J. A. Peck, in R. K. Viswanadham, D. J. Rowcliffe and J. Gurland (eds.), *Proc. 1st Int. Conf. on the Science of Hard Materials,* Plenum, New York, 1983, p. 873.
10 E. A. Almond and B. Roebuck, *Met. Technol., 5* (1978) 92.
11 B. Roebuck, E. G. Bennett and E. A. Almond, *Int. J. Refract. Hard Met., 3* (1984) 35.
12 H. E. Exner, *Int. Metall. Rev., 243* (1979) 149.
13 L. S. Sigl and H. E. Exner, *Metall. Trans. A, 18* (1987) 1299.
14 N. S. Stoloff, in C. T. Sims and W. C. Hagel (eds.), *The Superalloys,* Wiley, New York, 1972, p. 79.
15 H. Doi, in E. A. Almond, C. A. Brookes and R. Warren (eds.), *Proc. 2nd Int. Conf. on the Science of Hard Materials, Rhodes, September 23–28, 1984,* in *Inst. Phys. Conf. Ser., 75* (1986) 489.
16 B. Roebuck, E. G. Bennett and E. A. Almond, *J. Mater. Sci. Lett., 5* (1986) 473.
17 D. A. Grose and G. S. Ansell, *Metall. Trans. A, 12* (1981) 1631.
18 R. F. Decker and J. R. Mihalisin, *Trans. Am. Soc. Met., 62* (2) (1969) 481.

Sintering of Submicron WC–10wt.%Co Hard Metals Containing Nickel and Iron*

G. S. UPADHYAYA and S. K. BHAUMIK

Department of Metallurgical Engineering, Indian Institute of Technology, Kanpur 208016 (India)

(Received November 19, 1987)

Abstract

Although some results are available on the sintering behaviour and properties of WC–Co hard metals which are based on the micron range powders, almost no work has been reported on hard metals which is based on submicron size particles. With this in view, the present investigation covers sintering of submicron (0.78 μm) WC-based hard metals with 10 wt.% Co binder. The binder was partially or completely substituted by nickel or iron. In this study, we report mechanical properties such as hardness, transverse rupture strength and also oxidation resistance.

1. Introduction

From the introduction of cemented carbides in 1926, cobalt has been the traditional and predominant binder metal for tungsten carbide (WC), although nickel and iron are cited as alternatives. The reasons why nickel has not been used commercially are somewhat confused. Dawihl [1] indicated that cobalt is preferred to nickel because cobalt has superior comminution characteristics and hence a better distribution of cobalt on the WC particles is achieved during milling, which subsequently leads to the formation of a better skeletal structure during sintering. However, Dawihl [1] did indicate that nickel-bonded alloys equivalent to cobalt-bonded alloys can be obtained by using nickel oxide in place of nickel, since nickel oxide has comminution characteristics more akin to those of cobalt. Another reason given for the poor properties of the nickel-bonded alloys is that the solubility of WC in nickel differs from the solubility of WC in cobalt. At high temperatures, WC has a higher solubility in cobalt than in nickel [2] but at low temperatures the reverse is true [3]. Other researchers [4] disagree; they state that the solubility in cobalt at high temperatures is about the same as in nickel, particularly in the presence of excess carbon within the stipulated range for hard metal compositions. Nickel is said to be more prone than cobalt to the formation of brittle double carbides of the type $Ni_xW_yC_z$ [3, 5] particularly when the Ni–W–C system is deficient in carbon. Barlow [6] states that nickel-bonded WC alloys have similar properties to their cobalt-bonded equivalents.

Prakash [7, 8] studied the properties of WC hard metals containing a 20 wt.% Fe-Co-Ni alloy as binder. These hard materials were reported to be superior to conventional WC–Co hard metals in hardness and abrasion resistance, with slightly inferior transverse rupture strengths (TRSs). The carbon content was found to exert a critical influence on properties. An increase in the carbon content increased the strength of the martensite phase but at the same time stabilized the softer austenite phase. Agte [9] and Moskowitz *et al.* [10] showed that, by using excess carbon, Fe-Ni-bonded hard metals with optimum properties could be produced. Moskowitz *et al.* also found that the optimum nickel content in Fe–Ni-bonded hard metals was around 10 wt.%. Higher nickel contents tended to stabilize the austenite. Through the use of cobalt as an additional alloying element in the binder, the useful range of Fe–Ni–C binders could be enlarged. Nickel and carbon both depress M_s (the temperature at which the transformation of austenite to martensite starts during cooling; this is equivalent to stabilization of austenite). In addition to being a solid solution strengthener, up to 16 wt.% Co also raises M_s [11]. This means that the addition

*Paper presented at the 3rd International Conference on the Science of Hard Materials, Nassau, The Bahamas, November 9–13, 1987.

0921-5093/88/$3.50 © Elsevier Sequoia/Printed in The Netherlands

of cobalt increases the amount of martensite in the binder phase.

Fe–Ni-bonded WC materials have not yet been commercially applied, despite the fact that they are superior to the cobalt-bonded variety in TRS as well as in resistance to abrasive wear. The remarkably high strengths achieved in this system [12] (in the range 4120–4460 MPa) overshadow to some extent the system's superiority in abrasion resistance. However, the latter property may even be more significant than strength in terms of potential for industrial application.

The present study covers sintering of submicron (0.78 μm) WC–10wt.%Co hard metals, after partial or complete substitution of binder by nickel or iron. Apart from densification behaviour, the mechanical properties, grain size variation and oxidation resistance properties are reported.

2. Experimental procedure

The characteristics of various powders used in the study are given in Tables 1–4.

Wet milling of the carbide and binder powders was performed in a planetary ball mill using WC balls in acetone medium for 24 h. The milled powders were manually mixed for 1 h with 1% wax dissolved in benzene. The powders after ball milling were characterized in terms of particle size and size distribution using a Coulter counter. Rectangular green compacts of size 24.7 mm × 7.7 mm × 6.2 mm (approximately) from the waxed powders were prepared in a hydraulic compaction press at 350 MPa. Dewaxing of the green samples was carried out in a tubular furnace at 700 °C for 1 h under a dry hydrogen atmosphere. The pre-sintered compacts so obtained were kept in a graphite boat packed with alumina powder and sintered at temperatures of 1350 and 1400 °C for 1 h in dry hydrogen with a dew point of −35 °C. The sintered compacts were furnace cooled under hydrogen and their size, sintered densities, Vickers' hardness, TRS and magnetic properties were measured. From green, sintered and theoretical density data, densification parameters were also calculated.

The TRSs of as-sintered specimens were determined under three-point loading; the sintered compacts were placed on WC rollers of 3 mm diameter and kept 10 mm apart from each other. The coercivity and magnetic saturation were also measured.

A scanning electron microscope was used to observe the fracture surface of the failed samples.

An oxidation study in air on specimens of size 6 mm × 4 mm × 3 mm was carried out using a continuously recording semimicrobalance at a temperature of 800 ± 5 °C for 8 h. X-ray diffrac-

TABLE 1 WC powder

Average particle size	0.78 μm (Fisher subsieve size)
Size distribution	
0–1 μm	64.5 wt.%
1–2 μm	14.1 wt.%
2–3 μm	11.7 wt.%
3–4 μm	7.7 wt.%
4–5 μm	2.0 wt.%
Chemical analysis	
Mo	18 wt.ppm
Fe	61 wt.ppm
Cr	13 wt.ppm
Ni	2 wt.ppm
O	1036 wt.ppm
C (total)	6.13 wt.%
C (free)	0.04 wt.%

TABLE 2 Cobalt powder

Average particle size	2.3 μm (Fisher subsieve size)
Chemical analysis	
Ni	0.2 wt.%
Fe	0.04 wt.%
O	3500 wt.ppm
Co	Balance

TABLE 3 Nickel powder

Average particle size	3.7 μm (Fisher subsieve size)
Apparent density	1.6 g cm^{-3}
Surface area	0.3 m^2 g^{-1}
Chemical composition	
C	0.06 wt.%
Co	0.0003 wt.%
Fe	0.005 wt.%
O	0.05 wt.%
N_2	0.003 wt.%
S	0.0003 wt.%
Other elements	0.001 wt.%
Ni	Balance

TABLE 4 Iron powder

Average particle size	3–4 μm (Fisher subsieve size)
Chemical composition	
Fe	98.0 wt.% minimum
C	0.8 wt.% maximum
O	0.3 wt.% maximum
N	0.9 wt.% maximum

tion patterns were used to investigate the surface oxide. A general relation $(\Delta m)^n = kt$ for oxidation was presumed, where Δm is change in mass per unit surface area, k is the rate constant, t is the time and n is the power index.

3. Results

From Fig. 1 it is evident that the majority of WC–10wt.%(Ni–Fe) hard metals show higher sintered density values than do WC–10wt.%(Co–Ni) hard metals.

The reduction in sintered hardness values (Fig. 2) with increase in nickel concentration in the binder can be correlated using densification plots. An increase in sintering temperature from 1350 to 1400 °C resulted in an increase in hardness, although the pattern of variation remains unchanged.

As far as the variation in TRS is concerned, of all the compositions, hard metals containing a 50%Ni–50%Fe binder show a maximum TRS value, which is close to the TRS value of WC–10wt.%Co (Fig. 2).

The coercivity of WC–10wt.%(Co–Ni) hard metal decreases with increase in nickel content and ultimately reaches a zero value with the exception of the 50%Co–50%Ni binder composition. For WC–10wt.%(Co–Fe) the coercivity increases for substitution of up to 50% Co followed by a regular decrease. For WC–10wt.%Ni hard metals, the coercivity decreases with increase in iron content, followed by an increase. Magnetic saturation plots have a qualitative similarity to coercivity plots.

Grain size variation (Fig. 3) shows a slight increase in grain size for WC–10wt.%Co hard metal with increase in either nickel or iron in the binder. In contrast, the grain size remains almost the same for the WC–10wt.%(Ni–Fe) hard metals over the whole composition range.

The addition of nickel to the cobalt binder had a tendency to degrade the oxidation resistance (Fig. 4). From the double-logarithmic plots of mass gain *vs.* time (Fig. 4), the calculated index n values are 1.5–2. X-ray studies of the oxidized surface of the present hard metal system showed the presence of WC, WO_2, binder metal tung-

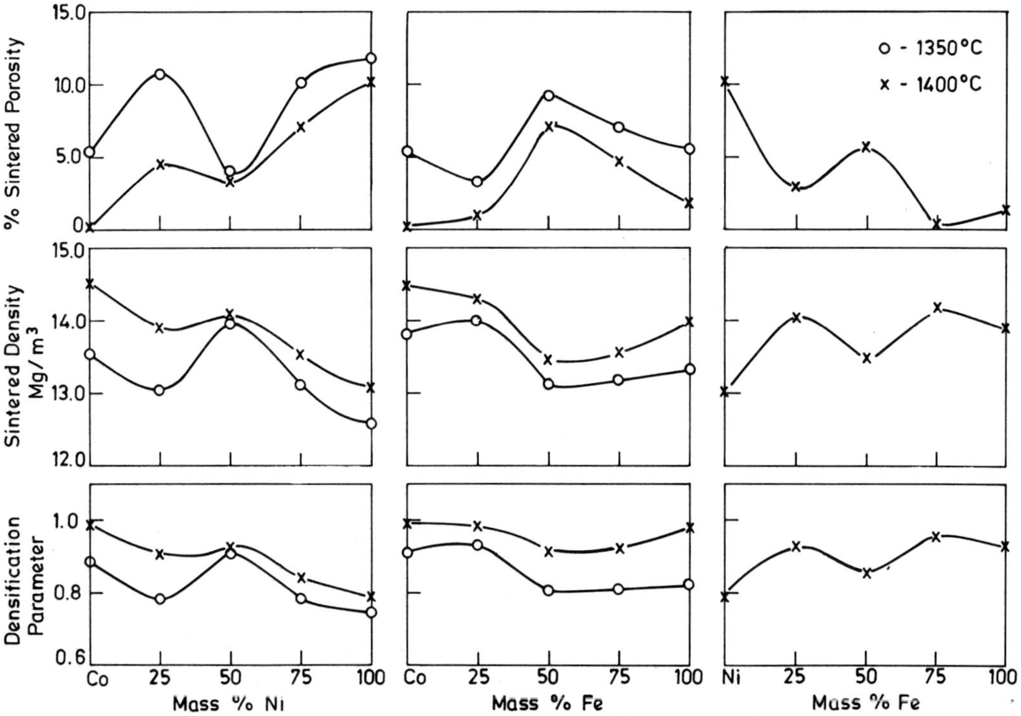

Fig. 1. Properties of W–10wt.%(Co–Ni–Fe) hard metals, sintered at 1350 and 1400 °C for 1 h in dry hydrogen.

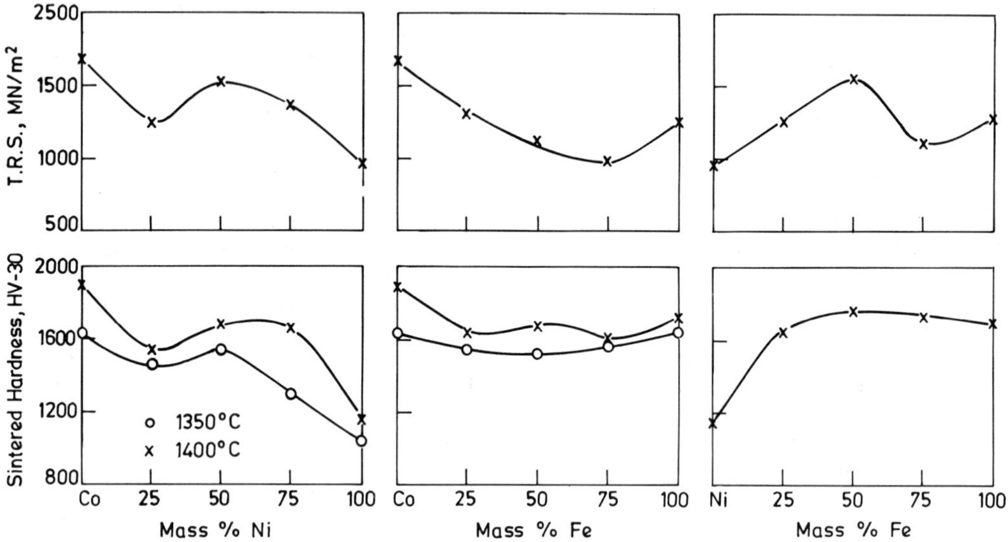

Fig. 2. Vickers' hardness and TRS plots of WC–10wt.%(Co–Ni–Fe) hard metals, sintered at 1350 and 1400 °C for 1 h in dry hydrogen.

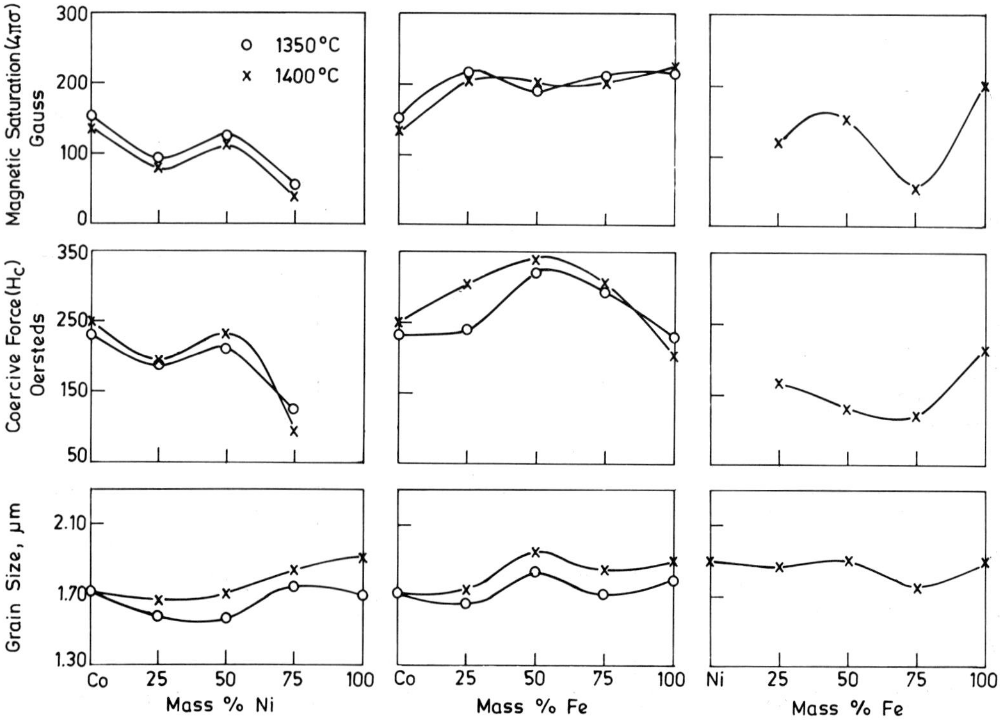

Fig. 3. Variations in grain size and magnetic properties of WC–10wt.%(Co–Ni–Fe) hard metals, sintered at 1350 and 1400 °C for 1 h.

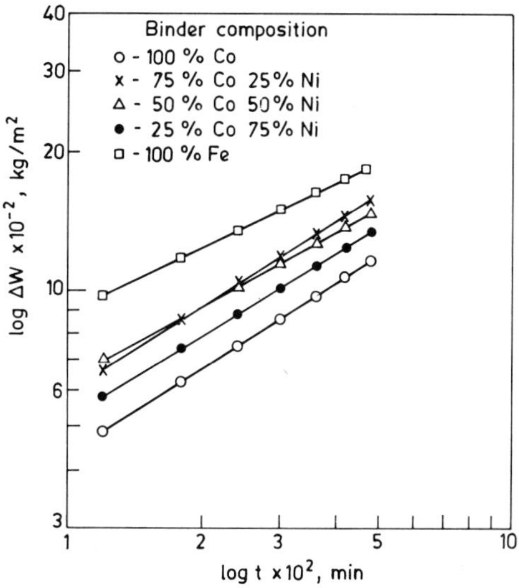

Fig. 4. Mass gain *vs.* time (logarithmic scale) plot of WC–10wt.%(Co–Ni) hard metals oxidized at 800 °C in air (sintered at 1400 °C for 1 h).

states (Fe, Ni)WO_4 and binder oxides (Fe, Ni)O and $(Fe, Ni)_2O_3$ for pure nickel or iron binder. Figure 5 shows some typical scanning electron fractographs of hard metals containing different binder compositions which had been sintered at 1400 °C for 1 h.

4. Discussion

The compatibility of the binder melt with the refractory carbide at the sintering temperature is an essential prerequisite to liquid phase sintering. WC, which is a group VI metal carbide, has excellent wettability with iron group transition metals, compared with group IV or V transition metal carbides. This has been explained in detail elsewhere on the basis of chemical bindings in the respective refractory carbides [13]. For example, TiC has the maximum share of the covalent sp^3 bond of all the refractory carbides and thus its contribution of non-localized electrons to the binder metal is minimal. The solubility of carbide phase in the binder melt, an important parameter for efficient liquid phase sintering, can also be explained on the basis of the above reasoning. Of all the binders selected in the present study, i.e. iron, cobalt and nickel, WC–Fe and WC–Co form pseudobinary systems, whereas WC–Ni does not. In addition, in the case of nickel the d band is more or less filled, explaining why WC–Ni hard metals are inferior.

Recently, German [14] has discussed the liquid phase sinterability criteria from the viewpoint of the nature of phase equilibria. The conditions proposed by him are as follows.

(1) The additive should have high solubility for the base component rather than vice versa.

(2) The temperature difference between the base component melting point and the eutectic should be as high as possible.

(3) The liquidus and solidus should decrease drastically in order to give rise to additive segregation at the interparticle interfaces, so as to enhance sintering.

Applying these criteria to the systems under investigation, cobalt (melting point, 1495 °C) dissolves 10–22 wt.% WC at the eutectic temperature of 1320 °C, whereas iron (melting point, 1535 °C) dissolves a much lower quantity, *i.e.* 7.5 wt.% WC, at the eutectic temperature (1143 °C). This suggests that the segregation effect would be greater in the WC–Fe system, where the difference between the binder melting point and the eutectic temperature is a maximum. The present results confirm that iron may be an efficient binder candidate provided that the range of WC solubility in it could be increased. This is evidenced by the fact that an addition of nickel to iron (up to 25%) enhances the sintered properties of the hard metals mainly because of the higher WC solubility in the alloy binder.

The liquid-phase-sintering process in cemented carbides is strongly dependent on the temperature of formation and the volume fraction of the liquid. Gabriel *et al.* [15] established the temperatures of formation of the liquid phase as 1286 °C, 1143 °C and 1326 °C for C–Co–W, C–Fe–W and Ni–C systems respectively, which suggests that the minimum liquid phase sintering temperature increases as the metal selected goes from iron to cobalt to nickel. As nickel provides a lower liquid fraction than do cobalt and iron, it can be expected that densification is more difficult with nickel than with cobalt or iron. Gabriel *et al.* [15] also established that, in order to obtain the same WC solubility in a nickel liquid phase, the temperature must be increased by 300 °C.

4.1. WC–10wt.%(Co–Ni) hard metals

Current experimental results, which are in agreement with those of Tracy and Hall [16], indicate that partial or complete substitution of

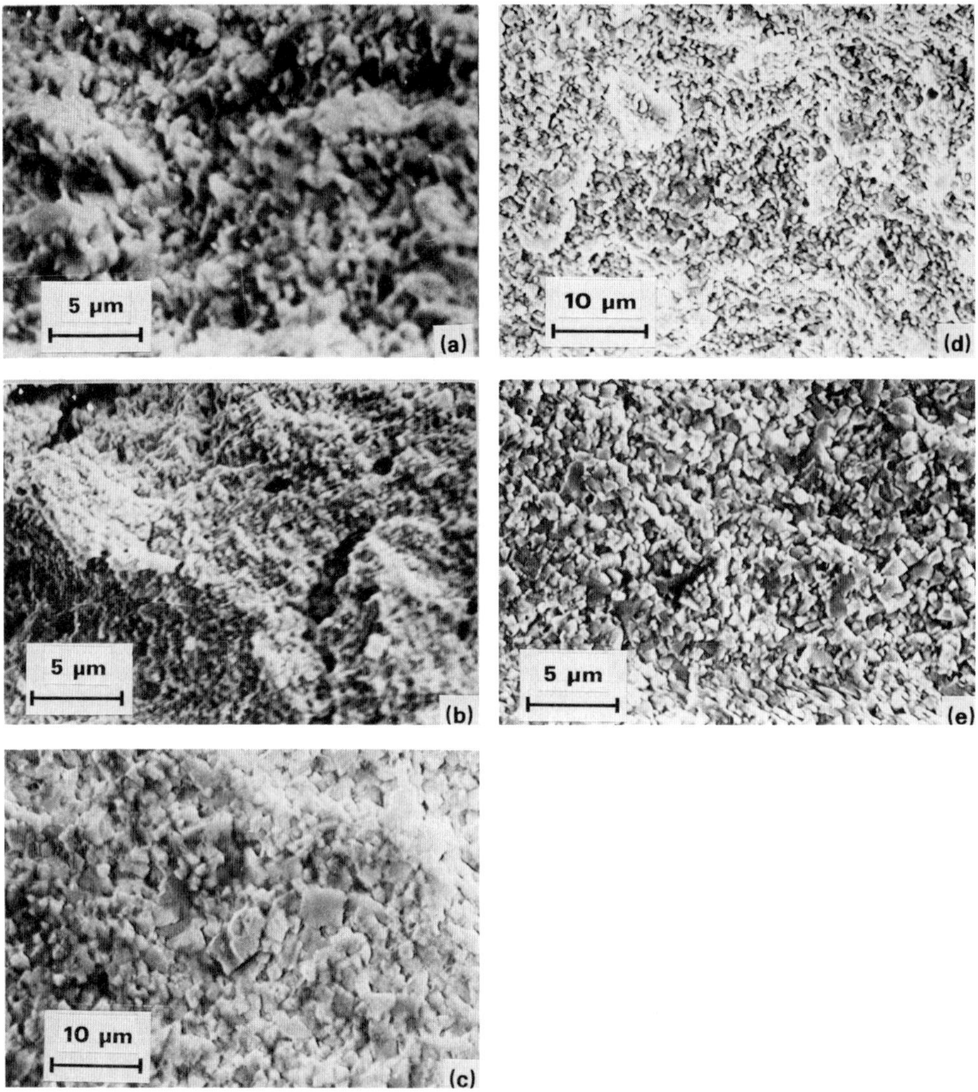

Fig. 5. Some typical scanning electron fractographs of WC–10wt.%(Co–Ni–Fe) hard metals with varying binder compositions (sintered at 1400 °C for 1 h) for the following binder compositions: (a) 100% Co; (b) 100% Ni; (c) 100% Fe; (d) 75%Co–25%Fe; (e) 25%Ni–75%Fe.

cobalt by nickel gives rise to inferior sintered properties compared with cobalt-bonded hard metals.

The decrease in TRS values with the addition of nickel may arise because nickel promotes free-graphite formation in the hard metal. The decrease in magnetic properties with increase in nickel content in the binder is followed by zero magnetic saturation in the case of complete nickel substitution. This confirms the findings of Schwarzkopf and Kieffer [17] and Brabyn *et al.* [18]. According to the former, the introduction of nickel leads to free carbon while, according to the latter, even slightly decarburized WC–Ni hard metals can show zero magnetic saturation and coercive force.

In the case of WC–Co hard metals, it is known that the stable h.c.p. (ε) structure is not present to a significant degree after sintering. It is thought that the metastable f.c.c. (β) phase is retained at room temperature. Brabyn *et al.* [18] indicate that the strength of WC–Co hard metals are dominated by the influence of the deformation-induced $\beta \rightarrow \varepsilon$ transformation via its effect as work hardening at the sites of concentration of plastic flow. In other words, such a transformation is promoted by the formation of h.c.p. lamellae during deformation [19].

4.2. WC–10wt.%(Co–Fe) hard metals

Enhanced densification of such hard metals in iron-rich compositions may be related to the relatively low eutectic temperature in the WC–Fe system (1143 °C). The drastic decrease in TRS with iron substitution is probably related to the lower solubility of WC in iron melt, as well as the poor wettability of WC by iron, compared with cobalt. It is thus obvious that such a feature is due to inadequate WC–binder bonding in the present binder system. Another possible reason may be the presence of η phase on the addition of iron, as suggested by Moskowitz *et al.* [12]. The result is at variance with that of Prakash [7, 8], who found comparable mechanical properties.

4.3. WC–10wt.%(Ni–Fe) hard metals

Results indicate that complete substitution of the cobalt binder by Ni–Fe enhances properties compared with the partial substitution by either nickel or iron. It has been reported [12] that the presence of iron in WC–Ni hard metals lowers the adverse effect of nickel, thus resulting in improved sintered properties. The lower solubility of WC in iron (7.5 wt.%) compared with that of nickel (15 wt.%) should promote carbide particle coarsening, although this was not observed in the present work.

4.4. Oxidation behaviour

Oxidation results characteristic of these hard metal systems run parallel to the variation in sintered porosity. Of all the binder compositions investigated, cobalt offers the maximum oxidation resistance. Cobalt forms three oxides cobaltous oxide (CoO), cobaltic oxide (Co_2O_3) and tricobalt tetraoxide (Co_3O_4). In contrast, nickel forms NiO, Ni_2O and Ni_2O_3. From the thermodynamic data, it is noted that the change in the free-energy formation of oxides at 1100 K for CoO and Co_3O_4 are -158 MJ kg^{-1} mol^{-1} and -517 MJ kg^{-1} mol^{-1} respectively, whereas the corresponding value for NiO is -139 MJ kg^{-1} mol^{-1}. These data confirm that the hard metals containing only cobalt as binder should be relatively more oxidation resistant. The stabilization of oxidation by the Co–Ni binary binder may be attributed to the oxidation resistance caused by the solute oxide.

5. Conclusions

(1) The densification and mechanical properties of submicron WC–10wt.%Co hard metals deteriorate on increase in the substitution level of binder by nickel.

(2) Complete cobalt substitution in WC–10wt.%Co hard metal by Ni–Fe binder imparts better properties compared with substitution by either iron or nickel.

(3) The grain size of the carbide phase increases with increase in the nickel content in the binder.

(4) The oxidation resistance of WC–10wt.%Co hard metal decreases on substitution of the cobalt binder by nickel.

References

1 W. Dawihl, *Handbook of Hard Metals*, H.M. Stationery Office, London, 1955, p. 68.
2 R. Edwards, T. Raine, *Proc. 1st Plansee Semin.*, Metallwerk Plansee, Reutte, 1953, p. 232.
3 S. Takeda, *Sci. Rep. Res. Inst. Tohoku Univ., Honda Anniv. Issue* (1936) 864.
4 I. N. Chaporova, V. I. Tretyakov, E. A. Shchetling and T. G. Makarenko, in S. I. Bashkirov (ed.), *Hardmetals Production Technology and Research in USSR*, Pergamon, Oxford, 1964, p. 172.
5 K. Whitehead and L. D. Brownlee, *Res. Rep. 5006*, 1954 (Metrovickers Electrical Co. Ltd.).
6 G. Barlow, *Proc. 4th Eur. Powder Metallurgy Symp., Grenoble, May 1975*, Société Française de Métallurgie, Paris.
7 L. Prakash, *KfK-Ext. 6/77-1*, 1977, p. 52 (Kernforschungszentrum Karlsruhe GmbH).
8 L. Prakash, *KfK-Ext. 6/78-1*, 1978, p. 87 (Kernforschungszentrum Karlsruhe GmbH).
9 C. Agte, *Neue Hütte, 2*, 1959, 537.
10 D. Moskowitz, in H. H. Hausner and P. V. Taubenblat, *Modern Developments in Powder Metallurgy*, Vol. 10, Metal Powder Industries Federation, Princeton, NJ, 1977, p. 543.
11 M. Tanaka, T. Suzuki and J. Komura, *Bull. Tokyo Inst. Technol.* (82) (1967) 65.
12 D. Moskowitz, N. J. Ford and M. Humenik, Jr., *Int. J.*

Powder Metall., *6* (4) (1970) 55.
13 G. S. Upadhyaya, in G. S. Upadhyaya (ed.), *Sintered Metal–Ceramic Composites*, Elsevier, Amsterdam, 1984, p. 41.
14 R. M. German, *Liquid Phase Sintering*, Plenum, New York, 1985.
15 A. Gabriel, H. Pastor, D. M. Deo, S. Basu and C. H. Allibert, in G. C. Kuczyuski, D. P. Uskokovic, H. Palmour III and M. M. Ristic (eds.), *Sintering '85*, Plenum, New York, 1987, p. 379.
16 V. A. Tracy and N. R. V. Hall, *Proc. Conf. on Recent Advances in Hardmetal Production, September 17–19, 1979*, Metal Powder Report, Shrewsbury, 1979.
17 P. Schwarzkopf and R. Kieffer, *Cemented Carbides*, Macmillan, London, 1960, p. 188.
18 S. M. Brabyn, R. Cooper and C. T. Peters, in H. M. Ortner (ed.), *Proc. 10th Int. Plansee Semin.*, Vol. 2, Metallwerk Plansee, Reutte, 1981, p. 675.
19 V. K. Sarin and T. Johannesson, *Met. Sci. J.*, *9* (1975) 472.
20 P. Rautellea and T. Norton, *J. Met.*, *4* (1952) 1045.

Investigations of Phase Equilibria in the Ti–N and Ti–Mo–N Systems*

WALTER LENGAUER and PETER ETTMAYER

Institute for Chemical Technology of Inorganic Materials, Technical University of Vienna, Getreidemarkt 9, A-1060 Vienna (Austria)

(Received November 9, 1987)

Abstract

As a part of a programme to develop hard materials with refractory binder metals the Ti–N and Ti–Mo–N systems were investigated.

In order to reassess the Ti–N system the following were prepared: arc-melted TiN–Ti mixtures; sintered TiN–Ti mixtures; diffusion couples. Special attention was paid to the area around 33 at.% N, where two new high temperature phases were found: η-Ti_3N_{2-x} and ζ-Ti_4N_{3-x}, both crystallizing in the space group $R\bar{3}m$ and being isostructural to ε-Hf_3N_2 and ζ-Hf_4N_3 respectively. The stability range and the formation of ε-Ti_2N were investigated by metallography and high temperature X-ray diffraction. ε-Ti_2N decomposes at 1332 ± 13 K. At this temperature it contains only 31.5 at.% N. The results of the metallographic study indicate that the formation of ε-Ti_2N from pure δ-TiN_{1-x} is retarded. In contrast, the formation of ε-Ti_2N is favoured by the presence of α-Ti(N).

In the case of the Ti–Mo–N system, arc-melted Ti–TiN–Mo mixtures were equilibrated at 1423 K and investigated by means of X-ray diffraction, metallography and electron probe microanalysis. The Ti–Mo–N system is characterized by the coexistence of δ-TiN_{1-x} with β-Ti_xMo_{1-x}, from $Ti_{0.70}Mo_{0.30}$ up to nearly pure molybdenum. The solubility of nitrogen in β-Ti_xMo_{1-x} is low (less than 1 at.%), as also is that of molybdenum in α-Ti(N) and δ-TiN_{1-x}. The composition of the β phase in equilibrium with α-Ti(N) + ζ-Ti_4N_{3-x} and ζ-Ti_4N_{3-x} + δ-TiN_{1-x} respectively was evaluated to be near $Ti_{0.70}Mo_{0.30}$. The width of the two-phase field ζ-Ti_4N_{3-x} + β-Ti_xMo_{1-x} is very narrow.

The alloy state of molybdenum as a binder phase is strongly dependent on the titanium activity in δ-TiN_{1-x}. If δ-TiN_{1-x} loses nitrogen during sintering, titanium diffuses into molybdenum. Since the titanium activity is a function of the nitrogen partial pressure, the latter should be kept as high as possible if a low alloy status of molybdenum is desirable.

1. Introduction

The mononitrides of the group IVA metals are known to exhibit exceptional properties such as high hardness and wear resistance, a high melting point and high thermal and electrical conductivities. Because of these properties, they offer great potential for use in cutting tools and wear-resistant parts. TiN, ZrN and HfN have already been successfully introduced as coatings on conventional hard metals to increase the cutting performance. In contrast, nitrides as components in sintered compositions with Ni–Mo binders have not been able to compete with conventional hard metals, and only molybdenum- and tungsten-containing carbonitrides of titanium with an alloyed nickel binder have had any commercial success. However, at least at high cutting speeds, nickel-containing binders have an insufficient hot strength compared with refractory metal binders.

As a part of a programme to develop such nitride-based refractory-metal-bonded hard materials, the Ti–N system and an isothermal section of the Ti–Mo–N system were investigated.

While it appears to be reasonably well established that the solid solubility of molybdenum and tungsten in the nitrides of the group IVA metals is negligible at all temperatures, the solubility of these metals in the binder is dependent on both temperature and nitrogen pressure. The investigation of the ternary system Ti–Mo–N should supply some of the necessary data for characterizing the interaction between the nitride phase and the binder alloy at various temperatures and nitrogen pressures.

Several studies have been carried out in the field of nitridation of Ti–Mo alloys (see for ex-

*Paper presented at the 3rd International Conference on the Science of Hard Materials, Nassau, The Bahamas, November 9–13, 1987.

0921-5093/88/$3.50 © Elsevier Sequoia/Printed in The Netherlands

ample ref. 1). If Ti–Mo alloys are nitrided, titanium nitride precipitates and leads to precipitation hardening of the surrounding molybdenum matrix. This improves the hot strength and the recrystallization temperature of the alloy. There is disagreement in the literature about the interaction of molybdenum with δ-TiN_{1-x}. While some researchers report changes in the lattice parameters of both molybdenum and δ-TiN_{1-x} during the sintering of powder mixtures [2, 3], others [4] state that there is no difference before and after annealing or sintering. These discrepancies can most probably be explained by changes in the composition of TiN_{1-x} as a result of differences in the sintering atmosphere. By choosing stoichiometric TiN and a nitrogen pressure sufficient to avoid nitrogen loss at the sintering temperature, there should be no observable change in the lattice parameters of either component.

Suganuma *et al.* [5, 6] performed high pressure sintering experiments using TiN and molybdenum with and without an intermediate layer of 60vol.%TiN–40vol.%Mo. Samples with an interlayer showed a higher thermal shock resistance in quenching experiments.

The interaction of TiN with molybdenum and tungsten (as well as the interaction of ZrN and HfN with molybdenum and tungsten) at very high temperatures and high pressures was investigated by Ettmayer and Hörmanseder [7]. They found that TiN and molybdenum form a pseudobinary eutectic with a melting point minimum at 2500 K at a composition of 25 mol.% TiN. Barabash and coworkers [8–10] and Psenichnaya and Verkhovodov [11] have also investigated the Ti–Mo–N system in the solid–liquid region, but their results differ considerably from those of Ettmayer and Hörmanseder. The Soviet authors apparently did not take into account the gas phase, which evidently has a crucial influence on the phase equilibria. So far no comprehensive studies of the phase diagram for the Ti–Mo–N system have been reported in the literature.

Because of the severely diverging opinions in the literature with respect to the Ti–N binary system, part of this system was reinvestigated. Several proposals for the phase diagram of the Ti–N system have been reported [12–18] but the agreement with respect to the homogeneity region of δ-TiN_{1-x} and the stability range of ε-Ti_2N is unsatisfactory. Values for the decomposition temperature of ε-Ti_2N have been reported as from 1300 to 2300 K [14–20]. Therefore, we focused our investigations of the Ti–N system on the area of around 33 at.% N.

The other binary boundary systems Ti–Mo and Mo–N appear to be reasonably well established. Titanium and molybdenum exhibit complete solubility above the α-Ti→β-Ti transformation temperature (1155 K) [21]. Nitrogen is slightly soluble in solid molybdenum, but lowers the melting point drastically [22]. Under the conditions prevailing in our experiments, molybdenum nitrides are not stable [23].

2. Experimental details

2.1. The Ti–N system

One part of the samples was prepared by arc melting pellets of Ti–TiN powder mixtures in an argon atmosphere. After a homogenization step at 1900 K for 10 h the samples were put into cylinders of molybdenum foil and sealed under argon at 10–30 kPa in silica tubes. The sample tubes were then subjected to heat treatments at 870–1470 K for up to several hundred hours. After this treatment the tubes were quenched in water. Because of nitrogen losses during arc melting, the equilibrated samples were chemically analysed for their nitrogen (by the Dumas method), titanium (by combustion) and oxygen (by vacuum hot extraction) contents. The oxygen content was found to be within the range 0.3–0.5 wt.%.

Some of the specimens were prepared by pressing mixtures of $TiN_{0.975}$ and titanium powders (grain sizes of 1.5 μm and 0.1 mm respectively) into pellets and sintering them in argon at 1700 K for 1–5 days. After this homogenization treatment, the samples were cut and heat treated in sealed silica tubes as described above.

Diffusion couples were prepared by heating titanium plates of various sizes in nitrogen. The process was carried out in an autoclave at pressures of 100–300 kPa. The reaction times were between 1 and 100 h. Since rapid cooling rates could not be obtained in the autoclave, some of these samples were annealed in silica tubes and quenched.

2.2. The Ti–Mo–N system

These samples were prepared by arc melting TiN–Ti–Mo mixtures in an argon atmosphere and then equilibrating them in silica tubes at 1423 K as described above. The chemical analyses for titanium and molybdenum were carried out by X-ray fluorescence after having dissolved the samples in an acid mixture. Nitrogen analysis was performed by helium carrier gas extraction and in some cases by the Dumas method. Oxygen was analysed by vacuum hot extraction and found to be in the range 0.25–0.35 wt.%. In order to determine the composition of the coexisting phases in the two-phase field δ-TiN_{1-x} + β-Ti_xMo_{1-x}, the lattice parameters of the phases were determined using the Debye–Scherrer technique. The relationships between the lattice parameter and the composition in the binary system Ti–Mo and within the homogeneity range of TiN_{1-x} were used to establish the tie lines within the two-phase field. The results for β-Ti_xMo_{1-x} were also confirmed by electron probe microanalysis. The detailed procedure for the measurements has been reported in a previous paper [24]. All samples (Ti–N and Ti–Mo–N systems) were characterized by X-ray diffraction. Parts of the samples were subjected to high temperature X-ray diffraction and metallographic investigations.

3. Results and discussion

3.1. The Ti–N system

In the Ti–N system, two new but possibly metastable phases were found [25, 26]. η-Ti_3N_{2-x} contains 29 ± 1 at.% N and appears to exist in the temperature range from 1340 to approximately 1390 K. ζ-Ti_4N_{3-x} has 31.5 ± 1 at.% N and exists in the temperature range from 1340 up to at least 1500 K. Crystal structure determinations from powder patterns showed η-Ti_3N_{2-x} to be isostructural with ε-Hf_3N_2 [27] and Ta_2VC_2 [28]. ζ-Ti_4N_{3-x} is isostructural with ζ-Hf_4N_3 [27] and the ζ-carbides of vanadium, niobium and tantalum [29]. The lattice parameters of the new phases (both crystallizing in space group $R\bar{3}m$) are as follows: η-Ti_3N_{2-x} has $a = 0.298\,09(4)$ nm and $c = 2.166\,42(85)$ nm (hexagonal) and $a = 0.742\,36(26)$ nm and $\alpha = 23.16(1)°$ (rhombohedral); ζ-Ti_4N_{3-x} has $a = 0.297\,95(1)$ nm and $c = 2.896\,55(47)$ nm (hexagonal) and $a = 0.980\,72(15)$ nm, $\alpha = 17.47(1)°$ (rhombohedral).

The structures of both phases are given in Fig. 1. The common feature of both phases is a close-packed metal atom arrangement of nine (η) and 12 (ζ) layers in the stacking sequences ABABCBCAC… and ABABCACABCBC… respectively. As far as can be judged from our powder X-ray diffraction patterns, the highly deficient nitrogen sublattices are disordered, *i.e.* the nitrogen atoms appear to be randomly distributed among the available interstitial voids. Analogous to the ζ-carbides [30–33] and hafnium nitrides [34], the structure of these phases can be derived from a modulation of the stacking sequence of δ-TiN_{1-x}. In order to obtain the structures of η-Ti_3N_{2-x} and ζ-Ti_4N_{3-x}, every third (η) or every fourth (ζ) layer of f.c.c. TiN_{1-x} has to be sheared. Figure 2 illustrates the shearing process. However, the ordering of the nitrogen atoms as implied by Fig. 2 could not be detected. The

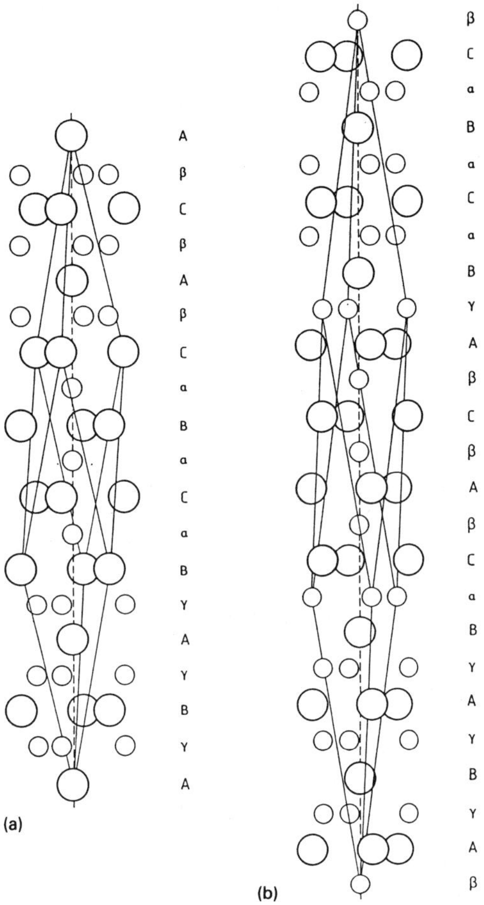

Fig. 1. Structure and rhombohedral unit cell of (a) η-Ti_3N_{2-x} and (b) ζ-Ti_4N_{3-x}. The stacking sequence is given in the margin. Large circles indicate titanium atoms, and small circles nitrogen atoms. The nitrogen sublattice is most probably randomly occupied.

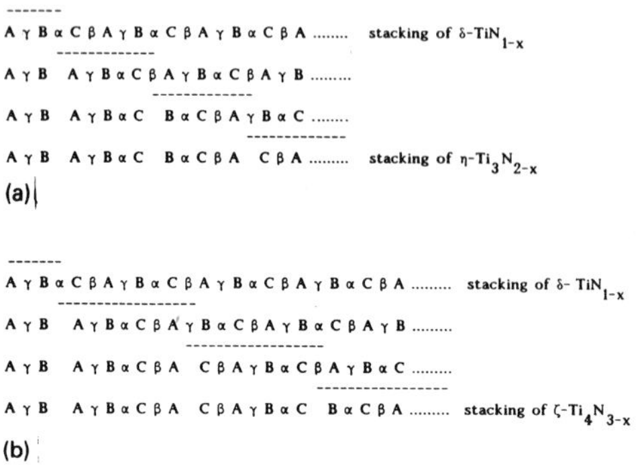

Fig. 2. Schematic model for the formation of (a) η-Ti_3N_{2-x} and (b) ζ-Ti_4N_{3-x} by periodic introduction of stacking faults in substoichiometric δ-TiN_{1-x}. The capital letters label the titanium layers, and the Greek letters label the nitrogen layers. The vacant nitrogen layers could not be verified by X-ray powder diffractometry.

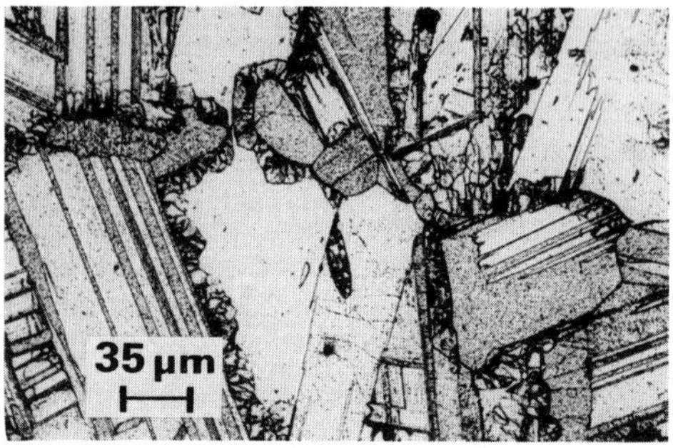

Fig. 4. Microstructure of an arc-melted sample containing α-Ti(N), ε-Ti_2N and η-Ti_3N_{2-x}, heat treated at 1360 K and quenched. The precipitation bands of the ε phase occurring during quenching are clearly visible around the dark α phase; the η phase appears bright. The sample was etched with H_2F_2–HNO_3.

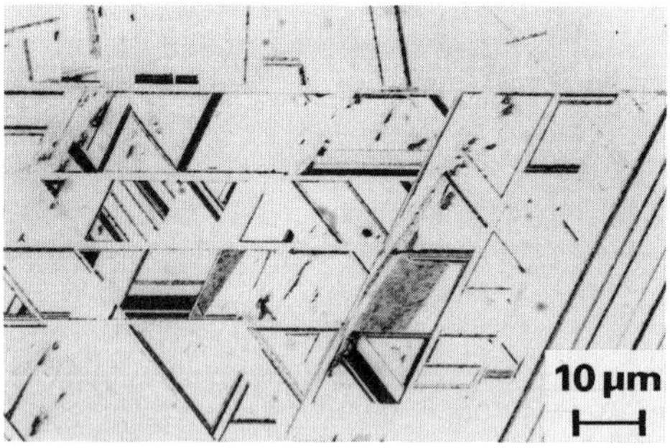

Fig. 3. Microstructure of a two-phase sample containing ζ-Ti_4N_{3-x} and small amounts of δ-TiN_{1-x} (etched with H_2F_2–HNO_3). The Widmannstätten-type pattern is clearly visible.

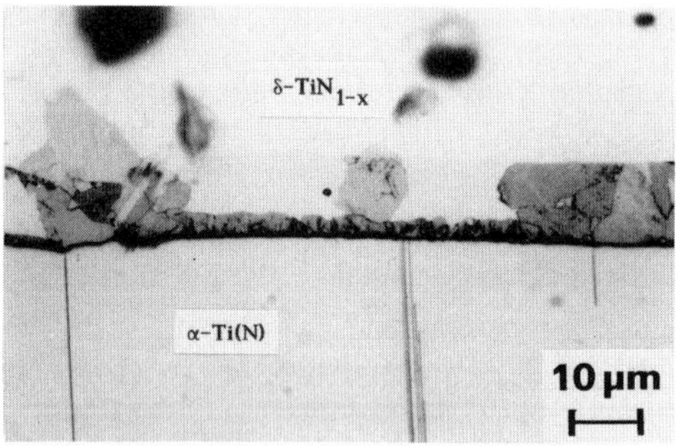

Fig. 5. Non-isothermal formation of ε-precipitation bands at the α-Ti(N) boundary during quenching of diffusion couples from 1420 K.

shearing process becomes visible in the microstructure of a two-phase sample containing ζ-Ti_4N_{3-x} plus small amounts of δ-TiN_{1-x} (Fig. 3). The lamellar appearance, which is similar to the Widmannstätten pattern, points to a twinning process occurring in δ-TiN_{1-x} in selected lattice directions. However, there is conflicting evidence against the metastable character of ζ-Ti_4N_{3-x} since, in rapidly quenched samples, this phase can be obtained in the form of distinct parallel diffusion zones having a thickness of about 6 μm at 1470 K [35].

Because of the strong disagreement about the stability limit of ε-Ti_2N, the decomposition temperature of this compound was measured by high temperature X-ray diffraction [36]. Several parallel runs yielded 1332 ± 13 K, in excellent agreement with a recent differential thermal analysis study [18]. The divergent data on the peritectoid decomposition of ε-Ti_2N are probably caused by the rapid formation of this compound during the cooling period. Apparently, however, its formation is fast only if α-Ti(N) is present in the sample, as the cooling of single-phase δ-$TiN_{0.50}$ from 1700 K to room temperature within 2 min did not result in any ε-Ti_2N precipitation. By contrast the precipitation bands of ε-Ti_2N in contact with α-Ti(N) could be observed in both arc-melted (Fig. 4) and diffusion (Fig. 5) specimens if both samples were quenched from temperatures higher than the decomposition temperature of ε-Ti_2N. As also observed by Hibbs *et al.* [37] for vapour-deposited TiN films, this fact can probably best be explained by the close structural similarity between α-Ti(N) and ε-Ti_2N, where the nitrogen atoms only have to move over short distances to transform α-Ti(N) into ε-Ti_2N. Therefore the α-Ti(N) phase boundaries may serve as nucleating centres for ε-Ti_2N. The structures of ε-Ti_2N and δ-TiN_{1-x}, however, are not as closely related. This difference between the structural arrangements of the atoms in ε-Ti_2N and δ-TiN_{1-x} is also most probably responsible for the formation of the metastable compound δ'-$TiN_{0.50}$ as an intermediate step in the phase transition δ-$TiN_{1-x} \rightarrow \varepsilon$-$Ti_2N$ at temperatures below 1147 K. The latter temperature value was obtained by the stepwise annealing of sintered TiN–Ti pellets [36]. δ'-TiN_{1-x} is a metastable compound and characterized by an ordered arrangement of nitrogen atoms with a shift of the positions of the titanium atoms towards the nitrogen atoms [38, 39]. This ordering results in a reduction in the lattice symmetry. These investigations gave evidence of a substoichiometric character of ε-Ti_2N at temperatures higher than 973 K. If δ-TiN_{1-x} with 33.3 at.% N is heat treated at temperatures above 973 K even for long periods, the sample contains significant amounts of δ-TiN_{1-x}. Since δ-TiN_{1-x} necessarily contains more nitrogen than ε-Ti_2N does, the latter phase must be substoichiometric. The deviation of ε-Ti_2N from ideal stoichiometry was also confirmed by chemical analysis. According to these results, the ε-phase has 46 at.% N. In Fig. 6 the experimental results of our investigations in the region with a nitrogen-to-titanium ratio of 0.2–0.6 are summarized. Because of the multitude of possibly metastable (even martensitic) phases within only a few atomic per cent of nitrogen around the decomposition point of ε-Ti_2N, it is not yet possible to offer a phase diagram in this area. This will be the object of future investigations.

3.2. *The Ti–Mo–N system*

In agreement with the results for the binary system Ti–N the ζ phase was also observed in ternary Ti–Mo–N alloys which had been heat treated at 1423 K. In Fig. 7, a tentative isotherm section through the phase diagram corresponding to nitrogen pressures lower than 100 kPa is shown. At 1423 K the two-phase field δ-TiN_{1-x} + β-Ti_xMo_{1-x} extends from pure molybdenum to $Ti_{\approx 0.70}Mo_{\approx 0.30}$ at the Ti–Mo boundary and from $TiN_{0.51}$ at low nitrogen potential to $TiN_{1.00}$ at high nitrogen partial pressure. The tie lines within the two-phase field δ-TiN_{1-x} + β-Ti_xMo_{1-x} have been included in Fig.

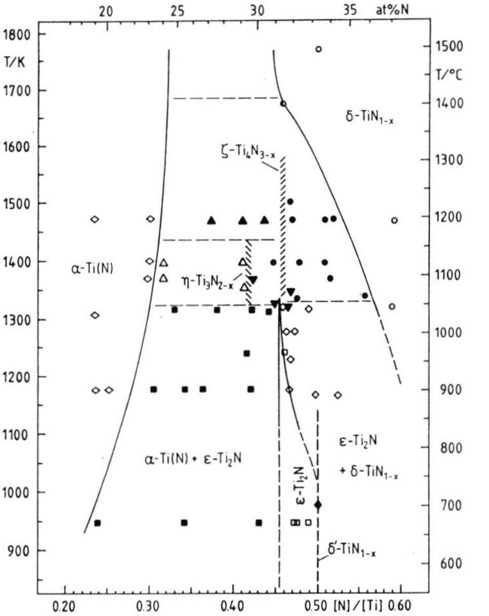

Fig. 6. Summary of results of experimental studies on the Ti–N system: ◇, α; ○, δ; □, ε; ■, $\alpha+\varepsilon$; ◇, $\delta+\varepsilon$; ▲, $\alpha+\zeta$; ●, $\delta+\zeta$; △, $\alpha+\eta+\zeta+(\varepsilon)$; ▼, $\eta+\zeta+\delta+(\varepsilon)$; ◆, δ'-$TiN_{0.50}$. The regions where η-Ti_3N_{2-x} and ζ-Ti_4N_{3-x} have been observed are shown as shaded. The ε phase is substoichiometric at higher temperatures.

7. Because of the small homogeneity range of ζ-Ti_4N_{3-x} and the obviously rather small stability difference between ζ-Ti_4N_{3-x} and the phase combination α-Ti(N) + δ-TiN_{1-x} at 1420 K, the two-phase field ζ-Ti_4N_{3-x} + β-Ti_xMo_{1-x} is rather narrow. The solid solubility of nitrogen in β-Ti_xMo_{1-x} is low, at least in the region where β-Ti_xMo_{1-x} is coexistent with δ-TiN_{1-x} and is estimated to be less than 1 at.% in that region. The solid solubility of molybdenum in the nitrogen-stabilized α-Ti(N) phase appears to be rather low; the positions of the tie lines in the two-phase field α-Ti(N) + β-Ti_xMo_{1-x} were not determined by experiments and therefore have not been included. Obviously the alloy status of molybdenum as a binder phase is strongly dependent on the titanium activity in δ-TiN_{1-x}, which in turn is a function of the nitrogen partial pressure. If δ-TiN_{1-x} loses nitrogen during (vacuum) sintering, titanium diffuses into molybdenum to form β-Ti_xMo_{1-x}. Thus the lattice parameter of TiN decreases whereas the lattice parameter of molybdenum increases. At higher temperatures the sensitivity to the nitrogen

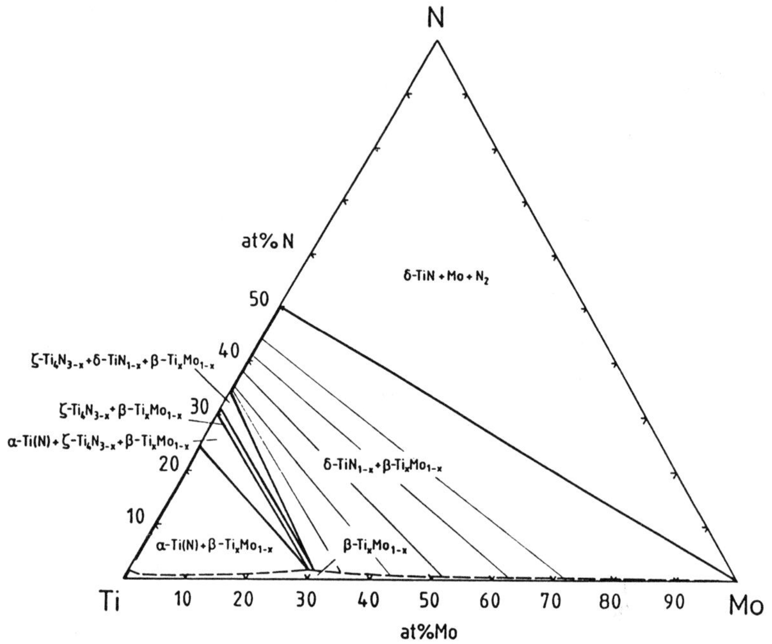

Fig. 7. Tentative phase diagram of the Ti–Mo–N system at 1420 K and pressures up to 100 kPa. In the two-phase field of δ-TiN_{1-x} + β-Ti_xMo_{1-x} the tie lines were determined experimentally. The phase boundaries at the end of the tie lines (not based on experimental data) are indicated by broken lines.

partial pressure becomes even more pronounced. If a low alloy status of molybdenum is desirable (because of its high hot strength) the nitrogen partial pressure should be kept as high as possible.

References

1 S. Yano, S. Morozumi and S. Koda, *Scr. Metall., 12* (1978) 467.
2 S. Morozumi, K. Kikuchi and S. Sugai, *Nippon Kinzoku Gakkaishi, 45* (1981) 184.
3 S. Takatsu and E. Ishimatsu, *Int. J. Refract. Hard. Met., 1* (1982) 75.
4 W. D. Kingery and F. A. Halden, *Ceram. Bull., 34* (1955) 117.
5 K. Suganuma, T. Okamoto, M. Shimada and M. Koizumi, *Commun. Am. Ceram. Soc.,* (1983) C-117.
6 K. Suganuma, M. Shimada and T. Okamoto, *Seramikkusu, 18* (1983) 112.
7 P. Ettmayer and W. Hörmanseder, *High Temp. High Pressures, 18* (1986) 161.
8 O. M. Barabash and K. A. Shurin, *Izv. Akad. Nauk S.S.S.R., Met., 4* (1978) 234.
9 O. M. Barabash and K. A. Shurin, *Metallofizika, 53* (1973) 101.
10 O. M. Barabash, V. F. Moiseev and E. P. Pechkovskii, *Fiz. Khim. Mekh. Mater., 12* (1976) 109.
11 O. V. Psenichnaya and P. A. Verkhovodov, *Porosk. Metall., 3* (1985) 26.
12 A. E. Palty, H. Margolin and J. P. Nielsen, *Trans. Am. Soc. Met., 46* (1954) 312.
13 L. A. McClaine and C. P. Coppel, *Rep. AFML-TR-65-299,* Part I, 1965 (U.S. Air Force Materials Laboratory, Dayton, OH).
14 N. R. McDonald and G. R. Wallwork, *Oxid. Met., 2* (1970) 263.
15 F. W. Wood and O. G. Paasche, *Microstruct. Sci., 2* (1974) 101.
16 M. P. Arbuzov, S. Ya. Golub and B. V. Khaenko, *Izv. Akad. Nauk S.S.S.R., Neorg. Mater., 13* (1977) 1779.
17 L. Wolff, G. Bastin and H. Heijligers, *Solid State Ionics, 16* (1985) 105.
18 E. Etchessahar, J.-P. Bars and J. Debuigne, *J. Less-Common Met., 134* (1987) 123.
19 E. H. Rennhack, W. C. Coons and E. A. Perkins, *Trans. Metall. Soc. AIME, 242* (1968) 343.
20 J.-P. Bars, E. Etchessahar and J. Debuigne, *J. Less-Common Met., 52* (1977) 51.
21 J. L. Murray, *Bull. Alloy Phase Diagr., 2* (1985) 185.
22 P. Ettmayer, *Monatsh. Chem., 101* (1970) 1720.
23 H. Jehn and P. Ettmayer, *J. Less-Common Met., 58* (1978) 85.
24 W. Lengauer, H.-J. Ullrich and P. Ettmayer, *Microchim. Acta, 1* (1987) 211.
25 W. Lengauer and P. Ettmayer, *J. Less-Common Met., 120* (1986) 153.
26 W. Lengauer, *J. Less-Common Met., 125* (1986) 127.

27 E. Rudy, *Metall. Trans., 1* (1970) 1249.
28 E. Rudy, *J. Less-Common Met., 20* (1970) 49.
29 K. Yvon and E. Parthe, *Acta Crystallogr., Sect. B, 26* (1970) 149.
30 A. Kelly and D. J. Rowcliffe, *Phys. Status Solidi, 14* (1966) K29.
31 J. L. Martin, B. Jouffrey and P. Costa, *Phys. Status Solidi, 22* (1967) 349.
32 J. L. Martin and B. Jouffrey, *J. Phys. (Paris), 29* (1968) 911.
33 D. J. Rowcliffe and G. Thomas, *Mater. Sci. Eng., 18* (1975) 231.
34 G. M. Demyashev, E. E. Koschug, N. N. Repnikov and R. K. Chuzko, *J. Less-Common Met., 107* (1985) 59.
35 E. Etchessahar, Institut National des Sciences Appliquées, Rennes, personal communication, 1987.
36 W. Lengauer and P. Ettmayer, *High Temp. High Pressures, 19* (1987) 673.
37 M. K. Hibbs, J. E. Sundgren, B. O. Johansson and B. E. Jacobson, *Acta Metall., 33* (1985) 797.
38 G. Lobier and J. P. Marcon, *C. R. Acad. Sci., Sér. C, 268* (1969) 1132.
39 A. N. Christensen, A. Alamo and J. P. Landesman, *Acta Crystallogr., Sect. C, 41* (1985) 1009.

Cemented Titanium Carbonitrides: Effects of Temperature and Carbon-to-Nitrogen Ratio*

D. MOSKOWITZ

Iscar Ceramics Inc., Livonia, MI 48150 (U.S.A.)

L. L. TERNER

Ford Motor Co., Detroit, MI 48239 (U.S.A.)

(Received November 9, 1987)

Abstract

Several cemented titanium-carbonitride-base materials containing from 0 to 0.6 weight fraction ratio of TiN to TiC + TiN were prepared and studied as a function of their carbon-to-nitrogen ratio. They all contained 22.5 wt.% Ni, plus 10 wt.% each of molybdenum and vanadium carbide. Vacuum sintering was carried out at temperatures of 1400, 1475 and 1550 °C. Scanning electron photomicrographs showed considerable inhibition of grain growth with increasing TiN content, especially at 1550 °C. Fracture toughness measurements were made using the short-rod method, and showed decreasing K_{Ic} with increasing TiN ratio, i.e. decreasing grain size. Machining tests were run, milling AISI 4140 steel of 230 Brinell hardness at 158 m min^{-1}. A maximum tool life of close to 200 min was obtained at a TiN fraction of 0.5 and a sintering temperature of 1550 °C.

1. Introduction

The present study is a continuation of an earlier investigation by the present authors and a coworker [1] into titanium-carbonitride-base materials. Significant changes in physical properties as a function of grain size were noted then but not investigated systematically. In the present work, seven increments of TiN-to-(TiC + TiN) ratios from 0 to 0.6 were all studied at three sintering temperatures, 1400, 1475 and 1550 °C. This allowed for a more detailed examination of the effects of the TiN ratio and sintering temperature on grain size and a study of some physical properties, such as fracture toughness, which are affected by grain size variation. Tanaka [2] and Doi [3], amongst others, have published data relating to the properties of TiC–TiN-base materials and the cutting performance of cermets based on them.

2. Experimental procedure

Cemented Ti(C, N) composites were prepared by ball milling TiC and TiN powders together with 22.5 wt.% Ni, 10 wt.% Mo and 10 wt.% VC. TiC, TiN and VC powders were all manufactured by the H. C. Starck Co. The nickel and molybdenum powders were produced by International Nickel and Amax respectively. Table 1 lists the chemical analysis of the as-received TiC and TiN powders. Their grain size prior to milling was in the range 3–4 μm. The compositions were ball milled for 4 days in Hastelloy B mills with TiC-base media under acetone. Compacts were pressed at 138×10^3 KPa using Carbowax 600 type of polyethylene glycol as the pressing lubricant–binder. Dewaxing was done under a dry 50% N_2–50% H_2 mixture at a temperature of 670 °C. Specimens were vacuum sintered for 1 h at 1400, 1475 and 1550 °C at a pressure of 0.13 Pa. Compositions were prepared at TiN-to-

TABLE 1 Chemical analysis of TiC and TiN powders (as received)

Powder	*Amount* (wt.%)			*Atomic fraction of interstitial element*		
	C	*N*	*O*	*C*	*N*	*O*
TiC	19.23	0.11	0.52	0.960	0.005	0.021
TiN	0.065	23.4	0.59	0.003	1.035	0.024

*Paper presented at the 3rd International Conference on the Science of Hard Materials, Nassau, The Bahamas, November 9–13, 1987.

0921-5093/88/$3.50 © Elsevier Sequoia/Printed in The Netherlands

(TiC + TiN) weight ratios of 0, 0.3, 0.20, 0.30, 0.40, 0.50 and 0.60. Scanning electron photomicrographs were taken using a model 1000A Amray scanning electron microscopy unit. The fracture toughness was measured by the short-rod method [4, 5]. Machining tests were run using a Cincinnati Milacron vertical milling machine to face mill AISI 4140 steel.

3. Results

Figures 1, 2 and 3 are scanning electron photomicrographs of specimens containing TiN-to-(TiC + TiN) weight ratios of 0, 0.30 and 0.60 respectively, which had been sintered at a temperature of 1550 °C. The marked effect of TiN content on inhibiting grain growth in this system is evident in these photomicrographs. This is especially true for additions in the range up to 0.30 TiN ratio. Scanning electron photomicrographs of the same compositions sintered at a temperature of 1400 °C show a similar effect. Although a decrease in grain size with increasing TiN ratio is evident here as well, the effect is not as marked as at the higher temperature.

Figure 4 is a plot of the densities of specimens sintered at 1400, 1475 and 1550 °C. The differences in measured density between 1475 and 1550 °C are believed to be within experimental error. It is evident from the data and from metallographic examination that the highest TiN content compositions, *i.e.* 0.50 and 0.60 TiN ratios, did not fully densify at 1400 °C.

This is also reflected in the Rockwell A hardness data plotted in Fig. 5, where specimens with

Fig. 1. Scanning electron micrograph of Ti(C, N)–22.5wt.%Ni–10wt.%Mo–10wt.%VC sintered at 1550 °C for 1 h (TiN-to-(TiC + TiN) weight ratio, 0).

Fig. 2. Scanning electron micrograph of Ti(C, N)–22.5wt.%Ni–10wt.%Mo–10wt.%VC sintered at 1550 °C for 1 h (TiN-to-(TiC + TiN) weight ratio, 0.30).

Fig. 3. Scanning electron micrograph of Ti(C, N)–22.5wt.%Ni–10wt.%Mo–10wt.%VC sintered at 1550 °C for 1 h (TiN-to-(TiC + TiN) weight ratio, 0.60).

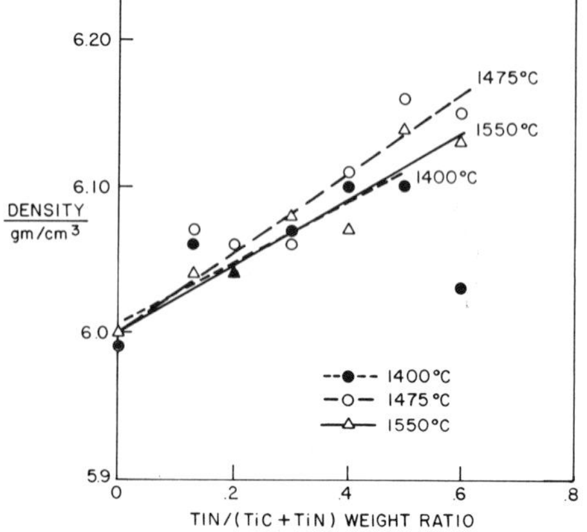

Fig. 4. Density *vs.* TiN content at sintering temperatures of 1400, 1475 and 1550 °C.

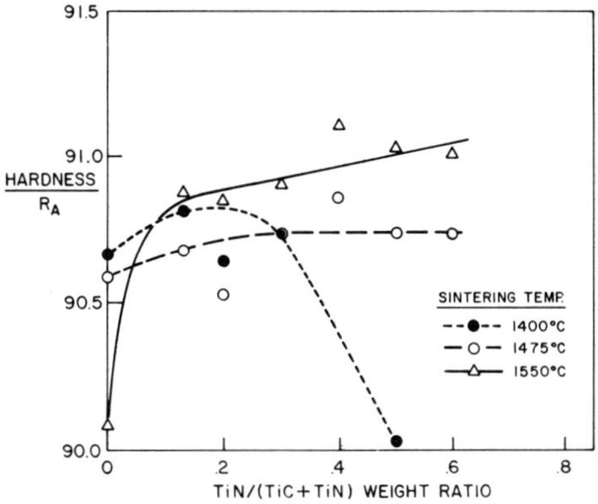

Fig. 5. Rockwell A hardness as a function of TiN-to-(TiC+TiN) ratio at sintering temperatures of 1400, 1475 and 1550 °C.

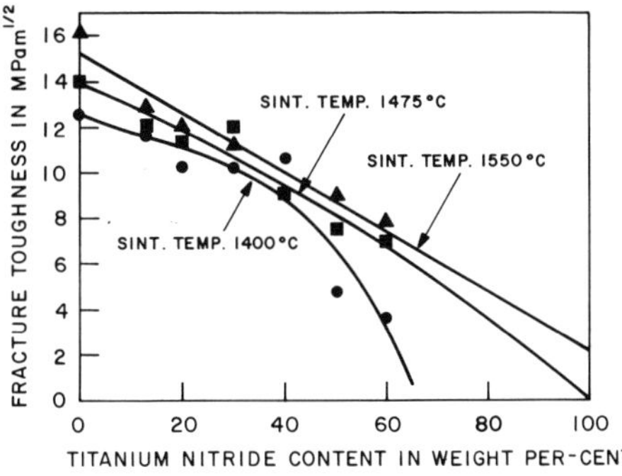

Fig. 6. Fracture toughness *vs.* TiN content at sintering temperatures of 1400, 1475 and 1550 °C.

TiN ratios of 0.50 or greater and sintered at 1400 °C had low hardness values. Maximum hardness values, in the neighborhood of Rockwell A 91.0, were obtained at a temperature of 1550 °C for TiN ratios of 0.40–0.60.

Figure 6 is a plot of K_{Ic} *vs.* TiN content, with sintering temperature as the parameter. There is a clear trend toward decreasing K_{Ic} with increasing TiN ratio at all sintering temperatures. This reflects the decreasing grain size with increasing TiN content, although both the hardening and the decreased K_{Ic} may be related to a solid-solution-strengthening effect. This effect has been noted previously in cemented carbides. The unusually low K_{Ic} values for the samples with 0.50 and 0.60 TiN ratios, which had been sintered at 1400 °C, arise because they are not fully dense. The K_{Ic} values of samples with TiN ratios in the range

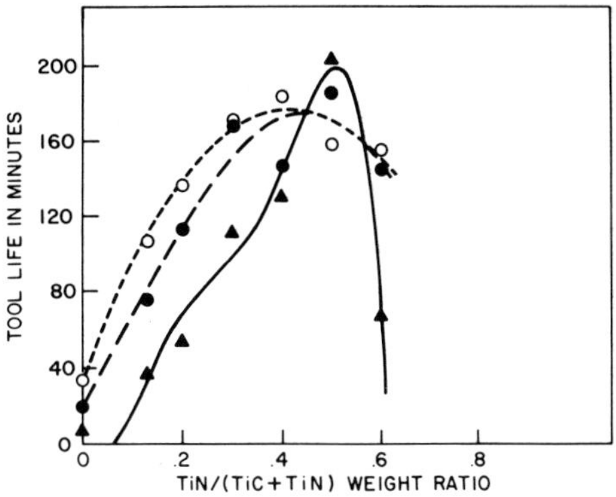

Fig. 7. Tool life of milling AISI 4140 steel at 158 m min^{-1} as a function of the TiN-to-(TiC+TiN) ratio at sintering temperatures of 1400 (○), 1475 (●) and 1550 °C (▲).

from 0 to 0.30, which had been sintered at 1475 °C and above, are from 16 MPa m$^{1/2}$ to 11 MPa m$^{1/2}$ respectively. This is in the same range as WC–Co-base materials of similar binder volume fraction.

Milling tests were run on AISI 4140 steel of 230 Brinell hardness at a speed of 158 m min^{-1}, a feed of 0.145 mm rev^{-1} and a depth of cut of 2.54 mm. A cutting edge chamfer of 0.15 mm and 20° was used in all these tests. These results are shown in Fig. 7. There is a sharp increase in tool life with increasing TiN content at all sintering temperatures. The peak value, in the neighborhood of 200 min, is reached for a TiN ratio of 0.50 sintered at 1550 °C. The marked increase in tool life with increasing TiN content, up to a TiN ratio of 0.50, is felt to be a reflection of both grain refinement and solid-solution-hardening effects. The decrease in tool life past this TiN content probably reflects their lower K_{Ic} values. It is especially interesting that tool life compares favorably with commercial coated WC-base grades at present in common use in steel milling and roughing applications. This indicates that Ti(C, N)-base materials have a promising future in the field of milling and rough machining of steel workpieces.

4. Summary

Some physical properties of cemented titanium carbonitrides were studied as a function of carbon-to-nitrogen ratio and sintering temperature. All compositions investigated contained 22.5 wt.% Ni, 10.0 wt.% Mo and 10.0 wt.% VC. Scanning electron microscopy examination showed marked grain growth inhibition with increasing TiN content, which was especially evident at the highest sintering temperature of 1550 °C. Short-rod fracture toughness measurements yielded values in the range from 11 MPa $m^{1/2}$ to 16 MPa $m^{1/2}$ for materials with TiN ratios from 0.30 to 0 respectively.

There was a decrease in K_{Ic} with increasing TiN content, reflecting the observed decrease in grain size. Milling tests run on AISI 4140 steel at 158 m min^{-1} showed an excellent tool life of approximately 200 min for a composition with a TiN ratio of 0.5, which had been sintered at 1550 °C. This type of material is expected to have commercial application as a steel milling and roughing grade.

Acknowledgments

The authors acknowledge the valuable work of R. L. Goss for the scanning electron photomicrographs in this study. The work of M. A. Yales in specimen preparation and of J. R. Dixon in obtaining the machining data is also gratefully acknowledged. The authors are indebted to C. O. McHugh and T. J. Whalen for proof reading the paper.

References

1 D. Moskowitz, L. L. Terner and M. Humenik, Jr., in E. A. Almond, C. A. Brookes and R. Warren (eds.), *Proc. 2nd Int. Conf. on the Science of Hard Materials, Rhodes, September 23–28, 1984,* in *Inst. Phys. Conf. Ser., 75* (1986) 605–617.

2 H. Tanaka, *Proc. Int. Conf. on Cutting Tool Materials, Fort Mitchell, KY, September 1980,* American Society for Metals, Metals Park, OH, 1981, pp. 349–361.

3 H. Doi, in E. A. Almond, C. A. Brookes and R. Warren (eds.), *Proc. 2nd Int. Conf. on the Science of Hard Materials; Rhodes, September 23–28, 1984,* in *Int. Phys. Conf. Ser., 75* (1986) 489–523.

4 L. M. Barker, *Engineering Fracture Mechanics,* Vol. 9, Pergamon, Oxford, 1977, pp. 361–369.

5 Standard test method for fracture toughness of cemented carbides, *ASTM Stand. B771-87,* 1987, Vol. 02-05.

The Sintering Characteristics and Properties of Hard Metal with Ni–Cr Binders*

R. COOPER, S. A. MANKTELOW and F. WONG

Hard Materials Research Inc., 980 Pacific Gate, Unit 15, Mississauga, Ontario L5T 1Y1 (Canada)

L. E. COLLINS

Physical Metallurgy Research Laboratories, CANMET, 568 Booth Street, Ottawa K1A OG1 (Canada)

(Received December 19, 1987)

Abstract

The substitution of an Ni–Cr binder for cobalt in conventional hard metals is known to enhance corrosion resistance in a variety of media. Chromium carbide is also noted for its grain growth inhibition properties in the production of abrasion resistant hard metals. However, in both cases, chromium has been associated with excessive porosity and reduced toughness in the sintered product.

An attempt has been made to characterize the effect of chromium on the sintering process, using dilatometry and resistivity determinations. The system studied was based on WC–10wt.%Ni, with up to 50% substitution by chromium in the binder to give WC–5wt.%Ni–5wt.%Cr.

Sintering activation energies were determined for each composition, the value increasing with increasing chromium content. While the straight nickel-bonded composition exhibited considerable shrinkage in the solid state (about 80% of total), increasing chromium levels delayed the onset of major shrinkage until the appearance of liquid phase. Isothermal treatments for extended times indicated that the liquid phase could appear at a temperature as low as 1210 °C.

It is proposed that dissolution of chromium in the nickel binder inhibits the tungsten carbide and nickel dissolution and that the formation of a low melting Ni–Cr–C eutectic results in most of the shrinkage in the high chromium compositions.

*Paper presented at the 3rd International Conference on the Science of Hard Materials, Nassau, The Bahamas, November 9–13, 1987.

1. Introduction

It has been demonstrated that the substitution of nickel for cobalt improves the corrosion resistance of hard metals in a variety of aqueous media. Further improvements are gained when chromium is added to the binder phase [1]. Chromium carbide is also well established as a grain growth inhibitor in the manufacture of submicron-grained hard metals [2].

However, during the development of corrosion-resistant and submicron grades of hard metals, it became apparent that the desired 100% densification could not be achieved by conventional techniques, the residual porosity resulting in reduced strength and toughness. Similar effects have been reported in the literature [3].

Little is known about the physical or chemical influence of chromium on the sintering process, and the present work was initiated to observe and to attempt to explain the effect of chromium additions to a WC–Ni-based hard metal composition. Both dilatometry and resistivity measurements have been successfully used to characterize sintering processes [4, 5]. These techniques were therefore adopted for this study, to support microstructural observations.

In order to quantify the effect of chromium on the energy requirements for initiation of shrinkage, the activation energy was determined for sintering hard metal compositions with various chromium contents, using linear shrinkage measurements during isothermal treatment.

2. Experimental procedure

Powder batches of the required compositions were prepared, using commercial tungsten

0921-5093/88/$3.50 © Elsevier Sequoia/Printed in The Netherlands

TABLE 1 Hard metal compositions investigated

Alloy	*Amount* (wt.%)		
	WC	*Ni*	*Cr*
WC–10Ni	90	10.0	—
WC–9.9Ni–0.1Cr	90	9.9	0.1
WC–9Ni–1Cr	90	9.0	1.0
WC–8Ni–2Cr	90	8.0	2.0
WC–5Ni–5Cr	90	5.0	5.0

carbide with a 4.5 μm nominal grain size, 1.5 μm nickel powder and 3.5 μm chromium carbide. In each case, the total binder content was 10 wt.%, and chromium carbide was added to obtain the required elemental chromium contents shown in Table 1. After conventional ball milling in acetone, drying, waxing and granulating, specimens of each composition were prepared by uniaxial pressing under 150 MPa to form cylinders 9.5 mm in diameter and 10 mm long. After drilling for thermocouple placement, dewaxing took place at 500 °C in flowing hydrogen.

Specimens were mounted in an MMC dilatometer and induction heated in vacuum (6 mPa) at a rate of 43 °C min^{-1} to 450 °C, and at 15 °C min^{-1} from 450 to 1500 °C. Argon was introduced at a pressure of 38 kPa at 1000 °C to minimize the binder evaporation, and specimens were quenched at about 200 °C s^{-1} on reaching 1500 °C.

For resistivity determinations, the lack of suitable means of on-line resistance and cross-sectional area determinations necessitated heating individual specimens, using the above cycle, to various temperatures between 700 and 1500 °C before quenching. The resistivity of each specimen was then established using a four-point probe and constant-current source.

Isothermal tests were performed at 50 °C increments from 1100 to 1400 °C. For these tests, specimens were heated from 1000 °C at 600 °C min^{-1} until the required isothermal temperature was reached. This temperature was held for 10^4 s while monitoring the change in specimen length.

3. Results and discussion

Figure 1 shows dilatometric curves, corrected for thermal expansion, for three of the compositions tested under constant heating. The plot for WC–9.9Ni–0.1Cr is omitted for clarity; it follows the straight nickel (WC–10Ni) composition, displaced 40 °C to the right. The WC–5Ni–5Cr specimens exhibited similar characteristics to the WC–8Ni–2Cr composition, but with the onset of shrinkage occurring some 50 °C earlier with the higher chromium content.

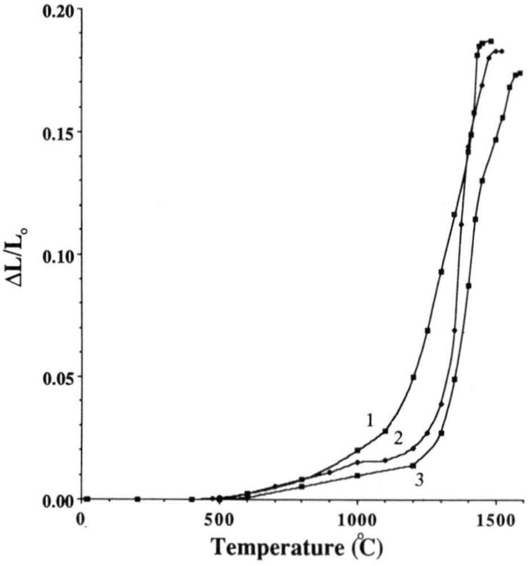

Fig. 1. Linear shrinkage under constant heating: 1, WC–10Ni; 2, WC–9Ni–1Cr; 3, WC–8Ni–2Cr.

Inspection of the WC–10Ni curve suggests that, once the pre-sintering temperature of 500 °C is passed, shrinkage occurs in the solid state at a steadily increasing rate until about 1100 °C, whereupon the rate increases sharply. The majority of the shrinkage is achieved at this new rate, until at 1410 °C the curve becomes vertical, signifying the appearance of a liquid phase. Although the equilibrium eutectic temperature for the Ni–W–C system is 1342 °C [3], it should be noted that continuous heating was being used, and it is unlikely that equilibrium conditions applied.

The fact that as much as 80% of the total shrinkage occurs in the solid state supports observations in the literature [4, 6, 7] which emphasize the importance of solid phase processes in liquid sintering practice.

Meredith and Milner [8] attributed the initial densification to the rapid formation of dense aggregates of carbide, as a result of increased surface and interfacial diffusion in the presence of the binder metal.

The WC–9Ni–1Cr, WC–8Ni–2Cr and WC–5Ni–5Cr compositions all exhibit a delay in the onset of significant densification of as much as

200 °C. However, the subsequent rate at which the curves rise suggests the presence of liquid phase at a lower temperature than for the straight nickel composition. The period associated with rapid solid state diffusion appears to be limited or absent.

The total linear shrinkage which the higher chromium compositions undergo is reduced, as shown in Table 2. This table also gives an estimate of the relative proportions of solid and liquid phase densification exhibited by each composition, as interpreted from the rate of densification. The decreasing contribution of solid state shrinkage with increasing chromium can be clearly seen.

The WC–9Ni–1Cr, WC–8Ni–2Cr and WC–5Ni–5Cr specimens all displayed a slowing of the shrinkage rate after about 13% linear shrinkage; this is thought to reflect a tendency to swell, associated with either a large contact angle between liquid and solid phases or an excessive volume of liquid [7].

Examples of the isothermal plots obtained are given in Figs. 2 and 3. Interpretation of these curves is similar to the rationale described above; a sudden vertical rise in shrinkage rate implies the presence of liquid. For WC–9Ni–1Cr the liquid phase was apparent as low as 1210 °C, provided that sufficient time is allowed for equilibrium to be attained. Vlasyuk *et al.* [9] described the presence of a eutectic in the Cr_3C_2–Ni system at 1100–1200 °C. Ramqvist [10] reported a eutectic melting at 1255 °C while, for the Ni–WC system, he did not observe liquid until 1453 °C.

It is therefore suggested that the shrinkage which takes place in the high chromium specimens is due to the formation of an Ni–Cr–C eutectic and that the dissolution of Cr_3C_2 in the nickel binder inhibits the reaction of the binder with WC. The solution of WC in the binder plays a major role as a rate controlling factor, even in the solid state [11]; hence a lack of densification was observed in the early stages of the cycle when appreciable chromium is present, as shown in Fig. 1. The restricted solubility of WC in the binder would also inhibit the important solution–reprecipitation phase of the sintering cycle, giving rise to the documented grain growth inhibition effects. The effect is shown clearly in Fig. 4 where, compared with the 10 wt.% Ni binder in

TABLE 2 Total shrinkage and estimated solid state contribution to shrinkage

Alloy	*Total linear shrinkage* (%)	*Solid state shrinkage* (estimated % of total)
WC–10Ni	18.7	80
WC–9.9Ni–0.1Cr	18.8	80
WC–9Ni–1Cr	18.4	30
WC–8Ni–2Cr	17.4	23
WC–5Ni–5Cr	17.9	15

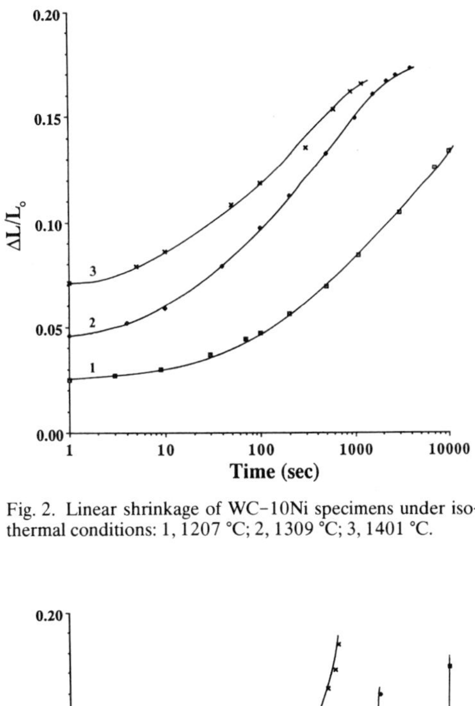

Fig. 2. Linear shrinkage of WC–10Ni specimens under isothermal conditions: 1, 1207 °C; 2, 1309 °C; 3, 1401 °C.

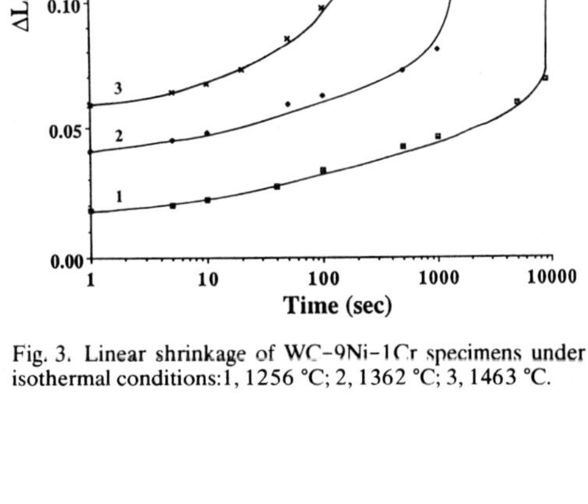

Fig. 3. Linear shrinkage of WC–9Ni–1Cr specimens under isothermal conditions: 1, 1256 °C; 2, 1362 °C; 3, 1463 °C.

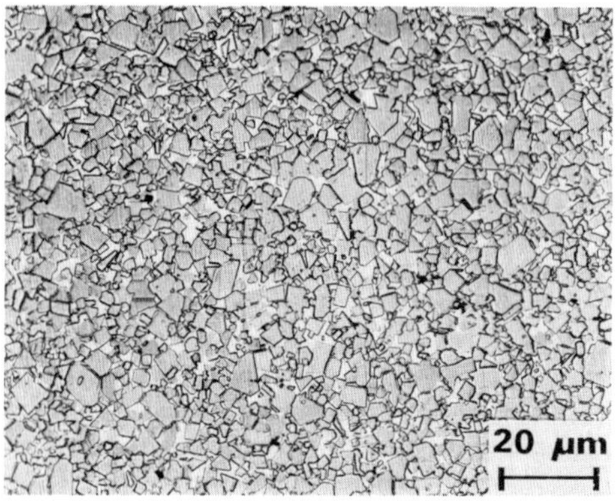

Fig. 4. Microstructure of a WC–9Ni–1Cr specimen, as sintered for 30 min at 1480 °C.

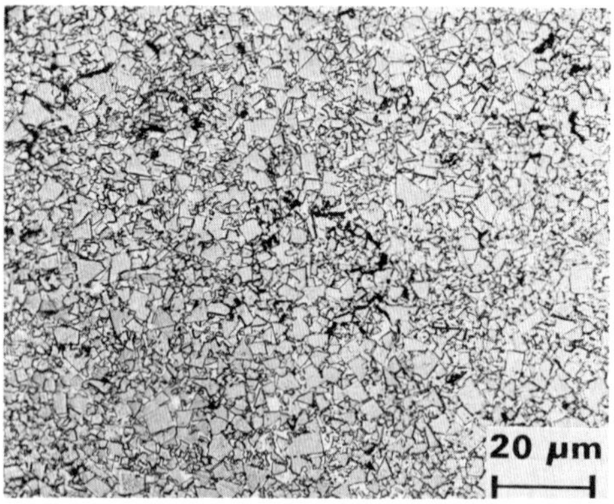

Fig. 5. Microstructure of WC–10Ni specimen, as sintered for 30 min at 1480 °C.

Fig. 5, there is an absence of grain growth and grain smoothing associated with the final two stages of sintering. The increased porosity is also evident in Fig. 4. Further work is proposed to observe the chemical composition of the binder phase, in order to confirm the mechanisms which may be operating.

The activation energy for sintering was calculated for each composition, using the Arrhenius equation of the form

$$\text{rate} = A\exp\left(-\frac{Q}{RT}\right)$$

where A is a constant, Q is the activation energy, R is the universal gas constant and T is the absolute temperature. Hence, by plotting $\ln(\Delta L/L_0)$ at a given elapsed time *vs.* reciprocal temperature, the slope of the curve yields Q/R. An isothermal holding time of 100 s was used for each data point, and a good fit to a straight line was found for every composition. The calculated values of Q are shown in Table 3, and these reflect the findings from the constant heating described above, as increased chromium levels require greater energy input to initiate densification.

The relationship of relative change in resistivity with temperature is shown in Fig. 6. It is interesting to note that, while no consistent effect of composition is apparent, the major change in resistivity for all cases occurs well before the onset of significant densification. It is suggested that sufficient solid state rearrangement has taken place to allow the development of a continuous conductive path, regardless of composition.

TABLE 3 Activation energies for the sintering of Ni–Cr-bonded WC

Alloy	*Activation energy* (kJ mol^{-1})
WC–10Ni	90.5
WC–9.9Ni–0.1Cr	98.8
WC–9Ni–1Cr	117.1
WC–8Ni–2Cr	126.7
WC–5Ni–5Cr	144.1

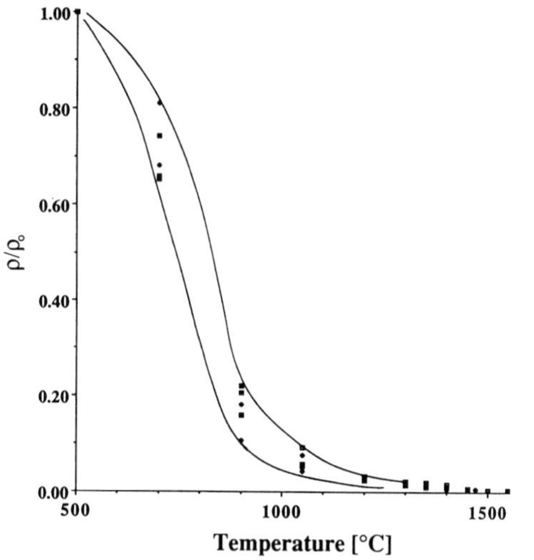

Fig. 6. Resistivity changes under constant heating: ⊡, WC–10Ni; ◆, WC–9.9Ni–0.1Cr; ◘, WC–9Ni–1Cr; ◇, WC–8Ni–2Cr; ■, WC–5Ni–5Cr.

It was also noted that the initial (as pre-sintered) values of resistivity increased with increasing chromium content. This effect decreased rapidly with increasing temperature and disappeared by 1200 °C, demonstrating the dissolution of Cr_3C_2 in the nickel matrix. The development of on-line resistivity determination would allow the measurement of isothermal resistivity changes and hence provide an alternative means of determining activation energies.

4. Conclusion

(1) Increasing chromium content in an Ni–Cr bonded hard metal results in an increased sintering activation energy, resulting in an increased temperature requirement to induce densification at normal heating rates.

(2) With no chromium addition to the nickel binder, most of the densification appears to occur in the solid state, whereas increasing chromium additions result in a decreasing solid state contribution to the total shrinkage.

(3) As-sintered structures indicate that the presence of chromium in the binder increases porosity and restricts the solution–reprecipitation stage of the sintering process.

(4) Resistivity measurements indicate that significant particle rearrangement below 900 °C takes place in all compositions tested. The composition effect on resistivity decreases as sintering proceeds and is eliminated between 1100 and 1200 °C.

References

1 E. Kny and L. Schmid, *Int. J. Refract. Hard Met., 6* (3) (1987) 145–148.
2 E. Lardner and S. Iggstrom, *Proc. 10th Plansee Semin.*, Metallwerk Plansee, Reutte, 1981, Paper HM17.
3 S. Ekemar, L. Lindholm and T. Hartzell, *Proc. 10th Plansee Semin.*, Metallwerk Plansee, Reutte, 1981, Paper HM8.
4 H. Ogawa and Y. Kataoka, *High Temp. High Pressures, 13* (1981) 481–494.
5 T. Piszczek and S. Stolarz, *Powder Metall. Int., 18* (1) (1986) 9–11.
6 R. F. Snowball and D. R. Milner, *Powder Metall., 11* (21) (1968) 23–40.
7 R. German, *Liquid Phase Sintering*, Plenum, New York, 1985.
8 B. Meredith and D. R. Milner, *Powder Metall., 19* (1) (1976) 38–45.
9 R. Z. Vlasyuk, I. D. Radomysel'skii and A. A. Sotnik, *Sov. Powder Metall. Met. Ceram., 24* (1985) 356–360.
10 L. Ramqvist, *Int. J. Powder Metall., 1* (4) (1965) 2–21.
11 H. E. Exner, *Recent Advances in Hardmetal Production*, Vol. 2, Metal Powder Report, London, 1979, Paper 13.

Materials Science and Engineering, *A105/106* (1988) 275-281

Residual Stress Distribution in Cermets*

A. D. KRAWITZ and M. L. CRAPENHOFT

Department of Mechanical and Aerospace Engineering, University of Missouri, Columbia, MO 65211 (U.S.A.)

D. G. REICHEL

Research Reactor Facility, University of Missouri, Columbia, MO 65211 (U.S.A.)

R. WARREN

Department of Engineering Metals, Chalmers University of Technology, Gothenburg (Sweden)

(Received December 7, 1987)

Abstract

The large difference between the coefficients of thermal expansion of the binder and carbide phases leads to substantial differential thermal residual stresses on cooling from the sintering temperature. The stress state is of the grain interaction or microstress type common in composite materials. In this study, neutron diffraction has been used to characterize the degree of strain variance and therefore stress distribution in the carbide phase for two types of cermet. Measurements of diffraction peak breadth in materials with rounded (NbC_x-Co) and angular ((NbC_x-TiC)-Co) carbide phases have provided evidence of greater variance of residual strain for the angular relative to the rounded carbide. This is attributed to a greater stress distribution due to the angular geometry of the carbide particles. Similar measurements on WC-Co cermets with a low and high binder content indicate that increased strain variance is present in the high relative to the low binder material. This indicates that a greater stress distribution may be present owing to the higher average residual stress level present in the high binder material. The magnitudes of the variance have been estimated for the WC phase and are found to be large, the halfwidth being two to three times the average compressive stress in the carbide. This implies that, assuming a normal (gaussian) stress distribution, in excess of 10% of the WC volume experiences tensile stress.

1. Introduction

Numerous studies of the differential thermal stress state in cemented-carbide composites have been performed using X-rays [1-10] and neutrons [11]. Although the quantitative validity of the X-ray studies has been questioned [12], they have generally shown the presence of compressive carbide and tensile binder stresses, in agreement with models [1, 13-18], as well as the expected variation with binder content. Furthermore, some of the more recent work has documented changes in the stress state due to cyclic and monotonic compression [10, 11], with attendant changes in Palmqvist crack length [10].

None of the experimental work to date has dealt with the variance or distribution of the residual stresses around their average values. In addition to stress gradients in the vicinity of binder-carbide interfaces as a consequence of the need to maintain continuity across boundaries, other contributors to the variance are possible. One is carbide morphology. If spherical carbide particles are present, a nearly uniform level of average hydrostatic residual stresses can be expected between the two phases. However, if angular carbide is present, it seems reasonable to expect more pronounced variation due to local geometrical configurations. Another potential source of broadening is the carbide particle size or mosaic. A smaller size is expected to cause broader diffraction peaks. Furthermore, it is anticipated that, as the average level of compressive carbide stress increases, the variance should also increase.

The present work is an attempt to test these notions by experimental determination of strain

*Paper presented at the 3rd International Conference on the Science of Hard Materials, Nassau, The Bahamas, November 9-13, 1987.

0921-5093/88/$3.50

© Elsevier Sequoia/Printed in The Netherlands

variances (stress distributions) through a study of diffraction peak breadths. The idea is to measure peak breadths in controlled samples and to compare them with each other and with those from standard samples which provide the smallest breadths obtainable from the experimental apparatus. Two sets of samples have been studied: NbC_x and WC cermets. The NbC_x samples are NbC_x–Co, with rounded carbide, and (NbC_x–TiC)–Co, with angular carbide. The WC samples are WC–6wt.%Co, with lower carbide stresses (due to reduced binder content), and WC–15wt.%Co, with higher carbide stresses. The diffraction analysis was based on the integral peak breadth approach to particle size–strain broadening. Details are presented below.

The general features of the differential thermal residual stress state may be seen from the formulation of Turner [14]:

$$\sigma_i = K_i(\alpha_c - \alpha_i)\,\Delta T \qquad (1)$$

where K_i is the bulk modulus of the ith phase, α_c is the volume coefficient of expansion of the cermet, α_i is the volume coefficient of expansion of the ith phase and ΔT is the difference between the measurement temperature and the set-up temperature. The value of α_c is calculated from

$$\alpha_c = \frac{\alpha_1 K_1 V_{f1} + \alpha_2 K_2 V_{f2}}{K_1 V_{f1} + K_2 V_{f2}} \qquad (2)$$

for a two-phase system. V_{fi} is the volume fraction of the ith phase. Equations (1) and (2) assume that both phases are elastic and that no cracks are present. They show that the carbide is in compression while the binder is in tension since $\Delta T < 0$ and $\alpha_{carbide} < \alpha_c < \alpha_{binder}$. Also, as $V_{fbinder}$ increases, $\alpha_c - \alpha_{carbide}$ increases so that the carbide stress level increases. This formulation does not take particle geometry into account or predict stress gradients at interfaces.

Relevant values for linear coefficients of thermal expansion α_{li} ($= \alpha_i/3$), as well as Young's modulus E, bulk modulus K and Poisson's ratio ν, are presented in Table 1.

2. Experimental details

2.1. Samples

The cermet samples used are summarized in Table 2. They had been studied on earlier occasions [18, 19]. In addition to these, pure NbC_x and WC powders were used. These reference powders were not from the same batches as the carbides in the cermets. This point is addressed in the discussion.

2.2. Neutron powder measurements

Neutron powder spectra were obtained using the 2XD powder spectrometer at the University of Missouri Research Reactor (MURR), a 10 MW light-water reactor. This instrument is equipped with a position-sensitive detector (PSD) consisting of three linear stacked PSDs. The PSD enables 25° ($= 2\theta$) to be collected at one detector setting, in 0.1° bins, thus greatly improving the rate of data collection compared with conventional single-slit detectors [21]. The wavelength of the thermal neutrons used was 1.29 Å.

The peaks were fitted assuming pure gaussian peak shapes. This is often observed to be substantially true for neutron powder pattern peaks. In fact, some lorentzian character was present in the peaks, as determined by initial use of a pseudo-Voigt fitting routine that used an optimized linear superposition of gaussian and

TABLE 1 Relevant physical properties of carbides and binder

Material	α_l ($\times 10^6 \cdot K^{-1}$)	*E* (GPa)	*K* (GPa)	ν
WC (*a* axis)	6.7	700	389	0.2
(*c* axis)	4.9			
NbC	6.7	338	≈ 188	≈ 0.2
Co	12	200	185	0.3

TABLE 2 Cermet samples used in the study

Sample	*Size* (mm)	*Binder content* (wt.% (vol. %))	*Mean carbide size* (μm)	*Carbide shape*	*Reference*
NbC_x–Co	5×6×7	21 (17.9)	11.8	Rounded	[19]
(NbC_x–5wt.%TiC)–Co	7×7×7	21 (16.1)	11.3	Angular	[19]
WC–Co	3×6×45	5.95 (10)	0.5	Angular	[20]
WC–Co	3×6×46	14.6 (23.9)	2.9	Angular	[20]

lorentzian functions with identical halfwidths. However, the lorentzian contribution was small (less than 20%) and it was found that the values of the full-width at half-maximum (FWHM) returned in either case were the same within statistical error. Hence the simpler gaussian was employed for this study. Peak intensities were in the 2000–20 000 count range, which ensured FWHM fitting errors of about 0.003° ($=2\theta$) for FWHM values of the order of 0.5° ($=2\theta$).

3. Results and discussion

3.1. *WC–Co*

The FWHM results for the WC–Co samples as well as the WC powder are shown in Fig. 1 and Table 3. The parabolic nature of the variation in FWHM for each sample is a result of the geometrical optics of the instrument [21, 22]. The FWHM values for both WC cermets are greater than for the powder. Furthermore, the FWHM values for the WC–15wt.%Co are somewhat greater than those for the WC–6wt.%Co. This is consistent with the notion that the material with the higher carbide stress (WC–15wt.%Co) contains a greater range of stresses than that with the lower stress (WC–6wt.%Co). A comparison of the FWHM values of the WC powder and a silicon powder standard is presented in Fig. 2. The silicon standard has peak breadths that are essentially representative of the instrument, *i.e.* contain little in the way of internal sources of peak broadening relative to the resolution of the instrument. The WC powder breadths are close to those of the silicon standard whereas those from the NbC_x powder, which is also included in Fig. 2, are clearly broader and are discussed below.

Table 2 shows that the 6wt.%Co material has a smaller grain size than the 15wt.%Co material has. Because a small grain size would tend to cause peak broadening, the fact that the broader peaks occur in the material with larger grains is even more significant. (Smaller grains are expected to cause broadening not only because the fraction of volume associated with geometri-

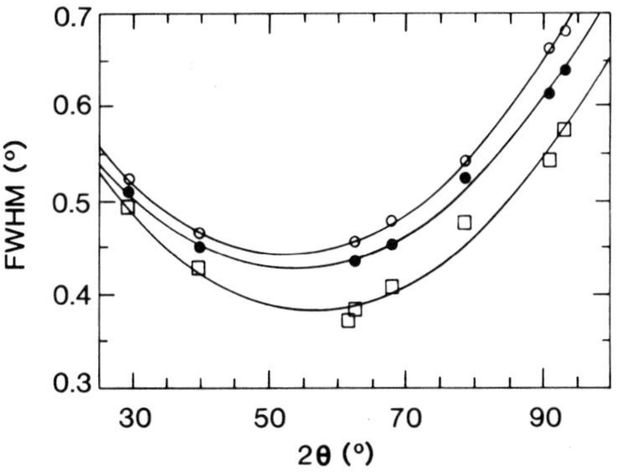

Fig. 1. FWHM values for the WC (□), WC–15wt.%Co (○) and WC–6wt.%Co (●) samples.

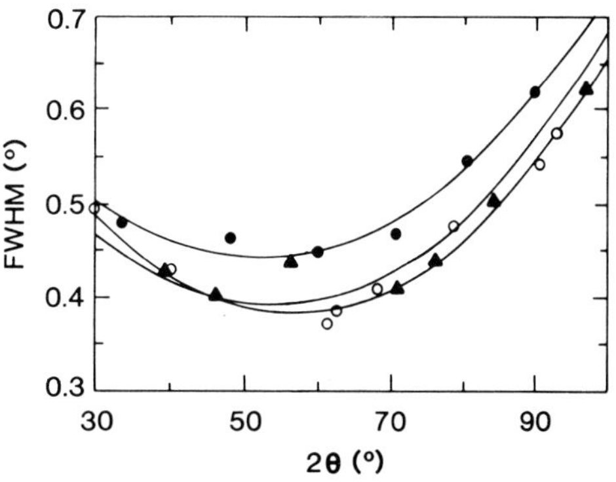

Fig. 2. FWHM values for silicon (▲), WC (○) and NbC_x (●) powders.

TABLE 3 Full width at half-maximum values for WC and WC–Co samples

hkl	2θ (deg)	*FWHM* 2θ (deg)		
		WC powder	*WC–6wt.%Co*	*WC–15wt.%Co*
100	29.6	0.494(3)	0.509(3)	0.523(2)
101	39.9	0.426(1)	0.450(1)	0.463(1)
102	62.8	0.383(3)	0.434(4)	0.455(3)
201	68.0	0.406(2)	0.453(2)	0.477(2)
112	78.7	0.474(3)	0.523(2)	0.543(2)
211	91.1	0.541(3)	0.614(2)	0.662(2)
103	93.3	0.575(5)	0.637(4)	0.683(3)

cal asperities is greater, a strain variance effect, but also because of the manner in which size affects breadth in any diffraction experiment. For this latter effect, it must be realized that the physical grain sizes in Table 2 would give no observable difference in breadth with a diffraction wavelength of 1.3 Å. Hence any particle size broadening is due to subgrain size, as discussed below in the particle size–strain analysis.) Thus the hypothesis that the 15wt.%Co material has broader peaks because of a greater strain variance appears to be supported even more strongly than the data suggest. However, in order to confirm this quantitatively, a particle size–strain analysis must be performed.

The breadth data were analyzed using the integral breadth method [23]. We summarize the basic aspects of this method here. The breadth B is the FWHM value of the sample after correction for instrumental breadth, *i.e.*

$$B^2 = B_{\text{meas}}^{\;2} - B_{\text{stan}}^{\;2} \tag{3}$$

where B_{meas} is the FWHM value measured for the sample and B_{stan} is the value for the standard used (see Table 3). This correction procedure assumes that all peak shapes are gaussian. The breadth B consists of a particle size component B_P and a strain component B_S. For the case where all sources of broadening are also gaussian, the appropriate equation is

$$B^2 = B_P^{\;2} + B_S^{\;2}$$

$$= \left(\frac{\lambda}{L\cos\theta}\right)^2 + (4\langle\varepsilon\rangle\tan\theta)^2 \tag{4}$$

where λ is the neutron wavelength, L is the particle size, θ is half the diffraction angle and $\langle\varepsilon\rangle$ is the variance of the strain in the material. The terms on the right-hand side of eqn. (4) correspond to particle size and strain respectively.

Rearranging eqn. (4) gives the following relationship:

$$\left(\frac{B\cos\theta}{\lambda}\right)^2 = \left(\frac{1}{L}\right)^2 + \left(\frac{4\langle\varepsilon\rangle\sin\theta}{\lambda}\right)^2 \tag{5}$$

From this we see that a plot of $\{(B\cos\theta)/\lambda\}^2$ *vs.* $\{(\sin\theta)/\lambda\}^2$ gives a line whose intercept and slope yield L and $\langle\varepsilon\rangle$ respectively.

For the cermet samples in this study the variance in strain is assumed to be due solely to strain distribution arising from carbide shape effects and is centered on the mean differential thermal stress level for the phase. In more conventional studies of metals and alloys, this quantity is associated with the variance in microstrain due to the presence of dislocations arising from plastic deformation.

The results for the WC–Co samples are shown in Fig. 3 and Table 4. The corresponding values for stress variance are also included in Table 4. The fact that the lines do not go through zero indicates that some of the peak breadth is due to particle size effects. The sizes derived from the analysis are much smaller than the actual grain sizes of the WC–Co samples. The result that the particle size of the 6 wt.% Co material is somewhat greater than for the 15 wt.% Co material even

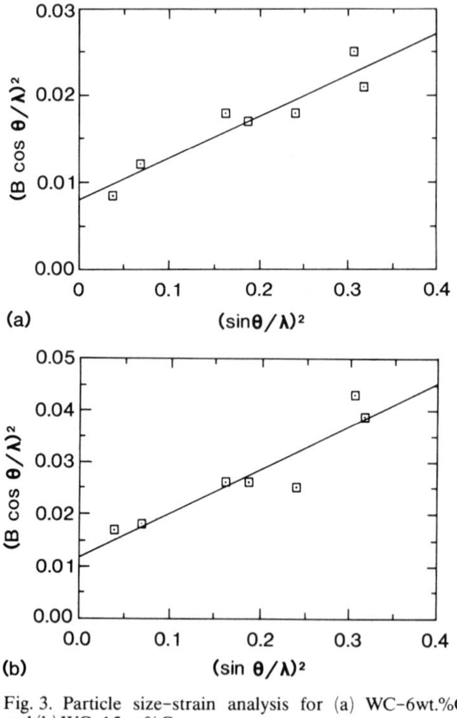

Fig. 3. Particle size–strain analysis for (a) WC–6wt.%Co and (b) WC–15wt.%Co.

TABLE 4 Results of particle size–strain analysis for WC–Co samples

Sample	*Particle size* (Å)	*Strain variance*	*Stress FWHM* (MPa)
WC–6wt.%Co	640(50)	$9.5(5)\times10^{-4}$	1110(60)
WC–15wt.%Co	530(40)	$12.6(7)\times10^{-4}$	1470(90)

though the grain size is considerably smaller is not as significant as the fact that both particle sizes are sufficiently large that size contributes little to the peak breadths. The important result is that the steep slopes indicate that the strain variance is large.

The relation of the stress distribution to the differential thermal stress state in the composites is now considered. Using eqns. (1) and (2) and the data in Table 1, values of average compressive stress in the WC phase of the 6 and 15 wt.% Co samples are calculated, neglecting relaxation, to be −280 MPa and −750 MPa respectively. (Owing to slight changes in instrumental parameters between runs, experimental values could not be determined.) The stress distributions are calculated from the values of strain variance by converting them to stress and using the gaussian distribution formula

$$y = \exp\left\{\frac{-(\sigma-\sigma_0)^2}{2s^2}\right\} \quad (6)$$

where y is normalized to 1 and s is related to the stress FWHM by

$$\langle\sigma\rangle = 2(2\ln 2)^{1/2}s \quad (7)$$

The stress variance is related to the strain variance, assuming hydrostatic stress, by

$$\langle s\rangle = 3K\langle\varepsilon\rangle \quad (8)$$

These distributions are shown in Fig. 4. The distribution is somewhat broader for the 15 wt.% Co material which has the larger strain variance. However, the important feature is that both distributions are very broad relative to the average stress values and range well into the compressive and tensile regions. From the properties of gaussian distributions, about 77% of the carbide volume lies in the stress range from about −840 to +270 MPa for the 6 wt.% Co material, and from −1490 to +10 MPa for the 15 wt.% Co material. Thus, about 12% of the carbide is under hydrostatic tensile stress in the range from about +270 to +1300 MPa, for the 6 wt.% Co material, and from +10 to +1400 MPa, for the 15 wt.% Co material. Presumably, the range of stresses is present within many different particles.

It is possible that the portion of WC under large tensile stress is more subject to crack initiation and/or propagation than are other regions. Also, the portions under extreme compression may tend to deflect propagating cracks.

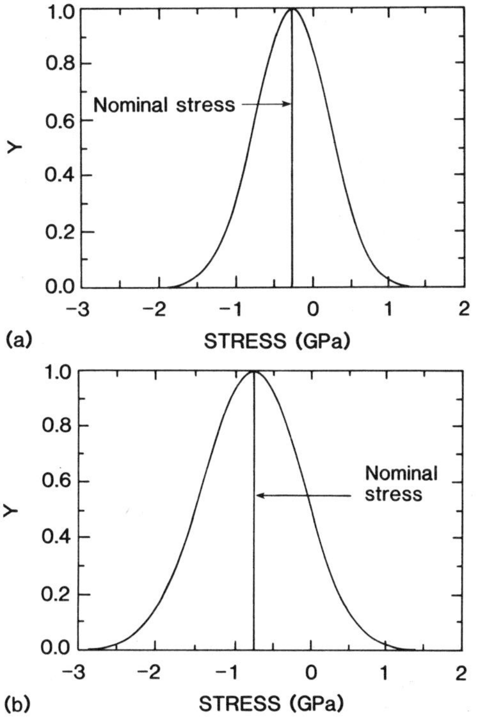

Fig. 4. Stress distributions for (a) WC–6wt.%Co and (b) WC–15wt.%Co.

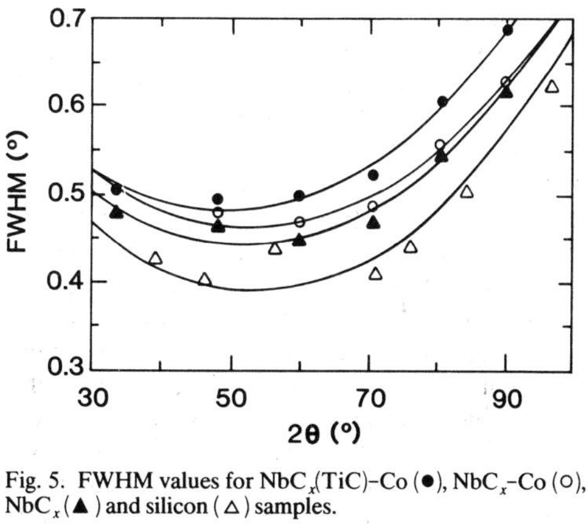

Fig. 5. FWHM values for NbC_x(TiC)–Co (●), NbC_x–Co (○), NbC_x (▲) and silicon (△) samples.

3.2. NbC–Co

The FWHM results for the NbC_x–Co samples are shown in Fig. 5 and Table 5. The data indicate that, as for WC–Co, the breadths of the NbC_x phase are broader in the cermet than in the loose

TABLE 5 Full width at half-maximum values for NbC_x and NbC_x–Co samples

hkl	2θ (deg)	*FWHM* 2θ (deg)		
		NbC_x powder	*NbC_x–Co*	*(NbC_x–TiC)–Co*
200	33.5	0.478(2)	0.5033(6)	0.5072(7)
220	48.2	0.463(2)	0.4794(6)	0.4940(7)
222	60.0	0.447(3)	0.4684(9)	0.498(1)
400	70.5	0.469(4)	0.487(1)	0.521(2)
420	80.4	0.543(3)	0.5543(6)	0.6067(9)
422	90.0	0.619(3)	0.6281(9)	0.686(1)

TABLE 6 Non-stoichiometry analysis for NbC_x in NbC_x powder, NbC_x–Co and (NbC_x–TiC)–Co

Peak ratios	*Value of x*		
	NbC_x	*NbC_x–Co*	*(NbC_x–TiC)–Co*
200/111	0.9027	0.9128	0.9529
200/311	0.8856	0.9322	0.9426
220/111	0.8988	0.9104	0.9508
220/311	0.8812	0.9300	0.9403
Average	0.89 ± 0.02	0.92 ± 0.02	0.95 ± 0.01

powder. Furthermore, the breadths for the TiC-bearing sample, in which the NbC_x is angular, are greater than for the NbC–Co sample, in which the NbC_x is rounded. An important difference for the NbC_x results is that the NbC_x powder gives markedly broader peaks than the silicon standard (and the WC powder). This is attributed to the non-stoichiometry of the carbide. In order to attempt to compare the breadths further, it is necessary to know the respective values of x, *i.e.* the degree of carbon deficiency, in the three samples.

The values of x were determined for each sample using a standard integrated intensity approach. NbC_x has the rocksalt structure for which the intensities from h, k, l all even peaks are proportional to $(b_{Nb} + b_C)^2$ and those from h, k, l all odd are proportional to $(b_{Nb} - b_C)^2$, where b is the neutron scattering length. The values of b for niobium and carbon are 0.7054×10^{-12} cm and 0.6648×10^{-12} cm, respectively. This closeness in the values of b makes the variation in intensity of the h, k, l all odd peaks very sensitive to the value of x for neutrons relative to X-rays. Thus the integrated intensity for the all odd peaks is proportional to $(b_{Nb} - xb_C)^2$. The ratio of this quantity for $x = 1$ and $x = 0.9$ is 7 so that the sensitivity is great. The peaks used in the analysis were 111, 311, 200 and 220. The results are shown in Table 6.

These results are consistent with the data in Fig. 5 and Table 5 in the following sense. As x deviates from 1, the compositional inhomogeneity of the carbide increases so that the peaks are expected to broaden. The NbC_x powder has the lowest of the x values while the TiC-bearing cermet has the highest value. This suggests that the differences between the FWHM values of the samples would actually be enhanced if corrections were made for non-stoichiometry. Such a correction is not possible without a calibrated series of NbC_x powders.

The particle size–strain analysis was not applied to the NbC samples because of the complication of stoichiometry effects.

4. Conclusions

Measurements of peak breadth for the carbide phases of WC–Co and NbC_x cermets were conducted. The results show that peak breadths are greater for carbides in cermets than in free powders. For the WC samples, WC peak breadths were greater than those of the powder for both 6 and 15 wt.% Co binder content cermets, with the 15 wt.% Co material having the greatest breadths. This increase in breadths over those of the free powder is attributed to a variance in stress due to the angularity of the carbide particles (since spherical particles are expected to be under uniform stress, which would only shift the peak positions but not increase the breadth). The 15 wt.% Co material has a greater variance owing to the higher average stress value associated with the increased binder content. Particle size–strain analysis indicates that the stress distribution is large, extending well into the tensile range if gaussian distributions are assumed.

The NbC_x samples contained rounded (NbC_x–Co) and angular ((NbC_x–TiC)–Co) carbide particles. The sample with angular carbide had broader peaks than the rounded material had. Stoichiometry effects prevented a more thorough analysis of the data from these samples.

Results from both the WC and the NbC_x samples appear to support the hypothesis that the differential thermal residual stress state in cemented-carbide composites is characterized by a distribution of stress in each phase which arises principally from the angular morphology of the carbide phase. The role of stress gradients at interfaces is unclear. In addition, it appears that the distribution is broader when the magnitude of the average stress increases, *i.e.* when the binder content increases. Quantitative analysis, assuming gaussian strain distributions, suggests that the stress distributions are very broad and that a significant fraction of the carbide volume is under a tensile stress.

References

1 J. Gurland, *Trans. Am. Soc. Met., 50* (1958) 1063–1071.
2 R. Bernard, *Jernkontorets Ann., 147* (1963) 22–36.
3 S. Ohman, E. Parnama and S. Palmqvist, *Jernkontorets Ann., 151* (1967) 126–159.
4 D. N. French, *J. Am. Ceram. Soc., 52* (1969) 267–271, 271–275.
5 A. Hara, M. Megata and S. Yazu, *Powder Metall. Int., 2* (1970) 43–47.
6 B. O. Jaensson, *Mater. Sci. Eng., 8* (1971) 41–53.
7 P. O. Snell and E. Parnama, *Planseeber. Pulvermetall., 21* (1973) 271–280.
8 J. Dubois, C. Mai and R. Riviere, *Z. Metallkd., 65* (1974) 130–134.
9 E. F. Drake and A. D. Krawitz, *Metal. Trans. A, 12* (1981) 505–513.
10 G. Vekinis and S. B. Luyckx, in E. A. Almond, C. A. Brookes and R. Warren (eds.), *Proc. 2nd Int. Conf. on the Science of Hard Materials*, in *Inst. Phys. Conf. Ser., 75* (1984) 591–600.
11 A. D. Krawitz, R. Roberts and J. Faber, in E. A. Almond, C. A. Brookes and R. Warren (eds.), *Proc. 2nd Int. Conf. on the Science of Hard Materials*, in *Inst. Phys. Conf. Ser., 75* (1984) 577–589.
12 A. D. Krawitz, *Mater. Sci. Eng., 75* (1985) 29–36.
13 F. Laszlo, *J. Iron Steel Inst., London, 147* (1943) 173–199.
14 P. S. Turner, *J. Res. Natl. Bur. Stand., 37* (1946) 239–240.
15 W. D. Kingery, *J. Am. Ceram. Soc., 40* (1957) 351–352.
16 J. Selsing, *J. Am. Ceram. Soc., 44* (1961) 419.
17 C. T. Liu and J. Gurland, *Trans. Am. Soc. Met., 58* (1965) 66–73.
18 D. Brooksbank and K. W. Andrews, *J. Iron Steel Inst., London, 210* (1972) 246–255.
19 R. Warren, *J. Mater. Sci., 3* (1968) 471–485.
20 B. Johannesson and R. Warren, in E. A. Almond, C. A. Brookes and R. Warren (eds.), *Proc. 2nd Int. Conf. on the Science of Hard Materials*, in *Inst. Phys. Conf. Ser., 75* (1984) 713–723.
21 C. W. Tompson, D. F. R. Mildner, M. Mehregany, R. Berliner and W. B. Yelon, *J. Appl. Crystallogr., 17* (1984) 385–394.
22 G. Caglioti, A. B. Pailetti and F. P. Ricci, *Nucl. Instrum. Methods, 35* (1958) 223–228.
23 C. N. J. Wagner and E. N. Aqua, *Adv. X-Ray Anal.*, 7 (1964) 46.

Atom Probe Microanalysis of the γ phase in Cemented Carbide Materials*

ULF ROLANDER and HANS-OLOF ANDRÉN

Department of Physics, Chalmers University of Technology, S-412 96 Göteborg (Sweden)

(Received November 9, 1987)

Abstract

Atom probe field ion microscopy and analytical electron microscopy were used to study the γ phase (cubic carbide or carbonitride phase) in 85wt.%WC–8wt.%(Ti, W)C–7wt.%Co and 87wt.%-(Ti, Ta, V, Mo, W)(C, N)–13wt.%(Co, Ni). In both materials, γ phase grains consisting of a core surrounded by a rim of the same structure but different composition were common. Microanalysis revealed that the cores are grains which have remained undissolved during sintering and that the rims were formed by epitaxial precipitation onto the cores from the binder phase. In both materials a carbide (or carbonitride) skeleton had formed with half a monolayer of binder phase atoms segregated to the carbide boundaries.

1. Introduction

During the past decade the interest in cemented carbide materials containing cubic carbides (γ phase) has grown rapidly. This is due both to economical and supply considerations and to the superior high temperature properties of TiC compared with WC [1, 2].

One severe problem when trying to understand the behaviour of the cubic carbides is the difficulty of obtaining accurate measurements of variations in the chemical composition through the carbides. Conventional microanalytical techniques lack either the required accuracy in chemical analysis or the required spatial resolution, to resolve, for example, variations in carbon, nitrogen and oxygen contents. Atom probe field ion microscopy (APFIM), however, has both the spatial resolution and the analytical accuracy to investigate fine-grained cemented carbide grades. This method has been described in detail previously [3].

Since atom probe analysis is rather time consuming, with respect both to specimen preparation and to actual analysis, the method is not very efficient for investigations on a broader scale. Therefore, much is gained by combining APFIM with analytical electron microscopy (AEM), using APFIM only for investigating well-defined and significant parts of the material. This, of course, also gives the advantage of being able to correlate results obtained by two rather different techniques.

2. Experimental details

In this investigation, two materials were analysed by both AEM and APFIM. The first material is a commercial 85.0WC–7.7(Ti, W)C–7.3Co grade (all compositions are given in weight per cent), produced by conventional liquid phase sintering. Titanium is in this material added in the form of $(Ti_{0.8}W_{0.2})C$ powder.

The second material is a 86.7(Ti, Ta, V, Mo, W)(C, N)–13.3(Co, Ni) cermet grade. Titanium is added in the form of carbide and nitride powders, whereas the other carbide formers are added as carbides only.

AEM was performed on a JEOL 2000FX transmission electron microscope–scanning transmission electron microscope equipped with a Link AN10000 energy-dispersive X-ray system. Thin foil specimens for electron microscopy were prepared using standard ion-beam-thinning techniques.

APFIM was performed at 90 K using 15% pulse fraction and a probe area of approximately 50 nm^2. The instrument used has been described in detail before [4, 5]. Needle-shaped specimens for atom probe analysis were made by electropolishing using 5% H_2SO_4 in methanol, cooled to −30 °C. In order to increase the length of the

*Paper presented at the 3rd International Conference on the Science of Hard Materials, Nassau, The Bahamas, November 9-13, 1987.

0921-5093/88/$3.50 © Elsevier Sequoia/Printed in The Netherlands

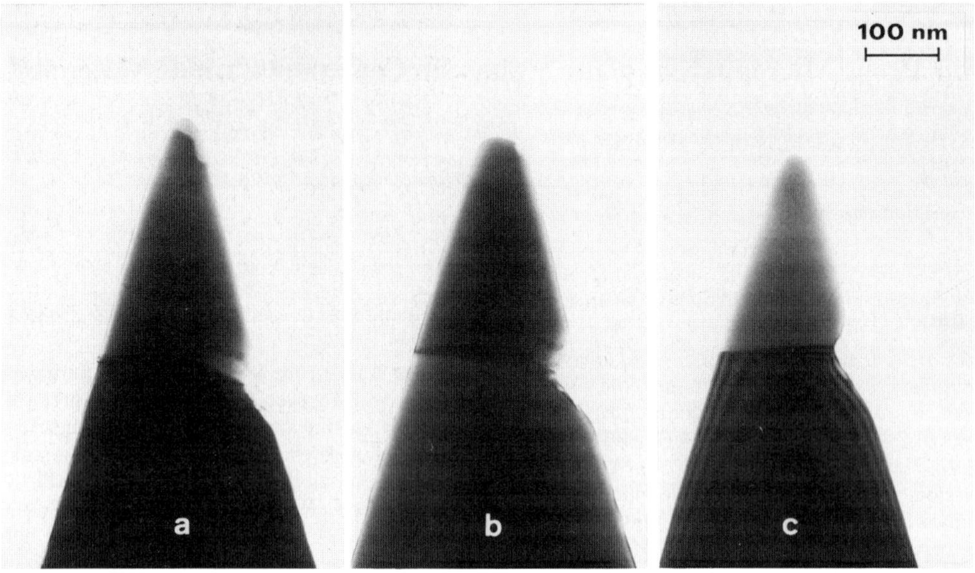

Fig. 1. By alternating atom probe analysis and controlled resharpening, fairly deep composition profiles can be made. A specimen containing a compositional boundary near the tip is shown (a) as prepared, (b) after the first analysis in the APFIM and (c) after resharpening before the next analysis.

atom probe composition profiles, use was made of a method to resharpen blunt specimens in a controlled way. This can be done by electropolishing, using very short polishing times [6].

Figure 1(a) shows a specimen containing a compositional boundary near the tip of the topmost grain. In Fig. 1(b) the specimen has been analysed by APFIM and then removed from the atom probe. The removal of a 30 nm surface layer has made the specimen too blunt for further analysis. In Fig. 1(c) the specimen has been resharpened by electropolishing at 10 V for 0.5 ms. It could then be re-inserted in the atom probe for a new analysis.

By alternating analysis and electropolishing, fairly deep (up to 0.5 μm) composition profiles could be obtained. As seen in Fig. 1, only approximately 5 nm in the axial direction of the specimen was removed during each step of electropolishing. It was thus fairly easy to plan the analyses so that no interesting boundary regions were lost. The specimens were imaged by transmission electron microscopy (TEM) and field ion microscopy (FIM) before and, when possible, after each atom probe analysis so that the profile length could be accurately determined.

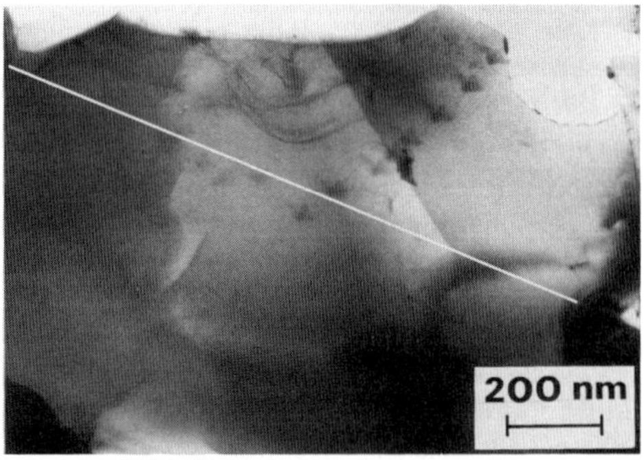

Fig. 2. TEM micrograph of a typical γ phase grain in the WC–(Ti, W)C–Co grade. An AEM composition profile along the straight line through the grain is shown in Fig. 3.

3. Results

Figure 2 is a TEM micrograph of a typical γ phase grain in the WC–(Ti, W)C–Co grade. The sharp compositional boundary can clearly be seen. An AEM composition profile consisting of point analyses along the straight line across the image is shown in Fig. 3. As can be seen, the core is rather homogeneous with a tita-

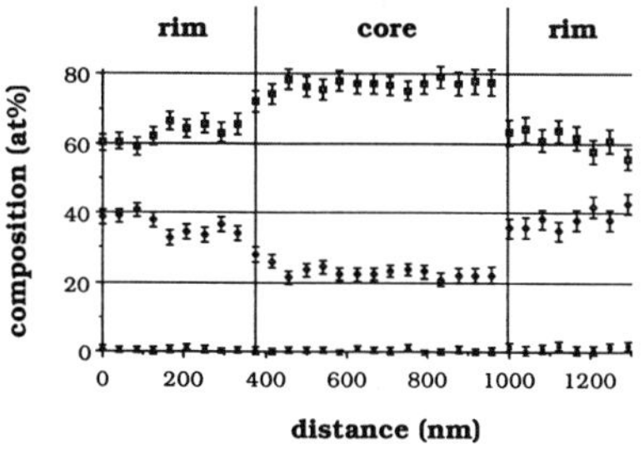

Fig. 3. AEM composition profile through the γ phase grain in Fig. 2. The titanium-to-tungsten ratio in the core agrees well with the composition of the (Ti, W)C raw material. The compositions are given in atomic per cent of the metal content; error bars are given as ± 1 standard deviation: ◘, titanium; ◆, tungsten; ■, cobalt.

nium-to-tungsten ratio which agrees well with the composition of the γ phase raw material. The core compositions of the different grains vary slightly. The rim regions are fairly homogeneous in the radial direction but the titanium-to-tungsten ratio both in the different grains and on different sides of the same grain vary slightly.

Compositional boundaries also gave a clear contrast in the FIM image, so that the area of analysis could be accurately selected. From the FIM image in Fig. 4 it appears that the boundary is sharp on an atomic scale and that the atomic planes are continuous across the boundary. Figure 5 shows part of an atom probe composition profile through a compositional boundary in the WC-based material. Atom probe analysis confirms the results obtained with AEM concerning the titanium and tungsten contents. The measured carbon content shown in the diagram is systematically too low (42.0 ± 0.9 at.%). This is due to the dead time of the ion detector. If the natural abundance of the ^{13}C isotope is used to correct the measured carbon content, a concentration of 44.7 ± 6.2 at.% is obtained.

It is interesting to note the rather high impurity level of nitrogen in the core which swiftly declines to a constant low level in the rim.

Our results from the cermet grade show several features which are similar to those described above. In this material, Ti(C, N) cores have been found. Figure 6 shows an atom probe profile through a grain containing a Ti(C, N) core. To measure this profile, the specimen had to be resharpened seven times, in the manner described above. As can be seen from the profile the core is free of alloying elements other than titanium. The rim regions are fairly homogeneous

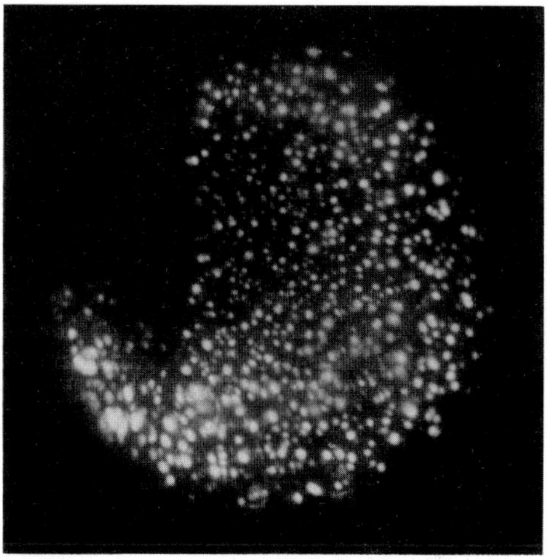

Fig. 4. FIM neon micrograph of a specimen containing a compositional boundary. The boundary between core (dark area) and rim (bright area) appears to be sharp on an atomic scale. The atomic planes (concentric rings) are continuous across the boundary (diameter of imaged area, approximately 60 nm).

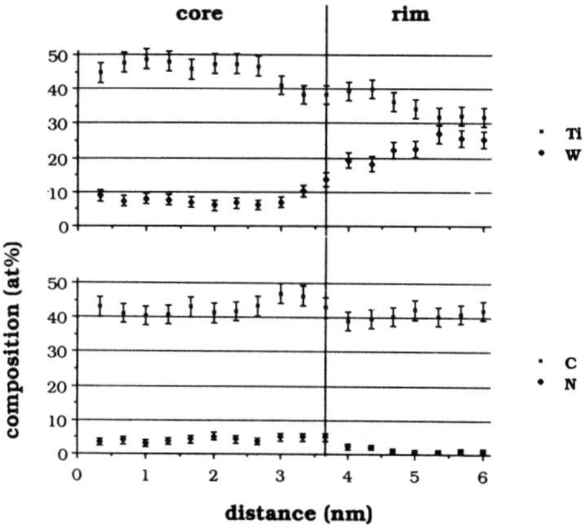

Fig. 5. Atom probe composition profile through a γ phase compositional boundary in the WC–TiC–Co grade. The rather large amount of nitrogen in the core which decreases rapidly in the rim should be noted. Nitrogen was not added intentionally to the material. The compositions are given in atomic per cent of the total content. Error bars denote ± 1 standard deviation.

in the radial direction but may differ somewhat both between different grains and on different sides of the same grain. The core also contains a rather large amount of oxygen (and should therefore in fact be designated as Ti(C, N, O)). AEM

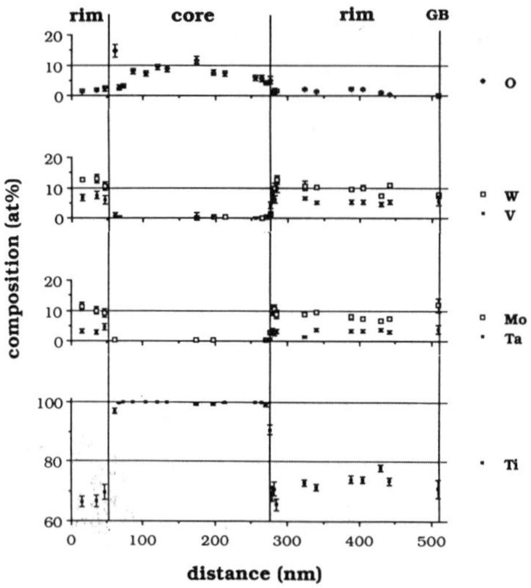

Fig. 6. Atom probe composition profile through a γ phase grain in the cermet grade, containing a Ti(C, N, O) core. GB indicates the position of the grain boundary. The oxygen content is in atomic per cent of interstitial content (carbon plus nitrogen plus oxygen); all other compositions in atomic per cent of metal content. Error bars are ± 1 standard deviation.

analysis confirms the results on the metal content obtained with APFIM.

The grain boundary composition of the WC–(Ti, W)C–Co grade has been reported previously [7]. It was shown that all carbide–carbide boundaries (WC–WC and γ–γ grain boundaries and WC–γ phase boundaries) contain approximately half a monolayer of cobalt. Atom probe analyses through grain boundaries in the cermet grade show that they contain approximately half a monolayer of binder phase atoms, in this case cobalt and nickel in roughly equal amounts.

4. Discussion

The good agreement between the metal contents of the cores in the two materials and the compositions of the raw material powders suggests that the cores are undissolved grains which act as precipitation sites during the sintering process. The rather high core content of impurity nitrogen in the WC–TiC–Co material, and oxygen in the Ti(C, N)-based material, supports this interpretation. The slight variations in the compositions of different cores can be explained by slightly inhomogeneous raw material powders.

The compositional variations in the rims obviously must be correlated to the sintering process. Since the rim composition of not only different grains but also of different sides of the same grain varies, it seems reasonable to believe that this is due to local variations in the binder phase composition when precipitation takes place.

The above reasoning together with the observation that the rims are fairly homogeneous in the radial direction imply that the precipitation is rather quick and occurs long before thermodynamic equilibrium is established. The precipitation obviously occurs in an epitaxial manner, so that there is a structural continuity across the compositional boundary.

In both materials a carbide skeleton has formed. Approximately half a monolayer of cobalt or, in the case of the cermet grade, cobalt plus nickel; has segregated to the grain boundaries.

5. Conclusions

(1) The γ phase of two different materials has been investigated with AEM and APFIM.

(2) In both materials, γ phase grains containing core to rim compositional boundaries are common.

(3) The good agreement between the core compositions and the compositions of the individual raw material powders suggests that the cores are in fact grains which were not dissolved during sintering.

(4) The rim compositions reflect the local composition of the binder phase from where the precipitation took place.

(5) In both materials a carbide skeleton has formed. Approximately half a monolayer of cobalt or, in the case of the cermet grade, cobalt plus nickel has segregated to the grain boundaries.

Acknowledgments

This work has been financially supported by the National Swedish Board for Technical Development (STU) and by AB Sandvik Coromant.

References

1 H. F. Fischmeister, in R. K. Viswanadham, D. J. Rowcliffe and J. Gurland (eds.), *Science of Hard Materials,* Plenum, New York, 1983, p. 1.

2 H. Doi, in E. A. Almond, C. A. Brookes and R. Warren (eds.), *Proc. 2nd Int. Conf. on the Science of Hard Materials, Rhodes, September 23–28, 1984,* in *Inst. Phys. Conf. Ser., 75* (1986) 489.
3 M. Hellsing, A. Henjered, H. Nordén and H. O. Andrén, in R. K. Viswanadham, D. J. Rowcliffe and J. Gurland (eds.), *Science of Hard Materials,* Plenum, New York, 1983, p. 931.
4 H. O. Andrén and H. Nordén, *Scand. J. Met., 8* (1979) 147.
5 H. O. Andrén, *J. Phys. (Paris), Colloq. C7, 47* (1986) 483.
6 U. Rolander, *J. Phys. (Paris), Colloq. C7, 47* (1986) 449.
7 A. Henjered, M. Hellsing, H. O. Andrén and H. Nordén, *Mater. Sci. Technol., 2* (1986) 847.

Materials Science and Engineering, *A105/106* (1988) 289–292

Binder Mean-free-path Determination in Cemented Carbide by Coercive Force and Material Composition*

R. PORAT and J. MALEK

ISCAR Ltd, Nahariya (Israel)

(Received November 9, 1987)

Abstract

In the present investigation, cemented carbide with different cobalt contents varying from 9 to 37 vol.% were prepared by conventional methods, and their mechanical and magnetic properties measured. Using optical microscopy and scanning electron microscopy analysis, the average WC grain size and binder mean free path were determined. These values were found to relate to the coercive force and cobalt volume percentage. K_{Ic} measurements using the "short-rod" test method, as a function of the binder mean free path calculated from the equation, were found to agree very well with K_{Ic} values for cemented carbide published by other investigators in the field.

1. Introduction

In metallurgical systems, the mean grain size and distribution of grain size have an important effect on material properties and performance. In cemented-carbide systems (WC–Co), properties such as hardness, toughness and strength depend strongly on the microstructure, and in particular on the mean free path of the cobalt binder.

The cobalt binder phase is ferromagnetic in hard metals and therefore the cemented carbide can be characterized by the parameters, the coercive force H_c and the specific magnetic saturation (SMS). The coercive force measures the ease with which the ferromagnetic domain walls are aligned by an induced magnetic field. The immobilization of the domain walls is caused by structural defects, particularly non-magnetic inclusions and material imperfections. Domain walls are attracted to these imperfections because the wall energy and the magnetostatic energy are thereby reduced. In cemented carbide, the total surface area of the WC grains (the WC being non-magnetic) is considered to be the main imperfection. Any process parameter which affects the microstructure of cemented carbide, such as the initial grain size of raw material, the milling time, the carbon balance and the sintering temperature, will directly influence the coercive force.

Many articles have been published [1–6] which show the existence of a strong relationship between the binder mean free path and fracture toughness of the material. In all the above-mentioned work the average carbide grain size was measured using the well-known method [7, 8] of line intersection on the sample cross-section. This method is tedious and requires many measurements to obtain statistically valid results.

The aim of our work was to find a better method of determining the mean free path by establishing a mathematical relationship between the coercive force (an easily measured parameter) and the cobalt mean free path.

The validity of our mathematical model was examined by measuring the K_{Ic} values for known compositions using the short-rod method [9], by relating these values to the calculated mean free path and by comparing the results with the data published by other researchers in the field of fracture toughness measurement.

2. Experimental details

2.1. *Sample preparation*

Samples of various materials were prepared according to the following compositional groups.

2.1.1. *Group A: WC + Co*

The only difference in the material in each group here is the WC grain size.

*Paper presented at the 3rd International Conference on the Science of Hard Materials, Nassau, The Bahamas, November 9–13, 1987.

2.1.2. Group B: WC + TaC–NbC + Co

This group has a coarse grain size.

2.1.3. Groups C and D: WC + TiC + TaC–NbC + Co

Group C differs from group D in the cobalt volume percentage. Other differences are in the grain size.

2.1.4. Groups E and F: WC + TiC + TaC–NbC + Co

Groups E and F differ from groups C and D in the percentages of TiC and TaC. Group E differs from group F in the cobalt volume percentage. In group F, various grain sizes were prepared.

All the materials were pressed as transverse rupture strength (TRS) specimens and vacuum sintered in the following manner: groups A and B at 1420 °C for 80 min; groups C–F at 1450 °C for 80 min. Table 1 is a summary of the compositions of the various groups.

2.2. Tests

All samples were tested for hardness, TRS, density, shrinkage, coercive force and specific magnetic saturation. The microstructure was examined according to ASTM specifications.

Table 2 summarizes some of the mechanical and magnetic results.

For the K_{Ic} measurements, samples of cylinder rods were made from the same composition groups as mentioned above. They were sintered together with the above samples. The cylinder rod samples were then centreless ground to the exact dimension as specified by the fractometer producer. The samples were then prepared for

TABLE 3 Summary of results on K_{Ic}, H_c and $\bar{\lambda}$

Material	H_c (kA m^{-1})	$\bar{\lambda}$ (μm)	K_{Ic} (MPa m$^{1/2}$)
A1	18.2	0.204	10.97
A2	14.2	0.279	13.06
A3	13.7	0.291	13.39
A4	10.0	0.430	14.90
B1	9.2	0.514	15.67
C1	17.6	0.208	10.29
C2	14.7	0.260	11.27
C3	13.9	0.279	11.50
D1	18.3	0.214	11.97
D2	11.0	0.389	13.72
E1	14.7	0.295	11.52
F1	13.1	0.364	12.21
F2	10.7	0.45	14.49

TABLE 1 Composition of the cemented carbide

Material	*Density* (g cm^{-3})	*Amount of Co* (vol.%)	*Amount of TiC* (vol.%)	*Amount of TaC* (vol.%)	*Amount of NbC* (vol.%)	*Amount of WC* (vol.%)
A	15.01	10.1	—	—	—	89.9
B	14.74	13.3	—	1.8	0.4	84.5
C	14.20	9.6	6.5	5.5	1.1	77.3
D	14.10	11.1	6.4	5.4	1.1	76.0
E	14.09	12.7	5.7	4.4	1.0	76.2
F	13.94	14.90	5.6	4.8	1.0	73.7

TABLE 2 Mechanical properties of the cemented-carbide materials

Material	H_c (kA m^{-1})	*Rockwell A hardness* (HRA)	*Vickers hardness (load, 30 kgf)* (HV30)	*TRS* (lbf in^{-2})	*TRS* (N mm^{-2})
A1	18.3	92.4	1750	320000	2250
A2	14.2	91.6	1600	330000	2300
A3	13.7	91.5	1570	350000	2450
A4	10.0	90.8	1460	330000	2300
B1	9.2	89.0	1250	400000	2800
C1	17.6	92.5	1650	255000	1800
C2	14.7	92.1	1570	270000	1900
C3	13.9	92.0	1550	290000	2050
D1	18.3	92.2	1600	270000	1900
D2	11.0	90.9	1420	290000	2050
E1	14.7	91.5	1500	320000	2250
F1	13.1	91.2	1450	340000	2400
F2	10.7	90.8	1400	330000	2300

the test according to the short rod method [9]. Table 3 summarizes K_{Ic} results.

The TRS samples were prepared for metallographic examination. A most representative sample area was finely examined with an optical microscope and a photograph was taken (Figs. 1 and 2).

The mean particle intercept distance was measured on each photograph by a microscope with a coordinate table, and a measurement fixture device capable of movement in the x–y direction with an accuracy of 1 mm. The distance length for measurement was 30 mm, and in each picture measurements were made at five cross-sections. This procedure was followed on 25 different compositions and various grain sizes. The coercive force for each sample was measured. Figure 3 shows the results of the d and H_c measurements.

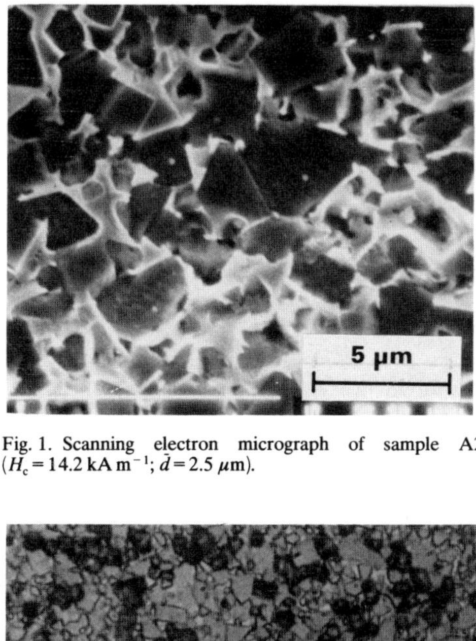

Fig. 1. Scanning electron micrograph of sample A2 ($H_c = 14.2$ kA m^{-1}; $\bar{d} = 2.5$ µm).

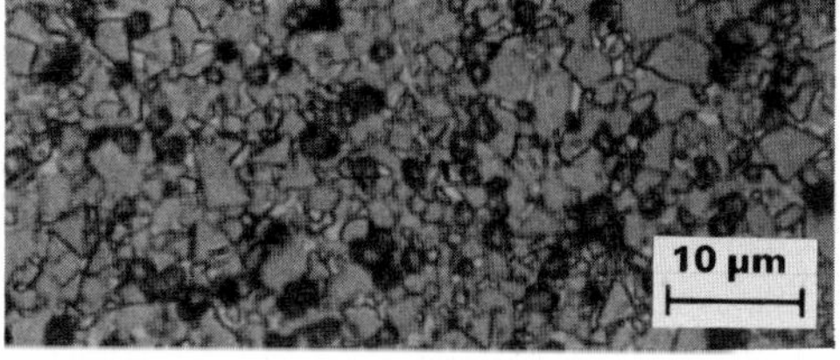

Fig. 2. Microstructure of sample F2 ($H_c = 10.7$ kA m^{-1}, $\bar{d} = 2.6$ µm).

3. Results and discussion

Figure 3 shows the mean intercept length of d, referred to as the mean WC grain size, as a function of coercive force H_c (kA m^{-1}) for various cobalt volume percentages on a log–log scale. It is observed that for a given d (mm) the coercive force decreases as the cobalt volume percentage increases. For a given cobalt volume percentage, the coercive force H_c increases as the grain size decreases.

From this graph, the following equation is deduced:

$$\bar{d} = 0.3\left(\frac{80}{H_c}\right)^{0.57(1/V_F)^{1/3}} \quad (\mu\text{m}) \qquad (1)$$

From the work of Underwood [7], the relationship between the mean intercept length and the mean free path of binder is

$$\bar{\lambda} = \bar{d}\,\frac{V_F}{1 - V_F} \ (\mu\text{m}) \qquad (2)$$

Equation (1) can be transformed into a more practical and meaningful relation, as follows:

$$\bar{\lambda} = 0.3\,\frac{V_F}{1 - V_F}\left(\frac{80}{H_c}\right)^{0.57(1/V_F)^{1/3}} \quad (\mu\text{m}) \qquad (3)$$

By introducing the weight percentage of cobalt instead of V_f using

$$V_F = \frac{X_{Co}}{8.9}\,\frac{\rho_{th}}{100} \qquad (4)$$

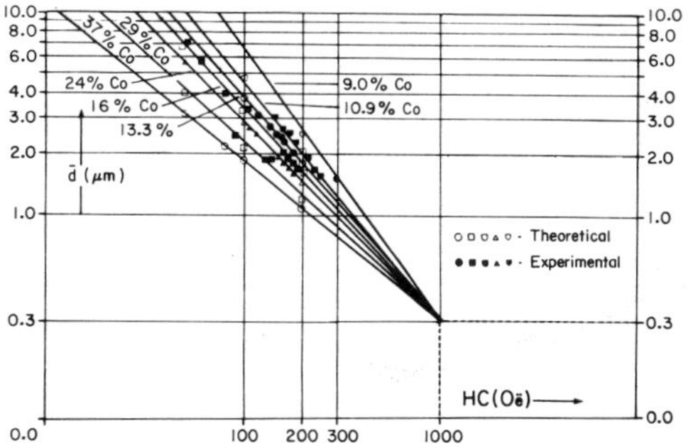

Fig. 3. Average grain size *vs.* coercive force for different cobalt volume percentages.

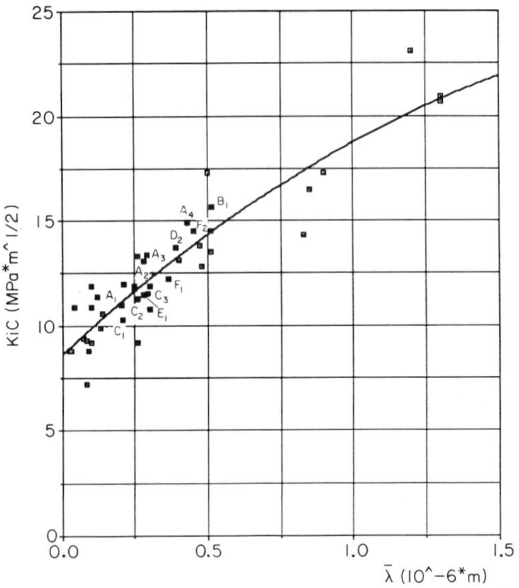

Fig. 4. Fracture toughness as a function of the mean free path of cobalt.

we obtain the following final expression for $\bar{\lambda}$:

$$\bar{\lambda} = \frac{0.3X\rho_{th}}{890 - X_{Co}\rho_{th}} \left(\frac{80}{H_c}\right)^{(164.822/X_{Co}\rho_{th})^{1/3}} \quad (\mu m) \qquad (5)$$

The validity of this equation was tested by measuring the fracture toughnesses K_{Ic} of known compositions, using the short-rod method. Table 3 shows the coercive force, the calculated $\bar{\lambda}$ (from eqn. (5)) and the measured K_{Ic} for the various materials. Figure 4 is a graph which shows K_{Ic} as a function of $\bar{\lambda}$ using published data [1–6]. Our measured K_{Ic} values, labelled by letters indicating the compositions, are seen to fit the distribution averages very well.

4. Conclusions

(1) Coercive force measurement makes it possible to calculate the average carbide grain size for cemented-carbide materials which differ in composition, grain size and cobalt content.

(2) Equation (5) gives the mean free path of cobalt, using the cobalt weight percentage, the coercive force and the theoretical specific weight.

(3) K_{Ic} results for the calculated mean free path are in agreement with other published data.

(4) Our method is a valuable, useful and non-destructive way of determining material toughness without requiring actual fracture toughness testing.

References

1 R. C. Leuth, *Ph.D. Thesis*, Michigan State University, 1972.
2 J. L. Chermant and F. Osterstock, *J. Mater. Sci., 11* (1976).
3 L. Lindau, in D. M. R. Taplin (ed.), *Fracture 1977*, Proc. 4th Int. Conf. on Fracture, Waterloo, Ontario, June 19–24, 1977, University of Waterloo Press, Waterloo, Ontario, 1977.
4 J. R. Pickens and J. Gurland, *Mater. Sci. Eng., 33* (1978) 135.
5 R. Porat and A. Ber, *CIRP Ann., 32* (1) (1983).
6 Z. Goffer and R. Porat, *Rep. EURO CVD 4*, 1983 (Eindhoven).
7 E. E. Underwood, *Quantitative Stereology*, Addison-Wesley, Reading, MA, 1970.
8 J. Gurland, *Trans. Metall. Soc. AIME, 236* (1966).
9 L. M. Barker, *Rep. TR77-91R*, 1977 (Salt Lake City, UT).
10 M. T. Laugier, *Powder Metall. Int., 5* (1986).

Appendix A: Nomenclature

$\bar{d}$	mean intercept length (mean grain size) of WC
H_c	coercive force (kA m^{-1}),
V_F	volume fraction of cobalt
X_{Co}	weight percentage of cobalt

Greek symbols

$\bar{\lambda}$	mean free pass of cobalt binder
ρ_{th}	theoretical density of the composition

Dislocation Microstructures in Explosively Deformed Hard Materials*

T. B. MERALA, H. W. CHAN and D. G. HOWITT

Department of Mechanical Engineering, University of California at Davis, Davis, CA 95616 (U.S.A.)

P. V. KELSEY

Aluminum Company of America, Alcoa Laboratories, Alcoa Center, PA 15069 (U.S.A.)

G. E. KORTH and R. L. WILLIAMSON

EG and G, Idaho, Idaho National Engineering Laboratory, Idaho Falls, ID 83415 (U.S.A.)

(Received November 27, 1987)

Abstract

The effects of shock loading at 23 GPa on the microstructures of Al_2O_3, TiB_2 and SiC were studied with special emphasis placed on the role of dislocation generation and localized transformations which could result in the absorption or conversion of kinetic energy by the material.

1. Introduction

The development and optimization of hard ceramics in systems which must function in high strain rate (HSR) environments require at least some qualitative knowledge of the associated mechanisms of failure. Very little is known about the micromechanical response of ceramics under HSR conditions. Material selection and optimization are based solely on experiment, and the correlation between microstructural relationships and performance is at best empirical. An important feature which has been generally ignored in ceramic failure analyses under HSR conditions is the role of the plastic deformation associated with fracture although observations of defect microstructures have been made [1]. In this paper, a quantitative determination of the dislocation microstructures in ceramics deformed at HSR is presented.

Hard crystalline ceramics derive their properties, high hardness and generally brittle fracture behavior, from relatively high bond strengths and from a substantial Peierls stress which inhibits slip. As a result, dislocation densities tend to be small and dislocation motion is impeded. Recent work [2–4], however, has shown that substantial dislocation generation (more than 10^{14} dislocations m^{-2}) can occur under high impact shock loading. Basal slip $\frac{1}{3}\langle 11\bar{2}0\rangle(0001)$ is observed in sintered α-Al_2O_3 as pyramidal $\frac{1}{3}\langle 10\bar{1}1\rangle\{11\bar{2}3\}$ and prismatic slip $\frac{1}{3}\langle 11\bar{2}0\rangle\{1\bar{1}00\}$ when it is explosively shocked (20 GPa; 4 μs pulse) at ambient temperature [3], and it has long been established by Kim and Palmour [5] and Kim *et al.* [6] that aluminas plastically deform under ballistic loading. In work by Ferguson and Rice [7], the ballistic performance of a wide variety of aluminas against a light-weight projectile was postulated to be a direct function of the potential for microplasticity, whilst other factors such as tensile strength and grain size were found to be of minor importance.

The crystal structures of α-Al_2O_3 [8], TiB_2 [9] and SiC [10–12] are well described in the literature.

2. Experimental procedure

Samples of Al_2O_3, TiB_2 and SiC with the geometry and dimensions shown in Fig. 1 were cut from larger 15.2 cm × 15.2 cm × 2.54 cm plates. The measured properties of the materials are shown in Table 1.

The samples were placed in a copper cup with a soldered lid as shown in Fig. 1. Grinding of the sample lid was performed to ensure a flush fit into the steel pedestal, shown in the schematic dia-

*Paper presented at the 3rd International Conference of the Science of Hard Materials, Nassau, The Bahamas, November 9–13, 1987.

0921-5093/88/$3.50 © Elsevier Sequoia/Printed in The Netherlands

gram in Fig. 2. The rectangular specimens were mounted in a fourfold offset array in a Nimonic 33 plate 22.9 cm × 22.9 cm × 6.7 cm. The plate was mounted on a type 316H stainless steel pedestal 36.8 cm in diameter and 10.2 cm high. The plate containing the specimens was surrounded by two steel layers of spall plates 0.64 cm thick. A copper cover plate 21.6 cm square by 0.159 cm thick was placed over the specimen plate to ensure a planar contact on shock initiation. The flyer plate 20.3 cm square by 2.5 cm thick, surrounded by double spall plates, was placed between the copper cover plate and the explosive. Shock loading of the specimens was by single-point initiation of the explosive Vibrogel (E. I. duPont deNemours and Co., Wilmington, DE 19898).

To measure the velocity of the impacting flyer plate, time-of-arrival pin gauges were utilized, which allowed the time between a known distance to be determined with a high speed digital storage oscilloscope. A sufficient initial offset between the flyer plate and the target was provided so that the flyer plate reached terminal velocity prior to impact. Multiple-velocity pin sets were used for each shot which provided redundant velocity information.

The pressure–time data were determined from a two-dimensional numerical model of the flyer plate and target assembly, using the CSQII hydrocode. Experimentally measured plate velocities were used for initial conditions and thus the calculations began at the instant of impact; the flyer plate was assumed to be perfectly planar at impact. All materials were modeled using an elastic, perfectly plastic constitutive model. Since the dynamic behavior of ceramic under shock loading is uncertain, the validity of this assumption remains somewhat uncertain, although the current activity indicates that plasticity is indeed occurring.

The samples of Al_2O_3, TiB_2 and SiC were reduced to fragments by the explosion. The larger fragments were carefully sectioned using a low speed diamond saw, polished with 6 μm diamond

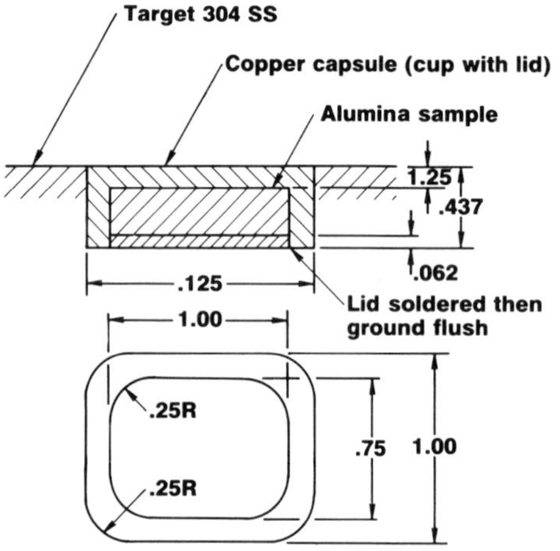

Fig. 1. Details of copper capsule.

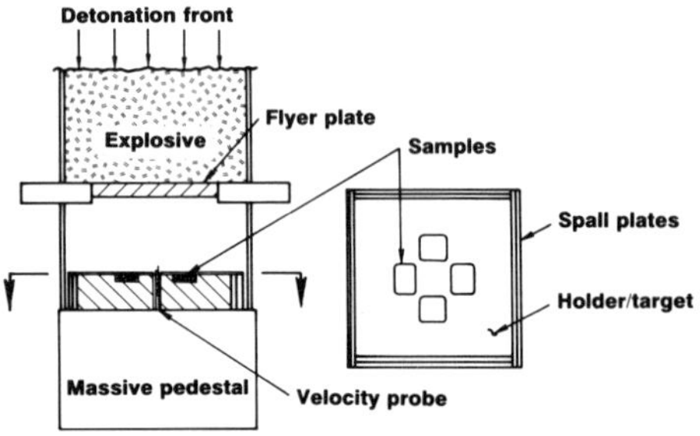

Fig. 2. Schematic diagram of explosive experimental apparatus.

TABLE 1 Properties of materials

Property structure (units)	*Al_2O_3, hexagonal*	*TiB_2, hexagonal*	*α-SiC, hexagonal*	*β-SiC, cubic*
Lattice parameters (nm)	$a = 47.58$ $c = 129.7$	$a = 30.2$ $c = 32.2$	$a = 30.7$ $c = 150.8$	30.7
Young's modulus (GPa)	376	551	417	425
Vickers microhardness (2 kgf load) (HV)	1950	2100	2642	2550
Melting point (°C)	2048	2871	3102	3102
Thermal conductivity (W m^{-1} K^{-1})	1–10	35–40	50–100	50–100

Manufacturers: Al_2O_3, A-99F sintered alumina, Alcoa; TiB_2, hot pressed, Eagle–Picher Industries; α-SiC, Hexallon SA, Sohio Engineered Materials; β-SiC, hot pressed, Eagle–Picher Industries.

paste and then ion beam thinned to perforation. The thinnest regions around the central hole were evaluated by transmission electron microscopy, and this technique was used to determine the extent and distribution of the plastic deformation that was induced.

3. Results and discussion

The predicted shock-loading profile is shown in Fig. 3. The peak pulse was approximately 23 GPa with measured midpeak duration of approximately 4.7 μs. An electron micrograph from the Al_2O_3 sample is shown in Fig. 4 and shows the formation of dense dislocation structures and cracks both at the grain boundaries and within the residual glass phases. Al_2O_3 is perhaps the most appropriate material to evaluate in detail because both the glide mechanisms and the morphology of the dislocation structures which form at high temperatures (*i.e.* greater than about half the melting temperature) and low temperatures have been evaluated (see for example Bretheau *et al.* [13]) as well as the deformation induced by very high pressures [14] and microindentation techniques [15]. A comparison of the microstructures formed from the HSR deformation shows a similarity to those which form at low temperatures. There is no evidence of dipole formation in this material, which is a common feature of the thermally activated glide processes observed for both basal and prismatic slip in sapphire at high temperatures [16], nor is there any evidence of partial dislocations. The microstructures are also quite similar to those observed in rapidly deformed metals and in some cases the intersecting glide plane activity is very similar to the deformation induced by diamond indentation in sapphire where the glide in both instances includes the $\frac{1}{3}\langle 10\bar{1}1\rangle\{11\bar{2}3\}$.

Another characteristic feature of the microstructure is the formation of shear bands, as shown in Fig. 5. These bands, which have been observed to form in explosively deformed metals of low stacking fault energy (see for example Murr [17]), have been attributed to a thermal softening mechanism brought about by an essentially adiabatic heating which is generated by either the intense slip activity itself or the friction between the sliding grains.

The α- and β-SiC samples were also essentially dislocation free prior to deformation although there is clear evidence of the formation of polytypes in both of these samples (Fig. 6). The deformed microstructure of the α sample clearly deforms in much the same way as the Al_2O_3, forming characteristic shear bands on

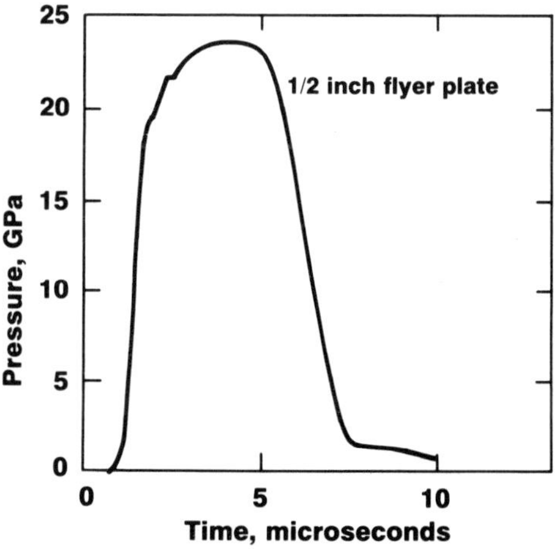

Fig. 3. Shock-loading profiles at the midpoint of ceramic samples.

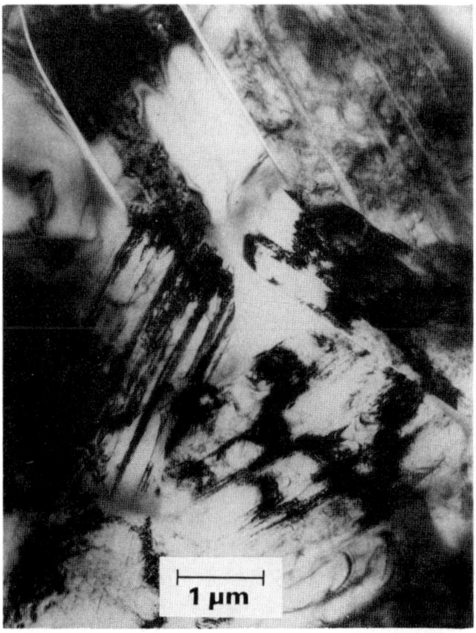

Fig. 4. A bright field micrograph showing the structure of deformed Al_2O_3. Regions of intense deformation appear as dark bands, and cracks are visible along the grain boundaries and in the glass phase.

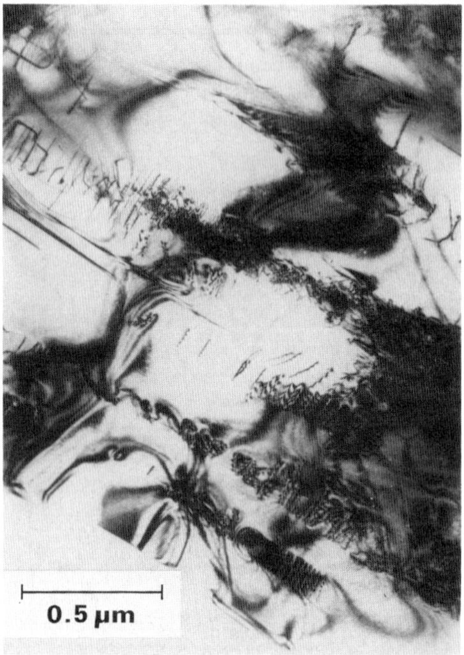

Fig. 5. Bright field micrograph of a shear band in the basal plane of Al_2O_3.

$(0001)\langle 11\bar{2}0\rangle$ (Fig. 7(a)), as well as some of the other close-packed planes. This basal slip system has been observed after both the indentation and frictional deformation of SiC although it appears that the $\{1\bar{1}00\}\langle 11\bar{2}0\rangle$ is the favored secondary system for the former and $\{1\bar{1}01\}\langle 11\bar{2}0\rangle$ for the latter [18, 19]. The cubic material, however, has a dislocation microstructure more closely resembling the dipoles characteristic of high temperature deformation, where the glide needs to be thermally activated (Fig. 7(b)).

In both of these materials, the local dislocation densities are extremely high and approach 10^{15} m^{-2} in the shear bands of the hexagonal material. The density of the dipoles is about half of this value and the overall dislocation density, like that of the Al_2O_3, is between 10^{13} and 10^{14}.

The dislocation density of the TiB_2 is almost as high although the microstructure is substantially different (Fig. 8), being highly strained and on an extremely fine scale, resembling in some ways that of an irradiated material. Nevertheless, the ion bombardment, which accompanies the specimen preparation technique, is not responsible

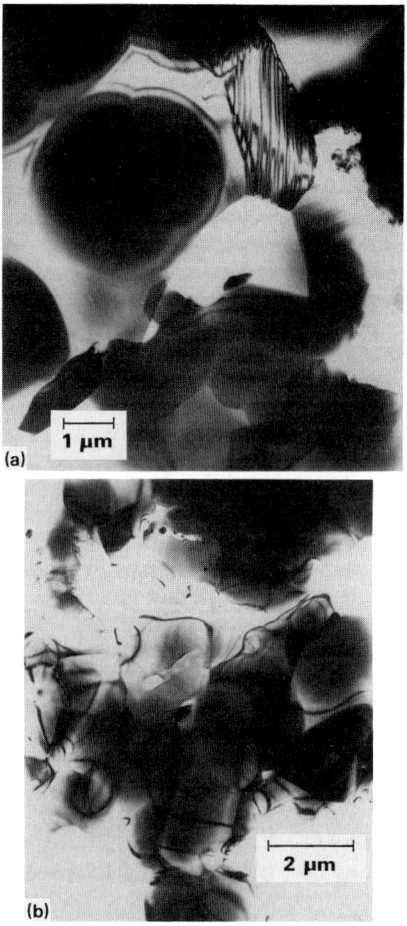

Fig. 6. The microstructures of the undeformed SiC (a) in the α-SiC hexagonal 6H form and (b) in the β-cubic form.

and the samples of material prepared prior to the deformation clearly confirm this (Fig. 9). The micrographs of the deformed material also reveal the development of a high concentration of cavities, visible as either dark or light spheres depending on their exact location in the foil. These cavities were not found in either the Al_2O_3 or the SiC although microcracking was observed in all the deformed materials.

It is possible to estimate the energy of the dislocation microstructures in these materials if the dislocation Burgers vectors have been distinguished [20]. All these materials have shear moduli in the range 100–200 GPa [21] and they have similar interatomic spacings. The lattice

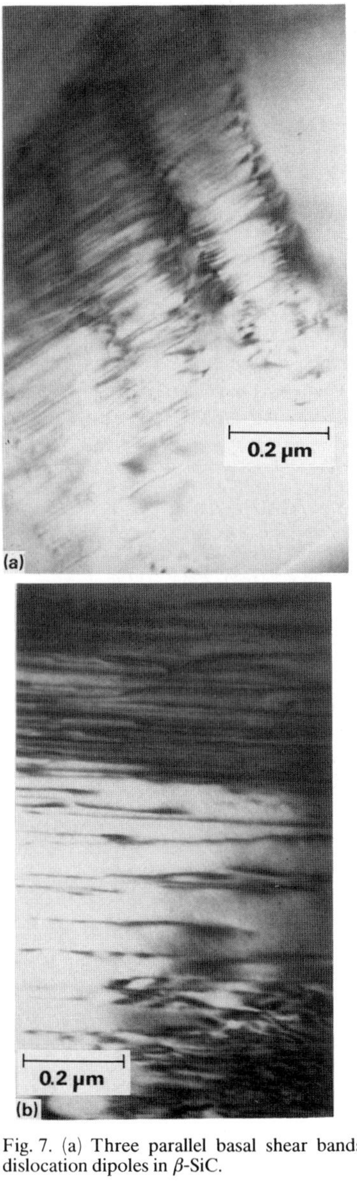

Fig. 7. (a) Three parallel basal shear bands in α-SiC; (b) dislocation dipoles in β-SiC.

parameters of the hexagonal unit cell of Al_2O_3 ($a = 47.5$ nm; $c = 129.7$ nm) and α-SiC ($a = 30.7$ nm; $c = 150.8$ nm) are substantially larger than those of TiB_2 ($a = 30.2$ nm; $c = 32.2$ nm) only because of the cation packing. These dislocation densities are associated with energies of about 5×10^6 J m^{-3} in Al_2O_3, 2×10^6 J m^{-3} in α-SiC and 1×10^6 J m^{-3} in β-SiC and TiB_2.

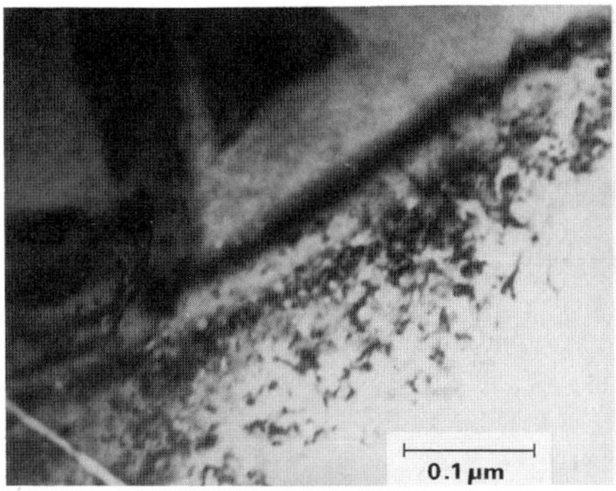

Fig. 8. Dislocations in TiB_2.

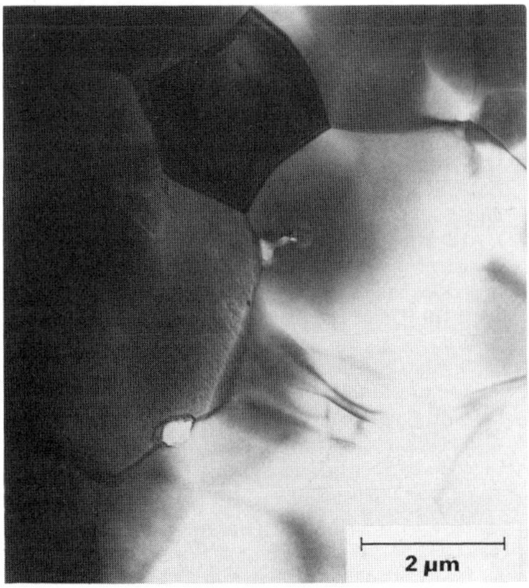

Fig. 9. The microstructure of TiB_2 prior to explosive loading.

4. Conclusions

The simple correlation of these results with the physical properties of these hard materials is not entirely obvious. The dramatic differences between the microstructures of the cubic and hexagonal forms of SiC must presumably be due to the greater number of equivalent slip systems in the former although it is difficult to explain the differences between the TiB_2 and the quite similar microstructures of α-SiC and Al_2O_3. The onset of slip in the $\{1\bar{1}00\}\langle 11\bar{2}0\rangle$ in preference to $\{11\bar{2}0\}\langle 1101\rangle$ is favored by both TiB_2 and SiC over Al_2O_3 [18, 19, 22, 23] and so it is unlikely that structural differences are responsible in this case. TiB_2 has a melting point, thermal conductivity and specific heat intermediate between the lower values determined for Al_2O_3 and those of α-SiC and only the values of the Knoop hardness (3000 for 100 gf, compared with 2000–2500) would seem to be substantially higher [21]. Although specific correlations of behavior cannot be universally made, it can be concluded that the role of plastic deformation for most ceramic materials under HSR conditions is, from an energy-absorptive perspective, extremely significant.

References

1 C. S. Yust and P. S. Sklad, in R. K. Viswoinadham, D. J. Rowcliffe and J. Gurland (eds.), *Science of Hard Materials,* Plenum, New York, 1983, pp. 911–930.

2 F. Longy and J. Cagnoux, *Proc. Int. Conf. on Impact Loading and Dynamic Behavior of Materials, Bremen, May 18–22, 1987,* to be published.

3 D. G. Howitt and P. V. Kelsey, *Proc. Int. Conf. on Impact Loading and Dynamic Behavior of Materials, Bremen, May 18–22, 1987,* to be published.

4 J. A. Brusso, D. E. Mikkola, J. E. Flinn and P. V. Kelsey, submitted to *Scr. Metall.*

5 C. H. Kim and H. Palmour III, *Tech. Rep. 70-8,* December 1970 (Office of Naval Research Contract NR064-504).

6 C. H. Kim, H. Palmour III, D. R. Johnson and C. E. Zimmer, *Tech. Rep. 69-5,* April 1969 (Office of Naval Research Contract NR064-504).

7 W. J. Ferguson and R. W. Rice, in W. W. Kniegel and H. Palmour III (eds.), *Ceramics in Severe Environments,* Plenum, New York, 1971, pp. 261–270.

8 M. L. Kronberg, *Acta Metall., 5* (9) (1987) 507–524.

9 T. Lundstrom, in V. I. Matkovich (ed.), *Boron and Refractory Borides,* Springer, Berlin, 1977, pp. 352–353.

10 P. T. B. Shaffer, *Acta Crystallogr. B, 25* (1969) 477–488.

11 A. H. Heuer, G. A. Fryburg, L. U. Ogbuji and T. E. Mitchell, *J. Am. Ceram. Soc., 61* (9–10) (1978) 406–412.

12 N. W. Jepps and T. F. Page, *J. Cryst. Growth, 7* (1983) 259–306.

13 T. Bretheau, J. Castaing, J. Rabier and P. Veyssiere, *Adv. Phys., 28* (1979) 829.

14 J. Castaing, J. Cadoz and S. H. Kirby, *J. Phys. (Paris), Colloq. C3, 42* (1981) 43.

15 B. J. Hockey, in R. C. Bradt and R. E. Tressler (eds.), *Deformation of Ceramic Materials,* Plenum, New York, 1975.

16 T. E. Mitchell, W. T. Donlon, K. P. D. Lagerlof and A. H. Heuer, in R. C. Bradt and R. E. Tressler (eds.), *Deformation of Ceramic Materials,* Plenum, New York, 1984, p. 125.

17 L. E. Murr, in M. A. Meyer and L. E. Murr (eds.), *Shock Waves and High Strain Rate Phenomena in Metals,* Plenum, New York, 1981, p. 607.

18 O. O. Adewoye and T. F. Page, *J. Mater. Sci., 70* (1981) 37.

19 O. O. Adewoye and T. F. Page, *Wear, 73* (1981) 247.

20 J. P. Hirth and J. Lothe, *Theory of Dislocations,* McGraw-Hill, New York, 1968.

21 J. F. Lynch, C. G. Ruderer and W. H. Duckworth, *Engineering Properties of Selected Ceramics,* American Ceramic Society, Columbus, OH, 1966.

22 S. A. Mersol, C. T. Lynch and F. W. Vahldiek, in F. W. Vahldiek and C. T. Lynch (eds.), *Anisotropy in Single Crystals of Refractory Compounds,* Vol. 2, Plenum, New York, 1968, p. 41.

23 A. H. Heuer and J. Castaing, *The Structure and Properties of MgO and Al_2O_3 Ceramics, Advances in Ceramics,* Vol. 10, American Ceramic Society, Columbus, OH, 1985, p. 238.

Compressive Deformation and Fracture in WC Materials*

DAVID J. ROWCLIFFE†

SRI International, Menlo Park, CA (U.S.A.)

VICKRAM JAYARAM‡, MARY K. HIBBS§ and ROBERT SINCLAIR

Stanford University, Stanford, CA (U.S.A.)

(Received December 8, 1987)

Abstract

Plastic flow and fracture have been studied in WC single crystals and in WC–Co materials. Deformation was introduced by several techniques which resulted in the development of high compressive stresses which encourage slip in WC. Optical and transmission electron microscopy studies show that plastic flow in the carbide phase always precedes fracture. A simple analysis shows that there are only four independent slip systems in WC. Consideration of the limitations for slip in WC lead to a mechanistic model for fracture initiation in WC–Co.

1. Introduction

Fracture processes in cemented WC–Co have been modelled extensively [1–6]. Many of these models are logically based on observations of fracture surfaces which give information on the fracture path. Most models emphasize the role of plastic deformation in the binder phase because much of the final fracture occurs through the binder phase or at carbide binder interfaces. However, Sarin and Johannesson [6] suggest that even small amounts of plastic flow in the carbide phase exert a significant influence on the deformation of WC–Co. Studies of single crystals [7–9] show clearly that WC can undergo plastic deformation relatively readily under certain stressing conditions. Although dislocation multiplication [10] has been reported in deformed WC–Co, there has been little attempt to relate single-crystal plasticity to specific deformation and fracture processes in WC–Co. This paper addresses this issue. Information is drawn from an extensive study of deformation and fracture of WC materials [8, 9, 11–15] which includes work on indentation [8, 9, 11, 12], isostatic compression [13] and dynamic impact using hard spheres [15]. These different stressing methods have in common a large component of compression in the presence of constraining forces which tend to restrict the development of high tensile stresses. This combination encourages accommodation of stress by plastic flow rather than by tensile fracture and, under these conditions, plastic flow in the carbide phase always precedes fracture. Using these considerations, this paper suggests a mechanistic model for fracture initiation in WC–Co based on limited slip in the carbide phase.

2. Deformation of single crystal WC

It is well established that the slip plane in WC is $\{10\bar{1}0\}$. The slip direction [8, 12, 15, 17] is $\langle 11\bar{2}3\rangle$ with slip occurring by the motion of identical $(a/6)[11\bar{2}3]$ partials [8, 12, 17]. Figure 1 illustrates schematically the slip traces S and some crack patterns for indentations on the (0001) and $\{10\bar{1}0\}$ planes. For indentations on the basal plane, cracks develop at intersecting slip bands on $\{11\bar{2}0\}$(A) and on $\{10\bar{1}0\}$(B). The former are more common and an example is shown in Fig. 2. Almond [18] observed similar cracks at intersecting slip bands in WC–Co but did not analyse them. Cracks of this type have been analysed by transmission electron micros-

*Paper presented at the 3rd International Conference on the Science of Hard Materials, Nassau, The Bahamas, November 9–13, 1987.

†Present address: Royal Institute of Technology, Stockholm, Sweden.

‡Present address: University of California, Santa Barbara, CA, U.S.A.

§Present address: Honeywell Inc., Bloomington, MN, U.S.A.

0921-5093/88/$3.50 © Elsevier Sequoia/Printed in The Netherlands

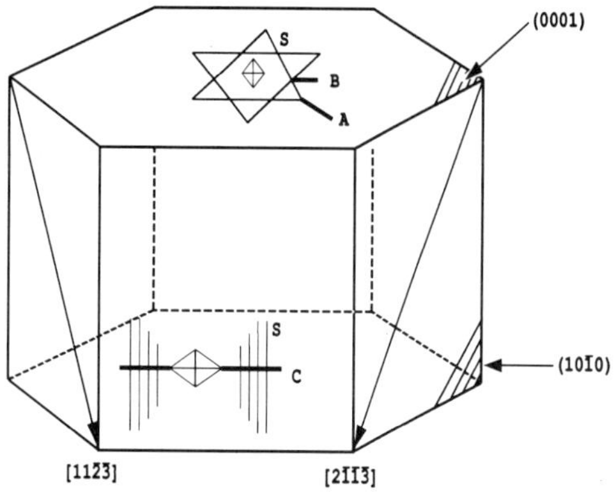

Fig. 1. Slip traces and crack patterns in WC.

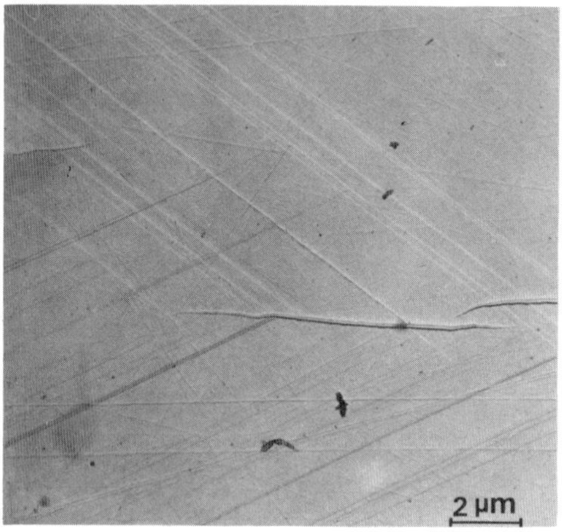

Fig. 2. Crack on {1120} formed at intersecting slip bands on the {0001} surface of WC.

copy [10]. It has been suggested that the cracks form by a mechanism analogous to that proposed by Cottrell for the nucleation of cracks in b.c.c. metals. In the case of WC the proposed interactions are as follows.

(1) For a crack on (1$\bar{2}$10)

$$\tfrac{1}{6}[\bar{2}11\bar{3}]+\tfrac{1}{6}[11\bar{2}3]=\tfrac{1}{6}[\bar{1}2\bar{1}0]$$

(2) For a crack on (10$\bar{1}$0),

$$\tfrac{1}{6}[2\bar{1}\bar{1}3]+\tfrac{1}{6}[11\bar{2}\bar{3}]=\tfrac{1}{2}[10\bar{1}0]$$

The reaction to form $\tfrac{1}{6}[\bar{1}2\bar{1}0]$ partials has been confirmed by Hibbs *et al.* [12] but the second reaction was not verified, owing to experimental difficulties. However, both dislocations are sessile and, under the action of an applied stress, coalescence of edge dislocations can be expected to open up wedge cracks of the type that are observed.

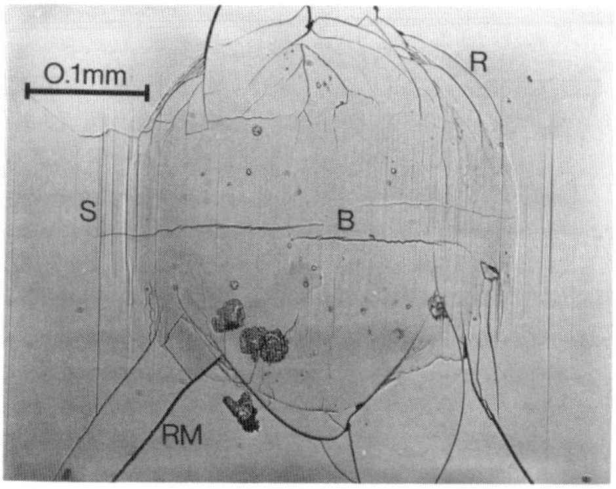

Fig. 3. Spherical indentation on the {1010} plane of WC: B, basal plane crack; S, slip steps; R, ring crack; RM, radial–median crack.

A second mechanism for fracture related to plastic flow has also been found. The basal plane crack C shown schematically in Fig. 1 has been observed at both the Vickers and spherical indentations made on {10$\bar{1}$0} and on {11$\bar{2}$0} planes and an example is shown in Fig. 3. The origin of this crack lies in the anisotropic nature of plastic flow in WC. During indentation on (10$\bar{1}$0), slip occurs on the two inclined {10$\bar{1}$0} planes, making a [0001] trace on the plane of indentation. The in-plane stresses normal to [0001] are readily relieved by plastic flow but relief of stress is not possible parallel to [0001] and the large tensile stress causes fracture. Inspection of Fig. 1 confirms that, with {10$\bar{1}$0}⟨1123⟩ slip, it is not possible to extend WC plastically along its *c* axis. Further, a formal strain tensor treatment [11] shows that there are only four independent slip systems in the {10$\bar{1}$0}⟨11$\bar{2}$3⟩ family rather than the five required for a general shape change in polycrystalline WC. This result has important consequences for WC–Co materials, as discussed below.

3. Intergranular fracture in polycrystalline WC–Co

The direct consequence of only four independent slip systems with the inability to extend WC in the *c* direction is that it is not possible to transmit slip across certain contiguous WC–WC boundaries. This leads directly to a mechanism to initiate intergranular microfracture and the WC–WC boundary configurations which are particularly susceptible can be readily proposed from the slip geometry.

The specific conditions are as follows.

(1) If the c axes of neighboring grains are nonparallel, there is a strain discontinuity across the boundary, leading to tensile (opening) stresses.

(2) The worst case (maximum tension) is when the c axes of neighboring grains are orthogonal and the c axis of one grain is normal to the boundary.

A further general requirement for fracture to occur preferentially at such boundaries is that the stress must be sufficiently high to exceed the fracture stress locally. Whereas this latter requirement does not follow directly from the considerations of limited slip systems, it arises implicitly because of the expected local inhomogeneity of strain in different carbide grains.

The two mechanisms which can initiate fracture in single crystals have been found to cause occasional transgranular carbide fracture in deformed WC–Co. However, intergranular fracture by the mechanism related to plastic incompatibility is much more common and the main cases are described in Section 4.

4. Analysis of grain boundaries in deformed WC–Co

More than 50 WC–Co grain boundaries in deformed WC–6wt.%Co and WC–10wt.%Co were analyzed by transmission electron microscopy. The details have been reported elsewhere [11, 13, 14] and only a few examples of the main types of boundary are considered here. Figure 4 illustrates a "worst-case" example, *i.e.* a mutual orientation of contiguous grains in which the c axes are orthogonal and the c axis of one grain is normal to the plane of the boundary. An intergranular crack has been initiated at a two-grain junction because of the local incompatibility. 35 such boundaries were analyzed in highly deformed samples and each was cracked as predicted. Similarly, many uncracked boundaries in nearby regions were also examined and it was found that they generally lay in orientations favoring transfer of slip.

Multigrain junctions are common in WC–Co and the need to accommodate strains in several neighboring grains increases the likelihood of finding orientations that cause a strain discontinuity. In Fig. 5, grains F and G are separated by a twist boundary about the c axis. According to the analysis presented here, this boundary should not crack, if this were a two-grain junction. However, the presence of the third grain, H, causes a strain discontinuity across the boundary with grain F and this discontinuity leads to the observed fracture.

Several different situations arise at junctions where a cobalt binder phase is present. It was generally found that, in multigrain junctions containing a compatible WC–WC boundary and neighboring cobalt pockets, both the carbide grains and cobalt pockets could be highly strained without fracture. It is likely that yielding in adjacent cobalt grains partially relieves some of the discontinuity difficulties arising from the limited number of slip systems in WC. In addition, very large amounts of strain can be accom-

Fig. 4. A cracked grain boundary containing a basal plane. The electron beam direction is close to [11$\bar{2}$0].

Fig. 5. Intergranular crack along a twist boundary between grains F and G. The crack initiated at the boundary between G and H (black arrow).

Fig. 6. Slip steps in WC adjacent to a cobalt pocket.

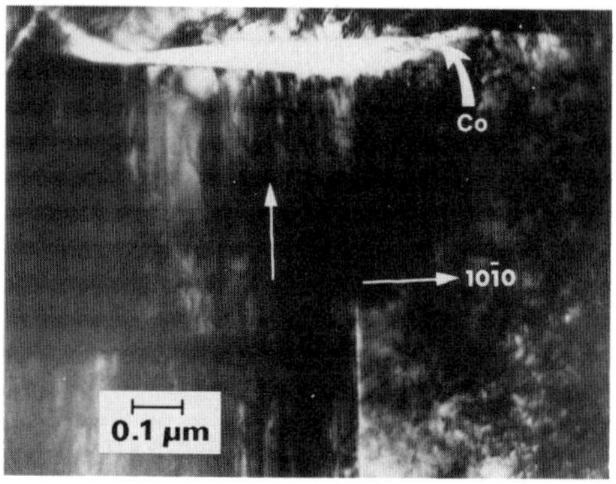

Fig. 7. Cracked (horizontal) WC–WC boundary containing cobalt and uncracked (vertical) boundary between contiguous carbide grains. The white vertical arrow denotes the *c* axis.

modated locally in the cobalt as illustrated in Fig. 6. The large slip steps in the WC grain are equivalent to about 5% plastic strain. Thus, local strains in deformed WC–Co can be expected to be much higher than the average strain, which should have important consequences for fracture initiation.

The presence of cobalt between WC grains and, where present, its effect on deformation and fracture has been the subject of much discussion. Our studies suggest that the presence of cobalt between WC grains does not ensure transfer of large strains if the WC grains are unfavorably oriented and the cobalt layer is relatively thin. A limited number of observations and an approximate calculation [14] suggest that cobalt layers 500 Å or less thick are insufficient to prevent fracture. Figure 7 shows an example of intergranular cracking at an unfavorably oriented WC–WC boundary containing cobalt, whereas an adjacent cobalt-free WC–WC compatible boundary remains uncracked even though both grains are heavily deformed.

5. Discussions and conclusions

In this paper, we reviewed evidence which points to limitations in plastic compatibility between contiguous WC grains as the main cause of the initiation of intergranular microfracture at such boundaries. The frequency of observation of this effect suggests that it could be the primary mode of failure initiation in WC–Co deformed under conditions of confined compression. It must be emphasized that these observations relate specifically to fracture nucleation, whereas macroscopic crack propagation clearly also involves fracture of cobalt ligaments. However, the major route to transfer strain is through cobalt pockets and the accommodation of large strains can be visualized as a percolation of slip through the cobalt along Co–WC–Co paths rather than over the more difficult route through the major WC constituent, in grades containing 6–10 wt.% Co. This view is consistent with the observation of local strain concentrations in the cobalt pockets and suggests the following possible sequence of events that lead to fracture in WC–Co.

(1) The applied stress exceeds the constrained yield stress for the cobalt binder phase which simultaneously transforms martensitically to the h.c.p. phase. Dislocations pile up in the cobalt setting up elastic stresses in the carbide which exceed the applied stress.

(2) As the applied stress is increased, the constrained yield stress of WC is exceeded in WC grains adjacent to deformed cobalt grains.

(3) Slip spreads across WC grains in the vicinity of cobalt.

(4) Slip transfers across favorably oriented WC–WC boundaries, but intergranular cracks form at other boundaries owing to plastic incompatibility.

(5) Additional slip systems operate in certain WC grains and transgranular cracks form at intersecting slip bands or owing to anisotropy of slip.

(6) Intergranular and transgranular cracks link up for macroscopic fracture.

Although most slip is probably transfered through the cobalt pockets, we emphasize the importance of WC–WC boundaries because this is the more difficult route which ultimately leads to failure in the low cobalt content grades. This view is consistent with the observation of Lea and Roebuck [19] who have shown that about 50% of the fracture propagates through WC–WC boundaries. As the cobalt content is increased, the difficulties due to only four independent slip systems may be overcome in microstructures containing sufficient WC–Co paths to circumvent incompatible WC–WC boundaries.

Acknowledgment

This work was supported by the National Science Foundation.

References

1 L. Lindau, in D. M. R. Taplin (ed.), *Fracture 1977, Proc. 4th Int. Conf. on Fracture, Waterloo, Ontario, June 19–24, 1977,* Vol. 2A, Pergamon, Oxford, 1977, p. 215.
2 J. L. Chermant and F. Osterstock, *J. Mater. Sci., 11* (1939) 1976.
3 J. R. Pickens and J. Gurland, *Mater. Sci. Eng., 33* (135) 1978.
4 M. Nakamura and J. Gurland, *Metall. Trans. A, 11* (141) 1980.
5 R. K. Viswanadham, T. S. Sun, E. F. Drake and J. A. Peck, *J. Mater. Sci., 16* (1981) 1029.
6 V. K. Sarin and T. Johannesson, *Met. Sci., 9* (1975) 472.
7 T. Takahashi and E. J. Friese, *Philos. Mag., 12* (1965) 1.
8 M. K. Hibbs and R. Sinclair, *Acta Metall., 29* (1981) 1045.
9 D. J. Rowcliffe, in R. K. Viswandham, D. J. Rowcliffe and J. Gurland (eds.), *Proc. 1st Int. Conf. on Science of Hard Materials*, Plenum, New York, 1983, p. 155.
10 T. Johannesson and B. Lehtinen, *Phys. Status Solidi A, 16* (1973) 615.
11 V. Jayaram, R. Sinclair and D. J. Rowcliffe, *Acta Metall., 31* (1983) 373.
12 M. K. Hibbs, R. Sinclair and D. J. Rowcliffe, *Acta Metall., 32* (1984) 941.
13 V. Jayaram, A. Kronenerg, S. H. Kirby, D. J. Rowcliffe and R. Sinclair, *Scr. Metall., 20* (1986) 701.
14 V. Jayaram, *Acta Metall., 35* (1987) 1307.
15 D. J. Rowcliffe, C. D. Craig and T. Cooper, in J. E. Field and J. P. Dear (eds.), *Proc. 7th Int. Conf. on Erosion by Liquid and Solid Impact, Cambridge, 1987,* Cavendish Laboratory, Cambridge, 1987, p. 59-1.
16 S. B. Luyckx, *Acta Metall., 18* (1970) 223.
17 S. Hagege, J. Vincens, G. Nouet and P. Delavignette, *Phys. Status Solidi A, 61* (1980) 675.
18 E. A. Almond, in R. K. Viswanadham, D. J. Rowcliffe and J. Gurland (eds.), *Proc. 1st Int. Conf. on Science of Hard Materials*, Plenum, New York, 1983, p. 517.
19 C. Lea and B. Roebuck, *Met. Sci., 15* (1981) 263.

Finite Element Modelling of Crack Propagation in WC–Co Hard Metals*

H. F. FISCHMEISTER, S. SCHMAUDER and L. S. SIGL

Max-Planck-Institut für Metallforschung, D-7000 Stuttgart (F.R.G.)

(Received January 13, 1988)

Abstract

According to a model which yields crack resistance values in good agreement with the experiments of Sigl and Fischmeister, crack propagation in WC–Co is opposed by the plastic stretching of binder ligaments and by the fracture energy of the carbides surrounding them. The size of the plastic zone is less than the mean free path in the binder, but very large strains occur in the ligaments. In contrast with this, a finite element model (FEM) study by Ljungberg et al. indicated a very large plastic zone. The present work shows that FEM calculations with a sufficiently fine net will model the localized deformation of the binder at the edges of the carbide cracks in good agreement with our experimental observations, confirming the proposed mechanism of ligament formation and crack propagation.

1. Introduction

Many models have been suggested for explaining the fracture toughness of cemented carbides. They have been reviewed by two of the present authors [1, 2]. Two problems are central to all approaches: (i) the precise quantitative determination of the different types of fracture paths in the structure; (ii) the size of the plastic zone.

Qualitatively, it has long been understood that the size of the plastic zone must be related to the geometry of the surrounding carbide structure or, in other terms, to the size of the binder phase regions, which is most usually represented by the mean free path in the binder.

Recently, we presented a model for the fracture toughness of cemented carbides [2–4] based on combining the specific energies of the four possible fracture paths in the ratios in which they occur in the fracture surface:

$$R = A_A{}^{C}\gamma^{C} + A_A{}^{C\text{-}C}\gamma^{C\text{-}C} + A_A{}^{B}\gamma^{B} + A_A{}^{B\text{-}C}\gamma^{B\text{-}C} \quad (1)$$

where γ is the fracture energy per unit area of crack and A_A is the area fraction, with indices as follows: C represents transgranular fracture in the carbide, C–C intergranular carbide fracture, B transgranular fracture in the binder and B–C fracture along the binder–carbide interface.

Fracture in the binder (type B) is of the ductile (dimple) type, and the same is true for type B–C where the crack runs parallel and close to the carbide–binder phase boundary. Decohesion at the phase boundary itself was never observed [5]. Paths B and B–C are responsible for the major part of the total fracture energy. Their specific energies were calculated from the plastic work expended in creating a dimple zone, using appropriate values of the *in situ* yield stress and work hardening of the thin binder layers [4].

In contrast with the assumptions of earlier models [6–8], microscopy observation of arrested cracks as well as measurements of characteristic distances in the microstructure before and after passage of the crack [2, 5] showed the plastic zone to be smaller than the mean free path in the binder. Using empirically determined area fractions and fracture energies for the carbide paths, and using the specific work of binder fracture calculated on the basis of this small plastic zone [4], a fully quantitative account could be given of the fracture toughness of WC–Co hard metals encompassing the whole range of industrial alloys [2, 3].

A finite element study of the deformation of WC–Co at a crack tip presented by Ljungberg *et al.* [9] indicated a plastic zone several times the size of the mean free path, which is much larger than the size indicated by the observations and calculations reviewed above. In a discussion of the paper of Ljungberg *et al.*, it was suggested [10]

*Paper presented at the 3rd International Conference on the Science of Hard Materials, Nassau, The Bahamas, November 9–13, 1987.

0921-5093/88/$3.50 © Elsevier Sequoia/Printed in The Netherlands

that the large plastic zone might be a consequence of the coarse finite element net used, and it was decided to carry out a finite element model (FEM) study with a finer net. The results are presented here.

2. The finite element model

The process zone of arrested cracks in WC–Co alloys has been found [2, 3, 5] to be of the multiligament type first described by Evans *et al.* [11]. The crack at first extends throughout the carbide phase, forming a continuous area in which the binder regions remain as isolated ligaments. As the crack opens, the ligaments are stretched until they rupture by dimple fracture. Figure 1 shows a portion of a crack in WC–Co at a stage where the ligaments have already been severed. Extensive plastic deformation is seen in the binder, but only in the immediate vicinity of the crack.

The finite element calculations were performed with a two-dimensional model (Figs. 2 and 3) copied from an actual micrograph taken after the passage of a crack (with correction for the crack extension), so that the crack paths were known. Only slight simplifications were made in adapting the finite element net (Fig. 3(a)) to the structure. In setting up the initial situation for the FEM calculations, intergranular cracks were assumed to have already formed in the carbide along B–C, D–E and F–R, creating a multiligament zone L1–L2–L3. The microstructural parameters of the model are given in Table 1. The discretized region is embedded in a homogeneous surrounding (Fig. 3(b)) with the average elastic properties of a WC–Co alloy of appropriate composition [12]. A crack has entered this surrounding and the discretized area from the left and has come to rest at *R*. Load is incrementally applied in the vertical direction in Fig. 3. It is

Fig. 1. Crack trace on the surface of a 10wt.%(16.4vol.%) WC–Co alloy with a coarse microstructure. Detectable plastic deformation is restricted to the ligaments.

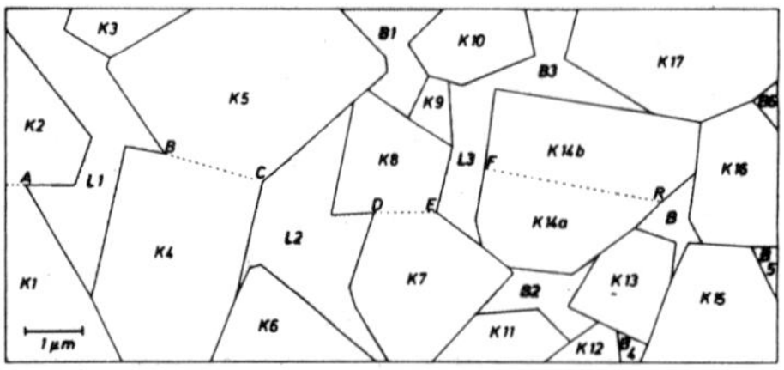

Fig. 2. Schematic view of the microstructure shown in Fig. 1 (uncracked): L, cobalt ligaments; B, binder region at the alloy crack tip; B1–B6, other binder regions; K, WC-grains; R, alloy crack tip; D–F, "local" crack tips.

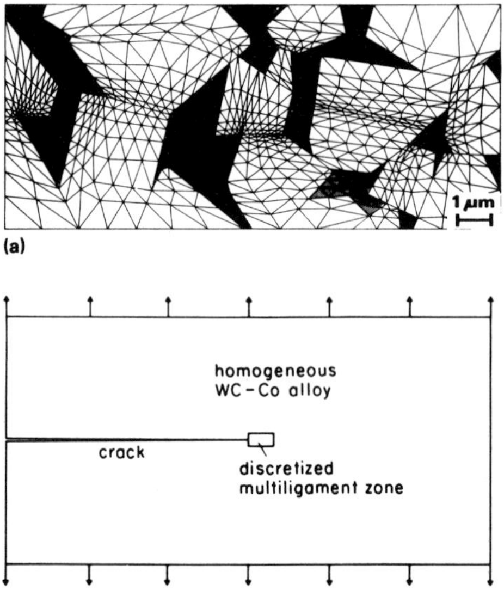

Fig. 3. The FEM: (a) net of the multiligament zone in which the bright areas represent carbide grains, and the binder areas are shaded; (b) the discretized region is embedded in a homogeneous surrounding (loading condition indicated).

TABLE 1 **Microstructural parameters**

	Volume fraction of binder (%)	*Mean carbide intercept length* (μm)	*Mean binder intercept length* (μm)
Model	22.8	1.98	0.96
Bulk alloy	16.4	2.18	0.85

evenly distributed over the horizontal boundaries of the homogeneous surrounding of the discretized region. The vertical boundaries remain unconstrained.

The deformation of the ligaments (L1–L3) and elsewhere in the binder (B1–B6) was modelled using von Mises' criterion with the associated rule of flow perpendicular to the yield surface. Strain hardening was described by Voce's equation

$$\sigma(\phi) = \sigma_y + (\sigma_S - \sigma_y)\left\{1 - \exp\left(\frac{-\phi}{\phi_c}\right)\right\} \tag{2}$$

where $\sigma(\phi)$ is the true stress, ϕ is the true strain, σ_y is the yield stress (which takes into account solid solution strengthening by dissolved tungsten and carbon and the limitation of slip length by the adjacent carbide crystals), σ_s is a saturation stress and ϕ_c is a characteristic strain which relates to the initial strain hardening of the material. The parameters to be adopted for eqn. (2) have been discussed against the background of available experimental data and hardening mechanisms in ref. 4. Table 2 summarizes the parameters used to describe the elasto-plastic behaviour of the binder phase.

The multiligament zone is subdivided into 1534 elements with typically more than 50 elements per ligament (Fig. 3). Calculations were performed for plane strain and plane stress loading using the small-strain finite element program ASKA [13] to simulate crack propagation at the surface and in the bulk respectively. Good convergence was obtained with less than 8 iterations per loading step.

3. Results

On loading in plane stress, the elastic–plastic boundary in the binder spreads as shown in Fig. 4. Initially, small-scale yielding plasticity is observed ahead of local crack tips in all ligaments. As the fracture load is approached, the plastic zone spreads across ligament L1 and a fully plastic band connecting the crack tips A and B develops, while ligaments closer to the outermost crack tip are still in the small-scale yielding regime. It should be noted that, at fracture load, most parts of the ligaments and all binder regions remote from the potential crack path are still in the purely elastic regime.

Plasticity as characterized by the equivalent plastic strain at fracture load is predominantly concentrated to the local crack tips at ligament L1 while ligaments L2 and L3 suffer less plastic deformation (Fig. 5). There is a general trend, however, for all ligaments to develop plasticity on a line linking opposite local crack tips in the WC skeleton, which leads to the formation of a band-like plastic zone approximately normal to the external load. This tendency is more pronounced where deformation is not constrained by the local geometry of carbide grains. It is seen from Fig. 5 that some carbide crystals (*e.g.* K2, K4 and K8) truncate part of the plastic zone that would have developed in their absence.

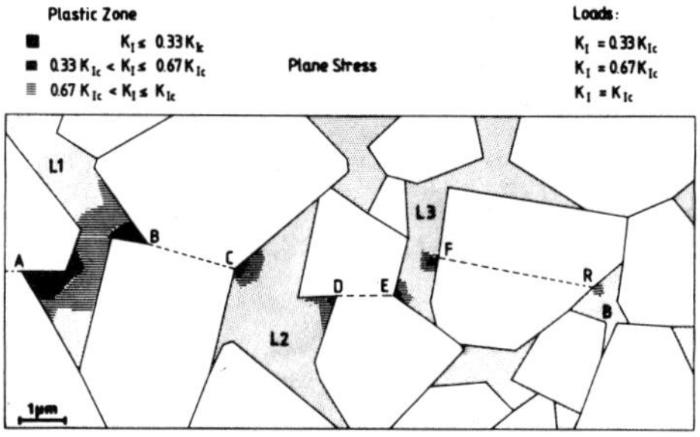

Fig. 4. Evolution of the elastic–plastic boundary in plane stress for three stress intensity levels. A plastic band forms across ligament L1, as the fracture load is approached.

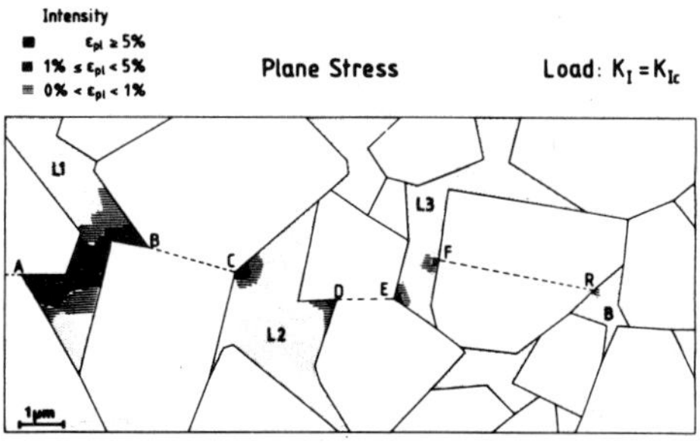

Fig. 5. Distribution of equivalent plastic strains in plane stress at fracture load.

TABLE 2 Elastic constants of binder, carbide and composite

	E (GPa)	ν
WC	714	0.194
Co	211	0.31
WC–10wt.%Co	595	0.22

The intensity distribution of equivalent plastic strain at fracture load is shown in Fig. 6 for ligament L1. Plastic strains higher than 5% are restricted to the local crack tips A and B at phase boundaries orthogonal to the direction of applied load.

In contrast with plane stress conditions, plane strain loading under a von Mises-type flow criterion forces plasticity to spread along carbide–binder interfaces approximately parallel to the applied load. This is evident from the shape of the elastic–plastic boundary (Fig. 7). Hence the plastic zone is truncated more severely by interfaces normal to the applied load than in plane stress. Furthermore, the gradient of equivalent plastic strain ahead of the local crack tips is steeper and regions with plastic strains of more than 10% are confined to phase boundaries approximately parallel to the applied load near the local crack tips A and B in ligament L1 (Fig. 8). Owing to additional constraint, the size of the plastified region is smaller than in plane stress. In either case, the mean size of the plastic zone does not exceed the mean intercept length in the binder.

As will be shown below, the degree of triaxiality of the stress state, defined [14] as

$$\zeta = \frac{\sigma_H}{\sigma_M} \tag{3}$$

where σ_H and σ_M represent the hydrostatic and the von Mises stress respectively, is important for the formation of voids in the ligament. Figure 9 shows a map of this quantity in ligament L1 at

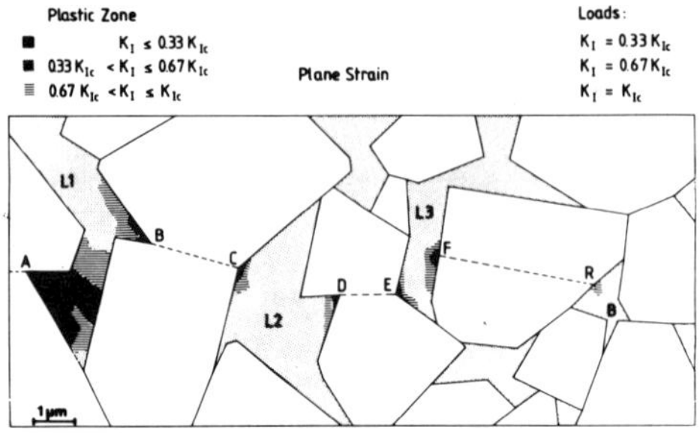

Fig. 7. Elastic–plastic boundary in plane strain loading.

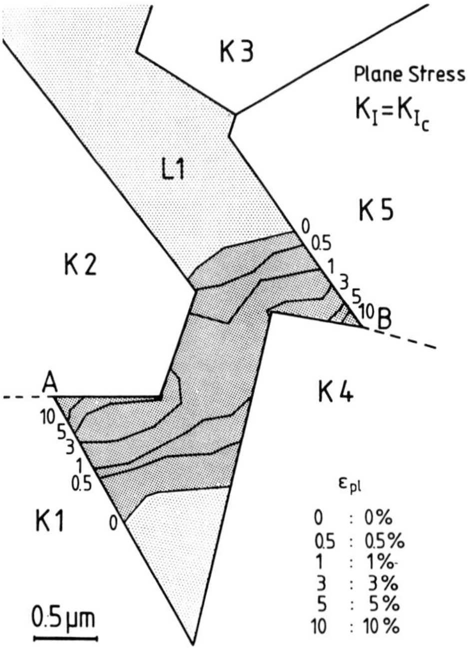

Fig. 6. Equivalent plastic strains in plane stress at fracture load in ligament L1. Large plastic deformations are localized at interfaces orthogonal to the direction of applied load.

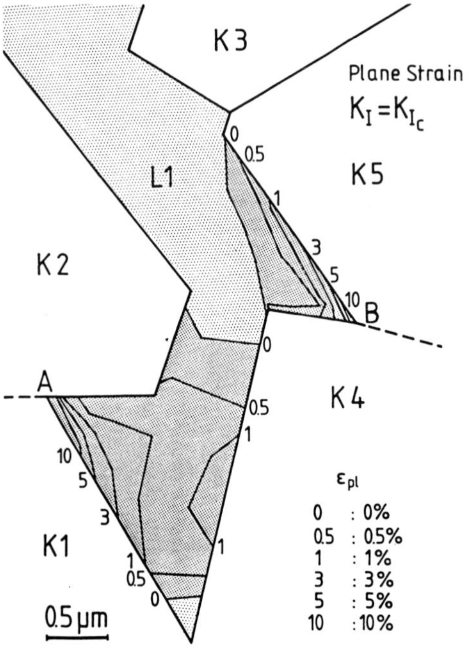

Fig. 8. Equivalent plastic strains in ligament L1 in plane strain loading at fracture stress. Large plastic deformations are concentrated at interfaces parallel to the direction of applied load.

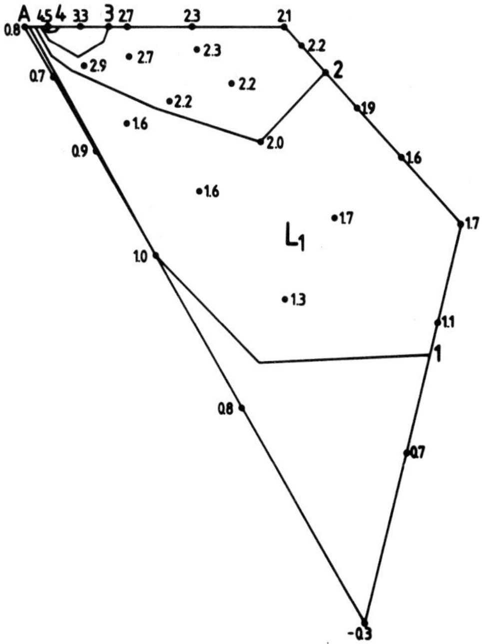

Fig. 9. Stress triaxiality at fracture load ζ in the lower part of ligament L1 under plane strain.

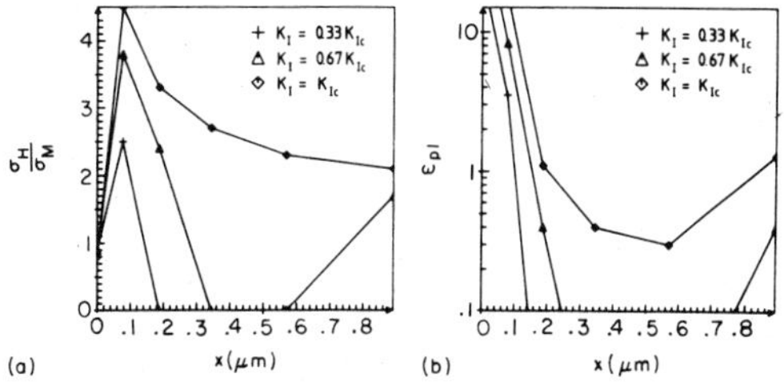

Fig. 10. (a) Stress triaxiality ζ; (b) equivalent plastic strain ε_{pl} at different loading steps in plane strain ahead of local crack tip A.

fracture load. Maximum triaxiality is observed on planes approximately normal to the applied load. This trend is most pronounced in L1 along the phase boundary ahead of crack tip A, but local maxima ahead of all local crack tips are observed throughout the multiligament zone. Hydrostatic stresses drop rapidly in the direction of the applied load, giving rise to inverse trends for triaxiality and plastic strain; while the former concentrates predominantly along interfaces nearly parallel to the applied load, the latter tends to be small in such regions. The variation in triaxiality and equivalent plastic strain along the interface ahead of crack tip A is plotted in Fig. 10 for three different load levels. It is seen that plastic strain falls off rapidly with increasing distance from the crack tip while the hydrostatic stress exhibits a pronounced maximum which increases on loading.

4. Discussion

The present study of crack growth in a multiligament zone is exploratory and has obvious limitations. The restriction to small-strain analysis implies that the results are pertinent only to the initial stages of ligament plastification. As soon as voids begin to form in the ligaments, the stress will be redistributed, and this brings the applicability of the present model to an end. However, the absence of detailed investigations on the mechanics of bridging zones makes it worthwhile to study the simplified and limited model presented here. Its mesh size is still too coarse to supply precise near tip solutions, but good predictions can be expected of the intermediate and far-field stresses and displacements and of the depth of plastification in the ligaments [15]. A finer net would not produce any significantly different results.

Confidence in the results is engendered by comparing the calculated elastic–plastic boundary in plane stress (Fig. 4) with scanning electron microscopy observations of the crack at the free surface (Fig. 1). The calculated limits of the plastic zone agree well with the observation [5] that plasticity is confined to the ligaments along a narrow band around the expected fracture path and that non-bridging binder regions deform purely elastically.

The good agreement for the plane stress case suggests that the results of the plane strain calculations should be significant as well and that they may serve as pointers to the way in which the ligaments fail. Fractography provides ample evidence that ligaments subject to plane strain fail by microvoid coalescence (Fig. 11). The resulting dimple population falls into two categories: (i) widely spaced deep dimples where the crack proceeds across a ligament; (ii) shallow and closely spaced dimples where the crack runs parallel and close to the carbide–binder interface.

In the absence of inclusions, the nucleation of voids in the binder will most likely occur at stacking fault intersections [2, 3, 5] in the highly plastified regions ahead of local crack tips. Substantial

Fig. 11. Dimpled fracture surface in a WC–10wt.%Co alloy. Deep dimples develop during fracture across a ligament; shallow dimples characterize crack paths parallel and close to binder–carbide interface.

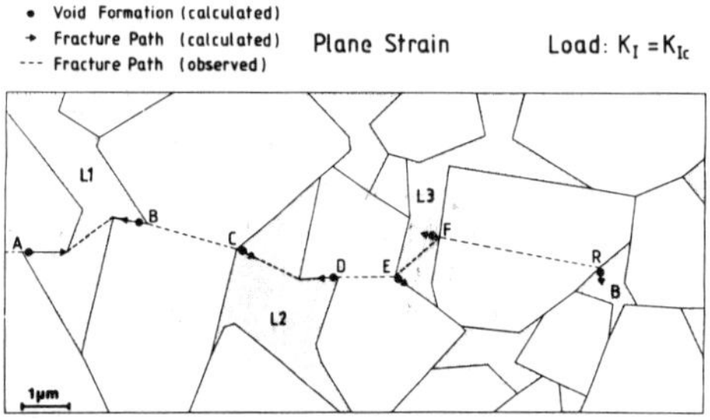

Fig. 12. Expected void formation sites are indicated by black bullets; the arrows indicate the most probable direction of crack advance through the ligament. The length of the arrows is scaled to the stress triaxiality ζ.

growth of such voids can only occur in regions of high hydrostatic stress [16]. Coincidence of high plastic strain and maximum triaxiality will therefore mark the sites where microvoid fracture may be expected to start.

The regions which satisfy both these conditions are marked with bullets in Fig. 12. As expected, they are all found where the carbide crack is about to enter the ligaments. It should be noted that maximum triaxiality is found slightly ahead of the local crack tip A, inside the ligament L1 (Figs. 9 and 10). This is, in fact, the place where void formation would be intuitively expected to start, and one would expect the first void to be followed by others along the plane linking the adjacent cracks in the carbide phase. This would give rise to a "perforation-tearing" manner of crack propagation where void formation and coalescence progresses across the ligament in the general macroscopic direction of the crack, as visualized in Fig. 13.

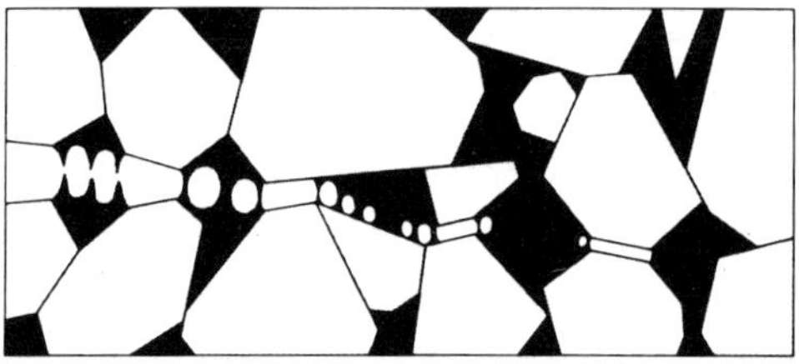

Fig. 13. New model for crack propagation in the multiligament zone; void growth increases with increasing distance from the matrix crack tip in each ligament.

TABLE 3 Hardening parameters of the constitutive law for the cobalt binder phase

	σ_y (GPa)	σ_s (GPa)	ϕ_c
Co	2.19	12.19	0.63

Although this process can no longer be described by the present model because of the redistribution of stress connected with the first voids, it is possible to extract trend indications for the likely direction of crack propagation by considering the distribution of high plastic strain and high triaxiality beyond the probable site of the first void. These directions are labelled by arrows in Fig. 12; the lengths of the arrows correspond to the degree of triaxiality as defined in eqn. (3). It should be noted that triaxiality is particularly intense close to the carbide–binder interfaces (Fig. 9). This suggests a qualitative explanation of the observed fine dimple structure of the C–B crack path segments (Fig. 11).

The behaviour of the FEM is, of course, strongly influenced by the constitutive law assumed for the binder phase, whose parameters are stated in Table 3. The good agreement of the FEM predictions with the microscopic observations [5] confirms the plausibility of the parameters, and this strengthens the arguments used in connection with the calculations of the work of fracture in the binder paths [4]. This confirmation is valuable because the *in situ* properties of the binder phase are not accessible by direct experiment.

When comparing the present results with those of the FEM study by Ljungberg *et al.* [9], it should first be noted that in that study the calculation of crack-opening displacement was based on the Westergaard solution [9] which is valid only for homogeneous materials. Secondly, Ljungberg *et al.* modelled the binder regions by only a few net elements, and the disposition of the elements ahead of the crack was such that highly localized strain concentrations such as those shown in the present study could not have been resolved. Thirdly, the yield stress assumed for the binder was only about one-half of the value adopted for the present study, because the large amount of binder strengthening by the densely spaced rigid B–C boundaries was not included. Also, the effect of strain hardening was assumed to be smaller than in our work. All these factors tend to produce larger values for the plastic zone size.

5. Conclusions

(1) The plastic zone size in the binder phase extends only through part of the binder regions directly intersected by the crack (*i.e.* those binder regions which will be stretched into ligaments when the crack opens). This is in agreement with microscopy observations [3, 5] and with calculations of the work of fracture in the binder path [4].

(2) Void formation in the binder can be predicted by a criterion combining intense strain and a large hydrostatic–deviatoric stress (stress triaxiality).

(3) The FEM suggests that the ligaments will fail by nucleation and growth of pores beginning at the points where the crack in the carbide enters a ligament and followed by the progress of void formation and coalescence across the ligament in the direction of macroscopic crack propagation in a manner similar to the tearing of a perforated sheet.

(4) The observed difference in dimple spacing between crack path segments across binder regions and those segments where the crack runs parallel and close to the carbide–binder interface can be understood on the basis of increased stress triaxiality in the zone adjacent to the interface.

References

1 H. Fischmeister, in R. K. Viswanadham, D. J. Rowcliffe and J. Gurland (eds.), *Science of Hard Materials*, Plenum, New York, 1983, p. 1.
2 L. S. Sigl, *Dissertation*, in *Fortschrittsber. VDI (Reihe 5) 104*, 1986 (Verein Deutscher Ingenieure, Düsseldorf).
3 L. S. Sigl, H. E. Exner and H. F. Fischmeister, *Proc. 2nd Int. Conf. on the Science of Hard Materials*, in *Inst. Phys. Conf. Ser., 75* (1986) 631.
4 L. S. Sigl and H. F. Fischmeister, *Acta Metall.*, (1988). in the press.
5 L. S. Sigl and H. E. Exner, *Metall. Trans. A, 18* (1987) 1299.
6 M. Nakamura and J. Gurland, *Metall. Trans. A, 11* (1980) 141.
7 T. Johannesson, *Proc. 4th Eur. Symp. on Powder Metallurgy, Grenoble, 1975*, Societé Française de Métallurgie, Paris, 1975, Paper 5-11, p. 1.
8 J. L. Chermant and F. Osterstock, *J. Mater. Sci., 11* (1976) 1939.
9 A.-B. Ljungberg, C. Chatfield, M. Hehenberger and B. Sundström, *Proc. 2nd Int. Conf. on the Science of Hard Materials*, in *Inst. Phys. Conf. Ser., 75* (1986) 619.
10 H. Fischmeister, *Proc. 2nd Int. Conf. on the Science of Hard Materials*, in *Inst. Phys. Conf. Ser., 75* (1986) 725.
11 A. G. Evans, A. H. Heuer and D. L. Porter, in D. M. R. Taplin (ed.), *Proc. 4th Int. Conf. on Fracture, Waterloo, Ontario, June 19–24, 1977*, Pergamon, Oxford, 1977, p. 529.
12 H. Doi, Y. Fujiwara, K. Miyake and Y. Oosawa, *Metall. Trans., 1* (1970) 1417.
13 J. H. Argyris and H. Kamel, *Res. Rep. 8*, 1965 (Institut für Statik und Dynamik, University of Stuttgart).
14 G. Rousselier, in S. Nemat-Nasser (ed.), *Three-dimensional Constitutive Relations and Ductile Fracture*, North-Holland, Amsterdam, 1981, p. 331.
15 J. R. Rice and A. M. Tracey, in S. J. Fenves, N. Perrone, A. R. Robinson and W. C. Schonobrich (eds.), *Numerical and Computer Methods*, Academic Press, New York, 1973, p. 585.
16 F. A. McClintock, *J. Appl. Mech., 35* (1968) 363.

Internal Friction in WC–Co Hard Metals*

R. SCHALLER and J. J. AMMANN

Institut de Génie Atomique, Ecole Polytechnique Fédérale de Lausanne, PHB-Ecublens, CH-1015 Lausanne (Switzerland)

C. BONJOUR

Stellram SA, CH-1260 Nyon (Switzerland)

(Received November 9, 1987)

Abstract

Internal friction measurements have been performed on various grades of WC–Co cemented carbides from room temperature to 1000 °C. These composite materials exhibit an internal friction spectrum which is mainly composed of a relaxation peak and a high temperature exponential background. The peak appears in the same temperature range where an increase in toughness has been observed and interpreted as being due to a brittle-to-ductile transition of the material. The exponential background can be associated with high temperature creep phenomena. The analysis of the results obtained shows the important role of the cobalt binder phase in the mechanical behaviour of WC–Co.

1. Introduction

WC–Co cemented carbides, used as cutting tools, are subjected to severe stress conditions at relatively high temperatures (800–1000 °C). During machining, the tool lifetime can be limited by various wear mechanisms such as abrasion, delamination, chemical dissociation and dissolution [1], but also by deficiency in the mechanical properties such as toughness or creep resistance [2]. It is important to obtain a better understanding of the fundamental mechanisms which are responsible for the mechanical properties, such as toughness, of these composite materials.

For simplicity, WC–Co can be considered as a two-phase material: WC hard particles and cobalt binder phase. It is evident that the mechanical properties of such a composite will strongly depend on the volumetric fraction and the morphology of the two phases [3, 4]. They are also greatly affected by structural defects such as dislocations, twin boundaries and point defects, which can play an important role in the plasticity of each phase [5, 6] or in grain boundary sliding [7].

Internal friction is a parameter which is very sensitive to the mobility of structural defects [8]. It results from material anelasticity. Anelastic deformation is due to the reversible movements of structural defects, as dislocations, around their equilibrium position. Such movements are generated by small applied stresses which do not give rise to any modification of the microstructure. Then, because of the very low strain amplitude investigated (10^{-8}–10^{-4}), internal friction can be considered as a non-destructive technique well suited to the study of the mechanisms of microplastic deformation which are responsible for the mechanical properties of hard and brittle materials exhibiting a limited plasticity.

In this sense, internal friction has been used to study the microscopic mechanisms which may be responsible for the mechanical behaviour of WC–Co cemented carbides from room temperature to 1300 K. The non-destructive character of the technique has revealed itself as very advantageous for measuring brittle specimens in a reproductive manner under various experimental conditions.

2. Internal friction technique

Anelasticity of solids [8, 9] occurs in the following simple experiment (Fig. 1(a)). A stress σ, of low intensity, is applied abruptly to a specimen

*Paper presented at the 3rd International Conference on the Science of Hard Materials, Nassau, The Dahamas, November 9–13, 1987.

0921-5093/88/$3.50 © Elsevier Sequoia/Printed in The Netherlands

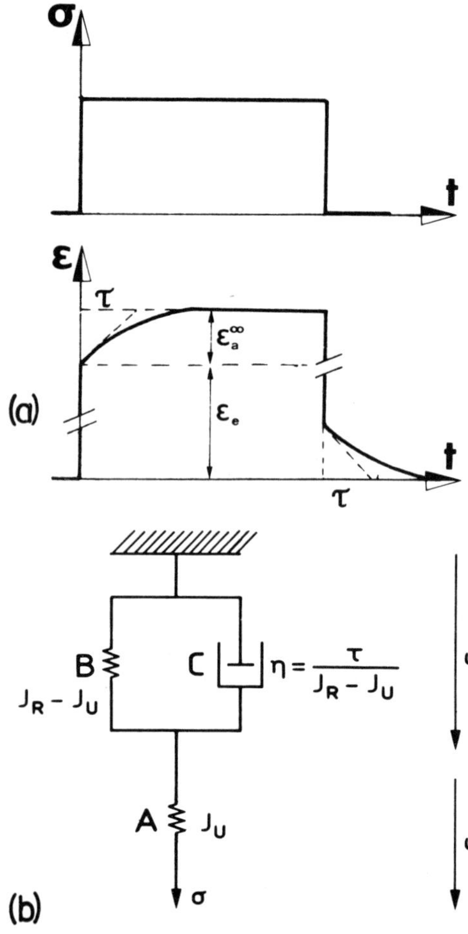

Fig. 1. Anelasticity of a solid: (a) elastic and anelastic strain ε of a solid submitted to stress σ; (b) rheological model corresponding to the anelastic behaviour in (a).

at time $t=0$ and held constant while the strain ε is recorded as a function of time. We observe the instantaneous elastic strain $\varepsilon_e = J_u\sigma$ (where J_u is the unrelaxed compliance) and the anelastic strain ε_a which increases with time from zero to an equilibrium value ε_a^∞. When equilibrium is reached,

$$\varepsilon = \varepsilon_e + \varepsilon_a^\infty$$
$$= J_r\sigma$$

where J_r is the relaxed compliance.

This evolution of the solid from one equilibrium state to a new state, under a mechanical applied stress, is called anelastic relaxation and may be defined by two parameters: the relaxation intensity

$$\Delta = \frac{\varepsilon_a^\infty}{\varepsilon_e} = \frac{J_r - J_u}{J_u}$$

and the relaxation time τ.

The total strain ε can be written as a function of time:

$$\varepsilon = J_u\sigma + (J_r - J_u)\sigma\left\{1 - \exp\left(-\frac{t}{\tau}\right)\right\} \tag{1}$$

If the stress is removed after a certain delay time t, we can observe the instantaneous recovery of the elastic strain and with the same relaxation time τ the recovery of the anelastic strain. In such an experiment, strain is completely recoverable.

This behaviour is representative of the standard anelastic solid, the equation of which can be derived from eqn. (1):

$$\varepsilon + \tau\dot{\varepsilon} = J_r\sigma + \tau J_u\dot{\sigma} \tag{2}$$

From a rheological point of view, the solid can be represented by the model shown in Fig. 1(b). The spring B in parallel with the dashpot C is essential to the recoverable nature of the anelastic strain. If this spring does not operate, creep behaviour is observed; anelastic strain does not reach an equilibrium value and is then no longer completely recoverable.

From a microscopic point of view, anelastic strain can be interpreted as being due to the movements of structural defects (elastic dipoles or dislocations) from one equilibrium position defined at $\sigma=0$ to another defined at $\sigma \neq 0$. The relaxation intensity Δ is then proportional to the concentration of defects which are relaxing, the relaxation time accounting for their mobility.

Measurements performed in the way shown in Fig. 1 are very delicate. Effectively, if the elastic strain amplitude is about 10^{-6}, anelastic strain values of 10^{-9}–10^{-8} must be obtained in order to detect anelastic relaxation phenomena in metals. For this reason, it is more convenient to use dynamic methods in practice for measuring the relaxation parameters.

In this case, an alternative stress $\sigma = \sigma_0 \exp(i\omega t)$ is applied to the system. The linearity of the stress–strain relationship ensures that strain is periodic with the same frequency:

$$\varepsilon = \varepsilon_0 \exp\{i(\omega t - \delta)\}$$

where δ is the phase lag of strain behind stress due to anelasticity.

Introducing these expressions for σ and ε into eqn. (2) leads to the following relationship:

$$\varepsilon = J^*(\omega)\sigma$$
$$= \{J_1(\omega) - iJ_2(\omega)\}\sigma \qquad (3)$$

where J_1 and J_2 respectively, are the real and the imaginary part of the complex compliance J^*:

$$J_1 = J_u + \frac{J_r - J_u}{1 + \omega^2\tau^2}$$

$$J_2 = (J_r - J_u)\frac{\omega\tau}{1 + \omega^2\tau^2} \qquad (4)$$

The internal friction Q^{-1} of the material is related to the loss angle δ by

$$Q^{-1} = \tan\delta = \frac{J_2}{J_1} \qquad (5)$$

Introducing eqn. (4) into eqn. (5) and assuming that $J_r - J_u \ll J_u$ gives

$$Q^{-1} = \Delta\frac{\omega\tau}{1 + \omega^2\tau^2}$$

$$\frac{\delta J}{J} = \frac{J_1(\omega) - J_u}{J_u} \qquad (6)$$

$$= \Delta\frac{1}{1 + \omega^2\tau^2}$$

where $\delta J/J$ is the variation of the dynamic modulus due to the anelastic relaxation.

Equations (6) show that the internal friction Q^{-1} has a maximum as a function of $\omega\tau$, centred at $\omega\tau = 1$. This internal friction peak gives us the relaxation intensity Δ (the height of the peak) and the relaxation time τ (the position of the peak on the $\omega\tau$ axis). If several independent relaxation mechanisms are activated, the superposition principle allows us to write

$$Q^{-1} = \sum_i \Delta_i \frac{\omega\tau_i}{1 + \omega^2\tau_i^2} \qquad (7)$$

where i refers to the mechanism. The material then exhibits an internal friction spectrum composed of internal friction peaks due to the various microscopic relaxation mechanisms.

When the relaxation mechanisms are thermally activated (the viscosity of the dashpot in Fig. 1(b) depends on the temperature), $\tau = \tau_0 \exp(E/kT)$, and it is possible to measure the internal friction as a function of temperature, keeping ω constant. Then,

$$Q^{-1} = \sum_i \Delta_i \left[2\cosh\left\{\frac{E_i}{k}\left(\frac{1}{T_{pi}} - \frac{1}{T}\right)\right\}\right] - 1 \qquad (8)$$

Relaxation i gives rise to a peak located at temperature T_{pi} (defined by $\omega\tau_0\exp(E_i/kT_{pi}) = 1$, where E_i is the activation energy), and the height of which is $\frac{1}{2}\Delta_i$. Therefore, we obtain an internal friction spectrum as a function of the temperature, which is characteristic of the material.

3. Experimental procedure

For metals, $\tan\delta$ (see eqn. (5)) is small and therefore it is generally difficult to measure the phase lag between strain and stress directly. In the low frequency range, it is easier to use resonant systems which vibrate at a natural frequency. The internal friction, Q^{-1} is deduced from the free decay of the vibrations. When $\tan\delta$ is small, it is possible to show that

$$Q^{-1} = \tan\delta$$

$$= \frac{1}{n\pi}\ln\left(\frac{A_i}{A_{i+n}}\right) \qquad (9)$$

where A_i and A_{i+n}, respectively, are the ith and the $(i+n)$th amplitudes in the free decay of the vibrations.

In the present case, the internal friction measurements were performed by means of an inverted torsion pendulum working in the low frequency range (0.2–5 Hz) under vacuum (10^{-3} Pa).

The specimens used for the measurements were sheets of size 0.5 mm × 2.5 mm × 110 mm. They were cut by spark machining into bars of size 2.5 mm × 20 mm × 110 mm. These initial bars of WC–Co, with varying amounts of cobalt and an initial carbide grain size of 3.6 μm, were then sintered.

The measurements of the internal friction and the dynamic elastic modulus were performed automatically during heating and cooling at a constant rate of 2 K mm^{-1}. The temperature range investigated was between room temperature and 1300 K. The maximum strain amplitude was approximately 5×10^{-6}.

4. Results

WC–Co cemented carbides exhibit a characteristic internal friction spectrum as a function of temperature (Fig. 2) [10]. It is mainly composed of a peak located at about 950 K in WC–11wt.%Co (vibration frequency, about 2.6 Hz). In the temperature range lower than the peak, *i.e.* from room temperature to 700 K, the internal friction is weak but constant. In the temperature range above the peak, the internal friction is high (approximately 15×10^{-3} at about 1100 K) and increases exponentially with the temperature. This part of the internal friction spectrum will be referred to as "the exponential background".

Associated with the peak, as well as with the exponential background, we observe anomalies in the decrease in the vibrational frequency as a function of temperature. These anomalies are due to the effects of the anelastic relaxations on the dynamic modulus (eqn. (6)).

The internal friction peak is effectively a relaxation peak, because it shifts in temperature when the frequency is modified. The activation energy, deduced from the Arrhenius plot, is 2.7 eV, which is very close to the self-diffusion energy in cobalt.

In contrast, the internal friction (mainly the exponential background component) increases with increasing cobalt content (Fig. 3). In addition, the relaxation peak shifts towards lower temperatures. These results tend to show that the internal friction phenomena, peak and exponential background, are associated with mechanisms which take place in the cobalt binder phase.

In order to localize more precisely the relaxation mechanisms, specimens of pure WC and pure cobalt were prepared. WC samples were sintered without binder at 2000 °C. Cobalt samples were obtained by melting cobalt industrial powders in an alumina crucible.

The internal friction results obtained from these specimens are compared to the spectrum for WC–11wt.%Co in Fig. 4. It clearly appears that the relaxation peak in WC–Co (curve a) is not due to a microscopic mechanism in the WC particles or at the WC–WC interfaces. The internal friction of pure sintered WC presents only a slight monotonic increase in this temperature range (curve b). Also, the peak is not due to the allotropic phase transition of cobalt which is responsible for the sharp peak located at about 700 K [11] (curve c). Pure cobalt exhibits a similar increase in the internal friction at high temperatures as WC–Co, but no significant peak appears in the temperature range at around 900 K, although the change in slope at about 1100 K is

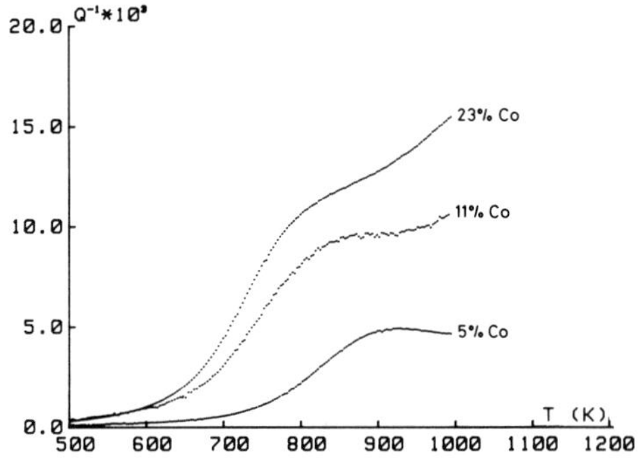

Fig. 3. Effects of the cobalt content on the internal friction of WC–Co cemented carbides.

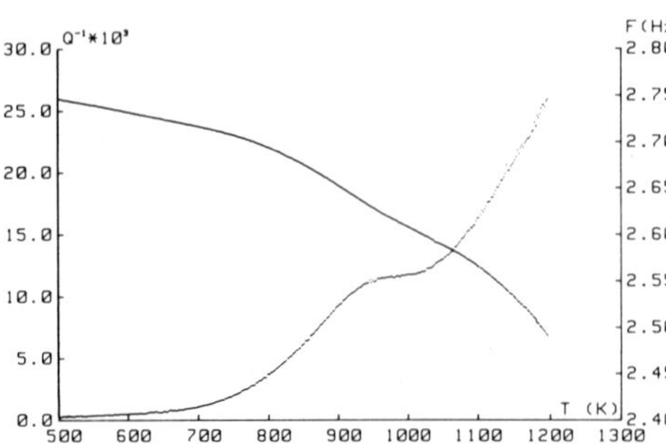

Fig. 2. Internal friction Q^{-1} and oscillation frequency F (giving approximately the dynamic elastic modulus) of WC–11wt.%Co as a function of temperature.

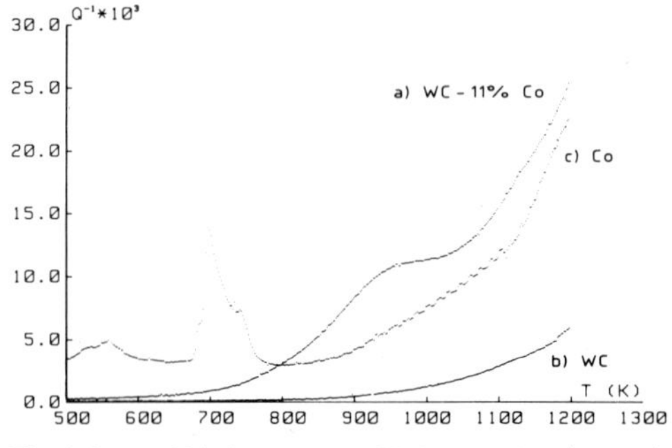

Fig. 4. Internal friction spectra of WC–11wt.%Co (curve a), pure WC sintered without binder (curve b) and a cobalt specimen obtained by melting cobalt industrial powders (curve c).

certainly due to the presence of a tiny peak located at about 1000 K.

From the results shown in Figs. 3 and 4, it is possible to conclude that the relaxation peak and the high temperature exponential background in WC–Co are due to the cobalt binder phase, but the relaxation mechanism is not related to the properties of pure bulk cobalt. The cobalt binder phase in WC–Co exhibits mechanical properties completely different from those of pure bulk cobalt. This also means that the behaviour of the cobalt phase is affected by the presence of the surrounding WC skeleton.

4.1. *Thermal treatments*

The internal friction spectrum of WC–Co is sensitively modified by heat treatment of the specimen (Fig. 5). In the just-sintered condition (curve 1), the internal friction peak is located at a relatively high temperature (about 900 K) and the exponential background is relatively high. After it has been annealed for 12 h at 1000 K, a decrease in the high temperature exponential background and a shift of the peak towards a lower temperature (about 800 K) are observed (curve 2). During subsequent aging at 1000 K, the peak temperature remains constant, but the peak height decreases somewhat. After 40 h at 1000 K, it reaches the stable state shown in curve 3. If the temperature is increased above 1000 K, a rapid increase in the internal friction with increasing temperature is observed (curve 3). After the specimen has been annealed at 1200 K for 1 h, curve 4 is obtained; this is very similar to curve 1. Annealing at 1200 K leads to a microstructural state similar to the sintered condition. The transition from curve 2 to curve 4 is reversible. Another anneal at 1000 K of the specimen exhibiting curve 4 leads to the reappearance of curve 2.

Such a behaviour is very close to that associated with precipitation and resolution processes. Effectively, other researchers [12] have detected, by high temperature X-ray analysis, the formation of Co_3W precipitates during aging at about 700 °C (973 K). Therefore, 1000 K anneals can activate the precipitation of tungsten atoms in a supersaturated Co–W solid solution, and 1200 K anneals lead to a resolution process because of an increase in the solubility.

If the solute atoms are controlling the mobility of defects, which are responsible for the anelastic relaxation in cobalt, it is easy to understand that changes in concentration of the solid solution by precipitation lead to shifts of the internal friction peak along the temperature axis during heat treatments where the relaxation time τ is modified.

4.2. *Changes in the binder*

The cobalt binder phase can be modified by the addition of some alloying elements. It is known that additions of ruthenium or chromium to WC–Co cemented carbides can reduce grain coarsening during sintering [13], but they can also modify the mechanical properties of the cobalt binder phase. In Fig. 6, the effects of chromium or ruthenium additions on the internal friction peak are presented. Each specimen was measured after the same heat treatment (after an anneal for 12 h at 1000 K). Curve a was obtained for the standard WC–11wt.%Co; curves b and c

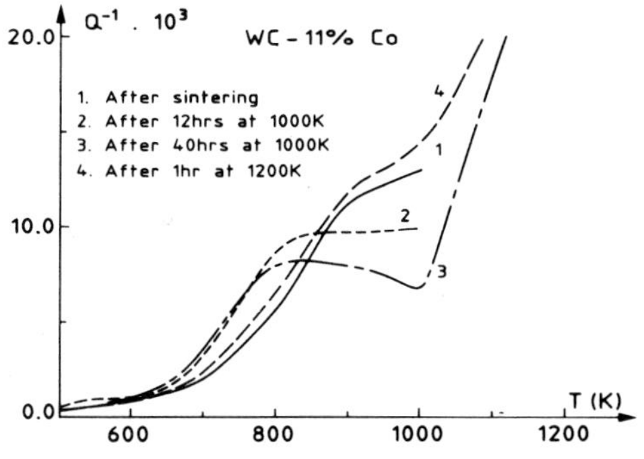

Fig. 5. Evolution of the internal friction spectrum of WC–11wt.%Co during subsequent heat treatments: curve 1, after sintering; curve 2, after annealing for 12 h at 1000 K; curve 3, after annealing for 40 h at 1000 K; curve 4, after annealing for 1 h at 1200 K.

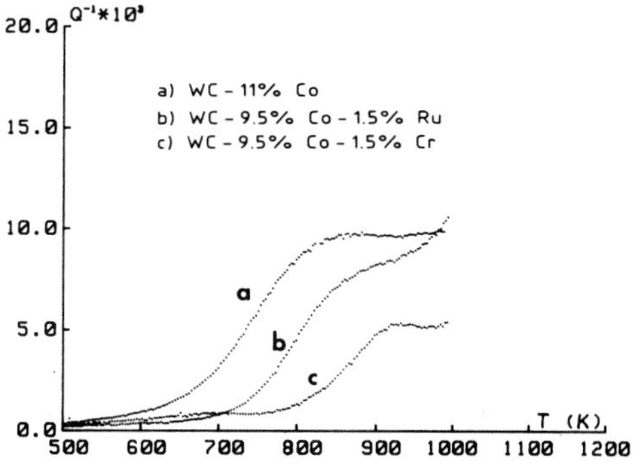

Fig. 6. Internal friction as a function of temperature, for WC–11wt.%Co (curve a), WC–11wt.%Co–Ru (curve b) and WC–11wt.%Co–Cr (curve c).

were obtained for specimens containing 1.5 wt.% Ru and 1.5 wt.% Cr respectively.

With respect to the internal friction of pure WC–11wt.%Co, the additions of ruthenium and chromium shift the curves towards higher temperatures. Moreover, a decrease in the peak height is observed. For WC–Co–Cr (curve c) the internal friction above the peak temperature is reduced by a factor of 2 in comparison with that for pure WC–Co. In addition, a large decrease in the peak height has been observed for the WC–Co–Cr system during anneals at 1000 K [10], accompanied by an increase in the elastic modulus. This can be a characteristic behaviour of precipitation hardening [14].

Even if the internal friction spectrum is rather sensitive to modifications of the cobalt binder phase, it does not change drastically when the nature of the binder phase is completely different (Fig. 7). Two other types of binder were tested. Figure 7 shows the internal friction spectra of WC–11wt.%Fe and WC–11wt.%Ni measured under the same experimental conditions as WC–11wt.%Co. These spectra are qualitatively similar. They are composed of a relaxation peak and a high temperature exponential background. The main differences are the position of the peak along the temperature axis and the intensity of the high temperature internal friction. When the peak temperature is lower, the high temperature exponential background is higher for a given temperature. It seems that the relaxation mechanism which gives rise to the peak is responsible for the rate of increase in the internal friction at high temperature. In Fig. 7, it is possible to note that the peak which appears for WC–Fe is 200 K

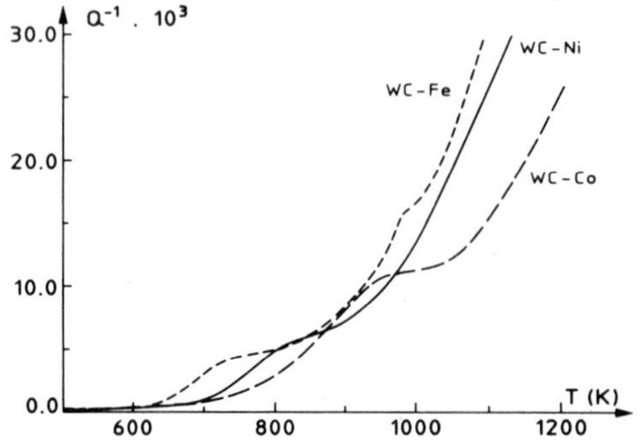

Fig. 7. Internal friction spectra of WC–11wt.%Fe and WC–11wt.%Ni compared with the spectrum for WC–11wt.%Co.

lower than for WC–Co. In the case of WC–Ni, the shift towards lower temperature is about 130 K.

The internal friction spectrum of WC–Fe exhibits a second peak at about 980 K. This peak is not a relaxation peak but a phase transition peak due to the allotropic transformation α-Fe to γ-Fe. The transition temperature T_c appears to be lower than in pure bulk iron ($T_c = 910$ °C). This could be due to the presence of carbon in the iron binder phase and also to the stresses created by the surrounding WC particles. Nevertheless, the detection of the peak at about 980 K assures us that the internal friction exponential background originates mainly in the binder phase.

5. Discussion

WC–Co cemented carbides exhibit a characteristic internal friction spectrum mainly composed of two components: the relaxation peak and the high temperature exponential background. These two components appear only when the WC skeleton is immersed in a ductile binder phase.

From a rheological point of view, a relaxation peak appears as the solution of a three-parameter model containing a Voigt unit (Fig. 1(b)) [8]. In such a model, the spring A of compliance J_u would be mostly due to the elasticity of the WC skeleton because the elastic modulus of WC–Co is much greater than the modulus of cobalt. However, the dashpot C is certainly relevant to the binder phase because the damping of WC, sintered without a binder, is weak (Fig. 4). Spring B of compliance $J_r - J_u$ is essential to the recoverable character of the anelastic deformation, and then to the appearance of a peak in the internal friction curve. If spring B does not work when dashpot C is activated, the anelastic deformation does not reach an equilibrium value. Creep behaviour appears. Then the internal friction does not reach an equilibrium value and an exponential increase as a function of temperature is observed. The restoring force associated with spring B, which limits the anelastic strain, could be due to the WC skeleton which limits the deformation of the binder or to pinning points in the binder which limit the movements of structural defects responsible for the relaxation.

It is interesting to compare the internal friction results with the mechanical properties of WC–Co cemented carbides measured by classical techniques. First, we can observe that the internal friction peak is located in the same temperature range in which an increase in the critical stress intensity factor has been observed [15, 16]. Effectively, in WC–6wt.%Co, K_{Ic} starts to increase at about 700 °C (about 973 K) and the relaxation peak in WC–5wt.%Co is centred at around 900 K (Fig. 3). Others [15, 16] have interpreted this increase in toughness as being due to a brittle-to-ductile transition of the material.

It is therefore possible to admit that the peak is associated with such a transition and may account for the increase in toughness due to a mechanism controlled by the cobalt binder phase. The increase in toughness results from the activation of the dashpot C (Fig. 1(b)) which dissipates a part of the energy used for crack propagation. This dissipation can be achieved by movements of structural defects, such as dislocations, under the effect of the applied stress.

It is known that two deformation paths may occur in the cobalt binder phase [6, 17]. At room temperature, the crystal structure of the cobalt binder phase is predominantly f.c.c. with a certain amount of h.c.p. faults. Effectively, the normal allotropic transition from f.c.c. to h.c.p. of pure bulk cobalt does not take place in WC–Co. This is confirmed by our results (Fig. 4) and can be interpreted as being due to the effects of carbide stresses and solutes acquired during sintering. Deformation is then achieved mostly by a stress-induced f.c.c.-to-h.c.p. martensitic phase transformation. As a consequence, the number of slip systems decreases during deformation, leading to macroscopic brittle behaviour. However, in this case, toughness is higher than in pure WC because of the binder ligament rupture mechanism behind the crack tip [18, 19].

As nickel is added to the cobalt, the stability of the f.c.c. structure at room temperature increases, and Vasel *et al.* [6] have shown that dislocation activity and twinning become the prevalent modes of deformation. However, also, the f.c.c. crystal structure of pure cobalt is stable at high temperatures. This suggests that the binder deformation can be activated by dislocation movements in the f.c.c. phase at high temperatures, giving rise to an increase in the ability of cobalt to relax strain in the composite material. As a consequence, toughness increases. In this way, the internal friction peak may account for the appearance of the second mode of deformation of the binder, *i.e.* by dislocation movements, and consequently for the increase in toughness as being due to an increase in the binder phase plasticity which may play an important role in front of the crack tip [20].

Above the peak temperature, the internal friction increases exponentially as a function of the temperature without reaching an equilibrium value. Such internal friction behaviour is normally observed when the restoring force (spring B in Fig. 1(b)) does not work, leading to creep of the material. Effectively non-negligible creep associated with pore formation along the WC–Co and WC–WC interfaces has been observed above 1100 K by Schmid *et al.* [21] in the same WC–11wt.%Co measured by internal friction.

Even if the exact nature of the internal friction mechanisms is not yet fully determined, it is possible to hypothesize that damping is due to dislocation movements in the binder phase. The mobility of these defects is limited by a viscous force which may be due to various interaction mechanisms with the solute atoms (pinning, depinnning and dragging) [9]. In fact, the diffusion energy of tungsten in cobalt is 2.95 eV atm^{-1} [22], which is close to the self-diffusion energy of cobalt (2.8 eV) or to the activation energy of the internal friction peak (2.7 eV). However, a peak in internal friction requires a restoring force opposed to the applied stress, which is essential to the recoverable nature of the relaxation (spring B in Fig. 1(b)). Here the restoring force could be due to the WC skeleton which prevents plastic deformation of the binder or to pinning points which limit dislocation movements. Pinning points on the dislocation loops can be precipitates in the binder or the WC–Co interfaces. When the pinning points are strong enough, the dislocations can move only around an equilibrium position and a peak is observed. When the pinning points become weaker, the dislocations can escape and creep is initiated. The weakening of the dislocation pinning points may be responsible for the exponential increase in the internal friction above the peak temperature. If the pinning points are WC–Co interfaces, the weakening may be due to grain boundary sliding. At high temperatures the amplitude of grain boundary sliding can be large enough to lead to cavitation [20].

Internal friction is good for the mechanical properties, because damping can increase toughness (for a discussion of toughening by martensitic transformation see ref. 23), but the exponential increase means that creep occurs. This behaviour is undesirable in cutting tools.

It is possible to reduce the high temperature exponential background, therefore reducing creep, by introducing hard pinning points (*i.e.* precipitates) on the dislocations. Our results (Fig. 5) show that heat treatments at about 1000 K, which can activate the formation of Co_3W precipitates, lead to a decrease in the high temperature exponential background. The shift of the peak towards lower temperatures due to such an annealing can be explained by a decrease in the solute atom concentration. These solute atoms (tungsten) leave the dislocations for the precipitates. As a consequence, the dragging force decreases and the peak appears at lower temperature. Annealing at 1200 K gives rise to the resolution of tungsten in the binder phase. The dragging force then increases and at high temperatures the restoring force decreases.

The dragging force depends on the nature of the binder. This may be the reason why the peak shifts with temperature when cobalt is replaced by nickel or iron (Fig. 7). For a rough approximation, it is possible to assume that the high temperature dislocation mobility is inversely proportional to the self-diffusion energy. In this sense, the position of the peaks along the temperature axis is understandable for the cases of WC–Ni and WC–Fe. Self-diffusion energy in nickel is higher than in iron, and consequently the relaxation peak in WC–Ni appears at a higher temperature than in WC–Fe.

If only the diffusion energies are considered, the peak in WC–Co would appear between the peaks observed in WC–Fe and WC–Ni. However, cobalt is a special case since, contrary to nickel, the fault energy in the cobalt f.c.c. structure is low at low temperatures. Then the dislocations can split into partials, limiting an h.c.p. fault. This means that the dislocation cores are extended and they can be decorated by tungsten atoms in a rather high concentration. (It seems that h.c.p. cobalt holds more WC in solution than f.c.c. cobalt does [24].) Then the dragging force which acts on dislocations is higher in the cobalt binder phase than in the nickel or iron phases.

Also, in f.c.c. cobalt, dislocations can move only when the h.c.p. structure is no longer stable. For instance, ruthenium stabilizes the h.c.p. structure. Consequently, the peak in WC–Co–Ru is shifted towards higher temperatures (Fig. 6). Chromium additions lead to a similar behaviour, but the high temperature exponential background is also weak in this case (Fig. 6(c)). This must result in a better creep resistance of the material.

6. Conclusions

WC–Co cemented carbides exhibit a characteristic internal friction spectrum mainly composed of a relaxation peak and a high temperature exponential background. The peak is located in the same temperature range in which an increase in toughness has been observed by other workers [15, 16] and is associated with a brittle-to-ductile transition of the material.

The analysis of the internal friction peak allows us to attribute the origin of this increase in toughness to an increase in the damping capacity of the binder phase.

In contrast, creep of the material seems to be connected with the high temperature exponential background. When the exponential background is high, the creep resistance is poor. It is possible to reduce the exponential background by introducing hard pinning points, such as precipitates, into the binder. The influence of creep would then be less.

Acknowledgments

This work was financially supported by the Swiss Commission pour l'Encouragement de la Recherche Scientifique.

References

1 N. P. Suh, *Wear, 62* (1980) 1.
2 Correlation of the mechanical properties of WC–Co with drilling performance in hard formations, *Final Rep.*, July 1982 (Terra Tek Inc., University Research Parka, Salt Lake City, UT).
3 E. Guyon and S. Roux, *Recherche, 18* (1987) 1050.
4 B. Roebuck and E. G. Bennett, *Metallography, 19* (1986) 27.
5 T. Johannesson and B. Lehtinen, *Phys. Status Solidi A, 16* (1973) 615.
6 C. H. Vasel, A. D. Krawitz, E. F. Drake and E. A. Kenik, *Metall. Trans. A, 16* (1985) 2309.
7 G. Wirkmark, C. Chatfield and G. L. Dunlop, in E. A. Almond, C. A. Brookes and R. Warren (eds.), *Proc. 2nd Int. Conf. on the Science of Hard Materials, Rhodes, September 23–28, 1984,* in *Inst. Phys. Conf. Ser., 75* (1986) 669.
8 A. S. Nowick and B. S. Berry, *Anelastic Relaxation in Crystalline Solids,* Academic Press, New York, 1972.
9 W. Benoit, G. Gremaud and R. Schaller, in B. Escaig and

G. G'sell (eds.), *Plastic Deformation of Amorphous and Semi-crystalline Materials,* Les Editions de Physique, Paris, 1982, pp. 65–91.
10 R. Schaller, J. J. Ammann, A. Kulik, C. Bonjour and W. Benoit, *J. Phys. (Paris), Colloq. C10, 46* (1985) 387.
11 J. E. Bidaux, R. Schaller and W. Benoit, *J. Phys. (Paris), Colloq. C10, 46* (1985) 601.
12 N. Bouaouadja, *Phil. Thesis,* Institut National des Sciences Appliquées, Lyon, 1986.
13 C. Bonjour, *Wear, 62* (1980) 83.
14 R. Schaller and W. Benoit, *J. Phys. (Paris), Colloq. C9, 44* (1983) 17.
15 G. Fantozzi, H. Si Mohand and G. Orange, in E. A. Almond, C. A. Brookes and R. Warren (eds.), *Proc. 2nd Int. Conf. on the Science of Hard Materials, Rhodes, September 23–28, 1984,* in *Inst. Phys. Conf. Ser., 75* (1986) 699.
16 B. Johannesson and R. Warren, in E. A. Almond, C. A. Brookes and R. Warren (eds.), *Proc. 2nd Int. Conf. on the Science of Hard Materials, Rhodes, September 23–28, 1984,* in *Inst. Phys. Conf. Ser., 75* (1986) 713.
17 L. Manlang, H. Xiaoying, D. Shitian, S. Dagin, C. Yumei and Y. Zhenmei, *Int. J. Refractory Hard Met., 2* (1983) 129.
18 L. S. Sigl, H. E. Exner and H. F. Fischmeister, in E. A. Almond, C. A. Brookes and R. Warren (eds.), *Proc. 2nd Int. Conf. on the Science of Hard Materials, Rhodes, September 23–28, 1984,* in *Inst. Phys. Conf. Ser., 75* (1986) 631.
19 H. G. Schmid, in E. A. Almond, C. A. Brookes and R. Warren (eds.), *Proc. 2nd Int. Conf. on the Science of Hard Materials, Rhodes, September 23–28, 1984,* in *Inst. Phys. Conf. Ser., 75* (1986) 473.
20 B. Aronsson and H. Pastor, Cobalt in cemented carbides, *Met. Powder Rep., 41* (1986) 499.
21 H. G. Schmid, D. Mari, W. Benoit and C. Bonjour, in B. Wilshire and R. W. Evans (eds.), *Proc. 3rd Int. Conf. on Creep and Fracture of Engineering Materials and Structures, Swansea, 1987,* Institute of Metals, London, 1987, p. 975; *Mater. Sci. Eng., A105/106* (1988) 343–351.
22 I. I. Kovenskii, *Sov. Phys.—Solid State, 3* (1961) 252.
23 A. G. Evans and R. M. Cannon, *Acta Metall., 34* (1986) 761.
24 H. Le Roux, in E. A. Almond, C. A. Brookes and R. Warren (eds.), *Proc. 2nd Int. Conf. on the Science of Hard Materials, Rhodes, September 23–28, 1984,* in *Inst. Phys. Conf. Ser., 75* (1986) 335.

The Failure of Hard Materials in Cyclic Compression: Theory, Experiments and Applications*

S. SURESH

Division of Engineering, Brown University, Providence, RI 02912 (U.S.A.)

(Received November 13, 1987)

Abstract

In this paper a summary is presented of recent developments in the understanding of mode I crack growth at room temperature in hard materials subject to uniaxial cyclic compressive stresses. Experimental observations of fatigue crack growth in cemented carbides, single-phase ceramics, transforming ceramics, ceramic composites and concrete are briefly reviewed. A detailed discussion of the "driving force" for the growth of a stable crack in a direction macroscopically normal to the far-field compression axis is presented. The "universal features" of cyclic compression fracture in a wide range of brittle and ductile solids are described. Some important applications of the compression fatigue phenomenon to the fracture toughness testing of brittle materials are highlighted. The overall framework for a general theory of fatigue failure under far-field cyclic compression is developed.

1. Introduction

The high compressive strength of hard brittle solids makes it appropriate to utilize them in a wide range of compressive load-dominated engineering applications. The mechanisms of failure of brittle solids subject to monotonic compressive loads have been the topic of many detailed investigations in the past (see for example refs. 1–3). There is, however, very little understanding of the fracture behavior of these materials under cyclic loading conditions which are often encountered in service.

Recent work by the present author and his coworkers [4–7] has shown that the application of uniaxial cyclic compressive stresses to notched plates of brittle solids can lead to stable mode I crack growth at room temperature. An interesting feature of this effect is that fatigue crack growth in hard materials has macroscopic characteristics which are very similar to those observed in ductile metallic materials under far-field cyclic compression. The mechanisms of fatigue failure in cyclic compression thus follow some universal features in a wide variety of brittle and ductile solids despite vast differences in their microscopic modes of deformation and fracture.

It is the objective of this paper to present a summary of recent developments and to build an overall perspective on the mechanics and micromechanisms of mode I fatigue fracture at room temperature in hard materials under cyclic compression. Firstly, experimental observations of fatigue crack growth in single-phase ceramics, transforming ceramics, cemented carbides, ceramic composites and concrete are briefly reviewed. Secondly, the "driving force" for the growth of a fatigue flaw in a direction normal to the far-field compression axis is examined. The origin of a zone of residual tension at the notch tip under far-field cyclic compression is discussed for a wide range of deformation modes. Finally, some important applications of the compression fatigue phenomenon to the fracture toughness testing of brittle solids are described.

2. Experimental observations

Figure 1 schematically illustrates the specimen geometry and loading conditions used in the uniaxial cyclic compression testing of hard materials. Figures 2–4 show examples of stable room temperature crack growth in single-edge-notched specimens of a variety of hard materials under cyclic compression. Figure 2 is an optical

*Paper presented at the 3rd International Conference on the Science of Hard Materials, Nassau, The Bahamas, November 9–13, 1987.

0921-5093/88/$3.50 © Elsevier Sequoia/Printed in The Netherlands

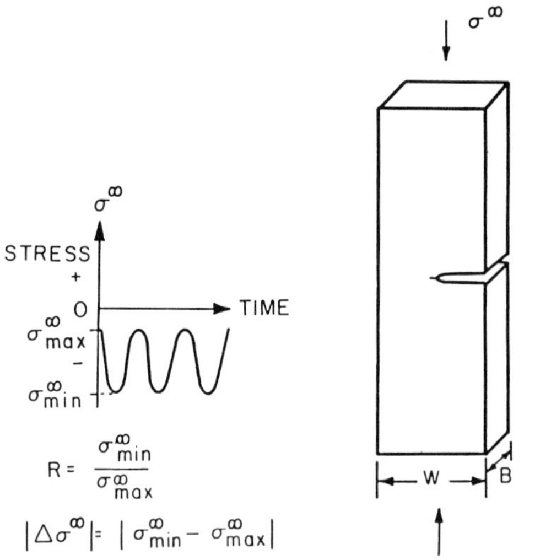

Fig. 1. A schematic diagram of the specimen geometry and loading conditions used in the uniaxial cyclic compression testing of hard materials.

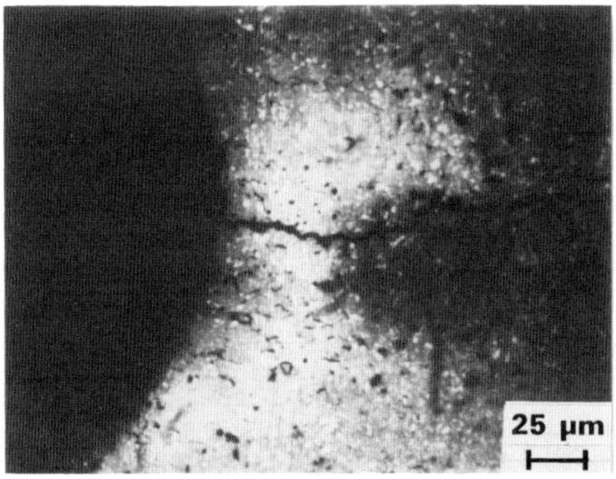

Fig. 2. Mode I crack growth in WC–5.6wt.%Co cermet under cyclic compression. The compression axis is vertical. The microcracks nucleated at the notch tip during compression loading should be noted.

micrograph of a mode I fatigue crack in a WC–5.6wt.%Co cermet subjected to a cyclic compressive load range from −60 to −660 MPa. Figures 3(a), 3(b) and 3(c) show examples of compression fatigue in polycrystalline aluminum oxide (Al_2O_3), hot-pressed silicon nitride (Si_3N_4) and fine-grained zirconia (ZrO_2) partially stabilized with 4.5 wt.% Y_2O_3 respectively. The ceramics were subjected to a compressive load range from −330 to −30 MPa. Figures 4(a) and 4(b) are optical micrographs of fatigue in cyclic compression in Al_2O_3–33wt.%SiC whisker com-

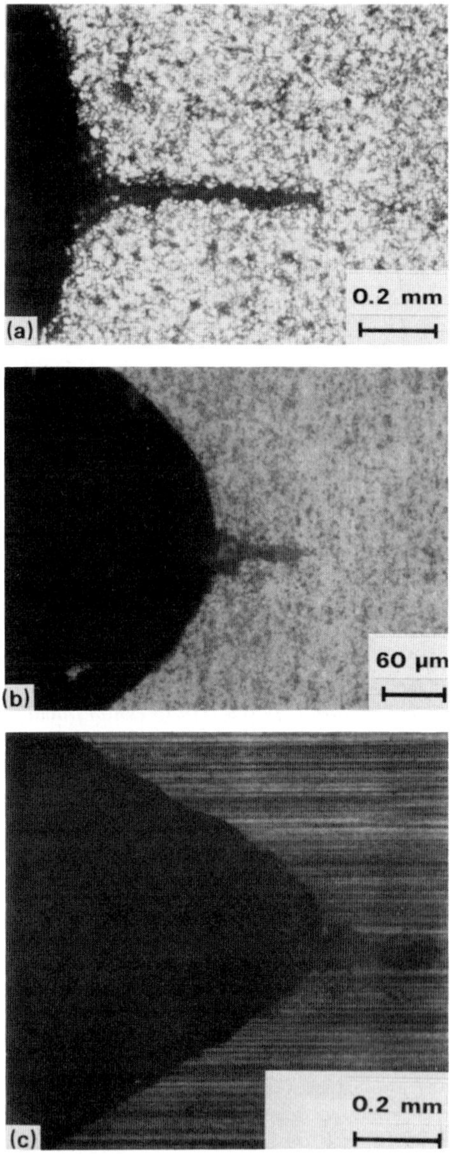

Fig. 3. Mode I fracture in (a) polycrystalline Al_2O_3, (b) hot-pressed Si_3N_4 and (c) fine-grained ZrO_2 partially stabilized with 4.5 wt.% Y_2O_3. The compression axis is vertical in all cases.

posite and Portland-cement-based concrete, respectively. The edge-notched specimens for all the cases had the following dimensions: height, 25–50 mm; width, 15–30 mm; thickness, 6–10 mm; notch-length-to-width ratio, 0.3–0.4; notch root radius, 0.15–0.75 mm. The cyclic frequency

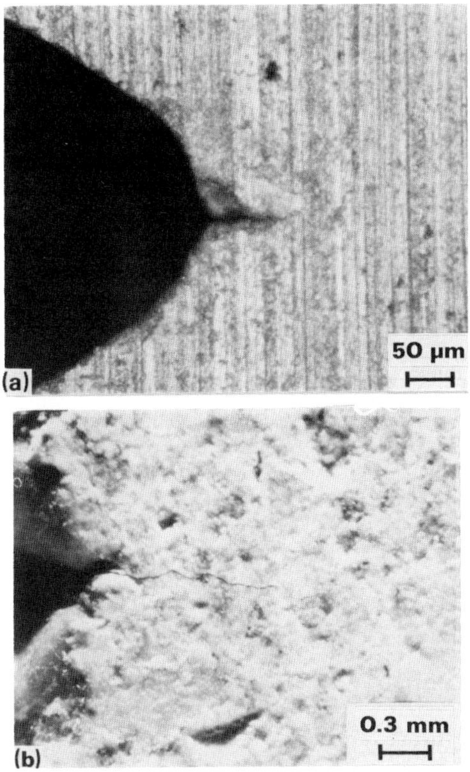

Fig. 4. Mode I fracture under far-field cyclic compression in (a) Al_2O_3 reinforced with 33 vol.% SiC whiskers and (b) concrete. The compression axis is vertical.

used for the fatigue tests was 1–10 Hz, with a sinusoidal waveform and a load ratio R of 10. For concrete a 75 mm × 75 mm × 15 mm specimen with a notch root radius of 1.5 mm was used. Further details of specimen design, loading conditions, material microstructure and experimental methods can be found elsewhere [5, 7].

For the materials in Figs. 2–4, mode I fracture was observed on unloading from the very first compression cycle. Further cycling led to crack advance at a progressively decreasing velocity until complete crack arrest was observed. In single-phase brittle solids, fracture occurs primarily along grain boundaries, while both intergranular and interfacial fracture is noticed in the composite microstructures. Our attempts to induce stable mode I fatigue fracture in notched single crystals of α-Al_2O_3 and in a soda–lime glass have thus far proved unsuccessful [5]. Only catastrophic splitting, parallel to the compression axis, is found in these materials. These results imply that contained damage in the vicinity of the notch tip is essential to induce mode I failure.

In single-phase hard materials, the principal damage mechanism is found to be the nucleation of grain boundary microcracks at the notch tip during compression loading [7]. Non-cubic ceramics and composites contain pre-existing microcracks due to thermal contraction mismatch between adjacent grains which leads to residual stresses during cooling from the processing temperature. Additionally, the introduction of the notch induces some microcracking. Our *in situ* microscopic observations and acoustic emission measurements indicate that significant levels of microcracking are induced at the notch tip during the loading portion of the very first compression fatigue cycle [6, 8]. Figure 5(a), for example, shows the damage induced at the notch tip during compression loading in a 99.9% pure Al_2O_3 with an average grain of about 18 μm. A higher magnification micrograph of the damage zone reveals grain boundary cracking (Fig. 5(b)). The microcracks are subject to local compressive and shear stresses under far-field compression. The frictional sliding of microcracks leads to tensile opening and possible microcrack nucleation at the adjacent grain boundaries. Quantitative analyses of microcracks on the specimen surface indicate that the average microcrack density β in alumina is 0.2. Here, $\beta = Na^3$, with N the number of microcracks per unit area and a the average microcrack length. A high density of microcracks was also nucleated at the WC–Co interface in the cemented carbide subject to cyclic compression. In this latter case, the non-dimensional microcrack density β was estimated to be 0.18; the presence of microcracks can be seen in the notch tip region in the optical micrograph shown in Fig. 2. As noticed in Fig. 2, a zone of microcracks is created at the notch tip during compression loading in the very first cycle. On unloading from the far-field compressive stress, a macroscopic mode I crack propagates through the microcrack zone. Similarly, in ceramics which undergo stress-induced martensitic transformation, compression loading creates a transformation zone at the notch tip. During the removal of the far-field compressive stress, a macroscopic flaw is initiated at the notch tip and is propagated within the transformation zone [7]. In ceramic–ceramic composites, there is evidence of both microcracking and interfacial sliding under far-field cyclic compression.

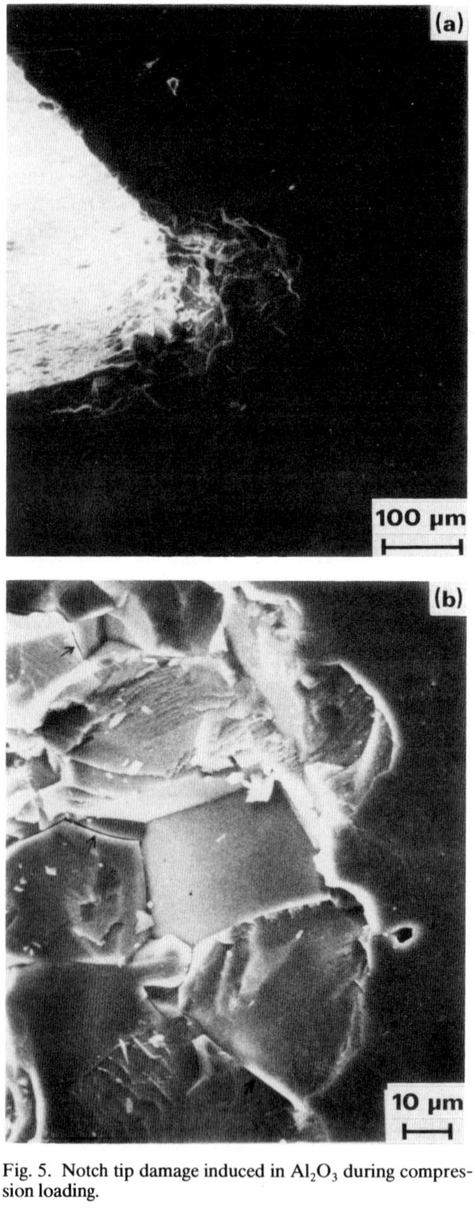

Fig. 5. Notch tip damage induced in Al_2O_3 during compression loading.

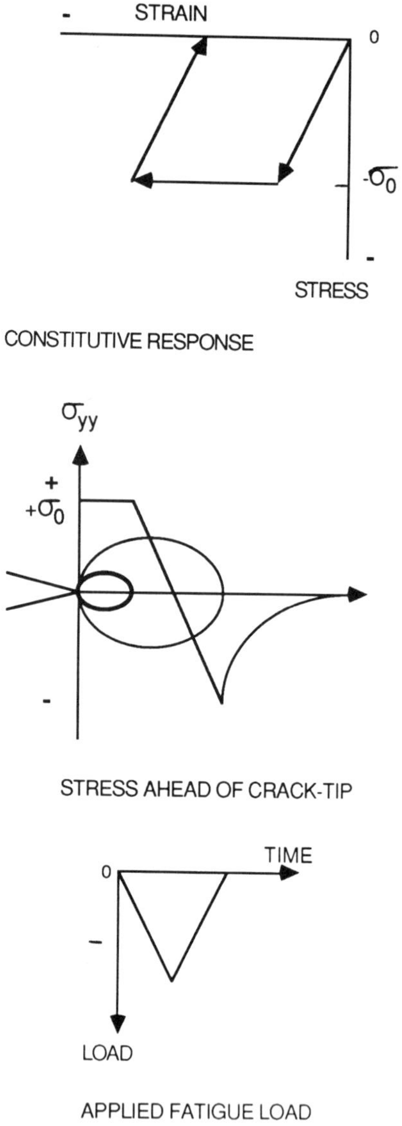

Fig. 6. Residual tensile stresses at the notch tip in an elastic, perfectly plastic solid subject to far-field cyclic compression.

3. Mechanics of failure under cyclic compression

The possibility of stable mode I crack growth from notches in metallic materials under far-field cyclic compression has been recognized for some time [9, 10]. Figure 6 schematically illustrates the residual tensile stresses generated at the tip of a "sharp" non-closing notch in an elastic, perfectly plastic solid which has been subject to one compression cycle. During the application of the maximum compressive stress, a monotonic compressive plastic zone is generated at the notch tip (Fig. 6, full curve). During unloading, as the elastically strained material surrounding the plastic zone relaxes, it subjects the notch tip to residual tension. The material at the notch tip yields in tension because of the high stress concentration, thereby creating a cyclic plastic zone (Fig. 6, broken curve). The residual tensile stresses, of magnitude equal to the flow stress in tension, are generated over a distance of the order of the cyclic plastic zone. At far field, residual compressive stresses exist to balance out the near-tip tension and to provide a net zero stress. The variation in the normal stress σ_{yy} as a function of the distance from the notch tip in an elastic, perfectly plastic solid subject to one compression cyclic as shown in Fig. 6. The notch tip residual tensile stresses constitute the "driving force" for mode I fatigue crack growth in ductile solids under cyclic compression.

As the presence of a plastic zone is generally considered to be essential for the inducement of mode I fracture under cyclic compression, this phenomenon has not been investigated in detail in brittle media which exhibit very little dislocation motion at room temperature. Recent studies, however, have clearly demonstrated that inelastic deformation at the notch tip, in the form of microcracking, martensitic transformation or creep, can also lead to the generation of residual tensile stresses upon unloading from the far-field compressive stress [7]. Figure 7 schematically illustrates possible constitutive behavior patterns in microcracking brittle solids (such as single-phase ceramics, ceramic composites and concrete) subject to cyclic compressive loads. During the loading portion of the fatigue cycle, microcracking at a threshold stress σ_0 promotes a nonlinear stress strain curve. The microcracks at the notch tip undergo frictional sliding under far-field compression. If the sliding displacement between the faces of the microcracks is completely reversible, no permanent strains will be observed on unloading. In this case, the unloading path will follow line B shown in Fig. 7. In contrast, when the pseudoplastic strains created during compression loading are fully irreversible, an elastic unloading behavior pattern (line C) will be observed. Most concrete materials are known to exhibit such a constitutive response. For conditions for partial closure of the microcracks, a situation in between these two, schematically shown by line D, is appropriate.

Recently, Brockenbrough and Suresh [6] have conducted a detailed finite element simulation of notch tip residual tensile stresses in single-phase ceramics using the constitutive model shown in Fig. 7 for the experimental conditions pertaining to the compression fatigue tests for alumina (Figs. 3(a) and 5). Figure 8 shows the variation in numerically estimated residual stress σ_{yy} as a function of distance from the notch tip for a polycrystalline Al_2O_3 containing randomly oriented penny-shaped flaws which are induced during compression loading. For the unloading situation corresponding to no permanent strains, $\lambda = 0$ (Fig. 7, line B), essentially no residual stresses are generated on unloading from a far-field compressive stress. In contrast, for the elastic unloading

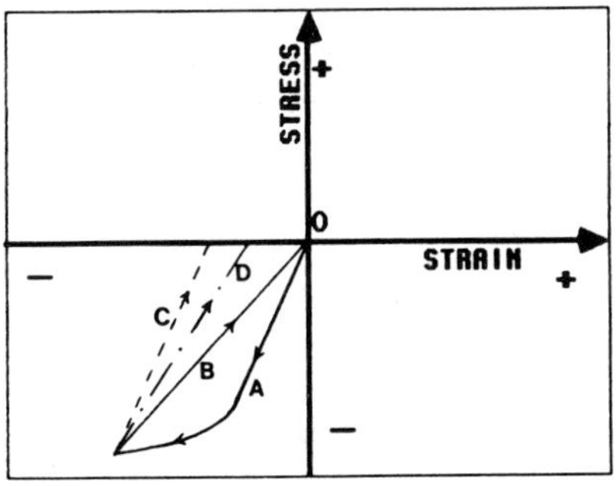

Fig. 7. Possible constitutive behavior patterns in a microcracking brittle solid subject to cyclic compression.

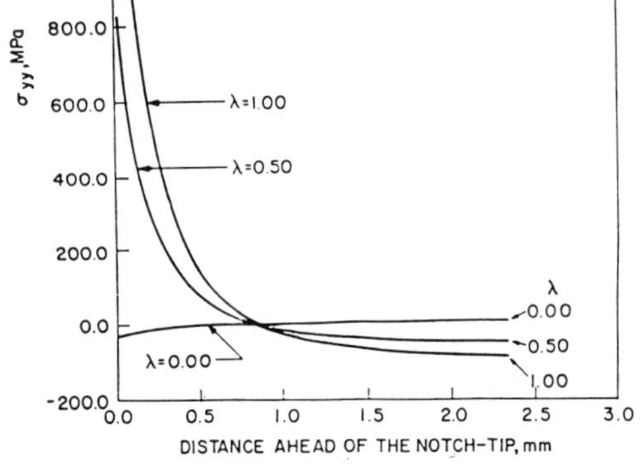

Fig. 8. Variation in numerically estimated σ_{yy} as a function of distance from the notch tip in a microcracking brittle solid subject to one compression cycle.

path $\lambda = 1$ (Fig. 7, line C), tensile stresses as high as 1600 MPa are generated at the notch tip. For $\lambda = 0.5$ (Fig. 7, line D), a tensile stress of up to 800 MPa is induced at the notch tip. As such residual tensile stresses significantly exceed the tensile strength (250 MPa) of the Al_2O_3, fracture occurs even after the very first compression cycle. The distance over which residual tensile stresses for $\lambda = 0.5$ exceed the tensile strength of Al_2O_3 closely correspond to the experimentally observed distance of crack growth after one cycle in this Al_2O_3 ceramic. The numerical analysis also reveals that the residual tensile stresses are generated in the last 25% of the unloading portion of the fatigue cycle.

For the case of transforming ceramics such as ZrO_2 partially stabilized with MgO, Suresh and Brockenbrough [7] have developed a generalized constitutive model which incorporates both dilational and shear strains induced by martensitic transformation in cyclic compression. A finite element analysis based on this model correctly predicts the experimentally observed crack length, crack shape and transformation zone size due to the application of one compression cycle.

An examination of the above results indicates that residual tensile stresses at the notch tip would be expected whenever the deformation mechanism induced during the application of compression loads leaves permanent strains on unloading. Since the zone of residual tension is fully embedded in material which is elastically strained in compression, stable fracture behavior is observed even in brittle solids. The continuation of compression loading beyond the first cycle leads to the exhaustion of the zone of residual tension, as the faces of the sharp fatigue flaw remain closed under far-field compression. There is, however, experimental evidence that the high frequency fretting contact between the faces of the fatigue crack leads to the formation of debris particles of the material within the crack [5]. This produces an apparently large crack-opening displacement even under far-field compression and may contribute to some crack advance during subsequent cyclic loading.

4. Applications

Crack growth under cyclic compression offers a novel possibility for pre-cracking brittle solids prior to fracture toughness tests. Our recent work has shown that, when fatigue cracks are introduced into notched specimens of cemented carbides and ceramics under low maximum compressive stresses and are propagated until complete crack arrest occurs, prior compression loading has no significant effect on subsequent tensile fracture toughness. Furthermore, the simple method of pre-cracking of single-edge-notched specimens under uniaxial cyclic compression offers marked advantages over currently available techniques such as the chevron-notched short rod or bar method [11], indentation techniques [12] and bridge compression procedure [13]. As pre-cracking in cyclic compression promotes a stable mode I through-thickness crack in a plate or a concentric flaw in a circumferentially notched cylindrical rod, the pre-crack shape and geometry are amenable to a direct and unambiguous characterization of subsequent tensile fracture using presently available fracture mechanics methods. Figure 9 shows an example of a fracture surface of polycrystalline Al_2O_3 specimen where a mode I pre-crack of length a was introduced at room temperature under uniaxial cyclic compression prior to a quasi-static four-point bend fracture test. Pre-cracking in cyclic compression has now been successfully used in the mode I and mode III fracture toughness testing of hard materials under both quasi-static and dynamic loading conditions [14] and in the fatigue threshold testing of ductile solids [10].

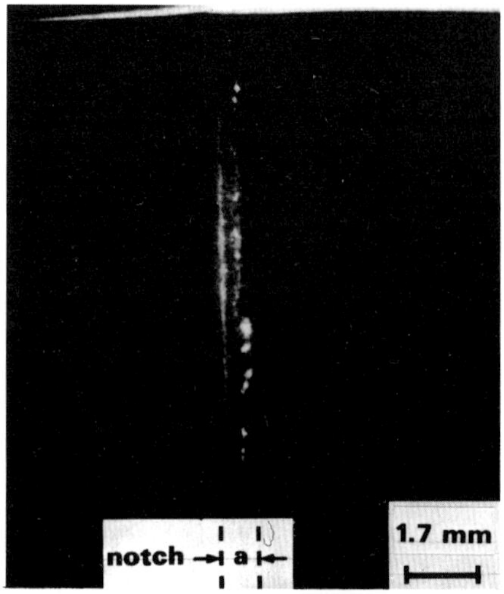

Fig. 9. An example of four-point bend fracture in Al_2O_3 following pre-cracking in cyclic compression.

5. Concluding remarks

Developing a thorough understanding of the mechanisms of fracture under cyclic compression is vital to the efficient and safe design of engineering components made of hard materials. Examples of applications where cyclic compressive loads are of prime interest include the following: cutting tools, metal-forming dies and rock-drilling tools made of cemented carbides or ceramic composites; civil engineering and transport applications involving concrete; ball-bearings, artificial human implants and potential engine structural units made of advanced ceramics and ceramic composites. The present work clearly shows that undesirable levels of subcritical crack growth can occur in hard materials with stress concentrations under far-field cyclic compression even when the maximum far-field (nominal) compressive stress is as low as one tenth of the unconstrained compressive strength.

The results presented here provide some insights into the key features of crack initiation and growth under cyclic compression in hard materials containing stress concentrations with experimental and theoretical results derived from a wide range of materials. A significant result of this work is the observation that hard materials comprised of very different microstructural units exhibit a macroscopically similar failure mode under far-field cyclic compression. Furthermore, it is clearly seen from our study that stable fatigue crack growth can occur even in the absence of plastic deformation, which is often considered to be a basic requirement for the occurrence of fatigue failure (see for example ref. 15). (This notion is engendered primarily by the conclusions drawn from the work on metallic materials over the past several decades and from the apparent paucity of fatigue in brittle solids.) The mechanisms of mode I fracture in cyclic compression are similar to those observed in quasi-static tensile failure of hard materials; however, this compression failure mode is an intrinsic fatigue phenomenon in that it is observed only under *cyclic* loads and that it is vastly different from the mode of fracture under monotonic compression loading conditions.

Acknowledgments

This paper is based on work carried out at Brown University with support from the U.S. Department of Energy and the National Science Foundation. The author is grateful to members of his research group, particularly to J. Brockenbrough and L. Ewart, whose results are included in the summary presented here. Thanks are also due to Dr. E. K. Tschegg for his help in obtaining the photograph shown in Fig. 4(b).

References

1 F. A. McClintock and J. B. Walsh, *Proc. 4th U.S. Congr. on Applied Mechanics, 1962*, Plenum, New York, p. 1015.
2 S. Nemat-Nasser and H. Horii, *J. Geophys. Res., 87* (1982) 6805.
3 M. F. Ashby and S. D. Hallam, *Acta Metall., 34* (1986) 497.
4 S. Suresh and L. A. Sylva, *Mater. Sci. Eng., 83* (1986) L7.
5 L. Ewart and S. Suresh, *J. Mater. Sci., 22* (1987) 1173.
6 J. R. Brockenbrough and S. Suresh, *J. Mech. Phys. Solids, 35* (1987), in the press.
7 S. Suresh and J. R. Brockenbrough, *Acta Metall., 36* (1988), in the press.
8 R. Godse, J. Gurland and S. Suresh, unpublished results, Brown University, 1987.
9 R. P. Hubbard, *J. Basic Eng., 91* (1969) 625.
10 S. Suresh, *Eng. Fract. Mech., 21* (1985) 453.
11 L. M. Barker, *Eng. Fract. Mech., 9* (1977) 361.
12 B. R. Lawn and T. R. Wilshaw, *J. Mater. Sci., 10* (1975) 1049.
13 R. Warren and H. Matzke, in R. Viswanadham and J. Gurland (eds.), *Science of Hard Metals*, Plenum, New York, 1984, p. 631.
14 S. Suresh and E. K. Tschegg, *J. Am. Ceram. Soc., 70* (1987) 726.
15 R. W. Hertzberg, *Deformation and Fracture Mechanics of Engineering Materials*, Wiley, New York (2nd edn.), 1983.

Materials Science and Engineering, *A105/106* (1988) 331–336

Applicability of the Critical Strain Fracture Criterion to WC–Co Hard Metals*

R. GODSE and J. GURLAND

Division of Engineering, Brown University, Providence, RI 02912 (U.S.A.)

(Received December 11, 1987)

Abstract

A generally accepted view that the fracture of hard metals needs to be modeled as a ductile event (their macroscopic brittle behavior notwithstanding) to account for the microplasticity of the binder phase, led to various attempts in the past to apply the Rice–Johnson model for ductile failure to the fracture of hard metals. The present work is an attempt to take these analyses a step further and is based on the concept of a crack tip process zone in which the contiguous carbide, since it is most severely stressed, is broken owing to microcracking and therefore does not contribute to the load-carrying capacity of the process zone. This leads to an expression relating fracture toughness and microstructural parameters of hard metals. The relationship contains one floating parameter for which reasonable values are obtained when the predictions of the present model are fitted to the available experimental data for WC–Co alloys with binder volume fractions approximately in the range 0.10–0.25.

1. Introduction

Commercial hard metal alloys possess unique combinations of mechanical properties on account of the composite nature of their microstructure, which consists of hard and brittle carbide particles cemented by a ductile binder phase. For example, relatively high values of fracture toughness K_{Ic} at high hardness levels are possible owing to the presence of the ductile binder which acts as a barrier to crack propagation through the carbide phase. Thus the fracture process in hard metals exhibits a dual character in that it is macroscopically a quasi-brittle phenomenon ($K_{Ic} < 20$ MPa m$^{1/2}$), but evidence of extensive plastic deformation in the binder phase shows it to be a microscopically ductile process.

It has been recognized for a long time that the mean free path λ_c in the binder phase exerts a large influence on mechanical properties of hard metals and several studies have been conducted to relate the fracture toughness of hard metals to this parameter (see ref. 1 for a review). The fracture toughness has been modeled analytically by considering the weighted contributions to the fracture energy of the composite from the various modes of crack propagation that have been observed to occur in the two-phase composite [2–6]. The microstructural dependence of the energy contributions yields a relation between the critical strain energy release rate G_{Ic} and microstructural parameters. A particularly elegant treatment of this concept was provided by Sigl *et al.* [7]. They were able to predict the fracture toughness of WC–Co hard metal alloys closely and thus validate the energy-based approach to modeling hard metal fracture.

Many researchers in the past have used an alternative stress-intensity-based model to describe the fracture in hard metals. In particular, a generally accepted view that fracture of hard metals needs to be modeled as a ductile event (their macroscopic brittle behavior notwithstanding) to account for the microplasticity in the binder phase has led to various attempts to apply the critical strain criterion, originally proposed by Rice and Johnson [8] for ductile failure, to hard metals [9–11]. These efforts suffered from a limited knowledge of the micromechanisms of fracture in hard metals. In recent years, significant improvements have been made in the understanding of these micromechanisms and this warrants another try at this approach.

*Paper presented at the 3rd International Conference on the Science of Hard Materials, Nassau, The Bahamas, November 9–13, 1987.

0921-5093/88/$3.50

© Elsevier Sequoia/Printed in The Netherlands

2. The critical strain criterion for ductile failure

The work of Rice and Johnson [8] demonstrated that, when the effects of large geometry changes associated with a deforming crack tip are incorporated in the analysis of stress and deformation fields, a zone of intense strain of size D is predicted in front of the crack tip. They suggested that this zone of intense strain should envelop a characteristic microstructural dimension l^* before ductile crack growth can commence. Thus the failure criterion can be expressed by the following relation:

$$D = l^* \tag{1}$$

Furthermore, Rice and Johnson showed that the crack-tip-opening displacement δ is comparable to D, so that at failure

$$\delta_c \approx D \tag{2}$$

where δ_c is the critical crack-tip-opening displacement. Equation (2) can be restated as

$$\delta_c = RD \tag{3}$$

where R is a constant of the same order as unity. Then, using eqns. (1) and (3), and the small-scale yielding expression for δ_c, namely

$$\delta_c = \frac{C_1 K_{Ic}^{\ 2}}{\sigma_0 E'} \tag{4}$$

we obtain

$$K_{Ic} = \left(\frac{Rl^* E' \sigma_0}{C_1} \right)^{1/2} \tag{5}$$

Here C_1 is a constant, σ_0 is the yield stress and

$$E' = \frac{E}{1 - \nu^2} \tag{6}$$

where E and ν are the elastic modulus and Poisson's ratio respectively.

The failure criterion of eqn. (1) can also be stated in terms of a critical strain ε_c that should be exceeded over the distance l^* before failure can occur. This alternative formulation is derived using the variation in strain ε with distance ahead of the crack tip [12–14]. The strain field can be written in the form

$$\varepsilon_c = \frac{C_2 \delta}{x}$$

where C_2 is a constant. At crack growth initiation, the strain at $x = l^*$ is ε_c and $\delta = \delta_c$, so that

$$\varepsilon_c = \frac{C_2 \delta_c}{l^*} \tag{7}$$

Then, by using eqns. (1), (3) and (7), we obtain

$$\varepsilon_c = C_2 R \tag{8}$$

This is the critical value of strain at which further straining causes accelerated unstable deformation, resulting in the initiation of macrocrack growth.

3. Application to WC–Co alloys

In the case of hard metals, the immediate region near the crack tip consists of an unbroken binder ligament followed by a carbide grain. Owing to the geometrical limitations imposed by the carbide grain which does not allow large strains to develop within itself, the maximum extent D of the large strain region is restricted to the ligament size. Hence the characteristic microstructural distance in hard metals is the binder mean free path λ_c. Thus, using $l^* = \lambda_c$, eqn. (5) can be written as

$$K_{Ic} = \left(\frac{R \lambda_c E' \sigma_0}{C_1} \right)^{1/2} \tag{9}$$

The adaptation of the critical strain model, which assumes the deforming medium to be an elastic–plastic continuum, to the heterogeneous microstructure of WC–Co alloys, requires the following considerations.

(1) Since a significant portion of the brittle carbide phase in the process zone is broken prior to the onset of crack propagation, the surviving material near the crack tip may be represented by a porous solid, with the microcracks constituting the pores. An approach originally advocated by Berg [15] to study the mechanical properties of porous solids by using a modified yield criterion could be useful in this respect. He suggested that, if the yield stress of the solid is appropriately modified to account for the presence of the pores, then the solid can be regarded as a continuum. In the present study, it is assumed that the only effect of the microcracks in the process zone is to cause a loss of load-bearing cross-section, so that the modified (effective) yield stress of the process zone can be represented as

$$\sigma_e = \sigma_B(1-f) \tag{10}$$

where σ_B is the yield stress of the binder phase and f is the volume fraction of the carbide phase that does not carry any load in the process zone owing to the presence of the microcracks.

(2) A rigorous estimate of f requires a knowledge of stress fields in the process zone, partitioning of stress and strain between the phases, and fracture stress of the carbide particles and carbide–carbide boundaries. In the absence of this type of detailed information, it is assumed here that the microcracks occur only in the partially connected contiguous portion of the carbide phase. This assumption can be justified on the basis of the generally accepted view that the skeleton-forming contiguous carbide is the principal load-bearing component in the hard metal microstructure (analogous to fibre composites where the stiff fibres carry most of the load) and is therefore also the most highly stressed. It has also been argued [16] that the relatively weak carbide–carbide boundaries present in the contiguous portion of the carbide phase are likely to be sites of stress concentration. Thus a combination of high stresses and weak sites in the contiguous carbide makes it the most likely component of the microstructure in which microcracking can occur. Hence the volume fraction f which carries no load in the process zone can be roughly estimated as the volume fraction V_c of the contiguous carbide. As has been shown by Lee and Gurland [16], V_c is given in terms of the total volume fraction V_{WC} of the carbide phase and its contiguity C as $V_c = CV_{WC}$. Thus,

$$f = CV_{WC} \tag{11}$$

With these assumptions, the effective yield stress σ_e of the binder can be used in place of σ_0 in eqn. (9) to obtain

$$K_{Ic} = \left\{\frac{R\lambda_c E'\sigma_B(1-CV_{WC})}{C_1}\right\}^{1/2} \tag{12}$$

Using the stereological expression

$$\frac{\lambda_c}{V_B} = \frac{d}{V_{WC}(1-C)} \tag{13}$$

where d is the mean intercept length of the carbide particles and V_B is the volume fraction of the binder, eqn. (12) can be written as

$$K_{Ic} = \left\{\frac{R(\lambda_c + d)\sigma_B V_B E'}{C_1}\right\}^{1/2} \tag{14}$$

In this manner, the critical strain criterion for ductile failure can be used to relate the fracture toughness K_{Ic} to the microstructural parameters λ_c, d and V_B.

4. Results and discussion

For the purpose of comparing the model with experimental data, eqn. (14) was written in the form

$$K_{Ic} = RP$$

where

$$P = \left\{\frac{(\lambda_c + d)V_B \sigma_B E'}{C_1}\right\}^{1/2} \tag{15}$$

This equation predicts that K_{Ic} varies linearly with the parameter P.

Numerical values of the terms in eqn. (15) were obtained as follows: the constant C_1 (=0.54) from McMeeking [17], E' for the binder from Doi *et al.* [18], σ_B from Sigl *et al.* [7], namely

$$\sigma_B = 480\ \text{MPa} + \frac{1550}{\lambda_c}\ \text{MPa} \tag{16}$$

and V_B, d, λ_c and K_{Ic} from Table 1 of Sigl [19].

Figure 1 shows the plot of experimentally measured K_{Ic} against the parameter P. The plot

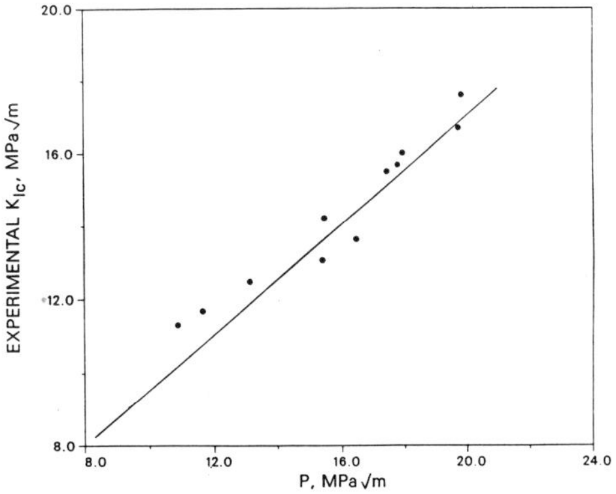

Fig. 1. Plot of experimental K_{Ic} *vs.* parameter P defined in eqn. (15): ——, linear least-squares fit to the data of Sigl [19].

demonstrates that the experimental data conform to the linear relation between K_{Ic} and P as predicted by the present model. From least-squares fitting of the data to a straight line, it was found that $R=0.78$. The measured and the calculated values of K_{Ic} are compared in Table 1. As can be seen, the agreement is good.

The above result ($R=0.78$) implies that initiation of crack growth occurs when the crack-tip-opening displacement is about $0.8\lambda_c$. Using eqn. (8), the critical strain at which crack growth initiation occurs is $\varepsilon_c=0.12$. This value is the strain at ligament instability and is smaller, as expected, than the average maximum strain of 0.5 at ligament fracture reported by Sigl *et al.* [7].

The expected values of δ_c for non-hardening materials (*i.e.* materials with strain-hardening exponent $n=0$) are given by $\delta_c=0.5l^*$, under the assumption that the region of intense strain ahead of the crack tip envelopes a characteristic microstructural dimension (*i.e.* $D=l^*$, eqn. (1)). For strain-hardening materials (with $n>0$) the value lies in the range $\delta_c=(0.5\text{–}2.0)l^*$, depending on the extent of strain hardening. Current understanding of the *in situ* deformation characteristics of the binder indicates that the binder exhibits an appreciable degree of strain hardening [7]. Thus the result $\delta_c=0.78\lambda_c$ lies within the expected range.

According to the present model, initiation of crack growth occurs when the region D of intense strain envelopes the characteristic microstructural distance λ_c, or alternatively when the strain at the distance λ_c from the crack tip reaches a critical value. An increase in toughness with an increase in λ_c results because the region D has to spread over a greater distance before crack growth can occur. Alternatively, the critical strain has to be exceeded over a greater interval as λ_c increases. The dependence of fracture toughness on V_B comes through the consideration of the effective yield strength of the binder in presence of microcracks. As can be seen from eqn. (14), an increase in V_B leads to an increase in the fracture toughness at constant values of other parameters.

It is of interest to note that eqn. (14) is similar in form to that derived by Budiansky [20] for the fracture toughness of a composite model involving crack bridging, namely

$$K_{Ic}=(EG_0+Bf\sigma_0ER)^{1/2} \tag{17}$$

where B is a constant. The last term in the equation above shows a dependence of K_{Ic} on the same parameters as in eqn. (14). This is not surprising since eqn. (17) involves the same mechanism in the last stages of fracture as in eqn. (14), *i.e.* rupture of ligaments.

The model of Budiansky [20] takes into account the direct contribution of the carbide cracking to the total fracture energy, as indicated by the first term on the right-hand side of eqn. (17). It is therefore applicable to a wide range of compositions. The present model assumes that the dominant contribution to crack growth resistance comes from the ductile binder. This assumption is justifiable for alloys with binder volume fractions roughly in the range 10%–25% [7]. For alloys with very low V_B, the above assumption does not hold true, since the intrinsic crack growth resistance of the carbide phase will also be a contributing factor. For example, in the extreme case where $V_B=0$, the model predicts zero fracture toughness for the WC–Co composite, whereas an experimental measurement

TABLE 1 Comparison of experimental and predicted K_{Ic} (data of Sigl [19])

Alloy	V_B	d (μm)	λ_c (μm)	P (MPa m$^{1/2}$)	*Experimental* K_{Ic} (MPa m$^{1/2}$)	*Calculated* K_{Ic}[a] (MPa m$^{1/2}$)
6F	0.100	0.66	0.24	10.87	11.31	9.74
6F/M	0.098	0.83	0.26	11.64	11.68	10.30
8M	0.129	1.43	0.51	15.46	14.21	13.69
8C	0.132	2.16	0.72	17.78	15.71	15.74
9F	0.148	0.68	0.31	13.13	12.48	11.62
9C	0.151	1.83	0.69	17.94	16.04	15.88
10M	0.161	1.50	0.58	17.43	15.52	15.42
10C	0.165	2.18	0.84	19.75	16.73	17.48
12F	0.204	0.69	0.35	15.40	13.06	13.63
12M	0.198	1.68	0.74	19.85	17.64	17.57
15F	0.247	0.64	0.42	16.46	13.63	14.58

[a]Calculated with $R=0.78$.

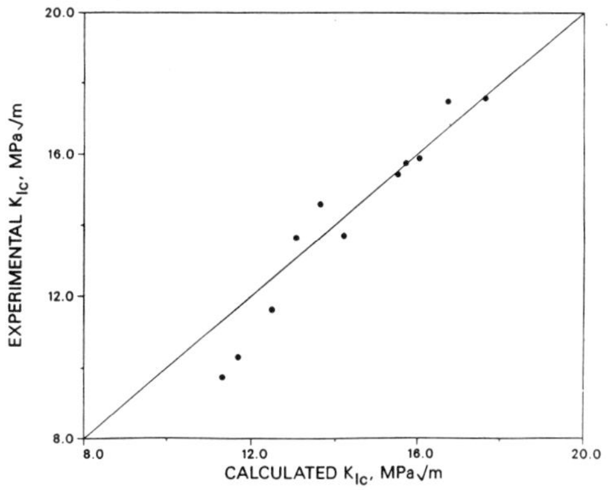

Fig. 2. Comparison of experimental K_{Ic} with that calculated from eqn. (15) with $R = 0.78$ using the data of Sigl [19].

would yield the toughness of the unbonded carbide. Thus the model predictions are expected to underestimate the true composite toughness in the low V_B (low P) range. This does seem to be the case, as can be seen from Fig. 2.

5. Summary

(1) A model relating the fracture toughness of hard metals to their microstructure has been developed. The model is based on the concept originally proposed by Rice and Johnson for ductile failure, that a critical level of strain must be exceeded over a characteristic microstructural dimension before crack growth can commence. Although the process of fracture in hard metals is macroscopically brittle, the above criterion for ductile failure is deemed applicable since the critical stage in the fracture process is the ductile rupture of the binder ligaments surviving in the process zone.

(2) In the present approach, the hard metal microstructure is viewed as being composed of three components: the contiguous carbide, the non-contiguous carbide and the binder. The partially connected contiguous carbide bears most of the load and therefore is the most severely stressed component when compared with the other microstructural constituents. Subsequent analysis is based on the concept of a crack tip process zone in which the contiguous carbide is broken owing to the elevated stresses and hence does not contribute to the load-carrying capacity of the process zone.

(3) The present model predicts a linear relationship between the fracture toughness and a parameter P that is dependent on microstructural variables. The experimental results of Sigl [19] corroborate this prediction fairly well.

(4) The relationship between fracture toughness and the microstructural variables contains one floating parameter R. When the present model is fitted to the experimental data, a value of R ($R = 0.78$) is obtained that lies in the expected range ($R = 0.5$–2.0).

Acknowledgment

This work was supported by the Materials Research Laboratory at Brown University.

References

1 E. Almond, in R. K. Viswanadham, D. J. Rowcliffe and J. Gurland (eds.), *Science of Hard Materials*, Plenum, New York, 1983, pp. 517–561.
2 T. Johannesson, *Proc. 4th Eur. Symp. on Powder Metallurgy, Grenoble*, 1975, Paper 5-11.
3 C. Chatfield, *Proc. 5th Eur. Symp. on Powder Metallurgy, Stockholm*, Vol. 2, 1978, pp. 57–62.
4 M. Nakamura and J. Gurland, *Metall. Trans. A, 11* (1980) 141.
5 R. K. Viswanadham, T. S. Sin, E. F. Drake and J. A. Peck, *J. Mater. Sci., 16* (1981) 1029.
6 E. A. Almond, in N. R. Comins and J. B. Clark (eds.), *Speciality Steels and Hard Materials*, Pergamon, Oxford, 1982, pp. 353–369.
7 L. S. Sigl, H. E. Exner and H. Fischmeister, in E. A. Almond, C. A Brookes and R. Warren (eds.), *Proc. 2nd Int. Conf. on the Science of Hard Materials*, in *Inst. of Phys. Conf. Ser., 75* (1986) 631–644.
8 J. R. Rice and M. A. Johnson, in M. F. Kanninen, W. F. Adler, A. R. Rosenfield and R. I. Jaffee (eds.), *Inelastic Behaviour of Solids*, McGraw-Hill, New York, 1970.
9 L. Lindau, in D. M. R. Taplin (ed.), *Fracture 1977, Proc. 4th Int. Conf. on Fracture, Waterloo, Ontario, June 19–24, 1977*, Vol. 2, University of Waterloo Press, Waterloo, Ontario, 1977, p. 215.
10 J. R. Pickens and J. Gurland, *Mater. Sci. Eng., 33* (1978) 135.
11 J. Hong, *Ph.D. Thesis*, Brown University, Providence, 1981.
12 R. K. Pandey and S. Banerjee, *Eng. Fract. Mech., 10* (1978) 817–829.
13 R. O. Ritchie, W. L. Server and R. A. Wallaert, *Metall. Trans. A, 10* (1979) 1557–1570.
14 K. H. Schwalbe, *Eng. Fract. Mech.*, 9 (1977) 795–832.
15 C. A. Berg, in M. F. Kanninen, W. F. Adler, A. R. Rosenfield and R. I. Jaffee (eds.), *Inelastic Behaviour of Solids*, McGraw-Hill, New York, 1970, p. 171.

16 H. C. Lee and J. Gurland, *Mater. Sci. Eng., 33* (1978) 125.
17 R. M. McMeeking, *J. Mech. Phys. Solids, 25* (1977) 357.
18 H. Doi, Y. Fujiwara, K. Miyake and Y. Oosawa, *Metall. Trans. A, 1* (1970) 1417.
19 L. S. Sigl, *Ph.D. Thesis*, Max-Planck-Institut für Metallforschung, Stuttgart, 1985.
20 B. Budiansky, paper presented at the *3rd Int. Conf. on the Science of Hard Materials, Nassau, The Bahamas, November 9–13, 1987.*

Pre-compression, Internal Stresses and Coercivity in WC–Co*

F. R. N. NABARRO and G. VEKINIS†

Condensed Matter Physics Research Unit, University of the Witwatersrand, Johannesburg, and National Institute for Materials Research, Council for Scientific and Industrial Research, Pretoria (South Africa)

S. BARTOLUCCI LUYCKX‡

Condensed Matter Physics Research Unit, University of the Witwatersrand, Johannesburg (South Africa)

(Received November 9, 1987)

Abstract

The results of X-ray stress and magnetic coercivity measurements on axially compressed samples of WC–6wt.%Co are interpreted using a model in which the cobalt and the contact regions of the carbide grains deform plastically under axial stresses of 3–4 GPa, while the bulk of the carbide grains deforms only under stresses of 4.5–5.5 GPa. Because the contact regions are very small and inhomogeneously stressed, the high stresses in them do not contribute appreciably to the measured position of the X-ray peak. The changes in coercivity are interpreted better in terms of anisotropy induced by internal stress than in terms of crystal magnetic anistropy.

1. Introduction

The preparation of the samples used in this work and the experimental procedures employed have already been described by Vekinis and Bartolucci Luyckx [1]. The material used was WC–6wt.%Co with a mean grain size of 2 μm. Cylindrical samples were cyclically compressed to various stresses by axial loads, fracture being avoided by enclosing the samples in shrink-fitted steel rings. The magnetic coercivity was measured, the samples were cut and polished on a plane through the axis, and X-ray stress measurements were made on this plane for both WC and cobalt in both axial and transverse directions.

2. Internal stresses measured by X-rays

The samples were prepared by hot sintering. The coefficients of thermal expansion are 5.9×10^{-6} K^{-1} for WC and 14.2×10^{-6} K^{-1} for cobalt [2]. After cooling, the WC is therefore in compression, measured to be about −80 MPa and, to maintain elastic equilibrium, the ductile cobalt (volume fraction about 0.10) must be in tension of about 720 MPa = 0.72 GPa. X-ray stress measurements are difficult, the stress system is distorted by the relaxation of the normal stress, and this relaxation cannot be calculated accurately for a composite of the elastically anisotropic phases. The ingenious calculation of Hanabusa *et al.* [3] is valid only if matrix and inclusions have the same elastic constants, a requirement which is not even approximately met in the present system. Nevertheless, we believe that the X-ray measurements provide a useful guide and, in particular, that fractional changes in the stresses have been measured reasonably accurately.

If σ represents a single cartesian component of stress, such as the axial tension, if the stresses in the two components are σ_{WC} and σ_{Co}, and if their volume fractions are f_{WC} and f_{Co}, then mechanical equilibrium requires that

$$f_{WC}\sigma_{WC}+f_{Co}\sigma_{Co}=0 \qquad (1)$$

The initial stress in the cobalt, before pre-compression, quoted above has been adjusted to be compatible with this requirement.

The relation between eqn. (1) and the internal stress measured by conventional X-ray tech-

*Paper presented at the 3rd International Conference on the Science of Hard Materials, Nassau, The Bahamas, November 9–13, 1987.

†Present address: Department of Engineering, University of Cambridge, Cambridge, U.K.

‡Present address: Schonland Centre for Nuclear Sciences, University of the Witwatersrand, Johannesburg, South Africa.

0921-5093/88/$3.50

© Elsevier Sequoia/Printed in The Netherlands

niques is not close. If σ is one of the cartesian stress components σ_{xx}, σ_{xy} or σ_{xz}, then eqn. (1) applies to the mean value of each of the components over any mathematical plane which is perpendicular to the x axis and divides the sample into two parts. By summing the contributions from an equally spaced set of parallel cuts, we see that eqn. (1) also applies to the mean value of the component over the whole volume of the sample. However, if we replace the mathematical cut by a physical cut perpendicular to the x axis, it is just these components σ_{xx}, σ_{xy} and σ_{xz} which relax to zero. The stress components σ_{yy}, σ_{yz} and σ_{zz} in the plane of the cut also relax. Their values measured by X-rays are different from their values before the cut was made. We believe that these differences, depending on Poisson's ratio, will not invalidate the measurements as qualitative guides to the stress in the original sample. However, let us suppose that we write eqn. (1) for the stress component σ_{yy}. It applies to the average over any mathematical cut perpendicular to the y axis, or to an average through the whole sample. It need not apply to an average taken over the actual physical cut surface perpendicular to the x axis. (Let us consider, for example, a heated bimetallic sheet with the surfaces of the sheet perpendicular to the x axis. One of the exposed x surfaces has σ_{yy} and σ_{zz} positive everywhere; the other has σ_{yy} and σ_{zz} negative everywhere.) Even if the measured relaxed stresses in the plane of the cut surface are reasonable estimates of the original unrelaxed stresses, their average value over the cut surface can be expected to obey eqn. (1) only if some restrictive conditions are satisfied.

In fact, the design of the experiments under consideration already implies that these restrictive conditions hold. What is measured is the stress pattern in a limited region of the cut surface. It is implied that these stresses are representative of those over the whole cut surface. The implied model, which we believe to be reasonable although unproved, is that, if macrostresses are present either before or after compression, they are small in comparison with the system of microstresses in which the stresses in a WC grain are balanced by the stresses in the surrounding cobalt layers. If this assumption is justified and, moreover, the microstress system is statistically homogeneous throughout the sample, then both eqn. (1) and the microstresses measured in a small region of the sample are representative of volume averages and must be compatible with one another. In the absence of fuller data, we shall discuss the observations on the assumption that the approximation which has been discussed will turn out to be justifiable.

The X-rays penetrate more deeply into cobalt than into WC. However, since the penetration depth is of the order of the WC grain size, and the volume concentration of cobalt is small, only over a small proportion of the surface will both incident and reflected beams travel solely through cobalt. The stresses in the two components are therefore measured at roughly the same depth.

3. The change in axial stress on pre-compression

Figures 1(a) and 1(b) show the changes in axial stresses in WC and in cobalt respectively after pre-compression to 100 cycles to various stresses. We initially concentrate attention on axial stresses. The main features are as follows.

(1) The internal stresses change rapidly when the applied stress is in the range 3–4 GPa, and the changes then saturate.

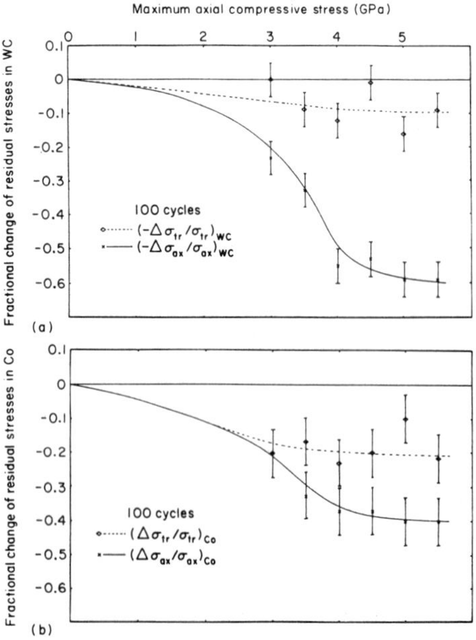

Fig. 1. (a) Fractional changes in the axial and transverse stresses in WC after compression of WC–Co by 100 cycles to various maximum stresses; (b) the corresponding fractional changes in cobalt.

(2) The fractional changes in stress are comparable, and the absolute changes in stress are therefore much greater in cobalt than in WC.

(3) Pre-compression reduces the measured tensile stress in cobalt and increases the measured compressive stress in WC, *i.e.* pre-compression renders the axial stresses less tensile both in WC and in cobalt.

The last observation is incompatible with relation (1). The observations of Krawitz *et al.* [4], using neutron diffraction, did not show this incompatibility. The reason for the discrepancy is not clear, and the neutron diffraction results, which sample the specimen to a much greater depth, could be more reliable. To explain our own observations, we consider the microstructure of the material shown in Fig. 2. There are many narrow regions of contact between WC grains, marked X in Fig. 3. Since Young's modulus is 717 GPa in WC and 211 GPa in cobalt [2], both internal and applied stresses are heavily concentrated in those narrow WC necks.

Fig. 2. Microstructure of a WC–Co sample.

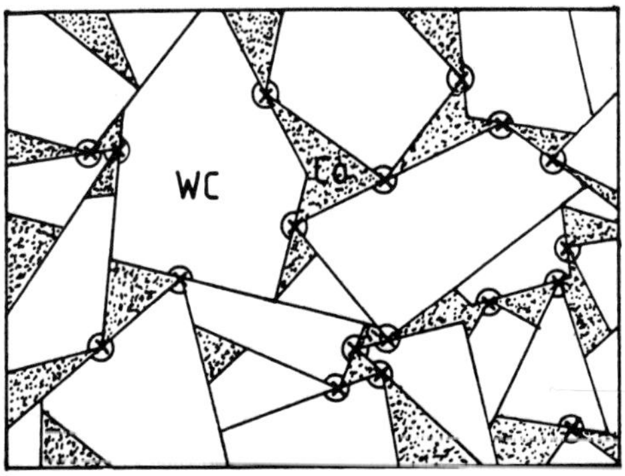

Fig. 3. Narrow regions of contact X between WC grains.

4. Interpretation of the measured initial axial stresses

We first consider the internal stresses produced by cooling after sintering. Cobalt, with a high coefficient of thermal expansion, shrinks onto the WC grains, which are almost completely rigid. The resulting stress system is best visualized by imagining that a rigid WC framework is expanding into an elastic cobalt matrix in which thermal effects need not be considered. Figure 4 shows two adjacent WC grains joined by a narrow neck, and embedded in cobalt. Whenever the cobalt is in contact with the WC, the stresses in the cobalt are of the order of the elastic modulus of cobalt multiplied by the uniform strain in the WC. At a point such as Y, the stresses in the WC are of the same order as the stresses in the cobalt. In the region of the neck X, however, the force exerted by the cobalt on an area of WC comparable with the total area of a WC grain is balanced by the stress in the narrow neck X. Because the area is small, the stress in the neck is large.

This large stress in the neck hardly affects the measured position of the WC X-ray peak. The neck regions are small, occupying perhaps less than 1% of the total value of the WC grains. They have linear dimensions of the order of 10–100 nm, and, in the real sample if not in our model, they are very inhomogeneously strained. They therefore contribute an X-ray peak of large width and small area. When this is superposed on the sharp high peak contributed by the bulk of the WC grain (Fig. 5), the position of the top of the

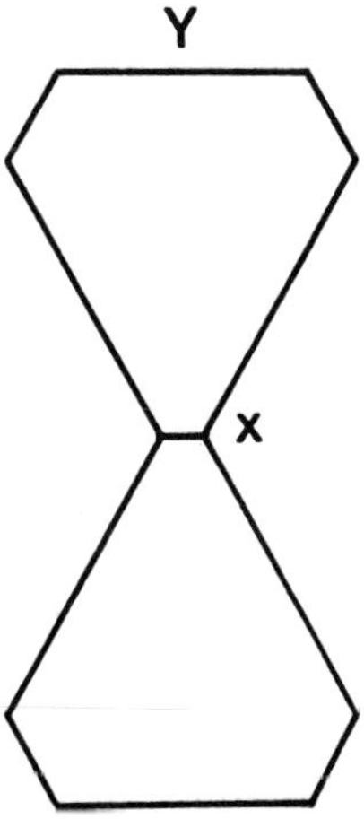

Fig. 4. Detail of narrow neck X between two WC grains.

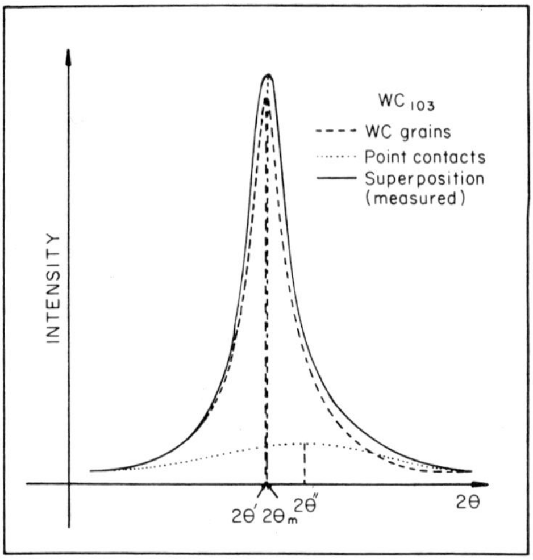

Fig. 5. The superposition of a broad displaced X-ray line of small area on a sharp line of large area, making only a small shift in the position of the maximum.

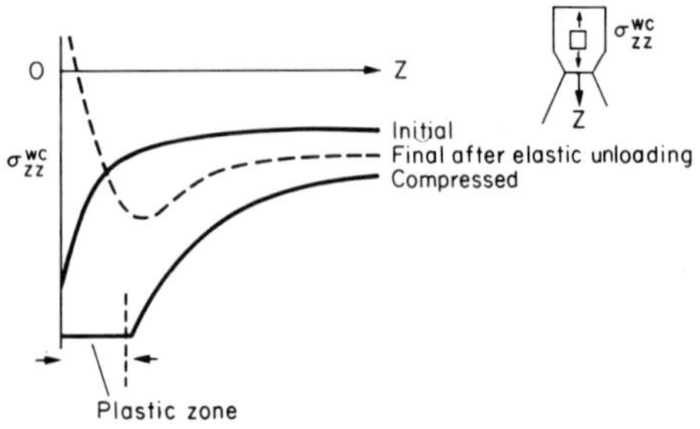

Fig. 6. Schematic representation of the axial tensile stress in the neighbourhood of the neck X in Fig. 4 as a function of distance in the XY direction, in the initial state, under maximum external load, and after removal of the external load.

peak is hardly shifted. Thus the X-rays do not measure the high compressive stresses present in the neck regions of the as-sintered material.

5. Interpretation of the measured changes in axial stress on pre-compression

When a sample is compressed axially, the applied stress is concentrated in the WC necks parallel to the axis of compression. Although the compressive strength of the material is about 4.9 GPa [5], the stress concentrations in the necks cause them to begin to yield at an applied stress of about 3 GPa. This transfers the load to the transverse layers of cobalt, which are now in axial compression to a stress equal to their compressive yield stress. On unloading, these cobalt layers go back into tension, but their axial elongation is restrained by the plastically compressed axial necks. Because of this restraint, they do not expand axially to their original thickness, and so they show reduced axial tensile stresses, as is measured. (We neglect for the time being the effects of the plastic strain in the cobalt, which tends to flow from transverse into axial channels.) This reduction in the tensile stress in the transverse cobalt layers must be compensated by a reduction in the mean axial compressive stress in the WC, whereas what is measured is an increase in this component. This is because the large compressive stresses in the axial necks, which the X-rays do not measure, have been relaxed or even reversed and, if reversed, are again not measured.

Figure 6 provides a schematic representation of the axial tensile stress in the neck X in Fig. 4 as a function of distance in the direction XY in the initial state, under maximum external load, and after removal of the external load.

6. Stresses transverse to the axis of pre-compression

The measured changes in stresses transverse to the axis of pre-compression are also shown in Figs. 1(a) and 1(b). The principal features are as follows.

(1) The transverse internal stresses change under the same applied stresses as do the longitudinal internal stresses.

(2) The changes are in the same directions as those in the axial stresses.

(3) The changes in transverse stresses are appreciably smaller than the changes in the axial stresses.

(4) Whereas the fractional change in the axial stress in WC exceeds the fractional change in axial stress in cobalt, the fractional change in transverse stress in WC is smaller than that in cobalt.

Observations (1) and (3) are in agreement with our model, in which the controlling mechanism is the plastic collapse of axial WC necks. The model also explains observation (4), *i.e.* that there is very little change in the transverse stress in WC. The model provides no obvious mechanism for such a change. To understand the reduction in transverse tension in cobalt in observation (2), we recognize that under axial pressure the transverse pools of cobalt are extruded outwards into the axial pools, thereby coming under transverse compression. When the axial load is removed, this process cannot completely reverse itself, because the axial WC necks have been plastically shortened, and so the transverse tension in the transverse cobalt pools is reduced.

7. Magnetic coercivity

Figure 7 shows the changes in the magnetic coercivity of the cobalt after loading to various stresses for 100 cycles [6]. In the stress range 3–4.5 GPa the change in coercivity is roughly proportional to the change in the mean axial stress in the cobalt, but in the range 4.5–5.5 GPa the coercivity falls steeply while the mean internal stress reaches a plateau. The lower stress range is that in which the plastic deformation is confined to the cobalt and the axial WC necks, while in the latter range the bulk of the WC grains is beginning to yield plastically. The cobalt grain size is much larger than the WC grain size [7], being characteristically 50–100 μm, or even 500–1000 μm [2], and we believe that the magnetic coercivity is controlled by the spatial variations in the internal stress in the cobalt; the energy of a Bloch wall is lower in some regions of the cobalt than in others, and therefore it cannot move freely. When plastic deformation occurs but is confined to small regions of the composite, the internal stresses in the cobalt are very non-uniform, and the coercivity is high. When general yield occurs,

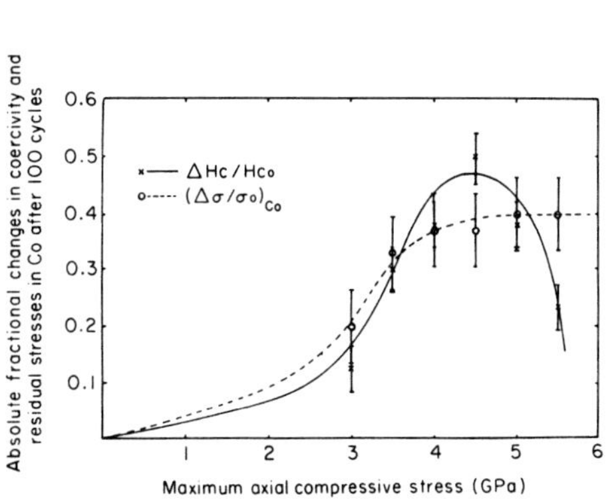

Fig. 7. Change in the magnetic coercivity of cobalt after compression of WC–Co by 100 cycles to various maximum stresses.

the stress throughout the WC is (if we neglect yield point phenomena and work hardening) equal to the flow stress of WC, and the stress throughout the cobalt is equal to the flow stress of cobalt. Since the stress in cobalt is uniform, the coercivity is low. Krawitz *et al.* [4] observed narrowing of the neutron diffraction peaks in some fatigued samples.

An alternative explanation of the changes in coercivity [8, 9] is based on the observation that the f.c.c. cobalt retained in the as-sintered sample is partly transformed into h.c.p. cobalt during plastic deformation. While little is known about the magnetic properties of f.c.c. cobalt at room temperature, there is little doubt that the presence of h.c.p. lamellae will enhance its coercivity. However, there does not seem to be any clear correlation between the extent of the phase change and the change in the coercivity. For a sample with 9.7 wt.% Co, Hara and Ikeda found no evidence of h.c.p. cobalt until the plastic strain reached 0.5%, the fraction of h.c.p. cobalt rising to about 15% at a plastic strain of about 1.7%. For the same material, the coercivity began to rise at a plastic strain of about 0.13%, although the rise usually began at 0.15–0.2%. While coercivity may be a more sensitive indicator of phase inhomogeneity than X-rays is, the discrepancy seems to be significant. In these experiments the coercivity continued to increase up to plastic strain of nearly 4%.

The observations of Miyake *et al.* on the fatigue of a high cobalt (20 wt.% Co) alloy to a varying number of cycles at a constant maximum stress of 100 kg mm^{-2} are difficult to reconcile with those of Hara and Ikeda. The X-ray peaks of h.c.p. cobalt become equal in intensity to those of f.c.c. cobalt, and, on a logarithmic scale of number of cycles, the major rise in the coercive force seems to occur after a substantial rise in the proportion of h.c.p. cobalt, and the coercivity continues to rise as the deformation proceeds.

In an alloy containing 17 wt.% Co, Vasel *et al.* [10] found 44.5 wt.% h.c.p. Co after a plastic strain of about 4.5%.

We may first attempt to explain these observations in terms of the crystal magnetic anisotropy of the unstrained material. We note first that the principal anisotropy coefficient of h.c.p. cobalt at room temperature [11] is $(40–53) \times 10^5$ erg cm^{-3}, much greater than the principal coefficients of cubic iron (4.2×10^5 erg cm^{-3}) or nickel (-0.34×10^5 erg cm^{-3}) at room temperature. There seem to be no data for f.c.c. cobalt at room temperature. The observed values decrease from -2.30×10^5 erg cm^{-3} at 500 °C to 0 at 1000 °C. The temperature dependence is far from linear, and extrapolation to room temperature is very difficult. The value could numerically exceed -15×10^5 erg cm^{-3} at room temperature and be comparable with the value for h.c.p. cobalt. Even so, h.c.p. cobalt with only two easy $\langle 0001 \rangle$ directions of magnetization and only 180° Bloch walls is likely to show a much higher anisotropy-based coercivity than is f.c.c. cobalt with eight easy $\langle 111 \rangle$ directions of magnetization and 70.5° Bloch walls. If f.c.c. domains of all four possible orientations are interleaved, the coercivity is likely to increase monotonically as the transformation from f.c.c. to h.c.p. progresses. If, however, the imposed stresses cause one of the four possible h.c.p. orientations to predominate in a particular region, the coercivity could be expected to pass through a maximum when about half of the f.c.c. cobalt had transformed to h.c.p. but not appreciably before this. Extrapolation of the curve of Hara and Ikeda suggests that a 50–50 composition would be reached only at very large strains. Vekinis and Bartolucci Luyckx did not observe any X-ray reflections arising from the h.c.p. phase. If the fall in coercivity in their experiments arose because the h.c.p. phase became predominant, they would almost certainly have observed the corresponding h.c.p. peaks.

In summary, it seems that the decrease in coercivity observed by Vekinis and Bartolucci Luyckx at high stresses and high cycles of fatigue can be explained on the crystal anistropy model only on the assumption that the cobalt had been converted almost entirely into the h.c.p. form, and that in each region a single h.c.p. orientation was strongly predominant, while the observations find a ready interpretation of the internal stress model.

Acknowledgment

We are grateful to a referee for constructive comments.

References

1 G. Vekinis and S. Bartolucci Luyckx, in E. A. Almond, C. A. Brookes and R. Warren (eds.), *Proc. 2nd In'. Conf. on the Science of Hard Materials, Rhodes, September 23–28, 1984*, in *Inst. Phys. Conf. Ser., 75* (1986) 591–600.
2 H. E. Exner, *Int. Metall. Rev., 24* (1979) 149–173.
3 T. Hanabusa, K. Nishioka and H. Fujiwara, *Z. Metallkd., 74* (1983) 307–313.
4 A. D. Krawitz, R. Roberts and J. Faber, in E. A. Almond, C. A. Brookes and R. Warren (eds.), *Proc. 2nd Int. Conf. on the Science of Hard Materials, Rhodes, September 23–28, 1984*, in *Inst. Phys. Conf. Ser., 75* (1986) 577–589.
5 H. Doi, *Elastic and Plastic Properties of WC–Co*, Freund, Tel Aviv, 1974, p. 66.
6 G. Vekinis and S. Bartolucci Luyckx, *Mater. Sci. Eng., 96* (1987) L21–L23.
7 V. K. Sarin and T. Johannesson, *Met. Sci., 9* (1975) 472–476.
8 K. Miyake, Y. Fujiwara and K. Nishigaki, *J. Jpn. Miner. Soc., 32* (1968) 1128–1131.
9 A. Hara and T. Ikeda, *Trans. Jpn. Inst. Met., 13* (1972) 128–133.
10 C. H. Vasel, A. D. Krawitz, E. F. Drake and E. A. Kenik, *Metall. Trans. A,, 16* (1985) 2309–2317.
11 Batelle Memorial Institute, *Cobalt Monograph*, Centre d'Information du Cobalt, Brussels, 1960.

Materials Science and Engineering, A105/106 (1988) 343–351

The Mechanical Behaviour of Cemented Carbides at High Temperatures*

H. G. SCHMID†, D. MARI and W. BENOIT

Institut de Génie Atomique, École Polytechnique Fédérale Lausanne, Lausanne (Switzerland)

C. BONJOUR

Stellram SA, Nyon (Switzerland)

(Received November 9, 1987)

Abstract

Specimens of WC–11wt.%Co were tested under three-point bending conditions at temperatures ranging between 20 and 1000 °C to investigate the effect of chromium and ruthenium additions on the mechanical behaviour of cemented carbides. Below 800 °C, a linear elastic brittle behaviour is observed. Above 800 °C, creep exponents of about $n=2$ and activation energies between 3 and 5 eV were measured. Pore formation and growth were observed during creep. These results agree well with models based on creep by grain boundary sliding and pore formation.

Displacement-controlled bend tests of chevron-notched specimens were performed to introduce stable crack growth. Above 800 °C, creep crack growth occurs owing to pore formation and growth. The pore distribution around the crack tip has been observed using a scanning acoustic microscope. It is shown that the equivalent strain distribution around a creep crack according to the models of Riedel agrees well qualitatively with the experimentally determined pore density. The transition between the linear elastic and viscoplastic behaviour (i.e. creep) is accompanied by a Portevin–Le Chatelier type of effect. If creep occurs in a tool made from this material, the edge blunts and is no longer useful for cutting. A characteristic time t_{ft} is introduced describing the transition between the brittle and creep behaviour. It is shown that cemented carbides with a ruthenium addition have longer transition times than those without. Therefore, cemented carbides with ruthenium are advantageous as cutting tool materials since they can be used for longer times at a high temperature.

1. Introduction

Cemented carbides are ceramic–metal composites which combine the high hardness of ceramics with a fracture toughness comparable with that of metals at room temperature [1–7]. At high temperatures (above 600 °C), rapid softening takes place [8–11]. Up to 1000 °C, creep associated with grain boundary sliding and pore formation and growth has been found to be responsible for the softening of WC–11wt.%Co [9, 10]. The cemented carbides are mostly used for cutting tools. If creep occurs, then the material blunts and will obviously lose its desirable properties as a cutting material. It has been shown that a transition time between the linear elastic brittle behaviour and the viscoplastic creep behaviour is able to characterize the time limit of a cutting tool in service at a high temperature [9, 10]. Empirical results obtained at Stellram SA have shown [25] that cemented carbide cutting tools with approximately the same weight percentages of cobalt but with a small amount of ruthenium give much longer lifetimes.

This work is the continuation of our investigations [9, 10] of the high temperature fracture of WC–11wt.%Co specimens, and the changes in the mechanical behaviour of WC–11wt.%Co cemented carbides with additions of chromium and ruthenium are studied.

2. Materials and procedures

The nominal specimen compositions were WC–11wt.%Co, WC–11wt.%Co–1.5wt.%Cr and WC–11wt.%Co–1.65wt.%Ru. The average grain

*Paper presented at the 3rd International Conference on the Science of Hard Materials, Nassau, The Bahamas, November 9–13, 1987.

†Present address: Département Matériaux, École Polytechnic Fédérale Lausanne, Lausanne, Switzerland, and Nordische Universität, Institut für Werkestofftechnik, Neumünster, FRG.

0921-5093/88/$3.50

© Elsevier Sequoia/Printed in The Netherlands

TABLE 1 The average bending fracture stress σ_{BB} and the Weibull parameter $m_{Weibull}$

Composite (wt.%)	σ_{BB} (MPa)	$m_{Weibull}$
WC–11Co	2458 ± 170	14
WC–11Co–1.5Cr	2497 ± 475	6
WC–11Co–1.65Ru	3478 ± 450	7

size of the powder of all specimens was 3.6 μm. The mean free path within the cobalt phase was about 1.1 μm for all specimens. The fracture values in Table 1 were obtained at room temperature for three-point bending tests with specimens of dimensions 35 mm × 7 mm × 3.5 mm.

The fracture tests were performed with a closed-loop testing machine (Instron 1361). The three-point system is surrounded by an induction furnace in a vacuum chamber [9, 10]. Three-point single-edge-notched bending specimens of the dimensions shown above, with a chevron notch were used to obtain stable crack growth. Precracking [2] to obtain straight crack fronts was not necessary since the measured quantities t_{ft} and $\bar{R}$ (see eqns. (2) and (3)) refer to the crack fronts of any geometry. Therefore, displacement-controlled tests were carried out without precracking where the displacement rates varied between 100 and 0.2 μm s^{-1}. The temperature was measured using a pyrometer and/or a Pt–Rh thermocouple.

The higher bending fracture stress of the WC–11wt.%Co–1.65wt.%Ru at room temperature may be one indication that there is an improvement in materials behaviour. However, the Weibull parameter is very low. Indeed, some of the measured bending fracture stresses for WC–11wt.%Co–1.65wt.%Ru have been below the highest values for WC–11wt.%Co. Furthermore, the longer lifetimes at elevated temperatures is also not explained by the room temperature bending stress values.

3. Measurement methods

In this work, we shall characterize the creep behaviour by Norton's law:

$$\dot{\varepsilon} = \frac{\dot{\sigma}}{E} + B\sigma^n \tag{1a}$$

$$B = A\exp\left(-\frac{\Delta Q}{kT}\right) \tag{1b}$$

where $\dot{\varepsilon}$ is the strain rate, σ is the applied stress, E is Young's modulus, A and B are constant factors, n is the creep exponent, ΔQ is the activation energy, k is the Boltzmann constant and T is the temperature.

The high temperature fracture test is characterized by the crack resistance R and the transition time t_{ft} which has been introduced in our preliminary work [9, 10]. The reasons for using these parameters are as follows.

(1) The concepts of linear elastic fracture mechanics (K_{Ic}, Weibull statistics, S–p–T diagram etc.) and elastic–plastic fracture mechanics (J integral, crack-opening displacement etc.) are not valid for describing the behaviour of elastic–viscoplastic materials [12–15]. Any arbitrary values of K_{Ic} and J_{Ic} can be measured.

(2) In the case of linear elastic or elastic–plastic fracture the critical values K_c or/and J_c characterizing the material behaviour at a given temperature have one value. In the case of viscous materials, C^* or the time t_f to failure depends on the temperature and the applied stress [14–17]. This situation makes it more difficult to distinguish between different qualities of materials.

Thus, we would like to have only one value which characterizes the behaviour of the viscous material at a given temperature. Therefore, we derived a transition time t_{ft} to failure from the time-to-failure concept [9, 10].

This transition time t_{ft} is defined as the time when the stress–strain field around the propagating crack tip changes from a linear elastic or elastic–plastic stress–strain field to a viscoelastic or/and viscoplastic stress–strain field [9, 10].

Our method for measuring t_{ft} is as follows. Chevron notched three-point bending specimens with the same geometry are tested with different constant displacement rates while a crack propagates from the initial crack length to fracture. The measured parameters are load P, displacement δ, displacement rate $\dot{\delta}$ and temperature T. The corresponding t_f is given by

$$t_f = \frac{\delta_{max} - \delta_0}{\dot{\delta}} \tag{2}$$

where δ_{max} is the maximum displacement at fracture and δ_0 is the displacement at crack initiation.

If the behaviour of the crack tip field is linear elastic, then the maximum displacement (deformation) δ_{max} is not dependent on t_f. If the crack

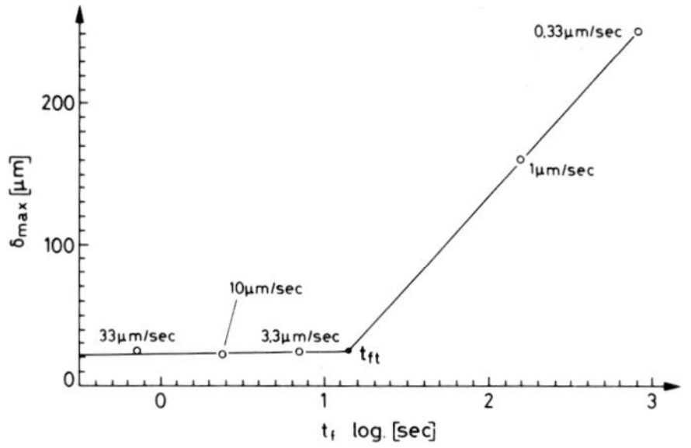

Fig. 1. Maximum displacement δ_{max} at different displacement rates *vs.* time to failure as $\log(t_f)$, to determine the transition time t_{ft} for WC–11wt.%Co tested at 870 °C. (After Schmid [10].)

tip field becomes viscoplastic, then the specimen deforms more and δ_{max} increases. Therefore a plot of δ_{max} *vs.* t_f provides t_{ft} (Fig. 1), where t_{ft} is given by the intersection of the constant $-\delta_{max}$ line at short lifetimes t_f and the increasing δ_{max} due to creep at longer lifetimes [9, 10].

A transition displacement rate $\dot{\delta}_t$ can be defined according to eqn. (2), which is equivalent to the definition of t_{ft}. The transition displacement rate $\dot{\delta}_t$ is the displacement rate which is sufficiently slow that viscoplastic effects take place and the linear elastic fracture behaviour changes to viscoplastic fracture behaviour. The t_{ft} or $\dot{\delta}_t$ value depends on the size and the geometry of the chosen specimens. The main goal of this work is to characterize the mechanical behaviour of specimens of the same geometry and the same microstructure but of different chemical compositions. Thus the proposed quantity t_{ft} satisfies this purpose.

The fracture toughness when linear elastic brittle behaviour predominates has been characterized by measuring the average crack resistance $\bar{R}$ [9, 10]; this is developed from the work of fracture [16]:

$$\bar{R} = \int_0^{\delta_{max}} \frac{P\,d\delta}{A_m} \tag{3}$$

where A_m is the crack surface created and P is the load.

4. Results

4.1. Controlled crack propagation tests

The preliminary studies have shown that above 800 °C the nature of the load–displacement curves for WC–11wt.%Co indicate elastic–viscoplastic behaviour (Fig. 2). Qualitatively, the same behaviour has been observed for the cemented carbides with chromium or ruthenium additions. Above 900 °C, instabilities occur in the load–dis-

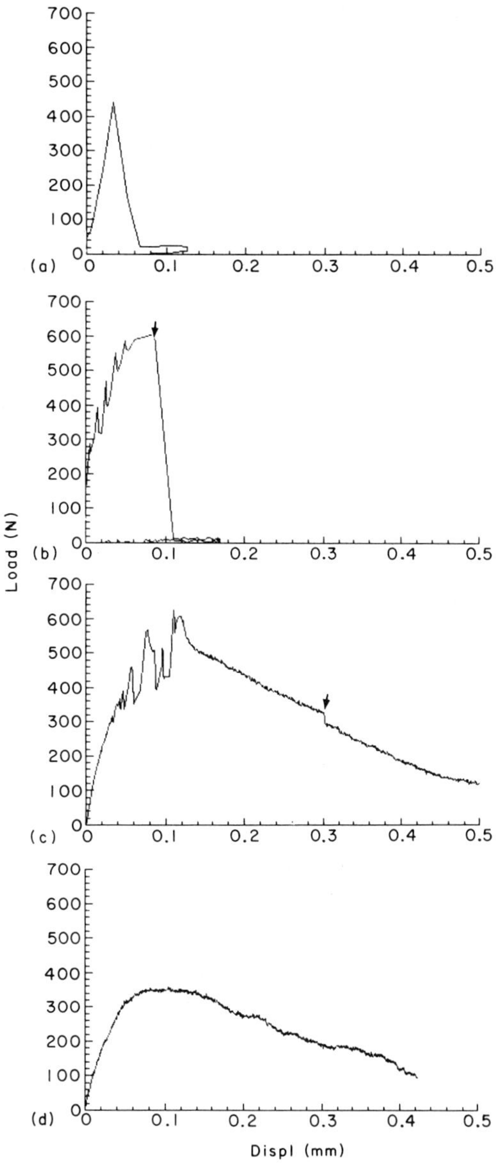

Fig. 2. Load–displacement curves of WC–11wt.%Co–1.65wt.%Ru tested at 940 °C with various displacement rates $\dot{\delta}$: (a) 50 $\mu m\ s^{-1}$; (b) 10 $\mu m\ s^{-1}$; (c) 3 $\mu m\ s^{-1}$; (d) 1 $\mu m\ s^{-1}$. The arrows show unstable crack propagation.

TABLE 2 Crack resistance values $\bar{R}$ for brittle behaviour

T (°C)	$\bar{R}$ (J m^{-2}) *for the following*		
	WC–11wt.%Co	*WC–11wt.%Co–1.5wt.%Cr*	*WC–11wt.%Co–1.65wt.%Ru*
20	400 ± 20	490 ± 80	600 ± 100
640			218
780		313	334
820	350		440
880	400		
910			450
950	500		

placement curves (Fig. 2). These instabilities appear just below the transition from the linear elastic behaviour (Fig. 2(a)) to the viscoplastic behaviour (Figs. 2(b) and 2(c)). If the displacement rate is slow, the instabilities disappear (Fig. 2(d)). The instabilities at increasing load can be compared by the Portevin–Le Chatelier effect [18]. If the load decreases owing to crack extension, instabilities of the crack propagation occur (Fig. 2(c)). It was occasionally not possible to introduce stable crack growth (Fig. 2(b)). The crack propagation was catastrophic. The crack resistance values obtained from the load–displacement curves for linear elastic behaviour ($t_f < t_{ft}$) are shown in Table 2.

4.2. *Creep tests*

The creep tests performed on specimens of WC–11wt.%Co with the same geometry as those used in fracture experiments indicate two distinct behaviours. Below 900 °C, a saturation creep strain is observed (Fig. 3(a)). At 900 °C for stresses higher than 100 MPa, and at 1000 °C for all stresses employed, we obtain "typical" creep curves (Fig. 3(b)) with a stationary state. The point of inflection of stress–strain curves was obtained in order to determine Norton's creep law parameters (eqn. (1)) (Table 3).

Constant-strain-rate tests were also performed. Instabilities of the Portevin–Le Chatelier type were observed for samples of WC–11wt.%Co at 900 and at 1000 °C. The stress–strain curves of hot-pressed pure WC and WC–11wt.%Co obtained under the same testing conditions indicate very different behaviours (Fig. 4). WC and WC–Co have very similar structures; in fact, voids in WC could be considered to be replaced by cobalt in WC–Co. The results obtained favour the idea that grain boundary sliding is predominant in WC–Co creep instead of the deformation of the WC grains. This is also confirmed by the

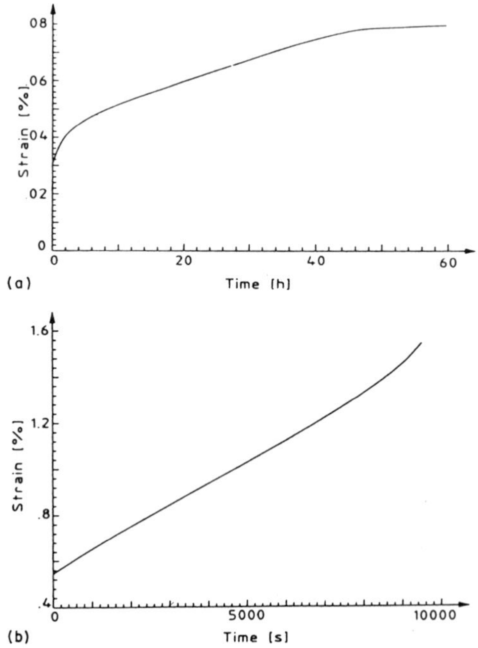

Fig. 3. (a) Strain–time curves for WC–11wt.%Co (a) at 800 °C under a constant stress of 110 MPa and (b) at 900 °C under constant stress of 300 MPa.

TABLE 3 Norton's creep law parameters

T (°C)	n	B (s^{-1} mm^{2n} N^{-n})	ΔQ (eV)
800	2.1	6.8×10^{-14}	4.8
900	2	9.6×10^{-12}	

formation of cavities in the cobalt matrix during creep (Figs. 5 and 6).

4.3. *Microscopy investigations*

When viscoplastic behaviour occurs (see Figs. 2 and 3), a bright zone around the crack tip in the

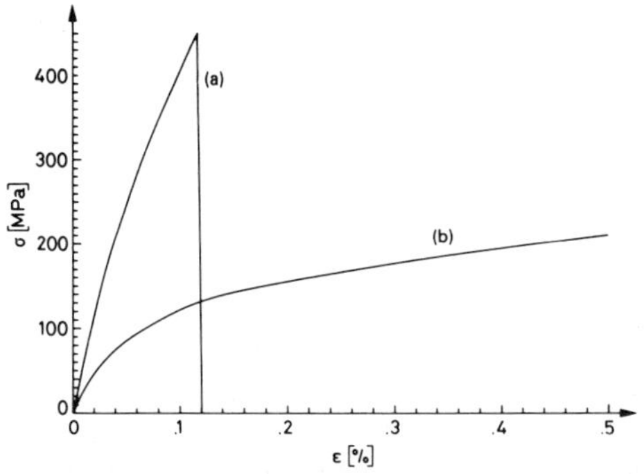

Fig. 4. Stress-strain curves for (a) pure WC and (b) WC–11wt.%Co at 1000 °C with a strain rate $\dot{\varepsilon}$ of $1.6\times10^{-5}\,s^{-1}$.

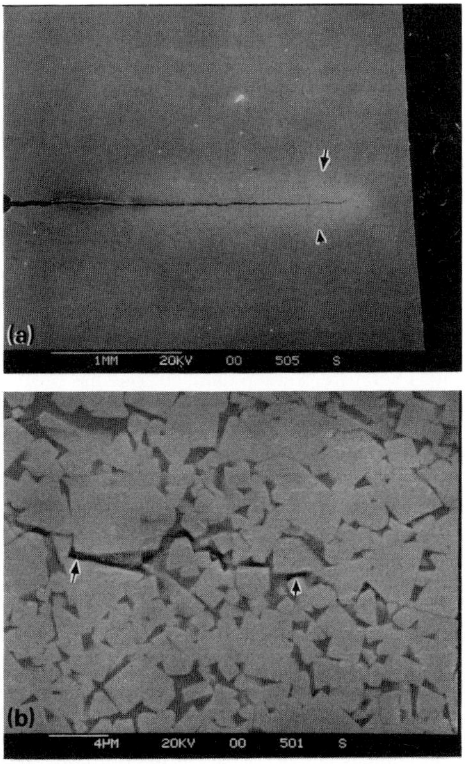

Fig. 5. Scanning electron micrographs of the surface of WC–11wt.%Co tested at 900 °C with $\dot{\delta}=1\ \mu m\ s^{-1}$: (a) the arrows show the bright creep zone; (b) the arrows indicate the pores.

scanning electron micrograph (in Fig. 5(a)) can be detected. Within this zone, creep pores are always observed along the WC–WC and WC–Co interfaces (Fig. 5(b)). In order to enhance the

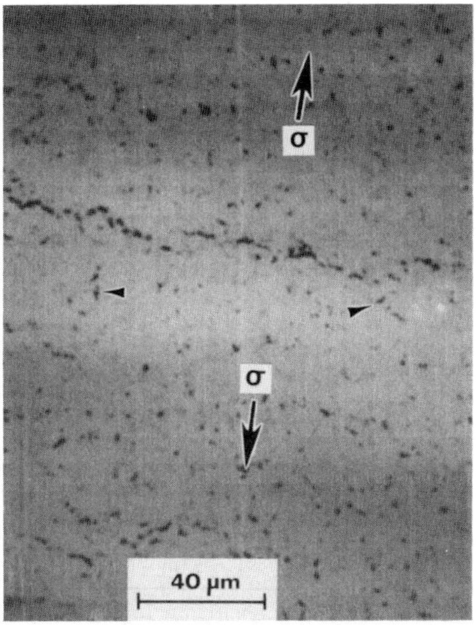

Fig. 6. Scanning acoustic micrographs of WC–11wt.%Co crept at 900 °C and a load of 200 MPa. The arrows show the pores.

contrasts of the small pores, an Ernst Leitz scanning acoustic microscope was chosen (Fig. 6). The black dots (particles) have much larger diffraction contrasts [9, 10] than the small pores do. This technique permits measurements of the density of the pores within large creep zones [9, 10].

5. Discussion

5.1. The load–displacement experiments

The load–displacement experiments show that three different types of mechanical behaviour exist (Fig. 2).

5.1.1. Linear elastic behaviour with stable crack growth at high displacement rates (Fig. 2(a))

The shape of the load–displacement curve is that of a linear elastic brittle material with stable crack growth. This shape is limited by the transition displacement rate $\dot{\delta}_t$. Linear elastic shapes were found at all temperatures for all materials tested. The crack resistance values (see Table 2) were obtained in order to check whether there exist strong deviations from the behaviour that has been investigated extensively by Fantozzi *et al.* [8]. The minimum values of the crack resistance occur at about 600 °C, as reported in ref. 8. This fact confirms the assumption of the model [9, 10] that the crack resistance below 600 °C is effected by the ligament rupture mechanism [2, 5, 7, 9, 10]. The plastic deformation of the unbroken cobalt ligament behind the crack tip controls the crack propagation. Above 500 °C the cobalt becomes weak and the contribution of the ligaments to the crack resistance decreases. Up to 1000 °C, the crack resistance reaches the room temperature values. This increase may be effected by the deformation of the WC skeleton [19, 20]. However, the mechanical behaviour of the materials tested should be investigated in more detail between 500 and 800 °C.

5.1.2. Unstable elastic–viscoplastic behaviour with intermitted stable crack growth or instantaneous fracture below $\dot{\delta}_t$

The instabilities at increasing load (the Portevin–Le Chatelier effect) occur above 900 °C and is observed in all three materials. Behaviour similar to the Portevin–Le Chatelier effect was always accompanied by steps of unstable crack propagation (Fig. 2(c), arrow) or instantaneous fracture (Fig. 2(b), arrow). The range of instabilities below the transition from linear elastic to viscoplastic increased with increasing temperature (Fig. 7). The instabilities were always found just below $\dot{\delta}_t$. Sometimes instabilities in the crack propagation were found without the Portevin–Le Chatelier effect. Between 800 and 900 °C a transition from linear elastic to viscoplastic takes place without the instability effects. Perhaps these instabilities always exist but the range near the point of transition is too small and the steps in displacement rates are too large for them to be detected.

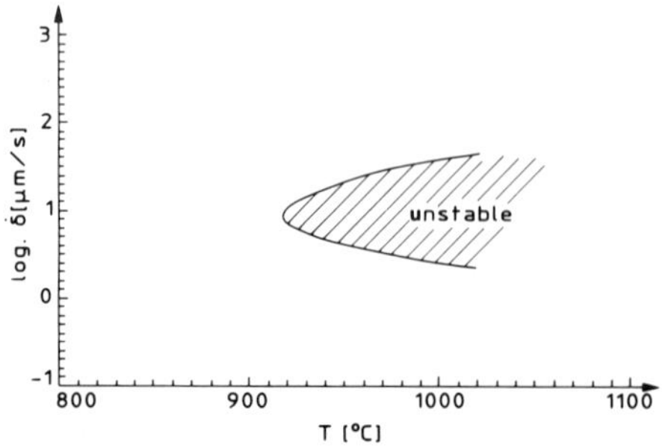

Fig. 7. Range of instabilities (the Portevin–Le Chatelier effect) as shown by the displacement rate as log $\dot{\delta}$ *vs.* temperature *T*.

The theories of cracks in elastic–non-linear viscous materials are almost all focused on the description of non-propagating cracks [14, 15, 21, 22]. Hui and Riedel [23] proposed a model for a propagating crack under uniaxial tension in a non-linear viscous material using Norton's material law (eqn. (1)). One result of this model is that, below a certain minimum growth rate, no stable steady state crack growth is possible if the creep exponent is less than 3. In our creep experiments on WC–11wt.%Co, a creep exponent *n* of 2 was determined. No stable steady state growth should be possible below a certain minimum crack propagation rate. Effectively, unstable crack growth was only found in an intermediate range (Fig. 7).

5.1.3. Stable steady state creep crack growth below $\dot{\delta}_t$

This type of creep crack growth (Fig. 2(d)) occurs if a transition from linear elastic to viscoplastic exists. The micrograph in Fig. 5(a) shows a creep crack which propagated over 90% of the total possible crack extension ($a_{tot} = 3$ mm) at 870 °C and a displacement rate $\dot{\delta}$ of 1 μm s^{-1}. The bright zone around the crack is the damage zone (*i.e.* the zone which remains plastically deformed and damaged by pores behind the creep zone). The thickness of this zone is nearly constant over 90% of the crack length *a* (= 2.7 mm). Therefore, a creep damage zone has propagated through the material with a nearly constant diameter, creating the creep crack.

In order to study the formation of the creep crack in detail, a micrograph obtained by scanning acoustic microscopy of the region around the crack tip of a specimen tested at 900 °C and 1 μm s^{-1} is shown in Fig. 8(a). The crack propagated over 2.7 mm and the displacement was halted. After 5 min the specimen was quenched. The white lines represent the contours of the zones of the same specimen which were obtained with backscattered electrons in a scanning electron microscope (see also Fig. 5(a)). The image enhancement of such micrographs has been reported in detail by Schmid and coworkers [9, 10]. This image shows the typical situation of crack growth by cavity formation and coalescence joining with the main crack, as has been postulated in the models of Riedel [14, 15] (Fig. 8(b)). The circle around the crack tip encloses the creep zone. Within the creep zone the pore density *D* has been measured [9, 10] parallel to the

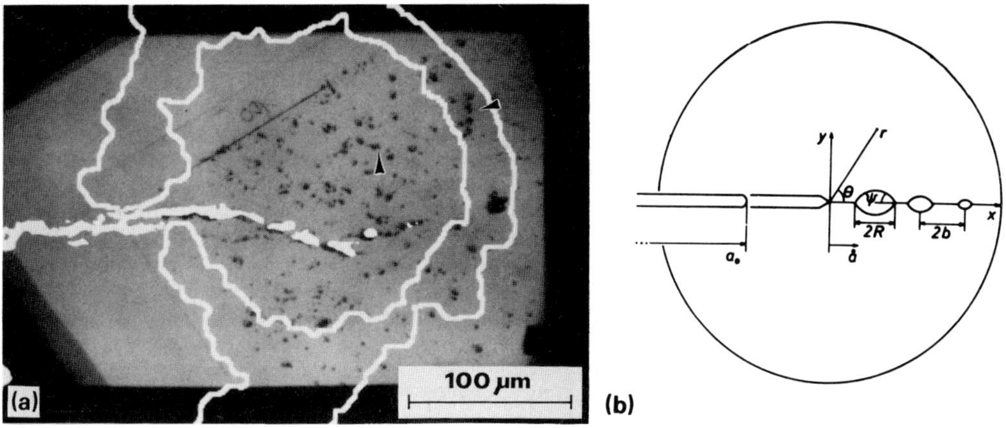

Fig. 8. (a) Enhanced scanning acoustic micrograph of the creep zone around the crack tip in WC–11wt.%Co with the contours of a scanning electron micrograph of the same region according to Schmid and coworkers [9, 10] ($T = 900$ °C; $\dot{\delta} = 1$ µm s^{-1}). The arrows show the pores. (b) Crack growth by cavity formation and coalescence joining with the main crack according to Riedel [14].

crack line at a distance of about 40 µm (Fig. 9(a)):

$$D = \frac{N_s}{S} \tag{4}$$

where N_s is the number of pores (cavities) measured within the surface area S.

The question arises of whether or not there exists a correlation between the pore density and the strain field. Therefore, we calculated the distribution of the equivalent strains $\bar{\varepsilon}$ in arbitrary units according to Riedel [21] along the line indicated in Fig. 9(a):

$$\bar{\varepsilon} \propto r^{-n/(n+1)} \tag{5}$$

where r is the distance to the crack tip.

The theoretical equivalent strain distribution agrees well qualitatively with the experimentally determined pore density (Fig. 9(a)). The continuous increase in the pore density suggests a continuous nucleation of the grain boundary cavities, as suggested by Riedel [22]. Dyson [24] evaluated data on cavity densities and concluded that the number of cavities is related approximately to the creep strain. He proposed an empirical law for the description of the nucleation rate J^*:

$$J^* = \hat{a}\dot{\varepsilon} \tag{6}$$

where $\hat{a}$ is a constant.

Our measurements confirm the empirical assumptions of Dyson [24] in that we can conclude that continuous nucleation and growth of

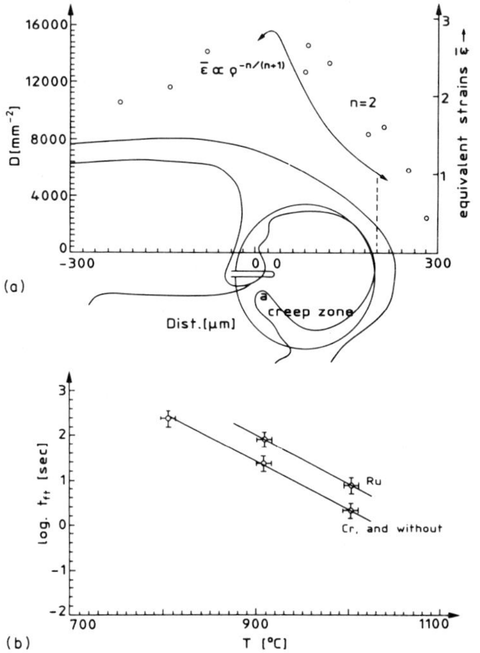

Fig. 9. (a) Density D of the creep pores (○) from the scanning acoustic micrograph (Fig. 8(a)) and the calculated equivalent strain field $\bar{\varepsilon}$ (——) according to Riedel [21] as a function of the position within the creep zone and the damage zone. The contours of the zones (black lines) were derived from a scanning electron micrograph from the same specimen shown by Schmid and coworkers [9, 10]. (b) Transition time to failure, as $\log(t_{ft})$, *vs.* temperature for WC–11wt.%Co, WC–11wt.%Co–1.5wt.%Cr and WC–11wt.%Co–1.65wt.%Ru.

grain boundary cavities caused the crack growth of the WC–Co hard metals tested.

5.2. The effect of ruthenium and chromium as additive

INCO Europe developed ruthenium as a carbide additive for cutting tools that provide faster metal removal [25]. An extensive series of evaluation trials [25] on WC–Co hardmetals with the Ru additive (the X44 and non-ruthenium tips) has been carried out. The main result is "that ruthenium type inserts can continue cutting for 50%–100% longer than equivalent non-ruthenium tips, before critical amounts of wear are reached". Stellram described X44 as an "extraordinary grade of great tenacity and high resistance to heat". Successful experiences with other carbides and ruthenium have been performed by Jackson and Warren [26].

The aim of this study is to describe the observed qualitative experience by physical quantities in order to understand the limiting mechanisms. The understanding of these mechanisms should give rise to the production of better materials for cutting tools. A cutting tool is only useful if no creep deformation can take place. Creep deformation is, of course, a type of wear which limits the use of cutting tools. Therefore the transition time t_{ft} to failure describes the length of time during which creep deformation does not contribute to the wear. Thus, t_{ft} gives an estimate of the time at which "critical amounts of wear are reached". The transition times for WC–11wt.%Co and WC–11wt.%Co–1.5wt.%Cr are similar (Fig. 9(b)), but at a given temperature the WC–11wt.%Co–1.65wt.%Ru remains linearly elastic three times longer. Within the scatter of the measurements, this fact agrees with the experience that "ruthenium-type inserts can continue 50%–100% longer than equivalent non-ruthenium tips" [25]. In terms of materials science, we can conclude that the ruthenium addition makes the WC–Co hard metal more resistant to creep.

Our microscopy investigations of the creep damage confirm the observations that creep is inhibited by the ruthenium additive. Creep pores were never observed in any specimens that were tested to shorter times than t_{ft}. At longer test times than t_{ft}, creep pore formation always takes place. The distribution of the creep pores (Fig. 9(a)) agrees well with the equivalent strain field distribution of elastic–viscoplastic materials (eqn. (5)) according to Riedel [21]. This observation confirms our definition of t_{ft} (see Section 3). The slope of an Arrhenius plot of the transition time *vs.* $1/T$ gives the activation energy ΔQ for the controlling process [10]:

$$t_{ft} = t_0 \exp\left(-\frac{\Delta Q}{kT}\right) \tag{7}$$

where t_0 is a constant factor, k is the Boltzmann constant and T is the temperature.

An activation energy ΔQ of 3.3 ± 1 eV was determined for all three materials. Thus, we suggest that the same creep mechanism is responsible for the creep behaviour of all three materials. However, the ruthenium addition reduces the creep rate.

6. Conclusions

The behaviour of the three materials tested show the following similarities.

(1) The microstructures of the three materials are the same.

(2) We observed the same rupture mechanism up to 600 °C with about the same crack resistances. The WC grains build up a linear elastic skeleton. This skeleton provides the conditions for the ligament rupture mechanism. This mechanism is controlled by the plasticity of the cobalt phase.

(3) If creep occurs, cavity formation and growth take place along the WC–WC and WC–Co grain boundaries within the cobalt phase.

(4) The density of the pores is closely related to the creep strain field according to the theory of viscoplastic solids.

(5) The creep crack growth is controlled by cavity formation joining the main crack.

(6) The activation energies of the creep mechanism obtained by creep tests and by transition time measurements are the same for the three materials.

(7) The development of large creep zones with large deformation can only take place if the skeleton is destroyed.

(8) In bending tests the plastic deformability of the WC below 1000 °C is much less than the observed deformations.

(9) Therefore only grain boundary sliding in WC–WC interfaces and/or decohesion can produce the large deformations observed. Grain

boundary sliding in several hard metals has been reported [3, 27].

In spite of all these similarities, the ruthenium addition significantly changes the high temperature behaviour. There is one possible reason for such a change. The ruthenium addition has reinforced the bonding of the WC–WC interfaces. The nature of the WC–WC interfaces in general and the type of interface which has been created by the ruthenium addition is in fact not yet well investigated. Probably the bonding of the WC–Co interfaces is also reinforced by the ruthenium addition. This could be the reason that the crack resistance of the material with the ruthenium addition is the highest of the three materials tested. Thus the nature and the mechanical behaviour of the interfaces in WC–Co composites should be investigated in the future.

Acknowledgments

We would like to thank Professor Ilschner for supporting the acquisition of the quantitative image analyser and our colleagues J. J. Ammann, Dr. Earthman, Dr. Eggeler, C. Mills, Professor H. O. K. Kirchner (Carnegie-Mellon University, Pittsburgh, PA, U.S.A.) and B. Senior for many hours of fruitful discussion and help with the experiments. This work was supported by Stellram, Nyon, Switzerland, and Commision pour l'Encouragement de la Recherche Scientifique Project 1304. Further support was given by Leitz Wetzlar, F.R.G.

References

1 H. E. Exner, A. Walter and R. F. Pabst, *Mater. Sci. Eng., 16* (1974) 231.
2 L. S. Sigl, H. E. Exner and H. F. Fischmeister, in A. Almond, C. A. Brookes and R. Warren (eds.), *Proc. 2nd Int. Conf. on the Science of Hard Materials, Rhodes, September 23–28, 1984,* in *Inst. Phys. Conf. Ser., 75* (1986) 631.
3 R. J. Gottschall, W. S. Williams and I. D. Ward, *Philos. Mag. A, 41* (1980) 1.
4 F. Osterstock, *Phil. Thesis*, Université de Caen, 1980.
5 H. G. Schmid, *18 Arbeitskreis Bruchvorgänge*, Deutscher Verband für Materialprüfung, Berlin, 1986, p. 71.
6 J. Hong and J. Gurland, in R. K. Viswandham, D. J. Rowcliffe and J. Gurland (eds.), *Science of Hard Materials*, Plenum, New York, 1980, p. 649.
7 V. D. Kristic and M. Komac, *Philos. Mag. A, 51* (1985) 192.
8 G. Fantozzi, H. Si Mohand and G. Orange, in A. Almond, C. A. Brookes and R. Warren (eds.), *Proc. 2nd Int. Conf. on the Science of Hard Materials, Rhodes, September 23–28, 1984,* in *Inst. Phys. Conf. Ser., 75* (1986) 699.
9 H. G. Schmid, D. Mari, W. Benoit and C. Bonjour, in B. Wilshire and R. W. Evans (eds.), *Proc. 3rd Int. Conf. on Creep and Fracture of Engineering Materials and Structures, Swansea*, Vol. 3, Institute of Metals, London, 1987, p. 975.
10 H. G. Schmid, *Mater. Forum, 10* (3) (1987) 184.
11 B. Johannesson and R. Warren, in A. Almond, C. A. Brookes and R. Warren (eds.), *Proc. 2nd Int. Conf. on Science of Hard Materials, Rhodes, September 23–28, 1984*, in *Inst. Phys. Conf. Ser., 75* (1986) 713.
12 A. Bornhauser, K. Kromp and R. F. Pabst, *J. Mater. Sci., 20* (1985) 2586.
13 J. R. Rice, P. C. Paris and G. Merkle, *Progress in Flow Growth and Fracture Toughness Testing, ASTM Spec. Tech. Publ. 536*, 1973, p. 241.
14 H. Riedel, *Advanced Semin. on Fracture Mechanics, Ispra, 1981.*
15 H. Riedel, *Arbeitskreis Bruchvorgänge*, Deutscher Verband für Materialprüfung, Berlin, 1980, p. 165.
16 A. F. C. Cocks and M. F. Ashby, *Prog. Mater. Sci., 27* (1982) 189.
17 M. F. Ashby and B. Tomkins, in K. J. Miller and R. F. Smith (eds.), *Proc. 3rd Int. Conf. on the Mechanical Behaviour of Materials, Cambridge, Cambridgeshire, August 20–24, 1979*, Vol. 1, Pergamon, Oxford, 1979, p. 47.
18 L. P. Kubin, K. Chihab and Y. Estrin, in D. Walgraef (ed.), *Patterns, Defects and Microstructure in Nonequilibrium Systems*, Nijhoff, Dordrecht, 1987, p. 220.
19 V. Jayaram, *Acta Metall., 35* (1987) 1307.
20 R. Greenwood, M. H. Loretto and R. E. Smallman, *Acta Metall., 30* (1982) 1193.
21 H. Riedel, *Z. Metallkd., 69* (1978) 755.
22 H. Riedel, *Z. Metallkd., 76* (1985) 669.
23 C. Y. Hui, H. Riedel, *Int. J. Fract., 17* (1981) 409.
24 B. F. Dyson, *Met. Sci., 10* (1976) 349.
25 K. Brookes, *Metalwork. Prod., 7* (1979) 77.
26 J. S. Jackson and R. Warren, *Powder Metall. Int., 17* (34) (1988) 255.
27 R. Arndt, *Z. Metallkd., 63* (1972) 274.

Subcritical Crack Growth and Plastic Deformation in the Fracture of Hard Metals*

BENGT JOHANNESSON and RICHARD WARREN

Department of Engineering Metals, Chalmers University of Technology, S-412 96 Göteborg (Sweden)

(Received November 8, 1987)

Abstract

The fracture behaviour has been studied of three commercial hard metals up to 1000 °C. The samples were in the form of pre-cracked single-edge-notched bend samples loaded in four-point bending in vacuum. Hard metals are characterized by a transition temperature range over which the load–deflection behaviour changes from linear elastic to non-linear. This non-linearity arises partly from plastic deformation and partly from subcritical crack growth. The proportions of these components depend on the temperature and the microstructure. These effects were examined for two of the alloys of the present study (coarse-grained WC–15wt.%Co and fine-grained WC–6wt.%Co) by carrying out interrupted fracture tests as well as constant-load creep tests followed by examination of crack profiles and measurement of permanent deformation.

For WC–15wt.%Co the transition temperature is around 500 °C. At this temperature, non-linear load–deflection behaviour is due largely to subcritical crack growth with relatively little plastic deformation at the crack tip. Crack propagation occurs by yield of the cobalt binder and early failure of the carbide skeleton ahead of the main crack tip. The apparent K_I value for crack propagation is relatively insensitive to testing rate. At higher temperatures the proportion of plastic deformation increases and is assumed to contain an increasing proportion of creep deformation.

For the WC–6wt.%Co alloy the transition temperature is around 750 °C. The non-linear deflection behaviour at lower K_I levels is almost entirely due to creep deformation. This deformation is extensive and leads to crack tip blunting and an apparent increase in K_I for unstable crack propagation. The onset of non-linearity is sensitive to loading rate. It is concluded that the more rigid carbide skeleton of this low cobalt alloy and the low value of the binder mean free path suppresses the yield of the binder phase; deformation first occurs and is controlled by creep of the carbide skeleton and this eventually leads to crack growth.

Nominally pure WC (containing 0.75 wt.% Co) remained linear elastic up to 1000 °C, thus demonstrating the intrinsic rigidity of the WC grain structure. A WC–γ-carbide–9wt.%Co alloy remained linear elastic to almost 1000 °C, confirming that the creep resistance of the γ-carbide skeleton is greater than that of the WC skeleton with a similar volume fraction.

1. Introduction

Until recently, studies of the fracture of hard metals at elevated temperature were rare. Moreover, the nature of these materials has made the choice of test method and the interpretation of results rather difficult. Braiden *et al.* [1] applied time-to-rupture tests in bending to WC–6wt.%Co and WC–16wt.%Co alloys at room temperature and 950 °C. It was found necessary to test a series of samples for each test condition to obtain statistically meaningful results. Their results for 950 °C indicated that creep deformation occurred. Ueda *et al.* [2] studied the failure in bending of a series of WC–Co alloys and were able to identify three failure mechanism regions with increasing temperature up to 1000 °C. Doi *et al.* [3] examined similar alloys in tensile creep. More recently, a number of studies have been made of the fracture of various hard metal compositions on pre-cracked specimens (*i.e.* K_{Ic} testing) (for example see refs. 4–7). In all these, a transition temperature is identified beyond which the fracture behaviour deviates significantly from linear

*Paper presented at the 3rd International Conference on the Science of Hard Materials, Nassau, The Bahamas, November 9–13, 1987.

0921-5093/88/$3.50 © Elsevier Sequoia/Printed in The Netherlands

elastic. The temperature of this transition seems to vary between about 400 and 1100 °C depending on the composition and microstructure of the alloy [4]. It also seems probable that, even at temperatures as low as 700 °C, time-dependent processes (*i.e.* creep) are involved in the failure. This has made it difficult to find a suitable quantitative parameter to express the fracture resistance of the material [5–7].

In this work, a study has been made of the fracture of three different hard metal types up to 1000 °C. Various modifications of the pre-crack bend test have been applied to provide a better understanding of the failure processes. In particular, a separation of plastic and subcritical crack growth contributions to the failure has been attempted. This work is a continuation of earlier investigations of the same three alloys [4].

2. Experimental details

2.1. Materials

The compositional and microstructural details of the three investigated alloys are summarized in Table 1. These are a fine-grained WC–6wt.%Co alloy, a medium-grained WC–15wt.%Co alloy (see Fig. 5) and a mixed-carbide grade (Fig. 1). Detailed microstructural analysis of the WC–15wt.%Co alloy has been described by Wasén and Warren [8].

2.2. Procedure

Fracture tests were performed by four-point bending in a resistance furnace inside a vacuum chamber. Details of the apparatus have been given in ref. 4. The sample dimensions were as follows: width $B=3$ mm; height $W=6$ mm. The outer and inner spans of the bend rig were 38 mm and 19 mm respectively. The pre-cracks were created by bridge indentation [4, 9, 10]. The pre-crack length was measured prior to testing by optical observation on the sides of the sample, which were polished to a 1 μm diamond paste finish for this purpose. When subcritical crack growth occurred before final fracture, it could normally be detected on the fracture surface and/or fracture profile after the test.

As a standard procedure the samples were loaded continuously to fracture at a cross-head speed of 60 μm min^{-1}. To provide more insight into the fracture mechanism, a number of modified tests were performed.

Fig. 1. Scanning electron micrograph of the mixed-carbide alloy; the crack was produced by bridge indentation at room temperature.

TABLE 1 The materials investigated

Parameter (units)	*Parameter value for the following nominal compositions*			
	WC–6wt.%Co[a]	*WC–15wt.%Co*	*WC–20wt.%TiC–15wt.%(Ta, Nb)C–9wt.%Co*	*Unbonded WC*
C (analysis) (wt.%)	5.81	5.21	—	—
Co (analysis) (wt.%)	5.95	14.6	—	0.75
Co (vol.%)	10	23.6	10.5	—
Binder phase[b] (vol.%)	7.5	24.5	9	—
WC phase[b] (vol.%)	92.5	75.5	6.5	—
γ-carbide[b] (vol.%)	—	—	84.5	—
Carbide grain size (mean linear intercept) (μm)	0.5	1.7	1.6 (γ) 1.1 (WC)	8
Hardness (load, 30 kgf) (GPa)	17.2	10.3	14.7	—
K_{Ic} (room temperature) (MN m$^{-3/2}$)	9±0.5	17±1	8.5±0.5	8–12

[a]Includes 0.2 wt.% NbC and 0.2 wt.% TaC.
[b]Measured by point counting on scanning electron micrographs of unetched sections; the binder fractions tend to be underestimated.

(1) Tests were interrupted before final fracture to permit assessment of the relative contributions of plastic deflection and subcritical crack growth to the total deflection. In these cases the subcritical crack growth could be measured from both the fracture profile and the unloading compliance.

(2) Samples were pre-loaded above the transition temperature and then tested to final fracture below the transition as a means of measuring the degree of crack blunting.

(3) Tests were carried out at various cross-head speeds between 6 and 600 μm min^{-1}.

(4) Pre-cracked samples were subjected to creep testing, *i.e.* were loaded to a nominally constant K_I below that giving fast fracture. In tests in which creep involved subcritical crack growth, K_I increased correspondingly during the test.

The permanent plastic deformation in both broken and unbroken samples was estimated by subtracting the crack growth deflection (obtained from the estimated or observed compliance) from the observed total deflection of the sample. The plastic deformation in the unbroken samples of interrupted fracture and creep tests was also measured on the samples. For example a region of permanent deformation could usually be observed around the crack tip on the polished sides of the sample (see also ref. 4). This had the form of a depression up to 1 μm in depth. The shape and extent of this depression were measured using a diamond stylus surface roughness instrument (Talysurf) with a vertical resolution of 0.05 μm. Linear profiles were taken across the depression at 0.25 mm intervals perpendicular to the crack direction. The permanent bend deflection of the samples was measured with a clock micrometer with a vertical resolution of 0.5 μm. This measurement was used to check the accuracy of the displacement measurement of the bend test equipment.

3. Results

3.1. General remarks

The results of this work confirm those of the earlier study [4], *i.e.* that a transition from linear elastic behaviour occurs in the WC–15wt.%Co alloy at around 500 °C, in the WC–6wt.%Co alloy at around 800 °C and in the mixed-carbide alloy at around 1000 °C. Consequently, the work described here concentrated on the fracture behaviour in these transition regions. The fracture of the binderless WC was linear elastic up to 1000 °C (K_{Ic} lying between 8 and 12 MN $m^{3/2}$ in this range), indicating that the transition in the other alloys is associated with the existence of the binder phase.

3.2. Creep

The results of creep tests on the three alloys are summarized in Fig. 2 and Table 2. The creep tests are presented here before the fracture tests since creep processes might be expected to play a significant role in the latter at high temperatures.

TABLE 2 Creep and interrupted tests giving unbroken samples (midpoint deflection refers to deflection over inner span)

Conditions				*Plasticity*		*Crack growth,*	*Midpoint deflection* (μm)		
Temperature (°C)	*Time* (min)	K_I (MN $m^{-3/2}$)	*Pre-crack length* a_0 (mm)	$2r_p$ (mm)	σ_{eff} (MPa)	Δa (mm)	*Total* y_{tot}	*Plastic* y_{pl}	*Crack growth* y_{cr}
WC–6wt.%Co									
650	23	6	1.6	0	—	0	0	0	0
750	20	10.6	1.5	5.2	83	0.05	4.7	4.7	0
850	12	10.8[a]	1.5	6.6	104	0.75	61.2	59.8	1.4
WC–6wt.%Co (interrupted fracture)									
850	—	9.1	2.9	0	—	0	1.1	1.1	0
850	—	16	2.8	2.3	188	0.15	10.6	8.3	2.3
WC–15wt.%Co									
450	19	13.7	3.5	1.2	223	0	1.5	1.5	0
536	28.5	12	3.1	2.0	151	0	8.4	8.4	0
650	34	14.9	2.9	1.5	217	0.05	4.8	4.6	0.2
Mixed carbide									
1000	17	7.5	1.9	4	67	0	1.9	1.9	0

[a]Initial value; K_I rose to 15 MN $m^{-3/2}$ at the end of the test.

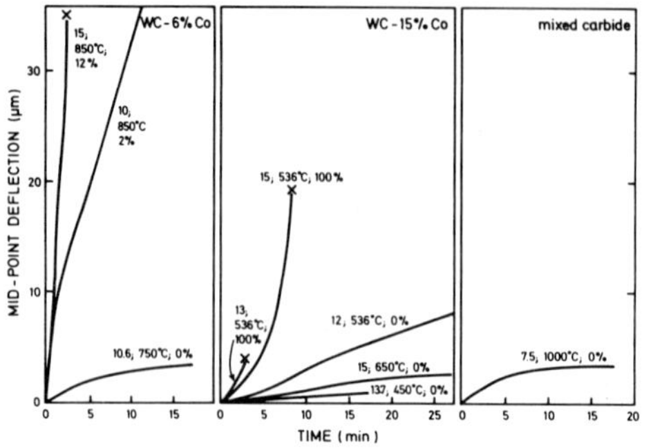

Fig. 2. Creep curves for the three investigated materials: ×, fracture. The numbers against each curve refer to the initial applied K_I, the temperature and the proportion of deflection at the end of the test due to subcritical crack growth. Midpoint deflection refers to the inner span.

Although limited in extent, the creep results demonstrate that, in the transition region of the WC–15wt.%Co alloy (500 °C) and the mixed-carbide alloy (1000 °C), the creep rates are negligible in relation to the deflection rates observed in a normal fracture test (the duration of such a test is generally less than 1 min). The creep that does occur in these two alloys is by plastic deformation without subcritical crack growth. Crack growth begins at stress intensities very close to the critical value and quickly leads to final fracture. It should be noted that for WC–15wt.%Co the creep rate is, rather unexpectedly, lower at 600 °C than at 536 °C. In the WC–6wt.%Co alloy, significant creep rates are observed in the transition region. Measurable proportions of the creep deflection are due to subcritical crack growth.

Table 2 includes the values of $2r_p$, the maximum width of the plastically deformed zone around the crack tip in the creep samples, measured on the side surfaces. From these values an effective creep stress at the edge of the zone has been estimated from the equation derived for plastic zone size [11]:

$$\sigma_{\mathrm{eff}} = \frac{K_I^{**}}{(2\pi r_p)^{1/2}} \tag{1}$$

where K_I^{**} is the stress intensity pertaining to the end of the test. These stress levels are consistent with those observed for measurable creep rates in tensile creep tests [3, 12].

3.3. *Fracture toughness*

The results of the fracture toughness measurements are summarized in Fig. 3; the behaviour of each alloy will be described in turn. The fracture of the WC–6wt.%Co alloy was linear elastic up to around 750 °C. The transition in behaviour above this temperature is marked by a deviation in linearity in the load–deflection curve. This is indicated in Fig. 3 by the curve for K_{pr} which is the K_I value at which non-linearity began. Interrupted tests at 850 °C (see Table 2 and ref. 4) revealed that the non-linearity began as plastic deformation at K_I around 7.5 MN m$^{-3/2}$; subcritical crack growth began only when the effective K_I reached about 11 MN m$^{-3/2}$. In this

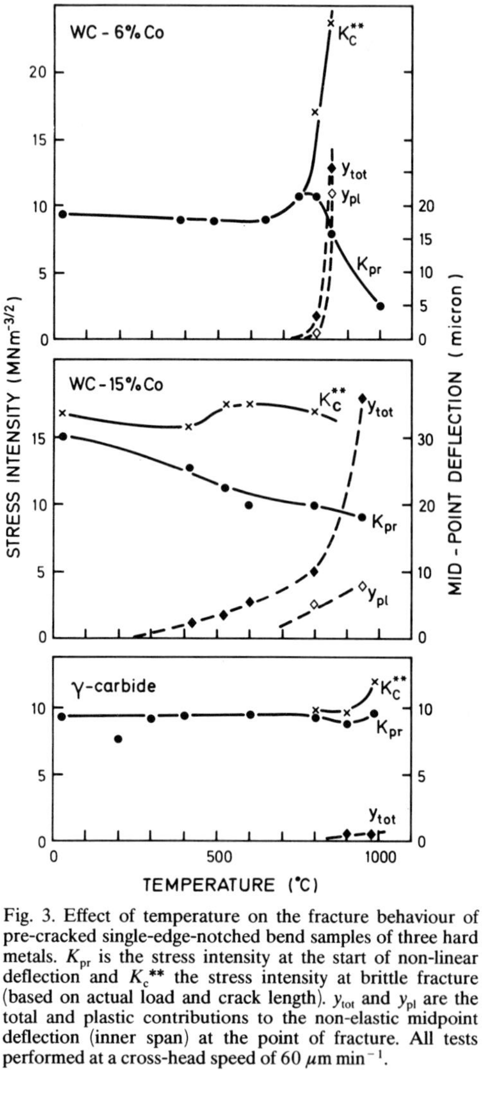

Fig. 3. Effect of temperature on the fracture behaviour of pre-cracked single-edge-notched bend samples of three hard metals. K_{pr} is the stress intensity at the start of non-linear deflection and K_c^{**} the stress intensity at brittle fracture (based on actual load and crack length). y_{tot} and y_{pl} are the total and plastic contributions to the non-elastic midpoint deflection (inner span) at the point of fracture. All tests performed at a cross-head speed of 60 μm min^{-1}.

region, loading can continue with combined plastic deformation and crack growth until brittle fracture occurs at a very high apparent K_I. This brittle fracture stress intensity is denoted K_c**, being estimated from the actual load and crack length at fracture. At the point of fracture the subcritical crack growth contributes a relatively small proportion of the total non-elastic midpoint deflection (y in Fig. 3). At 1000 °C, brittle fracture was not observed; fracture occurred entirely by subcritical growth. The creep studies described above indicate that observed plastic deformation is largely the result of creep processes. This is supported by the observed effect of testing speed on the apparent fracture toughness (Fig. 4); a significant increase is observed at low speeds at 750 °C.

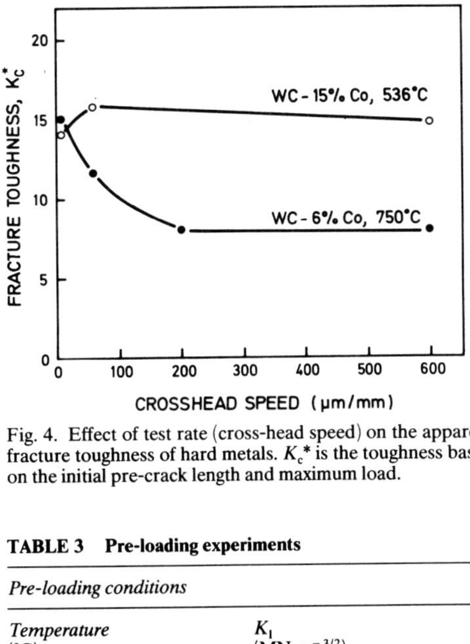

Fig. 4. Effect of test rate (cross-head speed) on the apparent fracture toughness of hard metals. K_c* is the toughness based on the initial pre-crack length and maximum load.

The apparently high values of fracture toughness K_c** are associated with crack blunting due to crack tip opening and crack branching. This is confirmed by microscopic observation of interrupted test samples and by pre-loading experiments (Table 3). Samples loaded into the plastic region, above the transition temperature retain a high K_{Ic} when tested to fracture at lower temperatures.

It can be concluded from the above combination of tests that this alloy resists all forms of plastic deformation up to about 750 °C but that at around 800 °C significant creep deformation occurs at a relatively low stress.

The WC–15wt.%Co alloy behaved differently in many respects. The first signs of a transition occur between 400 and 500 °C. Up to about 700 °C the onset of non-linearity (K_{pr}) is due almost exclusively to subcritical crack growth and not plastic deformation. As discussed above, the contribution of creep deformation to the observed deflection is negligible. This is supported by the observation of a low test speed sensitivity (Fig. 4).

The resistance K_{pr} to subcritical cracking decreases continuously with increasing temperature. At the same time, there is only a slight increase in K_c** for brittle failure. Pre-loading experiments confirmed that no significant crack blunting occurred (Table 3). Nevertheless, the fact that, in creep tests, a constant K_I above K_{pr} could not maintain subcritical crack growth indicates that some form of rising crack resistance develops above K_{pr}. Above 700 °C, plastic deformation becomes measurable, although crack growth remains as the dominant cause of midpoint deflection (y, Fig. 3). At 950 °C, subcritical crack

TABLE 3 Pre-loading experiments

Pre-loading conditions		*Test temperature* (°C)	K_{Ic} (MN m$^{-3/2}$)
Temperature (°C)	K_I (MN m$^{-3/2}$)		
WC–6wt.%Co			
—	—	20	9±0.5
800	2.5	200	10
800	2.5	650	12.2
870	2.5	400	9.9
870	9.6	220	16.3
800	15.3	20	>15.3 (interrupted)
WC–15wt.%Co			
—	—	20	17±1.5
536	14.8 (gave 0.12 mm subcritical crack growth)	20	15.9

growth extended through the sample. The crack was relatively straight and exhibited little sign of branching or blunting.

These results suggest that in the transition region the microstructure is subject to a subcritical deformation process which tolerates only a small strain to failure. Above 700 °C, creep deformation becomes significant but, in contrast with the WC–6wt.%Co alloy, deformation is predominantly by subcritical crack growth.

The mixed alloy exhibited a slight tendency to non-linearity at 950 °C as a result of subcritical crack growth. At 1000 °C, this is accompanied by an apparent increase in fracture toughness K_c**. In view of the negligible plastic deflection, it is unlikely that this is due to blunting.

4. Discussion

The results of this work together with observations and conclusions of previously reported studies permit the proposal of a qualitative model for the high temperature fracture of the hard metals of the type studied here. The fracture of the WC–15wt.%Co alloy up to 700 °C can be seen as an extension of its room temperature behaviour. In such an alloy with a relatively high cobalt content and a relatively low contiguity of the carbide skeleton, the contiguous boundaries suffer a high transfer of applied load. Ahead of a crack tip this results in the cracking of several carbide boundaries, the binder remaining intact, to produce a ligament zone (Fig. 5(a)) (see for example refs. 13 and 14). With increasing temperature the yield stress of the binder decreases and at sufficiently high temperature the binder begins to creep; consequently there occurs a corresponding increase in the stress transfer to the carbide skeleton with resulting growth of the ligament zone. Since presumably the failure stress of the cobalt also falls, the resistance K_{pr} to the initiation of crack growth will decrease. Nevertheless, if at elevated temperatures a longer ligament zone develops than that in the original pre-crack, an increasing fracture resistance from K_{pr} will be experienced. (The fact that the ligament zone length increases was indicated by examination of crack profiles on the polished sides of samples taken from interrupted tests (see for example Figs. 5(a) and 5(b)). However, quantitative measurements on profiles sectioned in the interior of samples have yet to be made.) Unstable fracture will begin at the true K_{Ic} when the ligament zone is developed. This K_{Ic} value can be estimated from a ligament model for fracture energy such as that of Schmid *et al.* [7] or Sigl *et al.* [14]. The onset of a proportion of plastic deformation at around 700 °C can be associated with an increased creep deformation. This was identified by Ueda *et al.* [2] as an increasing proportion of failure in the binder phase associated with microvoid formation assisted by grain boundary sliding of the carbide. This process can be expected to promote further subcritical crack growth.

The behaviour of the WC–6wt.%Co alloy is dominated by the highly contiguous carbide skeleton. This is able to sustain a larger stress transfer than that in the WC–15wt.%Co alloy. Moreover, because of its significantly lower mean free path, the binder experiences a larger constraint and therefore exhibits a higher *in situ* yield stress. Thus the early transition observed in the WC–15wt.%Co alloy is suppressed by the skeleton in the WC–6wt.%Co alloy. The transition from linear elastic behaviour at around 800 °C occurs when the carbide skeleton begins to yield by creep. The creep rate is significant down to relatively low stresses and consequently up to relatively large distances from the crack tip. This leads to the observed crack branching and crack tip blunting. A probable carbide creep process is grain boundary sliding.

The creep resistance of the WC carbide skeleton in these alloys may seem surprisingly low for such a high melting point compound. The effect can be explained partly by the very small grain size and possibly also by the presence of cobalt atoms segregated in the contiguous boundaries [15]. Cobalt can be expected to activate the grain boundary creep in a similar way to the activation of the creep of tungsten by segregated nickel atoms [16].

The above discussion can be extended to a more general description of the influence of composition and microstructure on fracture for WC–Co alloys. This is illustrated in Fig. 6 which is a schematic map showing the proposed effects of temperature and binder content on fracture processes in pre-cracked specimens. This shows that the transition, from near-linear elastic behaviour to the region of limited yield, subcritical crack growth and ligament zone growth, is expected to move to lower temperatures with increased binder content and grain size (*i.e.* increasing binder mean free path). It should be

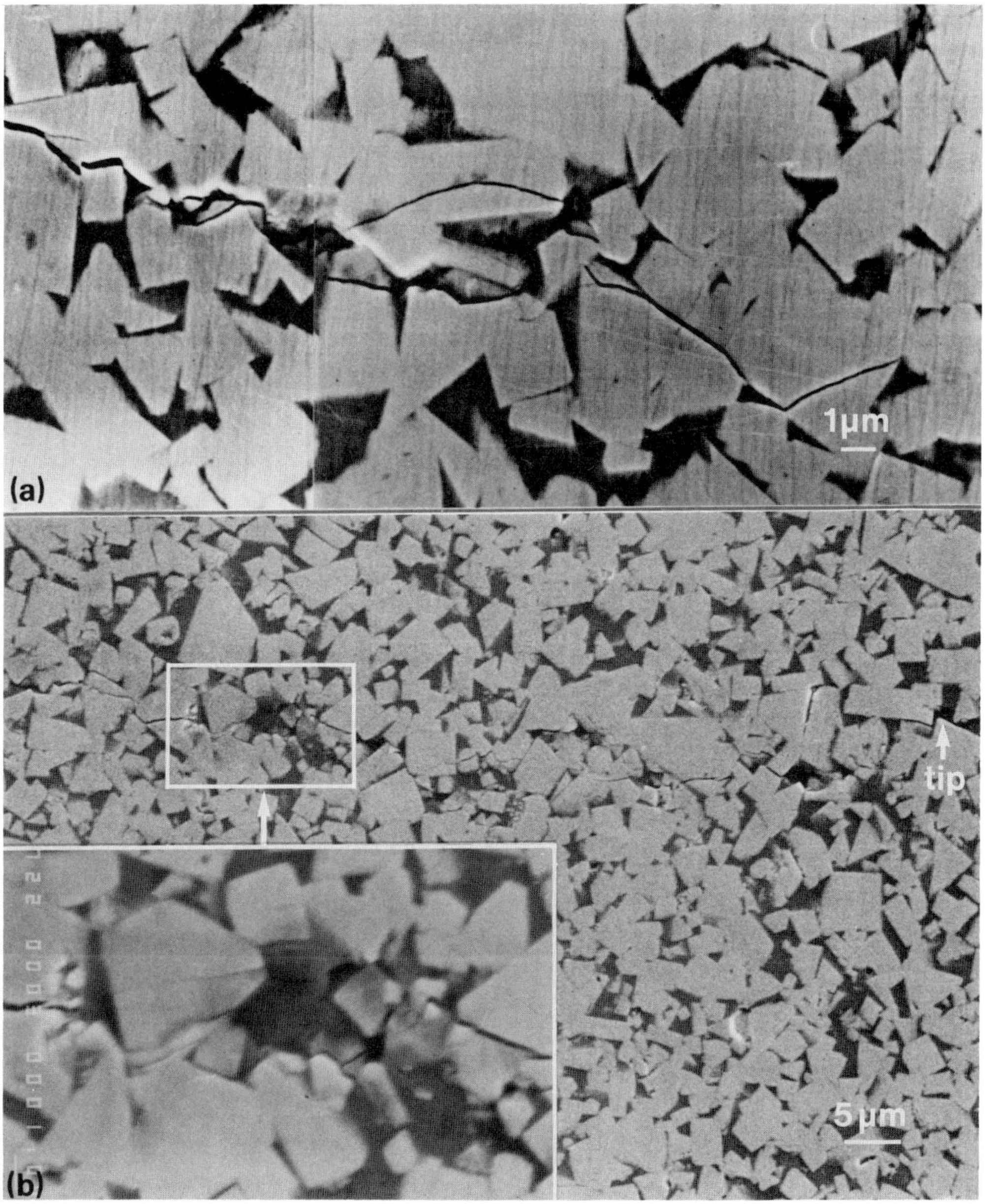

Fig. 5. Scanning electron micrographs of (a) crack tip produced by bridge indentation at room temperature in the WC–15wt.%Co alloy and (b) the crack tip zone in the WC–15wt.%Co alloy after loading to $K_I = 12$ MN m^{-32} at 536 °C. The enlarged inset in (b) shows the first binder ligament.

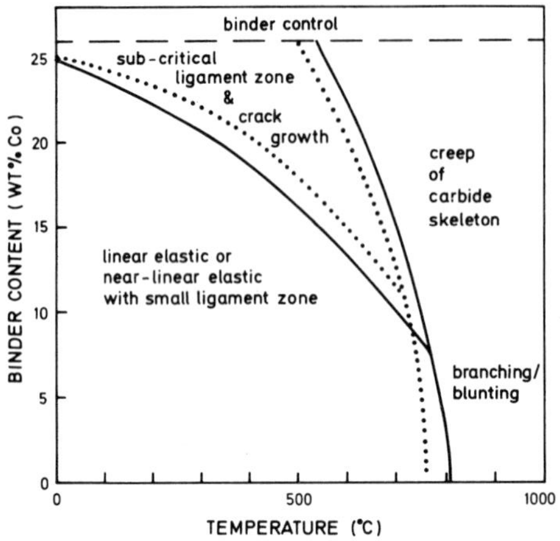

Fig. 6. Schematic map of fracture processes in pre-cracked WC–Co alloys: ·····, effect of reducing carbide grain size.

noted that the transition is expected to be gradual and not sharp as the figure might suggest. At low volume fractions, this transition is superseded by the transition caused by the creep of the carbide skeleton. A reduction in carbide grain size is expected to reduce the extent of the ligament zone and crack growth region, on one side because of a reduced binder mean free path and on the other because of an increased skeleton creep rate. In the skeleton creep region, the important boundary between crack branching–blunting and non-blunted crack growth is not shown since the relative influence of binder content and carbide grain size on this boundary is not clear.

The high transition temperature of the mixed-carbide alloy is explained by a very creep-resistant skeleton. The superior creep resistance of this alloy type has been demonstrated by Santhanam *et al.* [17].

5. Conclusions

Studies of the fracture behaviour of selected hard metals using various modifications of a pre-crack bend test have yielded information pertinent to an identification of fracture mechanisms in the alloys. The fracture behaviour of a fine-grained WC–6wt.%Co alloy undergoes a transition around 800 °C from linear elastic fracture to fracture dominated by creep of the carbide skeleton. This results in subcritical crack growth accompanied by significant crack blunting and branching. In medium-grained WC–15wt.%Co alloys the fracture begins to deviate from linear elastic from between 400 and 500 °C; this is a result of a decreasing resistance to subcritical crack growth which occurs with very little accompanying plastic deformation. It is proposed that in this region the cracking occurs by the formation of a ligament zone in which the binder phase has a yield stress decreasing with temperature. Above about 700 °C, creep deformation around the crack tip begins to become significant. However, the subcritical crack resistance continues to fall and crack propagation occurs with little blunting or branching up to 1000 °C.

A mixed γ-carbide–WC alloy with 9 wt.% Co remained linear elastic up to 950 °C, confirming the high creep resistance of this alloy type. At 1000 °C, a small amount of subcritical crack growth occurred.

Acknowledgments

Thanks are extended to the Swedish Institute of Silicate Research and Volvo for provision of testing machines. The hard metal samples were provided by Sandvik Hard Materials and Seco Tools. The work was supported financially by the Swedish Board of Technical Development.

References

1 P. M. Braiden, R. W. Davidge and R. Airey, *J. Mech. Solids, 25* (1977) 257–267.
2 F. Ueda, H. Doi, F. Fujiwara and H. Masatomi, *Trans. Jpn. Inst. Met., 18* (1977) 247–256.
3 H. Doi, F. Ueda, F. Fujiwara and H. Masatomi, *Int. J. Refract. Hard Met., 3* (1984) 146–148.
4 B. Johannesson and R. Warren, in E. A. Almond, C. A. Brookes and R. Warren (eds.), *Proc. 2nd Int. Conf. on the Science of Hard Materials, Rhodes, September 23–28, 1984,* in *Inst. Phys. Conf. Ser., 75* (1986) 713–723.
5 G. Fantozzi, H. Si Mohand and G. Orange, in E. A. Almond, C. A. Brookes and R. Warren (eds.), *Proc. 2nd Int. Conf. on the Science of Hard Materials, Rhodes, September 23–28, 1984,* in *Inst. Phys. Conf. Ser., 75* (1986) 699–712.
6 N. Bouaouadja, *Doctoral Thesis,* Institut National des Sciences Appliquées, Lyon, 1986.
7 H. G. Schmid, D. Mari, W. Benoit and C. Bonjour, in B. Wilshire and R. W. Evans (eds.), *Proc. 3rd Int. Conf. on Creep and Fracture of Engineering Materials and Structures, 1987,* Institute of Metals, London, 1987, pp. 975–988.
8 J. Wasén and R. Warren, *Abstracts 3rd Int. Conf. on Progress in Microstructures, 1987,* Rheinisch–Westfälische Technische Hochschule, Aachen, 1987, pp. 165–169.
9 R. Warren and B. Johannesson, *Powder Metall., 27* (1984) 25–29.

10 B. Johannesson, *J. Mater. Sci. Lett.*, *4* (1985) 299–301.
11 D. Broeck, *Elementary Engineering Fracture Mechanics*, Sijthoff and Noordhoff, Alphen, 1978.
12 G. Wirmark, C. Chatfield and G. L. Dunlop, in E. A. Almond, C. A. Brookes and R. Warren (eds.), *Proc. 2nd Int. Conf. on the Science of Hard Materials, Rhodes, September 23–28, 1984*, in *Inst. Phys. Conf. Ser.*, *75* (1986) 669–678.
13 E. A. Almond, in N. R. Comins and J. B. Clark (eds.), *Speciality Steels and Hard Materials*, Pergamon, Oxford, 1982, pp. 353–360.
14 L. S. Sigl, H. E. Exner and H. F. Fischmeister, in E. A. Almond, C. A. Brookes and R. Warren (eds.), *Proc. 2nd Int. Conf. on the Science of Hard Materials, Rhodes, September 23–28, 1984*, in *Inst. Phys. Conf. Ser.*, *75* (1986) 631–644.
15 A. Henjered, M. Hellsing, H.-O. Andrén and H. Nordén, *Mater. Sci. Technol.*, *2* (1986) 847–855.
16 R. Warren and C.-H. Andersson, in H. Lilholt and R. Talreja (eds.), *Fatigue and Creep of Composite Materials, Proc. 3rd Risø Int. Symp. on Metallurgy and Materials Science, 1982*, Risø National Laboratory, Roskilde, 1982, pp. 335–340.
17 A. T. Santhanam, K. P. Mizgalski and W. C. McCoy, in E. A. Almond, C. A. Brookes and R. Warren (eds.), *Proc. 2nd Int. Conf. on the Science of Hard Materials, Rhodes, September 23–28, 1984*, in *Inst. Phys. Conf. Ser.*, *75* (1986) 679–690.

Materials Science and Engineering, A105/106 (1988) 363–367

Elevated Temperature Properties of WC–Co Cemented Carbides*

M. T. LAUGIER

Materials Research Centre, National Institute for Higher Education, Limerick (Ireland)

(Received November 9, 1987)

Abstract

The Palmqvist toughness and erosion of WC–Co hard materials were characterized at elevated temperatures. The Palmqvist indentation technique was found to give reproducible cracking in suitably prepared cemented carbides. In earlier studies, Palmqvist crack extension was found experimentally to follow predictions of a half-penny crack model; an explanation for this surprising behaviour is now given. High temperature (750 °C) particle erosion of WC–Co composites was also investigated. Erosion models based on indentation fracture principles are found not to apply. Erosion in WC–Co materials may be described in terms of an indentation size effect.

1. Introduction

In most applications, cemented-carbide composites are subjected to elevated temperatures. Despite this, only limited data on their elevated temperature properties are available. In this paper, two critical properties, namely toughness and erosion, are discussed.

2. Palmqvist toughness

The cracks which form in brittle materials as a result of sharp indentation (usually with a Vickers diamond pyramid) have received considerable attention primarily because of their potential for toughness determination. In WC–Co composites, cracks are of the Palmqvist type (Fig. 1 illustrates crack geometry and shows a typical Palmqvist crack profile), and traditionally use has been made of the "crack resistance" parameter $g = P/l$ in the toughness assessment of WC–Co materials [1]. Indentation toughness in ceramics where the parameter $k = P/c^{3/2}$ has been found to be constant (for a given material), has been based on the half-penny crack analogy [2, 3] (the stress intensity factor for a centre-loaded half-penny crack is of the above form). However, the parameter k has also been found to be constant for a given WC–Co material [4–6]; an explanation has been provided in terms of the insensitivity of the surface stress intensity factor to marked changes in crack profile [7] (a geometrical explanation may also be given) [8]. The half-penny formalism has recently been calibrated for WC–Co materials [9].

Little attention has been given to the indentation behaviour of WC–Co materials at the elevated temperatures typical of cutting applications. In this work, preliminary results are presented on the indentation behaviour of a range of WC–Co materials at temperatures up to 1050 °C.

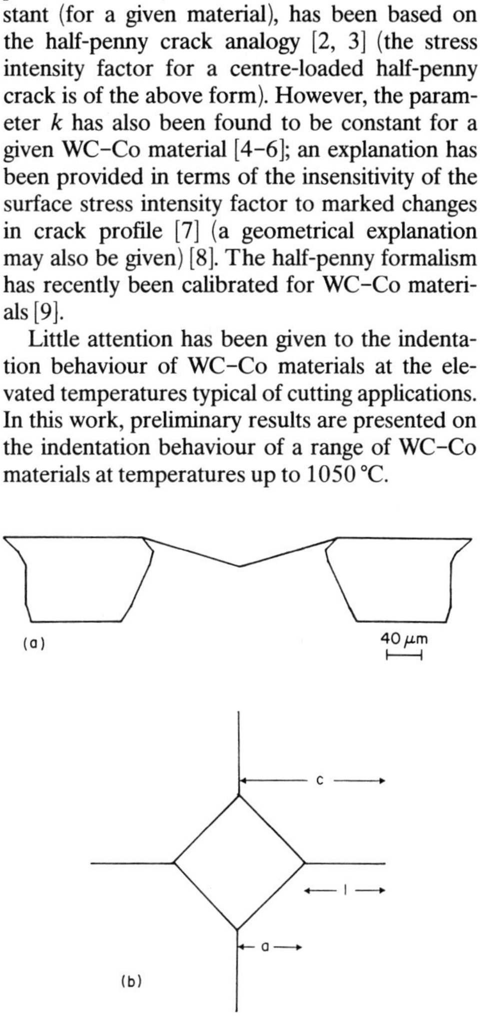

Fig. 1. (a) Palmqvist crack profile obtained by serial sectional in a WC–6wt.%Co material at $P = 50$ kgf and hardness (load, 30 kgf) 15 300 MPa; (b) illustration of indentation crack geometry.

*Paper presented at the 3rd International Conference on the Science of Hard Materials, Nassau, The Bahamas, November 9–13, 1987.

0921-5093/88/$3.50

© Elsevier Sequoia/Printed in The Netherlands

2.1. *Experimental details*

2.2.1. *Materials*

Specimens were commercial WC–Co tool materials in the form of broken compact tensile test pieces of approximate dimensions 10 mm × 10 mm × 3 mm. Surface preparation which is critical to reliable Palmqvist testing was performed by recording the time taken for the disappearance of polishing marks at each stage and then continuing polishing for an equal time interval under the same conditions. This procedure effectively removes subsurface damage from earlier polishing stages [6].

2.1.2. *Apparatus and procedure*

Specimens were mounted on an alumina pedestal and loaded via a double-beam system. Most elevated temperature Palmqvist indentations were performed at a load of 48 kgf, which required an applied load of 10 kgf. Load calibration was performed by comparing hardness impressions with impressions obtained using a standard Vickers hardness machine.

In the case of the elevated temperature Palmqvist tests, a 2.5 kgf pre-load was generally applied to the indenter prior to r.f. induction heating of the indenter–specimen assembly in a nitrogen atmosphere. The temperature was measured using an optical pyrometer. Indentation tests at intermediate temperatures were performed during cooling; temperature measurements were made using a thermocouple. Hardness measurements at a load of 5 kgf were made using a single beam with load directly applied vertically above the indenter. For the hardness measurements, the time under full load (the dwell time) was 10 s; in the high load Palmqvist tests, the load was applied gradually over 30 s. Observations of the impressions at elevated temperature were made using a simply constructed optical microscope at a working distance of 10 mm.

TABLE 1 Material properties and indentation parameters

Specimen	f_β^v	l_γ (μm)	T (°C)	P (kgf)	$2a$ (μm)	$2c$ (μm)	H (MPa)
1	0.1	0.6	1050	40	362		5550
1	0.1	0.6	1050	5	106		8090
1	0.1	0.6	450	48	257	526	13320
1	0.1	0.6	200	48	223	480	17360
1	0.1	0.6	20	50	230	560	17200
2	0.16	0.5	1050	5	106		8090
2	0.16	0.5	450	48	300	[a]	9700
2	0.16	0.5	200	48	240	452	15160
2	0.16	0.5	20	50	242	520	15530
3	0.1	1.5	1050	5	115		6750
3	0.1	1.5	450	48	296	[a]	10100
3	0.1	1.5	200	48	252	[b]	13970
3	0.1	1.5	20	50	254	470	14100
4	0.13	1.8	1050	5	144		4380
4	0.13	1.8	350	324			8520
4	0.13	1.8	200	48	274	[a]	11620
4	0.13	1.8	20	50	268	366	12830
5	0.17	1.6	1050	5	144		4380
5	0.17	1.6	450	48	320		8520
5	0.17	1.6	200	48	274	[a]	11620
5	0.17	1.6	20	50	270	340	12470

f_β^v is the cobalt volume fraction, l_γ is the WC mean linear intercept (Heyn's grain size) given by $l_\gamma = l_\beta^*(1-f_\beta^v)/f_\beta^v$ and [10] $l_\beta^* = (H/620)^{-5}$.
[a]Small cracks visible at a high magnification (5–20 μm).
[b]Non-symmetrical cracking; $2c \approx 375$ μm.

2.2. *Results*

Table 1 lists material properties and indentation parameters. Specimens are numbered in decreasing hardness. No cracking occurred at 1050 °C even in the hardest material (specimen 1) where hardness had fallen by a factor of more than 2 from its room temperature value. At 450 and 200 °C the crack length in the hardest material (specimen 1) was similar to the room temperature value. No crack extension could be observed during cooling. Very short cracks (5–20 μm) were observed in specimens 2 and 3 at 450 °C at a high magnification (400 ×). At 200 °C, cracking was observed in specimens 2 and 3 but cracking was non-symmetrical in the softer material; short cracks (5–20 μm) were observed at high magnification in the remaining specimens.

2.3. *Discussion*

Reliable Palmqvist toughness estimates in WC–Co materials require that $c/a > 1.3$ and, empirically, indentation cracking in WC–Co is difficult even at high loads when $H < 12\,000$ MPa. This is related to an indentation fracture size effect, which has recently been discussed [11]. According to this view, the toughness K_c is given by the formula

$$K_c = 0.072 P_c^{1/4} H^{3/4}$$

where P_c is the critical load required to initiate cracking. This result is applied to those cases where very short cracks were observed. The toughness K_c may also be computed from the formula of Anstis *et al.* [12], which was originally-developed for ceramics but recently calibrated for WC–Co materials. The modified Anstis *et al.* result is [9]

$$K_c = 0.011\left(\frac{E}{H}\right)^{1/2}\frac{P}{c^{3/2}}$$

(for ceramics the numerical term has a value of 0.016) where E is Young's modulus, and the other terms have been given earlier. In fact, broadly equivalent descriptions of indentation fracture in WC–Co materials may be given in terms of either $k = P/c^{3/2}$ (as above) or $g = P/l$ (the Palmqvist crack resistance), but a description in terms of k is preferable as it leads to less dispersion in the results [13]. Finally, the toughness may be calculated from a microstructural model [14]

$$K_c = 1.55G^{2.5}b_{Co}^{\ 0.5}\left(\frac{E}{H}\right)^{0.6} \times\left(1+\frac{0.012E}{H}\right)^{-0.6}\varepsilon_c^{\ 0.6}H^{-1.5}$$

recently validated [15] for $H > 10\,000$ MPa, where the expression $1.55G^{2.5}b_{Co}^{\ 0.5}$ takes the numerical value of 2.67×10^7 (G is the cobalt shear modulus, b_{Co} is the cobalt Burgers vector and $\varepsilon_c = 0.015$ is a critical composite strain a characteristic distance from the crack tip).

Toughness values computed from the formula given above are listed in Table 2, which also includes compact tensile toughness values obtained by Johansson [17]; pre-cracking was performed using a development of the wedge indentation technique of Ingelstrom and Nordberg [18]. There is a broad measure of agreement between computed toughness values, and the compact tensile results with little apparent change in toughness up to 450 °C. No cracking occurred in any case at 1050 °C and no estimate of toughness is available for those conditions.

3. Erosion

WC–Co composites are widely used in applications requiring good wear resistance and include dies for extrusion and wire drawing, punches, nozzles for concrete spraying and sand blasting, valves for coal conversion plants, and mining tools. In room temperature applications, the erosion performance of WC–Co generally improves with increasing hardness [19] and materials selection is usually made on this basis (however, in a recent study [20], no relation between WC–Co erosion and hardness was found).

The present work extends earlier studies of erosion in WC–Co materials [19–21] to temperatures in the region of 750 °C.

3.1. Models

Indentation fracture models have had some success in describing erosion of ceramics; material removal by microchipping is considered to

TABLE 2 **Toughness values for the experimental WC–Co materials**

Specimen	*T* (°C)	*E*[a] (×10^5 MPa)	K_c (MPa m$^{1/2}$)			
			Modified Anstis et al. formula[b]	*Ductile-to-brittle transition formula*[c]	*Microstructural model*[d]	*Compact tensile results*
1	450		8.4		10.8	
1	20	6.3	7.0		6.9	7.8
2	450			10.4		
2	20	5.9	7.7		7.8	9.0
3	450			10.7		
3	20	6.3	10.5		9.9	10.4
4	200			11.9		
4	20	6.1	15.4		11.6	15.2
5	200			11.9		
5	20	5.9	16.1		12.2	12.1

[a] From Doi *et al.* [16]; it is assumed that $E(20\,°C) \approx E(450\,°C)$.
[b] $K_c = 0.011(E/H)^{1/2}P/c^{3/2}$.
[c] $K_c = 0.072P_c^{1/4}H^{3/4}$.
[d] $K_c = 2.15\times10^6(E/H)^{0.6}(1+0.012E/H)^{-0.6}H^{-1.5}$.

occur when lateral cracks intersect the free surface. The semiempirical relation $c \approx (P/K_c)^{2/3}$ between the lateral crack radius c, the maximum load P resulting from impact and toughness K_c is used to obtain the volume loss per impact. Expressions [22, 23] for the rate of erosion as a volume loss per impact are

$$W \propto V^{2.4} R^{3.7} \rho^{1.2} K_c^{-1.33} H^{0.11} \quad (1)$$

$$W \propto V^{3.2} R^{3.7} \rho^{1.3} K_c^{-1.33} H^{-0.25} \quad (2)$$

where V, R and ρ are the particle velocity, radius and density respectively. Both expressions are found to underestimate the dependence of erosion on hardness and on toughness [24, 25].

3.2. Experimental details

3.2.1. Materials

A range of WC–Co materials with hardnesses in the range 15.58–10.09 GPa was investigated (Table 3). Specimens (about 1 cm × 1 cm × 0.2 cm) were cut from compact tensile test pieces. Erosion testing was performed using angular Al_2O_3 grit of size 350 μm (Fig. 2). The particle speed was measured using a two-disc time-of-flight method; discs were spun at 17×10^3 rev min^{-1} with a disc separation of 5 cm. Separation of the resulting erosion zones was 110 ± 15°, giving a particle speed $v = 46 \pm 6$ m s^{-1}.

Fig. 2. Scanning electron micrograph illustrating angular Al_2O_3 particles used for erosion testing.

3.2.2. Apparatus

Special apparatus was constructed for the high temperature erosion tests; grit (about 120 g) contained in a cylindrical body was forced vertically upwards through a 5 mm exit nozzle using compressed N_2 gas to provide an inert environment. Specimens were housed in a concentric cylinder above the exit nozzle. The whole assembly (dimensions: height, 20 cm; diameter, 10 cm) was placed inside a coil and heated by means of a 17 kW r.f. induction set. Temperature measurements were made by means of thermocouples.

3.3. Results

Experimentally determined erosion coefficients W are included in Table 3 (W is the mass of target removed divided by the mass of incident particles). Erosion rates increase significantly with increasing temperature under the present conditions but, although room temperature erosion is maximum at normal incidence, typical of brittle materials, erosion rates at 750 °C are quite similar for both normal and 45° incidence.

3.4. Discussion

Although WC–Co materials exhibit brittle erosion behaviour at room temperature and at 450 °C [21] (maximum erosion at 90° incidence), the elastic–plastic erosion models appear to be inapplicable as they predict increasing erosion rate with increasing hardness, contrary to observation [19] (hardness and toughness are inversely related in WC–Co; see for example ref. 14).

At 750 °C, there appears to be a change from purely brittle behaviour, necessitating further investigation.

Generally, fracture can only occur when the strain energy release rate is sufficient to provide

TABLE 3 Compositional and properties data together with erosion rates *W* for a range of WC–Co materials

Specimen	*H* (MPa)	f_β^v	l_β^* (μm)	*W*(90°, 20 °C) ($\times 10^{-4}$ g g^{-1})	*W*(45°, 20 °C) ($\times 10^{-4}$ g g^{-1})	*W*(90°, 750 °C) ($\times 10^{-4}$ g g^{-1})	*W*(45°, 750 °C) ($\times 10^{-4}$ g g^{-1})
1	15580	0.16	0.1	1.8	1.3	9.6	13
2	12740	0.13	0.3	4.1	1.9	17	20
3	12330	0.17	0.3	3.3	2.1	12	16
4	11045	0.24	0.6	3.7	2.0	13	22
5	10090	0.24	0.9	4.8	2.4	34	31

f_β^v is the cobalt volume fraction and l_β^* is calculated [10] from $l_\beta^* = (H/620)^{-5}$.

the energy associated with crack extension, leading to a size effect under conditions of inhomogeneous loading [26]. Palmqvist cracking in WC–Co composites loaded with a Vickers diamond indenter may be viewed as a ductile-to-brittle transition following [26], with the result that fracture is not predicted to occur for impression sizes smaller than about $10^3\ l_\beta^* \approx 100-1000$ μm [11] (see Table 3), where $l_\beta^* = l_\gamma f_\beta^{\mathrm{v}}/(1-f_\beta^{\mathrm{v}})$, l_γ is Heyn's grain size of WC (mean linear intercept) and f_β^{v} is the cobalt volume fraction. The maximum volume of an impression caused by particle impact is $mv^2/2H \approx 0.02 \times 10^{-12}$ m^3 ($R = 200$ μm; $\rho = 4 \times 10^3$ kg m^{-3}; $v \approx 50$ m s^{-1}; m is the mass; $H = H_{\mathrm{min}} \approx 10^4$ MPa) under quasi-static conditions with no work hardening, if all the energy is dissipated in plastic work. This corresponds to a crater depth of only about 6 μm and a diameter of about 100 μm, so that fracture initiation should be inhibited. In order that fracture may initiate, l_β^* must decrease significantly; this could occur by impact compaction and extrusion of cobalt. This process should proceed more readily at higher temperatures, consistent with the observed increase in erosion with temperature.

4. Conclusions

4.1. Toughness

Palmqvist testing of a range of WC–Co materials with loads up to 50 kgf and temperatures up to 1050 °C has been performed. No cracking occurred at 1050 °C and only the hardest material ($H = 17\,200$ MPa) exhibited cracking at 450 °C, although crack initiation was evident in some materials (specimens 2 and 3). At 200 °C, full Palmqvist cracking was still inhibited in the materials of lowest hardness (specimens 4 and 5) although crack initiation had occurred. Crack extension was not observed during cooling from 200 °C, suggesting that the resulting build-up of tensile stress is not significant in Palmqvist crack extension. Finally, toughness values computed by a range of methods showed broad agreement with compact tensile toughness values.

4.2. Erosion

Testing of a range of WC–Co composites at room temperature and 750 °C using 350 μm Al_2O_3 particles incident normally (90°) and at 45° confirmed the inverse hardness dependence of erosion in WC–Co materials.

Room temperature erosion was greatest at normal incidence and increased with temperature although, at 750 °C, erosion rates at normal and 45° incidence were similar.

Current elastic–plastic erosion models appear to be inapplicable as a result of a microstructure-controlled indentation size effect. It is suggested that the brittle fracture which occurs under the usual erosion conditions in WC–Co materials must be preceded by a critical level of compaction and associated cobalt extrusion.

References

1 S. Palmqvist, *Arch Eisenhüttenwes.*, *33* (1962) 629.
2 B. R. Lawn and E. R. Fuller, *J. Mater. Sci.*, *10* (1975) 2016.
3 A. G. Evans and E. A. Charles, *J. Am. Ceram. Soc.*, *59* (1976) 371.
4 E. L. Exner, J. R. Pickens and J. Gurland, *Metall. Trans. A*, *9* (1978) 736.
5 E. A. Almond and B. Roebuck, *Conf. on Recent Advances in Hardmetal Production, Loughborough University of Technology*, in *Met. Powder Rep.* (1979) 31-1.
6 M. T. Laugier, *J. Mater. Sci. Lett.*, *6* (1987) 897.
7 M. T. Laugier, in E. A. Almond, C. A. Brookes and R. Warren (eds.), *Proc. 2nd Int. Conf. on the Science of Hard Materials, Rhodes, September 23–28, 1984*, in *Inst. Phys. Conf. Ser.*, *75* (1986) 449.,
8 M. T. Laugier, *J. Mater. Sci. Lett.*, *4* (1985) 207.
9 M. T. Laugier, *Ceram. Int.*, in press.
10 M. T. Laugier, *Acta Metall.*, *33* (1985) 2093.
11 M. T. Laugier, *J. Mater. Sci. Lett.*, *6* (1987) 768.
12 G. R. Anstis, P. Chantikul, B. R. Lawn and D. B. Marshall, *J. Am. Ceram. Soc.*, *64* (1981) 533.
13 M. T. Laugier, *J. Mater. Sci. Lett.*, *4* (1985) 207.
14 M. T. Laugier, *Powder Metall. Int.*, *18* (1986) 330.
15 M. T. Laugier, *J. Mater. Sci. Lett.*, *6* (1987) 779.
16 H. Doi, Y. Fujiwara, K. Miyake and Y. Oosawa, *Metall. Trans.*, *1* (1970) 1517.
17 I. Johansson, Sandvik AB, personal communication, 1984.
18 N. Ingelstrom and H. Nordberg, *Eng. Fract. Mech.*, *6* (1974) 597.
19 M. T. Laugier, *J. Mater. Sci.*, *21* (1986) 3548.
20 H. Conrad, D. McCabe and G. A. Sargent, in R. K. Viswanadham, D. J. Rowcliffe and J. Gurland (eds.), *Proc. 1st Int. Conf. on the Science of Hard Materials*, Plenum, New York, 1983, p. 775.
21 M. T. Laugier, *Proc. 7th Int. Conf. on Erosion by Liquid and Solid Impact, September 7–10, 1987, Cambridge*, Cavendish Laboratory, Cambridge, 1987, p. 61-1.
22 A. W. Ruff and S. M. Wiederhorn, *Treatise Mater. Sci. Technol.*, *16* (1979) 69.
23 A. G. Evans, M. E. Gulden and M. E. Rosenblatt, *Proc. R. Soc. London, Ser. A*, *361* (1978) 343.
24 S. M. Wiederhorn and B. J. Hockey, *J. Mater. Sci.*, *18* (1983) 766.
25 M. T. Laugier, in P. Vincenzini (ed.), *Proc. 6th World Congr. on High Technology Ceramics, Milan, June 23–27, 1986*, in *High Tech Ceramics, Materials Science Monographs*, Vol. 38, Elsevier, New York, 1987, p. 1357.
26 K. E. Puttick, *J. Phys. D*, *13* (1980) 2249.

Materials Science and Engineering, *A105/106* (1988) 369–375

Fracture Toughness Measurement by Microindentation and Three-point Bend Methods*

MINYOUNG LEE and MILIVOJ K. BRUN

General Electric Research and Development Center, P.O. Box 8, Schenectady, NY 12301 (U.S.A.)

(Received January 13, 1988)

Abstract

The microindentation and the three-point bend test of the chevron-notched-bar methods were used to measure the fracture toughness of several common ceramic materials between room temperature and 1000 °C. The fracture toughness calculated from the maximum loads of chevron-notched beams was shown to be in good agreement with the values given in the literature.

Agreement of the results obtained by the two methods was excellent at room temperature. The two methods, however, gave opposite trends for the fracture toughness of silicon nitride as a function of temperature. Because of the fewer assumptions involved in the chevron-notched-beam method, it is considered preferable despite more complicated specimen preparation.

1. Introduction

The plane strain fracture toughness K_{Ic} of a material represents its inherent resistance to the propagation of crack-type defects. As such, it is useful for defining the critical conditions for brittle failure. The measurement of K_{Ic} generally requires ability to propagate cracks through the material in a stable manner and to measure the force necessary for their propagation. Because of their high modulus of elasticity and low values of K_{Ic}, achieving stable crack growth in ceramic materials is difficult and may require complicated specimen geometries and experimental set-ups. Many simplified methods have been proposed to avoid this difficulty. One method particularly attractive for routine engineering evaluation of materials is a microindentation technique developed by Evans and Charles [1].

Although the microindentation method can measure only an approximate value of K_{Ic}, it is a convenient technique for routine engineering applications to many brittle materials. Being based on a standard hardness test, the test is relatively simple to perform and requires only a standard hardness tester and a good measuring microscope. A small piece of material with a stress-free flat surface is sufficient as a test sample. The method, however, is not suitable for materials with a very low fracture toughness, significant ductility, large grain size or microstructural inhomogeneities. The method has been used at a high temperature but its validity has not been fully established. For example, the high temperature fracture toughness of silicon nitrides measured by the indentation method showed a small but consistent decrease with increasing temperature [2] while the "macroscopic" fracture toughness tests have often indicated an increase in the fracture toughness of similar materials at temperatures around 1000 °C [3].

The bend test of notched or pre-cracked beams is another relatively simple method which has been used for the measurement of fracture toughness [4]. Estimating the fracture toughness from the peak load of the bend test [5, 6] of chevron-notched beams is particularly attractive because it does not require pre-cracking of test bars or measurement of the pre-crack size after the test. The method also has a wider range of potential applications than the microindentation.

The chevron notch does not require pre-cracking, since the tip of the notch experiences the highest stress and a crack generally initiates at the tip. Because of the geometry of the notched section of beam, the propagating crack encounters a progressively wider crack front and its propagation is stable, provided that the fracture toughness remains constant during the crack growth (flat R curve). The load reaches a maxi-

*Paper presented at the 3rd International Conference on the Science of Hard Materials, Nassau, The Bahamas, November 9–13, 1987.

0921-5093/88/$3.50

© Elsevier Sequoia/Printed in The Netherlands

mum value when the crack reaches a critical length. The critical crack length is a parameter dependent on the geometry of specimen and it is essentially independent of material. The fracture toughness can be computed from the maximum load measured from following equation [5, 6]:

$$K_{Ic} = P_{max} \frac{Y_c(\alpha_0, \alpha_1)}{BW^{1/2}} \tag{1}$$

where P is the maximum load, Y_c is a parameter which is a function of the sample compliance, α_0 and α are a_0/W and a/W, respectively, and B and W are the dimensions of the sample, as shown in Fig. 1. The method allows calculation of the fracture toughness without having to follow the propagation of the crack through the sample, which is difficult to do experimentally. However, it requires determination of the sample compliance as a function of position of the crack throughout the test. Such measurements can be done either experimentally [6] or analytically [5, 7].

For an approximate measurement of the toughness, the specimen compliance calibration can be avoided. Since the compliance formula is essentially an equation of geometric parameters and has a very small dependence on material properties, a coefficient to estimate an approximate stress intensity factor from the peak load can be determined empirically for a fixed test specimen geometry using materials with known fracture toughness [8].

This study evaluated the fracture toughness of several ceramic materials between ambient temperature and 1000 °C by the indentation method and by the three-point bend test of chevron-notched bars. The applicability of both methods to high temperatures was assessed.

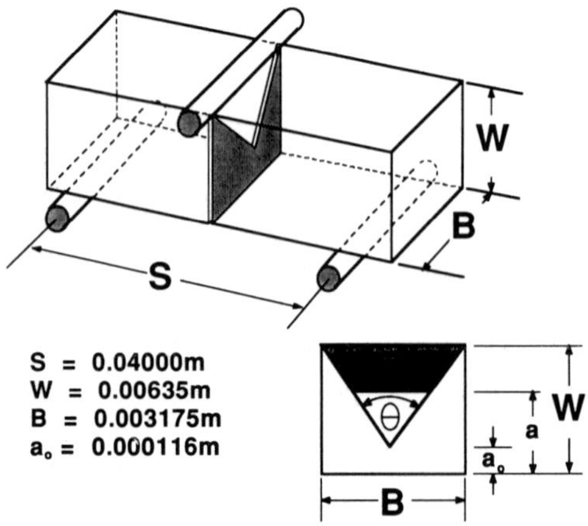

Fig. 1. Geometry of the chevron-notched bend bar.

2. Experimental details

The materials used are listed in Table 1. All materials were obtained as large billets, so that all the bend bars and the indentation samples could be machined from the same piece of material.

Microindentation fracture toughness measurements were performed by a procedure described by Evans [9] and the fracture toughness was estimated using the polynomial equation in that paper. Estimation of the toughness using the equation requires knowledge of the elastic modulus of the material. When the modulus could not be found in the literature, it was measured ultrasonically, using the time-of-flight method. A tungsten waveguide was used for the measurement of the elastic modulus at high temperatures.

Samples were polished by standard metallographic techniques, finishing with 1 μm diamond polish, to minimize residual surface stresses. Testing was done in an Akashi high temperature Vickers hardness tester in an argon atmosphere. Three different loads (1, 5 and 10 kgf) were used for each sample, to assure that the test was conducted in the load-independent regime. Samples were soaked at each test temperature to equilibrate the sample and the indenter.

The geometry of the three-point bend test specimen and the dimensions are shown in Fig. 1. The W/B ratio was higher than the optimum recommended for this type of test [5]. The maximum load occurred at relatively low values of $a/W = \alpha_{critical}$ of 0.43. All samples were ground by a commercial vendor (Bomas Machine Specialties Inc., Somerville, MA). The chevron notches were

TABLE 1 Materials used

Material	*Manufacturer*
Fused silica	General Electric Quartz & Chemical Product, Willoughby, OH
Sintered SiC	Experimental
Al_2O_3, Lucalox	General Electric Lamp Division, Cleveland, OH
Al_2O_3, McDanel	McDanel Refractory Porcelain Co., Beaver Falls, PA
Hot-pressed Si_3N_4	Norton Co., Worcester, MA
SiC whiskers–alumina	Experimental

cut with a fine (150 μm) diamond-tipped blade. All tests were conducted in a testing machine equipped with a high temperature furnace. Sintered silicon carbide fixtures were used for all tests. The fixtures were supported on a large-diameter alumina tube, and the load was applied with a sintered silicon carbide push rod, to provide maximum stiffness to the system. High temperature tests were done in a nitrogen atmosphere to prevent possible oxidation of the samples. All tests were run at a cross-head speed of 0.127 mm min^{-1}.

3. Results and discussion

The measurement of the fracture toughness by the indentation method requires development of a well-defined radial crack pattern around the indentation. The two aluminum oxide specimens and the fused silica developed very complex cracks around the hardness indenter even at the lowest load used. A reasonable estimate of the toughness from the indentation tests, therefore, was not possible for these three materials. The silicon carbide specimen was also very brittle, but several well-defined indentation marks were obtained at room temperature for this material. Only the silicon nitride specimen produced clean indentation cracks. The indentation fracture toughness of the silicon nitride specimen for room temperature, 600 °C and 1000 °C, and the room temperature toughness of the silicon carbide specimen, are shown in Table 2.

Two types of load–deflection curve were observed in the chevron-notched-bar tests, corresponding to stable and semistable crack propagation, and are illustrated in Fig. 2 as dotted and broken lines respectively. The stable crack propagation is indicated by a continuous and smooth load–deflection curve throughout the experiment. It was observed for both types of aluminum oxide, silica glass and hot-pressed silicon nitride. The semistable crack propagation is indicated by the sudden drop in load after the peak in the load–deflection curve. The sudden load decrease corresponds to unstable crack propagation, while the tail in the curve indicates that the crack was arrested before the complete sample failure. This type of curve was observed for silicon carbide and the silicon carbide whisker–alumina composite. The discontinuity was more prominent at elevated temperatures, probably owing to the lowering of the stiffness of the load train. With all the samples the experiment could be stopped at any time, and the extent of crack propagation observed with optical microscope.

TABLE 2 Experimental fracture toughnesses of the specimens

Material	*Test method*	*Fracture toughness* (MN m$^{-3/2}$) *Room temperature*	*600 °C*	*1000 °C*
Fused silica	Bend	0.7	0.8	0.9
	G/G_{RT}	1.0	1.13	1.09
Sintered SiC	Bend	2.3		
	Indentation	2.3		
Al_2O_3, Lucalox	Bend	3.0	2.5	2.1
	G/G_{RT}	1.0	0.88	0.76
Al_2O_3, McDanel	Bend	3.0	2.6	1.9
	G/G_{RT}	1.0	0.77	0.53
Hot-pressed Si_3N_4	Bend	5.6	5.7	6.2
	Indentation	5.9	5.1	4.8
	G/G_{RT}	1.0	0.89	0.86

The G/G_{RT} values are the areas under the load–deflection curves, normalized to room temperature value.

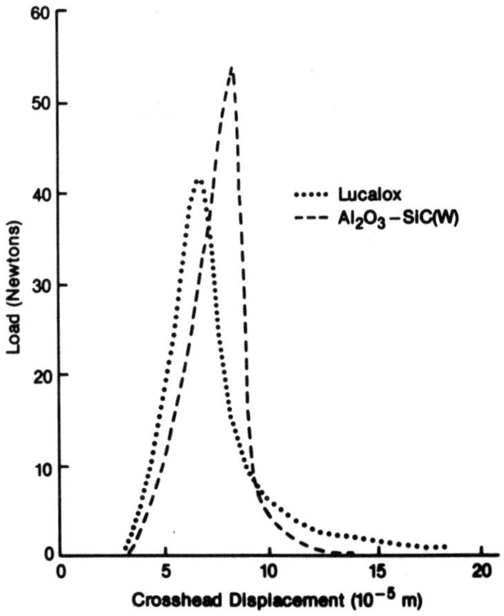

Fig. 2. Examples of a stable and a semistable load–displacement curve.

The fracture toughness values were calculated from the peak loads using eqn. (1). The values of Y_c were calculated by two methods. A complete analytical solution for a three-part bend geometry

TABLE 3 Calculated K_{Ic} values and K_{Ic} values from the literature

Material	*Peak load* P_{max} (N)	K_{Ic} (MN m$^{-3/2}$)			
		Wu [5]	*Munz et al. [7]*	*Other references*	*Calculated*
Fused silica	9.73	0.4	1.0	0.7 [10]	0.7
Sintered SiC	32.40	1.4	3.4	3.2 [11]	2.3
Al_2O_3, Lucalox	41.13	1.8	4.4	3.0 [12]	3.0
Al_2O_3, McDanel	41.13	1.8	4.4	3.0 [12]	3.0
Hot-pressed Si_3N_4	77.34	3.4	8.2	5.5 [2]	5.6
SiC whiskers–alumina	55.97	2.5	6.0	Not available	4.0

was given by Wu [5]. This equation was modified for the present geometry and solved to give $Y_c = 11.11$. The second value ($Y_c = 26.89$) was estimated from relations developed by Munz *et al.* [6] for the four-point bend test by setting the value of the inner span to zero (this is outside the range of variables which they considered). The values of the fracture toughness calculated from the two values at Y_c are shown in Table 3. Table 3 also shows the fracture toughness values of the materials given in the literature and they are very similar to those used in this study.

As the results show, the fracture toughness values calculated using the theoretically developed compliance relationship by Wu [5] are significantly lower than the previously known values for the materials, whereas the approximate toughness values estimated using the extrapolated relationship of experimental compliance equation by Munz *et al.* [6] are significantly higher than the previously known values. It is clear from the results that neither relation predicts the data satisfactorily.

The peak loads are plotted *vs.* the fracture toughness values in the literature in Fig. 3. It can be seen that all materials with the exception of silicon carbide fall on the straight line, as was predicted by eqn. (1).

From the slope of the line, we calculate a value of $Y_c = 18.20$ which is, as expected, intermediate between the values estimated from the two models. The values of K_{Ic} shown in the last column of Table 3 were calculated using the peak loads and the calculated value of Y_c. With the exception of silicon carbide, the agreement with the value in the literature is excellent as expected. The fracture toughness of silicon carbide measured by the bend test was considerably lower than the value in the literature for sintered silicon carbide. However, it was very close to the toughness of the material measured by indentation, shown in Table 2. It can be concluded, therefore, that the particular silicon carbide sample tested was inferior and that its K_{Ic} was correctly measured as being well below the values in the literature.

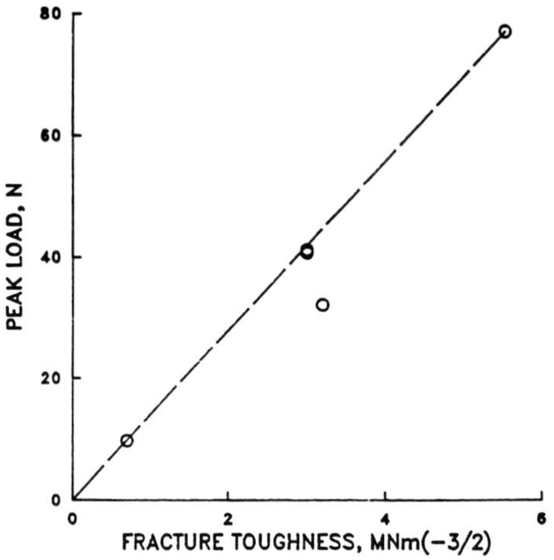

Fig. 3. Peak load *vs.* known fracture toughnesses of five ceramic materials.

As stated earlier, the specimen compliance equation has only a minor dependence on the material property through Poisson's ratio. Therefore, if the proportionality factor Y_c can be determined for a specific specimen geometry with a well-characterized material, it can be used for an approximate measurement of the fracture toughness of other materials, provided that the specimen geometry and the test conditions are unchanged. The relatively good agreement between the experimentally determined ratios of the known fracture toughness values to the peak loads supports this assessment. By the same argument, the method can be applied to high temperature measurements.

The load–deflection curves obtained at three test temperatures for the two aluminum oxides and for the hot-pressed silicon nitride are shown in Figs. 4, 5 and 6 respectively. The areas under the curves are proportional to the work G of fracture. The change in the work of fracture with temperature for the materials investigated is shown in Table 2, as the ratio of area at temperature, normalized with respect to the room temperature value.

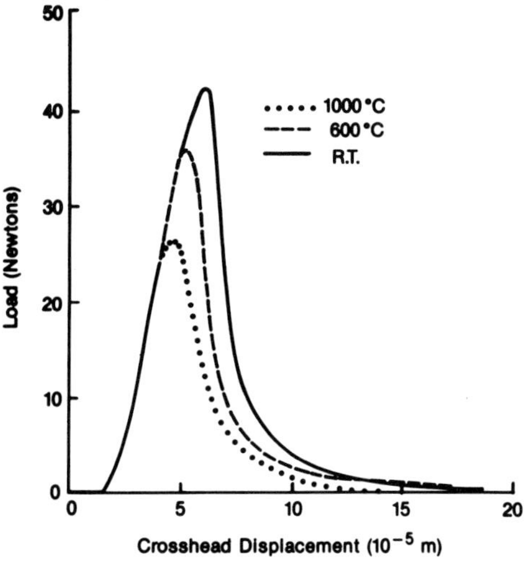

Fig. 4. Load–displacement curves of sintered alumina (McDanel 998).

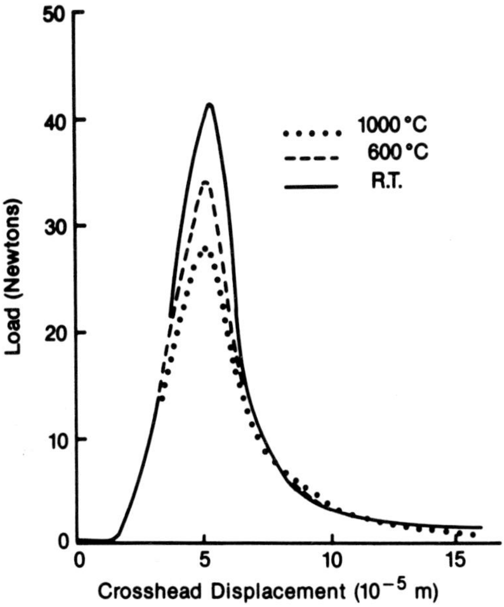

Fig. 5. Load displacement curves of sintered aluminum oxide (Lucalox).

The fracture toughness of aluminum oxide measured by the bend test showed a steady decrease with increasing temperature, in agreement with the microindentation results reported previously [2]. The decrease in the fracture toughness of aluminum oxide with increasing temperature can be explained by the general weakening of the bonds. Also, the internal stresses in the material which may arise during cooling from the processing temperature will be partially relieved by reheating to test temperature. If these stresses had the effect of raising the toughness of the material, their contribution will obviously decrease with an increase in test temperature. Figures 5 and 6 show that the general shape of the load–displacement curves of aluminum oxides did not change with temperature and that the total work of fracture decreased with the increasing temperature. It also shows some differences between the two materials. The modulus of McDanel alumina (the slope of the load–deflection curve) remains unchanged by temperature, while the modulus of Lucalox alumina decreases markedly with increasing temperature.

The fracture toughness of hot-pressed silicon nitride measured by the peak load showed essentially no difference between the value at room temperature and that at 600 °C. At 1000 °C, however, the value of the fracture toughness is

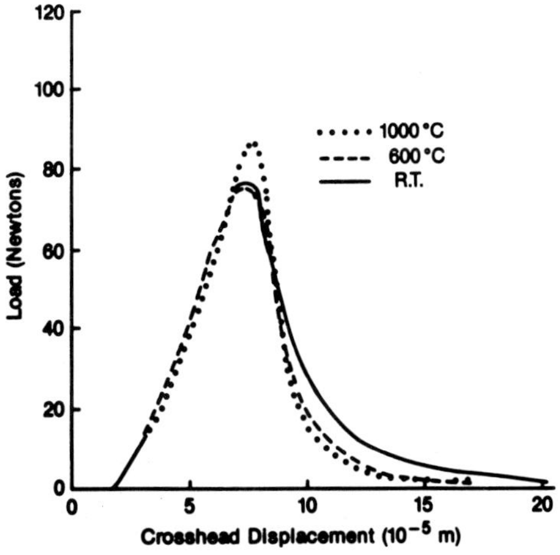

Fig. 6. Load–displacement curves of hot-pressed silicon nitride (Norton NCX-34).

clearly higher than at lower temperatures. The microindentation technique indicated an appreciable decrease in the fracture toughness with temperature. The work of fracture decreased with increasing temperature, but the decrease was very small.

The effect of temperature on the fracture toughness of silicon nitride has been reported by several investigators [3, 13] but the reported results over the temperature range up to 1000 °C are inconsistent. There have been observations of increasing fracture toughness [14] as well as decreasing fracture toughness [15] with increasing temperature. It has been reported that the size of test specimen [16] and loading rate [17] also have significant effects on the measured fracture toughness. In this work, since all the samples were prepared from the single piece of material, material variability cannot account for the difference in results obtained by the two different methods of measuring of fracture toughness. It is therefore reasonable to conclude that the observed difference in the trends of the fracture toughness change with temperature is caused by the two different test methods evaluated.

One possible explanation for the different trend observed by the two techniques may be the difference in the rate of crack propagation during the test. In the notched-beam test, crack propagation is relatively slow, while formation of indentation cracks is very rapid. If the material exhibits some degree of plasticity at the crack tip, or crack tip blunting or healing takes place, then the measured fracture toughness will be higher at slower propagation rates.

An alternative explanation would be that the differences are due to the measurement methods themselves. While the high temperature notched-beam test is not significantly different from the room temperature test, there are several aspects of the indentation test method which could be influenced by the high test temperature and could alter the results. These are possible annealing of the surface stresses, the change in friction between the sample and the indenter, an axial temperature gradient in the testing furnace and the assumption of limited plasticity in the derivation of the method.

To examine the possibility of annealing effect on the residual surface stress during the test, some tests were performed by making indentations at the highest test temperature first, so that the surface of the sample on which the low temperature indentations were made would have had the same thermal history. Also, some specimens were tested twice to evaluate the repeatability and the effect of thermal cycling on the toughness. None of these tests produced appreciably different toughness values.

It is also possible that the high temperature could have a significant effect on the friction between the specimen and the indenter compared with the room temperature case. The magnitude of that change could be different for silicon nitride and alumina ceramics, since the nature of the interactions (physical and chemical) of these materials with diamond is significantly different.

Although the indenters were carefully preheated, development of a slight temperature gradient at the specimen–indenter interface was unavoidable owing to the loading mechanism. This could have an effect on the local stress around the tip of indenter. The extent of any of these effects has not been evaluated in this study.

Finally, the derivation of the indenter method [1] assumes only a limited plasticity under the indenter tip. In materials which contain a vitreous phase which can soften at high temperature, such an assumption may not be satisfied.

Because of these possible complications, the applications of the indentation method at high temperatures seems to introduce some uncertainty. The magnitude of error, however, appears to be small and the technique is potentially useful for rapid evaluation of materials at moderately high temperatures (less than 1000 °C).

The bend test method of chevron-notched bars is also relatively simple to perform, although larger specimens are needed and preparation of samples is more complicated than for the indentation method. Consequently, if an accurate compliance formula and a rigid testing apparatus are available, the method is very useful for routine evaluation of materials. For a comparative evaluation of materials, the method can be used effectively, provided that a well-characterized standard material is used to determine the specimen compliance for a given specimen geometry, or the model is improved, so that reliable specimen compliance can be calculated.

4. Summary

The fracture toughness of several common ceramics was measured by the microindentation method and by the three-point bend test of chevron-notched bars at three different temperatures. Complete measurement by the indentation method was possible only for silicon nitride.

Stable crack propagation was observed during the bend tests of essentially all specimens. The two models suggested in the literature underestimated or overestimated the sample compliance appreciably. The ratios of the room temperature fracture toughness to the measured peak load were consistent and were used to estimate the sample compliance. The constant was used to measure the high temperature fracture toughnesses.

Both aluminum oxides showed a significant decrease in toughness with increasing temperature, while the fused silica showed a slight but consistent increase in fracture toughness with increasing temperature. The indentation fracture toughness of silicon nitride decreased at elevated temperatures, whereas the fracture toughness determined by the bend test showed an opposing trend. The magnitude of deviations, however, was small and the indentation technique appears useful for estimating toughness at high temperatures.

Acknowledgments

The authors are grateful to Louis P. Inzinna for help in sample testing and to Donna M. Merchant for preparation of the manuscript.

References

1 A. G. Evans and E. A. Charles, Fracture toughness determination by indentation, *J. Am. Ceram. Soc., 59* (7–8) (1976) 371.

2 M. Lee and M. K. Brun, The high temperature fracture toughness of SiAlON, *Ceram. Eng. Sci. Proc., 4* (9–10) (1983) 864.

3 T. T. Shih and J. Opoku, Application of fracture mechanics to ceramic materials—a state-of-art review, *Eng. Fract. Mech., 12* (1979) 479.

4 J. J. Petrovic and L. A. Jacobson, Controlled surface flaws in hot pressed SiC, *J. Am. Ceram. Soc., 59* (1976) 34.

5 S.-X. Wu, Compliance and stress-intensity factor of chevron-notched three-point bend specimen, in J. H. Underwood, S. W. Freiman and F. I. Baratta (eds.), *Chevron-notched specimens: Testing and Stress Analysis, ASTM Spec. Tech. Publ. 855*, 1984, p. 176.

6 D. Munz, R. T. Bubsey and J. L. Shannon, Jr., Fracture toughness determination of Al_2O_3 using four-point-bend specimens with straight-through and chevron notches, *J. Am. Ceram. Soc., 63* (5–6) (1980) 300.

7 D. Munz, J. L. Shannon, Jr., and R. T. Bubsey, Compliance and stress intensity coefficients for short bar specimens with chevron notches, *Int. J. Fract., 16* (1980) R137.

8 L. M. Baker, A simplified method for measuring plane strain fracture toughness, *Eng. Fract. Mech., 9* (1977) 361.

9 A. G. Evans, Fracture toughness: the role of indentation techniques, in S. W. Freiman (ed.), *Fracture Mechanics Applied to Brittle Materials, Spec. Tech. Publ. ASTM 678*, 1979, p. 112.

10 R. F. Cook and B. R. Lawn, A modified indentation toughness technique, *Commun. Am. Ceram. Soc., 66* (1983) C-200.

11 A. G. Evans and F. F. Lange, Crack propagation and fracture in silicon carbide, *J. Mater. Sci., 10* (1975) 1659–1665.

12 R. W. Rice, S. W. Freiman and P. F. Becher, Grain-size dependence of fracture energy in ceramics: I, Experiment, *J. Am. Ceram. Soc., 64* (1981) 345–350.

13 D. Munz, G. Himsolt and J. Eschweiler, Effect of stable crack growth on fracture toughness determination for hot-pressed silicon nitride at elevated temperatures, in W. S. Freiman and E. R. Fuller (eds.), *Fracture Mechanics Methods for Ceramics, Rocks, and Concrete, ASTM Spec. Tech. Publ. 745*, 1981, p. 69.

14 A. G. Evans and S. M. Wiederhorn, Crack propagation and failure prediction in silicon nitride at elevated temperatures, *J. Mater. Sci., 12* (1977) 212.

15 C. G. Annis and J. S. Cargill, Modified double torsion method for measuring crack velocity in NC-132 Si_3N_4, in R. C. Bradt, D. P. H. Hasselman and F. F. Lange (eds.), *Fracture Mechanics of Ceramics*, Vol. 4, Plenum, New York, 1977, p. 737.

16 M. H. Lewis and G. Smith, Fracture mechanics of Si–Al–ON ceramics, in D. M. R. Taplin (ed.), *Fracture Proc. 4th Int. Conf. on Fracture, Waterloo, Ontario, June 19–24, 1977*, Vol. 3, University of Waterloo Press, Waterloo, Ontario, 1977, p. 867.

17 M. G. Mendiratta, J. Wimmer and I. Bransky, Dynamic K_{Ic} and dynamic flexural strength in HS-130 Si_3N_4, *J. Mater. Sci., 12* (1977) 212.

Effects of Microstructure on Hardness and Palmqvist Fracture Toughness of WC–Co Alloys*

Y. SHIN† and W. CAO

Materials Science and Engineering Department, North Carolina State University, Raleigh, NC 27695-7907 (U.S.A.)

G. SARGENT

School of Engineering, University of Dayton, Dayton, OH 45469 (U.S.A.)

H. CONRAD

Materials Science and Engineering Department, North Carolina State University, Raleigh, NC 27695-7907 (U.S.A.)

(Received March 29, 1988)

Abstract

The effects of microstructure, specimen preparation and test procedure on the hardness and indentation fracture of commercial WC–Co alloys (4.5–20wt.%Co) fabricated by two companies were investigated. The measured hardness was in good accord with that calculated from the Lee and Gurland relations. The Palmqvist fracture toughness G_{Is} was proportional to the binder mean free spacing; moreover, G_{Is}^{-1} increased linearly with increasing hardness. By considering the variation in the bulk fracture toughness G_{Ic} (and of G_{Is}) with hardness, the two are related through

$$G_{Ic} = C\exp(-\beta G_{Is}^{-1})$$

Only a little difference in the hardness and G_{Is} values occurred for the two material sources, specimen preparation and test procedures considered.

1. Introduction

Of importance in the application of cemented WC–Co alloys is a knowledge of the effects of microstructure on their hardness and fracture toughness. A convenient method for obtaining both hardness and fracture toughness in WC–Co alloys is that proposed by Palmqvist [1], which consists of conducting a Vickers hardness test and measuring as a function of applied load the impression diagonal and the total length ($L_t = 4l$) of the cracks occurring at the four corners of the impression. For carefully prepared specimens, L_t has been found to increase linearly with increasing applied load [2], so that

$$L_t = \frac{P - P_c}{W} \qquad (1)$$

where W is a crack propagation resistance parameter [2] and can be considered [3] to be a surface fracture toughness parameter G_{Is}, similar to G_{Ic} for the bulk. P_c is generally small compared with the values of P normally employed, and in many cases a plot of L_t *vs.* P passes through the origin [2, 4], although this need not necessarily occur [3, 5, 6].

A number of correlations [7–11] have been proposed for the effects of microstructure on the hardness of WC–Co alloys. Of these, that by Lee and Gurland [10] has the firmest physical basis and is the least empirical. It gives for the Vickers hardness of an alloy

$$H_V = H_{WC} f_{WC} + H_{Co}(1 - f_{WC} C) \qquad (2)$$

where H_{WC} and H_{Co} are the *in situ* hardnesses of the WC and cobalt phases respectively, f_{WC} is the volume fraction of the carbide grains and C is their contiguity.

*Paper presented at the 3rd International Conference on the Science of Hard Materials, Nassau, The Bahamas, November 9–13, 1987.

†Present address: Korea Advanced Energy Research Institute, Taejun, Chungnow, South Korea.

0921-5093/88/$3.50 © Elsevier Sequoia/Printed in The Netherlands

The Palmqvist fracture toughness parameter W for cemented carbides has been found [4] to vary with hardness according to

$$W^{-1}=A(H_V-H_V^0). \quad (3)$$

This led Viswanadham and Venables [4] to conclude that W was proportional to the bulk fracture toughness parameter G_{Ic}. Peters [12] found that W increases linearly with increasing cobalt mean free path λ and deduced that $W=a+bG_{Ic}$ where a and b are constants. However, noting that G_{Ic} was related to H_V through

$$\ln G_{Ic}=c-gH_V \quad (4)$$

where c (=10.3) and g ($=3.3\times10^{-10}$ Pa) are constants, Perrott [3] showed the inadequacy of a simple linear relationship between W and G_{Ic} and derived a more appropriate, albeit more complex, relation between the two. Warren and Matzke [13] showed that, for a limited range of cobalt contents, W may be related to the stress intensity factor K_{Ic} through $K_{Is}=K_{Ic}=0.087(H_V W)^{1/2}$.

The major objective of the present investigation was to obtain additional data on the effects of microstructure on the hardness and Palmqvist fracture toughness of WC–Co alloys and to evaluate further the correlations presented above. A secondary objective was to compare results on specimens obtained from two fabrication sources, and which were prepared and tested by two procedures.

2. Experimental details

The WC–Co alloys used in this investigation were commercial materials containing 4.5–20wt.%Co (8–32vol.%Co) provided by the Multi-Metals Division of Vermont American Corporation and by Kennametal Corporation. The microstructural parameters of the materials determined by the linear intercept method [10] are presented in Table 1. The alloys from Multi-Metals have in general somewhat lower values for the WC grain contiguity factor C.

In one series of tests (series A), the as-received specimens were first ground on a 45 μm metal-bonded diamond wheel and then polished on cloth-covered wheels employing 30, 15, 6, 3 and 1 μm diamond sprays. After polishing, the specimens were annealed for 1 h at 900 °C in a vacuum furnace. The Palmqvist tests were then carried out with a commercial Vickers hardness tester (136° diamond indenter) at loads P of 0.2–40 kgf. The length of the impression diagonal and the Palmqvist cracks (which were an extension of the diagonals) were measured with an optical microscope at a magnification of 200–500×. In the second series of tests (series B), the as-received specimens were polished on a commercial vibrating polishing machine, starting with 15 μm and ending with 1 μm diamond paste. The Palmqvist tests were then carried out without an anneal using a standard 136° diamond Vickers indenter mounted on an Instron testing

TABLE 1 Microstructural parameters of the WC–Co alloys

Specimen	*Binder content* f_{Co} (wt.%(vol.%))	*Mean WC grain size* d_{WC} (μm)	*Mean Cobalt free path*† λ (μm)	δ (μm)	*Contiguity* C
4.5F	4.5 (8)	1.0	0.08	0.16	0.48
6FF	6.0 (10)	0.9	0.09	0.16	0.36
6F	6.0 (10)	1.1	0.12	0.20	0.28
6M	6.0 (10)	1.6	0.18	0.35	0.49
6MM	6.0 (10)	1.8	0.20	0.43	0.53
6CR	6.0 (10)	2.6	0.29	0.72	0.60
9MM	9.0 (15)	1.6	0.28	0.43	0.35
10M	10.5 (17)	1.7	0.35	0.51	0.32
10MM	10.5 (17)	1.8	0.37	0.53	0.30
13M	13.0 (21)	1.6	0.43	0.57	0.26
6A	6.4 (11)	0.7	0.08	0.18	0.55
6B	6.0 (10)	1.0	0.11	0.37	0.69
6C	6.3 (11)	3.3	0.35	0.93	0.63
10E	10.0 (16)	1.3	0.26	0.95	0.67

Specimens 6A, 6B, 6C and 10E were provided by Kennametal Corporation; the remaining specimens were provided by Multi-Metals.

† $\lambda=d_{WC}f_{Co}/(1-f_{Co})$: assumes presence of cobalt film between WC–WC interfaces. $\delta=\lambda/(1-c)$: assumes no cobalt film between WC–WC interfaces.

machine, which was operated at a cross-head speed of 2.1×10^{-5} cm s^{-1} to attain loads of 10–150 kgf. The lengths of the impression diagonals and associated cracks were measured as above.

3. Results

3.1. Measurements and fracture path

Typical behavior for test series A is shown in Fig. 1. It should be noted that the average length $2a$ of the impression diagonal increased with increasing $P^{1/2}$ over the entire load range, in accord with the Vickers hardness equation $H_V = 2P\sin(\theta/2)/(2a)^2$, and that H_V is independent of P. At a critical load P^*, cracks developed at each of the four corners of the impression. The average crack length $L = (l_1 + l_2 + l_3 + l_4)/4$ of the four cracks of length l_i increased with increasing P more rapidly than did the diagonal length. A plot of $L_t = 4l_i$ *vs.* P as a function of cobalt content is presented in Fig. 2, where it is seen that L_t increases linearly with P, giving a small intercept P_c (= 2–3 kgf) on the load axis. The slope of the lines in Fig. 2 increases and the intercept decreases as the cobalt content decreases and the WC grain size becomes smaller.

L_t *vs.* P plots as functions of cobalt content and of WC grain size d_{WC} for test series B are presented in Figs. 3 and 4 respectively. In view of the higher loads employed and the reduced scale for the load axis, the straight lines now appear to pass through the origin.

The fracture path in all cases was mostly along the cobalt phase, with only an occasional transgranular fracture of a WC grain.

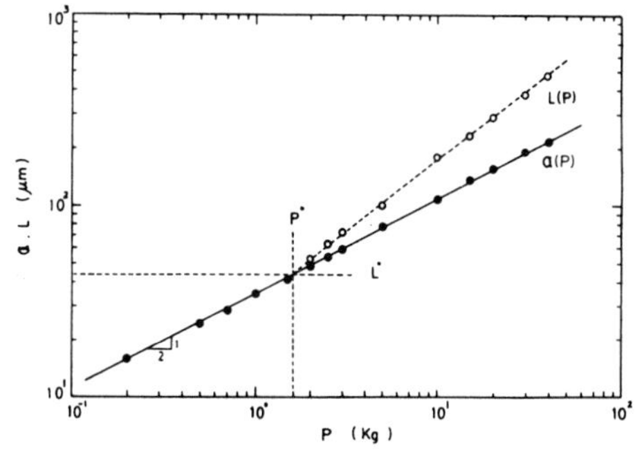

Fig. 1. Half-length a of indentation diagonal and average crack length L *vs.* applied load P for WC–6wt.%Co alloys in series A tests.

3.2. Hardness and fracture toughness

The Vickers hardness values obtained for the two tests series are given in Table 2; those for test series A tend to be slightly higher (average, 3%) than those for series B. In Fig. 5, the measured hardness for both test series is plotted *vs.* that calculated from the microstructural parameters according to Lee and Gurland [10]. As found by others [13], rather good agreement exists between the measured and calculated values.

The values of the fracture toughness $G_{Is} = dP/dL_t$ obtained for the two test series are included in Table 2. The values for test series A tend to be

TABLE 2 Hardness and fracture toughness of the WC–Co alloy specimens

Specimen	*Vickers hardness* (GN m^{-2})		G_{Is} (MJ m^{-2})	
	Series A	*Series B*	*Series A*	*Series B*
4.5F	17.38	16.83	0.681	0.686
6FF	16.17	16.39	0.782	0.815
6F	16.03	15.20		0.894
6M	14.62	14.36	1.186	1.390
6MM	14.21	13.94	1.381	1.265
6C	13.65	12.83	1.477	1.491
9MM	12.62	11.77	3.042	3.364
10M		12.43		3.844
10MM	12.16	12.16	2.635	2.630
13M	11.82	11.47	3.442	3.776
6A		17.59		0.722
6B		14.81		0.912
6C		11.69		2.589
6D		14.88		1.353
6E		12.26		2.707
14M		11.83		2.707
20M		10.23		10.230

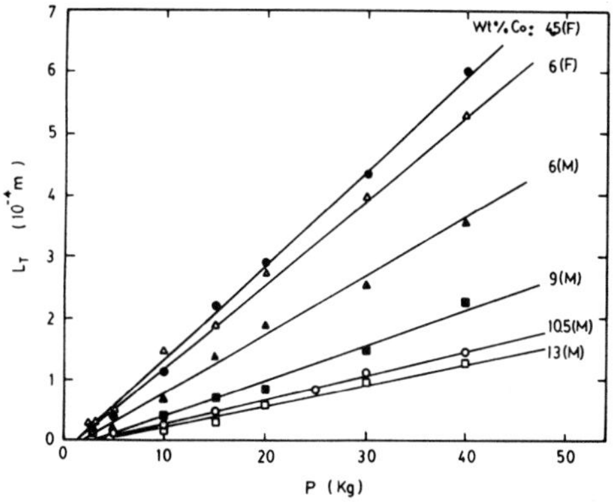

Fig. 2. Total crack length L_t *vs.* applied load as a function of cobalt content and WC grain size for WC–Co alloys in series A tests.

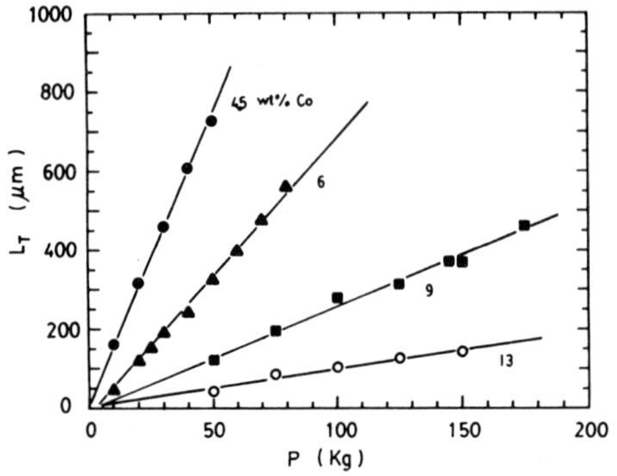

Fig. 3. Total crack length L_t *vs.* applied load as a function of cobalt content for WC–Co alloys in series B tests ($d_{WC} = 1–2$ μm).

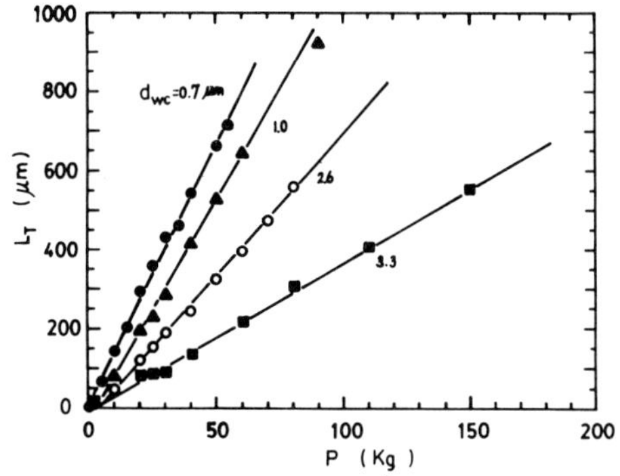

Fig. 4. Total crack length L_t *vs.* applied load for WC–6wt.%Co alloys as a function of WC grain size for series B tests.

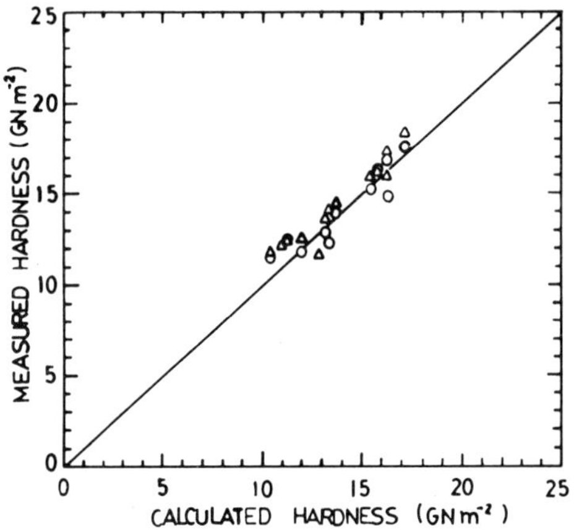

Fig. 5. Measured Vickers hardness *vs.* calculated hardness (△, series A; ○, series B).

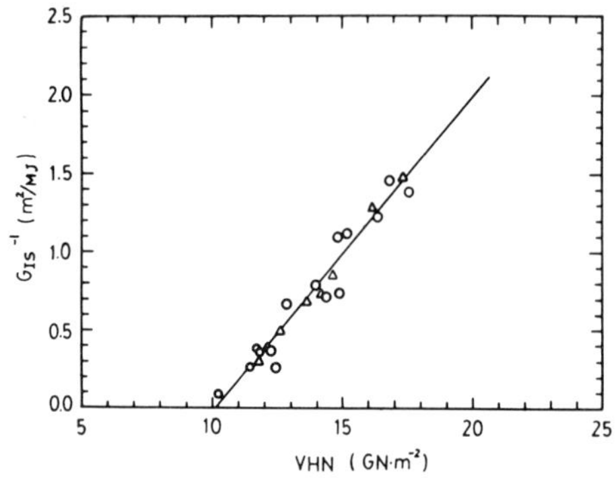

Fig. 6. G_{Is}^{-1} *vs.* hardness (△, series A; ○, series B).

slightly lower (average, 4%) than those for series B. Figure 6 shows that G_{Is}^{-1} increases linearly with increasing hardness in accord with eqn. (3), giving $A = 2 \times 10^{-16}$ m^2 N^{-2} and $H_V^0 = 10.26$ Pa; these values are in good accord with those reported by others [3, 4] for cemented carbides ($A = 1.85 \times 10^{-16}$ m^2 N^{-2} and $H_V^0 = 10.6$ GPa). Figure 7 shows that G_{Is} is proportional to cobalt mean free path, giving

$$G_{Is} = \beta\lambda \tag{5}$$

with $\beta = 7.6 \times 10^{12}$ J m^{-2}, which is in reasonable accord with the value of 5×10^{12} J m^{-2} obtained by Peters [12] for WC–Co alloys.

4. Discussion

The present results indicate reasonable agreement between the results for the various conditions considered here, namely two suppliers, two surface preparation procedures and two test methods. The slightly higher hardness and slightly lower G_{Is} values obtained in the series A tests compared with the series B tests probably result from a higher compressive stress existing in the WC grains (tensile in the cobalt) at the surface following the annealing treatment at 900 °C, which was employed in the former series and not in the latter. Exner [2] has shown that WC–Co specimens which had been given an annealing treatment exhibited slightly larger Palmqvist crack lengths (lower G_{Is}) than those which had been carefully polished. Moreover, the crack length was shown to increase with an increase in compressive stress in the WC grains (tensile in the cobalt) at the surface.

Of special interest is the relationship between G_{Is} and the bulk fracture toughness parameter G_{Ic}. Taking the data of Leuth [14] and Murray [15], Perrott [3] has shown that G_{Ic} decreases with increasing hardness according to eqn. (4) given above. Combining this relation with that between hardness and G_{Is} (eqn. (3)), we obtain

$$\ln G_{Ic} = \alpha - \beta G_{Is}^{-1} \tag{6}$$

$$G_{Ic} = C\exp(-\beta G_{Is}^{-1}) \tag{6a}$$

where $C = e^{\alpha}$, $\alpha = (c - gH_V^0)$ and $\beta = g/A$. A plot of log G_{Ic} (data of Leuth [14] and Murray [15]) *vs.* G_{Is} obtained in the present tests for the same hardness value is given in Fig. 8. It should be noted that reasonably good correlation is obtained, yielding $C = 8.70 \times 10^2$ J m^{-2} and $\beta = 1.86 \times 10^6$ J m^{-2}. In comparison, employing the values of the constants c, g, A and H_V^0 derived by Perrott [3], we obtain $C = 8.97 \times 10^2$ J m^{-2} and $\beta = 1.80 \times 10^6$ J m^{-2}.

5. Summary and conclusions

The effects of microstructure, specimen preparation and test procedure on the hardness and indentation fracture of commercial WC–Co alloys (4.5–20 wt.% Co) fabricated by two companies were investigated. The following is a summary of the results obtained and the conclusions derived therefrom.

(1) The effects of microstructure on hardness were in good accord with that calculated from the Lee and Gurland [10] relations.

(2) The Palmqvist fracture toughness parameter $G_{Is} = dP/dL$ was proportional to the binder free spacing and to the reciprocal of the hardness.

(3) The bulk fracture toughness G_{Ic} was related to the Palmqvist parameter G_{Is} through $G_{Ic} = C\exp(-\beta G_{Is}^{-1})$, where C and β are constants.

(4) Slightly higher hardness and lower G_{Is} values were obtained for specimens which had been annealed at 900 °C prior to testing compared with those which had not been annealed. This was attributed to a higher compressive stress in the WC grains in the annealed specimens.

Acknowledgments

The authors acknowledge and appreciate support of test series A of this research by the University of Kentucky Institute for Mining and Minerals Research and of test series B by the U.S. Department of Energy under Grant DE-FG05-84ER45115. They also wish to express their appreciation to Multi-Metals, a Division of Vermont American Corporation, and to Kennametal Corporation for providing the WC–Co alloys.

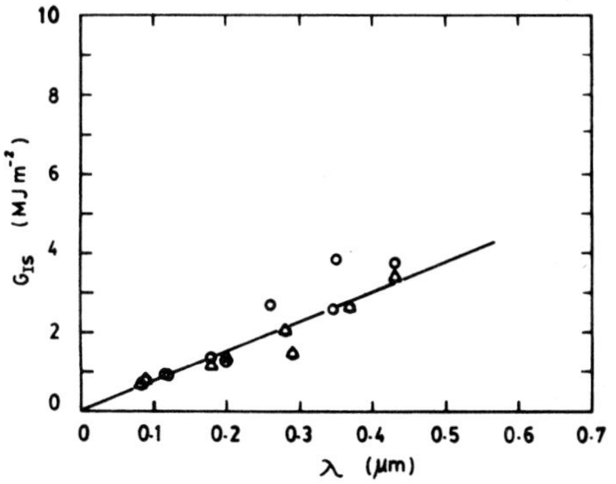

Fig. 7. G_{Is} *vs.* cobalt mean free path (Δ, series A; ○, series B).

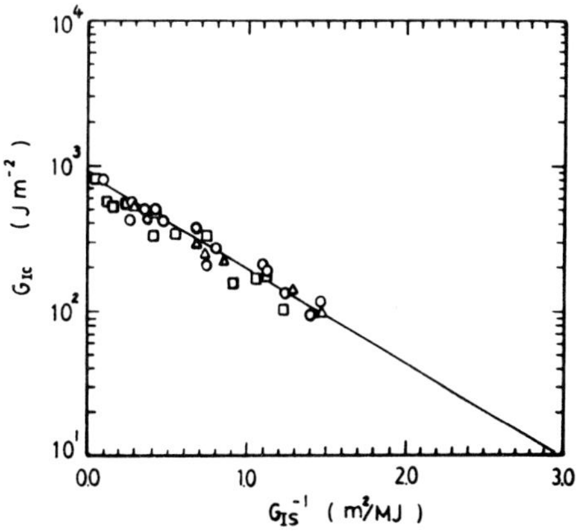

Fig. 8. Semilogarithmic plot of G_{Ic} *vs.* G_{Is}^{-1} (Δ, series A; ○, series B; □, Murray and Lueth).

References

1 S. Palmqvist, *Jernkonterets Ann., 141* (1957) 300; *Arch. Eisenhüttenwes., 33* (1962) 629.

2 H. E. Exner, *Trans. Metall. Soc. AIME, 245* (1970) 677.

3 C. M. Perrott, *Wear, 45* (1977) 293; *17* (1978) 82.

4 R. W. Viswanadham and J. D. Venables, *Metall. Trans. A 8* (1977) 187.

5 P. O. Snell and E. Pärnama, *Planseeber. Pulvermetall., 21* (1973) 17.

6 E. A. Almond and B. Roebuck, in R. K. Viswanadham, D. J. Rowcliffe and J. Gurland (eds.), *Science of Hard Materials*, Plenum, New York, 1983, p. 597.

7 J. L. Chermant, A. Deschanvres, G. Hautier, A. Iost and G. Manier, *Phys. Status Solidi, 15* (1973) 149.

8 G. Grathwohl and R. Warren, *Mater. Sci. Eng., 14* (1974) 55.

9 R. Warren and M. B. Waldron, *Powder Metall. Int., 7* (1975) 18.

10 H. C. Lee and J. Gurland, *Mater. Sci. Eng., 33* (1978) 125.

11 M. T. Langier, *Acta Metall., 33* (1985) 2093.

12 C. T. Peters, *J. Mater. Sci., 14* (1979) 1619.

13 R. Warren and H. Matzke, in R. K. Viswanadham, D. J. *Rowcliffe and J. Gurland (eds.) Science of Hard Materials*, Plenum, New York, 1983, p. 563.

14 R. C. Leuth, in R. C. Brandt, D. P. H. Hasselman and F. F. Lange (eds.) *Fracture Mechanics of Ceramics*, Vol. 2, Plenum, New York, 1974, p. 791.

15 M. J. Murray, *Proc. R. Soc. London, Ser. A, 356* (1977) 483.

Materials Science and Engineering, A105/106 (1988) 383–387

Effects of Residual Stresses in Fracture Toughness Testing of Hard Metals*

R. GODSE, J. GURLAND and S. SURESH

Brown University, Providence, RI 02912 (U.S.A.)

(Received November 11, 1987)

Abstract

Residual tensile stresses resulting from pre-cracking procedures, at the crack tip of cemented carbide specimens can affect the measured fracture toughness values. The nature of such stresses is studied in specimens pre-cracked by indentation methods or by cyclic compression. It is experimentally shown how the residual stress effects are reduced if the pre-crack is grown under tensile fatigue out of the zone of residual tension.

1. Introduction

Special techniques are needed for pre-cracking specimens used for fracture toughness measurements of hard metals, since conventional techniques such as fatigue pre-cracking in tension cannot be employed. Several methods have been advocated in this context in the past, but no single method has gained wide acceptance. In the present paper, two methods that have been advocated previously, *i.e.* bridge indentation [1] and cyclic compression [2], are dealt with. In particular, the paper is concerned with the influence of residual stresses arising from mechanical effects of the pre-cracking procedures on the measured values of fracture toughness.

2. Bridge indentation

This method was first described by Sadahiro and Takatsu [1] and was investigated in greater detail by Warren and Johannesson [3]. First a Vickers hardness indentation is made centrally on the top surface of the specimen. The specimen is then compressed with two rectangular punches placed on the two sides of the indentation while supported on an anvil. The geometry of such a compression produces a tensile stress which causes the cracks formed at the corners of the indentation (Palmqvist cracks) to extend outwards and downwards. This extended straight-through crack can then be used for K_{Ic} measurement with the single-edge-notched bend specimen.

The magnitude of the tensile stresses between the two punches depends on the dimensions d, s and w shown in Fig. 1. Analytical solutions for the stress distributions for various values of the above parameters are reported by Warren and Johannesson [3]. They have shown that the tensile stress at a given applied compressive load is maximum at the top surface and falls off with increasing depth into the specimen, eventually becoming compressive at a certain depth a'. A Palmqvist crack will grow if it lies in the tensile region. As the load is increased, the stress intensity at the Palmqvist crack tip increases until it reaches K_{Ic}, at which point the cracks parallel to the punch edges propagate unstably to form a pre-crack. This crack is arrested when it reaches the compressive region of the specimen.

For a given punch spacing and geometry, increasing the compressive load increases the stress levels in the specimen without altering the transition depth a'. However, a' increases when the punch spacing d or the punch sizes are increased. Therefore, longer pre-cracks are obtained from the Palmqvist cracks if larger punch spacings are used [3].

3. Experimental details

The tests were performed on a WC–5.5wt.%Co alloy supplied by Kennametal Inc.,

*Paper presented at the 3rd International Conference on the Science of Hard Materials, Nassau, The Bahamas, November 9–13, 1987.

0921-5093/88/$3.50

© Elsevier Sequoia/Printed in The Netherlands

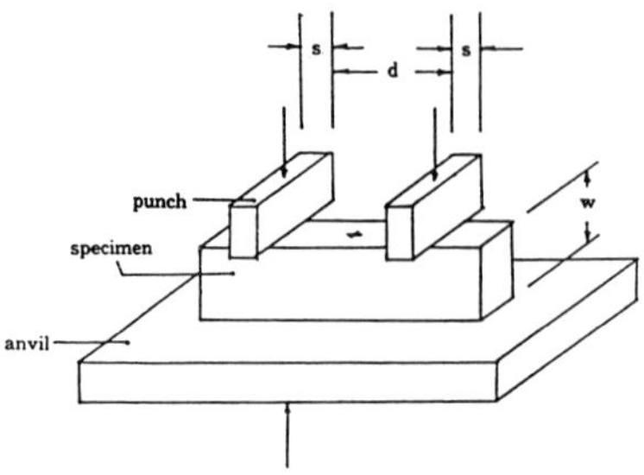

Fig. 1. Diagram showing main features of bridge compression.

TABLE 1 **Results from bridge compression tests**

Batch	P_V (N)	*d* (mm)	P_b (kN)	*a/w*
I	1471	8.0	20	0.35
II	981	10.0	25	0.60
III	589	12.0	26	0.68
IV	150	13.0	40	0.72

P_V, Vickers indentation load; P_b, load at unstable crack extension during bridge compression.

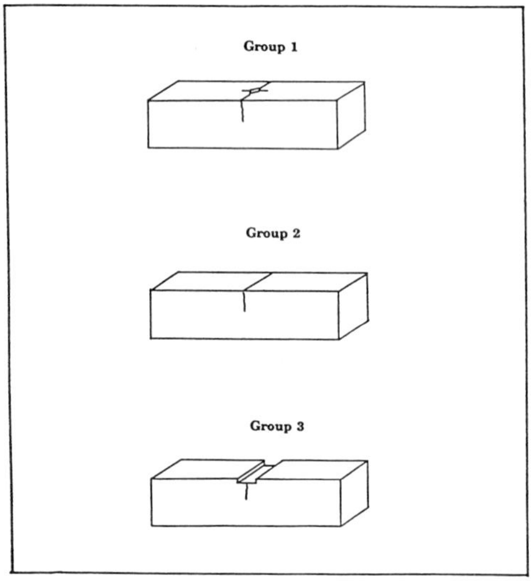

Fig. 2. Difference between specimens of groups 1, 2 and 3.

Latrobe, PA. The specimen dimensions were 3 mm × 6 mm × 30 mm. A Vickers indentation was placed centrally on the top surface. Four batches of specimens were prepared, and the load used for the Vickers indentation was varied as shown in Table 1. The specimens were then compressed as shown in Fig. 1 with various values of *d*. The punch size *s* was held fixed at 3.0 mm. The compression was carried out on a servohydraulic testing machine at a cross-head speed of 1.0 $\mu m\ s^{-1}$. The specimens were unloaded immediately after the unstable crack growth, audible in all cases, was detected.

To investigate the effect of the plastic deformation associated with the Vickers indentation, each specimen batch was further subdivided into three groups. In group 1, K_{Ic} was measured without removing the Vickers indentation and, in group 2, it was measured after grinding away the indentation on a 70 μm diamond wheel. In group 3, approximately 0.5 mm of material was removed from the area around the indentation with a diamond grinding wheel and the fracture toughness was measured subsequently (Fig. 2). The toughness values obtained are called K_{IQ}.

4. Results and discussion

Table 1 lists the pertinent results obtained from the bridge indentation experiments. Figures 3 and 4 summarize the overall results. As can be seen, increasing the punch spacing *d* decreases the load P_b at which unstable crack extension takes place and increases the length of the pre-crack obtained.

It seems that a low Vickers indentation load P_V should be used in order to reduce the effect of the plastic deformation around the indentation on K_{Ic} values. However, Palmqvist cracks associated with indentations made at low loads are short in length, and these require undesirably high compressive loads to obtain a pre-crack. Of course, the situation can be improved if a larger value of punch spacing *d* is chosen, but this cannot be done indefinitely. The length of the pre-crack increases with increasing *d*, and at very high *d* values the cracks obtained may be so long as to render the specimen useless. For example, batch IV specimens had to be compressed with a 13 mm punch spacing until a load of about 40 kN before a pre-crack could be obtained, whereas batch I specimens could be pre-cracked with a 8 mm punch spacing at a 20 kN load.

Figure 5 summarizes the main experimental results obtained from the fracture toughness tests performed. Two observations are important.

(a) In all the four batches, group 3 specimens showed the highest K_{IQ} values, while group 1 showed the lowest. This proves that residual ten-

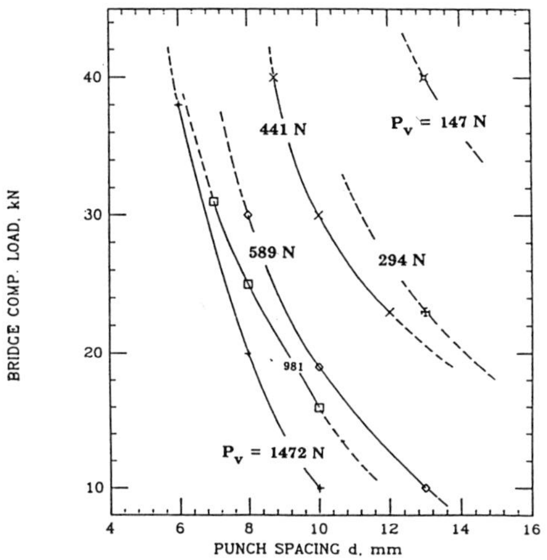

Fig. 3. Variation in P_b with d for various values of P_V.

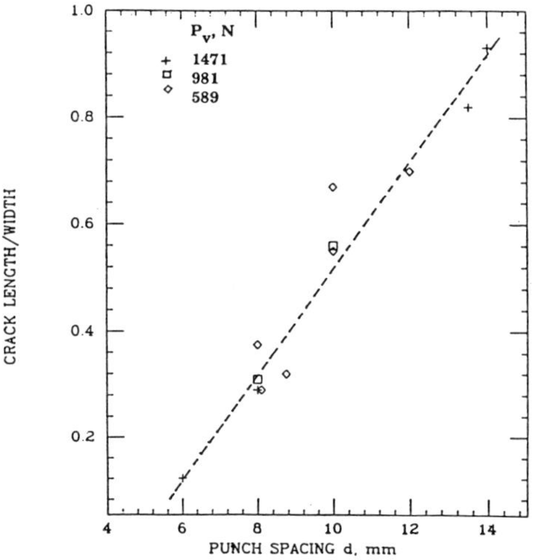

Fig. 4. Variation in normalized crack length with punch spacing d.

sile stresses exist at the pre-crack tip, arising most probably from the plastic deformation around the Vickers indentation. Also, these stresses are not effectively removed and the crack tip made stress free by merely grinding off the indentation, since the toughness in group 3 is consistently greater than that in group 2.

(b) The plot of K_{IQ} *vs.* P_V goes through a maximum for all the groups. The underestimation of K_{Ic} observed at the high P_V end (batch I) is

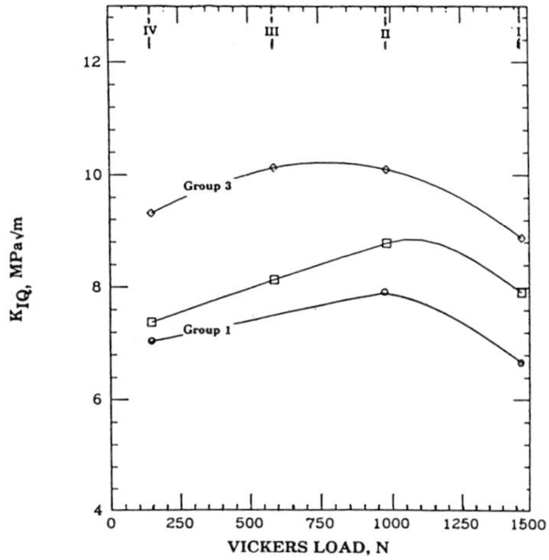

Fig. 5. Variation in K_{IQ} with indentation load P_V.

expected and is due to the high P_V values used in this batch, which introduce a large zone of plastic deformation. In contrast, the low K_{Ic} values obtained from batch IV seem to indicate that the high degree of compression used for pre-cracking in this batch may also affect the measurement of fracture toughness.

Thus the pre-crack obtained by bridge indentation may not be stress free. The residual stresses arise from the damage zone associated with the plastic deformation around the Vickers indentation. The results indicate that the loads used during pre-cracking by extension of Palmqvist cracks may also affect the measured fracture toughness.

5. Cyclic compression

The phenomenon of mode I fatigue crack growth in notched plates of cemented carbides subjected to uniaxial cyclic compression was first demonstrated by Sylva and Suresh [2]. The fatigue cracks propagate at a progressively decreasing rate and arrest after growth over a distance of the order of 1 mm. The driving force for this effect is the generation of a zone of residual tension at the notch tip on unloading from the far-field compressive stress. General features of the mechanics and micromechanisms of fatigue under cyclic compression in a variety of hard materials, including single-phase ceramics, transforming ceramics, cemented carbides and concrete, can be found in the review article by Suresh in this volume [4]. Recently, Godse *et al.* [5] have conducted a detailed investigation of the characteristics of compression fatigue in WC–Co and TiC–Ni composites. They found that the *R* ratio of the fatigue cycle (defined as the ratio of the minimum stress to the maximum stress of the fatigue cycle) and the phenomenon of crack closure affect the state of stress at the crack tip. These two factors therefore influence the fracture toughness measured using this pre-cracking technique.

6. Elimination of residual stresses

It can thus be concluded that the fracture toughness of hard metals measured by using indentation techniques and cyclic compression can be influenced by residual tensile stresses existing in a zone in front of the crack tip. The following discussion deals with a method by which the crack can be grown out of the residual stress zone and therefore be made approximately stress free. These tests were carried out on specimens pre-cracked by cyclic compression. Specimens were subjected to cyclic compression until crack arrest (saturation), following which the saturated crack was grown further under tensile fatigue in a bend configuration. The crack length was monitored as a function of the number of fatigue cycles. When a stage of crack arrest was detected, the imposed stress intensity range was incremented in steps of 0.25 MPa m^{-1}. Crack growth characteristics under these conditions are shown in Fig. 6. As can be seen, although the crack starts to grow at a particular value of the imposed stress intensity, its growth is arrested after some crack advance. A further increment in the stress intensity results in the resumption of crack growth and its eventual arrest. Finally, a stage is reached when no further arrest is detected.

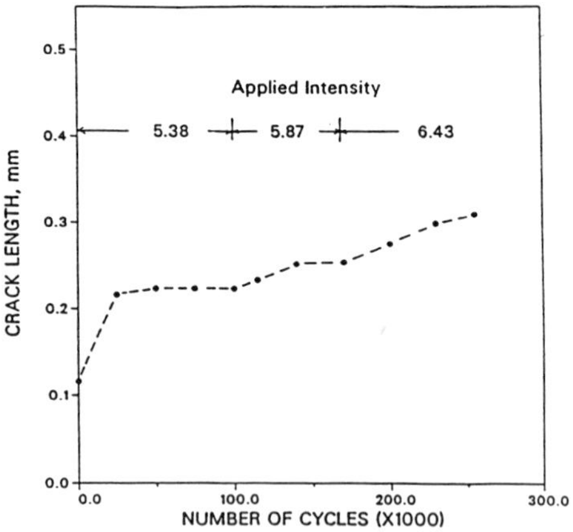

Fig. 6. Crack growth in tensile fatigue following saturation in cyclic compression. The applied stress intensity range is in megapascals per metre.

TABLE 2 Properties of multimetal alloys

Alloy	*Composition* (wt.%)	*Grain size*	*Rockwell A hardness* (HRA)
I	WC–6Co	Medium	89.9
II	WC–6Co	Coarse	88.0
III	WC–0.3TiC–6.3Co	Fine	90.9
IV	WC–10.2Co	Coarse	87.5
V	WC–4.9TiC–13.8TaC–8.1Co	Fine	91.8

The reduction in crack growth rates associated with the growth of the crack can be interpreted as resulting from the progressive reduction in residual tensile stresses at the crack tip (which assist fatigue crack growth) as the crack grows out of the zone of residual tension. Thus, when no further crack arrest is detected, a stage is reached when the applied stress intensity alone can cause the crack to grow. At this stage, the crack can be assumed to have grown out of the zone of resid-

TABLE 3 Results of fracture toughness tests of multimetal alloys and comparison with K_{IcSR} values

Alloy	*ASTM designation*	a_s (mm)	a_R (mm)	K_I^{max}/K_{IcSR}	K_{IcSR} (MPa $m^{1/2}$)	K_{IQ} (MPa $m^{1/2}$)
I	D	0.15	0.28	0.65	13.59	13.62
II	B	0.11	0.35	0.66	15.55	15.53
III	F	0.03	0.09	0.67	9.40	9.18
						8.48
IV	A	0.21	0.44	0.66	16.92	17.55
						17.94
V	E	0.06	0.22	0.68	10.19	9.10
						10.18

K_I^{max}, maximum stress intensity of the tensile fatigue cycle before termination of the test.

ual tension and therefore is approximately stress free. The fracture toughness measured under this condition is hence a valid measure of the true fracture toughness of the material. The above method (pre-cracking by cyclic compression followed by tensile fatigue to remove residual stresses) was used to measure the toughness of five WC–Co alloys (Table 2). Table 3 compares the K_{IQ} values evaluated by this method with those measured by the short rod-chevron notch specimen technique [6] on alloys with the same nominal composition. a_S refers to the saturation crack length obtained after cyclic compression, while a_R refers to the crack length after tensile fatigue. The K_{IcSR} values were obtained from an ASTM round-robin test series [7]. The agreement is fair and lends support to the present technique.

Although the present study has been concerned only with K_{Ic} measurement, the pre-cracking technique presented here can also be used to obtain stress-free sharp cracks which can be used in a variety of applications, such as fatigue crack growth rate measurements, R curve measurements, high temperature fracture and environmentally assisted fracture.

7. Conclusions

Residual tensile stresses at the crack tip in specimens pre-cracked by indentation methods or by cyclic compression can affect the measured toughness value. These stresses therefore need to be minimized before a valid measure of the material toughness can be obtained. This work presents a method by which the pre-crack can be grown out of the zone of residual tension and thus be made stress free. The results support the validity of the technique.

Acknowledgment

This work was supported by the Materials Research Laboratory at Brown University.

References

1 T. Sadahiro and S. Takatsu, *Mod. Dev. Powder Metall., 14* (1980) 561.
2 L. A. Sylva and S. Suresh, *Mater. Sci. Eng., 83* (1986) L7–L10.
3 R. Warren and B. Johannesson, *Powder Metall., 27* (1984) 25.
4 S. Suresh, *Mater. Sci. Eng., A105/106* (1988) 323–329.
5 R. Godse, J. Gurland and S. Suresh, unpublished results, Brown University, 1987.
6 L. M. Barker, *Eng. Fract. Mech., 9* (1977) 361.
7 ASTM Committee B09-06 for Standard, Method for short rod fracture toughness of cemented carbides, *ASTM Stand. B-771-87*, 1987.

Recent Observations of the Erosion of Hard Materials*

R. C. PENNEFATHER, S. E. HANKEY, R. HUTCHINGS and A. BALL

Department of Materials Engineering, University of Cape Town, Rondebosch (South Africa)

(Received November 8, 1987)

Abstract

A well-characterized set of tungsten carbide–cobalt cermets together with several engineering ceramics have been subjected to both cavitation erosion and particle impact erosion. The modes of erosion and the microstructural features which control material loss are described.

1. Introduction

Hard tungsten carbide (WC) cermets and ceramics are being used in a variety of engineering applications where their potential as wear-resistant materials is being utilized. For example, the development of hydraulically powered mining machinery for the South African gold mines requires reliable valves, seals and bearing surfaces. However, there is insufficient understanding of the various modes of wear under the wide variety of mechanical and environmental conditions encountered in these types of machine. Thus, in many instances, the most ideal material is not being used and there is scope for development or design of better materials for particular applications. We have therefore initiated a collaborative programme with industry which aims at the development of laboratory tests which will closely simulate the erosive conditions experienced in practice. This paper concerns the results obtained from tests which simulate the cavitation erosion and particle erosion experienced by valves and impellers. These tests are undertaken in order to evaluate and rank the erosion resistance of hard materials which are at present available. The close examination of the surface is undertaken by scanning electron microscopy in order to characterize the modes of deformation and fracture by which material is damaged and removed. The objective of this research is to use the information obtained in order to design the composition and microstructure of new materials which will provide better resistance to wear in given conditions.

2. Experimental details

The cavitation tests were carried out using an ultrasonic drill in the manner described by Heathcock *et al.* [1]. The disc-shaped specimens are placed in distilled water at a distance of 0.35 mm below the tip of the drill. Cavities are repeatedly created and collapsed in the liquid immediately above the specimen surface by the imposed high frequency oscillations. The coordinated collapse creates a high surface stress which damages the material and causes erosive wear. The specimens are removed at hourly intervals for the accurate determination of weight loss and examination in the scanning electron microscope.

The second experiment simulates solid particle erosion. Disc-type specimens were bombarded with 100 grit silicon carbide particles travelling at a velocity of 40 m s^{-1}. The weight loss was measured after repeated doses of 50 g of erodent. Scanning electron microscopy was also carried out during these intervals in the erosion test.

A well-documented set of WC–Co specimens were subjected to these two tests together with samples of alumina, Si–Al–O–N and a partially stabilized zirconia (PSZ). The 13 WC–Co specimens, with the volume of cobalt binder ranging from 7.13% to 44.2%, fell into two groups with grain sizes close to 1.8 μm and 2.8 μm respectively. All samples were polished to a surface finish of 0.25 μm prior to testing in order to gain information on the initiation of the erosion processes. Young's moduli and values of hardness and fracture toughness were determined for some of these specimens whilst available data for

*Paper presented at the 3rd International Conference on the Science of Hard Materials, Nassau, The Bahamas, November 9–13, 1987.

0921-5093/88/$3.50 © Elsevier Sequoia/Printed in The Netherlands

density, specific heat, thermal conductivity and melting point were collated from the literature.

3. Results and discussion

3.1. The cavitation and particle erosion resistance of WC–Co alloys

The rate of cavitation erosion was determined during steady state erosive conditions and the values of erosion rates and 5 h cumulative volume loss for the two grain size groups showed maxima at 12% Co and 24% Co respectively (Fig. 1). Specimens with cobalt contents less than these values showed continuous rigid skeletons of WC whereas specimens with cobalt contents above these maximum values had in general a discontinuous structure with the WC grains surrounded by the matrix. Scanning electron microscopy of the eroded surfaces indicated that the rigidity of the skeleton was resisting the erosion in the low cobalt specimens (Fig. 2). Thus the improvement in erosion resistance of these alloys with decreasing volume of binder is anticipated since the skeleton becomes more continuous and rigid. The erosion of the high cobalt alloys was pre-

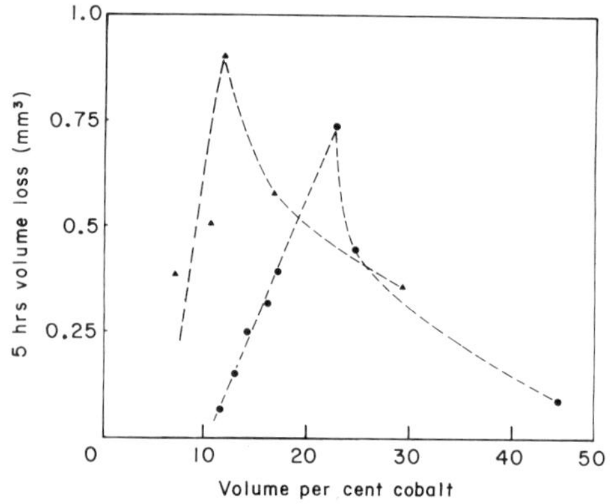

Fig. 1. Cavitation erosion performance of WC–Co cermets, where the 5 h cumulative volume loss is plotted as a function of binder content for specimens of the two grain sizes: ▲, 1.5 μm grains; ●, 2.8 μm grades.

Fig. 2. Scanning electron micrograph of the cavitation erosion damage of a WC–Co specimen containing 14 vol.% Co (grain size, 2.64 μm).

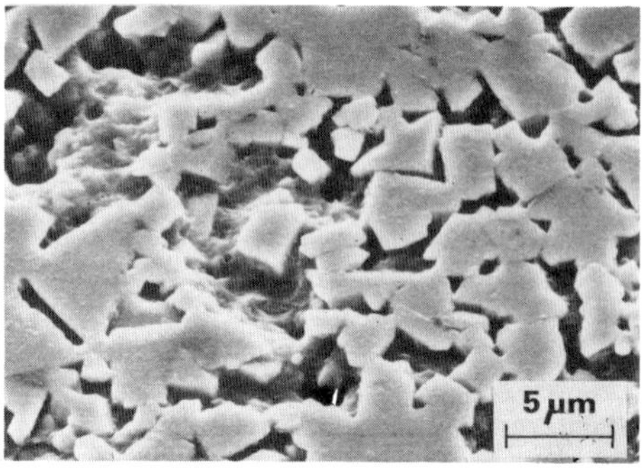

Fig. 3. Scanning electron micrograph of the cavitation erosion damage of a WC–Co specimen containing 44 vol.% Co (grain size, 2.17 μm).

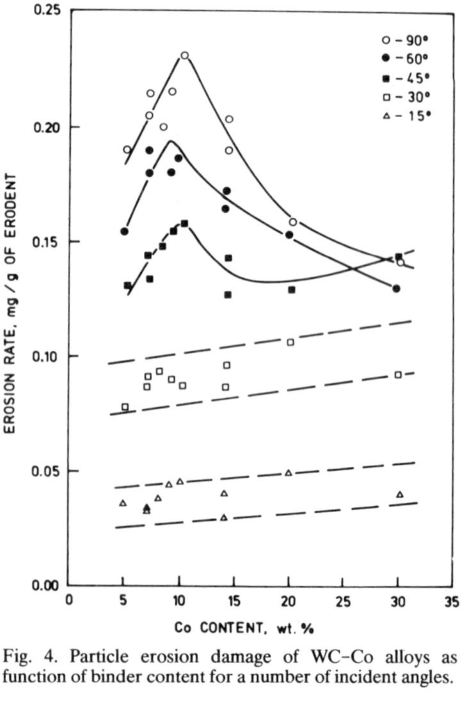

Fig. 4. Particle erosion damage of WC–Co alloys as a function of binder content for a number of incident angles.

dominantly occurring by the removal of cobalt binder phase (Fig. 3). The carbide grains play a passive role and are merely lost when the surrounding cobalt has been removed. Thus the properties of the cobalt matrix determine the overall erosion resistance. The transformation from f.c.c. to h.c.p. cobalt occurred during cavitation erosion and this transformation is capable of accommodating imposed strain [2]. Increased amounts of binder therefore afford increased strain accommodation. The above arguments are capable of explaining the observed maxima in erosion rates for the two grain size groups.

Particle impingement erosion studies on the same two groups of WC–Co specimens also displayed maxima at approximately 10 wt.% Co (Fig. 4). In this case, there was no discernible influence of grain size. These maxima were most pronounced when the specimens were placed at high incident angles to the stream of erodent. Scanning electron microscopy of the eroded surfaces after 90° erosion (Figs. 5 and 6) showed three modes of material removal, *i.e.* cobalt extrusion, brittle fracture of the carbide grains and ductile shear of the cobalt matrix. For specimens containing less than 10 wt.% Co, extrusion of the cobalt and fracture of the carbide grains were the predominant erosion mechanisms (Fig. 5) whilst specimens containing higher amounts of cobalt showed ductile shear to be the prominent mode of erosion (Fig. 6). It can be considered that the erosion of low cobalt alloys is controlled by the rigidity and toughness of the carbide skeleton while the transformation and deformation of the cobalt matrix controls the erosion of cobalt-rich specimens. Thus a maximum in the erosion resistance is again predicted.

These proposals have been substantiated [3] by an examination of the WC grains as revealed by leaching of the cobalt matrix subsequent to ero-

Fig. 5. Scanning electron micrograph of a 90° particle impact site on a WC–Co specimen containing 10 wt.% Co.

Fig. 6. Scanning electron micrograph of a 90° particle impact site on a WC–Co specimen containing 30 wt.% Co.

Fig. 7. Scanning electron micrograph of a WC–Co specimen containing 30 wt.% Co after the specimen had been subjected to steady state 90° particle erosion. The cobalt matrix had been chemically treated from the eroded specimen in order to reveal the unfractured WC–Co grains.

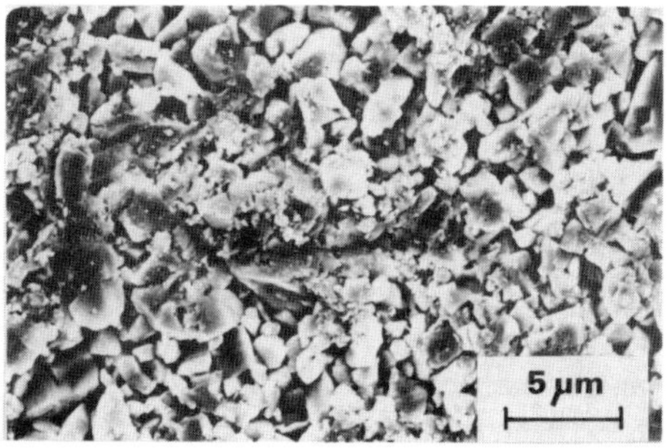

Fig. 8. Scanning electron micrograph of a WC–Co specimen containing 10 wt.% Co after the specimen had been subjected to steady state 90° particle erosion. The cobalt matrix had been chemically treated from the eroded specimen in order to reveal extensive fracturing of the WC–Co grains.

sion. For the high cobalt alloys and 90° erosion the tungsten carbide grains are unfractured (Fig. 7) whereas the tungsten carbide skeleton of the low cobalt alloys is severely fragmented by the erosion (Fig. 8).

The origin of maxima in the erosion rates as a function of volume percentage of the cobalt binder has been previously considered by Ball and Paterson [4], Pennefather [5] and Hankey [6]. Figure 9 which is taken from the work of Hankey depicts the microstructures above and below the maximum in erosion rates. The changes in mechanical properties which result from increasing contiguity of the WC skeleton or an increasing matrix mean free path are indicated. Thus, good erosion resistance is achieved with either a continuous strong skeleton of WC or with an amount of binder which provides an optimum compromise between strength and toughness.

3.2. The cavitation and particle erosion resistance of ceramic materials

The ceramic materials were tested under cavitation and particle erosion and their performances are given in Table 1. Under cavitation conditions the PSZ clearly outperforms the other materials with the Si–Al–O–N ceramic being a clear second. Under particle erosion conditions, Si–Al–O–N showed the best performance. Scanning electron microscopy of the eroded surfaces clearly shows that the modes by which material is removed are very different for the two erosive conditions. The initiation of cavitation

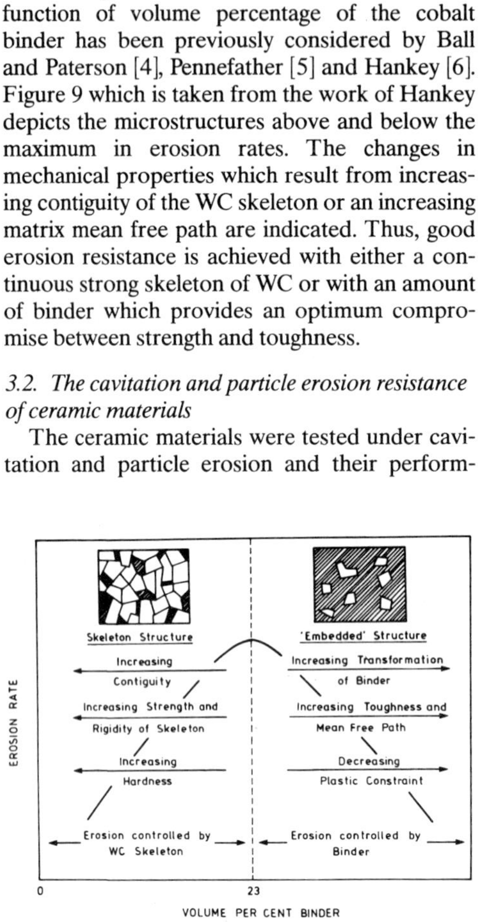

Fig. 9. A schematic summary of the erosion performance of WC–Co alloys as a function of binder content up to approximately 40 vol.% Co.

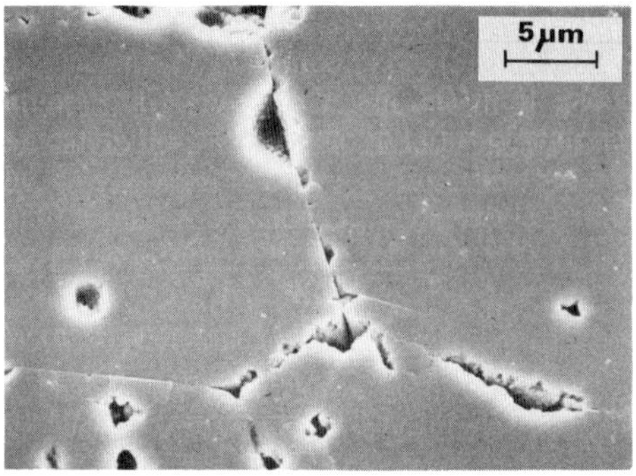

Fig. 10. Scanning electron micrograph of the initial cavitation erosion suffered by a PSZ specimen.

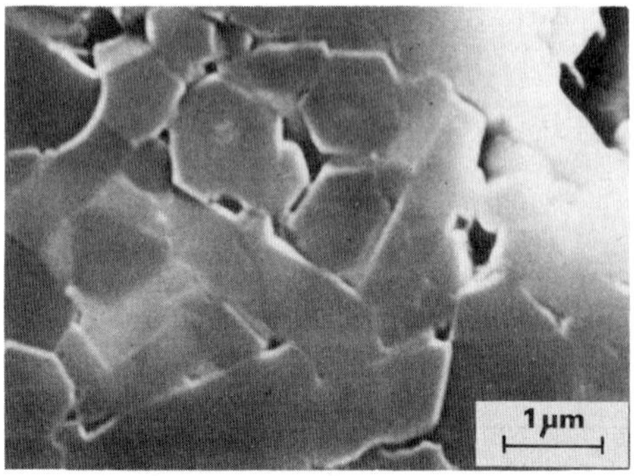

Fig. 11. Scanning electron micrograph of the initial cavitation erosion suffered by an Si–Al–O–N specimen.

TABLE 1 The cavitation and particle erosion performances of selected ceramic materials

	Density ($Mg\,m^{-3}$)	*Specific heat* ($kJ\,kg^{-1}\,K^{-1}$)	*Thermal conductivity* ($W\,m^{-1}\,K^{-1}$)	*Melting point* (°C)	*E* (GPa)	*Toughness K*		*Cavitation 5h volume loss* (mm^3)	*Particle erosion* ($mm^3\,g^{-1}\,cm^{-2}$)
						Literature	*Experiment*		
PSZ	5.7	0.70	2.0	2800	205	7.0	10.0	0.09	0.014
Si–Al–O–N	3.25	0.62	21.3	1830	288	7.7	4.6	0.25	0.008
Alumina	3.97	0.88	26.4	2072	380	3.6	2.8	1.21 (polished, 0.75)	0.015
WC–10wt.%Co	14.51	0.21	75–80	2870	579	10.3	8.2	0.581	0.014

erosion is extremely dependent on surface finish and structural defects. Damage is initiated at grain boundary porosities and proceeds in an intergranular manner (Figs. 10–12). As a consequence, the larger-grained PSZ is less susceptible to cavitation erosion. An added factor could be the strain accommodation provided by its phase transformation.

As cavitation erosion proceeds, the damage is predominantly intergranular and it is anticipated that the perfection and strength of the grain boundaries together with the grain size will control the erosion rate. It should be noted that a significant improvement in the initial erosion performance can be obtained by an improvement in the surface finish. For example, in Table 1 the values for the 5 h volume loss of alumina are given for specimens eroded with an as-sintered surface finish and that produced by a diamond polishing.

Particle erosion of the ceramic specimens occurs by individual strikes, and Figs. 13–15 show examples of damage after 90° impact. There is evidence of plastic flow, the formation of compacted platelets and brittle spallation. Localized heating and adiabatic shear have also been indicated. For the PSZ (Fig. 12), the particle impact has caused plastic shear and brittle spallation. Since the grain size of this material is much greater than the area of the strike, this spallation has occurred by a brittle concoidal mechanism.

For the Si–Al–O–N specimen, plastic compaction of the polyphase material is the predominant observation (Fig. 14). At some strikes, there is evidence of adiabatic shear and associated thermal degradation. The particle erosion of the alumina specimen (Fig. 15) also shows evidence of plastic compaction but the predominant mode of material loss is spallation by an intergranular fracture mechanism. It is not surprising from

Fig. 12. Scanning electron micrograph of the initial cavitation erosion suffered by an alumina specimen.

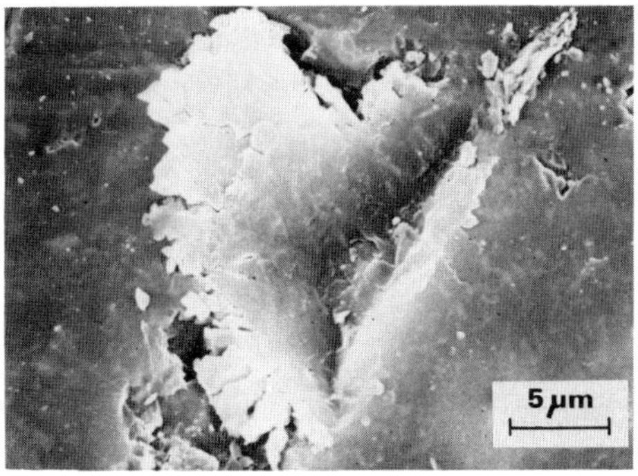

Fig. 14. Scanning electron micrograph of a 90° particle erosion strike on a Si–Al–O–N specimen.

Fig. 13. Scanning electron micrograph of a 90° particle erosion strike on a PSZ specimen.

Fig. 15. Scanning electron micrograph of a 90° particle erosion strike on an alumina specimen.

these observations that Si–Al–O–N has a superior resistance to particle erosion than either the PSZ or the alumina specimens (Table 1).

4. Conclusions

(1) Maxima in both cavitation and particle erosion loss have been measured for WC–Co cermets containing approximately 10% Co. It is hypothesized that the microtoughness of the WC skeleton controls the erosion at lower cobalt contents whilst the properties of the cobalt control the erosion at high cobalt contents.

(2) Microstructure has a very significant but different influence on the erosion of several ceramic materials. The stress pulses generated by cavitation cause grain boundary porosities to initiate intergranular fracture. Thus, grain size and defect density have an overwhelming influence. There is evidence of particle erosion at the individual impact sites for adiabatic plasticity. Brittle spallation occurs on relaxation.

Acknowledgments

The authors wish to thank the various industrial sponsors who have encouraged the work described in this paper and give thanks to Mrs. H. Bohm, Mrs. A. C. Ball and Mr. B. Greeves for their technical assistance.

References

1 C. J. Heathcock, A. Ball and B. E. Protheroe, Cavitation erosion of cobalt based stellite alloys, cemented carbides and surface treated low alloy steels, *Wear, 74* (1981–1982) 11.

2 C. J. Heathcock, B. E. Protheroe and A. Ball, Cavitation erosion of stainless steels, *Wear, 81* (1982) 311.

3 R. C. Pennefather, R. Hutchings and A. Ball, Erosion mechanisms in cemented carbides subjected to solid particle erosion, *Proc. 7th Int. Conf. on Erosion by Liquid and Solid Impact*, Cavendish Laboratory, Cambridge, p. 60-1.

4 A. Ball and A. W. Paterson, *Proc. 11th Int. Plansee Semin., Reutte, 1985*, Vol. 2, Metallwerk Plansee, Reutte, 1985, pp. 377–391.

5 R. C. Pennefather, The solid particle erosion of WC–Co alloys, *M.Sc. Thesis*, University of Cape Town, 1986.

6 S. E. Hankey, Cavitation erosion of WC–Co, *M.Sc. Thesis*, University of Cape Town, 1987.

Abrasive Wear of Some Titanium-carbonitride-based Cermets*

JORN LARSEN-BASSE

School of Mechanical Engineering, Georgia Institute of Technology, Atlanta, GA 30332-0405 (U.S.A.)

(Received November 9, 1987)

Abstract

A series of titanium-carbonitride-based cermets was tested in abrasion. The alloys had TiN-to-(TiC + TiN) alloy ratios of between 0 and 0.6. They had been cold pressed and sintered with a binder of 12.5wt.%Ni-10wt.%Mo-10wt.%VC.

The materials were abraded under three-body conditions by 85 μm SiC and quartz abrasives and by 1 μm diamond polishing paste. The wear mechanisms for the larger abrasives were similar to those described for WC-Co cermets, i.e. plastic indentation and microspalling for the hard abrasive (SiC), and fine-scale microspall formation for the relatively soft abrasive (quartz). The diamond polish gave the same wear mechanism as the SiC abrasives, only on a smaller scale.

The three abrasives all gave the highest wear rates for intermediate values of alloy hardness and toughness, at a TiN-to-(TiN + TiC) ratio of 0.2 and the lowest wear rate for the hardest alloy, which had an alloy ratio of 0.6. Zum Gahr's model for abrasion of low ductiltiy materials was used to evaluate the data. The model overestimates the contribution of brittle microfracture.

The wear data were also correlated with tool performance data in machining of gray cast iron. The correlation is quite poor; the need for additional work is discussed.

1. Introduction

Cutting tool wear represents a unique situation in the field of tribology because wear is the preferred failure mode for a tool (the alternatives cause more damage), and because the wear rate is two to three orders of magnitude greater than what is normally considered severe wear [1]. It is also unique (and difficult to study) because many different wear mechanisms may operate at the same time and on the same area of the tool [2].

Proposed tool wear mechanisms include abrasion, diffusion, delamination, fatigue, microchipping and thermal softening [1, 3-5]. Abrasion, which is the topic of this paper, is generally thought to be responsible for most flank wear and for general tool wear in the machining of sand-cast articles, cast iron, and alloys and composites which contain significant amounts of fibers or particles of a hard phase [6-8].

At the second conference in this series, Moskowitz *et al.* [9] presented physical and metal-cutting data for some titanium-carbonitride-based cermet materials. TiC and TiN are isomorphous and a continuous series of solid solutions can be prepared from them. The microhardness of the carbonitride increases with increasing carbon content. In contrast, for cermets based on this carbonitride and bonded with $Ni-Mo_2C-VC$ additions, there is an increase in resistance to plastic deformation (*i.e.* hardness) as the nitrogen content increases. Machining tests showed that TiN additions to these cermets increase tool life significantly in the machining of gray cast iron and heat-treated SAE 4140 and 4340 steels, while no clear effect was seen in the turning of normalized SAE 1045 steel.

The behavior observed in the machining of the hard steels was attributed to increases in thermal deformation resistance and in thermal conductivity as the TiN content increases. The shape of the curves for the machining of cast iron was somewhat different from that for the steels, and different between turning and milling operations. This was tentatively attributed to possible variations in abrasion resistance, which at that time had not been measured.

The present study was initiated in order to obtain abrasion data for these alloys under various test conditions and to compare them with the

*Paper presented at the 3rd International Conference on the Science of Hard Materials, Nassau, The Bahamas, November 9-13, 1987.

0921-5093/88/$3.50 © Elsevier Sequoia/Printed in The Netherlands

machining and other property data presented by Moskowitz *et al.* [9].

2. Experimental procedure

The materials were obtained from D. Moskowitz, Ford Motor Company. They were a series of TiC-based cermets prepared from solid solution carbonitride powders with different nitrogen contents. Values of the alloy ratio of TiN to (TiC + TiN) varied between 0 and 0.6. The powders had been cold pressed and sintered with a binder of 12.5wt.%Ni–11wt.%Mo, and they had a 10 wt.% VC addition, except for the nitrogen-free composition. The sintering temperatures had been adjusted somewhat, in order to obtain almost identical grain sizes, in the 1–4 μm range, for the various alloy compositions.

Values of the fracture toughness K_{Ic} were also obtained from Moskowitz while we measured the remaining properties of interest: the Vickers hardness at a 10.2 N load, the Palmquist cracks formed at the Vickers indentations at a 159 N load, and abrasion and polishing test data.

The abrasion tests were so-called three-body tests, with loose abrasives fed to the contact zone between a turning carbon steel wheel and the specimen. The abrasives used were 75–88 μm SiC commercial abrasives and 75–124 μm quartz abrasives crushed from 400 μm rounded beach sand. The steel wheel was 1.5 mm wide and 55 mm in diameter, and it operated with a surface velocity of 0.085 m s^{-1}. For the SiC abrasives a load of 10.9 N was used and 10 min running time was sufficient to generate significant mass loss. For the quartz abrasives the load for most tests was reduced to 5 N to minimize crushing of the abrasives, and an increase in the running period to 1 h was necessary. It was found that abrasion by quartz would not readily initiate on a smooth surface. These tests were therefore conducted in wear scars produced by the SiC tests. The measured mass loss data were converted to volumetric wear rates using density values obtained by the Archimedes method.

In the polishing tests a steel ball 19 mm in diameter was supported between the vertical sample surface and the rotating conical tips of two adjoining coaxial shafts. The ball was covered with 1 μm diamond paste before each test and the paste was replenished at intervals. The scar size was measured by optical microscopy after running for 30 min in this low load polishing test. Before testing, sample surfaces were finished with a 9 μm diamond polish.

3. Results and discussion

The mechanical property data are plotted in Fig. 1. They are the hardness and Palmquist crack susceptibility measured in this study, and the bulk K_{Ic} data obtained from Moskowitz *et al.* There is a significant increase in hardness with increasing nitrogen content and a concomitant decrease both in bulk toughness, as shown by the K_{Ic} values, and in local surface toughness, as shown by the susceptibility to formation of Palmquist cracks. The pure TiC-based cermet deviates slightly from the trend in the surface toughness curve, showing a greater Palmquist cracking susceptibility than expected from the trend. It is not clear at this stage whether this is significant and/or whether it is due to the absence of VC in this particular composition.

Abrasion and polishing data are shown in Fig. 2. The shapes of the curves are quite similar for the three very dissimilar abrasives, showing greatest wear at an alloy TiN-to-(TiN + TiC) ratio of 0.2 and decreasing with increasing nitrogen content above this value. The numerical values are somewhat different, of course, with the SiC abrading 30–35 times faster than the quartz and 700–1000 times faster than the diamond polish, or in the ratios 1:0.03:0.0015. Corrected to the same load and sliding distance the relative wear

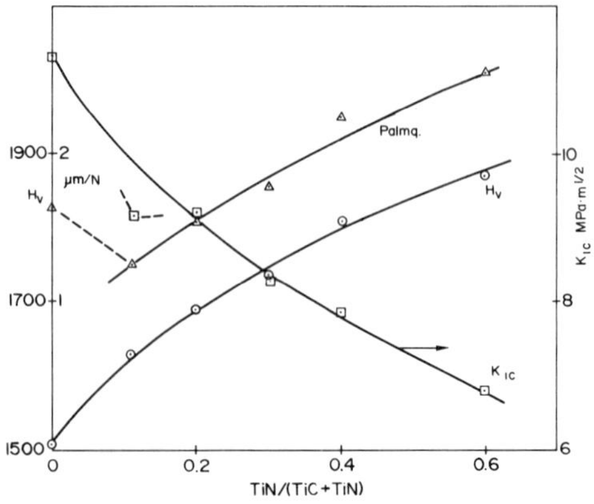

Fig. 1. The Vickers hardness and Palmquist cracking susceptibility (in micrometres per newton) measured in this study plotted with bulk K_{Ic} values (from Moskowitz *et al.*) *vs.* alloy ratio.

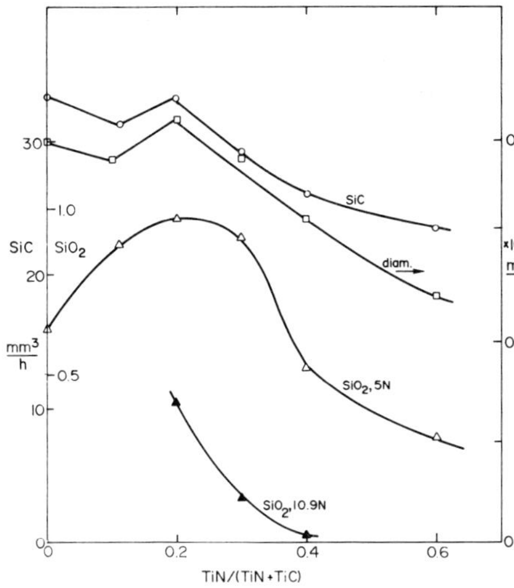

Fig. 2. Abrasion wear rates *vs.* alloy ratio.

rates become approximately 1:0.07:0.01 for SiC:quartz:diamond polish. This correction assumes that the wear per unit distance is proportional to the applied load and that the correction can be made without changes in wear mechanism. It is only for comparison purposes.

The few data shown for high load abrasion by quartz indicate that such an extrapolation is not permissible under operating conditions. For increased load the quartz grains crush and the wear rate decreases. For the regular 5 N test load, the wear rates by quartz are amazingly high, when it is considered that this abrasive has a Vickers hardness of only 1050–1100 HV and thus cannot directly indent any of the specimen materials during abrasion.

The wear rates all peak for the material with an alloy ratio of 0.2. Obviously, this alloy has the combination of relevant mechanical properties, such as hardness and surface microtoughness, which gives the lowest wear resistance. However, no serious quantitative reasoning can be forwarded at this point for what this combination of properties should be. It is quite possible that a grain size effect plays a role here and that wear is affected by grain size to a much greater extent than are the bulk mechanical properties. Such influence has been demonstrated for WC–Co alloys [10], whose abrasion resistance is partially related to $1/D^{1/2}$, where D is the WC grain size. In this case the WC grains fragment under the contact stresses when a significant amount of the cobalt binder has been preferentially removed and the fragments are lost from the surface as part of the wear loss. In the present case, as discussed below, the binder is not preferentially removed but part of the wear is due to spalling of the surface near the contacting points, with spall sizes equal to the grain diameter. It should be noted that, while the grain size of most specimens was around 1 μm, for the two alloys with alloy ratios of 0.2 and 0.3 (and showing the greatest wear) the grain size was 3–4 μm. This difference did not have any clear effect on the mechanical property values but could possibly be responsible for the lower wear resistance.

A detailed model for the abrasion of brittle materials has been developed by Zum Gahr [11]. In this model, material is removed by both plastic deformation in groove and crater formation and brittle cracking along the sides of the grooves. The predicted wear rate is

$$\dot{\omega} = \frac{L}{\beta H \tan\alpha} + \frac{h\delta}{\lambda^2}\frac{L^{3/2} D_a \mu^2 H^{1/2}}{K_{Ic}^{\,2}\sin(2\alpha)} \log\left\{\left(\frac{p}{p_{crit}}\right)^{1/2}\right\} \qquad (1)$$

where L is the applied load, β is a shape factor, H is the hardness, 2α is the aperture angle of the wear grooves, h is the groove depth, δ is the size of the microstructural features which favor crack propagation, λ is the mean free path between these features, D_a is the distance between active abrasive particles, μ is the coefficient of friction between the material and the abrasive, K_{Ic} is the fracture toughness, p is the applied pressure, and p_{crit} is a critical value for the applied pressure below which there is no contribution from brittle microcracking to the wear. p_{crit} is given by

$$p_{crit} = \frac{0.12\lambda K_{Ic}^{\,2}\sin(2\alpha)}{D_a^{\,2}\mu^2 H} \qquad (2)$$

The two terms in eqn. (1) are a "plastic" plowing term, which always is present when the abrasive can indent the surface, and a term for wear by brittle microcracking, which only appears when the contact stress reaches a certain level or the indentation a certain size.

Many of the factors in eqns. (1) and (2) are not well known for any system. However, for abrasion by a hard abrasive of constant grit size, many of the unknown parameters which depend only on abrasive type, shape and size can be lumped into one constant for each of the two terms in eqn. (1), *i.e.* for the "plastic" wear component and for the brittle wear component respectively. This was done for the case where only one abrasive grain carries the load at any one time, and the constants were determined by substituting the measured wear rates for SiC abrasives for alloy ratios of 0 and 0.2. These values were chosen because the samples had very different hardnesses and fracture toughnesses but almost identical wear rates. Results are shown in Fig. 3. There is a significant divergence between measured values and the calculated curve. The latter overestimates the contribution from brittle wear.

Several factors may be responsible for this deviation. Three of them are outlined here.

(1) It seems certain that the surface fracture toughness on the microscale of the indentations is somewhat different from the bulk value represented by K_{Ic}; as discussed above, it needs modification at least to take the observed grain size effect into account.

(2) The contribution to wear rate from brittle cracking around indentations is assured to be all the material penetrated by these cracks. However, since the cracks primarily run perpendicular to the surface, they need not necessarily result in immediate loss of material.

(3) The number of load-carrying abrasives may be substantially greater than the number of indenting abrasives. From scanning electron microscopy (SEM) observations, it is clear that, at any one time, only a small number of abrasive grains indent to cause wear. However, a significant number may be in only elastic contact with the surface [12], and this will decrease the effective load values to be used in eqn. (1).

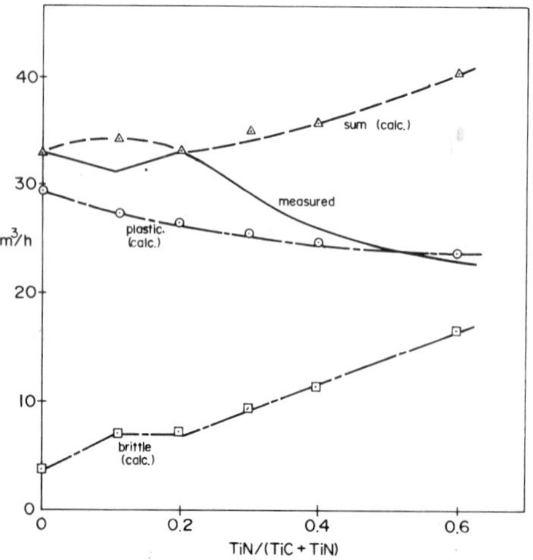

Fig. 3. Plastic and brittle wear components estimated according to Zum Gahr's [11] equations for abrasion by SiC. Values were adjusted to coincide with measured data for TiN-to-(TiC + TiN) ratios of 0 and 0.2.

Clearly, both model refinement and additional experimentation are needed before the model can accurately predict actual wear rates.

SEM observation of the abraded surfaces reveals features similar to those reported previously for WC–Co alloys [13]. The hard SiC abrasives indent the surface to form craters and minor surface cracking (Fig. 4). The extruded material is highly fragmented and some of it is smeared back onto the surface.

Surfaces abraded by quartz show a roughness which is much finer, of the order of the grain size (Fig. 5). The abrasives cannot indent the surface because of their relatively low hardness but hertzian stresses can apparently concentrate to produce spalls the size of a grain. It is interesting that the abrasion process would not initiate on a polished surface, as previously mentioned, probably because small sharp protrusions are needed to provide the requisite local stresses. In the few tests conducted at an 11 N load, the quartz grains crushed and produced a very smooth specimen. In these cermets there is no preferential removal of the binder such as is observed for WC–Co alloys [13].

Surfaces polished by 1 μm diamond paste also showed a fine roughness (Fig. 6). Material is removed primarily by microspalling, although

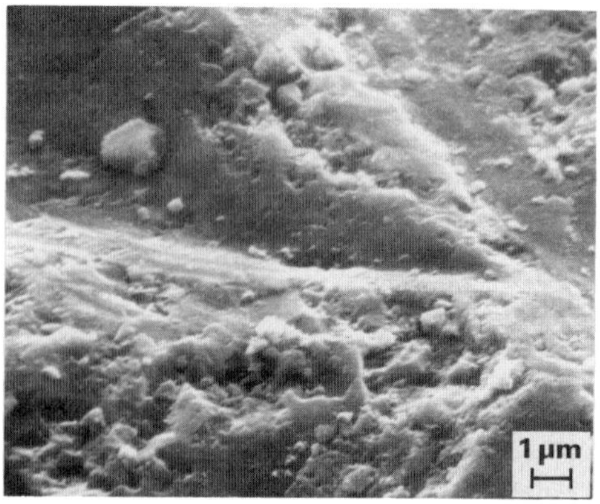

Fig. 4. Surface of a pure TiC cermet abraded by SiC.

Fig. 5. Surface of a cermet with a TiN-to-(TiC + TiN) ratio of 0.3 after abrasion by quartz.

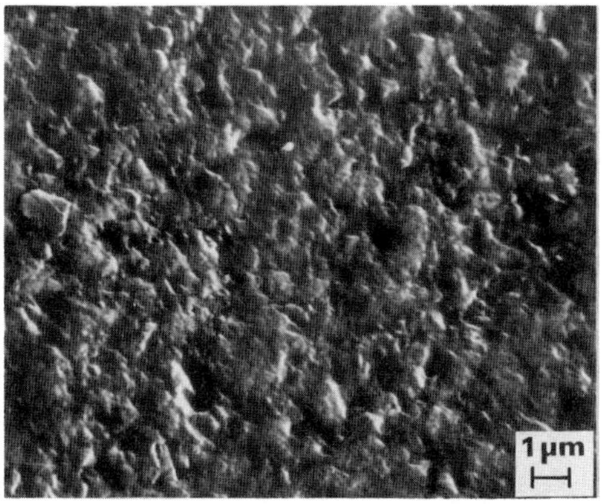

Fig. 6. Surface of a pure TiC cermet polished by 1 μm diamond paste.

some plastic craters and minor grooves are also seen. The spalls are up to the size of the grain diameter. The mechanism of polishing in general is not well understood. Both purely mechanical [14] and purely surface chemical [15] factors have been proposed. Under the present circumstances, mechanical factors obviously predominate, and the general mechanism is similar to abrasion by the other hard abrasive, SiC, even though the latter is much larger in size as is the scale of the surface deformation.

The machining data of Moskowitz *et al.* [9] for the milling and turning of gray cast iron have been converted to relative wear rates and are plotted in Fig. 7. Comparison with the wear data in Fig. 2 shows only a very tenuous connection between

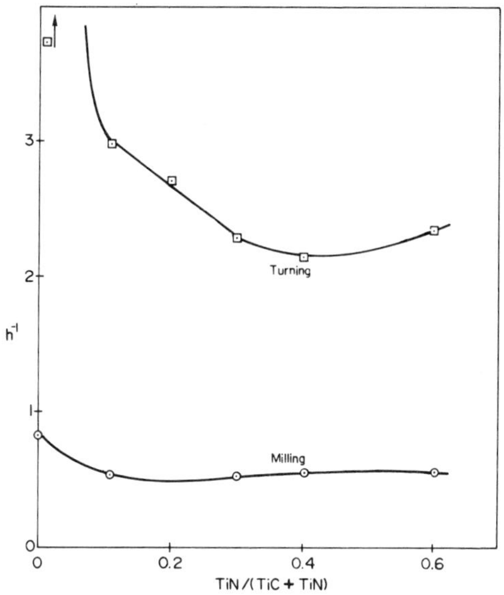

Fig. 7. Tool life data from Moskowitz *et al.* [9] for machining of gray cast iron, converted to relative wear rates and plotted *vs.* alloy ratio.

room temperature abrasion data, and wear in machining, at least under the conditions at hand. It is possible that the improvement in thermal resistance due to addition of the VC and the initial amount of TiN, rather than the room temperature abrasion resistance, is responsible for the great reduction in tool wear during turning. Little of this effect is seen in milling, where less tool heating takes place. It also appears that the contribution of brittle microfracture to wear is responsible for the slight increase in wear rate with increasing nitrogen content observed for both turning and milling; it begins at a lower alloy level for milling, which has a greater impact constituent.

Clearly, much work needs to be done before abrasion performance can be adequately predicted from microstructure and mechanical properties, and before tool material performance in cutting can be adequately related to wear mechanisms, let alone predicted from first principles.

4. Conclusions

The wear mechanisms in abrasion of relatively hard and brittle titanium carbonitride cermets by 85 μm SiC and 1 μm diamond polish are almost identical, except for scale. Material is removed by plastic indentation and microspall formation. The wear rates for the two abrasives show similar dependences on alloy composition and properties.

Abrasion by quartz is highly load dependent. Excessive loads lead readily to crushing of the abrasives, followed by fine polishing of the specimen surface. The degree of crushing is strongly dependent on specimen hardness. For loads under which crushing does not predominate, quartz abrasives remove material in the form of small spalls, of the size of the grain diameter or less. Since the relative hardnesses do not allow quartz to indent the specimen materials, these abrasives depend on surface microroughness to provide the high levels of local stress during contact, which are necessary for generation of the microspalls.

The currently available correlations between abrasive wear rates and mechanical properties, and between abrasive wear rates and tool wear in machining of gray cast iron are tenuous, at best. Both model refinement and experimental determination of a number of model parameters are needed.

Acknowledgments

The author is grateful to Mr. David Moskowitz, formerly of Ford Motor Company, for providing the specimen materials and some property values and for supplying general information on the machining behavior of these materials. Most of the wear data were collected by Ms. Marsha Inonye as part of a senior project in mechanical engineering at the University of Hawaii.

References

1 V. A. Tipnis, Cutting tool wear, in M. B. Peterson and W. O. Winer (eds.), *Wear Control Handbook,* American Society of Mechanical Engineers, New York, 1980, pp. 891–930.

2 D. G. Flom, R. Komanduri and M. Lee, High-speed machining of metals, *Annu. Rev. Mater. Sci., 14* (1984) 231–278.

3 E. M. Trent, Wear of metal cutting tools, *Treatise Mater. Sci. Technol., 13* (1979) 443–489.

4 N. P. Suh, New theories of wear and their implications for tool materials, *Wear, 62* (1) (1980) 1–20.

5 B. Kramer and N. P. Suh, Tool wear by solution: a quantitative understanding, *J. Eng. Ind., 102* (1980) 303–339.

6 S. Ramalingam, Wear mechanisms in machining, in N. P. Suh and N. Saka (eds.), *Fundamentals of Tribology,* Massachusetts Institute of Technology Press, Cambridge, MA, 1978, 663–675.

7 G. Faulring and S. Ramalingam, Oxide inclusions and tool wear in machining, *Metall. Trans. A, 10* (1979) 1781–1786.

8 M. K. Brun, M. Lee and F. Gorsler, Wear characteristics of various hard materials for machining SiC reinforced aluminum alloy, in K. C. Ludema (ed.), *Proc. Int. Conf. on the Wear of Materials, Vancouver, April 14–18, 1985,* American Society of Mechanical Engineers, New York, 1985, 539–544.

9 D. Moskowitz, L. L. Terner and M. Humenik, Jr, Some physical and metal-cutting properties of titanium carbonitride base materials, in E. A. Almond, C. A. Brookes and R. Warren (eds.), *Proc. 2nd Int. Conf. on the Science of Hard Materials, Rhodes, September 23–28, 1984,* in *Inst. Phys. Conf. Ser., 75* (1986) 605–617.

10 J. Larsen-Basse, Effects of hardness and local fracture toughness on the abrasive wear of WC–Co alloys, *Paper C229/87,* 1987, pp. 277–282 (Institution of Mechanical Engineers).

11 K.-H. Zum Gahr, Relation between abrasive wear rate and the microstructure of metals, in K. C. Ludema (ed.), *Proc. Int. Conf. on the Wear of Materials, Dearborn, MI, April 16–18, 1979,* American Society of Mechanical Engineers, New York, 1979, pp. 266–274.

12 J. Larsen-Basse, Influence of grit size on the groove formation during sliding abrasion, *Wear, 11* (1968) 213–222.

13 J. Larsen-Basse, Effect of composition, microstructure and service conditions on the wear of cemented carbides, *J. Met., 35* (11) (1983) 35–42.

14 R. L. Aghan and L. E. Samuels, Mechanisms of abrasive polishing, *Wear, 16* (1970) 293–301.

15 E. Rabinowicz, On the mechanisms of polishing with abrasives, *Wear, 18* (1971) 169–170.

Titanium-carbonitride-based Hard Alloys for Cutting Tools*

H. PASTOR

Ugicarb Morgon, 54 avenue Rhin et Danube, 38100 Grenoble (France)

(Received November 10, 1987)

Abstract

The methods of preparation and some intrinsic properties of titanium carbonitrides $TiC_{1-x}N_x$ *are described. Their thermodynamic stability as a function of temperature and nitrogen pressure has been studied; this is of great importance because of its influence on the sintering process. The fabrication of hard alloys* $TiC_{1-x}N_x$*–(Ni–Mo) and (Ti, Mo)*$(C_{1-x}N_x)$*(Ni–Mo) is described as well as the properties of these materials. Cutting tests of these alloys were performed and compared with corresponding tests on classical tungsten-carbide-based hard alloys, showing that carbonitride grades can easily compete with these classical cemented carbides.*

1. Introduction

The aim of this work was to study the preparation and main properties of titanium carbonitrides and cermets based on them, to enable cemented tungsten carbides to be replaced in some applications such as cutting tools and wear parts. A lower cost was expected because of the relatively low densities of these cermets and because of the abundance and low price of titanium dioxide (TiO_2).

2. Preparation of pure titanium carbonitrides

Five methods were investigated.

2.1. Synthesis, from TiC plus TiN, by high temperature diffusion

The synthesis was made according to Fig. 1.

2.2. Synthesis by nitridation of titanium plus TiC at a high temperature

The synthesis was made according to Fig. 2.

The method is shorter (only one heat treatment) and less polluting (only one ball milling) than the preceding method. It has been used throughout our investigations and allowed us to prepare (4 kg per batch) carbonitrides from $TiC_{0.1}N_{0.9}$ to $TiC_{0.8}N_{0.2}$. The nitridation temperature and the nitrogen pressure must be adapted to the required composition, according to the chemical stability, as explained in Section 5.2. The carbonitrides obtained are ball milled and sieved at 100 μm. The main impurities are WC (about 0.5–1.5 wt.% from the balls) and oxygen (about 0.5 wt.% from the starting titanium and TiC powders).

2.3. Preparation by carbonitridation of TiO_2

The expected reaction is as follows:

$$2TiO_2 + 2(3-x)C + xN_2 \rightarrow 4CO + 2TiC_{1-x}N_x$$

Starting from a TiO_2–C mixture with a 24.7 wt.% C content, nitridation tests were run for 2 h at temperatures between 1400 and 1800 °C, under a 600 l h^{-1} nitrogen flow. As a conclusion, we can say that, in order to obtain oxygen and free-carbon contents both lower than 0.2 wt.%, nitridation must be made at $T > 1600$ °C. However, according to the thermochemistry of the reaction (see Section 5.2), the composition of the carbonitride obtained depends mainly on the temperature: $TiC_{0.3}N_{0.7}$ at 1600 °C, $TiC_{0.4}N_{0.6}$ at 1700 °C, and $TiC_{0.5}N_{0.5}$ at 1800 °C.

The carbonitridation of TiO_2 is certainly economical in order to prepare carbonitrides with a formula between $TiC_{0.4}N_{0.6}$ and TiC. The preparation of high nitrogen content carbonitrides

*Paper presented at the 3rd International Conference on the Science of Hard Materials, Nassau, The Bahamas, November 9–13, 1987.

 © Elsevier Sequoia/Printed in The Netherlands

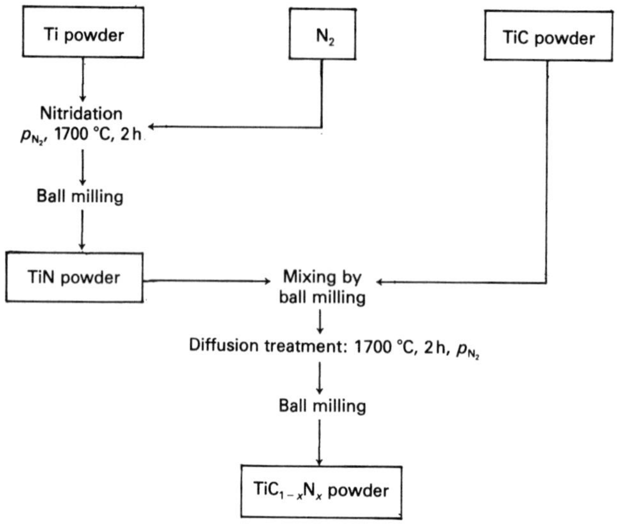

Fig. 1. Synthesis of $TiC_{1-x}N_x$ from TiC plus TiN.

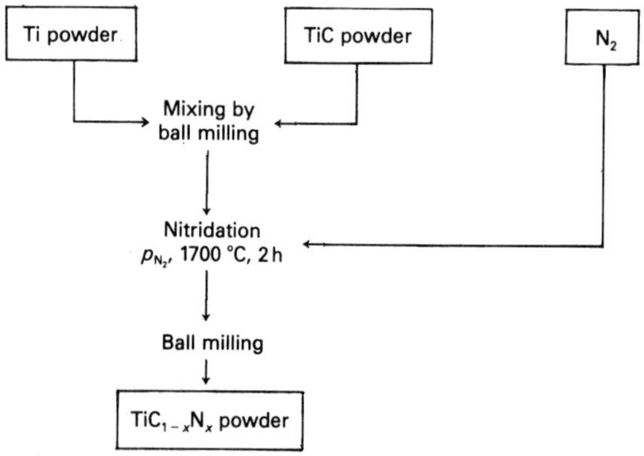

Fig. 2. Synthesis of $TiC_{1-x}N_x$ from titanium plus TiC.

needs low reaction temperatures and thus gives oxycarbonitrides contaminated with free carbon.

2.4. Preparation by thermal decomposition of titanium tetrachloride–amine (or nitrile) complexes

Titanium tetrachloride ($TiCl_4$) dissolved in carbon tetrachloride (CCl_4) is reacted with ethylenediamine ($C_2H_8N_2$) at room temperature. A brown ochre complex precipitates which is filtered under a protective nitrogen atmosphere. The complex is then decomposed in the following way: (1) from 20 to 600 °C, by heating (for 3 h) in a mixture of argon (50 l h^{-1}) and methylamine (CH_3NH_2) (3 l h^{-1}); (2) from 600 to 900 °C, by heating (for 45 min) under hydrogen (50 l h^{-1}) and a 2 h long plateau at 900 °C; (3) from 900 to 1200 °C, by heating (for 1 h) under argon (50 l h^{-1}) and a 2 h long plateau at 1200 °C.

Most of the precipitate is lost because it volatilizes at 200 °C. However, the remaining decomposition product has a composition $TiC_{0.48}N_{0.52}+2C$ and a high specific surface area (153 $m^2\ g^{-1}$).

Under the same conditions the yellow complex $TiCl_4$–acetonitrile (CH_3CN) gives a carbonitride with a 43 wt.% TiN content instead of the expected 33 wt.%.

The method seems to be more convenient for chemical vapour deposition (coatings and powders).

2.5. Preparation by carbonitridation of titanium powder by an organic gaseous compound

The method consists in reacting an oxygen-free organic gaseous compound (amine, nitrile or nitrogen-containing heterocyclic compound) with a fine and pure powder of metal or hydride at 800–1400 °C. The nitrogen-to-(carbon + nitrogen) ratio can be regulated by the nature of the organic compound and by the addition of an appropriate simple gas (nitrogen or CH_4).

In order to eliminate any possible formation of carbon black, cracking is prevented by diluting the organic gas with convenient amounts of hydrogen or argon.

The carbonitridation of titanium or TiH_2 powder, at 1200 °C, by an Ar–CH_3NH_2 mixture gives a $Ti(C_{0.5}N_{0.5})_{0.93}$ powder with a specific surface of about 1 $m^2\ g^{-1}$. $C_2H_8N_2$ also gives $TiC_{0.5}N_{0.5}$, while CH_3CN gives $TiC_{0.61}N_{0.39}$.

3. Crystalline structure of titanium carbonitrides

The method described in Section 2.2 was used to prepare batches of TiC, $TiC_{0.6}N_{0.4}$, $TiC_{0.4}N_{0.6}$, $TiC_{0.2}N_{0.8}$, $TiC_{0.1}N_{0.9}$ and TiN. The powders were sieved at 44 μm and carefully analysed (nitrogen by the Kjeldahl method and oxygen by neutron activation), and the cell parameter (cubic) was determined. The variation in the cell parameter is as a function of x; it seems that Vegard's law is applicable and the linear regression of our data gives the following formula (carbon plus nitrogen sublattice filled to more than 98%):

$$a\,(\text{Å}) = 4.305 - 0.070x \quad (r = 0.95)$$

4. Hot pressing of pure titanium carbonitrides

A study of the hot pressing of pure $TiC_{1-x}N_x$ was made in order to prepare cylindrical pellets which may be useful to investigate the wettability by liquid metals and alloys, the oxidation in air at high temperatures and the corrosion by acids and soda.

For each carbonitride, we determined the optimal parameters (grain size, pressure and temperature) in order to obtain sintered products with closed porosity (relative density, higher than 95%). The best conditions are as follows: powder sieved at 44 μm; $T = 1800$–2000 °C; $p = 200$ bar; $t = 5$–10 min.

5. Properties of pure titanium carbonitrides

5.1. Properties of pure TiN

The hot-pressed sample has the composition $TiN_{0.944}C_{0.006}O_{0.020}$ and a porosity of 17.4%.

The mean linear expansion coefficient was measured:

$$\bar{\alpha}_{25\,°C}^{1200\,°C} = 11.8 \times 10^{-6}\ °C^{-1}.$$

The measured electrical resistivity at room temperature ($\rho = 43\ \mu\Omega$ cm), corrected for the porosity ε (according to the formula, $\rho = \rho_0(1 - 1.5\varepsilon)$) gives $\rho = 33\ \mu\Omega$ cm.

Young's modulus ($E = 23\,200$ hbar), corrected for the porosity according to Eudier's formula $E = E_0(1 - 1.19\varepsilon^{2/3})$, gives $E_0 = 36\,900$ hbar.

5.2. Thermodynamic stability of $TiC_{1-x}N_x$

As one of the components of the carbonitrides is gaseous in its standard state at room temperature, the study of the thermodynamic stability is of great importance. At high temperatures the presence or absence of nitrogen in the pre-sintering or sintering atmosphere can have a great effect on the composition of the carbonitrides. We thus studied various possible equilibria using simple theoretical models and then compared the results with experience.

We studied the four following reactions (where $x < 1$):

$$(1 - x)\langle TiC\rangle + x TiN \rightleftharpoons \langle TiC_{1-x}N_x\rangle \qquad (1)$$

$$\langle TiC_{1-x}N_x\rangle + \frac{x' - x}{2}[N_2] \rightleftharpoons \langle TiC_{1-x'}N_{x'}\rangle + (x' - x)\langle C\rangle \quad (x' > x) \qquad (2)$$

$$\langle TiC_{1-x}N_x\rangle + \frac{1 - x}{2}[N_2] \rightleftharpoons \langle TiN\rangle + (1 - x)\langle C\rangle \qquad (3)$$

$$(1 - x')\langle TiC_{1-x}N_x\rangle \rightleftharpoons (1 - x)\langle TiC_{1-x'}N_{x'}\rangle + (x - x')\langle Ti\rangle + \frac{x - x'}{2}[N_2] \quad (x' < x) \qquad (4)$$

These reactions are very interesting from points of view of both synthesis and sintering of carbonitrides.

Equations (2) and (4) allow us to define the domain of existence of $TiC_{1-x}N_x$ in the ($\ln p_{N_2}$)–T–x diagram. We only need the free enthalpies of pure ⟨TiC⟩ and ⟨TiN⟩ [1]:

$$\Delta G^\circ_T\langle TiC\rangle = -44317 + 3.522T \text{ cal mol}^{-1}$$

$$\Delta G^\circ_T\langle TiN\rangle = -80638 + 22.604T \text{ cal mol}^{-1}$$

5.2.1. Free enthalpy of formation of $TiC_{1-x}N_x$ (reaction (1))

As the cell parameter of $TiC_{1-x}N_x$ solid solutions follows Vegard's law, we assumed that the solid solution is ideal; thus

$$\Delta G^\circ_T\langle TiC_{1-x}N_x\rangle = (1-x)\,\Delta G^\circ_T\langle TiC\rangle + x\,\Delta G^\circ_T\langle TiN\rangle + RT\{x\ln x + (1-x)\ln(1-x)\} \quad (5)$$

5.2.2. Free enthalpy of reaction (2)

$$\Delta G^\circ_T = \Delta G^\circ_T\langle TiC_{1-x'}N_{x'}\rangle - \Delta G^\circ_T\langle TiC_{1-x}N_x\rangle = -RT\ln p_{N_2}^{-(x'-x)/2}$$

and thus the minimum nitrogen pressure p_{N_2} in equilibrium over $\langle TiC_{1-x}N_x\rangle(x' \rightarrow x)$ is given by

$$\ln p_{N_2} = \frac{2(\Delta G^\circ_T\langle TiN\rangle - \Delta G^\circ_T\langle TiC\rangle)}{RT} + 2\ln\left(\frac{x}{1-x}\right) \quad (6)$$

In an x–$\ln p_{N_2}$ plot at constant T, the curve representing eqn. (6) defines two domains: one in which pure $TiC_{1-x}N_x$ is stable ($\ln p_{N_2} < \ln\{p_{N_2}(\text{equilibrium})\}$), and the other where $TiC_{1-x}N_x$ coexists with free carbon ($\ln p_{N_2} > \ln\{p_{N_2}(\text{equilibrium})\}$).

5.2.3. Free enthalpy of reaction (3)

Reaction (3) is in fact a particular case of reaction (2) for which $x' \rightarrow 1$. The equilibrium nitrogen pressure is here defined by eqn. (7):

$$\ln p_{N_2} = \frac{2(\Delta G^\circ_T\langle TiN\rangle - \Delta G^\circ_T\langle TiC\rangle)}{RT} - \frac{2}{1-x}\{x\ln x + (1-x)\ln(1-x)\} \quad (7)$$

In an x–$\ln p_{N_2}$ plot, at fixed T, the curve representing eqn. (7) defines two domains: one corresponding to $TiC_{1-x}N_x$ plus carbon, and the other to TiN plus carbon.

5.2.4. Free enthalpy of reaction (4)

Thermodynamics allows us to derive the minimum nitrogen pressure in equilibrium over $TiC_{1-x}N_x$:

$$\ln p_{N_2} = \frac{2}{RT}\Delta G^\circ_T\langle TiN\rangle + 2\ln x \quad (8)$$

In an x–$\ln p_{N_2}$ plot, at fixed T, the curve representing eqn. (8) defines two domains: one corresponding to pure $TiC_{1-x}N_x$ ($\ln p_{N_2} > \ln\{p_{N_2}(\text{equilibrium})\}$) and the other to the coexistence of $TiC_{1-x}N_x$ with free titanium ($\ln p_{N_2} < \ln\{p_{N_2}(\text{equilibrium})\}$), forming most probably $Ti(C_{1-x}N_x)_a$-$\square_{1-a}$ (where $\square$ is a vacancy).

5.2.5. Domain of existence of pure $TiC_{1-x}N_x$

Figures can be drawn to represent, at a given temperature the curves corresponding to eqns. (6)–(8) in an x–$\ln p_{N_2}$ plot. They mark the boundaries of the four following domains: $TiC_{1-x}N_x$ plus titanium; pure $TiC_{1-x}N_x$, $TiC_{1-x}N_x$ plus carbon; TiN plus carbon.

Figures can also be drawn to represent, at a given x value, the curves corresponding to eqns. (6)–(8) in a T–$\ln p_{N_2}$ plot, marking the boundaries of the same domains as above.

These figures are very useful because, for a given concentration x and temperature T, they allow us to determine the range of nitrogen equilibrium pressure permissible over $TiC_{1-x}N_x$ in order to prevent its decomposition.

5.2.5.1. Application 1. For instance, when we want to synthesize $TiC_{1-x}N_x$ at 1527 °C, starting from $(1-x)TiC + xTiN$, by diffusion at $p_{N_2} = 1$ atm, Fig. 3 shows that, if $0.2 < x < 0.625$, the carbonitride will be polluted by free carbon and, if $x < 0.2$, it is impossible to obtain a carbonitride and even TiC will be transformed into TiN plus carbon.

5.2.5.2. Application 2. Let us suppose that we want to sinter pure $TiC_{1-x}N_x$ at 2027 °C. Figure 4 shows that, if we work under a nitrogen pressure of 1 atm, we can sinter the carbonitrides with $x > 0.16$ without any problem. On the contrary, for carbonitrides with $x < 0.16$ and in particular TiC, during sintering a chemical reaction will occur with absorption of nitrogen and release of free carbon. If sintering is carried out under vacuum (nitrogen at 10^{-6} atm is equivalent to

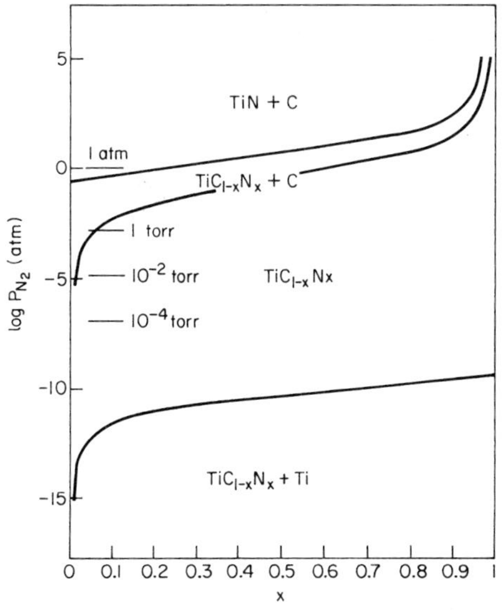

Fig. 3. Domain of stability of $TiC_{1-x}N_x$ at 1527 °C.

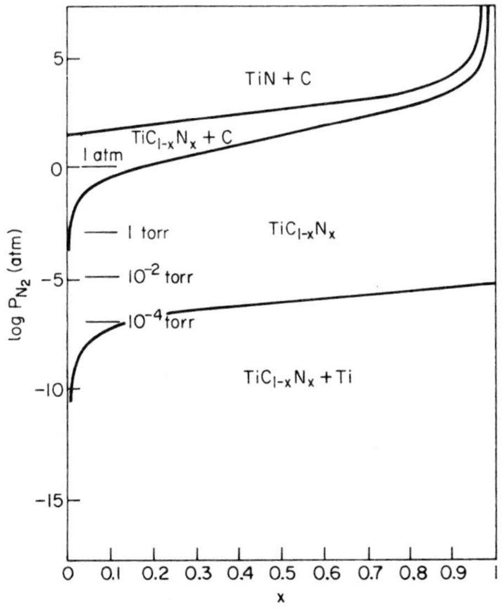

Fig. 4. Domain of stability of $TiC_{1-x}N_x$ at 2027 °C.

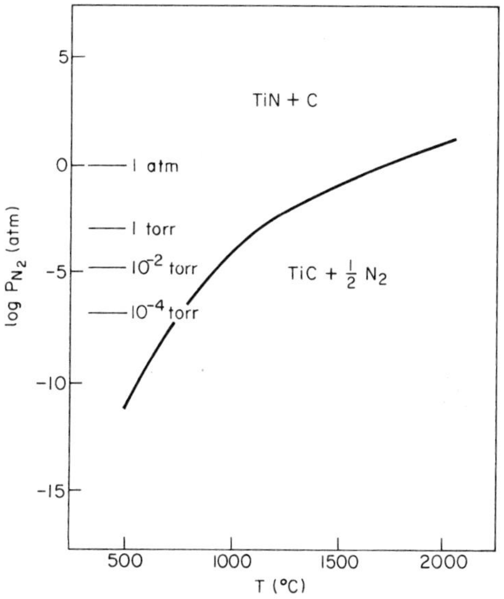

Fig. 5. Domain of stability of TiC in nitrogen atmospheres.

7.6×10^{-4} Torr), it will be possible to sinter safely all the carbonitrides with $x < 0.56$. Carbonitrides with $x > 0.56$ will lose some nitrogen and form free titanium or more probably $(TiC_{1-x}N_x)_a\square_{1-a}$.

5.2.5.3. Application 3. Figure 5 shows the stability domain of TiC in nitrogen or of TiN in the presence of carbon. We can see that, at $p_{N_2} = 1$ atm, TiC is stable only if $T > 1650$ °C. We can also say that, at $T > 1650$ °C, TiN is no longer stable in the presence of carbon.

5.2.6. Experimental data

For each test, 50 g of powder were put in a molybdenum crucible. Before heating, the furnace was pumped out and then filled with nitrogen; a slow nitrogen flow (bubbling) was maintained during the heating and heat treatment.

A first series of tests was made using a molybdenum resistor in order to prevent any carbon contamination by the furnace environment. The reaction treatment was 30 h at 1500 °C or 15 h at 1800 °C.

A second series of tests was made using a graphite resistor in order to verify possible carbon transport from the resistor to the powder by the presumed formation of cyanogen (C_2N_2), particularly at high temperatures.

After treatment the powder was carefully analysed and the cell parameter was measured. According to our results, at 1500 °C, carbonitrides with $x > 0.67$ are stable at $p_{N_2} = 1$ atm, confirming the thermodynamic results. Carbon transport through C_2N_2 does not seem to occur. At 1800 °C the carbonitrides with $x > 0.5$ are not

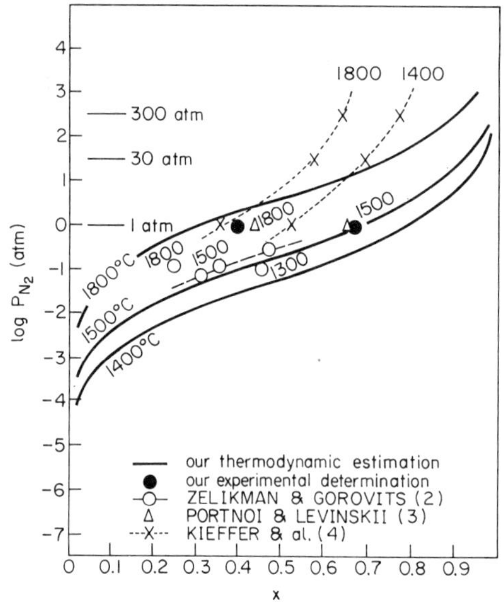

Fig. 6. Equilibrium nitrogen pressure over $TiC_{1-x}N_x$ *vs.* x.

stable at $p_{N_2} = 1$ atm and move to a limiting composition $TiC_{0.5}N_{0.5}$. The thermodynamic limit was calculated to be $TiN_{0.69}C_{0.31}$.

Carbonitrides with $x < 0.5$ are stable at $p_{N_2} = 1$ atm when a molybdenum resistor is used, but these carbonitrides are not stable when a graphite resistor is used. The carbon content increases and nitrogen is lost; carbon is most probably transferred from the resistor to the powder through gaseous C_2N_2 formed by the reaction at high temperatures of graphite with nitrogen.

Figure 6 shows the variation in the equilibrium nitrogen pressure over $TiC_{1-x}N_x$ as a function of x for various temperatures. Some results from other researchers are included and we can see that our results agree quite well with our experimental determinations and those of Soviet workers.

5.3. *Corrosion resistance of carbonitrides*

Corrosion tests were made on dense hot-pressed pellets of the six compounds quoted in Section 3. Six different media were tested: HCl (aqueous 40% solution), H_2SO_4 (93%), HF (aqueous 40% solution), H_3PO_4 (85%), HNO_3 (70%) and NaOH (aqueous 40% solution). The results can be summarized briefly as follows.

(1) Dense carbonitrides are not attacked by NaOH at room temperature, by HCl (385 h at 25 °C or 24 h at 110 °C), by H_2SO_4 (342 h at 25 °C), H_3PO_4 (200 h at 25 °C or 34 h at 170 °C) nor by HF (385 h at 25 °C).

(2) They are attacked by boiling HF; corrosion is greater for nitrogen-rich carbonitrides.

(3) They are attacked by HNO_3 (at 25 °C or boiling) and H_2SO_4 (white fumes) in various degrees.

5.4. *Oxidation by air at high temperatures*

Oxidation tests were made on dense hot-pressed pellets of the six compounds quoted in Section 3, at 800, 1000 and 1200 °C for 8–800 h.

At 1000 °C for instance, the oxidation kinetics of TiC and $TiC_{1-x}N_x$ are parabolic while the kinetics are cubic for TiN. Oxidizability passes through a minimum for the composition $TiC_{0.5}N_{0.5}$. The increase in mass per unit surface area was measured to be 2–7 mg cm^{-2} after 100 h and 0.5–1 mg cm^{-2} after 1 h (28 mg cm^{-2} for WC) according to the value of x.

5.5. *Wettability by molten metals and alloys*

A chip of nickel, cobalt, 75wt.%Ni–25wt.%Mo, 75wt.%Ni–25wt.%W or 18–8 stainless steel was placed on the polished face of a dense hot-pressed carbonitride sample. Wetting tests were made under the following conditions: $T = T_{melting} + 50$ °C, for 30 min; environment, argon, hydrogen or nitrogen (1 atm) or vacuum (10^{-2} Torr).

The best wetting was obtained with a 75wt.%Ni–25wt.%Mo alloy.

5.6. *Hardness of carbonitrides*

The measured Vickers microhardness (load, 100 gf) varied linearly from 2050 ± 200 kgf mm^{-2} for TiN to 3650 ± 350 kgf mm^{-2} for TiC.

6. Sintering of cemented titanium carbonitrides

The sintering process is given in Fig. 7.

Figures 8, 9 and 10 give the influence of the temperature and duration of sintering, the binder content and the composition of the binder respectively for $(100 - x)TiC_{0.2}N_{0.8} + x(yNi-(1-y)Mo)$.

Figure 11 shows the flexural strength *vs.* the Vickers hardness (load, 50 kgf) for the various cemented titanium carbonitrides investigated. We can see that the flexural strength of the cemented carbonitrides can attain a similar strength to

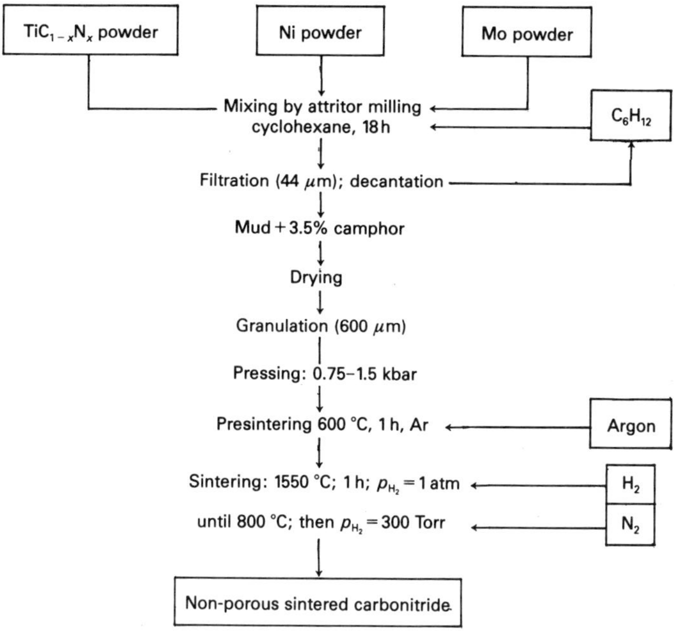

Fig. 7. Sintering process of cemented carbonitrides.

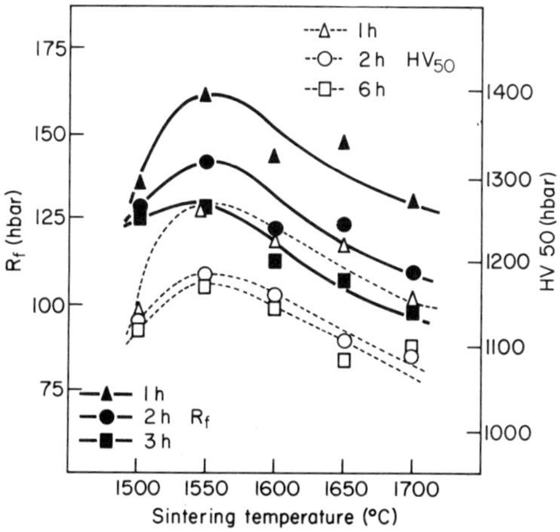

Fig. 8. Sintering of 80wt.%$TiC_{0.2}N_{0.8}$–20wt.% (75wt.%Ni–25wt.%Mo): influence of sintering temperature and duration on the flexural strength and hardness.

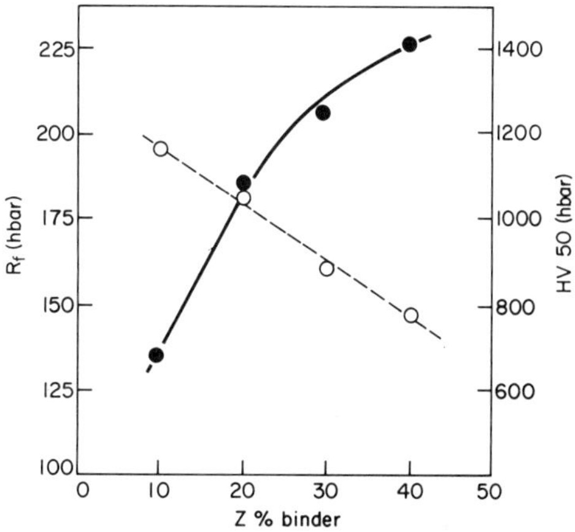

Fig. 9. Sintering of $(100-Z)TiC_{0.2}N_{0.8} + Z$(75wt.%Ni–25wt.%Mo) ($T = 1550$ °C; $t = 1$ h; $p_{N_2} = 300$ Torr): influence of the binder content on flexural strength (●) and hardness (○).

those of cemented tungsten carbides (P, M and K grades), but the hardness is lower. However, cutting phenomena are complex, and the flexural strength and the Vickers hardness are not the only important parameters; the chemical interaction and the friction coefficient between chip and tool are also very important and, from this point of view, carbonitrides may be a better material.

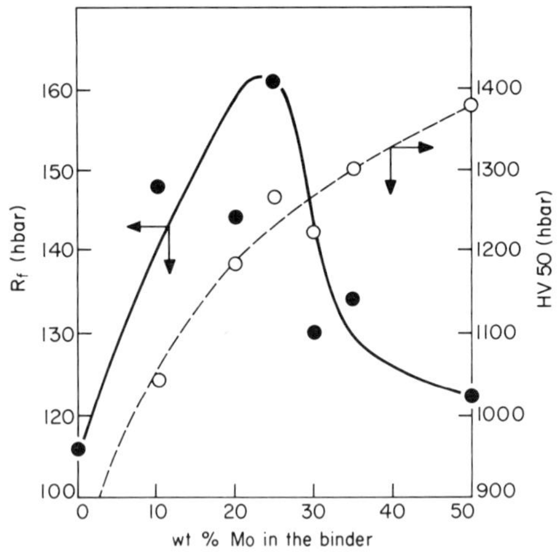

Fig. 10. Sintering of 80wt.%$TiC_{0.2}N_{0.8}$–20wt.%(Ni–Mo) (T=1550 °C; t=1 h; p_{N_2}=300 Torr): influence of the composition of the binder on the flexural strength (●) and hardness (○).

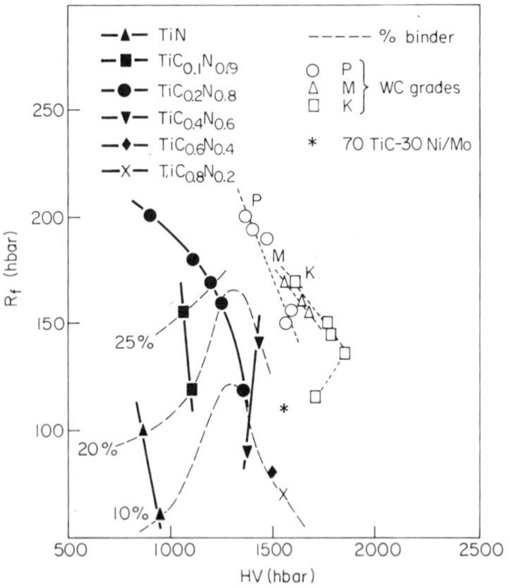

Fig. 11. Flexural strength *vs.* hardness of cemented carbo-nitrides.

TABLE 1 Samples selected for cutting tests

Composition (wt.%)	*Flexural strength R_f* (hbar)	*Vickers hardness* (HV 50)	*d* (g cm^{-3})
80$TiC_{0.30}N_{0.70}$–20$Ni_{4.9}Mo$	142	1225	5.86
90$TiC_{0.21}N_{0.79}$–10$Ni_{4.9}Mo$	108	1335	5.85
80$TiC_{0.39}N_{0.56}$–20$Ni_{4.9}Mo$	147	1380	6.00
90$TiC_{0.40}N_{0.57}$–10$Ni_{4.9}Mo$	120	1540	5.84

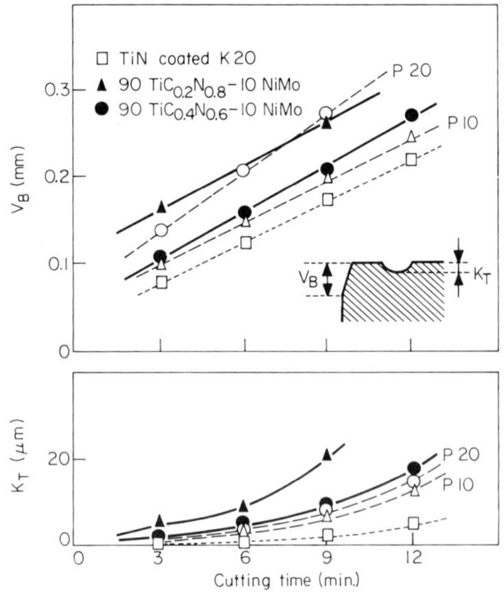

Fig. 12. Cutting test on steel XC 45 F (Brinell hardness, 192 HB; v_c=150 m min^{-1}; Feed F=0.5 mm; depth of cut, 2.5 mm).

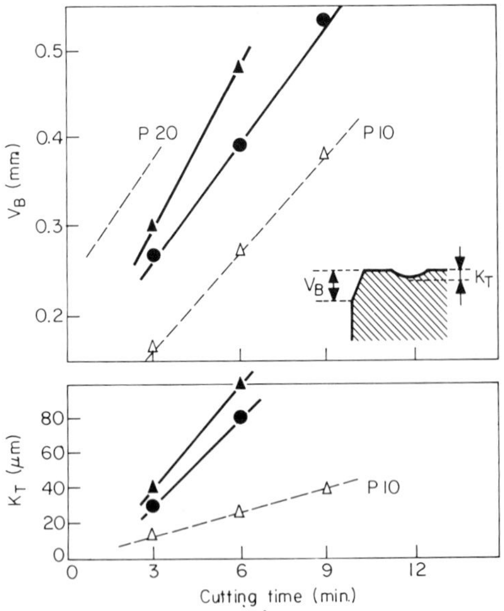

Fig. 13. Cutting test on steel XC 45 F (Brinell hardness, 192 HB; v_c=240 m min^{-1}; F=0.5 mm; depth of cut, 2.5 mm) for 90wt.%$TiC_{0.2}N_{0.8}$–10wt.%(Ni–Mo) (▲) and 90wt.%$TiC_{0.4}N_{0.6}$–10wt.%(Ni–Mo) (●).

7. Cutting tests

The four compositions given in Table 1 were selected for cutting tests.

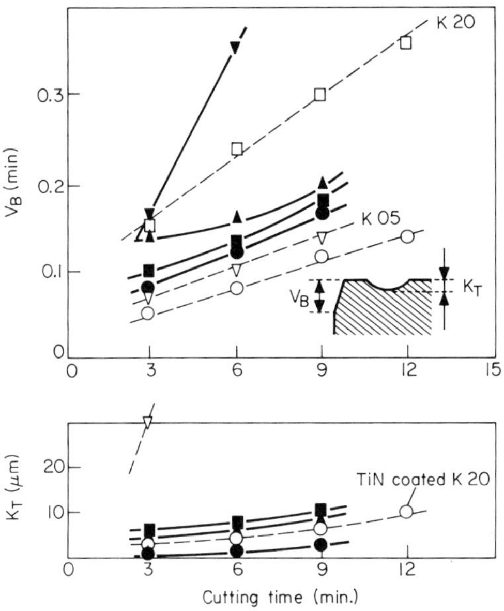

Fig. 14. Cutting test on cast iron FT 26 (Brinell hardness, 200 HB; $v_c = 80$ m min^{-1}; $F = 0.5$ mm; depth of cut, 2.5 mm); ▼, 80wt.%$TiC_{0.3}N_{0.7}$–20wt.%(Ni–Mo); ■, 80wt.%$TiC_{0.4}N_{0.6}$–20wt.%(Ti–Mo); ▲, 90wt.%$TiC_{0.2}N_{0.8}$–10wt.%(Ti–Mo); ●, 90wt.%$TiC_{0.4}N_{0.6}$–10wt.%(Ni–Mo).

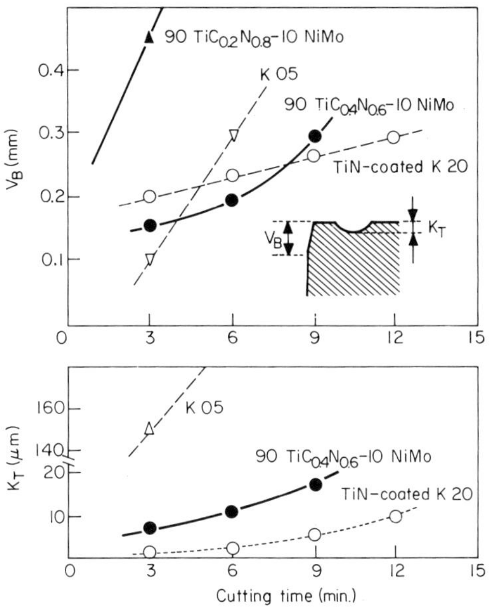

Fig. 15. Cutting test on cast iron FT 26 (Brinell hardness, 200 HB; $v_c = 180$ m min^{-1}; $F = 0.5$ mm; depth of cut, 2.5 mm).

Cutting tests were performed and the results compared with those obtained with cemented tungsten carbides (grades P10, P20 and TiN-coated-K20). The results are given in Figs. 12–15.

We can see that 90wt.%$TiC_{0.40}N_{0.57}$–10wt.%-$Ni_{4.9}Mo$ compares fairly well with classical carbides for the cutting of steel (comparable with grade P20 at a cutting speed v_c of 150 m min^{-1}) and for cast iron (comparable with TiN-coated-K20 at $v_c = 80$ and 180 m min^{-1}).

These preliminary results are very encouraging and studies are under way to improve the cutting qualities as well as for the use of cemented carbonitrides for wear applications.

8. Conclusions

We have described some methods of preparation and some properties of titanium carbonitrides $TiC_{1-x}N_x$ as well as their thermodynamic stability as a function of temperature and nitrogen pressure. Some examples clearly show the influence of thermodynamics on the preparation or sintering of these carbonitrides. Contrary to most papers published to date in which cermets of the type TiC–TiN–binder have been described, we have prepared Ti(C, N)–binder cermets, and the cutting tests performed with these materials appear to be very promising.

The present paper is a short summary of our work which will be published in more detail in the near future, together with a review paper on titanium-carbonitride-based cermets.

References

1 H. L. Schick, *Thermodynamics of Certain Refractory Compounds*, Vol. 2, Academic Press, New York, 1966.
2 A. N. Zelikman and N. N. Gorovits, Study of interaction between nitrogen and titanium carbide, *Zh. Prikld. Khim.*, *23* (1950) 689–695 (Henri Brutcher Translations, HB-2593).
3 K. I. Portnoi and Yu. V. Levinskii, High temperature equilibrium of the reaction between titanium nitride and carbon, *Zh. Fiz. Khim*, *37* (1963) 2627–2634 (*Russ. J. Phys. Chem.*, *37* (1963) 1424–1428).
4 R. Kieffer, H. Nowotny, P. Ettmayer and M. Freudhofmeier, Über die Beständigkeit von Übergangsmetallcarbiden gegen Stickstoff bis zu 300 at, *Monatsh. Chem.*, *101* (1970) 65–82.

Local Impact Damage and Erosion Mechanisms in WC–6wt.%Co Alloys

K. ANAND and H. CONRAD

Materials Science and Engineering Department, North Carolina State University, Raleigh, NC 27695-7907 (U.S.A.)

(Received November 9, 1987)

Abstract

The erosion by Al_2O_3 particles of WC–6wt.%Co alloys with WC grain size in the range 0.57–3.25 μm was determined as a function of impinging particle size (63–405 μm), velocity (35–93 m s^{-1}) and angle of impact (20°–90°). The angle at which the maximum in erosion rate occurred decreased from 90° as the particle size and velocity increased and the WC grain size decreased. Scanning electron microscopy observations revealed that the erosion mode depended on the relative size of the impact crater with respect to the WC grain size. When the crater encompassed less than 10 WC grains, the erosion mode was of the brittle type; when it encompassed more than 100 WC grains it had the characteristics of the ductile type. The principal material removal mechanism in the brittle mode was by cracking and crushing of the WC grains; that in the ductile mode by plastic deformation and fracture of the binder phase, the WC grains remaining relatively intact. Some plastic deformation of the WC grains occurred in both modes. The results for the ductile mode were in accord with a modification of the plastic indentation model, and those for the brittle mode with the elastic–plastic fracture models. Fragmentation of the impacting particles occurred and its magnitude and significance are discussed.

1. Introduction

Cemented WC–Co alloys, whose microstructure consists of a hard brittle WC phase in a ductile cobalt matrix, find application where resistance to wear or erosion is desired. A better understanding of the mechanism(s) by which these phenomena occur in these alloys is thus important from both technological and scientific viewpoints. The present investigation was undertaken to further our understanding of the effects of microstructure on the erosion of WC–Co alloys impacted by hard solid particles.

Previous work [1–5] on the erosion of WC–Co alloys by hard particles has consistently shown that the erosion rate increases with increasing cobalt content. However, some differences exist regarding the effect of the WC grain size. An increase in erosion rate with increase in WC grain size over the range of 1.5–10.6 μm was reported by Uuemyis *et al.* [2], whereas the opposite effect was found by Conrad *et al.* [4] at low impact velocities (45 m s^{-1} or lower) and carbide grain sizes in the range 0.9–5.1 μm. No effect of carbide grain size was noted by them at higher velocities. In a subsequent paper, Conrad *et al.* [5] suggested that the erosion rate increased with increasing WC grain size to about 1.5 μm and thereafter decreased with further increase in carbide grain size. It should be pointed out that the erosion conditions employed by the two laboratories were significantly different. Uuemyis *et al.* [2] employed large quartz sand particles (400–630 μm) as the erodent material and high impact velocities (75–330 m s^{-1}), whereas Conrad *et al.* [4, 5] employed 240 grit (30 μm) alumina particles at generally lower velocities (20–100 m s^{-1}).

A difference has also been reported for the impact angle dependence of the erosion rate. Employing quartz sand particles (100–200 μm) at a velocity of 507 m s^{-1}, Dankin [1] reported that the maximum erosion rate for WC–(6–20wt.%)Co alloys occurred at an impingement angle α of 60°. In contrast, Conrad *et al.* [4, 5] found that the maximum erosion rate always occurred at $\alpha = 90°$ for 30 μm Al_2O_3 particles impacting on WC–Co alloys containing 4.5–11.3 wt.% Co. A maximum erosion rate at normal impact is in accord with the behavior normally observed for brittle materials, whereas a

*Paper presented at the 3rd International Conference on the Science of Hard Materials, Nassau, The Bahamas, November 9–13, 1987.

0921-5093/88/$3.50

© Elsevier Sequoia/Printed in The Netherlands

maximum at 60° indicates a shift towards behavior more typical of ductile materials [6, 7].

In the paper by Anand *et al.* [8] on the erosion behavior of WC–Co alloys and of composites consisting of alumina fibers in stainless steel, it was proposed that scaling of the impact damage zone size with respect to the size of the constituents of the microstructure is an important factor in the erosion of multiphase alloys. When the impact damage zone size is of the order of the hard brittle phase, the erosion behavior is expected to include a contribution from the brittle phase and the angle for maximum erosion rate shifts towards 90°. In contrast, when the damage zone encompasses many brittle particles, a transition towards a more ductile-type erosion behavior should occur and the maximum erosion rate will then take place at an oblique angle. Direct evidence for the importance of scaling between the impact damage zone size and the microstructure size was provided by studies of (a) the damage produced by single impacts on WC–Co alloys [9] and (b) the erosion behavior of Al–Si alloys, which consist of a hard silicon phase in a ductile matrix [10, 11].

The objective of the present investigation was to examine further the effects of the microstructure on the erosion behavior of WC–Co alloys by hard solid particles, giving attention to (a) the angle dependence of the erosion rate, (b) the effect of carbide grain size, (c) the nature of the damage produced and the associated material removal mechanisms and (d) the influence of scaling between the impact damage zone size and the microstructure on (a)–(c). Scanning electron microscopy (SEM) observations of the craters produced by single-particle impacts and of the surface of specimens eroded by multiparticle impacts was an important aspect of the present studies.

2. Experimental details

2.1. *Microstructure*

The target materials were three commercial nominal WC–6wt.%CO alloys provided by the Kennametal Corporation containing a fine ($d_{WC}=0.57$ μm), a medium ($d_{WC}=1.10$ μm) and a coarse ($d_{WC}=3.25$ μm) WC grain size respectively. The microstructure of each alloy was examined by SEM and the parameters relating thereto are listed in Table 1. The cobalt mean free path λ_{Co} (or λ^c_{Co}), the mean linear intercept WC grain size d_{WC} and the WC grain continuity C were obtained using the method of Lee and Gurland [12]. The mean maximum WC grain diameter d_{WC}^{max}, the mean WC grain area A_{WC} and the shape factor δ_{WC} were determined using an Apple graphics tablet interfaced with an Apple II plus computer and a special software package [13]. The distribution of the WC grain areas in all three alloys was found to be log-normal. The principal difference between the alloys was in their WC grain size, and in turn the cobalt mean

TABLE 1 Microstructure parameters, hardness and fracture toughness of the WC–6wt.%Co alloys

Specimen	*Binder content* Amount (wt.%)	*Binder content* Volume fraction f_{Co}	*Binder mean free path* λ_{Co} (μm)	*Binder mean free path* λ^c_{Co} (μm)	*WC grain size* d_{WC} (μm)	*WC grain size* d_{WC}^{max} (μm)	*WC grain size* A_{WC} (μm)^2	*WC grain size* δ_{WC}	*WC contiguity* C	*Vickers hardness, Static* Load, 10 kgf (GPa)	*Vickers hardness, Static* Load, 250 gf (GPa)	*Vickers hardness* Dynamic[a], at load, 250 gf (GPa)	*Fracture toughness* G_{IS}[b] (MJ m^{-2})
A	6.0Co–0.4Cr	0.11	0.08	0.23	0.57	0.83	0.35	0.79	0.65	17.59	18.92	21.92	0.722
B	6.0Co	0.10	0.10	0.35	1.10	1.47	1.12	0.79	0.69	14.81	15.73	18.73	0.912
C	5.7Co	0.10	0.35	0.94	3.25	4.01	8.65	0.77	0.63	11.69	12.15	15.15	2.589

The following parameters were determined by the line intercept method on SEM micrographs employing the method of Lee and Gurland [12]: $d_{WC}=2f_a/(2N_{\alpha\alpha}+N_{\alpha\beta})$, mean linear intercept WC grain size; $\lambda_{Co}=d_{WC}f_a/f_\beta$, cobalt mean free path assuming the presence of a cobalt film between all WC grains; $\lambda^c_{Co}=d_{WC}f_a/f_\beta(1-C)$, cobalt mean free path assuming no cobalt film between contiguous grains; $C=2N_{\alpha\alpha}/(2N_{\alpha\alpha}+N_{\alpha\beta})$, fraction of the total surface area of the WC grains which is shared with other WC grains; $N_{\alpha\alpha}$ and $N_{\alpha\beta}$, the average number of intercepts per unit length of the test line with a WC–WC interface and a WC–Co interface respectively. The maximum WC grain diameter d_{WC}^{max}, the mean WC grain area A_{WC} and the WC shape factor $\delta_{WC}=4\pi A_{WC}/P_{WC}^2$ (P_{WC} is the perimeter) were obtained using an Apple graphics tablet connected to an Apple-II plus microcomputer using software provided by Hare and Russ [13].

[a] The increase in hardness resulting from dynamic loading in an erosion test taken from ref. 9.

[b] Obtained from Palmqvist indentation fracture tests.

free path. Also included in Table 1 are the Vickers hardness H_V of the alloys, which decreased with increase in WC grain size, and the Palmqvist fracture toughness parameter G_{IS}, which increased with increasing grain size.

2.2. Steady state erosion

The erosion tests were carried out with an air-stream-type tester [14] employing Norton Company angular alumina particles of grit sizes 240 to 60 whose average size D (equivalent spherical diameter) ranged from 63 μm to 405 μm respectively. A detailed statistical analysis of the size and shape of the particles was performed [15] using computerized image analysis of photographs taken of dispersions of the particles on a glass slide. The results of this analysis are given in refs. 9 and 15. The size distribution for each grit was a normal distribution. The alumina particles are estimated to have a Vickers hardness of about 25 GPa, being slightly harder than the WC grains [16]. The erosion tests were carried out with particle velocities v of 35–93 m s^{-1} determined by the paddle wheel method [17] and impact angles α of 20°–90°.

Prior to an erosion test, the as-received flat-ground specimen was ultrasonically cleaned in acetone and dried and the mass was determined to an accuracy of 0.1 mg. The specimen was then placed in a holder and impacted at a fixed angle by a known amount of erodent particles fed from a vibratory hopper at a controlled velocity and flow rate of about 0.033 g s^{-1} through a ceramic nozzle 3.2 mm in diameter. The resulting flux was sufficiently small that the rebounding particles are not expected to influence the erosion results significantly [14]. Following erosion, the specimen was again cleaned and dried and the mass determined. This test procedure was repeated a number of times to yield a plot of cumulative mass loss of the specimen *vs.* cumulative mass of impacting particles, which in all the cases was linear and passed approximately through the origin. The steady state erosion rate E_R (g g^{-1}) was taken as the slope of a linear least-squares fit to the data points. The surface damage produced by the multiparticle impacts in steady state erosion was examined by SEM.

2.3. Single-particle impacts

A number of tests were conducted to determine the size and nature of the damage produced by single impacts. For these tests, the specimen was first polished to a 1 μm finish using diamond paste and then impacted with a small amount of erodent (0.025 g) using the erosion tester mentioned above. The resulting single-impact craters were examined by SEM to determine their relative size with respect to the WC grains, and the nature of the damage produced.

2.4. X-ray diffraction studies

To obtain some insight into the amount of plastic deformation of the WC grains which occurs during steady state erosion, X-ray diffraction peak profiles were determined on specimens A and C prior to, and following, erosion. Both the peak position and its halfwidth were determined.

2.5. Fragmentation of erodent particles

To investigate the degree of fragmentation of the erodent particles which occurred on impact (and thereby influenced the erosion results), the distribution of erodent particle sizes before and after erosion was determined by sieve analysis employing a stack of standard ASTM sieves.

3. Results

3.1. Steady state erosion rate

3.1.1. Angle dependence

For a given WC grain size, the angle dependence of the steady state erosion rate depended on the severity of the erosion conditions. As reported previously [9], for a constant WC grain size the angle at which the maximum erosion rate occurred tended to decrease from 90° as the impacting particle size D and velocity v increased, *i.e.* the erosion rate shifts towards a ductile-type signature as D and v increase. Moreover, at a constant particle size and velocity, the shift towards the ductile mode was favored by a smaller WC grain size (Fig. 1). A shift towards the ductile-type erosion behavior could thus be obtained by independent variations in microstructure (the WC grain size) or in the severity of the erosion conditions (the erodent particle size and/or velocity).

3.1.2. Effect of particle velocity

The erosion rate varied with particle velocity v according to the generally observed power law relation $E_R = Kv^n$ for all erosion tests. Values of the velocity exponent n as a function of WC grain size, impacting particle size and impact angle are presented in Table 2. It should be noted that n

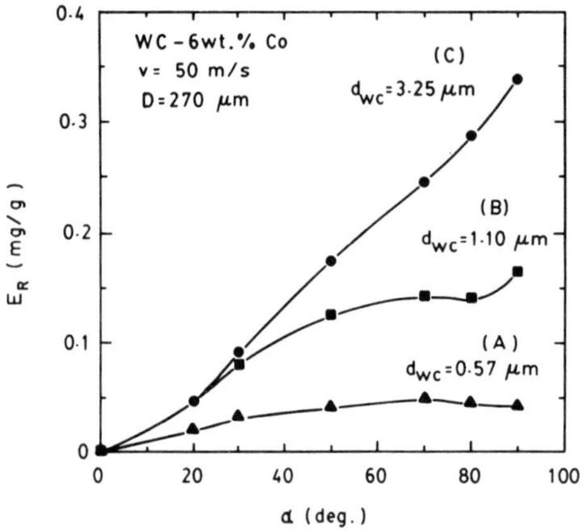

Fig. 1. Angle dependence of the erosion rate of the WC–6wt.%Co alloys as a function of WC grain size.

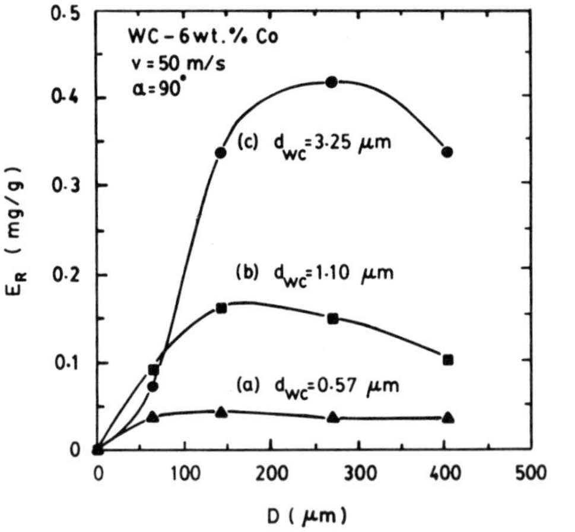

Fig. 2. Effect of impacting particle size *D* on erosion rate as a function of WC grain size.

TABLE 2 Velocity exponents as a function of WC grain size,impacting particle size and impact angle

Specimen	d_{WC} (μm)	*Angle of impact* (deg)	*Velocity exponents for the following impacting particle sizes*		
			270 μm	*143 μm*	*63 μm*
A	0.57	30	2.7	3.0	3.1
A	0.57	50	2.6	2.8	—
A	0.57	90	1.9	2.0	2.6
B	1.10	90	1.9	2.2	3.1
C	3.25	30	—	2.8	3.6
C	3.25	50	—	2.7	3.8
C	3.25	90	2.1	2.5	3.7

tends to increase for those conditions which favor a brittle-type erosion behavior and to decrease for those which favor a ductile-type behavior. In general, the velocity exponent *n* takes a value of 3.1–3.7 for pronounced brittle-type behavior and decreases to about 2.0 as the erosion behavior shifts to pronounced ductile-type behavior.

3.1.3. Effect of erodent particle size

The effect of impacting particle size *D* on erosion rate as a function of WC grain size is shown in Fig. 2. It is evident that E_R goes through a maximum with increasing *D*, the maximum occurring at a larger *D* as the WC grain size increases. The initial increase in E_R occurs for the brittle-type erosion mode; the subsequent decrease reflects the shift towards the ductile-type mode.

3.1.4. Effect of WC grain size

The erosion rate in the ductile regime increased monotonically with increasing WC grain size, whereas in the brittle regime E_R went through a maximum at $d_{WC} \approx 1.0$ μm.

3.2. Nature of the impact damage

3.2.1. Single-particle impacts

SEM examination of the nature of the damage by single-particle impacts revealed that, for those erosion conditions which produced a brittle-type signature, the impact craters encompassed only a few WC grains, and damage occurred mainly by brittle fracture of the WC grains, *i.e.* by cracking and crushing of the WC grains (Fig. 3). The cracks generally extended to the grain edges, where presumably WC fragments were ultimately torn loose. Although cracking occurred throughout the WC grains, separation often took place at grain corners or edges (Fig. 3(a)). The cracks ran both perpendicular and parallel to the surface, similar to radial and lateral cracks [18–21].

For those erosion conditions which produced a ductile-type signature, the crater encompassed a large number (more than 100) of WC grains and the damage consisted mainly of the extrusion ($\alpha = 90°$) or scooping out ($\alpha = 45°$) of material, leaving the WC grains relatively intact (Fig. 4). Plastic deformation of the binder now plays a dominant role and the craters resemble those observed for the impact of ductile metals [11, 22, 23].

Fig. 3. SEM micrographs of the damage produced in specimen C by single-particle impacts under conditions that give the fully brittle erosion mode for the following conditions: (a) $D=143$ μm, $v=50$ m s^{-1} and $\alpha=90°$; (b) $D=270$ μm, $v=50$ m s^{-1} and $\alpha=90°$.

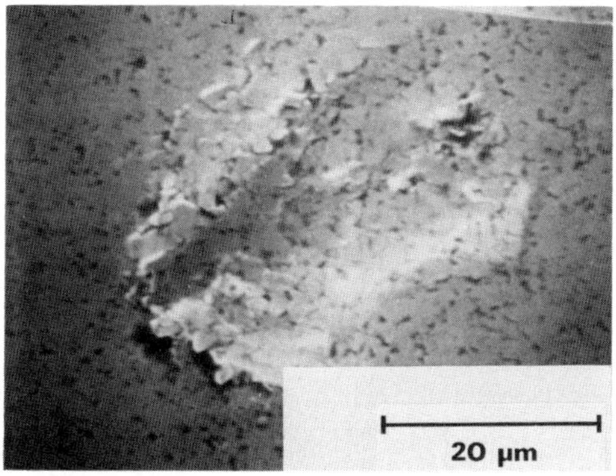

Fig. 4. SEM micrograph of the damage procedure in specimen B by a single-particle impact under conditions that give the fully ductile erosion mode (impact conditions: $D=405$ μm; $v=50$ m s^{-1}; $\alpha=90°$).

Fig. 5. SEM micrograph of the damage produced in specimen A by a single-particle impact under conditions that give an intermediate erosion mode (impact conditions: $D=63$ μm; $v=75$ m s^{-1}; $\alpha=90°$).

An example of a crater for impact conditions intermediate between the fully brittle (maximum E_R at $\alpha=90°$) and fully ductile (maximum E_R at $\alpha\approx45°$) modes is given in Fig. 5. Some carbide cracking, together with plastic deformation of the binder phase, now occurs. Extrusion of the binder is clearly evident.

3.2.2. Multiparticle impacts

The apperance of the surface of a specimen eroded in the fully brittle mode is shown in Fig. 6. The damage has features similar to those for single-particle impacts, namely the cracking and crushing of the WC grains. Again both radial and "lateral" cracks are evident, extending to the WC grain corners and edges.

The appearance of the surface of a specimen eroded in the fully ductile mode is illustrated in Fig. 7. Here also the damage features have characteristics of those observed for single-particle impacts, namely scooping out by plastic deformation and fracture of the binder phase of a volume containing a large number of carbide grains that remain relatively intact. Smearing of the binder phase over the WC grains has also occurred, giving the appearance of platelets [23].

3.3. X-ray studies of eroded surface

Changes in the X-ray diffraction profiles which occurred on eroding specimen A in the ductile mode and specimen C in the brittle mode are presented in Table 3. Only a little change in the

(111) peak angle and profile base width occurred for both modes; however, a significant decrease in the peak height resulted, giving a large change in the halfwidth. The change in the line halfwidth is slightly smaller for specimen C than for specimen A, indicating less plastic deformation of the larger WC grains in the brittle erosion mode. It should also be noted in Table 3 that the WC peak positions shifted slightly following erosion, indicating an increase in lattice parameter. This probably resulted from relief of the residual compressive stresses which are normally introduced during fabrication of cemented carbides. The observed lattice parameter increase represents a linear strain of about 0.2%.

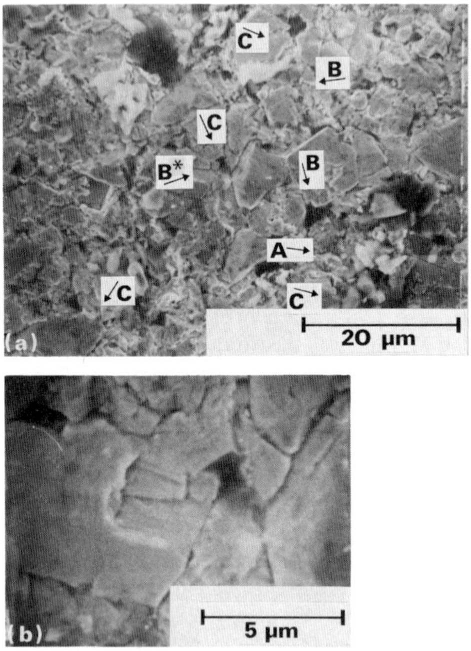

Fig. 6. SEM micrograph of the surface of specimen C eroded in the fully brittle mode (erosion conditions: $D=143$ μm; $v=50$ m s^{-1}; $\alpha=90°$): (a) lower magnification view with chipping at A, lateral cracks at B, and radial cracks at C; (b) higher magnification at location B*.

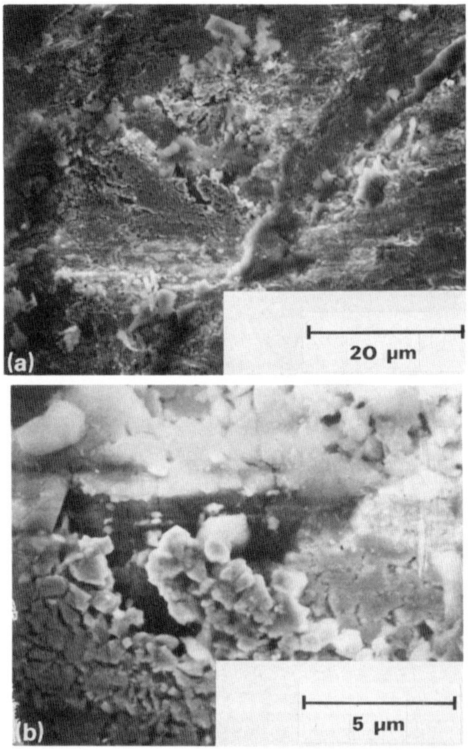

Fig. 7. SEM micrograph of the surface of specimen A eroded in the fully ductile mode (erosion conditions: $D=405$ μm; $v=50$ m s^{-1}; $\alpha=40°$): (a) lower magnification view; (b) higher magnification view.

TABLE 3 Effect of erosion on the (111) X-ray diffraction profile for a fine WC grain size (specimen A) and a coarse grain size (specimen C) WC–6wt.%Co alloy

Specimen	*Erosion conditions*			*Diffraction peak*	2θ (deg)	$\Delta(2\theta)$ (deg)	W (deg)	ΔW (deg)	Δd (Å)	$\varepsilon=\Delta d/d$
	D (μm)	v (m s^{-1})	α (deg)							
A, as received	—	—	—	111	73.04	—	0.23	—	—	—
A, eroded	270	50	90	111	72.86	0.18	0.77	0.54	0.0027	0.0021
C, as received	—	—	—	111	73.04	—	0.31	—	—	—
C, eroded	270	50	90	111	72.84	0.19	0.72	0.41	0.0029	0.0022

W is the halfwidth of the diffraction peak. The interplanar spacing d of (111) planes of the WC phase is taken to be 1.29315 Å.

3.4. Fragmentation of particles

Sieve analysis of 120 grit ($D=143$ μm) Al_2O_3 particles which had eroded WC–Co specimens indicated that only a slight change in the size distribution occurred for $v=50$ m s^{-1}, but significant fragmentation took place at $v=93$ m s^{-1}. The degree F_g of fragmentation was defined as

$$F_g = \frac{D_0^{\,s} - D_e^{\,s}}{D_0^{\,s}} \qquad (1)$$

where $D_0^{\,s}$ is the initial mean sieve particle size and $D_e^{\,s}$ is that following erosion. F_g was found to increase with increasing particle size and velocity [14]. However, the degree of fragmentation was relatively independent of target material (the present WC–Co alloys, 1018 steel and cast aluminum) and of impact angle (50°–90°).

4. Discussion

4.1. General comments

It was found that the angle dependence of the steady state erosion rate in WC–6wt.%Co alloys depends on the severity of the erosion conditions relative to the WC grain size. A brittle erosion signature was obtained for mild erosion conditions (a small particle size and/or a lower velocity) and a coarse WC grain size. This shifted towards a ductile signature as either the severity of the erosion conditions was increased (a larger impacting particle size and/or higher velocity) or as the carbide grain size was reduced. SEM observations of the craters produced by single-particle impacts revealed that, when the craters encompassed only a few (less than 10) WC grains, the damage consisted of brittle fracture of the WC grains. When the crater encompassed a large number (more than 100) of carbide grains, squeezing out ($\alpha=90°$) or scooping out ($\alpha=45°$) of material by plastic deformation and fracture of the binder phase occurred, retaining the WC grains as relatively intact.

The surfaces of eroded specimens exhibited damage features which resembled those for the single-particle impacts. Thus, material removal in the brittle mode occurred mainly by the cracking and crushing of WC grains and that in the ductile mode by the extrusion ($\alpha=90°$) or scooping out ($\alpha=45°$) of material by plastic flow and fracture of the binder phase. Although localized fracture at the WC grain edges ("the edge effect") was observed for both single-particle impacts and steady state erosion, it did not appear to be the dominant mechanism for material removal in either the brittle or the ductile erosion modes, in contrast with the finding of Kosel and Aptekar [24] for alloys in which the hard phase consisted of 50% or less of the volume fraction and the erodent particles were softer than the hard phase.

From the above, it is clear that the erosion behavior of the WC–Co alloys depends on the scaling between the damage zone size (crater) and the WC grain size. Therefore, any further evaluation of the erosion of these alloys should refer to the specific scaling size regime being considered. This will now be done in regard to (a) the effects of microstructure on the steady state erosion rate and (b) the erosion mechanism.

4.2. Ductile mode

Erosion in this regime is governed by the plastic deformation and fracture of the binder, which is constrained by the hard carbide grains. Hence erosion resistance might be expected to be proportional to $\lambda_{Co}^{-1/2}$ or that $E_R \propto \lambda_{Co}^{1/2}$. A plot of E_R *vs.* $\lambda_{Co}^{1/2}$ yielded a straight line with an intercept on the abscissa at $\lambda^0_{Co}=0.05$ μm. This critical value of the binder mean free spacing may represent the lower limit for erosion by the ductile mode.

The velocity exponent n for the ductile mode increased from about 2 at 90° to about 3 at 30°. The value of 2 is smaller than that (2.5–3.0) predicted by the theories [25, 26] which are based on the concept that material removal in this mode requires many repeated impacts leading eventually to a failure strain. The value of 2 obtained in the present tests for normal impact suggests that the erosion rate may be directly related to the kinetic energy of the impacting particle. Equating the kinetic energy of the particle to the plastic work required to form the resulting indentation, we obtain [27, 28]

$$pV_c = \frac{\alpha_1 m_p v^2}{2} \qquad (2)$$

where p is the pressure acting uniformly over the contact area and represents the force between the particle and the target, V_c the volume of the indentation (crater), α_1 the fraction of the kinetic energy contributing to the plastic deformation and m_p the mass of the impacting particle. If we take α_2 as the ratio of the volume V_d of the deformed material to the volume V_c of the crater (*i.e.* $\alpha_2 = V_d/V_c$) and α_3 as the ratio of the Vickers hardness H_V to the pressure p [29] (*i.e.* $\alpha_3 = H_V/p$), eqn. (2) gives

$$V_d = \frac{\alpha_1 \alpha_2 \alpha_3 m_p v^2}{2H_V} \tag{3}$$

Further, let us assume that the deformed material is removed when the binder phase has undergone a critical failure strain ε_f. This then gives, for the steady state erosion rate E_R,

$$E_R = \frac{\Delta\varepsilon}{\varepsilon_f} \frac{\alpha_1 \alpha_2 \alpha_3 \rho_t v^2}{2H_V} \tag{4}$$

where $\Delta\varepsilon$ is the plastic strain per impact and ρ_t is the density of the target.

The velocity exponent in eqn. (4) is in accord with that obtained experimentally at normal impact. Equation (4) further requires that E_R be proportional to $1/H_V$, all other parameters being constant. A plot of E_R *vs.* $1/H_V$ is given in Fig. 8. The static and dynamic values of H_V employed are those for a load of 250 g since the size of the hardness impression at this load is of the order of the size of craters obtained by single particle impacts [9]. Figure 8 shows that E_R increases linearly with $1/H_V$ beyond the critical value $H^0_V = 19.6$ GPa (static) or 22.7 GPa (dynamic), the former being in good agreement with that calculated from λ^0_{Co} mentioned above using the equation by Lee and Garland [12]. Thus, eqn. (4) is more appropriately written

$$E_R = \frac{1}{2} \alpha_4 \rho_t v^2 \left(\frac{1}{H_V} - \frac{1}{H^0_V} \right) \tag{5}$$

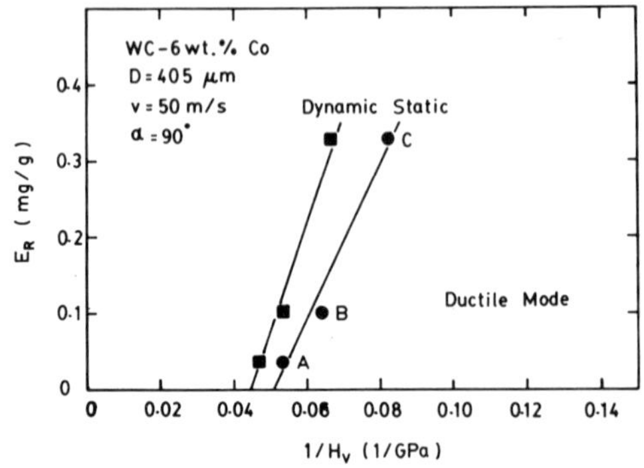

Fig. 8. Erosion rate in the ductile mode *vs.* the reciprocal of the dynamic and static hardness.

where $\alpha_4 = \alpha_1 \alpha_2 \alpha_3 (\Delta\varepsilon/\varepsilon_f)$. Figure 8 yields $\alpha_4 = 0.53$ and 0.74 for the static and dynamic hardnesses respectively. It is worthy of mention that a plot of $1/E_R$ *vs.* H_V yields $\alpha_4 = 1.36$ and $H^0_V = 12$ GPa and 15 GPa respectively.

The various constants in eqn. (4) can now be further evaluated in the following manner. In the previous investigation [9] the volume V_c of craters produced by single-particle impacts was measured directly employing a special SEM technique. Comparing the volume of such craters produced on specimen A (impact conditions: $D = 270$ μm; $v = 50$ m s^{-1}; $\alpha = 90°$) with the average volume V_i removed per impact calculated from steady state erosion under the same impact conditions gives $\alpha_5 = V_i/V_c = 0.77$. Substituting V_i for V_c in eqn. (1) gives

$$E_R = \frac{1}{2} \alpha_1 \alpha_3 \alpha_5 \rho_t v^2 \left(\frac{1}{H_V} - \frac{1}{H^0_V} \right) \tag{5a}$$

where now $\alpha_4 = \alpha_1 \alpha_3 \alpha_5$. Taking $\alpha_4 = 0.53$–0.74 and $\alpha_5 = 0.77$, we obtain $\alpha_1 \alpha_3 = 0.69$–0.96 and $\alpha_2 (\Delta\varepsilon/\varepsilon_f) = 0.77$. Making the reasonable assumptions that α_2 and α_3 are approximately unity, we obtain $\alpha_1 = 0.69$–0.96 and $\Delta\varepsilon/\varepsilon_f = 0.77$. The values for α_1 are quite reasonable, in that elastic losses associated with impact are expected to be of the order of 0.1 of the total kinetic energy [28] and fragmentation of the erodent particles upon impact to consume about 0.15 of the kinetic energy for the erosion conditions considered (9).

The above considerations thus lead to the conclusion that eqns. (2)–(5a) provide a reasonable description of the erosion of WC–Co alloys in the ductile regime, *i.e.* when the impact crater encompasses more than about 100 WC grains. The effects of target microstructure are then mainly through their influence on the hardness H_V (and H^0_V). Previous work [4, 9] has shown that the effects of microstructure on H_V are given reasonably well by the relations of Lee and Garland [12]. The value of H^0_V appears to be associated with a critical mean free binder spacing $\lambda^0_{Co} = 0.05$ μm.

The velocity exponent of 2 observed in the present tests for normal impact indicates that the ratio $\Delta\varepsilon/\varepsilon_f$ is independent of impact velocity. The larger values for the exponent at oblique angles suggests that either this ratio or one or more of the other constants of eqns. (2)–(5a) increases with increasing velocity. It should be mentioned in this respect that Hutchings *et al.* [30] found for steel balls striking mild steel specimens at 30° impact angle that the crater volume was proportional to $v^{2.4}$ and the erosion rate proportional to $v^{2.9}$.

Regarding a possible effect of erodent particle size on the erosion rate [31], the present data did not permit a clear valuation of the influence of D on E_R in the ductile mode.

4.3. Brittle mode

Erosion in this regime is governed mainly by the brittle feature of the WC grains, and therefore we anticipate that the erosion rate may vary with WC grain size. Figure 9 shows that E_R initially increases rapidly with increasing d_{WC}, reaches a maximum at $d_{WC} \approx 1$ μm and then decreases with further increase in WC grain size. The initial increase in erosion rate could result from one or more of the following causes: (a) the fracture strength of the WC grains is reduced as their size increases, for a particle generally becomes less perfect as it becomes larger, (b) it becomes easier to develop lateral cracks as the WC grain size increases or (c) the erosion mode at the smallest WC grain size is not completely brittle but contains a significant ductile component.

The decrease in erosion rate for $d_{WC} > 1$ μm was found to be in accord with the predictions of the effect of target properties on the erosion rate by the elastic–plastic indentation fracture models [22, 32–34] for the erosion of brittle materials. The dynamic impact model [32–34] predicts that $E_R \propto K_c^{-4/3} H_V^{-1/4}$, whereas the quasi-static impact model [22] predicts $E_R \propto K_c^{-4/3} H_V^{1/9}$. By taking $K_c \propto G_{IS}^{1/2}$ (Table 1), the erosion rates for specimen B ($d_{WC} = 1.10$ μm) and specimen C

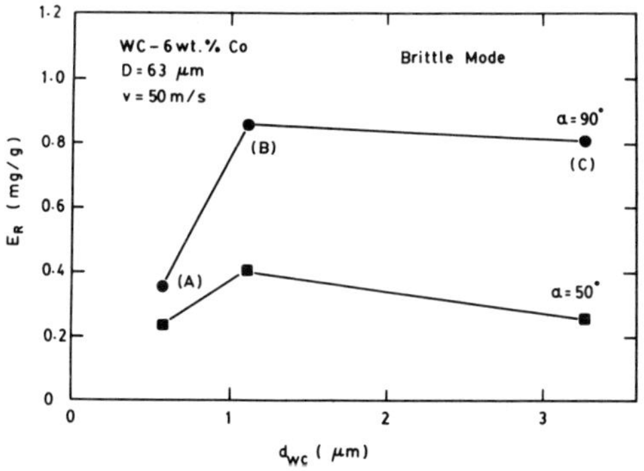

Fig. 9. Erosion rate in the brittle mode *vs.* the WC grain size at two impact angles.

($d_{WC} = 3.25$ μm) gave a reasonable fit to a straight line through the origin based on either the dynamic or the quasi-static relations for the effect of target properties on E_R. The agreement between the two results indicates that the hardness is not a sensitive indicator of the specific elastic–plastic fracture model which is operative, as had also been found by Wiederhorn and Hockey [35] for a wide range of monolithic ceramics.

Another possible contribution to the decrease in E_R with increase in d_{WC} beyond the maximum is that, as the particles become larger, their total perimeter length decreases, thereby reducing the contribution of the "edge effect" [24] to the erosion rate. Kosel and Aptekar [24] concluded that the decrease in erosion rate with increase in size of the hard second-phase particles in a 70wt.%Cu–30wt.%Al_2O_3 alloy resulted from reduction in material removal by the breaking off of the edges of the alumina particles, *i.e.* to an "edge effect".

The velocity exponent n for normal impact of the WC–Co alloys in the brittle regime ranged between 2.6 and 3.7, tending to increase with increasing WC grain size (see Table 2 for $D = 63$ μm). The elastic–plastic indentation fracture models for erosion of brittle materials based on dynamic impact conditions [32–34] lead to $n = 3.2$; those based on quasi-static impact conditions [22] give $n = 2.4$. The decrease in n with reduction in carbide grain size can be interpreted to indicate that, when the impact is fully absorbed by a WC grain, dynamic impact conditions prevail whereas, as the WC grain size is decreased, some of the impact is cushioned by the surrounding matrix and quasi-static conditions are approached. Moreover, as mentioned above, at the smallest WC grain size the erosion mode may contain a significant ductile component even for the smallest erodent particle size considered ($D = 63$ μm).

Regarding the influence of erodent particle size, both the dynamic and the quasi-static elastic–plastic fracture erosion models predict a $D^{2/3}$ effect of impacting particle size on the erosion rate, expressed in grams per gram, *i.e.* a particle size exponent $m \approx 0.7$. A log–log plot of E_R *vs.* D for specimen C in the brittle mode is given in Fig. 10. For $v = 50$ m s^{-1}, the erosion rate for the two larger particle sizes (140 and 270 μm) can be considered to follow the $D^{2/3}$ particle size effect. However, E_R for the 63 μm particles is

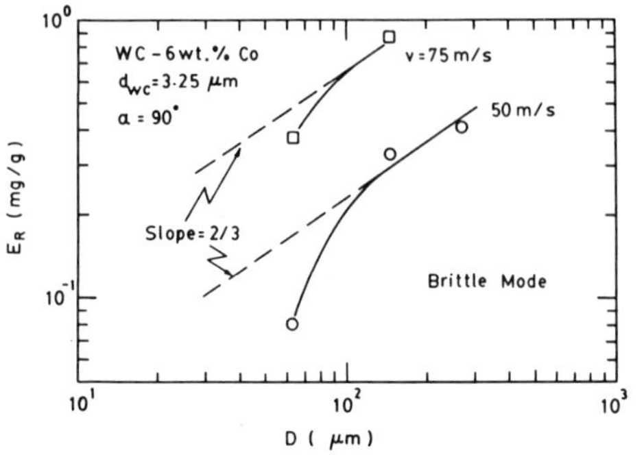

Fig. 10. Log–log plot of the erosion rate in the brittle mode *vs.* the impacting particle size D at two velocities.

lower than expected from an extrapolation based on the larger particles. The effect of particle size observed here is similar to that noted by Routbort and coworkers [36–39] for single-crystal silicon and reaction-bonded SiC. They interpreted the behavior in terms of a particle size threshold, which was being approached by the smaller erodent particles. They also noted that the velocity exponent for the erosion of silicon decreased from 3.6 to 2.5 as the impacting particle size was increased from 40 to 270 μm and proposed [38] that the decrease in n with increasing D may have resulted from localized heating on impact. The heating at the larger particle sizes was considered to promote the fully plastic deformation required for quasi-static indentation to occur. Hence, a contribution to the decrease in n with increase in D indicated in Table 2 for specimen C at $\alpha = 90°$ may have been the increase in localized heating on impact which occurs as D becomes larger.

4.4. X-ray diffraction results

The diffraction data in Table 3 are for the ductile erosion mode in specimen A and the brittle mode in specimen C. In both modes, plastic deformation of the WC grains occurred, although to a lesser degree in specimen C with the larger grains. Plastic deformation of the carbide grains (and the binder) reduced the residual compressive stress on the WC grains produced during fabrication, thereby making them more prone to cracking, which is especially important in the brittle erosion degree.

4.5. Fragmentation of erodent particles

In the ductile regime, fragmentation of the impacting particles was concluded to reduce the fraction of the kinetic energy available to form the impact crater. It can also lead to an increase in erosion rate with increase in particle size [31]. In the brittle regime, fragmentation may have contributed to the decrease in the velocity exponent n with increase in D.

Since the degree of fragmentation for a given set of impact conditions was independent of target material and microstructure, it is concluded that the changes in erosion behavior resulting from microstructural variations were not influenced by any fragmentation of the erodent particles which may have occurred.

5. Summary and conclusions

The following is a summary of the results obtained during this investigation into the erosion of WC–6wt.%Co alloys by Al_2O_3 particles and the conclusions derived therefrom.

(1) The erosion behavior depends on the relative size of the impact crater with respect to the WC grains. When the crater encompasses less than 10 WC grains, the behavior is of the brittle type; when it encompasses more than 100 WC grains, it shifts towards the ductile type.

(2) Material damage in the brittle mode is mainly by brittle fracture of the WC grains; that in the ductile mode is mainly by plastic deformation and fracture of the binder phase, the WC grains remaining relatively intact.

(3) The erosion rate in the ductile mode is in good accord with a modification of the model for ductile erosion based on plastic indentation; that in the brittle mode is in reasonable accord with elastic–plastic fracture indentation models.

(4) The effects of microstructure of the target on the erosion rate in the ductile mode is mainly through the hardness; that in the brittle mode is more complex and includes the fracture toughness and hardness.

(5) Some plastic deformation of the WC grains occurs in both erosion modes.

(6) Fragmentation of erodent particles increased with particle size and velocity, but was independent of target microstructure.

Acknowledgments

The authors gratefully acknowledge support of this research by the U.S. Department of Energy under Grant DE-FG05-84ER45115.

They also wish to thank the Kennametal Corporation for providing the WC–Co specimens. Thanks are also due to Dr. Y. W. Shin for his contributions in the early stages of this investigation.

References

1 A. A. Dankin, *Poroshk. Metall., 94* (1970) 73.
2 Kh. Uuemyis, L. Kleis, V. Tumanov and J. Tildermann, *Poroshk. Metall., 94* (1974) 185.
3 J. S. Hansen, in W. F. Adler (ed.), *Erosion, Prevention and Useful Applications, ASTM Spec. Tech. Publ. 664,* 1980, p. 148.
4 H. Conrad, D. McCabe and G. A. Sargent, in R. K. Vishanadham, D. F. Rowcliffe and J. Gurland (eds.), *Science of Hard Materials,* Plenum, New York, 1983, p. 175.
5 H. Conrad, Y. W. Shin and G. A. Sargent, in N. R. Cousins and J. B. Clark (eds.), *Specialty Steels and Hard Metals,* Pergamon, New York, 1983, p. 423.
6 G. L. Sheldon and I. Finnie, *J. Eng. Ind. B, 88* (1966) 387.
7 I. Finnie, in E. R. Nateson (ed.), *Corrosion–Erosion Behavior of Materials,* Metallurgical Society of AIME, Warrendale, PA, 1980, p. 118.
8 K. Anand, C. Morrison, R. O. Scattergood, H. Conrad, J. L. Routbort and R. Warren, in E. A. Almond, C. A. Brookes and R. Warren (eds.), *Proc. 2nd Int. Conf. on the Science of Hard Materials, Rhodes, September 23–28, 1984,* in *Int. Phys. Conf. Ser., 75* (1986) 949.
9 K. Anand and H. Conrad, Microstructural and Scaling Effects in the Damage of WC–Co Alloys by Single Impacts of Hard Particles, submitted to *J. Mater. Sci.*
10 S. K. Hovis, J. E. Talia and R. O. Scattergood, *Wear, 108* (1986) 139.
11 Y. W. Shin, G. A. Sargent and H. Conrad, *Metall. Trans. A, 18* (1987) 437.
12 H. C. Lee and J. Gurland, *Mat. Sci. Eng., 33* (1978) 125.
13 T. Hare and J. C. Russ, North Carolina State University, personal communication, 1986.
14 K. Anand, S. K. Hovis, H. Conrad and R. O. Scattergood, *Wear, 118* (1987) 243.
15 Y. Shin and H. Conrad, in H. Conrad and R. Scattergood, (eds.), Microstructural effects in solid particle erosion, *Prog. Rep. for the period June 1, 1984 to January 1, 1985,* 1985 (U.S. Department of Energy Grant DE-FG05-84ER45115).
16 R. Warren and H. Matzre, in R. K. Viswanadham, D. J. Rowcliffe and J. Gurland (eds.), *Science of Hard Materials,* Plenum, New York, 1983, p. 563.
17 A. K. Hovis, K. Anand, H. Conrad and R. O. Scattergood, *Wear, 101* (1985) 69.
18 A. G. Evans and T. R. Wilshaw, *Acta Metall., 24* (1976) 939.
19 I. M. Ogilvy, C. M. Perrott and J. W. Suiter, *Wear, 43* (1977) 239.
20 B. J. Hockey, S. M. Wiederhorn and H. Johnson, in R. Bradt, D. Hasselman and F. Lange (eds.), *Fracture Mechanics of Ceramics,* Vol. 3, *Flaws and Testing,* Plenum, New York, 1978, p. 379.
21 S. M. Widerhorn and Lawn, *J. Am. Ceram. Soc., 62* (1979) 66.
22 A. W. Ruff and S. M. Wiederhorn, *Treatise Mater. Sci. Technol., 16* (1979) 69.
23 A. L. Levy, in J. E. Field and N. S. Corney (eds.), *Proc. 6th Int. Conf. on Erosion by Liquid and Solid Impact, Cambridge, 1983,* Cavendish Laboratory, Cambridge, 1983, p. 39-1.
24 T. H. Kosel and S. S. Aptekar, *Corrosion '86,* March 17–21, 1986, Paper 113.
25 I. M. Hutchings, *Wear, 70* (1981) 269.
26 G. Sundararajan and P. G. Shewmon, *Wear, 84* (1983) 237.
27 D. Tabor, *The Hardness of Metals,* Clarendon, Oxford, 1951.
28 I. M. Hutchings, *Proc. 5th Int. Conf. on Erosion by Liquid and Solid Impact, Cambridge,* Cavendish Laboratory, Cambridge, 1979, p. 36-1.
29 E. A. Almond and R. Roebuck, in R. K. Viswanadham, D. J. Rowcliffe and J. Gurland (eds.), *Science of Hard Materials,* Plenum, New York, 1983, p. 597.
30 I. M. Hutchings, R. E. Winter and J. E. Field, *Proc. R. Soc. London, Ser. A, 348* (1976) 379.
31 G. P. Tilly and W. Sage, *Wear,* 16 (1968) 447.
32 A. G. Evans, M. E. Gulden and M. E. Rosenblatt, *Proc. R. Soc. London, Ser. A, 361* (1978) 343.
33 A. G. Evans, in R. C. Bradt, D. Hasselman and F. Lange (eds.), *Fracture Mechanics of Ceramics,* Vol. 3, Plenum, New York, 1978, p. 303.
34 A. G. Evans, *Treatise Mater. Sci. Technol., 16* (1979) 1.
35 S. M. Wiederhorn and B. J. Hockey, *J. Mater. Sci., 18* (1983) 766.
36 J. L. Routbort, R. O. Scattergood and H. P. L. Turner, *Wear, 59* (1980) 363.
37 J. L. Routbort, R. O. Scattergood and E. W. Kay, *J. Am. Ceram. Soc., 63* (1980) 635.
38 R. O. Scattergood and J. L. Routbort, *Wear, 67* (1981) 227.
39 R. O. Scattergood and J. L. Routbort, *J. Am. Ceram. Soc., 66* (1983) C-184.

Materials Science and Engineering, A105/106 (1988) 423–428

A Comparison of the Strength of WC–Co Measured by Ring and Transverse Rupture Strength Specimens*

F. R. VANDEPUT and N. MASTRANTONIS

De Beers Diamond Research Laboratory, Johannesburg (South Africa)

(Received November 9, 1987)

Abstract

The strength of WC–Co was measured using three-point bending specimens (transverse rupture strength) and ring specimens. The strength values obtained by the two tests were compared. The stress distribution in a ring specimen is similar to that in a three-point bending specimen, and a ring test can thus be regarded as a form of a transverse rupture test. However, tensile strength values obtained from ring specimens are lower than those obtained from three-point bending specimens and this discrepancy is more pronounced for specimens with a low cobalt content. This can be attributed largely to the size effect, the ring specimens having a much larger volume than the three-point bending specimens.

1. Introduction

Components made of cemented tungsten carbide (WC–Co) are used extensively in the synthesis of industrial diamond. The mechanical and thermal stresses in these components are quite high, often resulting in premature catastrophic failure. It is thus essential to develop reliable failure criteria in order to predict the service performance of these materials. The inherent brittleness of WC–Co necessitates the use of probabilistic design techniques. The Weibull analysis has been extended to multiaxial stresses but none of the available methods is capable of predicting the failure probability of WC components accurately [1, 2]. Numerous tests are now being conducted in an attempt to obtain a suitable failure criterion for WC–Co components. Tests have been conducted on Brazilian, biaxial flexure, ring, tube and three-point bending specimens. In the present paper, we compare the results obtained from the ring specimens with those obtained from conventional transverse rupture strength (TRS) bars (three-point bending test). The stress distribution in the ring specimen is similar to that in the TRS bars. However, the ring specimen has the advantage that failure occurs in the gauge section, this being a relatively large distance from the loading pins, whereas fracture in a TRS bar may be influenced by the position of the loading pin. Rolf and Weyand [3] carried out tensile tests on alumina ring specimens and compared the strength of these rings with the TRS obtained from three-point bending tests on specimens of the same material. It was decided to apply a similar exercise as part of a research project aimed at developing a suitable failure criterion for WC–Co.

2. Experimental techniques

Tests were conducted for seven WC–Co cemented-carbide grades with cobalt content of 3, 6, 9, 13, 15 and 20 wt.%. Test specimens were prepared from cold isostatically pressed slugs which were machined to shape in the green state. The samples were then vacuum sintered at a temperature between 1400 and 1530 °C depending on the cobalt content. The grain size was controlled between 2 and 6 μm. All test specimens were hot isostatically pressed prior to grinding to the final dimensions, with a surface finish with $R_a = 0.2$ μm. The cross-section of the TRS bars was 4 mm × 2 mm and the span length was 15 mm, except for the 6 wt.% Co grade for which specimens with a 40 mm span length were used. No chamfering was applied so that consistency could be retained with the sharp-edged ring specimens. The dimensions of the ring specimen are shown in Fig. 1.

*Paper presented at the 3rd International Conference on the Science of Hard Materials, Nassau, The Bahamas, November 9–13, 1987.

0921-5093/88/$3.50

© Elsevier Sequoia/Printed in The Netherlands

3. Results

Figure 2 shows Weibull plots of the TRS, *i.e.* ln[ln{1/(1 − *P*)}] *vs.* ln(TRS). The failure probability *P* is defined by the relation

$$P = \frac{i - 0.5}{N} \tag{1}$$

where i is the ranking number of the specimen when the specimens are placed in increasing order of strength and N is the total number of specimens tested. The TRS of the specimens is determined from the relation

$$\mathrm{TRS} = \frac{3FL}{2bh^2} \tag{2}$$

where F is the load at fracture, L is the span length, b is the width of the specimen and h is its height. The strength of the ring specimens is defined as the maximum tensile stress at failure. This maximum tensile stress occurs at the inner diameter of the ring in the narrowest part of the gauge section.

Median strength values obtained for different cobalt contents of the TRS and ring specimens are shown in Table 1 and Fig. 3. The TRS values given in Table 1 are slightly larger than the actual values because of the wedging effect resulting from the small span-to-height ratio used for the specimens [4]. Owing to the wedging effect, eqn. (2) overestimates the TRS by an amount

$$e = \frac{2F}{\pi bh} \tag{3}$$

This error amounts to about 2% for the long specimens (40 mm span length) and about 5.5%

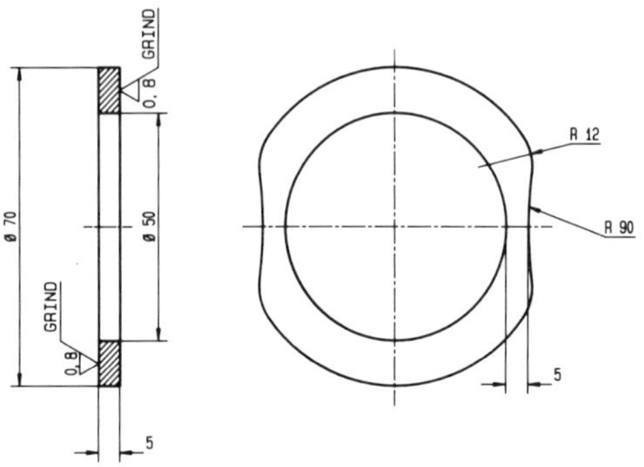

Fig. 1. Dimensions of the ring specimen.

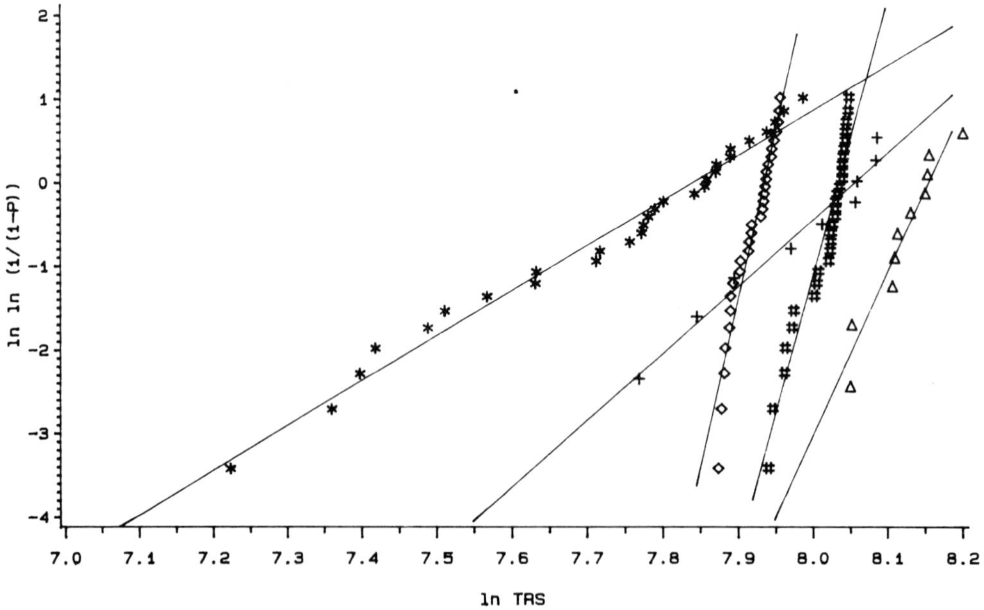

Fig. 2. Weibull plots of TRS for cobalt contents of 3 (✱), 6 (+), 9 (△), 15 (◇) and 20 (#) wt.%.

TABLE 1 Strength of WC–Co as measured by transverse rupture strength and ring specimens

Cobalt content[a] (wt.%)	*Failure stress in ring specimen*[b] ($N\ mm^{-2}$)	*Failure stress in three-point bending*[b] ($N\ mm^{-2}$)	*Size of TRS specimen* (mm × mm × mm)	*m*	σ_0[c] ($N\ mm^{-2}$)
3	—	2401.0	15×4×2	5.5	2529.0
6	1616.6	3015.0	40×4×2	8.4	3081.0
9	1964.8	3367.0	15×4×2	20.7	3449.0
13	3294.7	—	—	—	—
15	2636.8	2781.0	15×4×2	40.9	2791.0
20	2874.9	3069.0	15×4×2	33.5	3080.0

[a]Stoichiometric carbide (hot isostatically pressed).
[b]Median values.
[c]Preliminary values. Further calculations are being made at present and these results will be published at a later date.

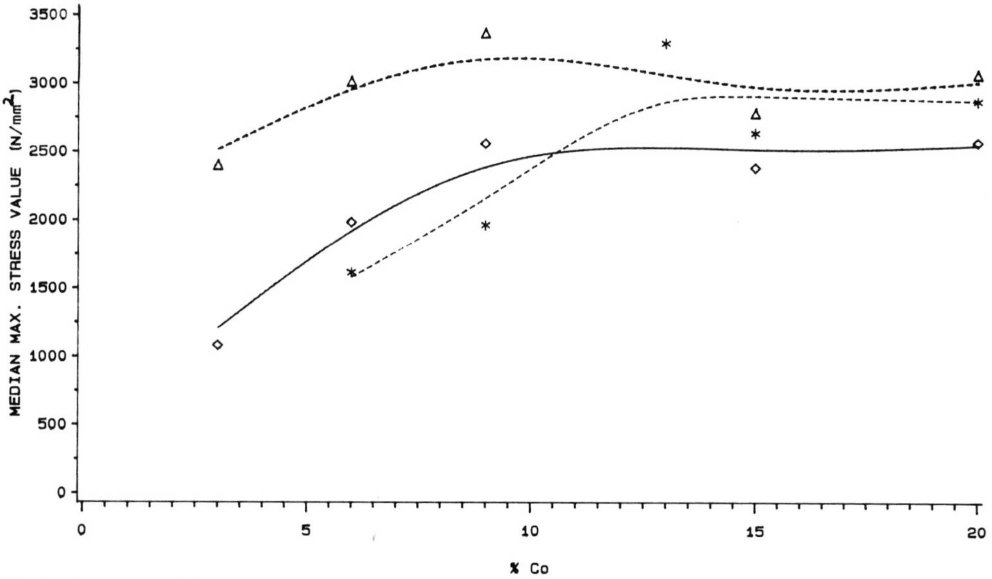

Fig. 3. Comparison of the strength of WC–Co measured by the three-point bending test and the ring specimen. The bold full line represents the calculated strength of the ring specimen. (△, TRS sample; ✱, ring sample (measured); ◇, ring sample (calculated).)

for the short specimens (15 mm span length). The Weibull modulus *m* and preliminary values for the characteristic strengths σ_0 derived from the TRS tests are also included in Table 1. It should be noted that longer specimens were used for the 6 wt.%Co specimens. A correction factor was thus used in order to compensate for size effects. The correcting relation can be expressed as [5, 6]

$$\frac{\sigma_1}{\sigma_2} = \left(\frac{V_2}{V_1}\right)^{1/m} \qquad (4)$$

where V_1 is the actual volume of the specimen, σ_1 is the measured strength and σ_2 is the corrected strength corresponding to a volume V_2. Thus the corrected value for the 6 wt.% Co TRS test is 3388 $N\ mm^{-2}$ and this is the result shown in Fig. 3. This volume correction is valid as post-fracture fractography revealed that volume flaws were the major cause of fracture. Figure 4 shows typical volume flaws at the fracture initiation site.

4. Discussion

The tensile strength of WC–Co measured by the ring specimens is lower than that obtained by the three-point bending specimens (Table 1 and Fig. 3). This discrepancy is very large for the specimens with a low cobalt content but is less

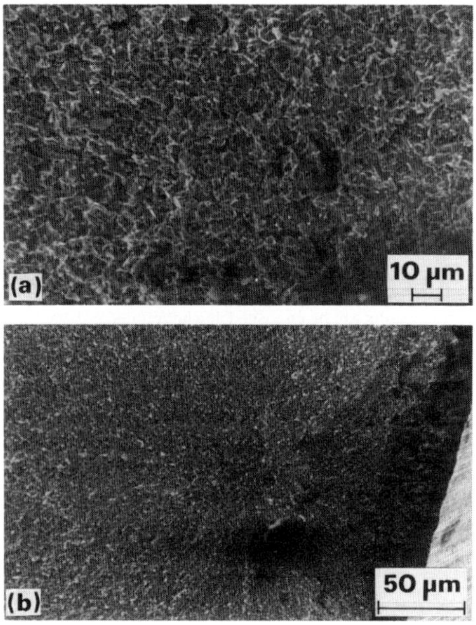

Fig. 4. (a) Pores at the fracture initiation site in the WC–15wt.%Co specimen; (b) η phase in the WC–3wt.%Co specimen.

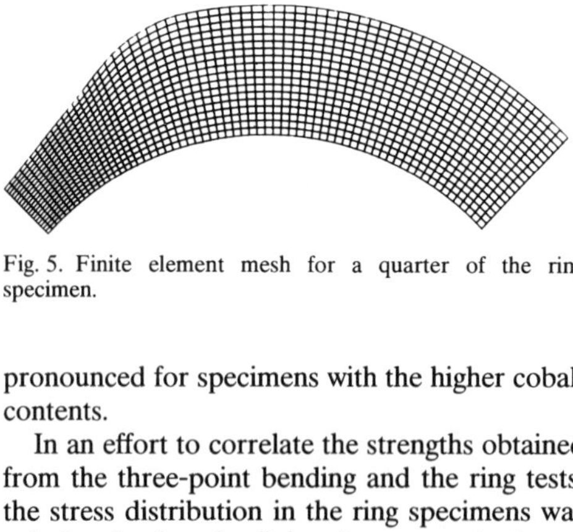

Fig. 5. Finite element mesh for a quarter of the ring specimen.

pronounced for specimens with the higher cobalt contents.

In an effort to correlate the strengths obtained from the three-point bending and the ring tests, the stress distribution in the ring specimens was calculated using the finite element method, and Weibull's formulation of the risk of rupture was used to predict the strength of the ring samples using data obtained from the three-point bending tests. The finite element mesh used in the stress analysis is shown in Fig. 5. A two-dimensional plane stress analysis was performed. Only a quarter of the ring specimen was modelled to take advantage of symmetry. The load applied by the loading pin on the specimen was modelled by a

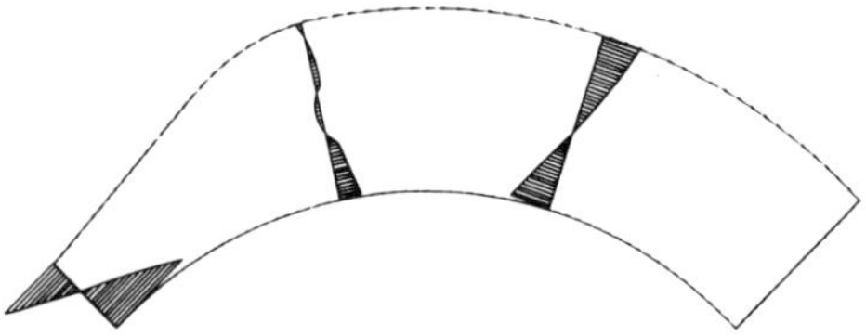

Fig. 6. Longitudinal stress distribution in the ring specimen.

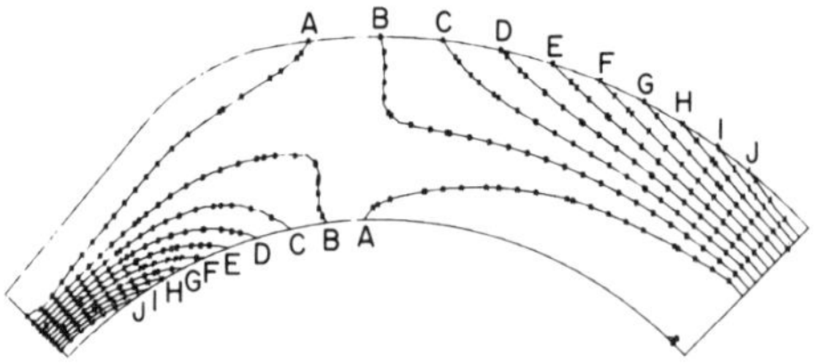

Fig. 7. Maximum principal stress distribution in the ring specimen for a load of 1000 N: A, 3.730; B, 14.90; C, 26.10; D, 37.40; E, 48.60; F, 59.80; G, 71.10; H, 82.30; I, 93.50; J, 104.0.

single point load. This leads to a localized stress concentration in the point where the load is applied but the overall stress distribution in the specimen is not affected. The stress analysis revealed that the stress distribution in the ring specimen is very similar to that in a three-point bending specimen, *i.e.* direct stresses acting in the longitudinal direction and a large stress gradient across the width of the specimen, with one side of the specimen under tension and the opposite side under compression (Figs. 6 and 7). The predominant stress component in both specimens is the stress in the longitudinal direction with only a small shear stress superimposed. The ring test can therefore be regarded as a form of a transverse rupture test.

To correlate the strength values obtained by the two tests, the one-dimensional Weibull [7] theory is used. In general, the probability of failure of a material that exhibits a scatter in its strength is given by

$$P = 1 - \exp(-R) \tag{5}$$

with R, the risk of rupture, defined as

$$R = \int_V \mathrm{d}V \int_0^\sigma g(\sigma)\,\mathrm{d}\sigma \tag{6}$$

where V is the volume of the specimen and $g(\sigma)$ is the strength distribution of the material [8, 9].

For uniaxial stress states, Weibull assumed a strength distribution given by

$$\int_0^\sigma g(\sigma)\,d\sigma = \left(\frac{\sigma}{\sigma_0}\right)^m \tag{7}$$

and the risk of rupture can then be calculated from

$$R = \int_V \left(\frac{\sigma}{\sigma_0}\right)^m dV \tag{8}$$

Several researchers have used the one-dimensional Weibull formulation to calculate the risk of rupture of three-point bending specimens [1, 5] using the stress in the longitudinal direction as the single variable. This longitudinal stress component is given by

$$\sigma(x, y) = \frac{6Fxy}{bh^3} \tag{9}$$

where x is the distance from the outer support and y is the distance to the neutral fibre. Substituting eqn. (9) into eqn. (8) and integrating over the volume under tension gives the following expression for the risk of rupture of a three-point bending specimen:

$$R = \frac{bhL}{2}\frac{1}{(m+1)^2}\left(\frac{\sigma_T}{\sigma_0}\right)^m \tag{10}$$

where σ_T is the maximum stress in the specimen as calculated from eqn. (2). In the derivation of eqn. (10), it was implicitly assumed that the longitudinal stress is the only stress component in the specimen, so that the uniaxial Weibull formulation can be used. This also implies that fracture of the specimen is governed by the value of the longitudinal stress. This is an approximation. In fact, a biaxial stress state exists in the three-point bending specimen. Apart from the longitudinal stress, there is also a shear stress present given by

$$\tau_{xy} = \frac{3F}{4bh^3}(h^2 - 4y^2) \tag{11}$$

The corresponding principal stresses are given by

$$\sigma_1 = \frac{\sigma(x, y)}{2} + \left\{\frac{\sigma(x, y)^2}{4} + \tau_{xy}^{\ 2}\right\}^{1/2} \tag{12}$$

$$\sigma_2 = \frac{\sigma(x, y)}{2} - \left\{\frac{\sigma(x, y)^2}{4} + \tau_{xy}^{\ 2}\right\}^{1/2} \tag{13}$$

It is clear from eqns. (11)–(13) that a uniaxial stress state only exists at the top and bottom face of the specimen, where $\tau_{xy} = 0$.

However, the tensile stresses are highest at the bottom face of the specimen where the uniaxial stress state exists and these stresses contribute most to the overall risk of rupture of the sample. At the neutral fibre, the two principal stresses have the same absolute value; so a fully biaxial stress state exists in the vicinity of the neutral fibre. However, these stresses are very small compared with the stress at the tensile face and their contribution to the risk of rupture is negligible. Therefore, it can be concluded that eqn. (10) gives a reasonably accurate estimation of the risk of rupture.

The strength of the ring specimens can now be predicted from the value of the TRS as follows. The risk of rupture of the three-point bending specimen is calculated for a given maximum stress σ_T. We then calculate the loading on the ring specimen which gives the same risk of rupture in the ring as in the TRS bar. The maximum stress in the gauge section corresponding to this load is the predicted strength in the ring. Rolf and Weyand [3] used the Barnett–Freudenthal approach for multiaxial stress states to calculate the risk of rupture of the rings. However, since the stress distribution in the rings is similar to that in the bending specimens, the same approximations can be made to calculate the risk of rupture, *i.e.* only the longitudinal stress component is taken into account. It is easy to verify that this longitudinal stress is the sum of the two principal stresses σ_1 and σ_2, which were calculated using the finite element method. In the case of the ring, the calculation of the risk of rupture, using eqn. (8), requires numerical integration and no closed formula can be derived. However, the risk of rupture is given by an expression of the form

$$R = f(\text{geometry}, m)\left(\frac{\sigma_R}{\sigma_0}\right)^m \tag{14}$$

where f depends on the geometry of the ring and m, and σ_R is the maximum tensile stress in the gauge section of the ring specimen. Equations (10) and (14) were used to predict the median values of the strength in the ring specimens from the median TRS. The predicted strength is shown in Fig. 3 and the values agree well with the measured strength for the low cobalt contents. For example, the TRS value of WC–6wt.%Co is almost twice as high as the strength measured by the ring specimens but the difference between the measured ring strength and the value predicted from the TRS value is only 18%. For higher cobalt contents the strength of the ring specimens is underestimated somewhat. However, in view of the relatively large scatter on the experimental data and the fact that the TRS and ring specimens were produced from different batches of powder, the agreement between the measured and predicted strength values for the rings is reasonable. Therefore, it can be concluded that the discrepancy between the strength values obtained by the two test methods is largely caused by the difference in volume, the ring specimens having a much larger volume under a high stress than the three-point bending specimens do.

In the previous derivations, it was implicitly assumed that failure is caused by volume flaws rather than by surface flaws. Hence the risk of failure was calculated by integration over the volume rather than over the surface area of the specimen. The fracture surface of each TRS bar was investigated and it was observed that fracture was initiated at internal flaws such as pores and inclusions or in the vicinity of the tensile face [10]. The fractures at or near the tensile face can still be regarded as volume failures on condition that fracture is caused by volume flaws. Figure 2 shows that there is only one population in the Weibull distribution, indicating that all samples fail in a similar way. The occurrence of fractures at or near the surface must therefore be attributed to the fact that the tensile face is the most highly stressed zone of the specimen rather than to the existence of surface flaws with a lower resistance to fracture than the volume flaws.

5. Conclusions

The strength of ring specimens was compared with the strength obtained from the more conventional TRS specimens. The strength values obtained by the two tests agree well for the samples with a high cobalt content (15 and 20 wt.% Co) but a large discrepancy exists for the lower cobalt contents. In order to correlate the strength values obtained by the two tests, Weibull analysis was used to predict the strength of the rings using the TRS value and the values of the Weibull modulus and the characteristic strength obtained by the three-point bending tests. The predicted values agree well with the measured values. The analysis underestimates the strength of the ring for the high cobalt contents. However, the large discrepancy between the strength values of the specimens with a low cobalt content can be explained adequately by the Weibull analysis. This suggests that the discrepancy can be attributed largely to the difference between the volumes of the TRS and ring specimens.

Acknowledgments

The authors are grateful to L. Smith for carrying out the tests and to Dr. S. Luyckx (University of the Witwatersrand, Johannesburg) for her scanning electron microscopy work and for many useful discussions.

References

1 D. K. Shetty, A. R. Rosenfield, W. H. Duckworth and P. R. Held, *J. Am. Ceram. Soc., 66* (1983) 36.
2 D. L. Hartsock, *Trans. ASME, 85-GT-100* (1985).
3 R. L. Rolf and J. D. Weyand, *Am. Ceram Soc. Bull., 64* (1985) 1360.
4 S. Timoshenko and J. N. Goodier, *Theory of Elasticity* McGraw-Hill, London, 1951.
5 P. M. Braiden, *Mater. Eng., 2* (1980) 73.
6 G. K. Bansal, W. H. Duckworth and D. E. Niesz, *J. Am. Ceram. Soc., 59* (1976) 472.
7 W. Weibull, *Ingenioersvetenskapsakad., Handl., 151* (1939) 45.
8 J. Lamon, *Rev. Int. Hautes Temp. Réfract., 22* (1985) 115.
9 A. G. Evans and R. L. Jones., *J. Am. Ceram. Soc., 61* (1978) 156.
10 S. Luyckx, personal communication, 1987.

Materials Science and Engineering, A105/106 (1988) 429–433

Microstructure and Wear of TiC–Cubic BN Tools*

R. M. HOOPER, J. I. SHAKIB and C. A. BROOKES†

Department of Engineering Science, University of Exeter, North Park Road, Exeter EX4 4QF (U.K.)

(Received December 7, 1987)

Abstract

Previous investigations of the wear of cubic boron nitride aggregate tools have shown that this is predominantly a chemical process involving the interaction of the tool with its environment and is restricted by the formation of protective layers on the exposed faces of the tool. Recently, the performance of such tools used in fine cutting operations has been improved by the addition of other components to the aggregate. The microstructure of these new composite tools has been investigated by selective etching and scanning electron microscopy. The wear features of a number of modified and unmodified tools used in fine cutting tests under identical conditions have been compared and the results have been interpreted in terms of the existing models for the wear of cubic-boron-nitride-based materials generally and the effects of the additives in the modified tools.

1. Introduction

The use of cubic boron nitride (BN) aggregates in the machining of hard materials is well established. Such tools retain high hardness at high temperatures [1] and hence permit the use of high stock removal rates on otherwise difficult-to-machine materials such as alloy steels and hard irons. This superior performance is not retained in fine or finishing cut operations; recently the wear resistance and performance of such tools under these conditions has been enhanced by the addition of other components to the aggregate at the synthesis stage. In this way the development of cubic BN tools parallels that of carbide tools in the fine tuning of their composition and microstructure to meet specific service criteria. In this paper, we present the results of an investigation of the microstructure and wear of these modified tools in comparison with existing tools based on cubic BN alone. The mechanism of their wear is also interpreted in terms of the existing models for the wear of cubic-BN-based materials.

2. Microstructure

Commercially available modified cubic BN tools were sectioned and polished using standard metallographic techniques to reveal the structure shown in Fig. 1. Light element energy-dispersive analysis of the phases observed showed that the dark areas were BN and the light areas were titanium carbide (TiC). In addition, some aluminium compounds were found to be present, mainly

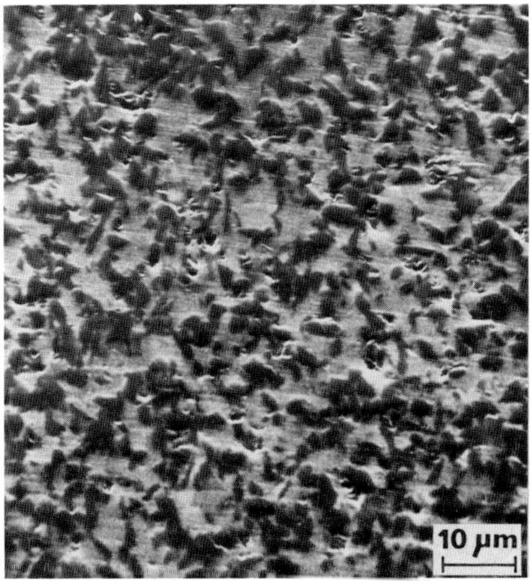

Fig. 1. Polished section of TiC–cubic BN aggregate tool.

*Paper presented at the 3rd International Conference on the Science of Hard Materials, Nassau, The Bahamas, November 9–13, 1987.

†Present address: Department of Engineering Design and Manufacture, University of Hull, Cottingham Road, Hull, HU6 7RX, U.K.

0921-5093/88/$3.50

© Elsevier Sequoia/Printed in The Netherlands

associated with the cubic BN particles. Traces of tungsten carbide (WC) were found in the TiC phase. This, together with X-ray diffraction analysis confirmed the composition to be 50 wt.% cubic BN, 40 wt.% TiC, 6 wt.% WC and 4 wt.% aluminium nitride and diboride. The size of the cubic BN particles was found to be of the order of 1–2 μm, in contrast with the 10 μm particle size of a conventional aggregate. The contiguity of the ultrahard phase was also generally lower, although in some cases the particles were observed to cluster together in groups. In order to investigate the structure of the individual components, the techniques of selective etching used in other aggregates of this type have been employed. In particular the cubic BN may be selectively etched by atmospheric oxidation [2] at approximately 900 °C. The results of such etching are shown in Fig. 2 where the cubic BN has been completely removed, revealing the oxidation-resistant structure of the TiC. Conversely, non-concentrated hydrofluoric or sulphuric acids may be used to leach out the TiC to reveal the structure of the cubic BN particles (Fig. 3). The morphology of these particles indicates that they have been much less deformed during synthesis than those in a conventional aggregate. Since the development of defect structures in the cubic BN can play a significant role in the wear process [3], the absence of evidence of plastic deformation, and the lower defect density in the modified material may therefore contribute to its increased wear resistance. These observations are summarized in Table 1.

4. Wear

The wear of modified materials and that of unmodified materials were compared by examination of tools used in machinability tests on a range of hard alloys including EN31, M2 and SKD11. Machining tests were carried out under identical "finishing" conditions in each case, as shown in Table 2.

After 20 min the tool inserts were removed and examined by scanning electron microscopy. The results illustrate the superior performance of

Fig. 2. Oxidation-resistant skeleton of TiC in modified tool revealed by oxidation etching of cubic BN.

Fig. 3. Cubic BN particles in TiC-modified tool exposed by extended etching.

TABLE 1 Microstructure of cubic BN aggregates

	Composition	*Particle size* (μm)	*Contiguity*	*Morphology*
Unmodified	Cubic BN	10	High	Deformed
Modified	Cubic BN + TiC	1	Low	Undeformed

TABLE 2 Fine cutting tests

Material	*Rockwell C hardness* (HRC)	*Composition* (wt.%)											
		C	*Si*	*Mn*	*Mo*	*W*	*V*	*Cu*	*Ni*	*Cr*	*S*	*P*	*Fe*
EN31	62	1	0.2	0.5	—	—	—	—	—	—	0.05	0.05	Balance
M2	62	0.9	—	—	5	6	2	—	—	—	—	—	Balance
SRD11	60	1.5	0.3	0.4	0.8	—	0.3	0.04	0.1	11.9	—	0.02	Balance

Surface speed, 120 m s^{-1}; time of cutting, 20 min; depth of cut, 0.25 mm; top rake; − 8°; dry cutting; feed, 0.1 mm rev^{-1}.

the modified tools under these conditions and a comparison of the wear scars on these and the unmodified tools leads to an interpretation of the wear process in terms of that proposed for cubic BN tools generally [2]. Essentially, the wear of cubic BN tools involves the following processes.

(1) 'Chemical wear' produced by a chemical interaction of the tool with its environment, including the workpiece and atmospheric oxidation. This may be exacerbated by the development of defect structures in the tool material. It has been shown [4] that, in common with other covalent materials, a threshold temperature (1200 K) exists for cubic BN, above which highly localized deformation introduced at lower temperatures expands to form extensive dislocation structures, the increased chemical activity of which contributes to the degradation of the tool by a process referred to as etching wear.

(2) Under favourable circumstances a thin protective layer of similar composition to the workpiece forms on the surfaces of the tool in contact with the work and chip.

(3) The stability of this protective layer then becomes the controlling influence on the wear of the tool. At lower temperatures and/or cutting speeds the layer becomes unstable with respect to the tool and is lost, exposing the surfaces of the tool to further chemical wear. The wear of the rake face is also increased by adhesion of the layer to the base of the chips, and hence enhanced wear by attrition as fragments of the tool are pulled away with the layer.

The operation of these processes is illustrated by the wear scars on the tools used in the cutting tests. Figure 4 shows an unmodified tool used in the fine cutting of EN31. No protective layer has formed and the chemical wear has produced severe cratering of the rake face and grooving of the nose and clearance faces. The chemical wear within the crater has produced an etched structure (Fig. 5 showing a detail of Fig. 4), revealing the cubic BN particles A and the filler phase B. In

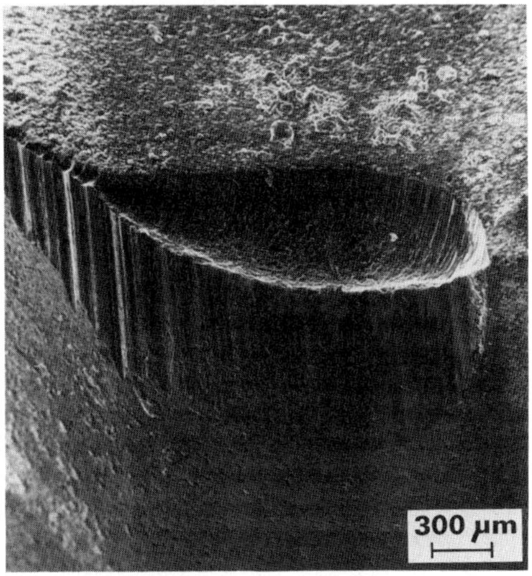

Fig. 4. Chemical wear of unmodified cubic BN tool used in fine cutting test on EN 31.

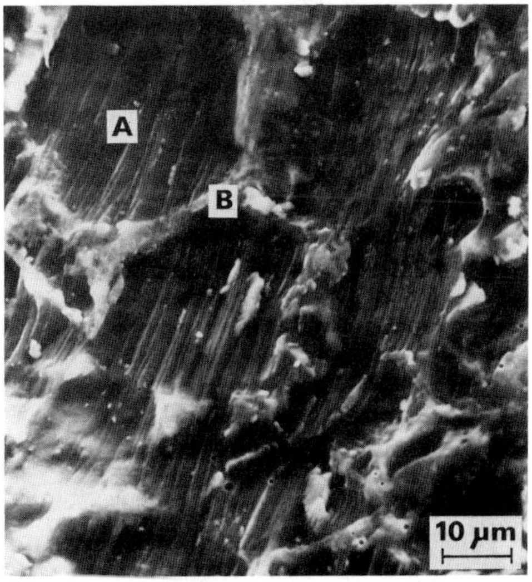

Fig. 5. Detail of crater in Fig. 4, showing etched cubic BN grains (A) and filler (B).

Fig. 6. Wear scar on unmodified aggregate tool used in fine cutting test on M2H.

Fig. 7. Wear scar on unmodified aggregate tool used in fine cutting test on SKD 11.

a similar tool used in the cutting of M2 and shown in Fig. 6 a layer has formed but was unstable, and its periodic removal by adhesion to the chip has contributed to the wear of the tool by attrition. Similar features can be observed on the tool used with SKD11 in Fig. 7. In contrast, examination of the modified tools from exactly similar operations (Figs. 8–10) shows greatly reduced wear scars and evidence of a stabilized protective layer structure.

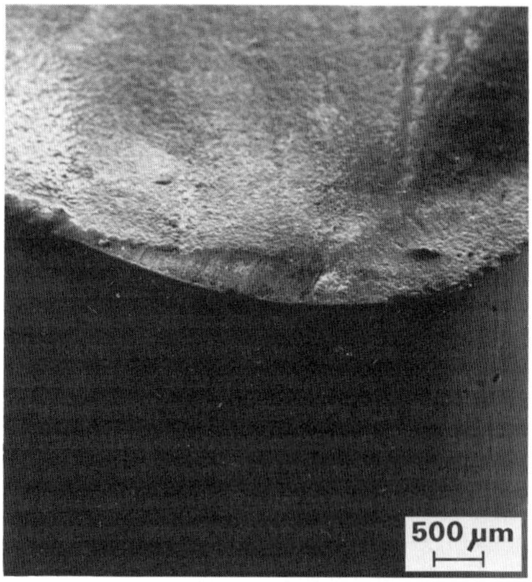

Fig. 8. Wear scar on TiC-modified cubic BN aggregate tool used in fine cutting test on EN 31.

Fig. 9. Wear scar on TiC-modified cubic BN aggregate tool used in fine cutting test on M2H.

4. Interpretation and conclusions

These observations lead to an interpretation of the wear processes in terms of the existing model for cubic-BN-based materials referred to and outlined above. Although detailed modelling of a particular cutting situation is extremely difficult, it has been shown [5, 6] that a reduction in the

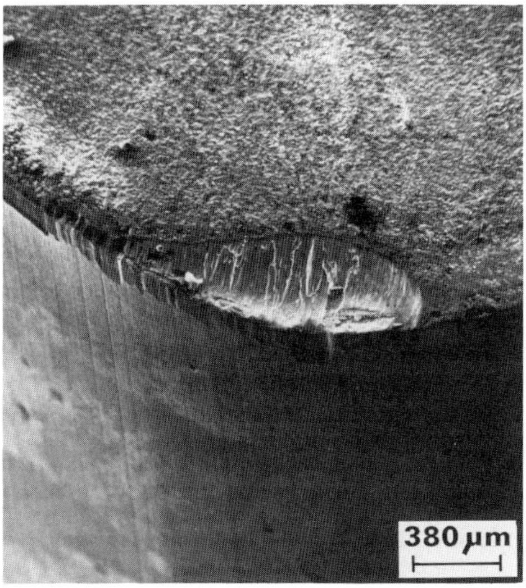

Fig. 10. Wear scar on TiC-modified cubic BN aggregate tool used in fine cutting test on SKD 11.

depth of cut leads to a corresponding decrease in the temperature of cutting. Thus, it is proposed that, in fine cutting operations with unmodified cubic BN tools, the temperatures developed in the tool, although high enough to cause chemical wear, are insufficient to stabilize the protective layer which, rather than protecting the tool from the chemical wear, can enhance it by a process of attrition as the layer breaks away with the chip. In the modified tools the cutting temperature is increased by the lower thermal conductivity obtained by the inclusion of the TiC (44 W m^{-1} K^{-1} as opposed to 100 W m^{-1} K^{-1} in unmodified material). This high cutting temperature then re-establishes the stability of the protective layer in a self-regulatory way.

In addition, the presence of the TiC increases the oxidation resistance of the material considerably and reduces its susceptibility to chemical wear processes generally.

References

1 C. A. Brookes, in E. A. Almond, C. A. Brookes and R. Warren (eds.), *Proc. 2nd Int. Conf. on the Science of Hard Materials, Rhodes, September 23–28, 1984,* in *Inst. Phys. Conf. Ser., 75* (1986) 207–220.

2 R. M. Hooper and C. A. Brookes, in E. A. Almond, C. A. Brookes and R. Warren (eds.), *Proc. 2nd Int. Conf. on the Science of Hard Materials, Rhodes, September 23–28, 1984,* in *Inst. Phys. Conf. Ser., 75* (1986) 907–917.

3 C. A. Brookes and R. M. Hooper, in P. Daniel (ed.), *Ultrahard Materials Application Technology,* Vol. 3, Hornbeam, London, 1984, pp. 41–48.

4 C. A. Brookes, R. M. Hooper, W. A. Lambert and J. D. J. Ross in P. Daniel (ed.), *Ultrahard Materials Application Technology,* Vol. 2, Hornbeam, London, pp. 72–80.

5 E. M. Trent, *Metal Cutting,* Butterworths, London, 1978.

6 M. C. Shaw, *Metal Cutting Principles, Oxford Series on Advanced Manufacturing No. 3,* Clarendon, Oxford, 1984.

Chemical Vapor Deposition for Surface Conditioning: Opportunities and Cautions*

JOHN M. BLOCHER, Jr.

915 Silvoor Lane, Oxford, OH 45056 (U.S.A.)

(Received January 13, 1988)

Abstract

The important role of chemical vapor deposition (CVD) in modern technology is reviewed in terms of its ability to form high purity (or controlled-composition) deposits, often of unique structure, with high throwing power. An overview of the factors influencing deposition rate and structure is given, with emphasis on the trade-offs that must be considered in achieving useful results most economically. Progress in the modelling of CVD reactors is cited. Characterization of CVD reactions by the rate-controlling factors has been enhanced in recent years by the van den Brekel technique of measuring and of interpreting the deposition profile on the wall of a narrow hole in the substrate, thus optimizing the deposition conditions for a given application. The increasing importance of two-phase CVD products is noted, with warnings of the limitations imposed by kinetic dissimilarity in the competing reactions. Further increase in the importance of CVD is foreseen through attention being paid to the fundamental aspects of this inherently complex technology.

1. Introduction

Chemical vapor deposition (CVD), the thermally activated deposition of solid materials by chemical reaction of gaseous precursors, has progressed in the last 50 years from a laboratory curiosity to an indispensable tool of modern technology, with an economic impact in the billions of dollars. The largest applications of CVD are found in the solid state microelectronics industry. However, wear-resistant coatings for cutting and forming tools are running a close second. CVD-generated optical fibers promise to supplant copper conductors in secure long-distance telephone communication and data transmission. Other applications of CVD abound.

Table 1 lists elements and compounds that can be conveniently deposited by CVD. We can add to this list a number of important compound semiconductors such as gallium arsenide. Table 2 gives representative CVD reactions with the temperatures required for activation and some suggestions for the use of catalysis to modify the course of the reactions. In the case of Al_2O_3 deposition, the water vapor is generally supplied by the reaction $CO_2+H_2 \rightarrow CO+H_2O$ so as to suppress the formation of gas-phase-precipitated Al_2O_3 "snow".

2. Advantages and limitations

The advantages of CVD include its versatility in the range of materials that can be deposited, and the fact that the precursors can be easily purified and their purity maintained to yield high purity deposits. Ability to control doping of such

TABLE 1 Important chemical vapor deposition materials[a]

Metals	Cu, Be, Al, Ti, Zr, Hf, Th, Ge, Sn, Pb, V, Nb, Ta, As, Sb, Bi, Cr, Mo, W, U, Re, Fe, Co, Ni, Ru, Rh, Os, Ir, Pt
Graphite, carbon and carbides	C, B_4C, SiC, TiC, ZrC, HfC, ThC, ThC_2, VC, NbC, Nb_2C, TaC, Ta_2C, CrC, Cr_4C, Cr_7C_3, Cr_3C_2, MoC, Mo_2C, WC, W_2C, VC_2, V_2C_3
Nitrides	BN, TiN, ZrN, VN, NbN, TaN, Si_3N_4
Boron and borides	B, AlB_2, TiB_2, ZrB_2, ThB_4, ThB, NbB, TaB, MoB, Mo_3B_2, WB, Fe_2B, FeB, NiB, Ni_3B_2, Ni_2B
Silicon and silicides	Si and various silicides of Ti, Zr, Nb, Mo, W, Mn, Fe, Ni, Co
Oxides	Al_2O_3, BeO, SiO_2, ZrO_2, Cr_2O_3, SnO_2

[a]Excluding compound semiconductors.

*Paper presented at the 3rd International Conference on the Science of Hard Materials, Nassau, The Bahamas, November 9-13, 1987.

0921-5093/88/$3.50

© Elsevier Sequoia/Printed in The Netherlands

TABLE 2 Representative chemical vapour deposition reactions

$$CrI_2(g) \xrightarrow{800\text{–}1000\ ^\circ C} Cr(s) + 2I(g)$$

$$WF_6(g) + 3H_2(g) \xrightarrow{350\text{–}1000\ ^\circ C} W(s) + 6HF(g)$$

$$WCl_6(g) + 3H_2(g) \xrightarrow{500\text{–}1100\ ^\circ C} W(s) + 6HCl(g)$$

$$TaCl_5(g) + \tfrac{5}{2}H_2(g) \xrightarrow{700\text{–}1100\ ^\circ C} Ta(s) + 5HCl(g)$$

$$(C_8H_{10})_2Cr(g) \xrightarrow{400\text{–}600\ ^\circ C^a} Cr(s) + 2C_8H_{10}(g)$$

$$Ni(CO)_4(g) \xrightarrow{150\text{–}200\ ^\circ C^b} Ni(s) + 4CO(g)$$

$$Al_2Cl_6(g) + 3H_2O(g) \xrightarrow{800\text{–}1400\ ^\circ C} Al_2O_3(s) + 6HCl(g)$$

$$BCl_3(g) + NH_3(g) \xrightarrow{500\text{–}1500\ ^\circ C} BN(s) + 3HCl(g)$$

$$TiCl_4(g) + CH_4(g) \xrightarrow{800\text{–}1100\ ^\circ C} TiC(s) + 4HCl(g)$$

$$Ga(g) + AsCl_3(g) + \tfrac{3}{2}H_2(g) \xrightarrow{750\text{–}900\ ^\circ C} GaAs(s) + 3HCl(g)$$

[a] With, for example, HI(g) as a catalyst to suppress codeposition of carbon.
[b] Deposition possible at 100 °C or lower with a trace of H_2S.

deposits in the parts per billion range is a crucial advantage of CVD in microelectronics. In other applications, the ability to form preferentially oriented crystalline structures is a key advantage. The high throwing power of CVD permits uniform coating in recesses and the impregnation of fiber-reinforced structures. Although operation at a reduced pressure improves the uniformity of deposition and the penetration of porous structures, many CVD processes can be operated at atmospheric pressure, which has the advantage of minimizing the capital investment required. The unique adaptability of CVD to coating particles or small parts suspended in a gas-fluidized bed is another significant advantage.

CVD is not without limitations, however. Many CVD coatings cannot be applied to heat-sensitive substrates.* Thermal expansion differentials between coating and substrate may lead to prohibitive stress as the system is cooled from the coating temperature or heated to a higher service temperature. Most of the precursors having appropriate volatility and thermal stability are corrosive, toxic or pyrophoric, requiring operation in closed systems with carefully engineered precautions. Further, many of the feed materials may be expensive. Despite these limitations, a wide variety of applications have emerged for which CVD is the most economical technique, or in some cases the only suitable technique available.

3. Examples of chemical vapor deposition surface modification

Examples of the use of CVD for surface modification include the application of TiN, TiC and Al_2O_3 coatings to cemented-carbide tool inserts or other tool surfaces [1]†. The presence of a few microns of TiC or TiN on the tool surface during machining decreases adhesion between the binder of the cemented carbide and the emerging chip, thus decreasing "pull-out" and resultant crater wear. Alumina coatings are effective for high speed machining with certain alloys where higher chip temperatures are encountered. Underlayers of TiC or TiN are used to increase the adhesion of the alumina coating by increasing nucleation density in the first stages of its deposition.

Wear-resistant CVD coatings of pyrolytic carbon strengthened with codeposited silicon carbide were developed nearly 20 years ago by Bokros [2] as the basis of a new generation of medical protheses such as artificial heart valves, whose use continues to increase.

An interesting example of CVD surface conditioning for emissivity control was reported by Cuomo *et al.* [3], who deposited a forest of surface-perpendicular preferentially oriented ⟨111⟩ tungsten needles to generate a "hohlraum" black-body coating. Another approach to selective spectral absorption based on coating structure in the two-phase Mo–MoO_3 ("black molybdenum") system has been developed by workers at Arizona State University [4].

CVD coatings as diffusion barriers or bond control layers between fibers and matrices, as oxidation-resistant or corrosion-resistant coatings and as decorative coatings (particularly gold-colored TiN) round out the surface-conditioning applications.

Several useful reviews are already available in the literature which contain additional examples

*The use of a plasma discharge to decrease the deposition temperature is feasible in some applications but is frequently limited by the codeposition of unwanted impurities that would normally vaporize at the higher temperature of the thermally activated process.

†Many of the references in this paper are given as examples only, without an attempt to be comprehensive or to reflect priority.

of surface-conditioning and other applications of CVD [5–8], to which the reader is referred. The present review is directed to the participants at the present conference and those interested in the subject who may be contemplating the use of CVD for new surface modification applications and should be made aware of some of the complexities of CVD technology, thus being better prepared to achieve useful results.

4. Anatomy of chemical vapour deposition processing

In coating by CVD, both the rate of deposition (of economical importance) and the coating structure (of technological importance) depend on the outcome of competition between the rate of arrival of precursors or reaction intermediates at the surface, and the rearrangement by surface diffusion of adsorbed species, before, during or after reaction to give the product species. Ordered large-grained structures are thus favored at higher temperatures and at lower reactant concentrations (lower deposition rates). Fine-grained disordered structures are favored conversely [9]. Thus, we often face a trade-off between rate of deposition and the ability to obtain the desired structure, a problem which increases with increasing thickness of the coating required. In any event, intelligent choices of the optimum chemical reaction and optimum conditions for CVD processing require an uncommonly intimate understanding of the process.

Except in rare cases where the thermodynamic yield or the rate of nucleation at the substrate surface is limiting, coating uniformity, coating structure and the presence or absence of gas phase precipitates in the usual laminar flow CVD reactor depend on the degree to which the rate of deposition is limited (a) by the rate of reactant and reaction by product transport through a transition or boundary layer between the bulk gas stream and the substrate surface or (b) by kinetic processes either in the gas phase or at the surface.

For the purpose of obtaining reasonably efficient conversion of the reactants (or "precursors"), CVD processing is generally carried out at low Reynolds number (less than 2500) where flow-friction-induced turbulence is not a factor and streamline flow predominates. However, the flow patterns in CVD reactors can be complicated by eddy currents where the reactants are introduced, and a fully developed streamline flow may not be present over an appreciable length of the reactor at the entrance end. Distortion of streamline flow by irregular substrates is a further complication. Other complications may be encountered in cold-wall reactors where thermal gradients lead to natural convection currents superimposed on the forced convection of the processing gas. In a horizontal reactor, this can lead to "convection rolls" in the gas stream above the substrate [10]. In a vertical downflow reactor, the resultant direction of reactant flow at the heated substrate may even be opposite to that of the forced convection [11]. Further, where large molecular weight differences exist between the gaseous species and steep thermal gradients occur in the gas phase, thermal diffusion, known as the "Soret effect" where light molecules diffuse preferentially up the temperature gradient, should be taken into account. However, in cases where these complications can be minimized, the mass transfer equivalent of the simple boundary layer model of heat transfer from a flowing gas to a heated surface has been a useful tool in interpreting in an elementary way the characteristics of many CVD reactors and providing guidance for their development and scale-up.

The simplest case is one in which the transport through the boundary layer occurs by interdiffusion of reactants and reaction products, the kinetic limitation occurs at or very near the substrate, and the composition of the bulk gas stream changes in the downstream direction in accordance with the amount of material deposited upstream. The thermal decomposition of silane (SiH_4) is a widely treated example [12–14], although Coltrin *et al.* [15] have shown that limiting silicon deposition to the surface reaction of SiH_4 is an oversimplification. The formation of SiH_2 in the gas phase is an important correction.

If we write expressions for diffusion flux through a boundary layer and for mass transport by simple first-order chemical reaction at the surface and then equate the two, eliminating the unknown interfacial concentration [16], we obtain an expression such as

$$J=\frac{1}{RT}\,\frac{P_{\mathrm{b}}-P_{\mathrm{eq}}}{\delta/D+1/k} \qquad (1)$$

where J is the flux of deposition species to the surface, $P_{\mathrm{b}}-P_{\mathrm{eq}}$ is the driving force for the reaction, *i.e.* the difference between the reactant partial pressures at equilibrium and in the bulk gas stream, δ is the boundary layer thickness, D is the diffusivity of deposition species (roughly inversely proportional to pressure), k is the mass transfer (surface reaction rate) coefficient, R is the gas constant and T is the absolute temperature. Equation (1) has the form of the familiar equation for conduction in a system having two impedances in series. When the diffusional impedance δ/D is large relative to the kinetic impedance $1/k$, the reaction is said to be diffusion controlled. When the reverse is true, the reaction is said to be kinetically limited. As the substrate temperature is raised, the deposition rate commonly reflects the transition from the kinetically limited regime to the less temperature-sensitive diffusion-controlled regime as shown for different partial pressures of inert carrier gas in Fig. 1.

The ratio $k\delta/D$ of diffusional impedance to kinetic impedance is the Sherwood number of chemical engineering, a useful guide to CVD reactor performance that will be discussed later.

Many boundary layer models based on eqn. (1) have appeared in the literature for use in explaining or improving the performance of CVD reactors, primarily those used for microelectronic processing. They have met with moderate success despite the fact that the boundary layer in such reactors is ill defined and can be used only as a useful approximation for the region through which interdiffusion occurs and in which a concentration gradient exists. The fact that the diffusional impedance is given by δ/D, and D is usually not precisely known allows us to make an approximate choice of δ (*e.g.* half the diameter of a tubular reactor or half the distance between wafers stacked vertically) and to adjust the parameter δ/D to the reactor performance data without seriously compromising the available diffusivity data.

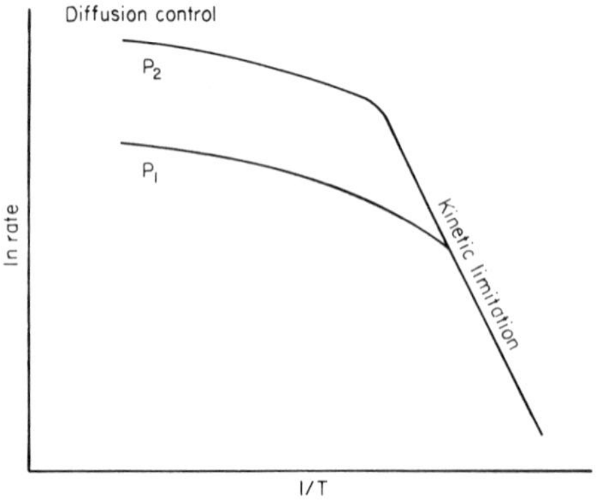

Fig. 1. Deposition rate as a function of temperature at different partial pressures of inert carrier gas ($P_1 > P_2$).

Caution is required with respect to a too literal interpretation of the boundary layer, as was made in an otherwise very useful modelling of a horizontal silicon deposition reactor by Eversteijn *et al.* [13]. In this case, flow patterns in the reactor containing TiO_2 smoke revealed a particle-free zone above the heated susceptor with a gas-mixing region above. The thickness of the particle-free zone was taken to be the boundary layer thickness in the subsequent treatment. However, the effect of thermophoresis in separating the smoke particles from the surface was overlooked. Thus the thickness of the particle-free region can at best be proportional to the boundary layer thickness. This caution applies to all smoke flow pattern visualization studies of CVD reactors in which steep temperature gradients exist in the gas phase.

Despite their limitations, boundary layer models of CVD reactors have the advantage of simplicity and are useful in optimizing the overall performance of a given reactor but break down in detail and on generalization. More revealing, but mathematically more complex, are the finite element analyses which have appeared in recent years [17–20]. In such analyses, continuity equations for momentum balance, mass balance, chemical species balance and energy balance are solved at nodes of, for example, a two-dimensional grid representing the reactor geometry. Data required are the heat capacities, viscosities, thermal conductivities, diffusivities and molecular weights of the gas components, and rate expressions for reactions occurring in the gas phase and at surfaces. The parameters include the substrate temperature, wall temperature, pressure, inlet gas composition and flow rate, inlet gas temperature and outlet gas temperature. With boundary conditions established, the computer iterates solutions of all equations node by node until all constraints are satisfied. The results include flow, temperature and composition profiles in the gas phase and deposition rates as a function of location on the substrate surface and position in the reactor. Although these treatments give highly improved results, they have been limited to rather simple reactor–substrate geometries such as those encountered in silicon wafer processing and can at best serve as a guide in the development of the more complex reactor–substrate geometries encountered generally in surface-conditioning applications. A particularly useful review of CVD reactor modelling has been given by Hess *et al.* [21].

5. Uniformity of deposition

Two of the major goals in CVD processing are (a) to obtain uniform surface coverage (particularly of microfeatures in electronic applications) and (b) to avoid spurious growth of protrusions from the coating surface. van den Brekel and Jansen [22] and Jansen and van den Brekel [23] have shown quantitatively how uniformity of deposition is promoted by operation in the kinetically limited deposition regime and how surface irregularities which penetrate the concentration gradient of the boundary layer in the diffusion-controlled regime are self-perpetuating. Such protrusions can be initiated as dust particles from a "dirty" gas stream or as particles of gas-phase-precipitated reaction product (favored in the diffusion-controlled regime). These facts, plus the knowledge that maximum penetration of porous structures by CVD occurs in the kinetically limited regime, make it desirable to characterize CVD processes as to the degree to which diffusional transport and kinetics are important. A useful technique for obtaining the Sherwood number (the ratio of diffusional to kinetic impedance) in a single run has been proposed by van den Brekel *et al.* [24], which is so simple as to recommend its use in the characterization of any CVD process in its developmental stage.

To use this approach, we simply measure the deposit thickness profile on the wall of a blind hole ($L/D \geqslant 15$) in the substrate surface. For a first-order reaction (approximately)

$$\frac{d_z}{d_0} = \exp\left\{-\left(\frac{z}{R}\right)(2K)^{1/2}\right\}$$

where d is the deposit thickness at position z from the entrance 0 down the "deep" blind hole of radius R.

We plot $\ln(d_z/d_0)$ *vs.* z/R and derive the slope $-(2K)^{1/2}$. K/R equals k/D, the ratio of the first-order reaction rate constant to the diffusivity. Multiplication by an estimated boundary layer thickness δ for the reactor (*e.g.* half the diameter of a tubular horizontal reactor, or half the distance between wafers in a vertically stacked reactor) gives the approximate Sherwood number $\mathrm{Sh} = k\delta/D$, the ratio of diffusional to kinetic impedance in the system.

The Sherwood number is also a measure of the concentration gradient in the gas phase adjacent to the substrate:

$$\mathrm{Sh} = \frac{C_B - C_S}{C_S - C_E}$$

where C_B, C_S and C_E are the concentrations of the rate-determining reactant in the bulk stream, at the substrate surface and in equilibrium with the substrate respectively.

To obtain maximum utilization of reactants in a CVD reactor and yet to maintain uniformity, reactant depletion downstream may be offset in the kinetically limited regime by imposing a positive substrate temperature gradient in that direction or, in the diffusion-controlled regime, by constricting the reactor downstream to increase the flow rate and to decrease the boundary layer thickness.

6. Grain refinement and multiphase deposits

Grain refinement in CVD coatings has long been of concern and currently receives increasing attention, one objective being to avoid the poor strength parallel to the substrate surface of large-grain columnar CVD coatings formed by an uninterrupted heterogeneous growth mechanism. To interrupt the mechanism and to nucleate new grains requires that a competing mechanism be introduced. This can take the form of (1) mechanically contacting the substrate during deposition to distort the growth front as was done by Holman and Hugel [25] for tungsten and W–Rh alloys, (2) codepositing a second-phase material, either by adjusting reaction conditions to give a two-phase deposit where possible or by introducing a parallel CVD reaction, (3) codepositing gas-phase-precipitated material or (4) condensing droplets of high molecular weight reaction intermediates which decompose *in situ*, as proposed by Holzl [26].

The potential for codepositing second-phase material is inherent, for example, in depositing SiC from the Si–C–H–Cl system (*e.g.* from methyltrichlorosilane) where SiC, SiC + carbon, and SiC + silicon are equilibrium phases, and where non-equilibrium silicon deposition can easily occur. An example of the addition of a parallel CVD reaction is illustrated in Fig. 2 where pyrolytic carbon was codeposited with beryllia on nuclear fuel particles to suppress the deleterious growth of beryllia grains at the elevated temperatures potentially used in gas-cooled nuclear reactors [27].

A word of caution is in order for those undertaking to exploit parallel CVD reactions to obtain two-phase or multiphase deposits; whereas uniformity of deposition is assured by rapid recirculation of particles in a fluidized-bed CVD reactor, uniformity in the normal flow-through reactor for larger parts presents problems. Since the kinetics of the parallel reactions are generally not equivalent, compensating for reactant depletion downstream by the imposition of a temperature gradient is ineffective. We must use kinetically highly favorable parallel CVD reactions and operate in the diffusion-controlled regime with special attention to gas flow dynamics to achieve uniformity.

Obtaining the desired multiphase deposits from parallel CVD reactions can encounter other problems, as reviewed recently by Lackey *et al.* [28]. For example, in seeking to deposit the oxidation-resistant and thermal-shock-resistant dispersed-phase composite containing 26 mol.% SiC in HfB_2, we cannot simply use the stoichiometric ratio of BCl_3, $HfCl_4$ and methyltrichlorosilane in hydrogen, shown as point A in the corresponding CVD deposition diagram (Fig. 3). The reactant composition required to produce the thermodynamically stable HfB_2–SiC two-phase region is shown to lie far toward the methyltrichlorosilane corner of the diagram. Although such thermodynamic analyses are useful as guidance, we must be cautious about relying too much on their predictions, particularly in this case at temperatures as low as 1200 K where non-equilibrium silicon is known to deposit in the methyltrichlorosilane–hydrogen system, apparently owing to kinetic restrictions on the mobility of adsorbed carbon [29]. The experimental study of the multiphase CVD system shown in Fig. 3 has yet to be done. The results should be interesting.

7. Conclusion

As a bottom line, it is concluded that CVD processes are very useful in surface-conditioning applications. However, they can involve complex factors whose successful control can best be accomplished by informed attention to the process fundamentals.

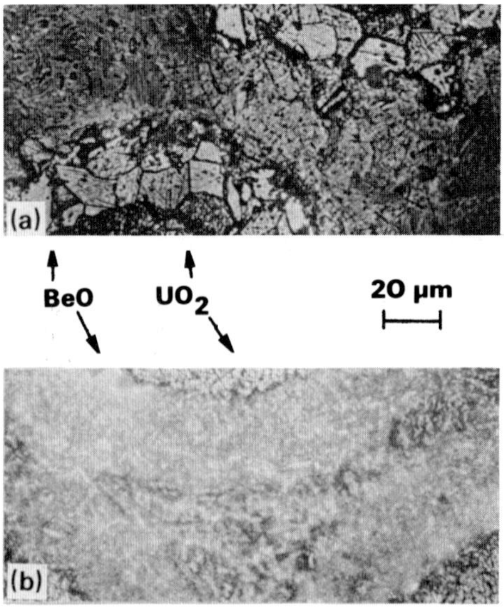

Fig. 2. BeO coatings on UO_2 nuclear fuel particles after an anneal for 1 h at 1600 °C in air: (a) initially pure BeO; (b) BeO initially containing 8 wt.% pyrolytic carbon [27].

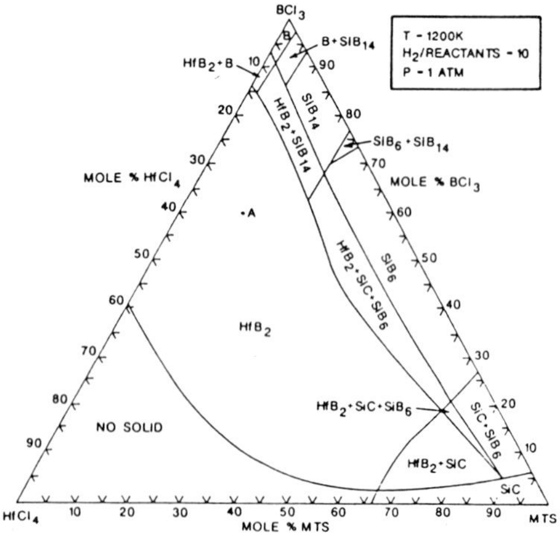

Fig. 3. CVD deposition diagram showing the thermodynamically stable solid phases as a function of initial feed gas composition consisting of 1 part of $HfCl_4 + BCl_3$ + methyltrichlorosilane in 10 parts of H_2 at 1200 K and 1 atm [28].

References

1 H. E. Hintermann, *Metall, 35* (1981) 213.
2 J. C. Bokros and R. J. Akins, Applications of pyrolytic carbon to artificial heart valves; a status report, *Proc. 4th Buhl Int. Conf. on Materials*, Carnegie Press, Carnegie-Mellon University, Pittsburgh, PA, 1971, p. 243.
3 J. J. Cuomo, J. F. Ziegler and J. M. Woodall, *Appl. Phys. Lett., 26* (10) (1975) 557–559.
4 E. E. Chain and B. O. Seraphin, The optical properties of black molybdenum, in D. Adler *et al.* (eds.), *Physics of Disordered Materials*, Plenum, New York, 1985, p. 811.
5 J. H. Oxley *et al.* (eds.), *Vapor Deposition*, Wiley, New York, 1966.
6 W. A. Bryant, *J. Mater. Sci., 12* (1977) 1285–1306.
7 K. K. Yee, *Int. Metall. Rev., 1* (1978) 19–42.
8 J. L. Vossen and W. Kern (eds.), *Thin Film Processes*, Academic Press, New York, 1978.
9 J. M. Blocher, Jr., *J. Vac. Sci. Technol., 11* (4) (1974) 680–686.
10 R. Takahashi, Y. Hoga and K. Sugawara, *J. Electrochem. Soc., 119* (1972) 1406–1412.
11 T. Nyce and F. Rosenberger, in G. W. Cullen (ed.), *Proc. 10th Int. Conf. on Chemical Vapor Deposition*, Electrochemical Society, Pennington, NJ, 1987, pp. 53–60.
12 F. C. Eversteijn and H. L. Peek, *Philips Res. Rep., 25* (1970) 472–481.
13 F. C. Eversteijn, P. J. W. Severin, C. H. J. van den Brekel and H. L. Peek, *J. Electrochem. Soc., 117* (1970) 925–931.
14 F. W. Dittman, *Adv. Chem. Ser., 133* (1974) 463–473.
15 M. E. Coltrin, R. J. Kee and J. A. Miller, *J. Electrochem. Soc., 131* (1984) 425–434.
16 J. H. Oxley, Transport processes, in J. H. Oxley *et al.* (eds.), *Vapor Deposition*, Wiley, New York, 1966, pp. 102–125.
17 C. Houtman, D. B. Graves and K. F. Jensen, *J. Electrochem. Soc., 133* (1986) 961–970.
18 K. F. Jensen, in McD. Robinson *et al.* (eds.), *Proc. 9th Int. Conf. on Chemical Vapor Deposition*, Electrochemical Society, Pennington, NJ, 1984, pp. 3–20.
19 C. Houtman, H. Moffat and K. F. Jensen, *Proc. 5th Eur. Conf. on Chemical Vapor Deposition*, Uppsala University, Uppsala, 1985, pp. 73–87.
20 F. Rosenberger, in G. W. Cullen (ed.), *Proc. 10th Int. Conf. on Chemical Vapor Deposition*, Electrochemical Society, Pennington, NJ, 1987, pp. 13–22.
21 D. W. Hess, K. F. Jensen and T. J. Anderson, *Rev. Chem. Eng., 3* (2) (1985) 97–185.
22 C. H. J. van den Brekel and A. K. Jansen, *J. Cryst. Growth, 43* (1987) 364–370.
23 A. K. Jansen and C. H. J. van den Brekel, *J. Cryst. Growth, 43* (1987) 371–377.
24 C. H. J. van den Brekel *et al.*, in J. M. Blocher, Jr., *et al.* (eds.), *Proc. 8th Int. Conf. on Chemical Vapor Deposition*, Electrochemical Society, Pennington, NJ, 1981, pp. 116–127, 131–141, 142–156.
25 W. R. Holman and F. J. Hugel, in A. C. Schaffhauser (ed.), *Proc. 1st Int. Conf. Chemical Vapor Deposition of Refractory Metals, Alloys, and Compounds*, American Nuclear Society, Hinsdale, IL, 1967, pp. 127–148.
26 R. Holzl, *Wear, 48* (1978) 305–308.
27 W. J. Wilson, M. F. Browning, V. M. Secrest and J. M. Blocher, Jr., Preparation and irradiation of beryllia-coated fuel particles, *BMI-1718, UC-25 Metals, Ceramics and Materials* (TID-4500), 38th edn., 1965.
28 W. J. Lackey, A. W. Smith, D. M. Dillard and D. J. Twait, Codeposition of dispersed phase ceramic composites, in G. W. Cullen (ed.), *Proc. 10th Int. Conf. on Chemical Vapor Deposition*, Electrochemical Society, Pennington, NJ, 1987, pp. 1008–1027.
29 J. Chin, P. K. Gantzel and R. G. Hudson, *Thin Solid Films, 40* (1977) 57–72.

Mechanical Properties, Structure and Performance of Chemically Vapor-deposited and Physically Vapor-deposited Coated Carbide Tools*

D. T. QUINTO, A. T. SANTHANAM and P. C. JINDAL

Philip McKenna Laboratory, Kennametal Inc., Greensburg, PA 15601 (U.S.A.)

(Received November 10, 1987)

Abstract

The relative merits of TiN coatings produced by physical vapor deposition (PVD) and chemical vapor deposition (CVD) on cemented carbide tools are examined by a comparison of microstructure, mechanical properties and metal-cutting performance of two hard metal alloys in the uncoated condition and after coating with either type. Intrinsic microstructural differences between PVD and CVD coatings are attributed to the differences in deposition process conditions. It was found that the transverse rupture strength (TRS) of the ground substrates decreased after CVD coating or high temperature treatment but remained relatively unchanged after PVD coating. The TRS values are correlated with the residual stress states of the TiN coating and the WC phase of the underlying substrate. The optimum condition for high TRS and minimum edge chipping sensitivity was obtained in the ground and PVD coated hard metal where the TiN coating and the WC phase were under relatively high residual compression. The best metal-cutting performance in both turning and milling tests was obtained with the substrate exhibiting high deformation resistance and the PVD TiN coating which had a higher microhardness and beneficial compressive residual stress.

1. Introduction

The advantages of applying hard coatings on cemented carbide cutting tools by chemical vapor deposition (CVD) are well established [1, 2] and have led to commercial application for almost two decades. These coatings have evolved from a single TiC layer to more common triple layers of TiC–TiCN–TiN and TiC–Al_2O_3–TiN to as many as 12 thin multilayer variants [3, 4]. In the last 5 years, TiN produced by physical vapor deposition (PVD) has been demonstrated as an effective hard coating for high speed steel tools. More recently, PVD TiN-coated carbide tools were commercially introduced after their superiority over CVD counterparts were realized in interrupted cutting such as milling of steel [5]. It is of interest to compare the relative merits of PVD and CVD coatings on cemented carbide tools because both technologies are now available. This comparison can be best made using the same cemented carbide substrates coated by both CVD and PVD processes to the same coating thickness. In this paper, we describe a comparison of two carbide substrates coated by PVD and CVD TiN in terms of their microstructures, mechanical properties and metal-cutting performance.

2. Experimental methods

The hard metal compositions employed in this study were the same as those used in a previous study [6]. The cobalt content and grain size were kept constant, but the amount of cubic carbides was varied. Alloy A had a very small amount of cubic carbides (1.6 vol.%) whereas alloy B had 50 vol.% of cubic carbides. The alloys were processed by conventional powder metallurgical techniques. The chemical composition, physical and mechanical properties of the two alloys are presented in Tables 1 and 2.

Samples 18.75 mm × 5.0 mm × 5.0 mm in size were prepared for transverse rupture testing. SNG 433 and SPG 433 style inserts were used for metal-cutting tests. To study the effect of variations in surface conditions, samples were either ground or polished to remove 30 μm of surface

*Paper presented at the 3rd International Conference on the Science of Hard Materials, Nassau, The Bahamas, November 9-13, 1987.

0921-5093/88/$3.50 © Elsevier Sequoia/Printed in The Netherlands

layer. A few transverse rupture bars were annealed in a hydrogen atmosphere at 1000 °C for 1 h to simulate partially the thermal effects of the CVD coating.

Conventional practice was used to deposit CVD TiN and PVD TiN (ion-plating process). Thicknesses were controlled to 5–6 μm in both types of coating for all specimens, except those used for microhardness measurements in which the coating thickness was 10 μm. The heavier thickness was required to eliminate possible substrate effects for the 50 gf indentation load used [7, 8].

Substrate and coating microstructures were examined by optical and scanning electron microscopy. Transmission electron microscopy was also used to characterize the coating microstructures.

Transverse rupture strength (TRS) values were determined at room temperature with the standard three-point bending test. The high temperature compressive yield strengths of alloys A and B were determined from 800 to 1100 °C. The microhardness of the CVD and PVD coatings were evaluated from room temperature to 1000 °C with a high temperature microhardness tester [9].

Residual stress measurements were made at room temperature on the uncoated substrates and on coated samples with the standard X-ray $\sin^2\psi$ method [10]. The stresses in the TiN coatings using {422} or {511} reflections and in the WC phase of the substrate using the {30.0} reflection were measured using monochromatic Cu Kα radiation at angles of 0°, 21°, 30°, 38° and 45°. Young's moduli used in the calculations were 640 GPa for TiN and 700 GPa for WC [11].

Metal-cutting tests consisted of turning AISI 1045 steel (speed, 230 m min^{-1}; depth of cut, 2.5 mm; feed rate, 0.25 mm rev^{-1}) and flycut milling AISI 4140 steel (speed, 150 m min^{-1}; depth of cut, 2.5 mm; feed rate, 0.13 mm rev^{-1}). The carbide tool was considered to have reached its end of life if examination of metal-cutting wear on the used cutting edge exceeded any of the following criteria: for the turning test, a crater depth of 0.10 mm or flank wear of 0.40 mm; for the milling test, uniform flank wear of 0.40 mm, edge chipping of 0.40 mm or localized flank wear of 0.50 mm.

TABLE 1 Composition of hard metals

	Amount (wt.%)			*Amount* (vol.%)		
	Co	*TaC*	*TiC*	*Co*	*Cubic carbides*	*WC*
Alloy A	6.0	0.5	0.3	10.0	1.6	88.4
Alloy B	7.2	20.8	8.9	10.0	50.0	40.0

3. Results

3.1. Microstructure

Figures 1 and 2 are optical micrographs of polished cross-sections of the coated samples. The mean linear intercept carbide grain size was 0.7 μm in both substrates. Both CVD and PVD coated samples have nearly the same TiN thicknesses, although the former has a rougher surface morphology. Little or no η phase formation is seen underneath the CVD TiN layer in the optical micrographs. However, there is an indication of some carbon diffusion from the substrate because of a slight discoloration of the TiN coating layer adjacent to the coating–substrate interface.

Transmission electron micrographs of the coatings, where the thin foils were prepared parallel and close to the substrate interface, are shown in Figs. 3 and 4. As described in our previous studies [10], there are distinct differences between the two coating types. The CVD TiN grains, which are formed under thermal equilibrium, have well-defined large-angle grain boundaries, with grain sizes of about 200–400 nm. In contrast, the PVD TiN grains are extremely fine grained; the regions of dark contrast in the bright field image represent clusters of fine grains which

TABLE 2 Properties of hard metals

	Density (kgf m^{-3})	*Room temperature hardness* (HV 30)	*Coercivity* (kA m^{-1})	*Magnetic saturation* (T m^3 kg^{-1})	*Carbide grain size* (μm)
Alloy A	14.97×10^3	1639	17.7	112×10^{-7}	0.68
Alloy B	12.67×10^3	1646	14.5	128×10^{-7}	0.71

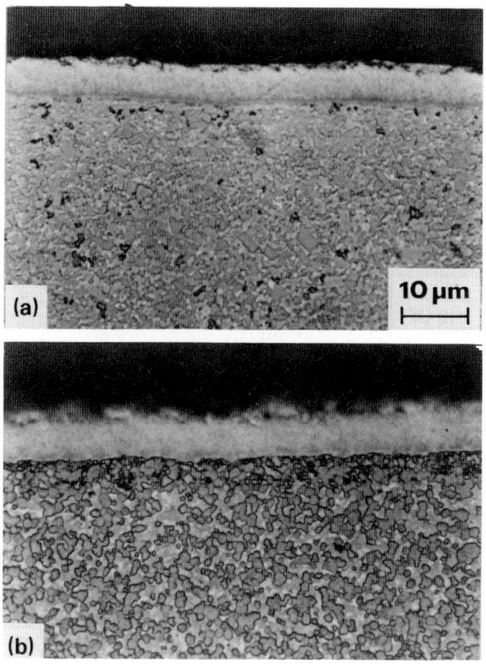

Fig. 1. Optical micrographs of CVD TiN coating on (a) alloy A and (b) alloy B.

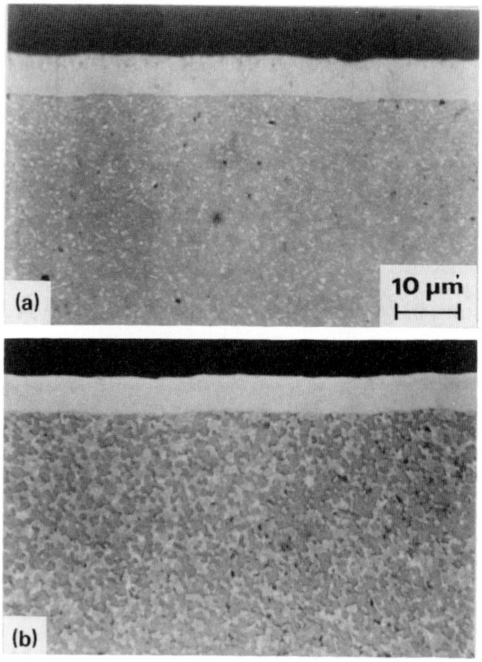

Fig. 2. Optical micrographs of PVD TiN coating on (a) alloy A and (b) alloy B.

are in close crystallographic orientation, the smallest of which may be approximately 50 nm as seen from the corresponding dark field micrograph. The electron diffraction rings of the PVD sample are broad and continuous relative to the discrete spotted rings of the CVD sample. These indicate both a finer grain size and a higher lattice distortion in the PVD sample, which is a consequence of the low deposition temperature and non-equilibrated condition associated with energetic particle bombardment during PVD.

3.2. *Mechanical properties*

3.2.1. *Deformation resistance*

High temperature compression test data of alloys A and B are presented in Fig. 5. It should be noted that the compressive yield strength of alloy B is significantly higher than that of alloy A over the range of test temperatures.

Coating microhardness data were generated for similarly processed but thicker (10 μm) CVD and PVD TiN coatings and are shown in Fig. 6. The PVD coating is consistently harder than the CVD coating, although the difference of about 500 kgf mm^{-2} at room temperature decreases with temperature, to less than 100 kgf mm^{-2} at 1000 °C.

3.2.2. *Transverse rupture strength*

The TRS values for alloys A and B in various conditions are presented in Fig. 7. The following features are shown.

(a) The TRS of alloy A is always higher than that of alloy B.

(b) Annealing lowers the TRS of uncoated samples significantly.

(c) After CVD coating, the TRS levels are at a minimum, although not significantly lower than those after annealing.

(d) After PVD coating, the TRS values are slightly, but not significantly, higher than those of the ground uncoated samples.

(e) Annealing of the PVD coated samples reduces TRS values to levels comparable with those of the CVD coated samples.

3.2.3. *Residual stress in TiN coatings*

The measured residual stresses in the CVD and PVD TiN coatings are graphically shown in Fig. 8. Parallel behavior is seen in both alloys.

Fig. 3. Typical TEM (a) bright field and (b) dark field micrographs of CVD TiN together with (c) selected area diffraction pattern.

Fig. 4. Typical TEM (a) bright field and (b) dark field micrographs of PVD TiN together with (c) selected area diffraction pattern.

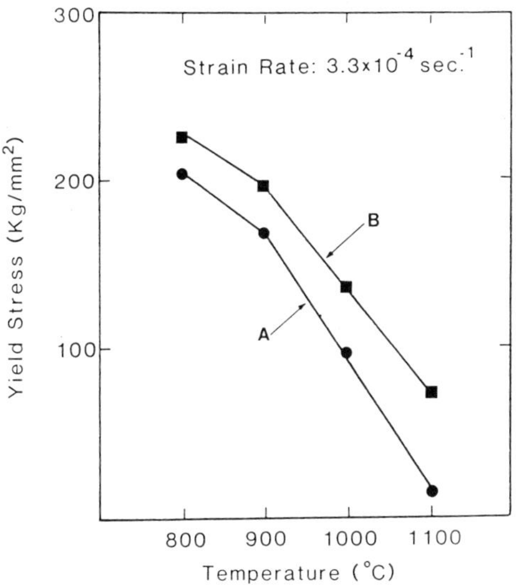

Fig. 5. Compressive yield stress (0.2% offset) of alloys A and B as a function of temperature.

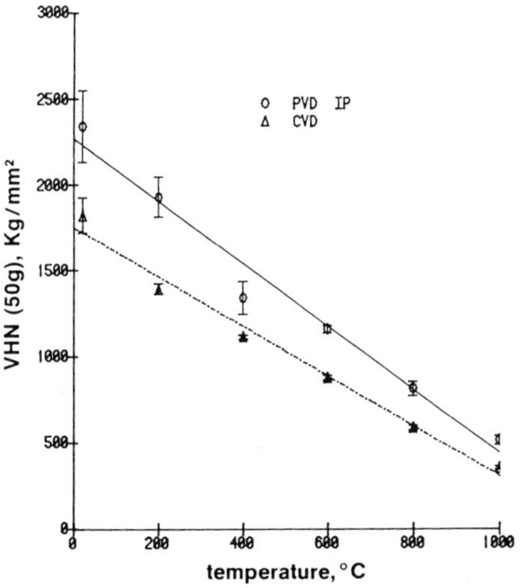

Fig. 6. Vickers microhardness of CVD and PVD TiN as a function of temperature.

(a) The CVD TiN coatings are under a relatively low tensile residual stress, +0.66 GPa on alloy A, and +1.2 GPa on alloy B.

(b) The PVD coatings are under a high compressive residual stress, regardless of whether the substrate was initially ground or polished. Residual stress values ranged from −3.2 to −4.2 GPa

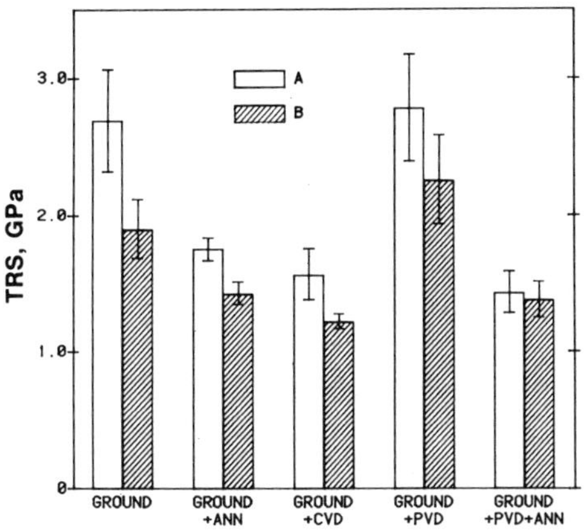

Fig. 7. TRS of uncoated and coated alloys.

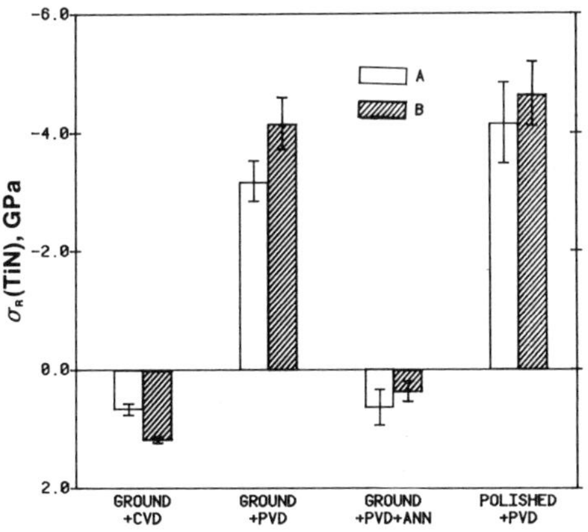

Fig. 8. Residual stress in PVD and CVD TiN coatings.

on alloy A, and from −4.2 to −4.7 GPa on alloy B.

(c) Annealing removed the compressive residual stresses in the PVD TiN coatings and converted them to low tensile stresses of +0.66 GPa on alloy A and +0.37 GPa on alloy B.

3.2.4. Residual stress in the WC phase

The measured residual stresses in the uncoated ground samples and after coating (with the coating intact, X-ray penetration through the coating was sufficient to yield diffracted intensity for these measurements) are illustrated in Fig. 9. Within the error of measurement, it is again evident that similar trends are shown in both alloys with varying treatments.

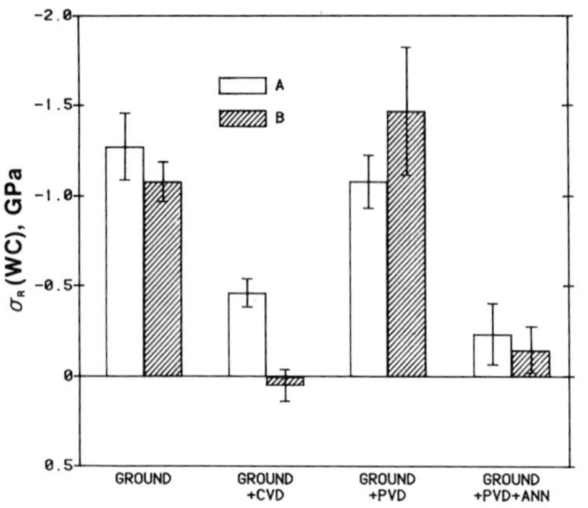

Fig. 9. Residual stress in the WC phase of the substrate for uncoated and coated alloys A and B.

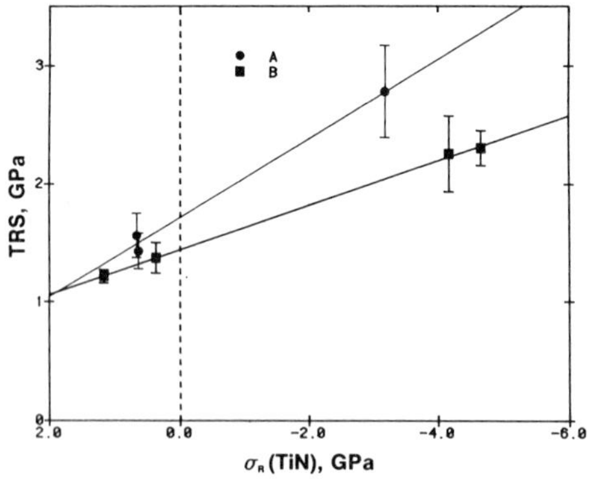

Fig. 10. Correlation between TRS and residual stress in TiN of coated alloys A and B.

(a) Grinding causes compressive residual stresses in the WC phase of both alloy A and alloy B.

(b) After the CVD cycle, these compressive stresses are decreased significantly.

(c) The PVD cycle slightly changes the levels of the compressive stress in the WC phase.

(d) Annealing after the PVD cycle again causes a significant decrease in the compressive residual stresses in the WC phase.

3.2.5. Correlations of the transverse rupture strength and residual stress

The TRS values are plotted against the residual stresses in the TiN coatings (Fig. 10) and in the WC phase of the substrate (Fig. 11). There is

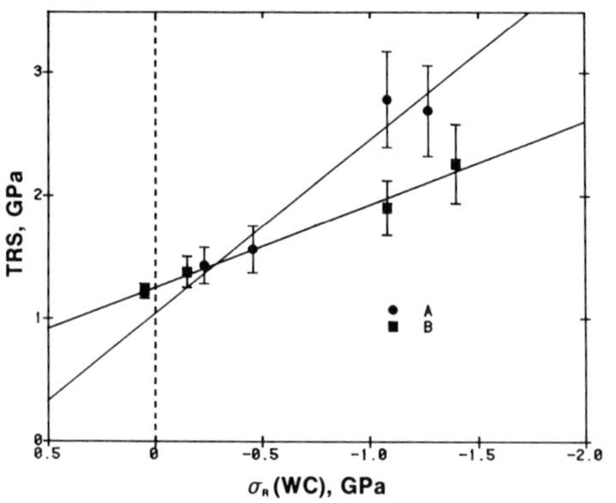

Fig. 11. Correlation between TRS and residual stress in the WC phase of the substrate for uncoated and coated alloys A and B.

a linear correlation between these two sets of variables in both alloys. The linear fit is particularly good in alloy B ($R^2 = 0.99$ and $R^2 = 0.98$ in Fig. 10 and Fig. 11 respectively). It is also apparent that the dependence of TRS on the residual stresses in the TiN coating or in the WC phase of the substrate is stronger in alloy A than in alloy B.

3.3. Metal-cutting results

The average wear curves for three cutting edges of each uncoated, CVD TiN-coated and PVD TiN-coated inserts for alloy A are given in Fig. 12(a). The failure criterion for either flank or crater wear is represented by the horizontal broken line. It is seen that the tool life of 0.5 min for the uncoated tool is increased by CVD coating and PVD coating (4 min and 9 min respectively). Whereas crater wear is the failure mode for both uncoated and CVD coated inserts, the PVD coated insert wears by both crater and flank wear modes. The SEM micrographs in Fig. 13 illustrate these forms of wear. A similar pattern is observed for alloy B, shown in Fig. 12(b). Both crater wear and flank wear curves are again displaced to longer times, thus increasing the tool life of 1 min for the uncoated alloy by CVD coating and PVD coating to 10 min and 17 min respectively.

The results for the milling tests for alloy A are shown in Fig. 14. The CVD coating increases the tool life of the uncoated tool only slightly; a far greater tool life increase is obtained by PVD coating. Uniform flank wear is the wear mode for the uncoated alloy, 100% edge chipping for the CVD coated alloy, and 67% uniform flank wear for the

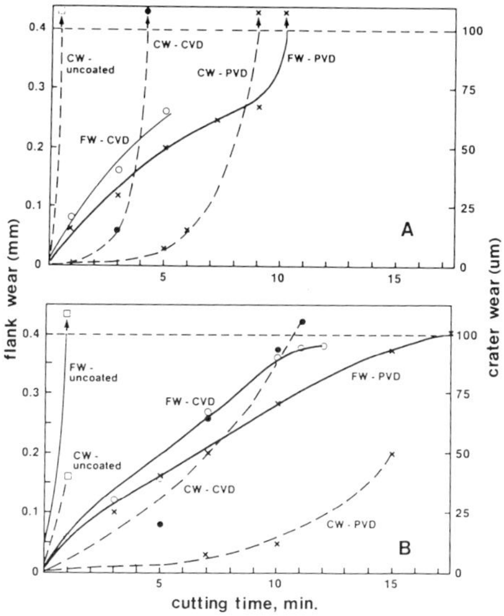

Fig. 12. Flank and crater wear curves for alloy A and alloy B in turning AISI 1045 steel (speed, 230 m min^{-1}; depth of cut, 2.5 mm; feed rate, 0.25 mm rev^{-1}).

PVD coated alloy. In the milling test for alloy B (Fig. 15), the increase in tool life over that of the uncoated alloy is about the same for both CVD and PVD TiN coatings. However, it is noted again that the failure mode of the CVD coated alloy is 100% edge chipping (as in alloy A), while it is 100% uniform flank wear for the PVD coated alloy. These are illustrated in the SEM micrographs in Fig. 16.

4. Discussion

The superior performance of the PVD coated tools in both turning and milling tests can be explained in terms of the structure and property relationships implied by the data.

First, it is noted that flank wear resistance is promoted by increased hardness in the protective coating and higher resistance to plastic deformation of the substrate at the metal-cutting temperature. In the turning test, this is obviously optimized by the choice of alloy B over alloy A substrate, and having the PVD TiN coating rather than the CVD TiN coating. Thus the best combination is alloy B with PVD TiN coating.

It is interesting to observe, in Fig. 12, that the resistance to both flank wear and crater wear is improved by the PVD coating in both alloy A and alloy B. It is generally believed that crater wear is governed by chemical interaction or solubility considerations between the workpiece and the tool material [12, 13] at the tool–chip interface. Since the TiN coating material is the same, whether processed by CVD or PVD, it is inferred that there must be an abrasive wear component in the crater wear process and that the higher hardness of PVD over CVD TiN accounts for this increase in crater wear resistance for a given alloy.

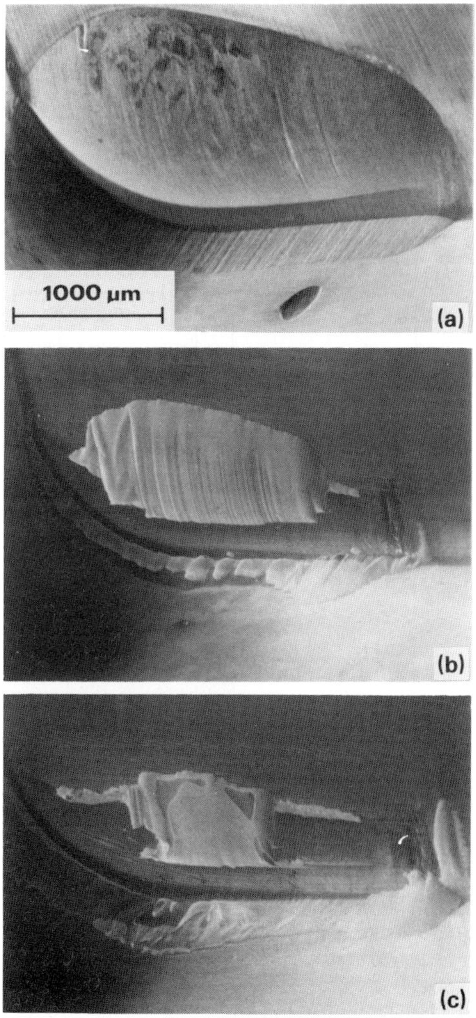

Fig. 13. Typical appearance of (a) uncoated, (b) CVD coated and (c) PVD coated inserts at the end of life in turning AISI 1045 steel.

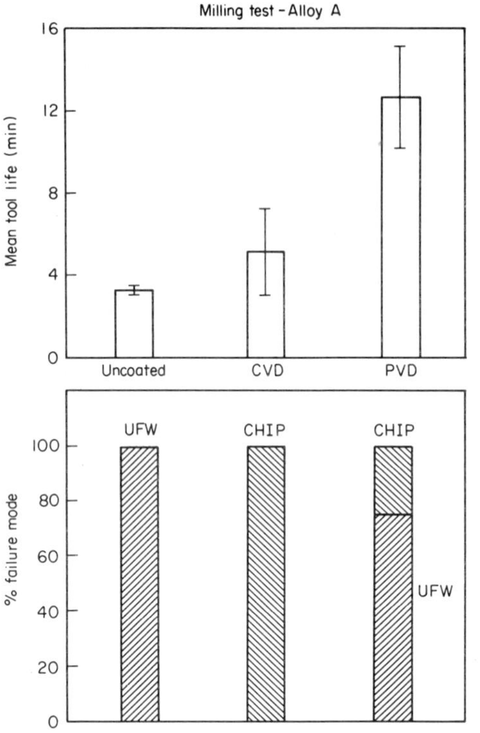

Fig. 14. Mean tool life and the associated failure modes of uncoated and coated alloy A in milling AISI 4140 steel (speed, 150 m min^{-1}; depth of cut, 2.5 mm; feed rate, 0.13 mm rev^{-1}).

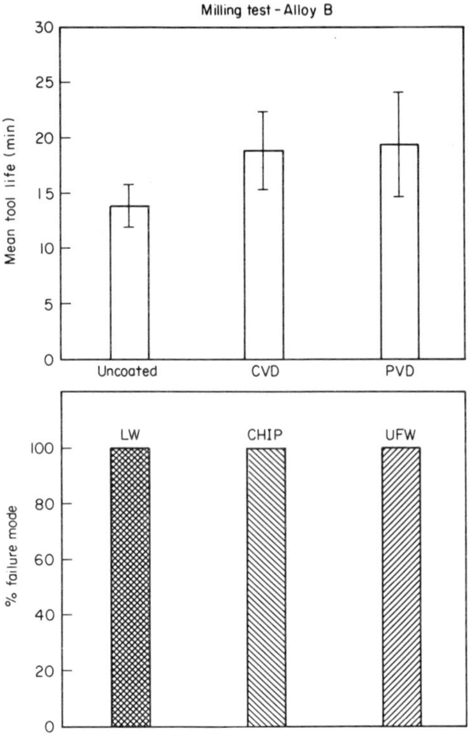

Fig. 15. Mean tool life and the associated failure modes of uncoated and coated alloy B in milling AISI 4140 steel.

In the milling test, the tool tip is subjected to the interrupted cutting process associated with thermal and mechanical fatigue stresses. Other properties besides abrasion wear resistance are required of the tool; its thermal shock resistance and edge fracture strength should be increased to minimize thermal cracking at the cutting edge and the subsequent growth of these cracks, which lead to edge chipping. It is remarkable that the CVD TiN-coated alloys A and B both failed by chipping while the corresponding PVD TiN-coated alloys failed predominantly by uniform flank wear. (This observation was made in a similar comparison of milling performance of PVD and CVD coated tools [5].) It is apparent that the high temperature CVD process increases tool edge chipping sensitivity, while the PVD coating preserves the edge strength of the uncoated ground tool; this is mirrored by lower TRS values after CVD, and relatively unchanged TRS values after PVD as shown in Fig. 7.

The measured TRS values reflect edge chipping sensitivity, which in turn is proportional to surface residual stress. It was observed in Figs. 10 and 11 that the TRS values correlated linearly with measured residual stresses in the TiN coatings and within the WC grains of the substrate in both alloys. It is not clearly understood, however, why alloy A seems to have a stronger dependence of TRS on residual stress than in alloy B.

Grinding introduces compressive residual stress which inhibits surface microcrack initiation. The PVD TiN coating, owing to its intrinsic growth stress [10, 14], is invariably under residual compression (Fig. 8). Because of a low deposition temperature, the stress state of the substrate is unaffected by PVD. Hence the best configuration for high compressive residual stress in the surface and minimum crack sensitivity is obtained in the PVD-coated ground carbide tool. Conversely, high temperature treatment in the CVD cycle anneals out the compressive residual stress of the ground substrate. In addition, the TiN coating is

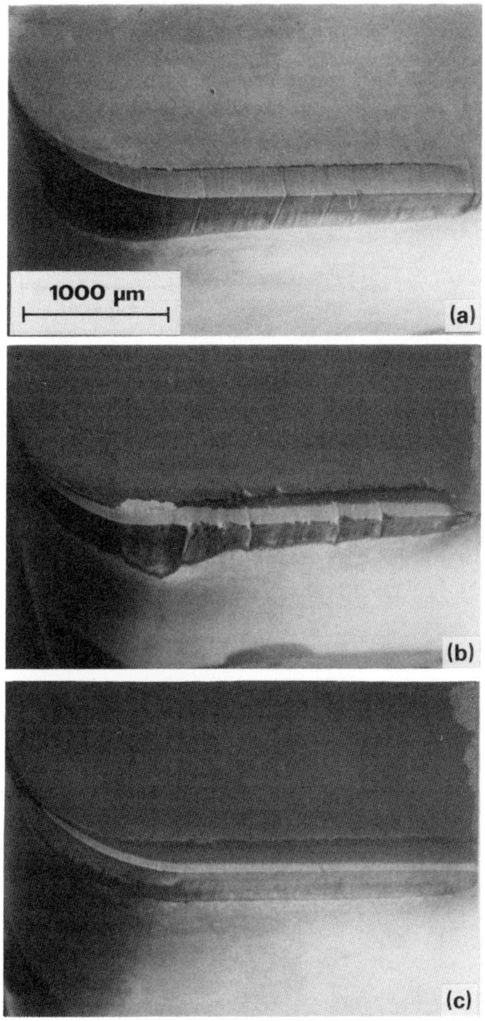

Fig. 16. Typical appearance of (a) uncoated, (b) CVD coated and (c) PVD coated inserts at the end of life in milling AISI 4140 steel.

in residual tension. This combination makes the CVD coated tool very susceptible to fatigue cracking of the cutting edge.

This work endorses the idea that compressive residual stresses within the plane of the coating, inherent in various PVD processes, are beneficial to the surface fracture strength which translates to better performance, particularly in milling. Since there was little η phase formation in the CVD coating evaluated here, its role as a detrimental factor in edge chipping has probably been overestimated previously, although it is likely to contribute to tool life inconsistency of CVD coated milling tools.

The present work agrees with most of the results of Kamachi *et al.* [15] who carried out a similar analysis of PVD and CVD coated carbide tools. A point of departure, however, was their finding that the flank wear resistance was lower in the PVD coated than in the CVD coated tools. Unfortunately, they did not report the microhardness of their coatings, and the level of compressive residual stress obtained in their PVD TiN coating (thickness, 2 μm) is significantly lower than the measured values reported here. It is speculated that their PVD TiN coating, produced by a different PVD process from ours, may have had a lower hardness, which explains its poor performance relative to CVD TiN, contrary to our present finding. It has been clearly demonstrated that a range of microhardnesses and corresponding residual stress values can be obtained in TiN produced by various PVD processes [10]; hence, generalizations on the performance of PVD TiN coatings should be made with caution.

5. Conclusions

(1) The TRS of a given hard metal alloy is influenced by its surface condition; surface grinding and subsequent coating with (low temperature) PVD TiN result in a high TRS, while coating with (high temperature) CVD TiN or heat treatment decrease the TRS.

(2) The residual stress state of the coating and the underlying substrate affect the TRS of the coated alloy. Compressive residual stress developed in the PVD TiN coating, coupled with the compressive stress in the WC phase due to grinding, offer an optimum combination in increasing the TRS of the coated tool and its resistance to edge chipping.

(3) The application of a CVD TiN hard coating to a hard metal tool is effective in increasing the tool life in turning, but it is not so effective in milling because of its tensile residual stress state and its effect in lowering the TRS of the coated tool and its resistance to edge chipping.

(4) PVD TiN is superior to CVD TiN coating in increasing the metal-cutting performance of cemented carbide tools owing to the former's greater resistance to abrasive wear and its associated higher surface fracture strength. These

follow from the PVD TiN coating microstructure and beneficial compressive residual stress both of which contribute to higher microhardness.

(5) The TRS of alloy B is consistently lower than that of alloy A (in both the uncoated and the coated conditions). However, the higher deformation resistance of alloy B results in superior metal-cutting performance over that of alloy A. This demonstrates the synergism between coating and substrate in determining overall tool life.

References

1 W. Schintlmeister and O. Pacher, *Metall, 28* (1974) 690–695.
2 V. K. Sarin, in G. Y. Chin (ed.), *Advances in Powder Technology*, American Society for Metals, Metals Park, OH, 1982, pp. 253–288.
3 J. N. Lindstrom and R. T. Johannesson, *J. Electrochem. Soc., 123* (1976) 555–559.
4 N. Reiter and H. Kolaska, *Proc. 11th Plansee Semin.*, Metallwerk Plansee, Reutte, 1985, pp. 335–376.
5 G. J. Wolfe, C. J. Petrosky and D. T. Quinto, *J. Vac. Sci. Technol. A, 3* (1986) 2747–2754.
6 A. T. Santhanam, K. P. Mizgalski and W. C. McCoy, in E. A. Almond, C. A. Brookes and R. Warren (eds.), *Proc. 2nd Int. Conf. on the Science of Hard Materials, Rhodes, September 23–28, 1984*, in *Inst. Phys. Conf. Ser., 75* (1986) 679–690.
7 P. K. Mehrotra, D. T. Quinto and G. J. Wolfe, *Proc. 9th Int. Conf. on Chemical Vapor Deposition*, Electrochemical Society, Pennington, NJ, 1984, pp. 757–771.
8 P. K. Mehrotra and D. T. Quinto, *J. Vac. Sci. Technol. A, 3* (1985) 2401–2405.
9 P. K. Mehrotra and D. T. Quinto, *Proc. 11th Plansee Semin.*, Metallwerk Plansee, Reutte, 1986, pp. 639–657.
10 D. T. Quinto, G. J. Wolfe and P. C. Jindal, *Thin Solid Films, 153* (1987) 19–36.
11 E. Torok, A. J. Perry and L. Chollet, *Thin Solid Films, 153* (1987) 37–43.
12 B. M. Kramer and P. K. Judd, *J. Vac. Sci. Technol. A, 3* (1985) 2439–2444.
13 B. M. Kramer, *J. Vac. Sci. Technol. A, 4* (1986) 2870–2873.
14 J. E. Sundgren, *Thin Solid Films, 128* (1985) 21–44.
15 K. Kamachi, T. Ito and T. Yamamoto, *Surf. J. Int., 1* (1986) 82–86.

Mechanical and Tribological Properties of Chemically Vapour-deposited Boron Carbide Coatings*

M. OLSSON and B. STRIDH

Department of Technology, Materials Science Group, Uppsala University, P.O. Box 534, S-751 21 Uppsala (Sweden)

U. JANSSON and J.-O. CARLSSON

Department of Chemistry, Thin Film and Chemistry Group, Uppsala University, P.O. Box 534, S-751 21 Uppsala (Sweden)

S. SÖDERBERG

AB Sandvik Coromant, P.O. Box 42056, S-12612 Stockholm (Sweden)

(Received November 10, 1987)

Abstract

Boron carbide coatings of different compositions and thicknesses were deposited on TiC-coated cemented carbide by chemical vapour deposition. Rhombohedral boron carbide ($B_{13}C_2$) with a carbon content ranging from 9 to 17 at.% and amorphous boron carbide with 22 at.% C were deposited. $B_{13}C_2$ exhibits a nodular morphology with crystal facets corresponding to the grain size of the coating. The nodules are nucleated on cobalt-rich outgrowths originating from the interface. Transmission electron microscopy examination indicates the presence of high local stresses in the coating.

The Vickers hardness of the $B_{13}C_2$ coatings was estimated to range from 4000 to 4500 HV, the higher value corresponding to the coating with the highest carbon content. An attempt to evaluate the adhesion of the coatings, using the scratch test, failed owing to the high hardness of the coatings. However, the adhesion of $B_{13}C_2$ to TiC appears to be comparable with or even superior to that of TiC to cemented carbide. In contrast, the adhesion of the amorphous coating to TiC is very poor.

The solid particle erosion resistance of $B_{13}C_2$ was found to be far superior to that of TiC. The best performance was obtained for the high carbon content coating. In contrast, the amorphous boron carbide gave very poor performance owing to its poor adhesion. The sliding wear resistance was investigated by a simulated tool wear test (a modified pin-on-ring test with continuous introduction of a fresh counter-material surface). Using quenched-and-tempered steel and austenitic stainless steel as counter materials, no improvement in properties compared with those of TiC was obtained. The results are discussed in relation to the wear mechanisms identified.

1. Introduction

Boron carbide has many interesting properties. For example, the high hardness and high temperature stability suggest a promising potential as a wear-resistant material [1]. Industrial components made from bulk boron carbide are available today, but the number of applications has been limited owing to the extreme brittleness of the material [2]. This problem may be possible to avoid if the boron carbide can be applied as a surface coating. The most promising method for the deposition of dense boron carbide coatings appears to be chemical vapour deposition (CVD) [3–5]. In the present work, the potential of CVD coatings of boron carbide for tribological applications has been explored. Coatings of boron carbide have been deposited and characterized with respect to composition, microstructure, mechanical properties and wear resistance. The work focuses on the influence of composition and thickness on coating properties.

*Paper presented at the 3rd International Conference on the Science of Hard Materials, Nassau, The Bahamas, November 9–13, 1987.

0921-5093/88/$3.50

© Elsevier Sequoia/Printed in The Netherlands

2. Experimental details

2.1. Boron carbide deposition

Boron carbide coatings were deposited in a hot-wall reactor from a reaction gas mixture of boron trichloride (purity, 99.99%), methane (purity, 99.998%) and hydrogen (purity, 99.9997%). All coatings were deposited at 1325 K, at a total pressure of 6.7 kPa. A linear gas flow velocity of 0.47 m s^{-1} was used.

The ratio of the partial pressure of hydrogen to that of boron trichloride was kept constant at 10 in all experiments. The partial pressure of methane and the deposition time were the two parameters used to control the composition and thickness of the coating. In Table 1, the different deposition conditions used in the present investigation are summarized.

Titanium-carbide(TiC)-coated Cemented-carbide pins (diameter, 5 mm) and indexable inserts (12 mm × 12 mm × 3 mm) were used as substrate materials. The chemical composition of the cemented carbide was 77.0 wt.% WC, 6.1 wt.% TaC, 4.0 wt.% TiC, 1.9 wt.% NbC and 11.0 wt.% Co. The TiC coating was deposited by CVD to a thickness of 6.5 ± 0.3 μm and 4.9 ± 0.3 μm for the pins and inserts respectively using the Sandvik Coromant standard procedure [6]. Examination of cross-sections revealed that the TiC coating consisted of two layers: an inner layer with fine equiaxed grains and an outer layer with larger columnar grains. By energy-dispersive X-ray analysis, it was found that cobalt was transported to the TiC surface during the TiC deposition. This results in a surface concentration of 0.5 at.% Co on the TiC surface. In addition, the TiC displays a network of thermal cracks perpendicular to the surface and porosity at the TiC–cemented-carbide interface. At the upper region of the substrate a non-continuous layer of η phase, was observed. All these observations are in agreement with earlier findings in the literature [7, 8].

TABLE 1 Deposition parameters for the boron carbide coatings

Specimen	P_{CH_4} (kPa)	*Deposition time* (min)
1	0.06	18
2.1	0.24	12
2.2	0.24	24
2.3	0.24	36
3	0.45	70

2.2. Characterization of boron carbide coatings

The boron carbide coatings were characterized with respect to phase and chemical composition by X-ray diffraction and Auger spectroscopy respectively. The Auger analysis was performed in a scanning Auger microscope. The experimental procedure has been described previously [9]. The thickness of the boron carbide coatings was determined from cross-sections obtained by scanning electron microscopy (SEM). SEM was also utilized to characterize the as-deposited morphology of the coatings. Finally, transmission electron microscopy (TEM) on thin foils in cross-section was used to investigate the microstructure of the coatings. The specimen preparation technique for the TEM work has been presented earlier [10].

2.3. Hardness, adhesion and wear testing

The hardness of the boron carbide coatings was measured directly on the surface of the coating after careful polishing. In order to account for the influence of the underlying substrate on the hardness readings, the technique proposed by Vingsbo *et al.* [11] was applied. Their method requires the determination of a series of microhardness readings at different normal loads for both the coated specimen and the substrate. The experimental data are then fitted to a model from which the "true" hardness value of the coating can be estimated.

Adhesion evaluation of the coatings was attempted by scratch testing *in situ* in the scanning electron microscope with a spherical diamond stylus (radius, 25 μm). The applied normal load was increased continuously during scratching, while monitoring the surface damage caused by the moving stylus. The critical load at which spalling of the coating was first induced was determined. It has been proposed in the literature that this value is a measure of adhesion [12].

The wear resistance of the boron carbide coatings was evaluated using two different types of test equipment. The erosion resistance was tested in a centrifugal solid particle erosion tester ("erofuge") [13] using high purity alumina (99.7 wt.% Al_2O_3) particles as erodent. The particle size range was 88–106 μm and normal impingement at a particle velocity of 66 m s^{-1} was used in the tests. A maximum of 18 specimens can be eroded simultaneously in the erofuge. This allowed all specimens in the present investigation to be eroded in one test series.

In the determination of the erosion rates of the coatings, the characteristic dose variation over the specimen surface in the erofuge was utilized. The dose variation was determined accurately by surface profilometry on an eroded glass target. The boron carbide specimens were then eroded until complete coating removal had been produced at the point of maximum erosion. The particle dose corresponding to complete coating removal was then determined from the backscattered electron image of the specimen surface in a scanning electron microscope equipped with an image analysis system.

A simulated tool wear test was used to evaluate boron carbide as a candidate material for machining applications. The test is based on a conventional pin-on-ring type of geometry but modified to allow the pin to be continuously fed into a fresh counter-material surface during the test. This is accomplished by mounting the ring material in a lathe. The normal force is applied using a spring-loading system and the resulting frictional force is measured with strain gauges. Conventional pin-on-ring testing can be performed in the same equipment simply by disengaging the feed control of the lathe. The equipment has been described in more detail earlier [14] and it has been demonstrated that this laboratory wear test is capable of closely reproducing the wear processes at the tool–workpiece interface in machining.

Wear tests were performed using two different counter materials, one quenched-and-tempered steel (AISI 4337: composition, Fe–0.36wt.%C–1.4wt.%Cr–1.4wt.%Ni–0.2wt.%Mo–0.65wt.%Mn; hardness, 310 HV) and one titanium-stabilized austenitic stainless steel (AISI 321: composition, Fe–0.08wt.%C–18.0wt.%Cr–10.0wt.%Ni–0.4wt.-%Ti; hardness, 160 HV). The sliding speed was 250 m min^{-1} for AISI 4337 and 200 m min^{-1} for AISI 321. The normal force was 500 N and the sliding distance 300 m in both cases.

Conventional wear rate determination by mass loss measurement cannot be applied in the present case owing to the very low wear rates of the investigated coatings. Instead, image analysis of the backscattered electron image in the scanning electron microscope was used to determine the area of exposed TiC and cemented carbide respectively. Prior to the analysis, the specimens were immersed in hot hydrochloric acid in order to dissolve adhered counter material.

3. Results

3.1. Thickness and composition

In Table 2 the thickness values and the phase and chemical compositions of the investigated coatings are summarized. It can be seen that all coatings, except the extreme high carbon content coating, exhibit a rhombohedral structure which corresponds to the phase $B_{13}C_2$ [15]. Specimens 1, 2.1 and 3 were selected to study the influence of chemical composition and specimens 2.1, 2.2 and 2.3 to study the influence of coating thickness on the mechanical and tribological properties of boron carbide. Specimen 4 corresponds to the substrate material and specimen 5 is the uncoated cemented carbide. These two materials were used as reference materials in the tribological tests.

A comparison between Tables 1 and 2 shows that the deposition rate is strongly dependent on the methane partial pressure. In particular, the deposition rate becomes very low at conditions corresponding to deposition of amorphous boron carbide. In an introductory study, it was found to be impossible to grow boron carbide directly on uncoated cemented carbide, except at a very low methane partial pressure in the gas. It is believed that this is due to an inhibition of the reaction by absorbed hydrocarbons on the surface [9].

3.2. Morphology and microstructure

The surface morphology of the coatings was found to be strongly dependent on the carbon

TABLE 2 Phase composition, chemical composition and thickness values of boron carbide coatings

Specimen	*Phase*	*Carbon concentration* (at.%)	*Thickness of pins* (μm)	*Thickness of inserts* (μm)
1	$B_{13}C_2$	8.8 ± 0.3	2.9 ± 0.6	3.9 ± 0.6
2.1	$B_{13}C_2$	15.6 ± 0.3	2.6 ± 0.4	3.4 ± 0.4
2.2	$B_{13}C_2$	15.2 ± 0.3	5.9 ± 0.6	6.8 ± 0.6
2.3	$B_{13}C_2$	12.7 ± 0.3	11.5 ± 0.6	11.9 ± 0.6
3	Amorphous	22.5 ± 0.3	2.9 ± 0.6	3.5 ± 0.6
4	TiC coated cemented carbide		6.5 ± 0.3	4.9 ± 0.3
5	Uncoated cemented carbide			

Fig. 1. Surface morphology of the boron carbide coatings for (a) specimen 1; (b) specimen 2.1 and (c) specimen 3. (d) Fractured cross-section of boron carbide coating (specimen 1) with smooth boron carbide (at A), and columnar (at B) and equiaxed TiC (at C).

content. From Figs. 1(a) and 1(b), it can be seen that the rhombohedral boron carbide coatings display a nodular morphology with a microcrystalline fine structure. The nodule size is independent of carbon content but the size of the crystal facets on the nodules decreases significantly with an increasing carbon content (see Figs. 1(a) and 1(b)). The extreme high carbon content, amorphous coating (Fig. 1(c)) exhibits a perfectly globular morphology with no signs of crystallinity. In fractured cross-sections, all boron carbide coatings yield a smooth surface (Fig. 1(d)). Further, as a result of the boron carbide deposition an increased porosity at the TiC–cemented-carbide interface was obtained.

Examination of the rhombohedral boron carbide by TEM reveals that the coatings have a spherical growth structure which generates the nodular morphology observed on the surface (Fig. 2(a)). Electron diffraction and dark field imaging reveal that each nodule is composed of several grains and that there is a direct relation between the grain size and the crystal facets observed on the surface. The nodules are generally nucleated on cobalt-rich outgrowths from the interface (A in Fig. 2(b)). This observation indicates that trans-

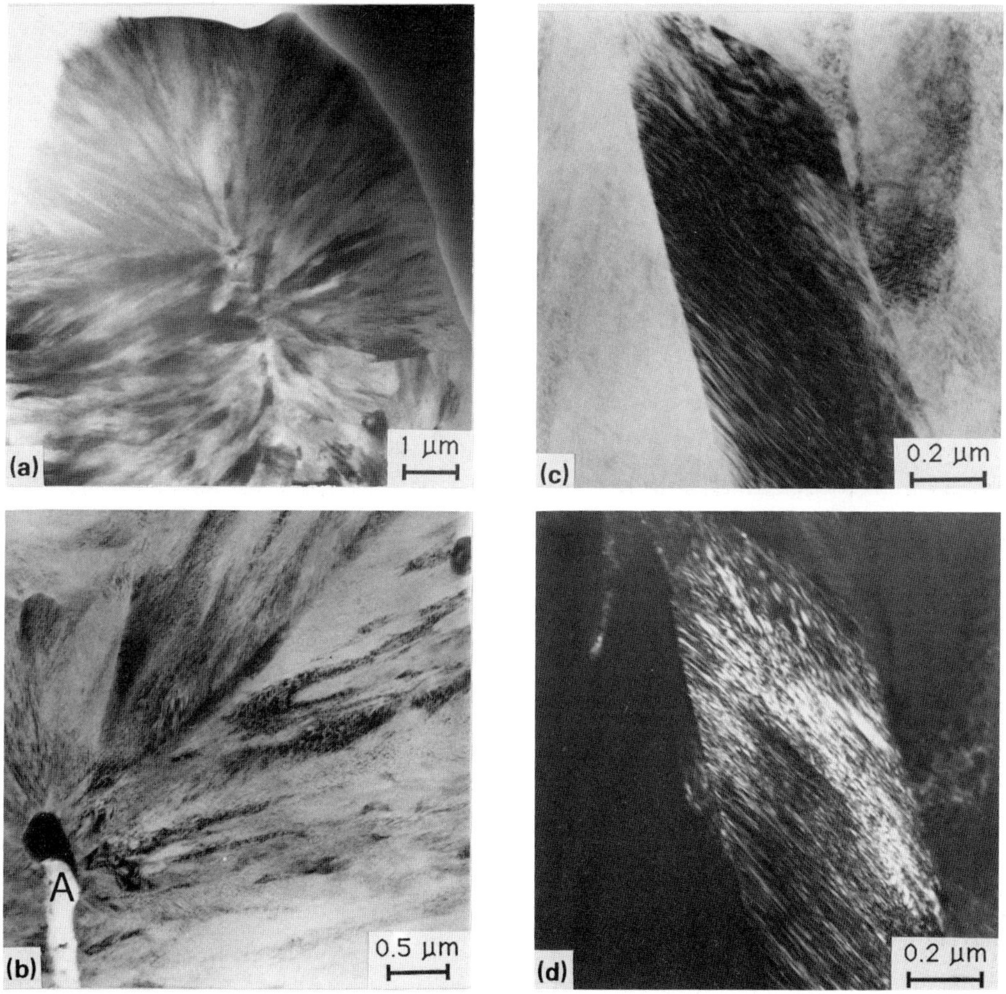

Fig. 2. TEM micrographs from rhombohedral boron carbide (specimen 1): (a) nodule in cross-section with spherical growth structure; (b) centre of a nodule with nucleation at cobalt-rich particle (at A); (c) detail of needle-like fine structure within the individual grains; (d) dark field micrograph corresponding to (c).

port of cobalt from the underlying substrate through the TiC coating occur during the deposition process. The observed porosity at the TiC–cemented-carbide interface is therefore believed to be due to diffusion of cobalt from the cemented carbide near the interface.

X-ray diffraction reveals that TiB_2 is formed at the boron carbide–TiC interface. Further, the formation of borides at the interface was supported by the observation of a number of four to five weak additional diffraction lines. Three of these lines were in agreement with the strongest lines expected from CoB. The X-ray line intensities show that the thickness of the interfacial layer increases with increasing deposition time and, in the case of the amorphous boron carbide, the layer is visible directly on fractured cross-sections.

Within the boron carbide grains, a fine structure resulting in a needle-like contrast is observed. The diameter of the needles is less than 1000 Å (Figs. 2(c) and 2(d)). The origin of this contrast is not fully understood yet, but it is believed to be related to local variations in the intrinsic stress field. This hypothesis is supported by the absence

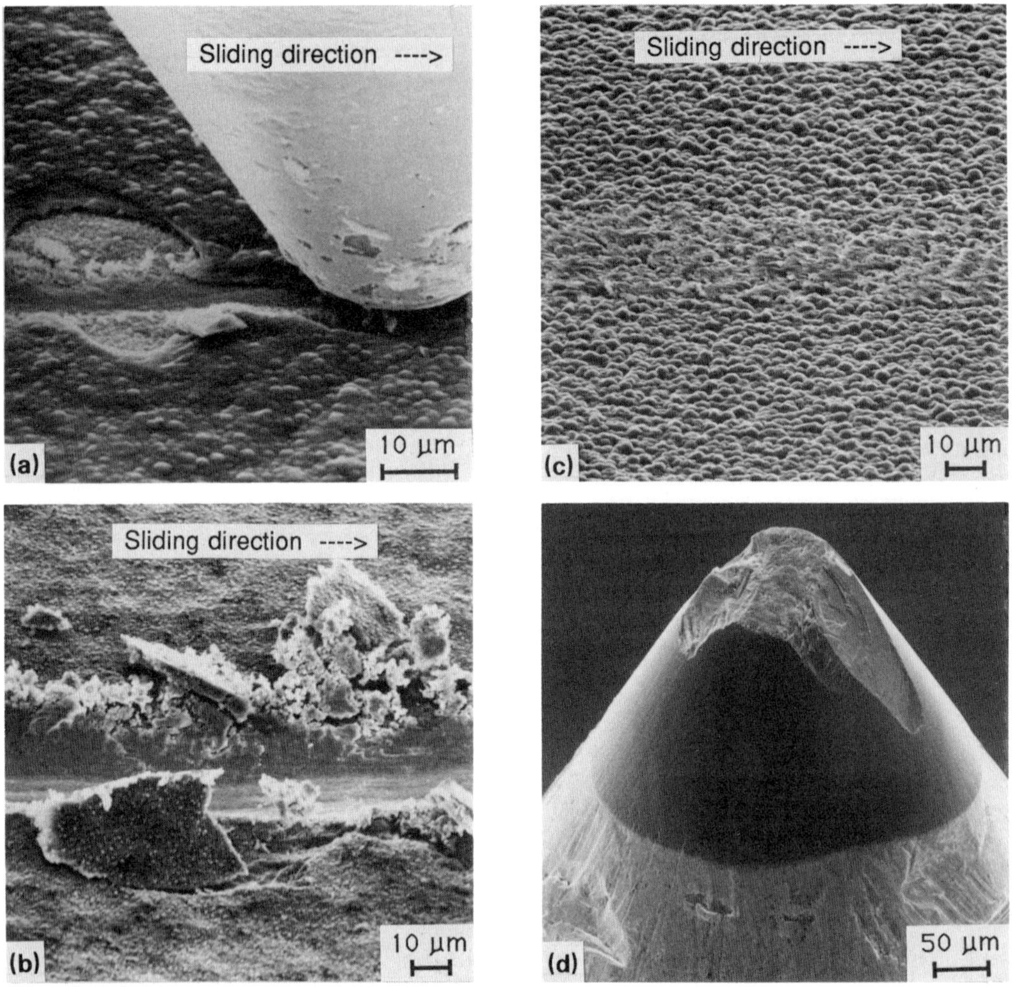

Fig. 3. (a) Spalling of the amorphous boron carbide at 0.9 N normal load in the scratch test; (b) spalling of TiC at 7.5 N normal load in the scratch test; (c) groove on $B_{13}C_2$ after scratching at 10 N normal load; (d) diamond stylus after scratch testing on rhombohedral boron carbide.

of Kikuchi lines in the diffraction pattern from thicker parts of the foil.

3.3. *Hardness and adhesion*

The hardness of thin coatings can only be measured directly on the coating surface if the depth of indentation is considerably less than the coating thickness [16]. This criterion was only fulfilled for specimen 2.3 in the present work which shows a Vickers hardness value of 4500 ± 300 HV (load, 1.0 N). For the other specimens, a method proposed by Vingsbo *et al.* [11] was applied. On the assumption of a crack geometry corresponding to $C = 0.5$ in their model, Vickers hardness values from 4000 to 4500 HV were obtained for the $B_{13}C_2$. The highest hardness value was recorded for the high carbon content $B_{13}C_2$ coatings (specimens 2.1 and 2.2), in agreement with the results of Niihara *et al.* [17]. The same technique yields a hardness value for the TiC coating of 2400 ± 100 HV. This value is considerably lower than the value of 3000 ± 200 HV which is measured directly on the surface of thick TiC coatings [18]. The discrepancy between the two hardness values for TiC indicates that the recorded values for $B_{13}C_2$ may be lower than the true values.

In the case of the amorphous boron carbide, a hardness value could not be established since the indentation induced spalling of the coating even at very low loads. This indicates extremely poor adhesion of the amorphous boron carbide to TiC.

In the scratch test, the critical load value for incipient spalling of the amorphous boron carbide was found to be 0.9 N (Fig. 3(a)). This value should be compared with a value of 7.5 N for the TiC coating (Fig. 3(b)). For the $B_{13}C_2$ coatings, it was impossible to induce spalling even at the maximum load (10.0 N) available. Figure 3(c) demonstrates that the surface damage is almost negligible at this high load. Examination of the diamond stylus after extended scratch testing of the rhombohedral boron carbide reveals that the tip has been severely damaged by chipping (Fig. 3(d)).

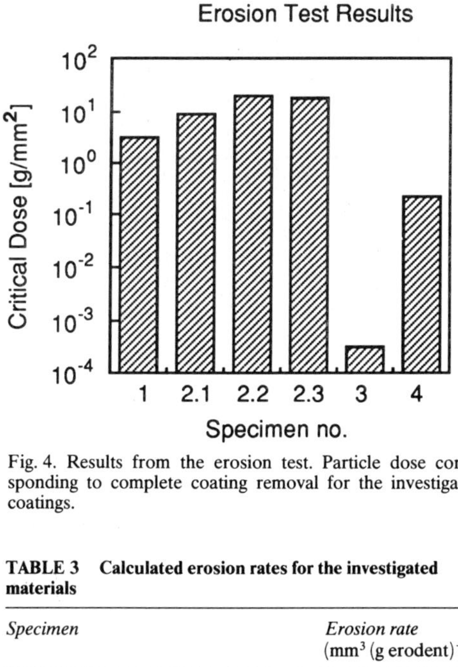

Fig. 4. Results from the erosion test. Particle dose corresponding to complete coating removal for the investigated coatings.

TABLE 3 Calculated erosion rates for the investigated materials

Specimen	*Erosion rate* (mm^3 (g erodent)$^{-1}$)
1	1.2×10^{-3}
2.1	3.8×10^{-4}
2.2	3.7×10^{-4}
2.3	6.5×10^{-4}
3	1.2×10^{1}
4	2.1×10^{-2}
5	2.3×10^{-2}

3.4. *Erosion resistance*

In the erosion tests the critical particle dose corresponding to complete coating removal was determined for the different coatings. The results are summarized in Fig. 4. From these experimental values, the corresponding erosion rates of the coatings were calculated (Table 3). The results clearly demonstrate the extremely high erosion resistance of the $B_{13}C_2$ compared with TiC and uncoated cemented carbide. Further, the results indicate that there exists an optimum coating thickness for erosion prevention and that the erosion resistance of rhombohedral boron carbide increases with increasing carbon content. The latter result agrees well with the observed increase in hardness at higher carbon content. Figure 5(a) shows the characteristic appearance of the eroded $B_{13}C_2$. A fairly smooth surface is produced with occasional signs of microchipping. Fragments of the alumina erodent can also be found on the surface.

Penetration of the $B_{13}C_2$ is first observed as incipient microchipping on the surface. At a certain critical particle dose, extensive spalling begins to develop rapidly until the boron carbide is completely removed (Figs. 5(b) and 5(c)). Exposed TiC is almost never found on the spalled areas, which indicates that spalling mainly takes place along the TiC–cemented-carbide interface, *i.e.* not along the boron carbide–TiC interface. This indicates that the TiC–cemented-carbide interface is the weakest of the two interfaces. This is supported by the observation that the particle dose required to induce spalling of TiC without a boron carbide coating is much lower than for $B_{13}C_2$.

In contrast with the crystalline rhombohedral boron carbide, the amorphous coating displays an inferior erosion resistance, even compared with TiC. This is not unexpected, considering the poor adhesion observed earlier. Figure 5(d) illustrates how the coating is removed by spalling already after a few single impacts on the surface.

3.5. *Wear resistance in simulated tool wear test*

The results from the simulated tool wear test are summarized in Fig. 6. The results are disappointing with respect to boron carbide as a potential tool material. For both counter materials, all boron carbide coatings tested yield higher wear rates than the TiC-coated substrate. There is a tendency towards an increasing wear rate with an increasing coating thickness. The amorphous

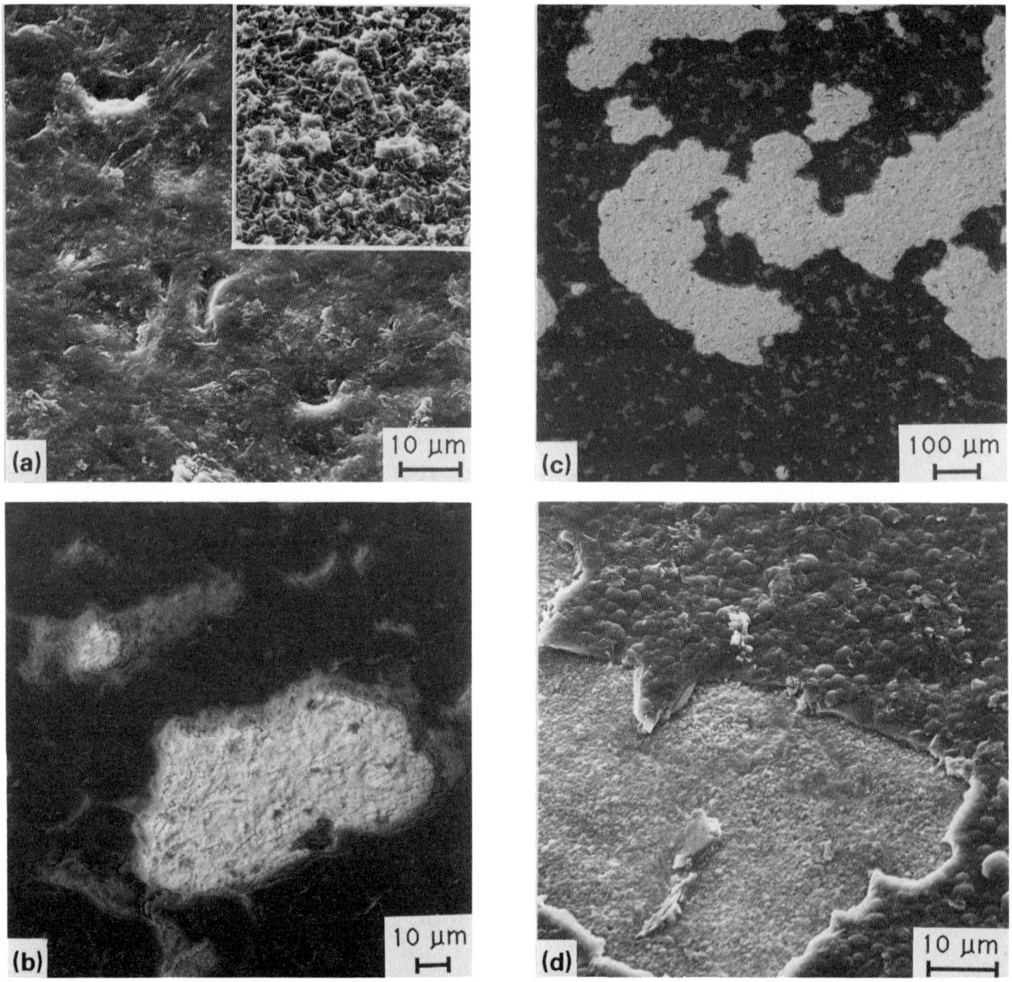

Fig. 5. Characteristic features of eroded surfaces: (a) smooth surface of $B_{13}C_2$ with signs of microchipping and alumina fragments (the inset shows the as-deposited surface morphology); (b), (c) spalling of the $B_{13}C_2$ and the underlying TiC once the critical particle dose has been exceeded (SEM; backscatter electron mode); (d) early spalling of the amorphous boron carbide.

boron carbide yields the least reduction in wear resistance of all the tested coatings. This result is surprising, considering the extremely poor erosion resistance and adhesion of the amorphous material. Therefore, a careful metallographic investigation of the worn specimens was performed to identify the active wear mechanisms.

Figure 7(a) gives an example of the characteristic wear topography of boron carbide from the simulated tool wear test against quenched-and-tempered steel. It is seen that a smooth topography is produced, resembling that of a polished surface. This could be interpreted as a mild abrasive wear mechanism. However, abrasive wear is unlikely in view of the extremely high hot hardness of boron carbide; for example ref. 19 gives a value of 2170 HV at 900 °C. Instead, it is proposed that the topography is an effect of wear by chemical dissolution of the boron carbide into the counter material. This hypothesis is supported by theoretical estimates from the available models in the literature [20, 21]. Using the equations given by Kramer and Suh [21] the solution wear resistance of B_4C was computed to be approxi-

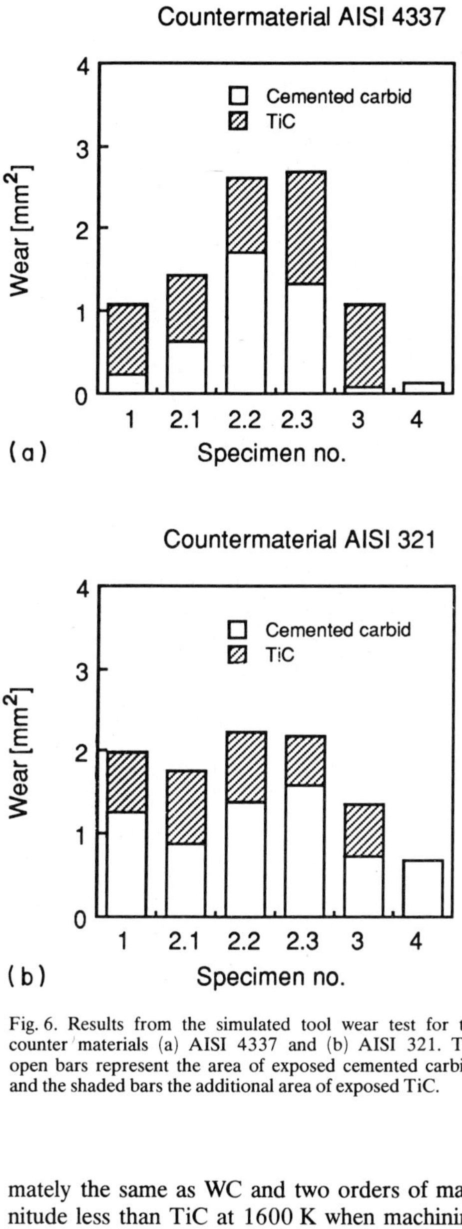

Fig. 6. Results from the simulated tool wear test for the counter materials (a) AISI 4337 and (b) AISI 321. The open bars represent the area of exposed cemented carbide and the shaded bars the additional area of exposed TiC.

mately the same as WC and two orders of magnitude less than TiC at 1600 K when machining α-Fe. In agreement with these predictions, solution wear is also observed for the cemented carbide (WC) while the TiC is worn by localized surface shear and fracture (Fig. 7(b)).

In addition to solution wear, spalling of the boron carbide is also observed. Spalling may occur along the boron carbide–TiC interface, but spalling along the underlying TiC–cemented-carbide interface is also frequently observed. Spalling is predominantly found on the entrance side of the wear scar.

For the austenitic stainless steel counter material, spalling becomes the dominant wear mechanism (Figs. 7(c) and 7(d)). The tendency of the spalling to extend down into the cemented-carbide interface is even more pronounced with this counter material. This suggests that the higher wear rate of the boron carbide compared with TiC is not an effect of poor adhesion to the TiC. This is also seen from the fact that the amorphous boron carbide yields the best results in spite of its strong tendency of spalling. Instead, it is believed that the addition of an outer layer of boron carbide results in a decrease in fracture toughness. This result is in good agreement with the findings of Stjernberg [22] on fracture toughness of TiC coatings.

4. Discussion and conclusions

The present investigation has shown that it is possible to deposit $B_{13}C_2$ of high hardness on TiC-coated cemented carbide. Dense coatings with good adhesion characteristics can be obtained by control of the deposition conditions. However, further work will be required in order to gain a better understanding of the nucleation and growth processes during deposition and how these processes influence the properties of the coating.

The rhombohedral boron carbide shows great potential as an erosion-resistant material. However, the deposition temperature is too high for most of the substrate materials used in erosive environments. The results from the simulated tool wear test do not indicate any beneficial effect from boron carbide in the machining of steel. The wear mechanism analysis suggests that the boron carbide does not have the required chemical stability in contact with steel at the high temperatures generated at the tool–workpiece interface during machining. This result may not apply to other types of work material. For example, theoretical estimates indicate that boron carbide may be more suitable in the machining of titanium [23]. The very high erosion–abrasion resistance of boron carbide may also make it a tool material candidate for machining alloys with a high content of abrasive inclusions. Therefore, additional tribological tests should be carried out.

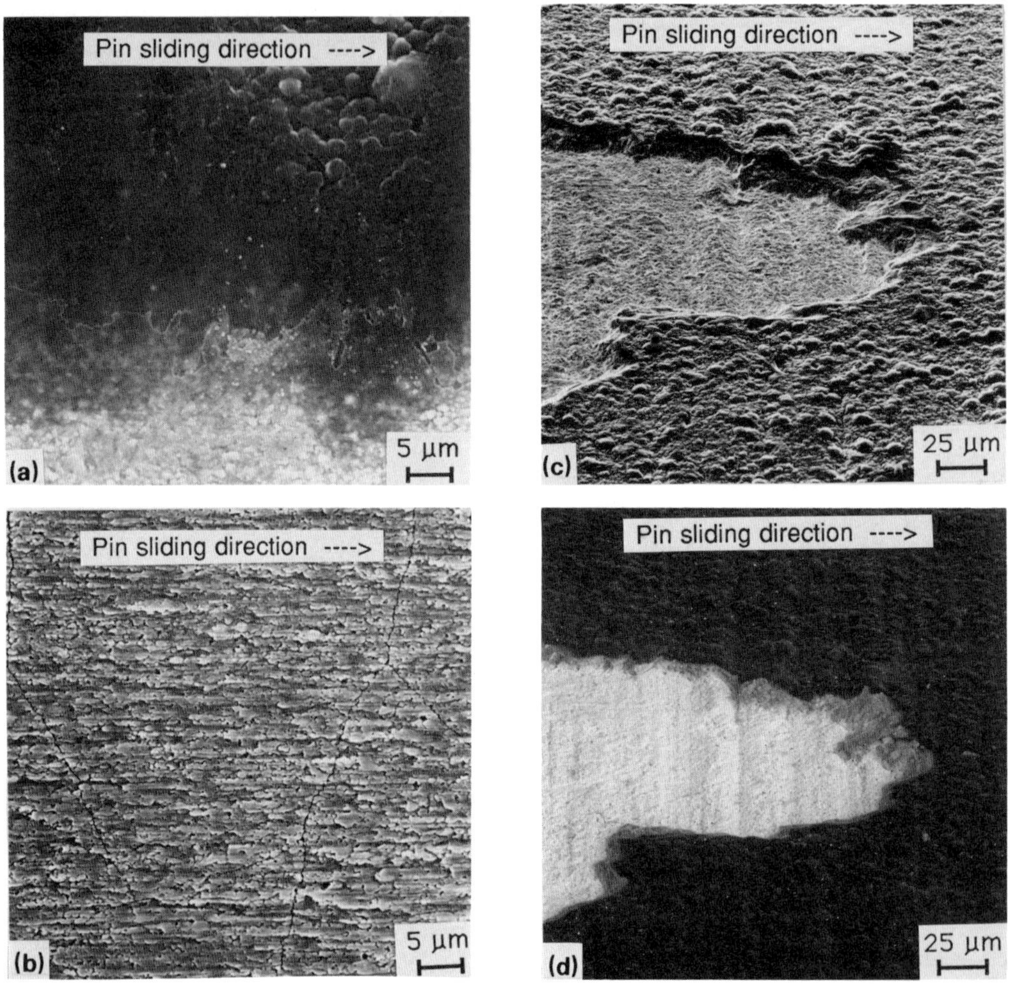

Fig. 7. Characteristic wear damage of coatings from the simulated tool wear test: (a) smooth wear topography of boron carbide suggesting solution wear (unaffected TiC is seen where the boron carbide has been completely removed) (counter material, AISI 4337); (b) surface shearing of TiC (counter material, AISI 4337); (c), (d) spalling of boron carbide and TiC at the rear side of the wear scar (counter material, AISI 321), in the SEM secondary-electron mode and SEM backscatter electron mode.

Acknowledgments

The financial support of the National Swedish Board for Technical Development (STUF) is acknowledged. The authors also wish to express their gratitude to Mr. Per Hedenqvist for valuable help in the experimental work.

References

1 H. Holleck, *Chem. Ztg., 106* (1982) 213–224.
2 D. Lange, Verschleissfeste Werkstoffe auf Borcarbidbasis, *Thesis,* Universität Karlsruhe, Kernforschungszentrum Karlsruhe, KfK 4009, 1985.
3 L. G. Vandenbulcke, Theoretical and experimental studies on the chemical vapour deposition of boron carbide, *Ind. Eng. Chem. Prod. Res. Dev., 24* (1985) 568–575.
4 U. Jansson and J.-O. Carlsson, *Thin Solid Films, 124* (1985) 101–107.
5 S. Lartigue, D. Cazajous and G. Male, Depots chimiques en phase vapeur de carbure de bore sur carbure de tungstène fritte et sur acier, *J. Phys. (Paris), Colloq. C1,* (1986) 197–202.
6 M. E. Sjöstrand and A. G. Thelin, *Proc. 8th Int. Conf. on*

Vacuum Metallurgy, Linz, September 30–October 4, 1985, Vol. 3, Eisenhutte Österreich, Vienna, 1985, pp. 321–338.
7 V. K. Sarin, *Proc. 7th Int. Conf. on Chemical Vapour Deposition,* Electrochemical Society, Princeton, NJ, 1979, p. 476.
8 W. D. Sproul and M. H. Richman, *J. Vac. Sci. Technol., 12* (1975) 842.
9 B. Stridh, M. Olsson, S. Söderberg, U. Jansson and J.-O. Carlsson, *Rep. UPTEC 8745 R,* 1987 (Uppsala University).
10 B. Stridh, M. Olsson, S. Söderberg, U. Jansson and J.-O. Carlsson, *Rep. UPTEC 8744 R,* 1987 (Uppsala University).
11 O. Vingsbo, S. Hogmark, B. Jönsson and A. Ingemarsson, in P. J. Blau and B. R. Lawn, (eds.), *Microindentation Techniques in Materials Science and Engineering, ASTM Spec. Tech. Publ. 889,* 1986, p. 257.
12 J. Valli, U. Mäkälä and A. Matthews, *Surf. Eng., 2* (1986) 49–53.
13 S. Söderberg, S. Hogmark, U. Engman and H. Swahn, *Tribol. Int., 14* (1981) 333–343.
14 M. Olsson, B. Stridh, S. Söderberg and U. Jansson, in K. C. Ludema (ed.), *Proc. Int. Conf. on Wear of Materials, Houston, TX, 1987,* Vol. 2, American Society of Mechanical Engineers, New York, 1987, pp. 505–516.
15 E. Amberger, M. Druminski and K. Ploog, *J. Less-Common Met., 23* (1971) 43–52.
16 A. I. Il'inskii and G. E. Lyakh, *Ind. Lab. (Engl. Transl.), 44* (1978) 1719.
17 K. Niihara, A. Nakahira and T. Hirai, *J. Am. Ceram. Soc., 67* (1984) C13–C14.
18 B. Stridh, P. Hedenqvist, M. Olsson and S. Söderberg, *Proc. 7th Int. Conf. on Erosion by Liquid and Solid Impact, Cambridge, Cambridgeshire, 1987,* Cavendish Laboratory, Cambridge, 1987, p. 19/1–8.
19 *Gmelin Handbook of Inorganic Chemistry, Boron,* Supplement Vol. 2, Springer, Berlin, 8th edn., 1981.
20 P. A. Dearnley, *J. Eng. Mater. Technol., 107* (1985) 68–82.
21 B. M. Kramer and N. P. Suh, *J. Eng. Ind., 102* (1980) 303–309.
22 K.-G. Stjernberg, *Met. Sci., 5* (1980) 189.
23 E. D. Whitney, *Powder Metall. Int., 6* (1974) 73.

Materials Science and Engineering, A105/106 (1988) 465–480

Chemical Vapour Deposition Coating of Crystalline Si_3N_4 on a Quartz Crucible for Nitrogen-doped Czochralski Silicon Crystal Growth*

H. DOI, N. KIKUCHI and Y. OOSAWA

Central Research Institute of Mitsubishi Metal Corporation, Omiya (Japan)

(Received November 10, 1987)

Abstract

A dense crystalline α-Si_3N_4 layer was formed on a quartz substrate by chemical vapour deposition (CVD) from an $SiCl_4$–NH_3–H_2 gas mixture under the following reaction conditions: NH_3-to-$SiCl_4$ ratio, 0.7–0.9; temperature, 1310 °C; pressure, 5–9 Torr; the Si_3N_4 deposition rate is 30–40 μm h^{-1}. A Si_3N_4-coated crucible remained almost intact after immersion in reactive molten silicon for 24 h; the rate of erosion of Si_3N_4 layer was as small as 5–6 μm day^{-1}. On the basis of these test results an α-Si_3N_4 approximately 300 μm thick layer was formed by CVD on the inner surface of a 7 in quartz crucible. Czochralski silicon crystal pulling tests were carried out using the coated crucible. The result indicates that solution of α-Si_3N_4 from the coated crucible and precipitation of β-Si_3N_4 whiskers disturbed single-crystal growth and led to polygonization of the grown crystal. The concentrations of nitrogen and oxygen in the Czochralski crystal were $(5-12)\times10^{15}$ cm^{-3} and $(5-16)\times10^{16}$ cm^{-3} respectively.

1. Introduction

Chemical vapour deposition (CVD) from a number of gaseous precursors has been well documented and the relevant CVD phase diagrams have been calculated [1]; deposition at $T<1500$ K results in amorphous deposits and deposition at $T<1500$–1650 K in general results in crystalline deposits of the α form. Niihara and Hirai [2] showed that crystalline Si_3N_4 up to 4.6 mm thick can be deposited on a hot graphite substrate using an $SiCl_4$–NH_3–H_2 mixture, the relevant deposition rates being as high as 0.7–1.2 mm h^{-1}. Si_3N_4 in a bulk form, such as in a crucible form, can be deposited on graphite; the graphite may be later burned off. Watanabe *et al.* [3] made such a crystalline Si_3N_4 crucible using a graphite substrate and investigated Czochralski silicon growth; the Si_3N_4 crucible enables a nearly oxygen-free crystal to be grown, which is useful for some device applications. For such an objective, oxygen contamination during CVD of Si_3N_4 should be strictly avoided.

Now, the use of quartz as a substrate for thick deposition of crystalline Si_3N_4 from an $SiCl_4$–NH_3–H_2 mixture appears to be attractive for certain applications such as Czochralski silicon growth on account of the very high purity of the product and comparatively low material and processing costs. CVD conditions in this case may differ from those for deposition on the hot graphite substrate. Si_3N_4–quartz thermal expansion mismatch and possible softening of quartz at the CVD temperature may be a problem. To the present authors' knowledge, no systematic investigations have been reported on the CVD coating of dense crystalline Si_3N_4 on a quartz crucible to be used for Czochralski silicon growth.

The purpose of the present work is to deposit a thick layer of crystalline Si_3N_4 on the inner surface of a quartz crucible. It will be shown that a CVD-Si_3N_4-coated quartz crucible remained intact during Czochralski silicon growth.

2. Experimental details

The CVD apparatus used in this experiment is shown in Fig. 1. The reactor was constructed by utilizing a part of a Czochralski silicon single-crystal puller. A 10 or 7 in quartz crucible was placed on a rotating graphite susceptor. The reactor was first evacuated by a mechanical booster and a rotary pump to about 10^{-3} Torr;

*Paper presented at the 3rd International Conference on the Science of Hard Materials, Nassau, The Bahamas, November 9–13, 1987.

0921-5093/88/$3.50 © Elsevier Sequoia/Printed in The Netherlands

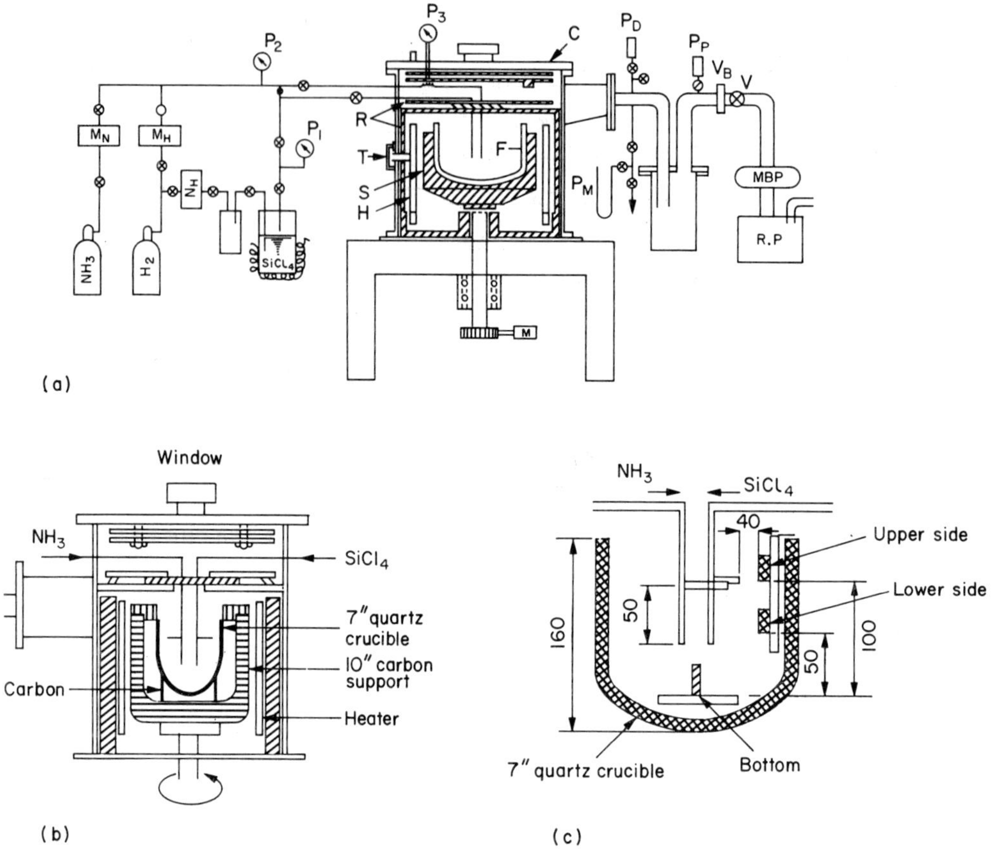

Fig. 1. (a) Outline of the CVD equipment for Si_3N_4 coating on a quartz crucible: C, reaction chamber 400 cm in diameter and 560 cm high; F, quartz crucible; H, carbon heater 288 cm in diameter; M, motor; M_N, M_H, mass flow meters; MBP, mechanical booster pump; P_1, P_2, P_3, pressure gauges; P_D, diaphragm-type vacuum gauge; P_B, Pirani-type vacuum gauge; P_H, mercury manometer; R, carbon thermal shield; R_p, rotary pump; S, split-type 10 in carbon support; T, thermocouple (Pt–(13wt.%Rh–Pt)); V_B, butterfly valve. (b) Schematic drawing showing the geometrical arrangement for CVD Si_3N_4 coating on the inner surface of a 7 in crucible. (c) Schematic drawing showing the mode of introduction of reaction gases into a 7 in quartz crucible and the location of the quartz specimen on the inner surface for the Si_3N_4 deposition test.

then the quartz crucible was heated to 800 °C in 2 h with a graphite resistance heater and thereafter quickly heated to 1400 °C in another $\frac{1}{2}$ h. Before CVD tests, the temperature distribution inside the crucible was measured using a Pt–(13wt.%Rh–Pt) thermocouple contained in an evacuated quartz tube.

Si_3N_4 was deposited from the vapour phase by a reaction of $SiCl_4$ with NH_3 under a reduced pressure on a transparent amorphous quartz (referred to as a standard grade quartz) substrate 25 mm × 35 mm × 2 mm in size and placed near the bottom of a quartz crucible. The reactant gases were fed into the reactor with H_2 as a carrier gas through two separate quartz pipes. Mixing of the reactant gases occurs inside the rotating crucible before Si_3N_4 deposition takes place. The distance between the gas outlet and the substrate was varied by moving the gas outlet position vertically. The NH_3-to-$SiCl_4$ ratio, reaction temperature and pressure, and gas-outlet-to-substrate distance were varied to find an optimum reaction condition. X-ray diffraction analyses of Si_3N_4 deposits were made under various CVD reaction conditions and scanning electron microscopy (SEM) observations of the free and fracture surfaces of the deposits were carried out. Since Si_3N_4 is an insulator, SEM observations were made after deposition of a thin gold film on the surfaces.

The resistance of a Si_3N_4-coated quartz specimen against erosion by reactive molten silicon was evaluated under static and dynamic conditions. A static erosion test means dipping the specimen for 24 h in a stationary molten silicon pool, while in a dynamic test the specimen is dipped in molten silicon which is moving in a rotating quartz crucible. Both standard grade quartz substrate and another type of substrate cut from a quartz crucible primarily used for Czochralski silicon crystal pulling (referred to as a crucible grade quartz) were utilized as the substrate materials for Si_3N_4 coatings to be used for the erosion tests, for comparison. Erosion and morphological changes in the Si_3N_4 deposit made on quartz substrate by molten silicon were examined.

Si_3N_4 deposits as thick as 250–350 μm were obtained on the inner surface of a 7 in quartz crucible by placing it inside a 10 in graphite susceptor as shown in Fig. 1(b). Czochralski silicon crystal pulling tests were carried out using a standard-type silicon crystal puller; the mass of initial molten silicon was 3–4 kg. The structural and morphological changes in the Si_3N_4 deposit on the quartz crucible were examined. Moreover, the oxygen and nitrogen contents in the Czochralski silicon crystal incorporated from the Si_3N_4-coated quartz crucible were determined through Fourier transform infrared spectroscopy measurements.

3. Experimental results

3.1. Temperature distribution inside the quartz crucible and the chemical vapour deposition conditions

The temperature distribution inside a 10 in quartz crucible which is rotating at a speed of 3–5 rev min^{-1} is shown in Fig. 2. During the temperature distribution measurements, a constant stream of mixed gases consisting of H_2 at 1.0 l min^{-1} and N_2 at 0.4 l min^{-1} were fed into the reactor under a pressure of 40 Torr. By placing several graphite circular plates as thermal shields in the upper part of the reactor, a comparatively uniform temperature distribution inside the crucible was established. When a regulating temperature near the graphite heater was set at 1400 °C, temperatures at various positions inside the crucible were found to be within 1310 ± 10 °C as shown in Fig. 2; position 4 corresponds to a typical gas outlet position. As a result of preliminary deposition tests, the appropriate regulating temperature to achieve a dense crystalline Si_3N_4 deposit was found to be between 1390 and 1400 °C. CVD coating tests were carried out under a NH_3-to-$SiCl_4$ ratio of 0.4–1.8 and a pressure of 4–40 Torr. Another important variable which affects the properties of the Si_3N_4 deposit is the distance from the reactant gas outlet to the quartz substrate; it was varied between 60 and 125 mm. Table 1 shows the reaction conditions for Si_3N_4 CVD coating on a quartz substrate and the thickness of Si_3N_4 deposit.

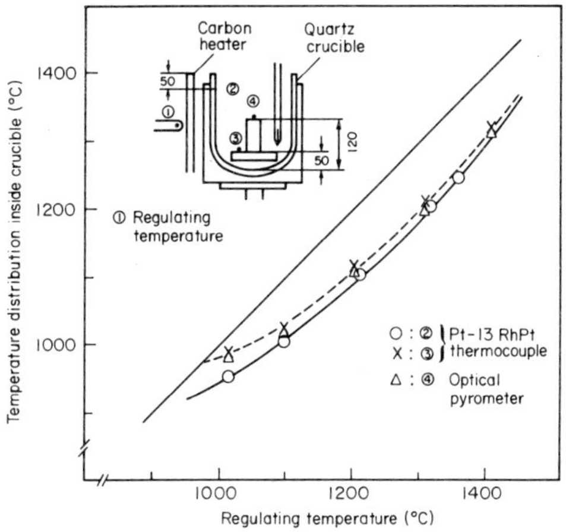

Fig. 2. Variation in temperature at various points inside the crucible with increase in the regulating temperature.

3.2. Morphology and structure of Si_3N_4 deposits made under various chemical vapour deposition conditions

Figure 3 shows the surface and fracture morphology of Si_3N_4 deposits made under various reaction pressures. It should be noted that the surface morphology of Si_3N_4 deposited under a comparatively high pressure of 15 Torr is characterized by globular growth morphology (specimen 3). Furthermore, the fracture surface of the deposit is porous and amorphous like. When the reaction pressure was reduced to 7 Torr, the surface morphology of the Si_3N_4 deposit changed to a fairly flat surface consisting of well-developed crystalline facets (specimen 16). Deposition under an intermediate pressure gave a morphology which bears both features concurrently. A further reduction in the reaction pressure results in the growth of comparatively thick whiskers from the crystalline surface. These results indi-

TABLE 1 Chemical vapour deposition reaction conditions for deposition of Si_3N_4 on a quartz substrate

Specimen	*Reaction temperature* (°C)	*Reaction pressure* (Torr)	*Reaction time* (h)	*Gas flow rate* (ml min^{-1})			*NH_3-to-$SiCl_4$ ratio*	*Substrate-to-nozzle distance* (mm)	*Film thickness* (μm)
				$SiCl_4$	*NH_3*	*H_2*			
1	1320	40	2	220	100	500	0.45	60	—
2	1320	70	2	110	80	500	0.73	60	—
3	1300	15	2	110	80	500	0.73	70	50
4	1300	12	2	110	90	500	0.82	90	140
5	1300	8 ~ 9	2	110	90	500	0.82	125	8
6	1310	6	2	220	90	600	0.41	125	0
7	1310	6	2	220	90	700	0.41	120	15
8	1310	10	2	220	200	700	0.91	100	110
9	1310	6	2	165	80	500	0.48	100	20
10	1310	4–5	4	110	90	400	0.82	80	30
11	1310	7	4	110	80	400	0.73	80	100
12	1310	35	4	110	80	400	0.73	80	140
13	1310	7	4	110	200	400	1.82	90	90
14	1310	6	3	110	60	350	0.55	90	60
15	1310	30	3.5	110	90	350	0.82	90	50
16	1310	7	6	110	90	350	0.82	90	90
17	1310	9	4	110	80	350	0.73	125	150
18	1310	8	8	160	100	450	0.63	125	200
19	1310	5	4	110	80	450	0.73	120	140
20	1310	6	4	135	80	450	0.59	90	150

TABLE 2 Chemical vapour deposition reaction conditions for deposition of Si_3N_4 on a quartz substrate for corrosion tests

Specimen	*Reaction temperature* (°C)	*Reaction pressure* (Torr)	*Reaction time* (h)	*Gas flow rate* (ml min^{-1})			*NH_3-to-$SiCl_4$ ratio*	*Substrate-to-nozzle distance* (mm)	*Film thickness* (μm)
				$SiCl_4$	*NH_3*	*H_2*			
21	1310	7	8	160	100	450	0.63	120	350
22	1310	9	6	160	100	450	0.63	120	400
23	1310	9	8	160	100	450	0.63	120	250
24	1310	7	7.5	160	100	450	0.63	120	190
25	1310	8	8	160	100	450	0.63	120	185
26	1310	8	8	180	100	450	0.56	120	180
27	1310	9	7	150	110	470	0.73	120	180
28	1310	11	8	160	110	600	0.69	120	250
29	1310	8	8	160	110	500	0.69	120	130

cate that optimum reaction pressures for depositing dense crystalline Si_3N_4 with a comparatively smooth surface are 6–9 Torr. X-ray diffraction analyses of Si_3N_4 deposits obtained under various reaction pressures reveal that a comparatively high reaction pressure gives a poorly crystallized α-Si_3N_4 deposit with weak diffraction lines. Lowering the pressure increases the crystallinity of the deposit and leads to the occurrence of preferred orientations such as (102) and (002). The most dominant preferred orientation depends on the reaction pressure: (102) at 7 Torr and (002) at 5 Torr.

Figure 4 shows the effect of the NH_3-to-$SiCl_4$ ratio on the morphology of free and fracture surfaces of Si_3N_4 deposits. A dense deposit with well-developed crystalline facets was obtained at a NH_3-to-$SiCl_4$ ratio of 0.73. Higher ratios led to a globular surface morphology, a situation which is very similar to Si_3N_4 deposition under comparatively high reaction pressures (see Fig. 3). Lower ratios also led to roughened surfaces although the Si_3N_4 deposits remain dense and crystalline. Thus the NH_3-to-$SiCl_4$ ratios between 0.7 and 0.9 were found to give a dense crystalline Si_3N_4 deposit with a comparatively smooth surface.

3.3. Observations of the erosion of an Si_3N_4 deposit on quartz substrates by molten silicon

On the basis of the systematic Si_3N_4 deposition experiments mentioned in Section 3.2, Si_3N_4 deposits as thick as 130–250 μm were obtained on both standard grade and crucible grade quartz

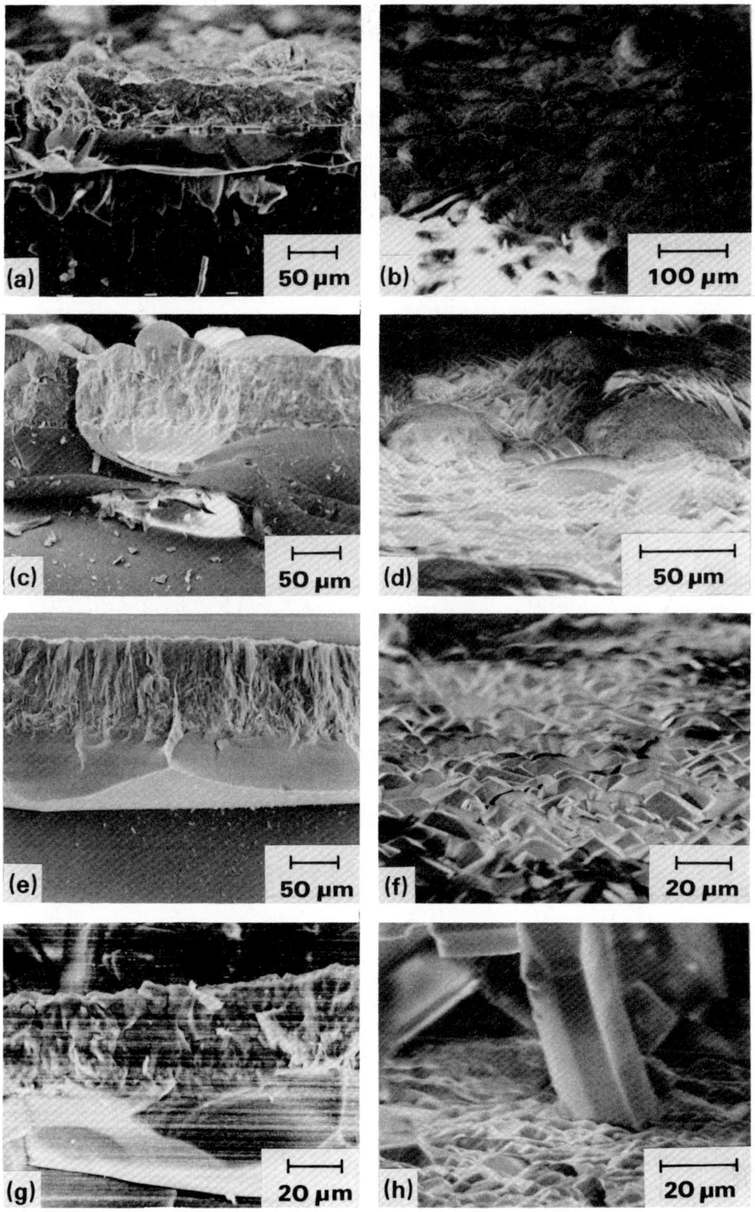

Fig. 3. Effect of reaction pressure in the CVD Si_3N_4 coating conditions on the surface and fracture morphology of Si_3N_4 deposits on a quartz substrate: (a), (b) run 3, 15 Torr; (c), (d) run 5, 9 Torr; (e), (f) run 16, 7 Torr; (g), (h) run 10, 5 Torr.

substrates. The conditions for CVD deposition are shown in Table 2. In many cases the deposit was highly dense and characterized by a relatively smooth surface morphology with well-developed crystalline facets. However, some deposits were characterized by a globular growth morphology. The very large scatter in the thicknesses of the deposits obtained under nearly identical reaction conditions should be noted; this appears to correspond to the effect of small fluctuations in some

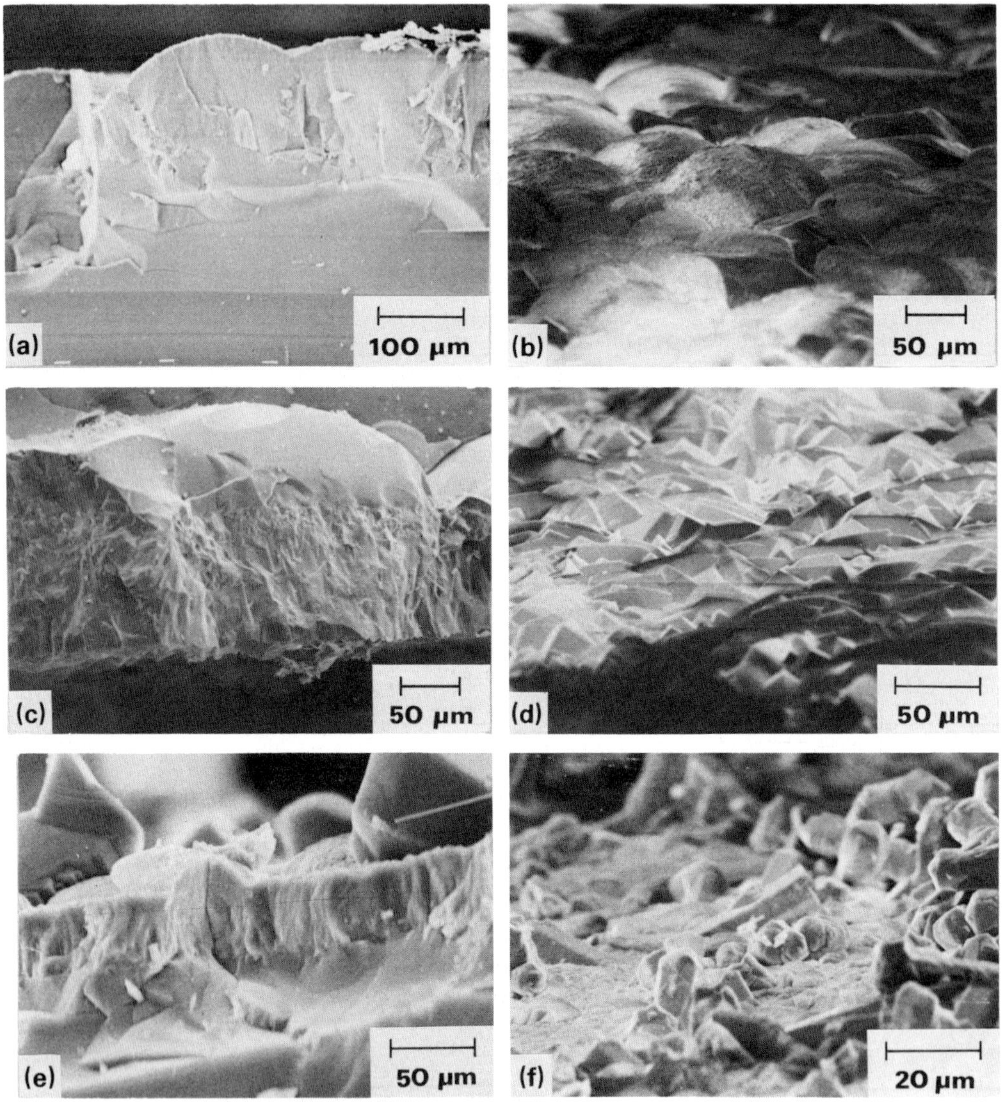

Fig. 4. Effect of the NH_3-to-$SiCl_4$ ratio in the CVD Si_3N_4 coating conditions on the surface and fracture morphology of Si_3N_4 deposits on quartz substrates: (a), (b) run 13, NH_3-to-$SiCl_4$ ratio of 1.8; (c), (d) run 17, NH_3-to-$SiCl_4$ ratio of 0.73; (e), (f) run 7, NH_3-to-$SiCl_4$ ratio of 0.41.

of the experimental parameters. Si_3N_4-deposited quartz specimens were subjected to static and dynamic erosion tests in a Czochralski silicon crystal puller.

3.3.1. Static erosion test results

Standard grade and crucible grade quartz specimens with coatings 300–350 μm thick (Table 2, specimen 21) were dipped for 24 h in a stationary molten silicon pool, the temperature of which was estimated to be 1340–1350 °C. Figure 5 shows the surface morphology of the specimens after the static erosion test. It should be noted that cracks are visible on the Si_3N_4 deposit on the standard grade quartz substrate. In comparison, the Si_3N_4 deposit on the crucible grade quartz substrate remained almost intact. Investigations of the origin of cracking in the former case disclosed that, in the near-surface region of the standard grade quartz substrate, extensive generation of voids occurred and, moreover, crystal transformation from the amorphous to the crystalline state (cristobalite and/or tridymite) took place. The resulting thermal expansion mismatch is believed to have generated cracks near the interface region and throughout the Si_3N_4 deposit.

Fig. 5. Effect of static erosion on the appearance of Si_3N_4-coated standard grade quartz specimen (right) and crucible grade quartz specimen (left).

The erosion-tested specimens were then immersed in a solution consisting of 1 part of 50% HF to 3 parts of concentrated HNO_3 to dissolve the silicon attached on the surface. The change in the mass of the specimen before and after the static erosion test is shown in Table 3. The mass loss due to erosion of the Si_3N_4 deposit for the standard grade quartz substrate was as high as 8%. In contrast, for the crucible grade quartz substrate, the mass loss was only 2.5%. X-ray diffraction analysis before and after the static erosion test indicated that the crystal structure is largely unchanged. However, microstructural observations reveal that anomalous growth of comparatively thick Si_3N_4 faceted crystals had preferentially taken place from the region near the silicon melt surface, as shown in Fig. 6; these faceted crystals were found to be of β-Si_3N_4 type structure. In contrast, in other regions which correspond to positions at considerable distances

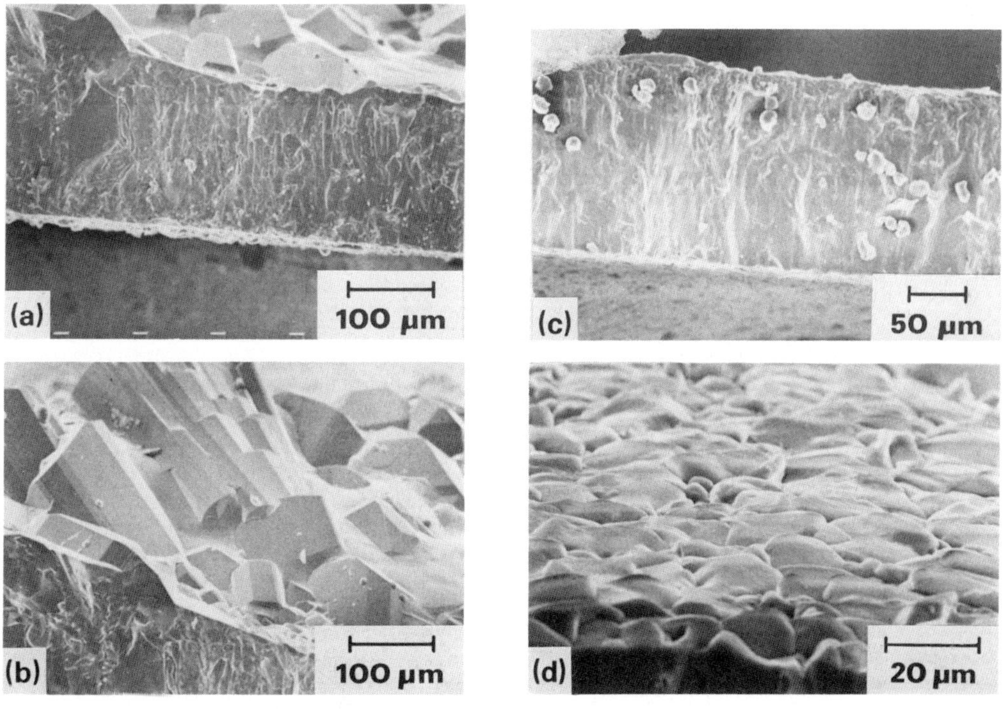

Fig. 6. (a), (b) growth of β-Si_3N_4 thick whiskers due to static erosion from the inner surface of Si_3N_4-coated crucible near the molten silicon surface; (c), (d) the morphology changes to a rounded hilly appearance towards the region away from the surface. The regions of observations are indicated by arrows in Fig. 5.

TABLE 3 Mass change of Si_3N_4 deposits on a quartz substrate due to static erosion tests

	Standard grade quartz substrate	*Crucible grade quartz substrate*
Mass before Si_3N_4 coating (g)	4.1464	11.6194
Mass after Si_3N_4 coating (g)	5.6714	13.4080
Mass of Si_3N_4 layer before static erosion test (g)	1.5250	1.7886
Mass of Si_3N_4 layer after static erosion test (g)	1.4118	1.7447
Amount of eroded Si_3N_4 (%)	8.1	2.5

TABLE 4 Dynamic erosion tests of Si_3N_4-coated quartz specimens simulating Czochralski silicon crystal growth steps

Target in simulation test	*Temperature of Si melt* (°C)	*Erosion time* (h)	*Crucible rotation speed* (rev min^{-1})	*Specimen*	*Thickness of Si_3N_4 layer* (μm)
Si crystal pulling temperature	1340	24	10	27	100–130
				28	120–200
Si ingot melting temperature	1450	1.5	3	28	120–200
Intermediate temperature between melting and pulling	1390	24	10	25	100–180
Effect of temperature change (melting–pulling)	1450–1340	25.5	3–10	25	100–180
				29	80–130

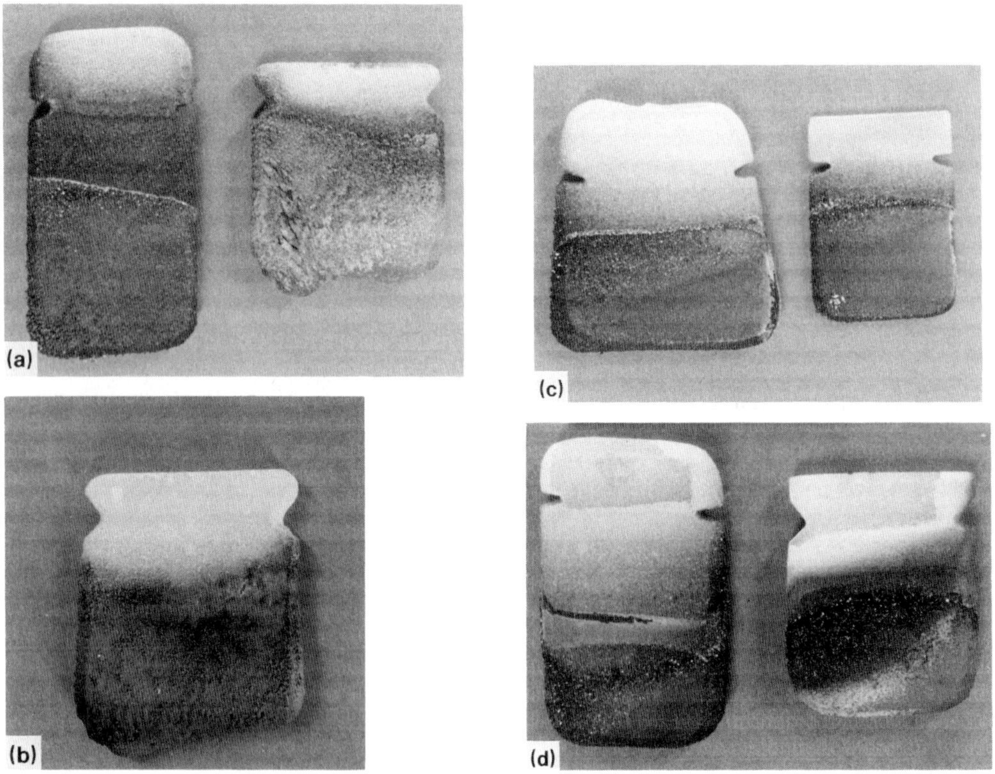

Fig. 7. Effect of the temperature of molten silicon on the appearance of the Si_3N_4-coated crucible grade quartz specimens subjected to dynamic erosion. The temperatures of the molten silicon near the crucible are (a) 1340 °C, (b) 1450 °C, (c) 1390 °C and (d) 1450–1340 °C.

from the melt surface, the original well-developed crystalline facets changed to a much more rounded hilly appearance. This seems to indicate that the α-Si_3N_4 deposit gradually dissolves from the region below the melt surface and precipitates preferentially on the region near the melt surface as β-Si_3N_4.

3.3.2. Dynamic erosion test results

In order to simulate the erosive environment prevailing during a Czochralski silicon crystal pulling process in which the quartz crucible is exposed, dynamic erosion tests were carried out. It should be noted that, although the average melt temperature during Czochralski silicon crystal pulling is 1340 °C, a higher temperature of about 1450 °C was attained in the melting step of the polycrystalline silicon material. As shown in Table 4, Si_3N_4 deposits on crucible grade quartz specimens were subjected to a dynamic erosion test in which the erosion time, temperature and crucible rotation speed were adapted to those corresponding to (a) a melting step, (b) a crystal pulling step and (c) a combination of melting and crystal pulling steps.

Figure 7 shows the morphological changes due to dynamic erosion of the Si_3N_4 deposits on quartz specimens. The crucible grade quartz changed colour from white to light brown. In one case, decohesion of the Si_3N_4 deposit was noted along the line which approximately corresponds to the silicon melt surface. The thickness of the deposit there was much thinner than in other regions well below the melt surface. This indicates the severeness of the erosive attack in such a region. Thus, a sufficiently thick deposition of Si_3N_4 must be made on the crucible grade quartz substrate in order to keep it intact as much as possible during the entire Czochralski crystal pulling process.

The morphological features of the free and fracture surfaces of the Si_3N_4 deposit on crucible grade quartz specimens after the dynamic erosion test indicate that erosive attack due to molten silicon was especially aggressive during the melting step; the precipitation and growth of new Si_3N_4 crystals occur extensively at the crucible wall region near the silicon melt surface. Figure 8

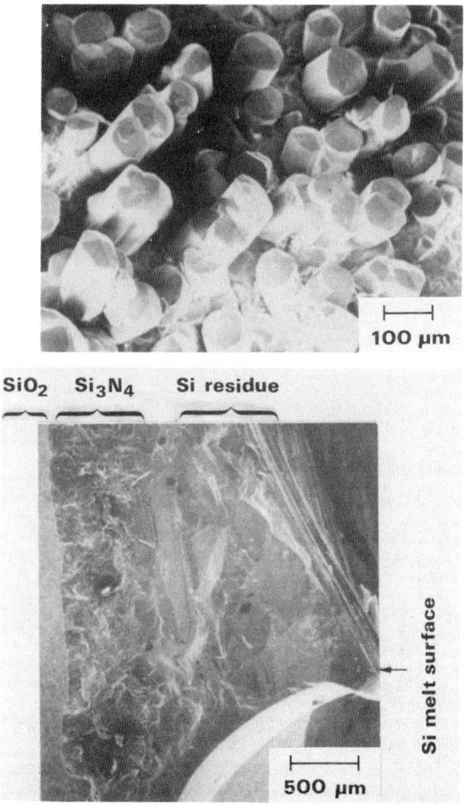

Fig. 8. Columnar growth of β-Si_3N_4 whiskers as a result of dynamic erosion from an Si_3N_4-coated crucible (Table 2, specimen 2a) (see Fig. 7(d) along the molten silicon surface).

TABLE 5 Summary of the results of dynamic erosion tests simulating Czochralski silicon crystal growth of Si_3N_4-coated quartz specimens

Target in simulation test	*Mass of quartz substrate* (g)	*Mass of coated quartz substrate* (g)	*Mass of Si_3N_4 layer* (g)	*Amount of eroded Si_3N_4* (g (wt.%))	*Mass of coated quartz substrate after dynamic erosion* (g)
Si crystal pulling temperature	17.2138 9.5175	19.1142 12.1030	1.9004 2.5855	0.1492 (7.9) 0.1015 (3.9)	18.9650 12.0015
Si ingot melting temperature	10.8195	13.3941	2.5746	0.1182 (4.6)	13.2759
Effect of temperature change (melting–pulling)	16.7455 18.1429	18.8053 19.6010	2.0598 1.4581	0.2006 (9.7) 0.1995 (13.7)	18.6047 19.4015

shows more clearly the columnar growth of Si_3N_4 whiskers from the crucible wall region along and very near the silicon melt surface. The mass losses of the Si_3N_4 deposits due to dynamic erosion under various temperature and time schedules were measured and are given in Table 5. It is clear that erosion proceeds more rapidly at higher temperatures. In the melting step which lasted for only $\frac{1}{2}$ h, erosive attack proceeded most rapidly and this governed the extent of erosion during the entire Czochralski crystal pulling process.

In order to obtain information on the mutual diffusion of atoms through the Si_3N_4 deposit–molten silicon interface, Si_3N_4-deposited crucible grade quartz specimens which had been subjected to dynamic erosion at 1390 °C for 24 h were cut by a diamond wheel and line analyses of such elements as silicon, oxygen and nitrogen were made along the thickness direction of the cross-section by means of electron probe microanalysis (EPMA) (Fig. 9). It is evident that nitrogen has diffused considerably from the deposit into the molten silicon, whereas almost no diffusion of oxygen has taken place from the quartz substrate into the molten silicon. This indicates that Si_3N_4-deposited quartz crucible remains almost intact during Czochralski silicon crystal pulling although the incorporation of a very large amount of nitrogen into the molten silicon is inevitable, as mentioned above.

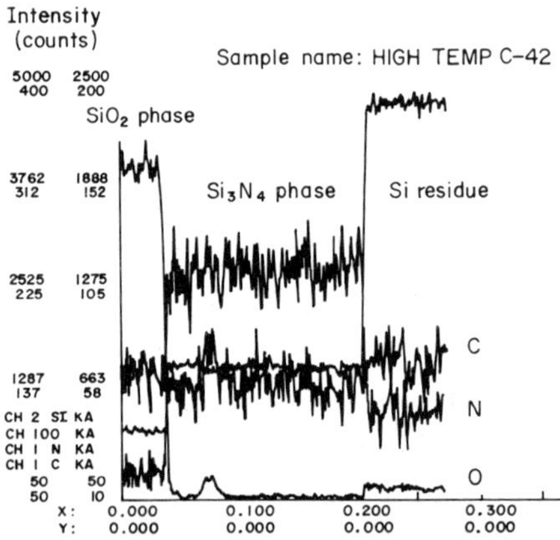

Fig. 9. EPMA line analysis of silicon, oxygen, nitrogen and carbon along the thickness of the Si_3N_4 deposit on crucible grade quartz substrate (Table 2, specimen 24) after a dynamic erosion test.

3.3.3. Preliminary experiments on Si_3N_4 coatings on the inner surface of a 7 in crucible by chemical vapour deposition

In order to establish the optimum conditions for depositing a dense crystalline CVD Si_3N_4 coating as uniformly as possible on the inner surface of a 7 in crucible, the effect of the CVD conditions on the variation in morphology and thickness of Si_3N_4 deposits on substrates placed at various positions was investigated. Figure 1(c) shows schematically the position of the quartz specimen as a substrate placed on the inner surface of the crucible and the positions of the outlets of each reactant gas. Both types of reactant gases were fed in the same direction(s), *i.e.* either in one direction towards the bottom of the crucible or in two directions as shown in Fig. 1(c). Table 6 shows the reaction conditions for Si_3N_4 deposition on the crucible grade quartz specimen. Figure 10 shows a typical example of the surface and fracture morphology of various Si_3N_4 deposits.

The crystallinity of the Si_3N_4 deposit is worse in the lowest position than in the side wall position as can be inferred from the appearance of the globular growth morphology at the former posi-

TABLE 6 Chemical vapour deposition reaction conditions for Si_3N_4 deposition on 7 in quartz crucible

Specimen	*Reaction temperature* (°C)	*Reaction pressure* (Torr)		*Reaction time* (h)	*Gas flow rate* (ml min^{-1})			*NH_3-to-$SiCl_4$ ratio*	*Substrate-to-nozzle distance* (mm)	*Film thickness* (μm)
					$SiCl_4$	*NH_3*	*H_2*			
30	1320	9[a]	7[b]	3	200	170	615	0.85	28	170[c]–50[d]
31	1320	5	4	3	250	220	710	0.88	40	250–110
32	1320	5	4	3	200	170	615	0.85	40	220–180
33	1320	5	3	1.5	300	265	755	0.88	90	170–100
34	1320	8	2	10	200	170	615	0.85	90	—

[a]Value measured with diaphragm-type vacuum gauge.
[b]Value measured with manometer-type vacuum gauge.
[c]Film thickness in bottom region.
[d]Film thickness in side region.

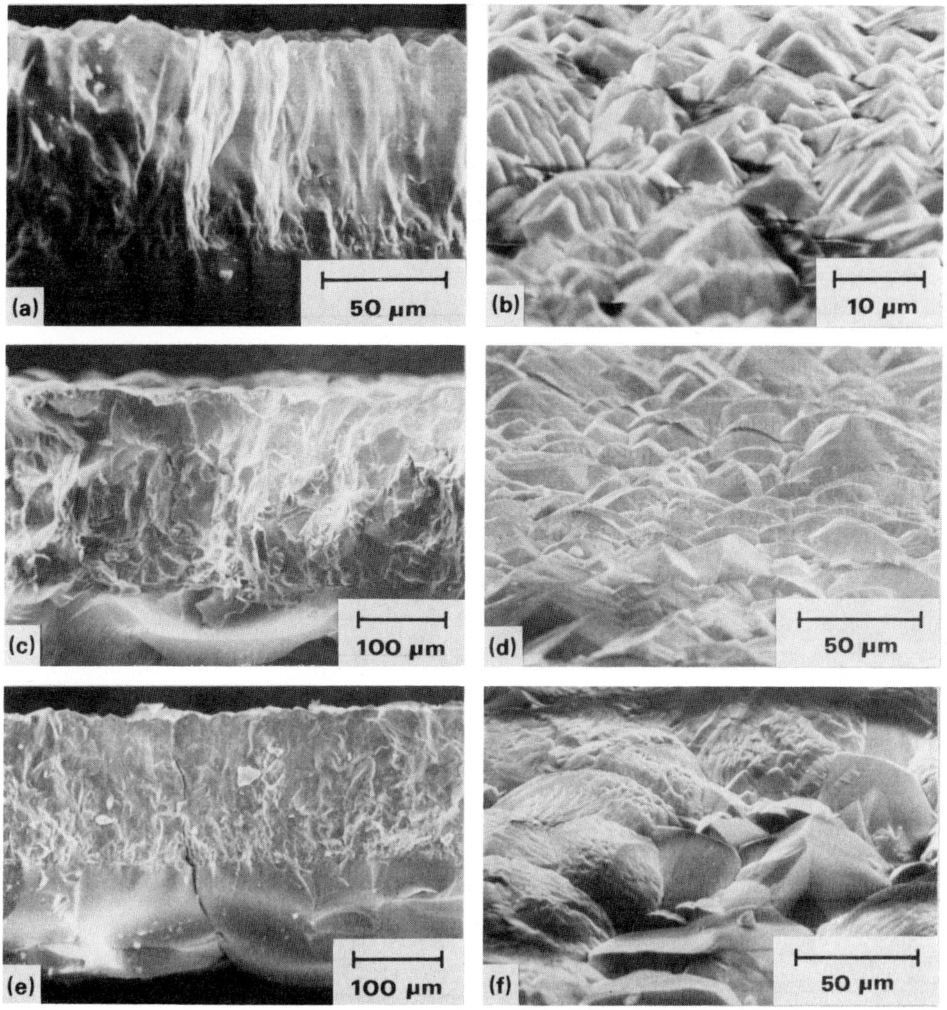

Fig. 10. Effect of the location on the inner surface of an Si_3N_4-coated 7 in crucible on the surface and fracture morphology of Si_3N_4 deposits (Table 6, specimen 32): (a), (b) the upper side; (c), (d) the lower side; (e), (f) the bottom. These regions are as indicated in Fig. 1(c).

tion. As shown in Fig. 11, when the concentrations of the reactant gases were increased by 50% while the NH_3-to-$SiCl_4$ ratio was kept constant, the Si_3N_4 deposit in the lowest position became completely amorphous and this tendency towards amorphism is visible even in the side wall position. In Fig. 12, the rates of Si_3N_4 depositions at two positions were plotted against $SiCl_4$ concentration. It is noteworthy that the deposition rate differs considerably between the two positions; at the lowest position, it increases rapidly with $SiCl_4$ concentration whereas, at the middle side wall position, it soon reaches saturation towards a level of 40 $\mu m\ h^{-1}$. When the effect of the $SiCl_4$ concentration on the crystallinity of the Si_3N_4 deposit is taken into account (see Figs. 10 and 11), it can be concluded that an optimum $SiCl_4$ concentration is about 200 $cm^3\ min^{-1}$. With regard to the reaction pressure, comparatively low pressures such as 5–7 Torr are desired since

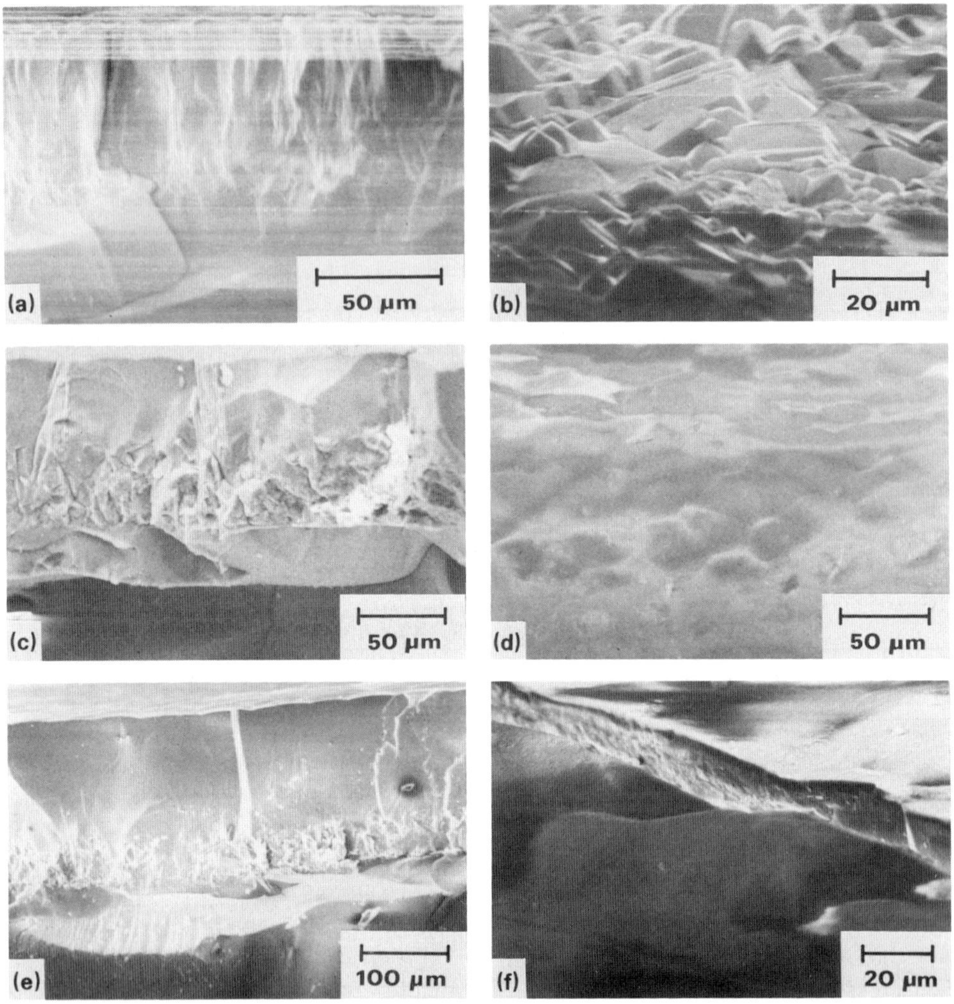

Fig. 11. Effect of increases in the concentrations of reaction gases on the surface and fracture morphology of Si_3N_4 deposit (Table 6, specimen 33): (a), (b) the upper side; (c), (d) the lower side; (e), (f) the bottom. These regions are as indicated in Fig. 1(c).

they lead to better crystallinity of the Si_3N_4 deposit, although the deposition rate necessarily decreases with decrease in reaction pressure.

It must be added that introduction of reactant gases in two directions was found to improve slightly the deposition rate at the upper side wall position. However, for Si_3N_4 deposition on the inner surface of 7 in crucible, this is not necessary since, even in the case of one-way introduction, very good uniformity in the crystallinity of the Si_3N_4 deposit can be attained by the choice of an optimum reaction condition for CVD.

3.3.4. Si_3N_4 coating by chemical vapour deposition on the inner surface of the 7 in crucible and Czochralski silicon crystal pulling using the coated crucible

On the basis of the information obtained from the preliminary experiments, thick dense Si_3N_4 deposition on the 7 in crucible was attained by CVD. The relevant experimental scheme is shown in Fig. 1(b). Table 7 gives the CVD conditions adopted for the deposition. With reference to the CVD condition for specimen 35, the reactant gases (NH_3 and $SiCl_4$) were fed into the CVD

TABLE 7 Chemical vapour deposition coating conditions used for Si_3N_4 deposition on 7 in crucible for Czochralski silicon crystal growth

Specimen	*CVD reaction conditions*			*Gas flow rate* (ml min^{-1})			*NH_3-to-$SiCl_4$ ratio*	*Crucible rotation speed* (rev min^{-1})	*Type of gas inlet nozzle*	*Thickness of Si_3N_4 layer* (μm)
	Temperature (°C)	*Time* (h)	*Pressure* (Torr)	*$SiCl_4$*	*NH_3*	*H_2*				
35	1320	12	5–6	110	100	500	0.91	4	Two way	250–350
36	1320	10	10–15	150	130	615	0.87	3	One way (downwards)	300–350

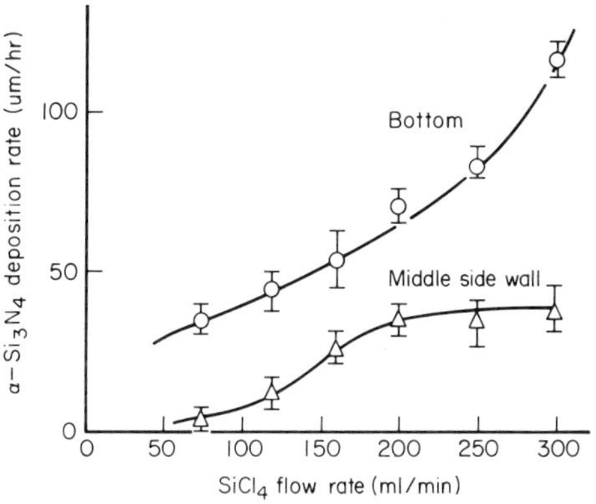

Fig. 12. Deposition rate in CVD Si_3N_4 coating plotted against $SiCl_4$ flow rate.

Fig. 13. Effect of the high temperature reached during the melting of silicon on the surface morphology of an Si_3N_4 deposit on a 7 in quartz crucible along the silicon melt surface.

reactor through quartz pipes with two-way outlets. In contrast, for specimen 36, quartz pipes with only a one-way outlet were utilized. The thickness of the deposition was as large as 300 ± 50 μm.

Czochralski silicon crystal pulling experiments were carried out utilizing Si_3N_4-coated 7 in crucibles. High purity polycrystalline silicon material, 3–4 kg in mass, was put in the coated crucible and the temperature was raised to 1450 °C to melt the silicon. The first silicon crystal pulling was then attempted at a slightly lowered temperature (about 1340 °C). However, the crystal pulling test was soon stopped because fluffy precipitates, which will be mentioned later, were generated from the silicon melt, and these floated on the surface and drifted towards the seed crystal; it was assumed that their existence might disturb the growth of the single crystal. A very large volume expansion during solidification of the molten silicon caused the Si_3N_4-coated crucible to crack. Erosion of the coated crucible by molten silicon during the Czochralski crystal pulling test was examined by EPMA and SEM analyses.

EPMA line analyses with respect to silicon, oxygen, nitrogen and carbon on the cross-section of the coated crucible after the pulling test are essentially the same with Fig. 9; nitrogen has diffused considerably from the surface of Si_3N_4 deposit towards silicon melt, whereas oxygen diffusion from the quartz crucible towards the melt has been suppressed owing to the existence of Si_3N_4 deposit. Figure 13 shows the morphology of the inner surface of the coated crucible near the silicon melt surface. The surface was preliminarily etched with an HF–HNO_3 solution to dissolve the attached silicon. Extensive columnar growth of Si_3N_4 whiskers was noted near the silicon melt surface. Many of the columnar whiskers were found to be hollow shaped, as shown in Fig. 13.

The second silicon crystal pulling test was carried out utilizing Si_3N_4-coated quartz crucible from run 36. The crucible containing 3 kg of polycrystalline silicon was heated at 1460 °C for $\frac{1}{2}$ h in argon for melting. Slowly decreasing the temperature to a level appropriate for seeding again caused extensive precipitation of fluffy whiskers on the surface; these soon became visible on the entire surface of the melt. Since the Czochralski silicon single crystal will be very disturbed because the whiskers drift towards and collide incessantly with the seed crystal, these whiskers were dissolved away by raising the temperature. Several temperature cycling tests were made around a seeding temperature in the hope of finding an optimum temperature of seeding at which these precipitates did not form. However, these efforts were in vain; the precipitation of whiskers was inevitable and, as a result, the pulled crystal became polycrystalline. The pulled Czochralski silicon polycrystal was 60 mm in diameter and 260 mm long. Fluffy precipitates which had floated on the silicon melt surface during the seeding step were photographed after solidification and slightly etched off the attached silicon with HF solution, as shown in Fig. 14. The X-ray diffraction analysis indicated that the precipitates had a β-Si_3N_4-type crystal structure.

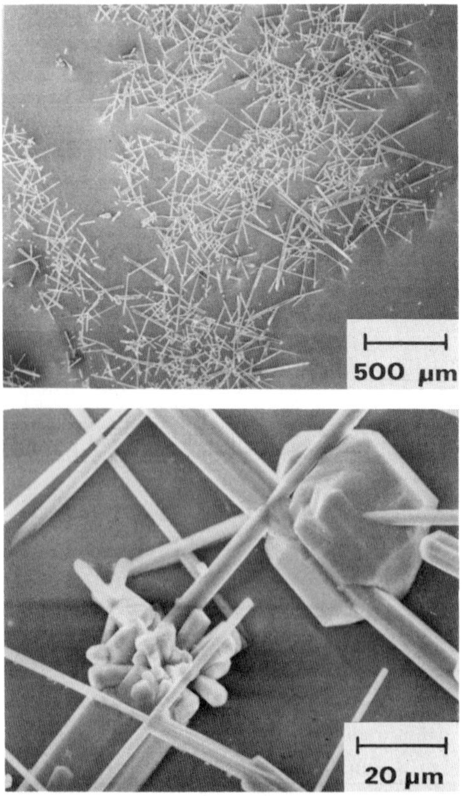

Fig. 14. Morphology of the fluffy Si_3N_4 precipitates on the surface of the molten silicon, which were collected on the silicon seed crystal.

The appearance of the Si_3N_4-coated 7 in crucible after Czochralski silicon crystal growth indicates that the surface morphology of the Si_3N_4 deposit in a region above the silicon melt surface remained almost unchanged. In contrast, the growth of comparatively long whiskers from the region of the Si_3N_4 deposit which had been in contact with the silicon melt surface during the melting step of silicon polycrystals was noted as shown in Fig. 15. These long whiskers can be recovered by etching off the attached silicon on the surface. The growth of extensive columnar Si_3N_4 whiskers was noted from the region of the Si_3N_4 deposit in contact with the remnant silicon melt surface just before solidification of the melt. X-ray diffraction analysis showed that the columnar Si_3N_4 whiskers had a β-Si_3N_4-type crystal structure. It is evident that β-Si_3N_4 whiskers have precipitated on the melt surface as fluffy precipitates or on Si_3N_4 deposit in the form of extensive columnar growth as the temperature was lowered to levels appropriate for seeding or Czochralski crystal pulling.

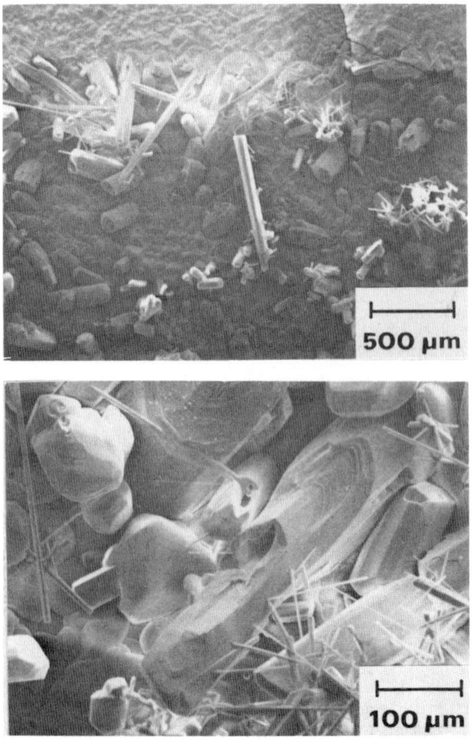

Fig. 15. Growth of β-Si_3N_4 whiskers from the inner surface of an Si_3N_4-coated 7 in quartz crucible along the molten silicon surface.

Oxygen, nitrogen and carbon contents at various positions of the Czochralski silicon crystal were analysed by Fourier transform infrared spectroscopy. The oxygen and nitrogen contents in the cylindrical part are $(5-16)\times 10^{16}$ cm^{-3} and $(5-12)\times 10^{15}$ cm^{-3}, respectively. Typical values of measured oxygen contents in a Czochralski silicon single crystal using an uncoated quartz crucible are $(1.5-1.8)\times 10^{18}$ cm^{-3}. Therefore, coating the quartz crucible with a dense crystalline Si_3N_4 layer by CVD is effective for suppressing oxygen incorporation from the crucible and reducing the oxygen atom content in the grown Czochralski crystal to less than 1/10.

4. Discussion

The present work showed that a thick (250–350 μm) deposit of crystalline α-Si_3N_4 can be formed on the inner surface of a 7 in quartz crucible from an $SiCl_4$–NH_3–H_2 gas mixture. The Si_3N_4-coated crucible remained almost intact in a highly corrosive environment during Czochralski silicon growth, although we failed to grow a single crystal owing to precipitation and drift towards the seed of β-Si_3N_4 whiskers. The relevant temperature T and total pressure P_t for depositing such crystalline Si_3N_4 are 1310 °C and 5–9 Torr respectively. In contrast, according to Niihara and Hirai [2], deposition at $T=1300$ °C and $P_t<40$ Torr or $P_t>70$ Torr resulted in amorphous Si_3N_4 deposits; crystalline deposits were formed at $P_t=40$–75 Torr. The disagreement with our result is considerable. In our case, crystalline Si_3N_4 was deposited on the quartz substrate under outside heating of the reaction system instead of deposition on the hot graphite substrate. During Si_3N_4 deposition, the intermediate species which can be formed in the gas phase before condensation on the substrate to form crystalline deposits might be somewhat different with our CVD apparatus from those with the apparatus used by Niihara and Hirai and this might have caused such a discrepancy.

A brief mention is made of the observed incorporation of nitrogen into the grown silicon crystal from the CVD-Si_3N_4-coated quartz crucible. The observed content ($(5-12)\times 10^{15}$ cm^{-3}) was of the same order as that reported by Watanabe *et al.* [3], *i.e.* 6.5×10^{15} cm^{-3}. Now, the solubility limit of nitrogen in molten silicon is also of the same order, *i.e.* $(4.5\pm 1.0)\times 10^{15}$ cm^{-3} [4]; it is evident that, on account of this limited solubility, the precipitation of Si_3N_4 whiskers from molten silicon occurred when the temperature of molten silicon was lowered to the levels appropriate for seeding, *i.e.* 1340 ± 30 °C from temperature of melting (1450 °C). With reference to the carbon content of the Czochralski crystal, it is of the same order as that in a Czochralski crystal which utilized an uncoated quartz crucible for crystal growth.

5. Summary of experimental results and conclusions

The results obtained on CVD Si_3N_4 coatings on quartz substrates and the effect of erosion by reactive molten silicon environment on the morphology and structure of the Si_3N_4 deposits can be summarized as follows.

(1) Under a deposition temperature of 1280–1320 °C and a reaction pressure of 5–30 Torr, a crystalline layer having an α-Si_3N_4 structure was formed. Above a temperature of 1350 °C, deposition of the β-Si_3N_4 phase which overlapped with the α-Si_3N_4 phase was noted.

(2) Si_3N_4 deposition made under a comparatively low reaction pressure of 5–9 Torr led to a relatively flat dense layer which was characterized by well-crystallized growth facets. In contrast, when deposition was made under a pressure higher than 15 Torr, surface morphology showing globular growth appeared between fairly flat regions consisting of well-crystallized growth facets; these globules were largely amorphous and porous, and the area of the globular growth surfaces increases with increase in the reaction pressure.

(3) A typical deposition rate of a dense crystalline Si_3N_4 layer on a 7 in quartz crucible under an optimum CVD condition was 30–40 μm h^{-1}.

(4) For a dense Si_3N_4 up to 200 μm thick deposited on the inner surface of a quartz crucible, the coated crucible remained almost intact when immersed in a highly erosive environment for 24 h. The rate of erosion of the Si_3N_4 deposit was as low as 5–6 μm h^{-1}.

(5) As a result of the erosion of the α-Si_3N_4 deposit on a quartz crucible, nitrogen was inevitably incorporated into molten silicon and precipitated in the form of β-Si_3N_4 whiskers on the side wall regions or on the free surface of the melt. An unusual growth morphology involving hollow or thick whiskers was frequently observed in the side wall regions.

(6) Czochralski silicon single-crystal pulling tests were attempted, utilizing the Si_3N_4-coated crucibles. However, extensive precipitation of β-Si_3N_4 whiskers during the seeding step disturbed single-crystal growth and resulted in polygonization of the pulled crystal.

(7) A Fourier transform infrared spectroscopy analysis of oxygen and nitrogen in the pulled crystal revealed that oxygen atom incorporation was greatly suppressed to a level of less than 1/10, whereas the nitrogen content was $(5-12) \times 10^{15}$ cm^{-3} which was considerably larger than the levels usually added by other doping methods.

References

1 A. I. Kingon, L. J. Lutz and R. F. Davis, *J. Am. Ceram. Soc., 66*(1983) 551.
2 K. Niihara and T. Hirai, *J. Mater. Sci., 11*(1976) 593.
3 M. Watanabe *et al.*, in H. R. Huff *et al.* (eds.), *Semiconductor Silicon 1981*, Electrochemical Society, Pennington, NJ, 1981, p. 126.
4 Y. Yatsurugi, N. Akiyama, Y. Endo and T. Nozaki, *J. Electrochem. Soc., 120*(1973) 975.

Materials Science and Engineering, A105/106 (1988) 481–488

The Structure and Composition of Ti–Zr–N, Ti–Al–Zr–N and Ti–Al–V–N Coatings*

O. KNOTEK

Lehrstuhl für Werkstoffkunde B, Rheinisch-Westfalische Technische Hochschule Aachen, D-5100 Aachen (F.R.G)

M. BÖHMER

National Physical Laboratory, Teddington, Middx. (U.K.)

T. LEYENDECKER

Cemecoat G.m.b.H. Technologiezentrum, D-5100 Aachen (F.R.G.)

F. JUNGBLUT

Lehrstuhl für Werkstoffkunde B, Rheinisch-Westfalische Technische Hochschule Aachen, D-5100 Aachen (F.R.G)

(Received December 14, 1987)

Abstract

In this paper, we deal with physically vapour-deposited hard coatings in the ternary system Ti–Zr–N and in the quaternary systems Ti–Al–Zr–N and Ti–Al–V–N on various substrate materials such as cold-worked steels, high speed steels and cemented carbides. The reactive gas high performance sputtering technique has specific advantages in the deposition of these innovative complex coating materials with extremely different melting points because there is no intermediate liquid phase in the coating process.

All the compound coatings investigated crystallize in a cubic face-centred TiN lattice with reduced or enlarged lattice parameters depending on the amount and type of the "foreign" atoms present. There was no real correlation found between wear properties, foreign metal type and content, and lattice distortion.

1. Introduction

A consideration of current development perspectives in coating technology shows further improvements of the coating process (reproducibility, coating of parts of complicated geometry etc.) and must take particular account of the introduction of innovative coating materials and the ability to work at lower temperatures.

One path to the further development of conventional hard material thin films lies therefore in depositing compounds with a complex structure, *i.e.* ternary or quaternary mixed phases. By incorporating additional metals (aluminium, vanadium or zirconium) or metalloids (carbon or oxygen), coatings consisting of ternary or quaternary mixed phases with improved properties can be obtained [1–4].

In order to deposit multicomponent films, high rate sputtering or sputter ion plating seems to be the most appropriate technique. In contrast with other physical vapour deposition (PVD) processes the transformation of the original material into a gaseous state does not involve an intermediate liquid phase but instead occurs directly as a result of an atomistic momentum transfer at the target. The danger of decomposition, especially of multicomponent alloys, is not as serious as with vaporization induced by an electron beam [5]. Reactive sputtering also makes it possible, by appropriate variation in the partial pressure ratio of the reactive to the processing gases, to achieve the required composition of multicomponent alloys and refractory materials even at low temperatures.

*Paper presented at the 3rd International Conference on the Science of Hard Materials, Nassau, The Bahamas, November 9–13, 1987.

0921-5093/88/$3.50

© Elsevier Sequoia/Printed in The Netherlands

2. New hard compound films

To realize coatings with a more complicated structure, first aluminium as a partial substitute of titanium was embedded in a Ti–N composition. Investigations of the composition carried out by X-ray analysis and electron probe microanalysis (EPMA) showed no H phases of the form Ti_2AlN as originally expected. Instead, interference bands of a cubic face-centred TiN structure were determined by diffraction examination. Figure 1 shows the well-known TiN X-ray pattern and the shifted peaks of the (Ti, Al)N solid solution. Apparently, titanium atoms in the TiN lattice are replaced by aluminium atoms, and compositions of the form (Ti, Al)N arise (Fig. 2). Because of the smaller aluminium atoms the TiN cell becomes increasingly reduced in size as the aluminium content increases (Fig. 3).

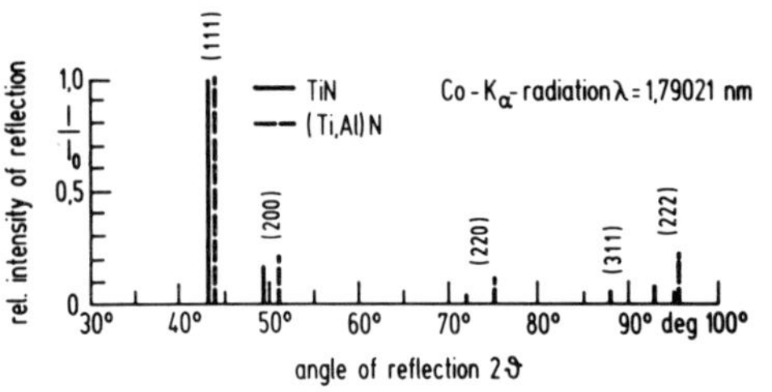

Fig. 1. X-ray pattern of (Ti, Al)N coatings in comparison with TiN coatings.

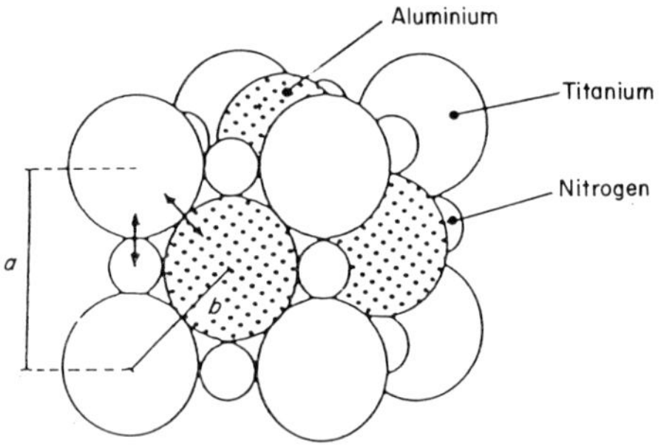

Fig. 2. TiN monocell (NaCl type) with embedded aluminium atoms.

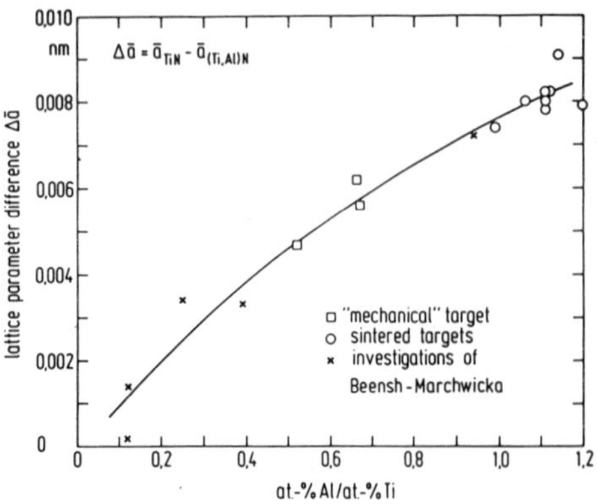

Fig. 3. Lattice parameter difference in relation to the aluminium-to-titanium ratio.

Aluminium was chosen as an additional element in the development of new wear-resistant coatings because the formation of stable aluminium oxide layers on top of the (Ti, Al)N coating could be expected on exposure to elevated temperatures. These thin oxide films should be able to reduce diffusion out of or into the layer substantially. Accordingly, diffusion wear (one of the major wear mechanisms on cutting tools) would be reduced. The improved oxidation behaviour of the hard coatings obtained by incorporating aluminium into the deposited films has been studied by carrying out step stress experiments at elevated temperatures [6]. Coated samples were heated up in steps of 100 or 50 °C for 1 h for each step. The mass gain caused by oxidation was determined from observations of the mass change. Figure 4 shows that (Ti, Al)N films deposited from targets with a titanium-to-aluminium ratio of 50 to 50 start to oxidize at temperatures from 700 to 750 °C. Storage experiments at elevated temperatures revealed that (Ti, Al)N-coated samples with the same composition exhibited a mass gain as low as 20 μg cm^{-2} after a storage time of 400 h at 700 °C. Obviously, a protective film of amorphous Al_2O_3 was formed which prevented the sample from further oxidation [7].

Investigations of the Ti–Zr–N system were based on similar considerations as explained above. Zirconium also gives stable oxides. The phase equilibrium of the isotypic phases TiN and ZrN is characterized by complete miscibility of the solid state [8, 9]. The stability of nitrides decreases from the refractory elements in the fourth

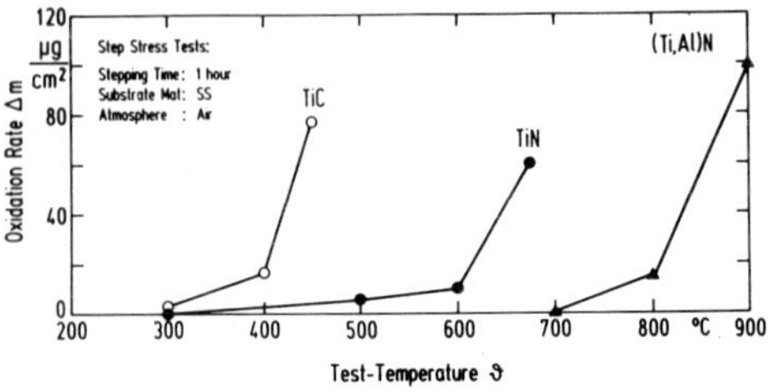

Fig. 4. Oxidation rate of hard coatings in step stress tests.

group. Coatings of the Ti–Zr–N system will have a stability comparable with that of TiN.

In order to construct an even more complicated unit cell, aluminium and zirconium or aluminium and vanadium have been added as alloying agents in the coatings [10]. These additions, embedded in an f.c.c. structure, might cause further stabilization as three metallic components of different atomic radii and different electron configurations lead to conditions which can be observed in solid-solution-hardened materials. In only a few publications have the results for Ti–Al–Zr–N or Ti–Al–V–N coatings been reported [11, 12]. Some researchers have discovered two phases after deposition by sputtering.

3. Experimental details

The new multicomponent hard coatings were deposited with a Leybold–Heraeus Z 400 magnetron sputtering installation working in the d.c. reactive gas mode.

The composition of the coating compounds was easily altered by changing the composition of the targets and varying the dominant process parameters, such as the reactive gas partial pressure, sputtering power, substrate bias and substrate temperature.

Targets of different materials were used. These targets were of a metallurgical nature based on a standard aircraft engineering alloy (Ti–6wt.%Al–4wt.%V) as a low-priced target material, a sintered Ti–Zr target with about 30 at.% Zr and various "mechanical" targets formed by inserting bolts made of the alloying agent materials (aluminium or vanadium) in the anticipated sputter zone of the target.

Sintered hard metals of ISO group M15 (82wt.%WC–11wt.%(Ti, Ta, Nb)C–7wt.%Co), two types of high speed steel (S 6-5-2/1.3343 and S 10-4-3-10/1.3207), a high alloyed cold-worked steel (X 210 CrW 12/1.2436) and a medium alloyed cold-worked steel (90 MnCrV 8/1.2842) were used as the substrate materials.

Micrographic, X-ray and scanning electron microscope tests were used to determine the coating properties. Coatings were also examined by EPMA and Auger electron spectroscopy (AES). Annealing tests at 1000 °C in a vacuum (less than 10^{-4} mbar) furnace provided data on the stability of the sputtered compounds. Subsequent wear testing (tool life turning test) was carried out by Krupp Widia G.m.b.H. (Essen).

4. Ternary Ti–Zr–N compound coatings

The Ti–Zr–N coatings were deposited by reactively sputtering a sintered Ti–30at.%Zr target. The Vickers microhardness (load, 0.05 kgf) of the coatings was 3000 HV 0.05 although the hardnesses of pure TiN and of ZrN are lower. This corresponds to the values reported by Holleck [13], who also found a hardness maximum for materials of the TiN–ZrN system. The microhardness values are constant over a wide range of partial pressure ratios of the nitrogen to argon gases.

X-ray analysis showed an f.c.c. TiN spectrum shifted towards smaller reflection angles. This confirms the results of Kieffer *et al.* [9], who found that the lattice parameters of comparable compositions increased because of the larger atom radius of zirconium. The composition seems to be stable as annealing tests caused no major decrease in the lattice parameters (Fig. 5). Only in coatings deposited at low nitrogen partial pressures did the solid solutions decompose to ZrO, TiN, ZrN and (Ti, Zr)N after heat treatment.

In Fig. 6 and Fig. 7 the cutting performance of coated M15 cemented-carbide tips is plotted *vs.* the performance of usual carbide tips coated with TiC–TiN by chemical vapour deposition (CVD) and is compared with other PVD coatings. The (Ti, Zr)N coatings are characterized by improved wear resistance. The wear properties do not change with a higher partial pressure ratio. Annealing the coated inserts at 1000 °C for 24 h had no influence on the cutting performance. Only after a cutting time of 10 min did the

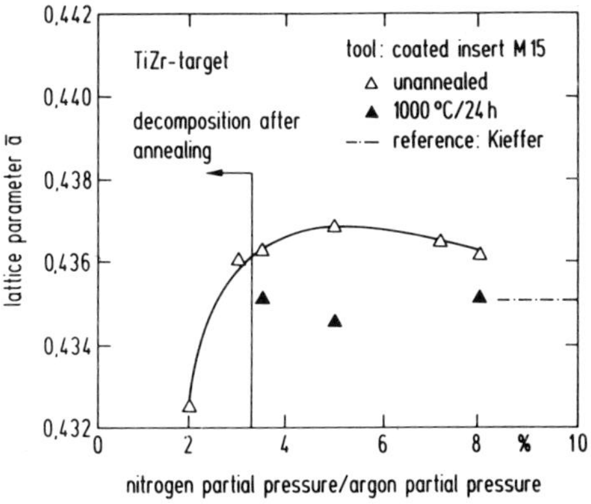

Fig. 5. Lattice parameter in relation to the partial pressure ratio.

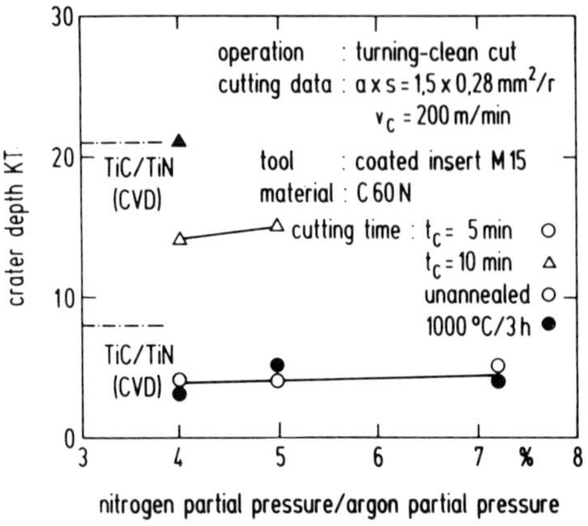

Fig. 6. Cutting performance tests of (Ti, Zr)N-coated carbide inserts.

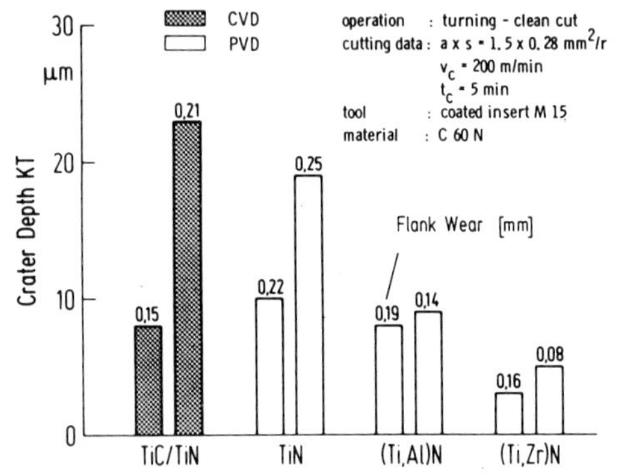

Fig. 7. Cutting performance tests of (Ti, Zr)N-coated carbide inserts.

annealed sample show a performance comparable with the CVD coated standard.

The improved cutting performance of the (Ti, Zr)N coatings could be due to the stabilization effect of zirconium in the f.c.c. TiN monocell. This improved performance is also related to the formation of very thin zirconium oxide layers on the top of the coating in the crater area. After cutting tests, TiN and (Ti, Zr)N surfaces were analysed by means of AES in the crater area as well as in the unworn areas. The oxygen peaks on the unworn TiN and (Ti, Zr)N surfaces were comparable, but a lower oxygen content in the crater area of the TiN-coated inserts contrasts with a significant oxygen peak on the (Ti, Zr)N-coated insert, indicating the formation of oxide layers which probably hamper further diffusion and accordingly reduce diffusion wear; this latter

	Ti	Al	Zr	V	N	O	
at. - %	24	17	10	-	47	2	(Ti, Al, Zr) N
at. - %	41	5	-	2	50	2	(Ti, Al, V) N

Fig. 8. Composition of quaternary (Ti, Al, Zr)N coatings and (Ti, Al, V)N coatings.

is the main wear mechanism under the chosen cutting conditions.

5. Quaternary Ti–Al–Zr–N and Ti–Al–V–N compound coatings

Previous investigations of (Ti, Al)N coatings have demonstrated the good wear behaviour of this coating used on cutting tools. Therefore an attempt was made to combine the advantages of (Ti, Al)N and (Ti, Zr)N coatings by the development of quaternary (Ti, Al, Zr)N coatings. In addition, the effect of vanadium was examined by adding small amounts of vanadium to the (Ti, Al)N.

Interesting microhardness results were obtained with both compositions. The maximum values of 2700 HV 0.05 were measured on (Ti, Al, V)N coatings and about 3500 HV 0.05 on (Ti, Al, Zr)N coatings. The adhesion of both compounds was good, with critical loads of up to 70 N on M15 cemented carbides.

EPMA was used to quantify the composition of these quaternary compounds. Figure 8 depicts two typical coating compositions. The stoichiometry is of interest. (Ti, Al, V)N coatings with a high microhardness and good adhesion exhibit metalloid-to-metal ratios of more than 1.0. By contrast, efficient (Ti, Al, Zr)N coatings are stoichiometric or slightly substoichiometric. The oxygen content is in the 1–3 at.% range. The lattice parameter of (Ti, Al, V)N differs only slightly from that of the TiN monocell; additions of aluminium and zirconium enlarge the monocell significantly (Figs. 9 and 10). Cutting performance tests using (Ti, Al, V)N and (Ti, Al, Zr)N coated carbide tools were carried out at cutting speed of 224 m min^{-1}, representing cutting parameters tougher than those in previous tests. The cutting properties of CVD TiN were used as a basis of comparison. Figure 11 indicates the superiority of complex PVD (Ti, Al, V)N coatings. The appearance of flank wear marks after uninterrupted cutting for 10–15 min is attributable to

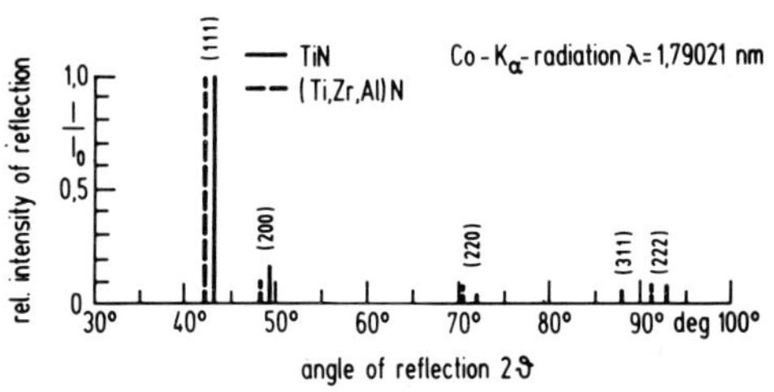

Fig. 9. X-ray pattern of (Ti, Zr, Al)N coatings in comparison to TiN coatings.

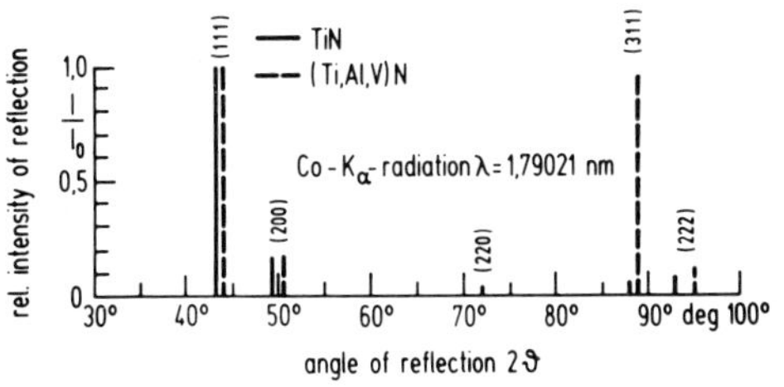

Fig. 10. X-ray pattern of (Ti,Al,V)N coatings in comparison with TiN coatings.

poor coating conditions on the flank because the inserts were not turned during the coating process. Nevertheless, in cutting tests the tool life of the (Ti, Al, Zr)N-coated tools failed to reach CVD standards. The reason for this unexpected result is hard to explain since the microhardness and adhesion tests showed significant increases.

A metallurgical Ti–6wt.%Al–4wt.%V target was used for the development of complex low temperature coatings. These developments in low temperature (less than 200 °C) coatings form a particular focus of interest because they allow temperature-sensitive substrate materials such as cold-worked steels to be coated without a hardness decrease and distortion. In order to reduce process temperatures to below 200 °C, extensive modification of the coating processes was necessary. Special attention needed to be paid to the important process step of ionic etching, which causes severe temperature loading of the substrate due to argon bombardment.

Also the sputtering power and the bias in particular have to be varied for the actual coating process. By suitable choice of all the process parameters, it proved possible to achieve low temperature coatings at economic coating rates and high critical load values in the scratch test. The low temperature coatings had a lower microhardness (about 2000 HV 0.05) than did the

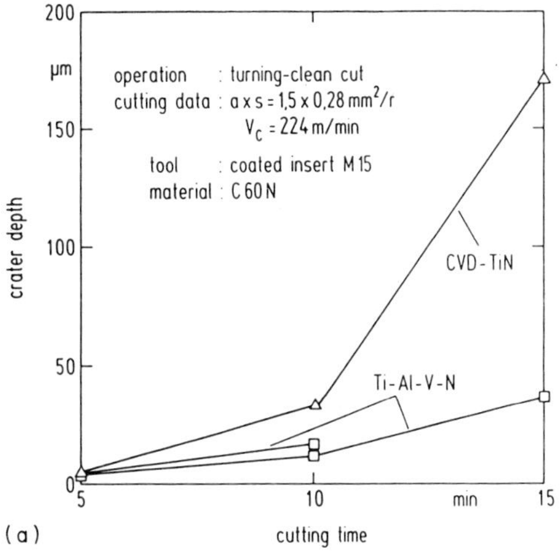

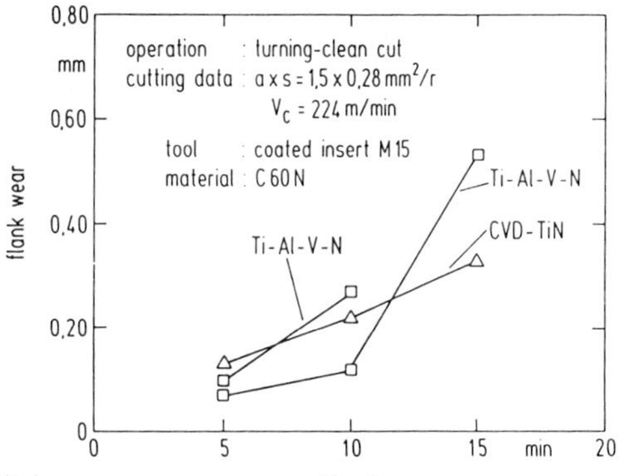

Fig. 11. (a) Crater depth and (b) flank wear of (Ti, Al, V)N-coated inserts.

coatings deposited at higher temperatures. This inferior microhardness is attributable to the coarser columnar growth of the low temperature coating (Fig. 12).

The significant influence of the alloying agents aluminium and vanadium on the physical and chemical properties of multicompound hard (Ti, Al, V)N coatings for cemented carbides was a special point of further investigation. By using "mechanical" targets for this coating material development and varying the aluminium and vanadium contents, the reactively sputtered films fell into the aluminium-rich and vanadium-rich (Ti, Al, V)N ranges. The aluminium-rich films had a typical violet (Ti, Al)N colour and the vanadium-rich films a yellow gold colour similar to that of TiN. Scratch tests revealed high critical

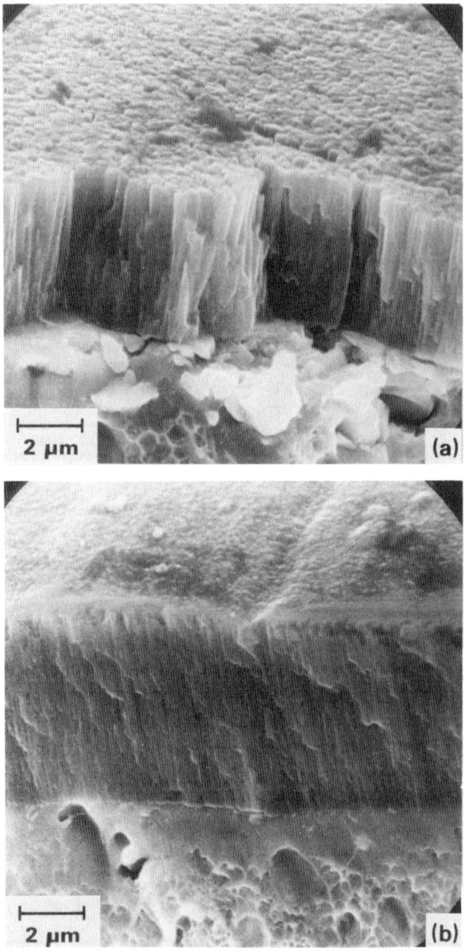

Fig. 12. Rupture structure of (Ti–6wt.%Al–4wt.%V)N coatings deposited at different process temperatures: (a) 200 °C; (b) 550 °C.

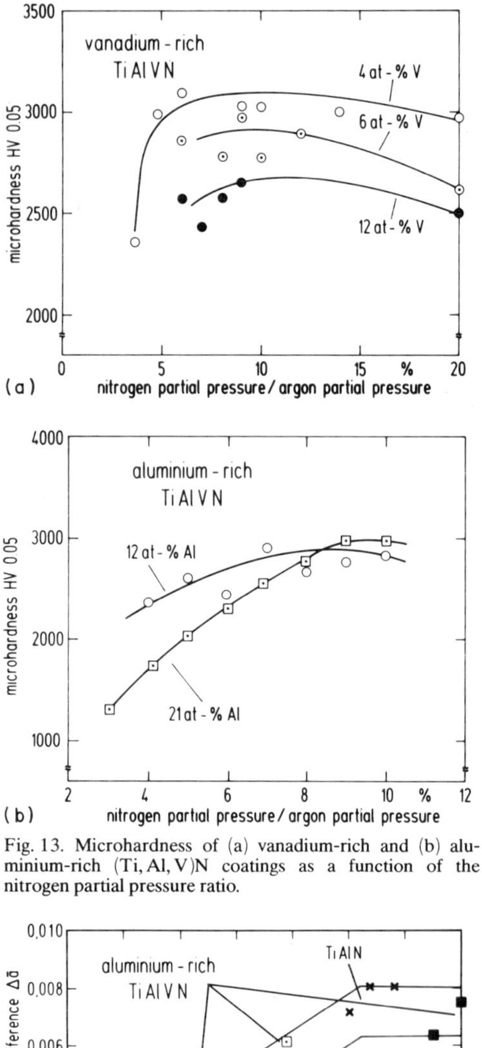

Fig. 13. Microhardness of (a) vanadium-rich and (b) aluminium-rich (Ti, Al, V)N coatings as a function of the nitrogen partial pressure ratio.

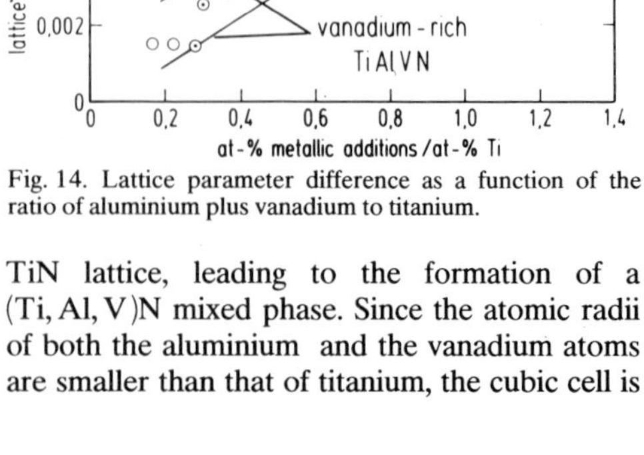

Fig. 14. Lattice parameter difference as a function of the ratio of aluminium plus vanadium to titanium.

loads above 75 N for all coatings, no notable composition-dependent differences being observable. The microhardnesses of both the aluminium-rich and the vanadium-rich (Ti, Al, V)N coatings were between 2600 and 3000 HV 0.05 (Fig. 13). X-ray microstructural analysis of the quaternary (Ti, Al, V)N coatings indicates a cubic face-centred lattice structure of TiN. Under the influence of the aluminium and vanadium alloying agents the X-ray spectrum, however, is displaced towards larger angles, with the (Ti, Al, V)N cell much reduced in size (Fig. 14).

Titanium atoms thus appear to be partly substituted for aluminium and vanadium atoms in the TiN lattice, leading to the formation of a (Ti, Al, V)N mixed phase. Since the atomic radii of both the aluminium and the vanadium atoms are smaller than that of titanium, the cubic cell is reduced in size when these alloying elements are incorporated (Fig. 14). This confirms the results of previous investigations into (Ti, Al)N and (Ti, Al, V)N coating structures. At high aluminium contents, amorphous AlN appears to be present together with the (Ti, Al, V)N cell, although X-ray analysis revealed no second phase alongside the cubic (Ti, Al, V)N.

Aluminium-rich and vanadium-rich (Ti, Al, V)N coatings respond differently to annealing in the vacuum (less than 10^{-4} mbar) furnace at 1000 °C. In the obviously metastable aluminium-rich coatings, AlN is demonstrated as a second phase after annealing. The lattice parameter also increases to a certain extent, owing to decomposition of the (Ti, Al)N solid solution. Distinctly more stable thermal behaviour is evident in vanadium-rich (Ti, Al, V)N coatings. No other phases apart from the previously determined (Ti, Al, V)N mixed phase were observed after annealing.

The wear behaviour of the newly developed (Ti, Al, V)N coatings was examined in tool-life turning tests (Figs. 15 and 16). The different influences of vanadium and aluminium were confirmed. It was shown that the addition of vanadium increases the wear resistance on the first face but tends also to increase the brittleness of the material, which is manifested as flank wear. By contrast, coatings with added aluminium possess good wear resistance even on the tool flank. This emphasizes the ductility of the coating. It is evident that further improvements are still possible in the direction of maximum wear resistance.

6. Conclusions

All these results show the great potential of wear-resistant thin film coatings and also demonstrate the great potential which there still is in the development of new and highly efficient coating

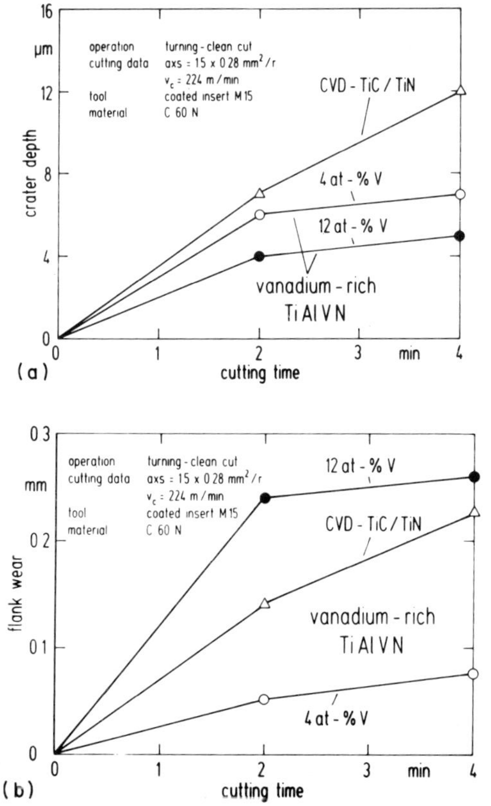

Fig. 15. (a) Crater depth and (b) flank wear of hard metals with vanadium-rich (Ti, Al, V)N coatings.

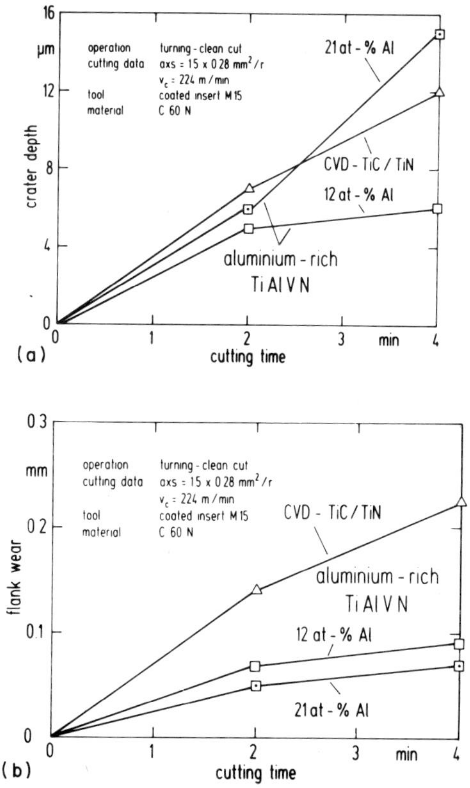

Fig. 16. (a) Crater depth and (b) flank wear of hard metals with aluminium-rich (Ti, Al, V)N coatings.

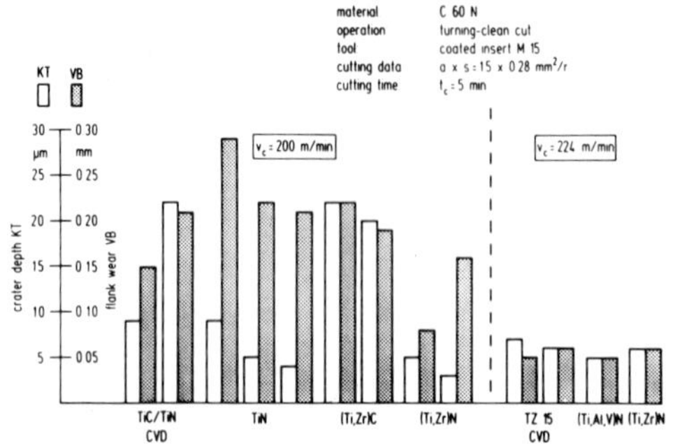

Fig. 17. Improvements in cutting test results through complex coatings shown by the increased wear resistance at higher cutting speeds.

materials. Figure 17 clearly shows the great progress in the field of coated hard metal cutting inserts made by new complex coatings, not only with PVD but also with CVD [14]. The newly developed compound coatings give better results under harder cutting test conditions.

For the future, new coating materials will appear which are specially tailored to special demands. For example it seems to be logical and necessary that high cutting speeds require different coatings as well as different machined materials. In addition, the PVD coating technology sputtering sources leave open the choice of coating materials.

7. Summary

In this paper, we report the development of innovative hard materials in the ternary Ti–Zr–N system and the quaternary Ti–Al–Zr–N and Ti–Al–V–N systems, using magnetron sputter ion plating. All compound coatings investigated crystallized in a cubic face-centred TiN lattice with reduced or enlarged lattice parameters depending on the amount and radius of foreign atoms. There was no clear correlation between wear properties, foreign metal content and lattice distortion. Although all these ternary and quaternary compound films showed high hardness and good adhesion, not all combinations had good wear behaviour. Cutting performance tests proved the outstanding wear properties of (Ti, Zr)N and some (Ti, Al, V)N coatings. According to these results, the development of the coating material itself will be one major point of interest in the future.

References

1 O. Knotek and W. Bosch, *Met. Powder Rep., 39* (1984) 406.
2 T. Leyendecker, *Doctoral Thesis,* Technische Hochschule Aachen, 1985.
3 O. Knotek, W. Bosch and T. Leyendecker, *Proc. 11th Plansee Semin.*, Metallwerk Plansee, Reutte, 1985, pp. 677.
4 W. Bosch, *Doctoral Thesis,* Technische Hochschule Aachen, 1985.
5 R. F. Bunshah, *Deposition Technologies of Films and Coatings,* Noyes, Park Hill, NY, 1982.
6 O. Knotek, W. D. Münz and T. Leyendecker, *J. Vac. Sci. Technol. A, 5* (4) (1987) 2173.
7 W. D. Münz, *J. Vac. Sci. Technol. A, 4* (1986) 2721.
8 P. Duwez and F. Odell, *J. Electrochem. Soc., 97* (1950) 229.
9 R. Kieffer, H. Nowotny, P. Ettmayer and G. Dufek, *Metall, 26* (1972) 701.
10 O. Knotek, M. Böhmer and T. Leyendecker, *J. Vac. Sci. Technol. A, 4* (1986) 2699.
11 M. A. Kuzenkova, P. S. Kislyi and O. V. Pshenichaya, *Izv. Akad. Nauk S.S.S.R., Neorg. Mater., 12* (1976) 430.
12 K. Wasa and S. Hayakawa, *Thin Solid Films, 10* (1972) 367.
13 H. Holleck, *Binäre und ternäre Carbid- und Nitridsysteme der Übergangsmetalle,* Bornträger, Berlin, 1984.
14 H. van den Berg, U. König and N. Reiter, in R. Porat (ed.), *Proc. 6th Eur. Conf. on Chemical Vapour Deposition, Jerusalem, 1987,* pp. 114–119.

Friction and Wear of Ion-implanted TiB_2*

C. S. YUST, C. J. McHARGUE and L. A. HARRIS

Oak Ridge National Laboratory, P.O. Box X, Oak Ridge, TN 37831 (U.S.A.)

(Received November 9, 1987)

Abstract

The unlubricated sliding of diamond and cobalt-bonded tungsten carbide (WC) against nickel-ion-implanted titanium diboride (TiB_2) have been investigated. The diamond sliding is characterized by low friction (less than 0.02) and the formation of a carbon-bearing debris layer on the wear path. WC slides on TiB_2 with a stick–slip motion and coefficients of friction ranging from 0.2 to 1.0 and forms wear grooves. The effect of implantation on sliding wear rates has been explored by sequential wear tests which indicate a limitation of the wear groove depth to that of the modified lattice.

1. Introduction

The inherent high hardness of titanium diboride (TiB_2) and the significant increase in hardness produced by the implantation of energetic ions into the surface makes this material an item of continuing interest for friction and wear applications. We have previously reported the properties of ion-implanted TiB_2 surfaces and the results of TiB_2 wear experiments using diamond as a counterface material [1, 2]. The results of further studies utilizing both diamond and cobalt-bonded tungsten carbide (WC) as counterface materials, and including ion-beam-mixing effects as well as ion implantation, are reported in this paper.

Studies at this laboratory [3–7] and others [8, 9] have shown that ion implantation of ceramics may produce highly defective but still crystalline structures with increased hardness or amorphous phases which are softer than the unimplanted material. The modified surface layer depth is a function of the implantation energy, implanted ion and host lattice and is typically 200–400 nm. Our earlier results on TiB_2 showed an increase in hardness of 70% due to implantation of 10^{17} Ni^+ ions cm^{-2} (1 MeV) [1]. Padmanabhan and Sorensen [10] reported that implantation of 4×10^6 Kr^+ ions cm^{-2} into sputter-deposited TiB_2 films caused an increase in hardness of about 50% [10]. It has long been recognized that hardness is not the only property which determines wear resistance, and further evidence can be found in these results. Diamond, which is harder than the ion-implanted TiB_2, forms a deposit on the wear track and does not remove a significant amount of surface material. Conversely, WC, which is not as hard as the ion-implanted diboride, does form wear grooves in TiB_2, although the WC is not significantly worn. The wear process depends on many parameters of the wearing system, including, among others, the composition of the bodies in contact, the microstructure of those bodies, the stress applied at the interface, the chemical environment and the temperature and velocity at the interface. In ion-implanted specimens, the analysis of wear phenomena becomes more complicated by the creation of a near-surface zone, 0.1–0.5 μm in thickness, in which both physical and chemical changes are introduced into the lattice. Changes in surface wear mechanisms may result from these microstructural variations, but the test and test conditions must be selected with care to achieve a meaningful evaluation of the implantation effects. The work presented in this paper concerns the evaluation of very-near-surface wear effects and the problems associated with the measurement of these effects.

2. Materials and experiment

The wear experiments were performed on dense polycrystalline TiB_2 obtained from PPG Industries Inc. The sintered body was prepared

*Paper presented at the 3rd International Conference on the Science of Hard Materials, Nassau, The Bahamas, November 9–13, 1987.

0921-5093/88/$3.50 © Elsevier Sequoia/Printed in The Netherlands

from powder produced in an arc plasma reactor from vapor reactants. The powder was blended with paraffin as a binder, cold pressed without additional sintering aids and fired under vacuum for 1 h in the temperature range 1900–2275 °C. The grain size is 20–30 μm and the density is 99.4% of the theoretical density. The surfaces of the TiB_2 specimens were polished with 0.5 μm diamond paste to provide a finish permitting the observation of small degrees of surface wear.

The Oak Ridge National Laboratory (ORNL) 5 MV Van de Graaff facility was used to implant nickel ions at 1 MeV. A fluence of 1×10^{17} Ni^+ ions cm^{-2} was used, giving a calculated nickel-to-titanium ratio of 0.12 at the position of maximum nickel concentration. A portion of the sample surface was protected from the ion beam by a metal mask to provide a reference unimplanted region. The specimen was nominally at room temperature during the implantation. In addition, a specimen was prepared at a lower implantation energy, 170 keV, at the ORNL surface modification and characterization facility.

The role of a carbon film on the wear response of TiB_2 was also part of this study. For this purpose, a carbon layer 60 nm thick was sputter deposited onto the sample surface prior to ion implantation. The samples were masked during ion bombardment so that four surface regions were obtained: (a) unimplanted and uncoated TiB_2, (b) nickel-ion-implanted TiB_2, (c) a surface carbon layer ion beam mixed into the surface and (d) a surface carbon layer without implantation. These four areas were worn by both diamond and WC counterfaces. The diamond counterface was a Rockwell C indenter with a spherical tip radius of 200 μm. A cobalt-bonded WC sphere (5.5–6.5 wt.% Co and the balance WC; grain size, 3–4 μm) was held in a collet-like retainer.

Electron spectroscopy for chemical analysis (ESCA) was accomplished by a study of photoelectron peaks obtained using a Phi ESCA system 5100 and Mg Kα radiation. Raman spectroscopy was performed on a micro-Raman system, using a spot size approximately 5 μm in diameter.

The wear test system is a pin-on-disc apparatus constructed in our laboratory and provides either circular or linear motion. In the tests discussed here, wear was accomplished by unlubricated reciprocating sliding along a straight path 7 mm in length. The linear motion is driven by an eccentric shaft; consequently the velocity varies sinusoidally along the path length. Dead-weight loading is used. The pin fixture bears against the end of a strain-gauged cantilever beam which deflects as frictional forces are generated at the sliding interface. A computer-controlled data acquisition system records data at rates as high as 1000 readings s^{-1} at programmed time intervals. Typically, tangential force readings were taken at the start of sliding, 2 and 5 min after sliding, and at other intervals distributed throughout the duration of the test. Tests were run at room temperature in either laboratory air of approximately 50% relative humidity or in a dry nitrogen atmosphere. The loads and test durations were selected to produce small but measurable increments of wear. The average sliding velocity was 2.5 mm s^{-1}.

3. Results and discussion

A summary of the tests performed and test conditions used is presented in Table 1. The test runs include both diamond and WC sliding against each of four surface conditions.

3.1. Diamond sliding on TiB_2

The unlubricated sliding of diamond on TiB_2 consistently results in the formation of an adhered layer of debris on the wear track as well as loose debris adjacent to the track. Even though diamond has a higher hardness than TiB_2, the tip of the diamond pin wears and forms a surface layer on the wear track. This result is consistently obtained on each of the four regions of the carbon-coated and implanted surface. We have previously reported this phenomenon, confirmed by surface profilometry, and we observed it again during the present work [1]. With sufficiently high levels of stress and/or prolonged test duration, failure by fracture of individual grains in the TiB_2 surface occurs. This behavior has been previously reported by Yust [11] and shown to be the consequence of the accumulation of lattice strain within the grains. Figure 1 illustrates both the adhered debris layer on the wear path and the initiation of single-grain failure.

The structural form and the composition of the adhered diamond debris layer have been investigated by Auger electron spectroscopy, ESCA and micro-Raman spectroscopy. The Auger study yielded distribution maps for carbon, oxygen, titanium and boron which confirmed that the

TABLE 1 Unlubricated sliding wear experiments on TiB_2 disc surfaces

Pin material	*Disc condition*	*Applied normal force* (gf)	*Test duration* (h)	*Sliding distance* (m)	*Test atmosphere*	*Comments*
Diamond	Unimplanted	100	24	216	N_2	Friction coefficient, 0.02
Diamond	Ion implanted (Ni, 1 MeV)	100	24	216	N_2	Friction coefficient, 0.03
Diamond	Carbon coating	100	24	216	N_2	Friction coefficient, 0.01
Diamond	Ion-beam-mixed carbon coating (Ni, 1 MeV)	100	24	216	N_2	Friction coefficient, 0.02
WC	Unimplanted	300	24	216	N_2	Stick–slip motion; friction coefficient extreme values, 0.2–1.0
WC	Ion implanted (Ni, 1 MeV)	300	24	216	N_2	
WC	Carbon coating	300	24	216	N_2	
WC	Ion-beam-mixed carbon coating (Ni, 1 MeV)	300	24	216	N_2	
WC	Ion implanted (Ni, 1 MeV)	300	46	414	N_2	Stick–slip motion; friction coefficient extreme values, 0.2–1.0
WC	Ion-beam-mixed carbon coating (Ni, 1 MeV)	300	46	414	N_2	
WC	Unimplanted	300	45	405	N_2	
WC	Ion implanted (Ni, 1.7 keV)	100	5.6	50.4	N_2	Maximum wear groove depth, 200.0 nm
WC	Ion implanted (Ni, 1.7 keV)	50	2.8	25.2	N_2	Maximum wear groove depth, 200.0 nm
WC	Ion implanted (Ni, 1.7 keV)	50	4.5	40.5	Air (relative humidity, 50%)	Maximum wear groove depth, 200.0 nm
WC	Ion implanted (Ni, 1.7 keV)	50	4.5	40.5	Air (relative humidity, 50%)	Maximum wear groove depth, 175.0 nm (5 m increments)
WC	Ion implanted (Ni, 1.7 keV)	50	G(0.55)	—	N_2	Eight sequential grooves
WC	Unimplanted	50	G(0.55)	—	N_2	

G, groove number.

debris and the adhered layer were primarily carbon with some incorporated oxygen [2].

ESCA was performed on a relatively large diamond debris area, approximately 4 mm × 10 mm in extent, prepared by placing several reciprocating sliding test wear paths adjacent to each other. The resultant broad-scan ESCA spectra of the as-received and sputter-etched wear tracks are shown in Figs. 2 and 3 respectively. The spectra show significant amounts of carbon and oxygen in addition to titanium and boron, and trace amounts of silicon, calcium, sodium and nitrogen. Sputter etching for 8 min reduced the amount of carbon and oxygen. High resolution spectra obtained by ESCA were also obtained specifically for titanium, boron, oxygen and carbon. The positions of the photoelectron peaks for these elements are shown in Table 2. The binding energies for titanium and boron determined for the as-received surface indicate the presence of TiO_2 (458.7 eV for titanium, and 530.4 eV for oxygen) and B_2O_3 (191.9 eV for boron, and an unknown value for oxygen). However, after the sputter etch the amount of these oxides present was significantly reduced, and the binding energies determined are in excellent agreement with values for TiB_2 in the literature (454 eV for titanium, and 187.2 eV for boron).

The presence of a carbon photoelectron peak at 283.45 eV was also noted, this energy being equivalent to the binding energy of carbon in graphite. In order to determine the exact form of carbon present, the sample was examined using micro-Raman spectroscopy. This technique yields spectra for diamond and graphite which are distinguishable from one another. Figure 4 is the spectrum obtained from clumps of material present in the wear track and is similar to the spectrum for low temperature carbon materials and it is distinctly different from the spectrum obtained from the clean diamond stylus (Fig. 5).

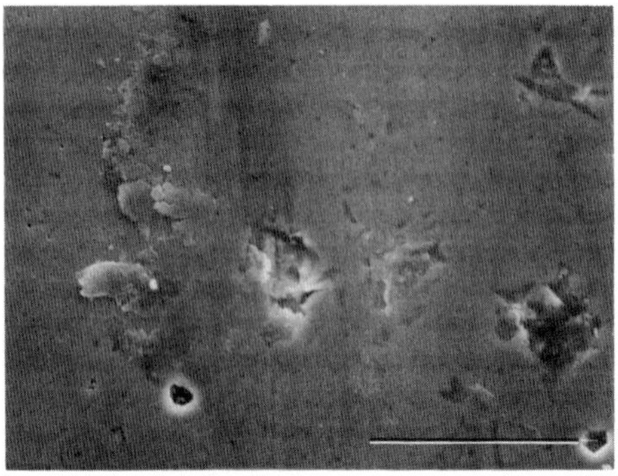

Fig. 1. Scanning electron photomicrograph of debris produced by the unlubricated sliding of diamond on a nickel-ion-implanted TiB_2 surface. The edge of the debris layer is seen at the left of the field of view, and the initiation of grain failures are seen at the center and right. The reference bar equals 30 μm.

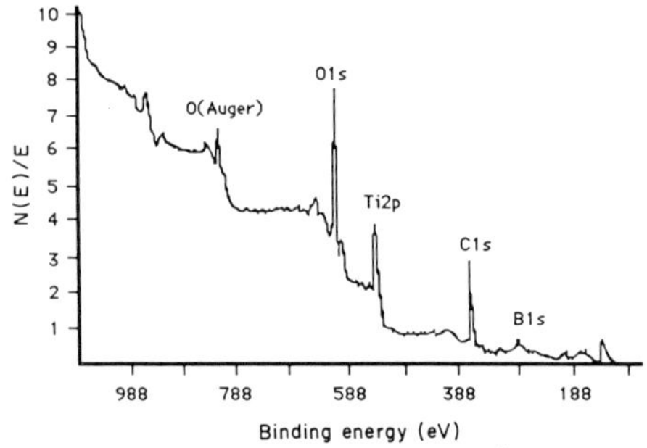

Fig. 2. Broad-scan ESCA spectrum for the wear track debris in the as-tested condition. Carbon, oxygen and titanium are indicated as the principal constituents of the debris layer.

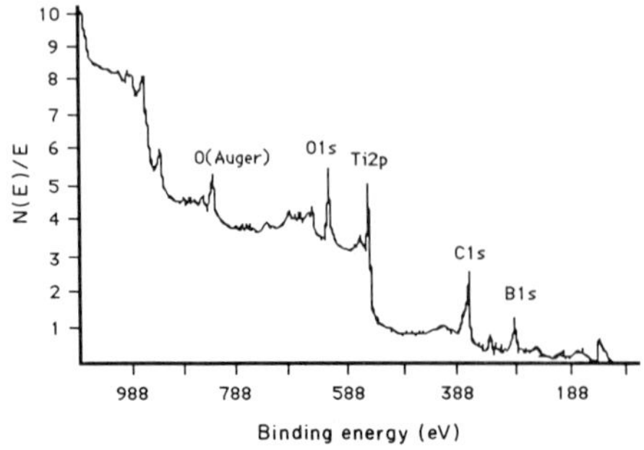

Fig. 3. Broad-scan ESCA spectrum for the debris layer after sputter etching for 8 min. The carbon and oxygen peaks are less prominent than before etching, and the titanium and boron peaks become more significant as the debris layer becomes thinner.

TABLE 2 Results for wear tracks obtained by high resolution electron spectroscopy for chemical analysis

Surface conditions	*Binding energy* (eV)			
	C 1s	*O 1s*	*Ti 2p*	*B 1s*
As received	284.6	530.4 532.4	458.7	187.1 191.85
Sputtered	283.45 284.6	530.6	454.05	187.0

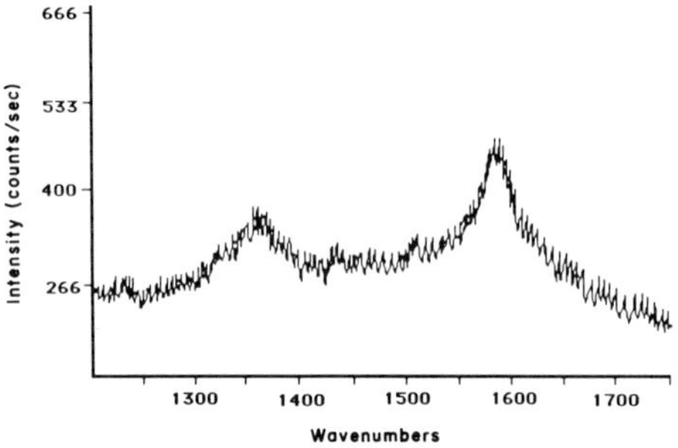

Fig. 4. A Raman spectrum obtained from material on the wear track. The form of the spectrum is similar to that obtained from low temperature carbon materials.

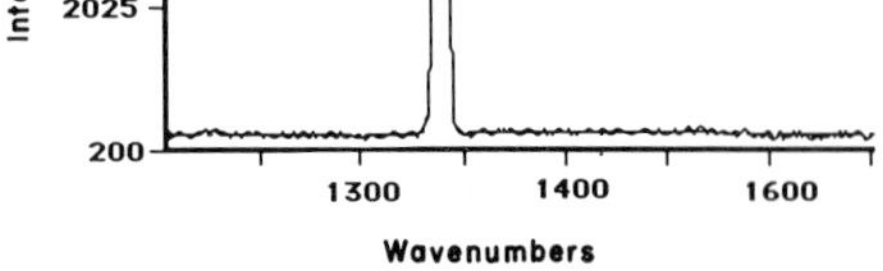

Fig. 5. A Raman spectrum obtained from the diamond stylus used as the slider on TiB_2.

The Raman spectrum at lower wavenumbers obtained from the wear track layer contains several peaks similar in intensity and position to those from TiO_2 (Figs. 6 and 7). These results suggest that wear of diamond against TiB_2 causes formation of a carbon layer incorporating some TiO_2. The carbon debris forms a compacted and adhered surface layer during subsequent passages of the stylus which has the carbonized form indicated by the Raman peak at 283.45 eV.

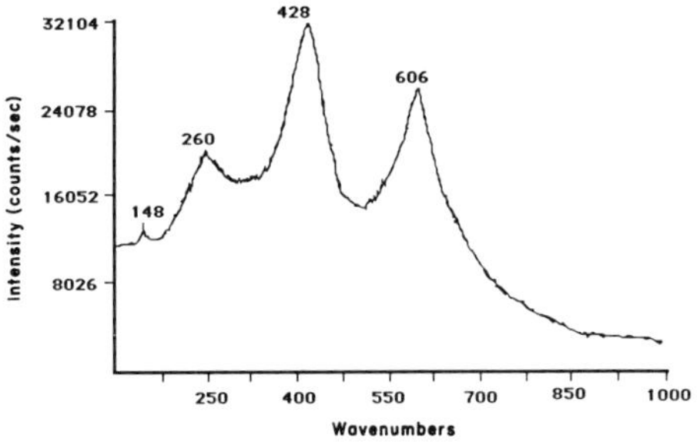

Fig. 6. A low wavenumber Raman spectrum from the wear track debris. The peaks are indicative of the presence of TiO_2.

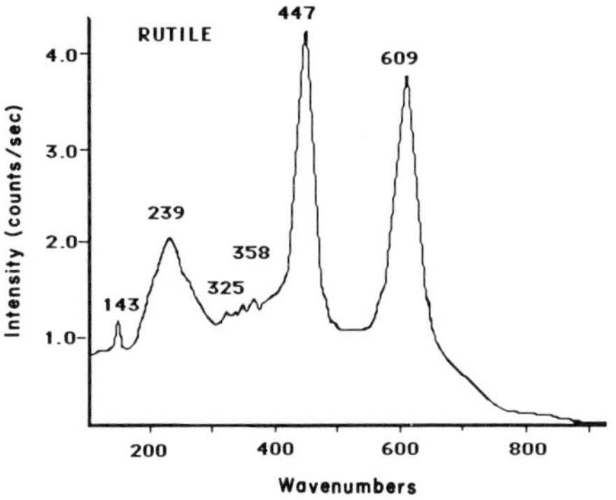

Fig. 7. A Raman pattern for TiO_2, confirming the existence of TiO_2 in the wear debris.

It is interesting to note that Thornton and Wilks [12] have reported that the wear of diamond against steel is primarily the result of graphitization of the diamond. They observed that the graphitization reaction is not thermally activated and occurs at velocities as low as 0.02 m s^{-1} in air at 0.1 Torr. The present experiments involve an even lower velocity than any used by Thornton and Wilks, but the analytical result reported here, the formation of a carbon-rich debris layer, is consistent with the operation of a graphitization mechanism comparable with that described by these researchers.

The magnitude and variation in the friction coefficient during reciprocating sliding tests of diamond on each of the TiB_2 surface regions were determined. At the start of sliding on the unimplanted surface, the friction coefficient is 0.02, but this decreases to less than 0.01 after sliding for 24 h. On the ion-implanted surface, the initial friction coefficient is 0.04, and this value also decreases to less than 0.01 after sliding for 24 h. On the ion-beam-mixed carbon region, the initial friction coefficient is relatively high, about 0.4, and the value fluctuates widely. After sliding for only 2 min, however, the friction coefficient is reduced to a value of about 0.01. On the sputtered-carbon-coated but unmixed surface, the friction coefficient value is initially small (less than 0.01) and remains at that level for the duration of the test. In each of these tests, the initial period of sliding involves the interaction of the diamond surface with the disc surface. In the first few passes, the diamond is observed to form a debris deposit layer which persists as the diamond subsequently slides on that layer. The initial stage of sliding against the ion-implanted surface results in a slightly increased tangential force compared with that observed on the unimplanted surface, possibly owing to microscale roughening of the surface during the implantation process. Similarly, the ion-beam-mixed surface shows a very high friction for a short initial time period, possibly as a result of a residual partially ion-beam-mixed carbon layer on the surface as well as some degree of surface roughening. Within 2 min, however, the tangential force value drops to the value characteristic of the sliding of diamond on the adhered debris layer on TiB_2. For the sputtered carbon layer, the friction coefficient is smaller than for each of the other surfaces, even though the carbon is readily stripped from the surface and appears to become entrained in the adhered debris layer. The decreased friction coefficient may be the result of lubrication by the carbon film or its residual particles but does not imply an improved wear resistance of that surface.

3.2. *WC sliding on TiB_2*

In contrast with diamond, cobalt-bonded WC as a sliding counterface results in the removal of surface material from the TiB_2, but essentially no removal of material from the WC ball. Unlubricated sliding experiments using a WC sphere as a counterface on each of the four areas of the TiB_2 specimen resulted in wear coefficient values which varied by a factor of only 3. Measurable wear flats were not observed on the WC spheres, but track widths on the TiB_2 surface were approximately 250 μm for an applied force of 300 gf and approximately 125 μm for an applied force of 50 gf. The applied normal force was 300 gf and the test duration was 24 h (216 m).

Significantly, the groove depth for each of these tests is less than 200 nm, while the peak of the concentration gradient for nickel ions implanted at 1 MeV is 330 nm. In these experiments, therefore, the wear process did not progress deeply enough into the bulk to reach the peak concentration of implanted ions. The near-surface volume in which the wear process was active did not differ chemically in the implanted and unimplanted specimens. However, the unimplanted and implanted regions do differ physically, since the implanted surface contains lattice damage resulting from the implantation process. Transmission electron microscopy has shown that the near-surface form of the damage in TiB_2 is the development of coarse dislocation tangles, followed by fine-scale damage, primarily point defect clusters extending to a depth about twice that of the implanted ions [13]. An increased dislocation content would not be expected to influence the mild wear removal process at the WC–TiB_2 interface since the material removal mechanism would involve only small volumes at the asperity tips. Consequently the similarity of wear rates within 200 nm of the surface indicated by the foregoing experiments is reasonable.

The experiments on the unimplanted and nickel-ion-implanted surfaces were repeated for longer test durations, 48 h or about 400 m to extend the wear process into the range of implanted ions. These extended tests did result in deeper penetration of the stylus into the bulk of the TiB_2; depths of approximately 400 nm were attained. The wear depth for the unimplanted material was found to be slightly greater than that for the implanted region, implying an increased wear resistance in the region of peak implanted ion concentration.

3.3. Incremental sliding distance tests: WC on TiB_2

In view of the above results, a further series of tests was performed in an effort to measure directly the influence of the implanted layer of ions and the associated damage layers on the wear process. The objective of the tests was gradually to wear away the surface layers by a mild wear process and to determine variations in wear rate with depth as the wear groove was deepened. In this instance, tests were performed both in a dry nitrogen atmosphere and in a laboratory humid air environment (50% relative humidity). The data from the test series are shown in Fig. 8. The TiB_2 specimen used in these tests had been

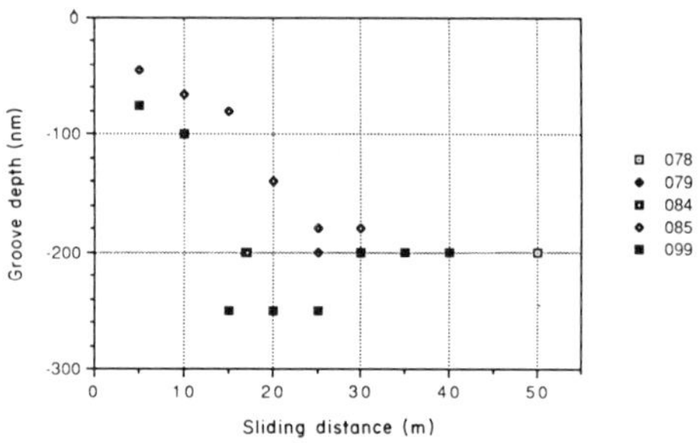

Fig. 8. Wear groove depths produced by the unlubricated sliding of WC against TiB_2. Tests 085 and 099 represent sequences of tests, as discussed in the text.

implanted with low energy ions (170 keV), and the resultant peak of the nickel ion concentration was 70 nm below the surface. Several tests (077, 078 and 079) were run as a single sliding increment. For test 077 (not shown in Fig. 8 since it would distort the scale of the plot for most of the data) using an applied load of 300 gf and sliding for 100 m over 11.1 h, the total groove depth was almost 400 nm, far deeper than the 70 nm range of the implanted ions. Test 078, using 50 gf and sliding 50 m, resulted in a groove depth of 200 nm, while test 079 also yielded a groove depth of 200 nm. These results demonstrated that very small wear increments would be required to permit an evaluation of possible wear rate variations at the 70 nm depth.

Experiment 084, sliding for 17 m under a 50 gf force in laboratory air (50% relative humidity), also resulted in a 200 nm groove depth. At this point, the results appeared to be inconsistent, indicating a constant groove depth independent of sliding distance and load. Because of this inconsistency, an additional test series was formulated to progress stepwise through the near-surface material, including the implanted layer. Dry nitrogen was again utilized as the test atmosphere to reduce the possibility of tribo-oxidation effects. Test 085 was performed in 5 m sliding intervals, the groove depth was measured after each interval, and the stylus was replaced on the wear track for an additional 5 m sliding increment. For the increments through 15 m, only small increases in groove depth are recorded while, for the subsequent increments through 30 m, much larger increases in groove depths are measured. These results seem to indicate a reduced wear rate at the level of the peak concentration of nickel ions.

The final increment 25–30 m, however, did not yield a further deepening of the groove.

The test procedure was then modified to avoid repetitive sliding increments in a single groove. The tests were repeated by preparing a set of grooves in which the sliding distance in each successive groove was increased by 5 m. Eight grooves were prepared by sliding the WC stylus on the TiB_2 in nitrogen under a fixed load of 50 gf. Measurements of the groove depths for this series of tests, identified as test 099, are also shown in Fig. 8. The groove depth progressed in the first 15 m of sliding to a depth of about 200–250 nm and remained at about that level for sliding distances as great as 40 m. The series of groove depths plotted in the figure strongly suggest a limitation of wear groove depth at 200–250 nm. It should be noted that this depth is about the same as that attained in the earlier single-increment tests at varied sliding distances and loads.

The transmission electron microscopy analysis of damage produced in nickel-ion-implanted TiB_2 also shows that fine-scale damage, *i.e.* point defect clusters, are found to exist to a depth about twice that of the ion-implanted species as a result of recoil of some of the host lattice atoms when struck by the implanting ions [13, 14]. In this instance, the fine-scale damage would extend to a depth of approximately 150–160 nm. These results can be interpreted as indicating that the wear rate is significantly affected by the fine-scale damage layer and/or a combination of the implanted ion species and the fine-scale damage layer. Although the wear depth is limited to about 200 nm, the width of the wear groove gradually increases with increasing sliding distance. On the assumption that the width reflects the diameter of the contact surface on the pin tip, the increasing groove width corresponds to a decreasing applied stress at the apparent contact area. An alternative interpretation of the observed limitation of groove depth is that the wear rate decreases significantly at apparent stress values lower than some critical value; in this case that value corresponding to the apparent stress when a depth of 200 nm is first attained. Since wear of the surface is continuing as the groove width increases, however, the process does not seem to be stress dependent.

A final set of wear tests was done on the unimplanted surface. The sliding distance in each successive groove was again increased by 5 m. It was anticipated that on this unmodified surface the groove depths would increase with increased sliding distance. While groove formation is observed, the debris produced by the early stage of wear is retained in the wear groove more readily than in the case of implanted surfaces. The result is the formation of a protective layer which inhibits the contact between the pin and the disc. The debris layer is generally not continuous over the entire length of the wear groove, and some wear does take place. The maximum groove depth, however, is only 100 nm. Further experiments to elucidate the mechanisms responsible for differences between implanted and unimplanted TiB_2 are planned.

4. Conclusions

(1) The mechanisms for unlubricated sliding wear in diamond–TiB_2 couples and WC–TiB_2 couples are markedly different. For the diamond–TiB_2 pair, the diamond pin wears, leaving an adhered debris layer on the wear path, even though the diamond has a greater hardness than the TiB_2. Surface analysis of the adhered layer indicates that it contains carbon and TiO_2. The initial sliding motion forms the layer on which the diamond later slides with very low friction. Prolonged sliding ultimately leads to failure by fracture of some surface grains of TiB_2.

(2) WC sliding against TiB_2 causes removal of material from the diboride surface, but little visible wear of the WC. A discontinuous debris layer may form in the wear path. The sliding friction is characterized by stick–slip behavior with extreme values of approximately 0.2–1.0.

(3) Progressive removal of surface layers by WC sliding on TiB_2 suggests a greatly lowered rate of material removal in the fine-scale damage layer just below the peak implanted species concentration depth. Corroborating experiments in an unimplanted region of the same specimen were complicated by debris accumulations, and further experiments are required to establish more firmly the effect of implantation.

Acknowledgments

The author wishes to acknowledge with thanks the contributions of C. E. DeVore, who performed the wear tests in a meticulous manner, and P. J. Blau and T. N. Tiegs, who reviewed the manuscript and provided constructive discussion and criticism. This research was sponsored by the Division of Material Sciences, U.S. Department of Energy, under Contract DE-AC05-84OR21400 with Martin Marietta Energy Systems Inc.

References

1 C. J. McHargue, C. S. Yust, P. Angelini, P. S. Sklad and M. B. Lewis, in E. A. Almond, C. A. Brookes and R. Warren (eds.), *Proc. 2nd Int. Conf. on the Science of Hard Materials*, in *Inst. of Physics Conf. Ser., 75* (1986) 803–812.

2 C. S. Yust, C. J. McHargue, C. L. White, P. Angelini and M. B. Lewis, *Eurotrib 85*, Elsevier, New York, 1985, pp. 1.4.8(1)–(9).

3 C. J. McHargue, M. B. Lewis, B. R. Appleton, N. Naramoto, C. W. White and J. M. Williams, in D. J. Rowcliffe, R. K. Viswanadham and D. J. Gurland (eds.), *Proc. 1st Int. Conf. on the Science of Hard Materials*, Plenum, New York, 1983, pp. 451–465.

4 C. J. McHargue and J. M. Williams, in S. T. Picraux and W. J. Choyke (eds.), *Metastable Materials Formation by Ion Implantation*, Elsevier, New York, 1982, pp. 303–309.

5 C. S. Yust and C. J. McHargue, in R. F. Davis, H. Palmour III and R. L. Porter (eds.), *Emergent Process Methods for High Technology Ceramics*, Plenum, New York, 1984, pp. 533–547.

6 C. J. McHargue and C. S. Yust, *J. Am. Ceram. Soc., 67* (1984) 117–23.

7 C. J. McHargue, P. S. Sklad, P. Angelini and M. B. Lewis, *Nucl. Instrum. Methods B, 1* (1984) 246–252.

8 T. Hioki, A. Itoh, S. Noda, H. Doi, J. Kawamoto and J. Kamigaito, *Mater. Sci. Lett., 3* (1984) 1099–1101.

9 P. Burnett and T. F. Page, *J. Mater. Sci., 19* (1984) 3524–3545.

10 K. Padmanabhan and G. Sorensen, *Thin Solid Films, 81* (1981) 13–19.

11 C. S. Yust, in K. C. Ludema (ed.), *Proc. Int. Conf. on Wear of Materials, Reston, VA, April 11–14, 1983*, American Society of Mechanical Engineers, New York, 1983, pp. 167–173.

12 A. G. Thornton and J. Wilks, *Wear, 53* (1979) 165–187.

13 P. S. Sklad, P. Angelini, M. B. Lewis and C. J. McHargue, in G. K. Hubler, O. W. Holland, C. R. Clayton and C. W. White (eds.), *Ion Implantation and Ion Beam Processing of Materials, Materials Research Society Symp. Proc.*, Vol. 27, Elsevier, New York, 1984, pp. 407–412.

14 M. B. Lewis, *Nuc. Instrum. Methods B, 7–8* (1985) 530–534.

Hard Transition Metal Nitride Films Deposited by Triode Ion Plating*

A. S. KORHONEN, J. M. MOLARIUS, I. PENTTINEN and E. HARJU

Helsinki University of Technology, Department of Materials Science and Mining, Vuorimiehentie 2 A, 02150 Espoo (Finland)

(Received December 3, 1987)

Abstract

The principles of reactive triode ion plating will be briefly described and examples of new nitride coatings deposited on cutting tools will be given. The coatings discussed include TiN, (Ti, Al)N and ZrN. Their relative order of performance in turning tests will be discussed and microstructural details will be studied by means of scanning electron microscopy (SEM) and transmission electron microscopy (TEM). It will be shown that, although the best coatings may appear rather uniform and hardly any structural details can be distinguished by SEM, cross-sectional TEM studies always reveal that a very-fine-grained columnar morphology exists in the coatings.

1. Introduction

Since the beginning of the 1980s, gold-coloured titanium-nitride(TiN)-coated high-speed steel (HSS) cutting tools have become generally available for various cutting applications ranging from drilling to gear hobbing. So far, most work has concentrated on the improvement of the quality and performance of the stoichiometric TiN coating and much less attention has been devoted to alternative coatings.

There are a number of different physical vapour deposition (PVD) processes which can be used to coat tools. They differ mainly in the type of evaporation technique used to produce the metal (titanium) vapour. The three main classes are sputtering and arc evaporation (which both use solid metal targets) and electron beam evaporation (which is based on the melting of the metal source). Often the generic term "ion plating" is used to describe all three, although it most commonly refers to electron beam evaporation onto negatively biased substrates.

(Ti, Al)N coatings have recently evolved as an alternative to TiN. The basic idea is to form a protective aluminium oxide layer on top of the coating, which can then improve the wear resistance further. So far, the coatings have been deposited mostly by sputtering [1, 2] although both arc evaporation [3] and electron beam [4, 5] evaporation have also been used.

The Ti–N system itself has been studied intensively [6, 7]. Although zirconium is a neighbouring element of group IV, the Zr–N system is not well understood and may behave quite differently from the Ti–N system [8, 9]. According to some reports, ZrN coatings may in fact outperform TiN in cutting [10]. The third material which has been studied in order to find better nitride coatings is hafnium. However, it is much more expensive than either titanium or zirconium and, although good results have been reported on the chemical vapour deposition of HfN coatings since the early 1970s, the PVD applications have not been extensive.

In a recent survey [4], it was found that both (Ti, Al)N and ZrN coatings deposited by reactive triode ion plating could outperform standard stoichiometric TiN coatings in turning. In what follows the results will be discussed in more detail.

2. Experimental details

2.1. Deposition of the coatings

The coatings were deposited on negatively biased steel substrates by means of reactive triode ion plating based on electron beam evaporation of metal into a nitrogen-containing atmosphere. The glow discharge within the chamber was intensified by emitting electrons from resistively heated tungsten filaments. The details of the present deposition technique have been described

*Paper presented at the 3rd International Conference on the Science of Hard Materials, Nassau, The Bahamas, November 9–13, 1987.

0921-5093/88/$3.50 © Elsevier Sequoia/Printed in The Netherlands

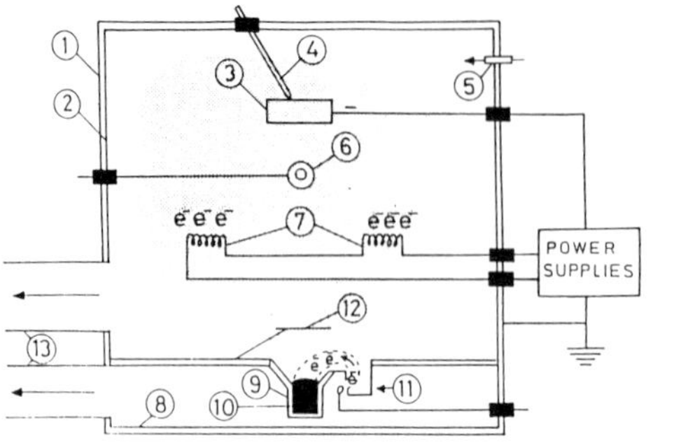

Fig. 1. Schematic diagram of the reactive triode ion plating system: 1, coating chamber, grounded; 2, thermal shield; 3, substrates; 4, thermoelement; 5, gas inlet; 6, quartz crystal sensor; 7, heated electron-emitting filaments; 8, chamber for electron beam gun; 9, water-cooled copper crucible; 10, material to be evaporated; 11, electron beam gun; 12, shield; 13, diffusion pump.

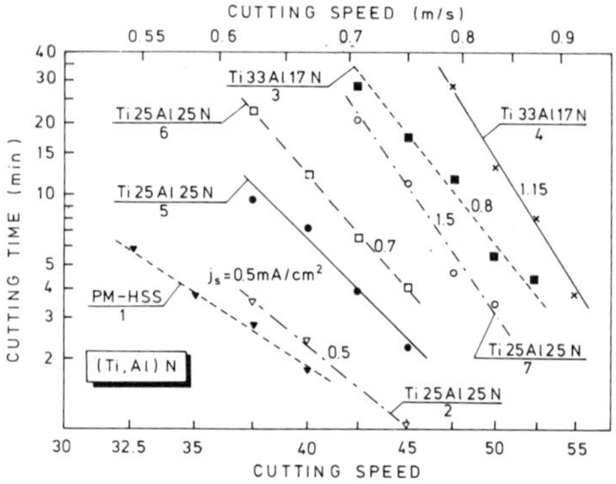

Fig. 2. Cutting performance of (Ti, Al)N coatings. The effect of increasing the substrate current density j_c is also shown. (Curves 3–7 taken from ref. 4.)

thoroughly elsewhere [4, 6, 11] and they will therefore not be discussed here. Figure 1 shows a schematic view of the system used.

2.2. Materials

Substrates were either samples cut from austenitic stainless steel sheet 0.8 mm thick of type AISI 304 or HSS inserts produced by powder metallurgy (PM) and having a nominal composition of 1.3 wt.% C, 4.3 wt.% Cr, 3.8 wt.% Mo, 3.2 wt.% V, 9.8 wt.% W and 10.5 wt.% Co. The cutting geometry of the inserts was in accordance with our previous study [4].

Titanium and zirconium sources were formed by cutting slices of metal rods of nominal purities 99.15% and 99.5% respectively. Various (Ti, Al)N compositions were then obtained by the mixing and pressing of titanium and aluminium powders and by subsequently pre-melting the source by means of an electron beam gun or by using slices cut from Ti–6Al–4V rod as a source. Evaporation was carried out in a pressure of 0.6 Pa by simultaneously feeding in argon and nitrogen gases having nominal purities 99.99% and 99.998% respectively.

2.3. Testing

The cutting tests were carried out by dry turning quenched-and-tempered steel OVAKO MoC 410 M, which corresponds approximately to AISI 4140 but was calcium treated for improved machinability. The feed and depth of cut were 0.25 mm rev^{-1} and 2.5 mm respectively. The cutting speed was varied between 0.5 and 1.0 m s^{-1}. The other details of the testing procedure have been reported elsewhere [4] and will therefore not be repeated here.

The structure of the coatings was studied by means of both scanning electron microscopy (SEM) and transmission electron microscopy (TEM), and also X-ray diffraction and nuclear-resonance-broadening (NRB) depth profiling for both nitrogen and aluminium [5]. In the following, the results from structural studies will be discussed in more detail.

3. Results and discussion

The Taylor curves of various (Ti, Al)N coatings are compared with each other and with uncoated PM HSS inserts in Fig. 2. It can be seen that the performance of the coatings can vary greatly. As shown previously [4], increasing the substrate current density tended to improve the performance of the (Ti, Al)N coatings and the corresponding Taylor curves moved from left to right in Fig. 2. The best composition corresponded approximately to the composition 33 at.% Ti, 17 at.% Al and 50 at.% N. Because of the greatly varying vapour pressures of aluminium and titanium, some non-uniformity in the aluminium distribution was evident in the coatings, as discussed elsewhere [5]. However, evaporation from alloy sources worked reasonably well and a two-source system having separate sources for aluminium and titanium was therefore not considered. It should be noted that especially for the ion plating of aluminium a resistively heated continuous wire feed system has recently been developed [12].

Fig. 3. Scanning electron micrographs showing the surface morphologies of the (Ti, Al)N coatings corresponding to (a) curve 2, (b) curve 3 and (c) curve 4 in Fig. 2.

There was some variation in the performance of the coatings even with the same nominal composition when a low substrate current density was used (see for example Ti–25Al–25N, coatings 2 and 5, in Fig. 2). Coating 2 yielded the lowest performance and was only slightly better than the bare PM HSS substrate. It was grown at a much higher rate (108 nm min^{-1}) than coating 5 (36 nm min^{-1}). It was markedly columnar as can be seen in Fig. 3, where the surface morphologies of the various (Ti, Al)N coatings are compared. It can be seen by comparing Figs. 2 and 3 that, when the

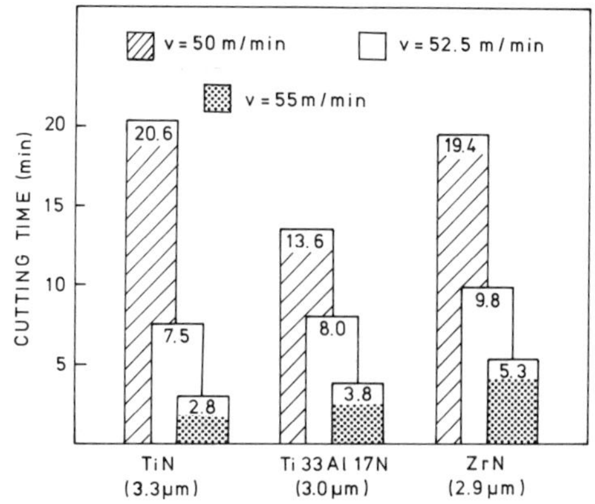

Fig. 4. Comparison of triode-ion-planted TiN, Ti–33Al–17N and ZrN coatings at three different cutting speeds. (Data from ref. 4.)

coating became smoother and less columnar, the cutting performance generally improved. The uniformity of the coating structure generally improved with increasing substrate current density both with TiN and (Ti, Al)N coatings [4]. However, while the performance of the best (Ti, Al)N composition was very good, especially at the highest cutting speeds, the structures were still not nearly as uniform as with TiN or ZrN coatings. The best coatings from the aforementioned three nitride groups, *i.e.* TiN, (Ti, Al)N and ZrN, are compared in Fig. 4 where the tool lives at three selected cutting speeds are shown. It can be seen that at the highest speed (55 m min^{-1}) the ZrN coatings last the longest while Ti–33Al–17N and TiN come second and third respectively. When the cutting speed decreases, the relative positions of TiN and (Ti, Al)N change and TiN becomes the best of the three coatings, although the difference between it and ZrN is not large. It has been suggested that the behaviour at the highest cutting speeds could be explained on the basis of the oxide layer which forms on the top of the coating [13]. Our results seem to support this view.

As already mentioned, the gradual improvement in the surface morphology tended generally to improve the cutting performance. Despite the considerable improvement in morphology with (Ti, Al)N coatings with increased substrate current densities and lower growth rates, the structure did not become quite as uniform as with TiN or ZrN coatings.

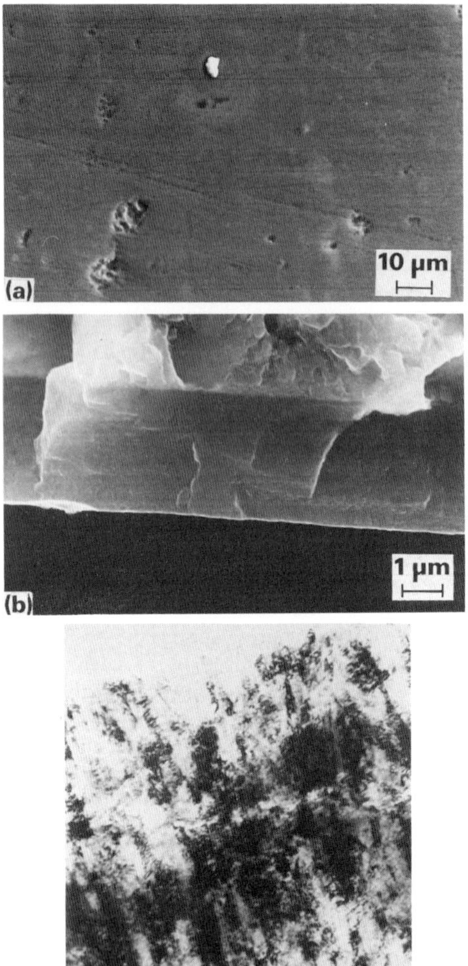

Fig. 5. (a) Plan-view SEM, (b) cross-sectional SEM and (c) cross-sectional TEM micrographs of a stoichiometric TiN coating.

Figures 5(a), 5(b) and 5(c) are two SEM micrographs and a cross-sectional TEM micrograph respectively of a good-quality TiN coating. It can be seen that, although hardly any details can be distinguished by SEM (Fig. 5(b)), the cross-sectional TEM micrograph in Fig. 5(c) reveals a fine-grained columnar growth pattern. The column

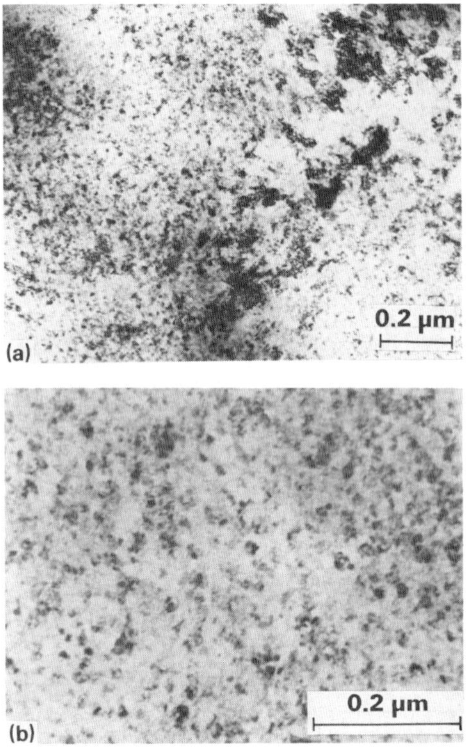

Fig. 6. Plan-view TEM micrographs of (Ti, Al)N coatings showing (a) the two-zone grain structure and (b) an enlargement of the fine-grained zone. (Magnifications: (a) 60 000 ×; (b) 120 000 ×.)

diameter is generally small, in this case about 0.04 μm. The grains are elongated in the through-thickness direction, indicating the general growth pattern of the coatings.

The structure of the (Ti, Al)N coatings as seen by SEM was generally slightly more columnar and less uniform than that of the mononitrides, *i.e.* TiN or ZrN. Figures 6(a) and 6(b) are plan-view TEM micrographs of the (Ti, Al)N coating at two different magnifications. Figure 6(a) indicates a two-zone structure having different grain sizes up to 0.1 μm. Figure 6(b) shows a view of the small-grain-size zone, where the grain size is about 0.02 μm. It should be noted that the existing two-zone structure in these coatings could also be confirmed by analysing the aluminium distribution by the NRB technique [5].

The most uniform structure was obtained with ZrN coatings. As reported earlier [4], the cutting performance tended generally to improve with

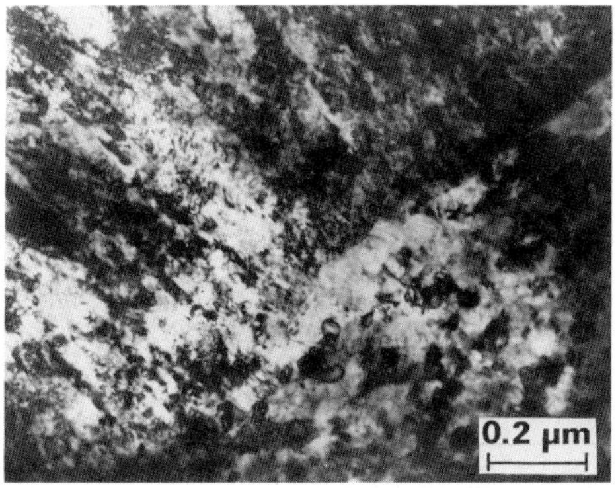

Fig. 7. A cross-sectional TEM micrograph of a ZrN coating.

decreasing substrate current density, which was contrary to the case with both TiN and (Ti, Al)N coatings. On the basis of SEM micrographs, it became difficult to distinguish any structural details. However, again by means of TEM, a fine columnar growth pattern could be seen, as shown in Fig. 7. The grains are elongated and the grain size is approximately 0.04 μm.

4. Summary and conclusions

The basic principles of reactive triode ion plating were briefly discussed and examples of the deposited coatings were given. It was shown that the ion plating of HSS cutting tools with various nitride coatings can readily be accomplished. Turning tests indicate that new nitride coatings including (Ti, Al)N and ZrN may even outperform standard stoichiometric TiN in some cutting applications. The cutting performance of the coatings generally tends to improve with improving uniformity although, in turning, even slightly columnar coatings may yield good performance. Even the best (Ti, Al)N films were slightly columnar as observed by SEM. In contrast, hardly any details could be distinguished in either TiN or ZrN coatings. However, even here a very-fine-grained columnar morphology could be found in cross-sectional TEM micrographs.

Acknowledgments

The authors are grateful to the Technology Development Centre (TEKES) and the Academy of Finland for financial support for this work.

References

1 O. Knotek, M. Bohmer and T. Leyendecker, *J. Vac. Sci. Technol. A, 4* (6) (1986) 2695.

2 W.-F. Münz, *J. Vac. Sci. Technol. A, 4* (6) (1986) 2717.

3 H. Randhawa and R. Cunningham, Characterization of ternary nitrides, *34th American Vacuum Society Symp., Anaheim, CA, November 2–6, 1987.*

4 J. M. Molarius, A. S. Korhonen, E. Harju and R. Lappalainen, Comparison of cutting performance of ion plated NbN, ZrN, TiN and (Ti, Al)N-coatings, *Surf. Coatings Technol.*, in the press.

5 I. Penttinen, J. M. Molarius, A. S. Korhonen and R. Lappalainen, Structure and composition of ZrN and (Ti, Al)N coatings, *34th American Vacuum Society Symp., Anaheim, CA, November 2–6, 1987.*

6 J. M. Molarius, A. S. Korhonen and E. O. Ristolainen, *J. Vac. Sci. Technol. A, 3* (6) (1985) 2419.

7 E. Niemi, A. S. Korhonen, E. Harju and V. Kauppinen, *J. Vac. Sci. Technol. A, 4* (6) (1986) 2763.

8 K. Salmenoja, A. S. Korhonen, E. Erola and J. M. Molarius, *Appl. Phys. Lett., 49* (9) (1986) 505.

9 E. O. Ristolainen, J. M. Molarius, A. S. Korhonen and V. K. Lindroos, *J. Vac. Sci. Technol. A, 5* (4) (1987) 2184.

10 M. Valkhoni and S. Chandrashekhar, *Int. J. Adv. Manuf. Technol., 2* (1) (1987) 75.

11 J. M. Molarius, *Doctor of Technology Thesis*, Department of Materials Science and Mining, Helsinki University of Technology, Espoo, 1987, 83 pp.

12 I. Pyyppönen and U. Mäkelä, Diode ion plating—aluminium coatings as an application, *Res. Notes 480*, 1985, 32 pp. (Technical Research Centre of Finland.)

13 B. Kramer, *Ph.D. Dissertation*, Massachusetts Institute of Technology, Cambridge, MA, 1979.

Microstructural Study of Ion-implanted WC–Co*

A. D. ANDERSON and M. H. LORETTO

Department of Metallurgy and Materials, University of Birmingham, Edgbaston (U.K.)

G. DEARNALEY

Atomic Energy Research Establishment, Harwell, Didcot, Oxon (U.K.)

(Received November 9, 1987)

Abstract

In this paper, we present the initial results of work aimed at observing the microstructure produced by nitrogen ion implantation into WC–6wt.%Co cemented carbide. Implantation is reported to increase the wear resistance of many tools but little is actually known about how such a shallow implanted layer can so affect the wear resistance. Techniques have been developed to allow the implanted microstructure to be examined in cross-section by transmission electron microscopy. The results show that implantation produces fine dislocation arrays in the carbide phase in the implanted layer to a depth of 0.2 μm and produces a lattice strain of the order of up to 0.5% in that phase. Even after annealing at 1000 °C the carbide is still damaged although voids or nitrogen bubbles are seen to form in the implanted layer.

1. Introduction

Ever since ion implantation of nitrogen into WC–Co composites and other materials for tribological applications has been investigated, there has been, and still is, much controversy as to whether any significant effects related to increased wear resistance are produced by implantation. While many reports suggest that there is no effect (see for example ref. 1), others indicate that significant increases in wear resistance do occur (see for example ref. 2), particularly in specific applications where wear rates are moderate to low, such as drill bits for printed-circuit boards and wire-drawing dies [3]. Increases of up to 10 times have been reported for drill bits and two to three times for wire-drawing dies [4], which can represent substantial savings in both tool costs and tool change-over costs. The possibility that such savings can be made is of such importance that research into the effects of ion implantation on wear resistance cannot be ignored until all related phenomena are explored and understood.

This field of research has, until fairly recently, been hampered by the fact that observed changes in mechanical properties have not been able to be related to microstructural changes because of the extreme difficulty in preparing useful samples, as the depth of these changes from the surface is very small, typically 0.2 μm. The first microstructural observations were reported in 1981 by Greggi [5], which gave some insight into the effects of implantation in cemented carbides.

The results presented here are our first observations in a project aimed at characterizing any microstructural changes produced by implantation and relating these to changes in mechanical properties.

2. Techniques

Suitable samples were implanted at the Atomic Energy Research Establishment, Harwell, and were prepared for transmission electron microscopy (TEM) by gluing together two implanted surfaces of "straight grade" WC–6wt.%Co (1.4 μm carbide grain sized), supplied by Sandvik U.K. Ltd., with a thin layer, of about a micron, of epoxy resin. The glued block was then sectioned to produce thin slices of around 200 μm which were then mechanically polished on both sides, the final slice being 80–100 μm thick. 3 mm discs were then ultrasonically cut from the slice with

*Paper presented at the 3rd International Conference on the Science of Hard Materials, Nassau, The Bahamas, November 9–13, 1987.

0921-5093/88/$3.50

© Elsevier Sequoia/Printed in The Netherlands

the interface across the centre of the disc. Discs were then ion beam thinned to produce electron-transparent samples. TEM was performed using a JEOL 4000FX microscope operated at 400 keV.

X-ray diffraction was performed on bulk samples to examine whether any elastic strain was produced in the surface by implantation. Microhardness testing was carried out on implanted surfaces with a Leitz "miniload" Vickers indenter using 20 and 50 gf loads. The indents were 4–6 μm in size and were measured using scanning electron microscopy (SEM).

3. Results and discussion

3.1. Transmission electron microscopy

Diffraction contrast studies using a 400 kV medium-voltage electron microscope of samples prior to implantation shows that the WC phase contains a few dislocations and stacking faults but is otherwise featureless, whereas the f.c.c. cobalt phase, which is known to contain a significant amount of tungsten and carbon in solution [6], is heavily faulted and shows some evidence of damage from the ion beam thinning necessary to prepare the samples. These facts, together with the preferential thinning and possible loss of cobalt due to sputtering effects during implantation, make analysis of any influence of nitrogen implantation on the cobalt very difficult.

The use of cross-sectional TEM specimens has allowed the depth-dependent microstructure of implanted material to be examined and the depth of any effects to be accurately measured. Samples implanted with nitrogen to a dose of 3.5×10^{17} N^+ ions cm^{-2} at 150 keV and 400 °C showed fine dislocation tangles and finer-scale damage in the carbide phase in the implanted layer to a depth of 215 nm (2150 Å) (Fig. 1). The only evidence, to date, of implantation damage in the cobalt is the appearance of broad parallel multiple faults or laths of hexagonal cobalt in surface regions (Fig. 2). The formation of these laths may be caused by the intense surface stress generated by implantation. This stress may be accommodated in the cobalt by faulting which would produce hexagonal cobalt which is the stable room temperature form of cobalt. Such a transformation has also been suggested by Greggi and Kossowsky [7]. When implanted samples are annealed at 1000 °C, voids or bubbles are seen to form. It is known that voids form in preference to or in addition to dislocation loops when gas atoms are present [8]. The maximum in void density occurs at the peak of the damage and nitrogen profiles, at approximately half the implanted depth (Fig. 3). It was also observed that the implantation-affected layer in this particular sample increased in depth by 80 nm (800 Å) over that observed and expected in other samples. No explanation can as yet be made for this observation.

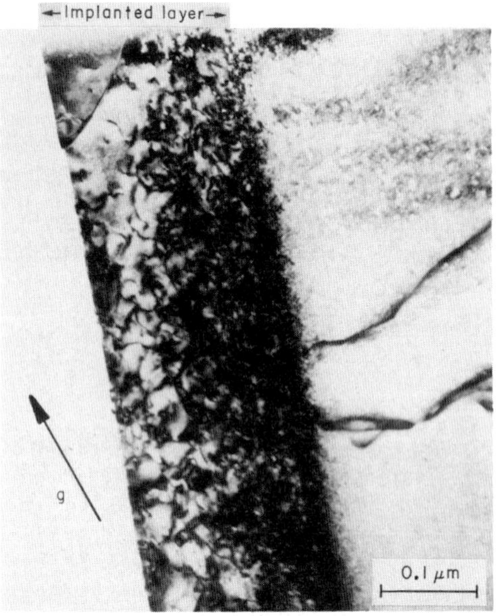

Fig. 1. Cross-sectional TEM micrograph of a sample implanted at 400 °C with 3.5×10^{17} N^+ ions cm^{-2} at 150 keV, showing small dislocations produced from the implantation damage in the implanted layer. $\boldsymbol{g} = \bar{1}101$.

A sample implanted at a higher dose of 5.8×10^{17} N^+ ions cm^{-2} at 90 keV and 300 °C showed striations of amorphous material in the near-surface half of the implanted layer while the deeper half of the layer showed dense fine damage (Fig. 4). The crystalline material surrounding the amorphous striations, although still damaged, did not appear to contain as much damage as the rest of the layer. Thus the severe lattice disturbance appears to have caused a certain amount of self-annealing in the crystalline regions of the near-surface half of the layer.

Thin window energy-dispersive X-ray analysis (EDXA) and electron energy loss spectroscopy (EELS) have been used in an attempt to determine the nitrogen composition profile, but the experiments are not straightforward and no evidence for the presence of nitrogen has been

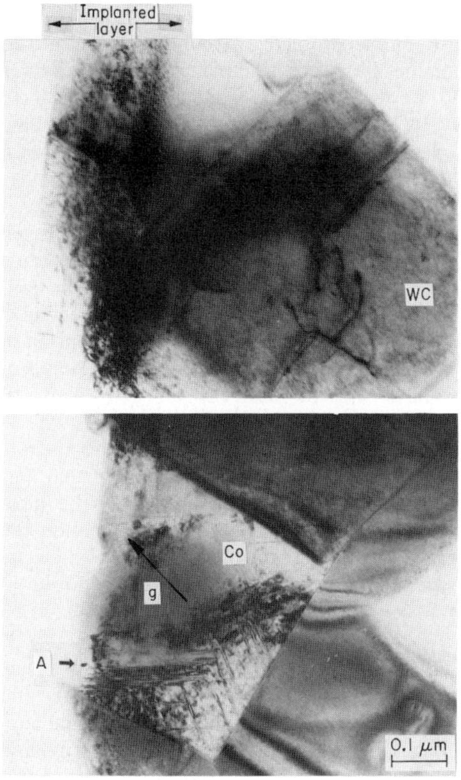

Fig. 2. Cross-sectional TEM micrographs of a sample implanted at room temperature with 3.5×10^{17} N^+ ions cm^{-2} at 150 keV, showing broad bands of faulting or laths in the cobalt binder phase at the surface (feature A, arrowed) which represent a phase change from the metastable cubic to the stable hexagonal form. This possibly occurs as a stress relief mechanism due to the surface compressive stress produced by implantation. It also appears that erosion of the cobalt has occurred because of sputtering effects either during implantation or during sample preparation by ion beam thinning. $\boldsymbol{g} = 0\bar{2}0$.

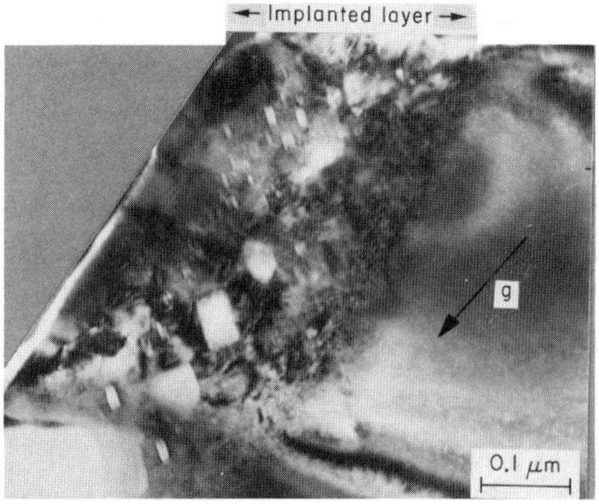

Fig. 3. Cross-sectional TEM micrograph of a sample implanted at room temperature to 3.5×10^{17} N^+ ions cm^{-2} at 150 keV and then post-implantation annealed at 1000 °C for 16 h, showing formation of voids or nitrogen bubbles in the implanted layer. $\boldsymbol{g} = 4\bar{2}\bar{2}2$.

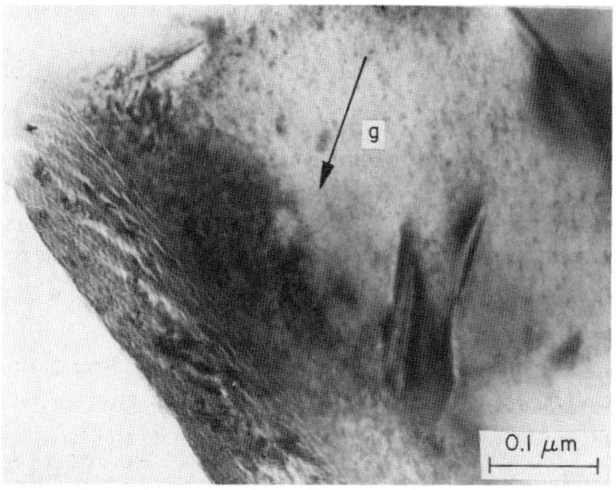

Fig. 4. Cross-sectional TEM micrograph of a sample implanted to a dose of 5.8×10^{17} N^+ ions cm^{-2} at 90 keV and 300 °C, showing amorphous striations (white) in the near-surface half of the layer. $\boldsymbol{g} = 01\bar{1}\bar{1}$.

obtained using these techniques. In the case of the EDXA work the absorption of the relatively low energy nitrogen X-rays in the heavy WC is expected to reduce greatly the ability to detect the presence of the nitrogen. For EELS, not only is the nitrogen edge superimposed on a large background caused by low energy events in the WC, but also the probability that electrons suffer a double-scattering event in these samples is very high, so that the nitrogen edge is mixed with plasmon loss events; the edge is then not visible. Only if far thinner samples can be prepared is it going to be possible to obtain useful data from analytical microscopy and to combine good spatial resolution with compositional information. In the further work outlined below the techniques of Auger electron spectroscopy (AES) and nuclear reaction for nitrogen depth profiling will be used to provide data on the distribution of nitrogen but these data will clearly have significantly inferior spatial resolution to that which would be available with EDXA or EELS data from thin foils, *i.e.* these two techniques are complementary to the analytical electron microscopy.

3.2. *X-ray diffraction*

X-ray diffraction results for the carbide phase in a sample implanted at 400 °C showed no change in lattice parameter but did show peak broadening, indicating a strain of the order of 0.04% in the implanted layer. The strain was calculated by relating the increase in peak width to an increase in lattice parameter. However, the X-ray penetration for the chromium radiation used was 2 μm compared with an implanted depth of 0.2 μm, but calculations show that only about 25% of the X-ray intensity would come from the implanted layer and the remaining amount would come from the substrate. Manual subtraction of an unimplanted carbide peak from the results indicated that the actual elastic strain in the implanted layer could be as high as 0.5%. The composite can accommodate up to 1% elastic strain; so, with a modulus of 600 GN m^{-2} (600 GPa) [9], this gives a stress of around 3000 MN m^{-2} (3 GPa). The value of this stress appears to be in agreement with work done by Burnett and Page [10] on implanted sapphire.

3.3. *Microhardness*

The low load Vickers microhardness tests on implanted surfaces showed no change in hardness due to implantation. However, Knoop microhardness testing, which was unavailable at the time, is reported to show increases in hardness of up to 40% [11] because the ratio of contact area to penetration depth is greater than that for the Vickers indentation.

4. Conclusion

The results of the work show that the effects of implantation can now be effectively examined by TEM and thus microstructural changes can be related to any change in mechanical properties. There is little other published work to compare the results with as the cross-sectional TEM technique has not been applied to this material before; however, the microstructural observations for the carbide phase do appear to agree with those of Greggi [5]. No conclusive observations have been made of any microstructural changes in the cobalt phase caused by nitrogen implantation, because of the small amount present in the composite and the difficulty of retaining it in the TEM specimens, although a possible cubic-to-hexagonal transformation has been observed. The elastic lattice strain of up to 0.5% observed by X-ray diffraction seems to be in agreement with that obtained by Burnett and Page [10].

Work is soon to be under way to overcome some of the problems mentioned above and to examine the effect of implantation temperatures and post-implantation annealing from 200 to 1100 °C on the microstructure within the implanted layer for both the composite material (WC–6wt.%Co) and a simulated cobalt binder phase alloy. The microstructure will be observed by cross-sectional analytical TEM and nitrogen distribution assessed by AES and, by implanting with ^{15}N, the nuclear reaction N(p, α) can be used as an independent nitrogen-profiling technique. X-ray diffraction will be used to assess strain in the implanted layer and low load Knoop microhardness will be used as an assessment of mechanical properties.

By the use of these implantation parameters and analytical techniques, it is hoped that this continuing work will shed some light on the question of whether implantation does increase the wear resistance of cemented carbides and on the underlying mechanisms by which it may do so.

Acknowledgments

Gratitude is expressed to the U.K. Science and Engineering Research Council for financial support and the Department of Metallurgy and Materials, University of Birmingham, for the provision of research facilities. The support of Sandvik is gratefully acknowledged and partial funding from the Commission of the European Community BRITE Project 1015.

References

1 D. W. Oblas and V. K. Sarin, Ion implantation and its application to WC–Co cemented carbides, *Carbide Tool J., 18* (4) (1987) 15–19.

2 G. Dearnaley and F. J. Minter, Microhardness and nitrogen profiles in ion implanted tungsten carbide and steels, *Nucl. Instrum. Methods B, 7–8* (1985) 188–194.

3 G. Dearnaley and S. J. B. Charter, Ion implantation of cemented carbide cutting and forming tools, *Eng. Dig., 42* (3) (1981) 35–40.

4 G. Dearnaley, Ion implantation for improved resistance to wear and corrosion, *Mater. Eng. Appl., 1* (1978) 28–41.

5 J. Greggi, TEM observations of defect structures in nitrogen ion-implanted WC–Co, *Scr. Metall., 17* (1983) 765–768.

6 G. Wirmark and G. L. Dunlop, Phase transformations in the binder phase of Co–W–C cemented carbides, *Proc. 1st Int. Conf. on the Science of Hard Materials*, Plenum, New York, 1981, pp. 311–328.

7 J. Greggi and R. Kossowsky, A STEM microanalytical

investigation of nitrogen implanted cemented WC–Co, *Proc. 1st Int. Conf. on the Science of Hard Materials*, Plenum, New York, 1981, pp. 485–497.

8 R. Bullough, B. L. Eyre and K. Krishnan, *Proc. R. Soc. London, Ser. A, 346* (1975) 91.

9 V. Thompson, personal communication, Sandvik Hard Materials (U.K.) Research Centre, Coventry, 1987.

10 P. J. Burnett and T. F. Page, An investigation of ion implantation-induced near-surface stresses and their effects in sapphire and glass, *J. Mater. Sci., 20* (1985) 4624–4646.

11 G. Dearnaley, unpublished, 1987.

Wear Mechanisms Modification Induced by Ion Implantation in Cemented Tungsten Carbide*

S. BARTOLUCCI LUYCKX and J. P. F. SELLSCHOP

Wits Council for Scientific and Industrial Research, Schonland Research Centre, University of the Witwatersrand, Johannesburg (South Africa)

(Received November 9, 1987)

Abstract

The effects of ion implantation on the wear resistance of cemented-tungsten-carbide components are found to depend on the wear mechanisms of the components. Maximum improvement in wear resistance after implantation is expected to occur when the controlling wear mechanism is cobalt extrusion followed by pull-out of carbide grains.

1. Introduction

The effect of ion implantation on the wear resistance of cemented tungsten carbide (WC-Co) is controversial. According to some investigations the wear resistance of WC-Co components is improved by ion implantation. The components tested in these investigations are as follows: round wire-drawing dies, rectangular wire dies, punches and dies for steel, blades for rotary scrap choppers [1], printed-circuit board drills [2, 3], dental burrs, slitters for cutting synthetic rubber and precision punches for electronic parts [2].

According to other investigators, WC-Co components do not benefit from ion implantation. These investigators tested steel cutting tools, tools for turning operations, components subjected to erosion and printed-circuit board drills [4].

The components which appear in the first list do not appear in the second, with the exception of printed-circuit board drills. In general, therefore, the effects of ion implantation on wear resistance appear to vary from component to component, which suggests that they vary with the wear mechanisms of the component.

In this paper, we summarize results obtained on the modifications induced by ion implantation in the surface of WC and cobalt, discuss the effects of these modifications on the wear mechanisms of WC-Co components and finally clarify the apparent disagreements found in the literature on this topic.

2. Experimental techniques and results

2.1. WC

{0001} faces of WC single crystals were implanted at 100 keV to doses ranging from 5×10^{16} to 2×10^{17} N^+ ions cm^{-2}. The {0001} faces were indented by means of a Vickers micro-indenter before and after implantation and were found to be embrittled by implantation [5]. Transmission electron microscopy of WC-Co thin foils before and after implantation at 50 keV and to the same doses as above revealed that the dislocation density in WC grains increases dramatically after implantation [6] and that the embrittlement can thus be explained in terms of decreased dislocation mobility. The observed embrittlement should imply an increase in hardness [7]. However, this could not be confirmed by means of standard hardness tests because the surface layer affected by implantation in the present experiments was only about 0.1 μm thick and standard indenters penetrated the substrate even at the lowest indenting loads. Fremlin and Askouri [8], however, proved that ion implantation hardens WC-Co surfaces by implanting surface layers as thick as 140 μm, whose hardness could be tested by standard methods.

Single-pass scratch tests with a diamond stylus on WC-Co implanted and unimplanted surfaces

*Paper presented at the 3rd International Conference on the Science of Hard Materials, Nassau, The Bahamas, November 9-13, 1987.

0921-5093/88/$3.50

© Elsevier Sequoia/Printed in The Netherlands

were carried out by applying to the stylus a load of 0.3 kgf and by moving the stylus at a speed of approximately 25×10^{-6} m s^{-1}. These tests showed that along the scratches on unimplanted surfaces a large number of WC grains exhibited slip lines and hardly any grains exhibited cracks (Fig. 1) [9]. The opposite was observed along scratches on implanted surfaces (see for example Fig. 2).

Since the WC grains along the scratches are oriented randomly, these results indicate that all WC surfaces (not only the {0001} surfaces) are embrittled by ion implantation.

2.2. *Cobalt*

The scratches described above were examined by scanning electron microscopy after the scratched surfaces had been tilted by up to 75°. Under these conditions, it was possible to observe the rod-like features indicated by arrows in Fig. 1. These features were identified as extruded cobalt because the energy-dispersive X-ray analysis (EDXA) gave a cobalt-to-tungsten ratio significantly higher than that obtained from the analysis of the rest of the surface (Fig. 3). These rod-like features were found more easily along scratches on unimplanted surfaces than on implanted surfaces. Thus the high local compressive stresses exerted by a scratching stylus on the adjacent cobalt (Fig. 4) extrude unimplanted cobalt more easily than implanted cobalt.

In the thin WC–Co foils described above the same cobalt areas were examined by TEM before and after N^+ ion implantation. After implantation the dislocation density increased and electron diffraction rings were recorded, consistent with the *d* values of Co_2N precipitates [9]. These observa-

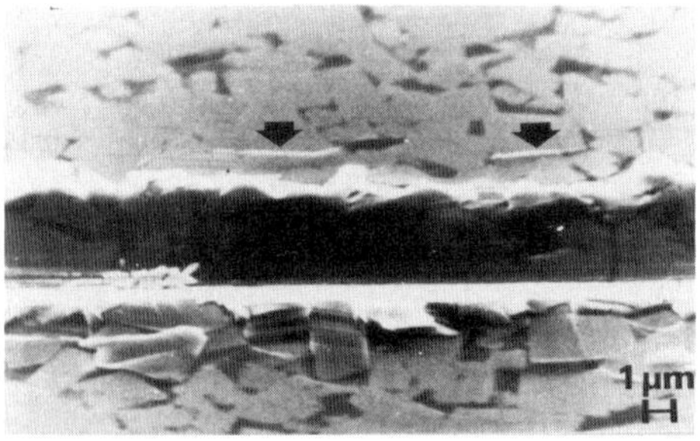

Fig. 1. A scratch produced by a diamond stylus on an unimplanted WC–Co surface at a load of 0.3 kgf. A large number of WC grains exhibit slip lines. The arrows point to rod-like features which were identified as extruded cobalt. The WC–Co surface was tilted by 60°.

Fig. 2. A scratch produced by a diamond stylus at a load of 0.3 kgf on a WC–Co surface implanted to a dose of 2×10^{17} N^+ cm^{-2} at 100 keV. The grains indicated by arrows exhibit cracks.

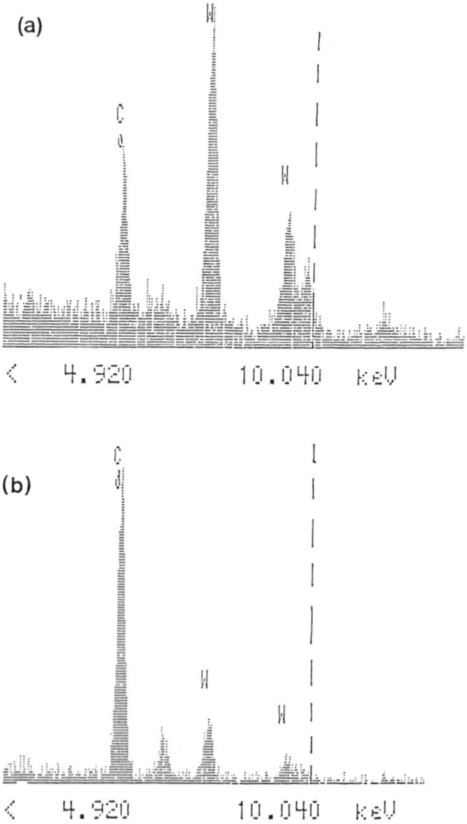

Fig. 3. EDXA spectra from (a) the whole area in Fig. 1, showing a cobalt-to-tungsten peak ratio of less than 1 and (b) from the rod-like features indicated by arrows in Fig. 1, showing a cobalt-to-tungsten ratio of more than 1.

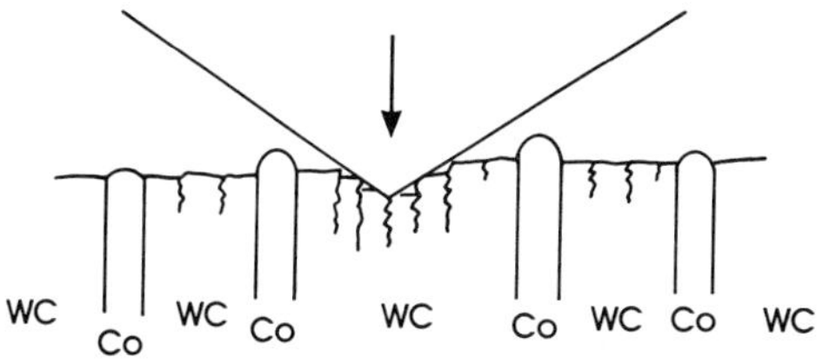

Fig. 4. Schematic representation of the effects of a diamond stylus loaded into a WC–Co surface. Compressive stresses parallel to the free surface of the material cause the extrusion of cobalt.

tions suggest that ion implantation hinders the extrusion of cobalt by hardening the cobalt.

2.3. *Adhesion tests*

Preliminary adhesion tests were carried out by reciprocating steel ball-bearings on implanted and unimplanted WC–Co surfaces under the load of 1 kgf, in a Cabris reciprocating wear-testing machine. The amount of seizure occurring after implantation was found to be less than before implantation (Fig. 5). This disagrees with Dearnaley's [10] results but may be explained in terms of the hardening of the cobalt phase. These results, however, may be influenced by the fact that implanted surfaces are smoother, on account of sputtering.

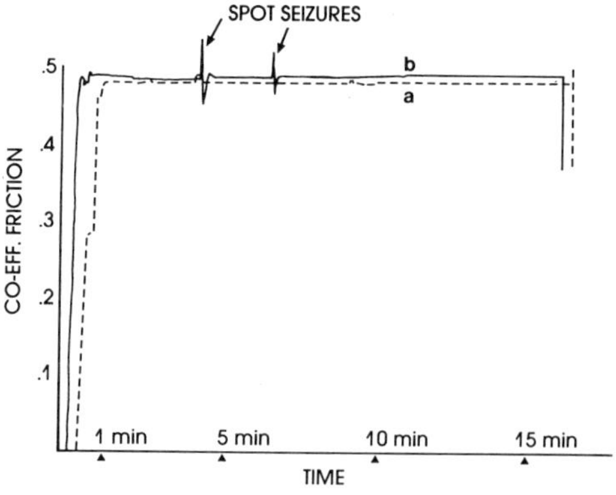

Fig. 5. Results from adhesive wear tests on an implanted (line a) and an unimplanted (line b) WC–Co surface. The coefficient of friction did not change appreciably but the amount of seizure decreased after implantation.

3. Discussion

WC–Co components degrade by abrasion, adhesion or fracture.

The mechanism controlling sliding abrasion and low stress grinding abrasion is the selective removal of the cobalt phase followed by pull-out of WC grains [11]. Since the extrusion and removal of cobalt appear to be hindered by implantation through precipitation hardening and dislocation multiplication, the resistance to sliding and low stress grinding abrasion should be improved by ion implantation. This is indeed the case, as shown by results from tests on wire-drawing dies, extrusion dies and punches [1] and from low stress laboratory tests [12].

High stress grinding abrasion and scratching abrasion may involve indentation-induced fracture as well as cobalt extrusion. In this case the benefits from ion implantation should be moderate because implanted WC grains fracture more easily than unimplanted grains. This should offset the advantages of a harder surface and of less cobalt extrusion, as it was indeed shown by high stress laboratory tests carried out by Fayeulle and coworkers [12].

Adhesion plays a role in the degradation of a large number of WC–Co cutting tools, where localized welding occurs followed by tearing near the welds. Preliminary results from adhesion tests suggest that the adhesion resistance of WC–Co increases after ion implantation. However, at the high temperatures involved in metal cutting, ions are expected to diffuse out of the implanted surface [4] with no extension of the life of the tool, as was found in field tests carried out by Oblas and Sarin [4].

Fracture is a frequent degradation mechanism in WC–Co components subjected to high stresses, thermal shocks or impact (*e.g.* in rock-drilling components). Since the ductility of both WC and cobalt is decreased the toughness of implanted WC–Co surfaces is reduced by ion implantation. The implanted components therefore should be more susceptible to fracture than unimplanted components. This may contribute to the lack of improvement in the implanted cutting and turning tools and in the components subjected to erosion which were tested by Oblas and Sarin [4].

4. Conclusions

The modifications induced by ion implantation in the surface of WC and cobalt can explain the apparently contradictory results reported in the literature on the effects of ion implantation on the wear resistance of WC–Co components. The modifications are the following: the carbide phase is hardened but embrittled and the cobalt phase is hardened so that its extrusion is hindered. As a result, components which degrade by abrasion and adhesion benefit from the increased hardness and reduced cobalt extrusion, but components which degrade by microfracture suffer from the reduced toughness of the carbide phase.

In conclusion, the maximum improvement in wear resistance after implantation is expected to occur in components which wear by sliding and low stress grinding abrasion.

References

1 R. E. Fromson and R. Kossowsky, *Carbide Tool J., 16* (1984) 3.

2 J. K. Hirvonen, in G. K. Hubler, O. W. Holland, C. R. Clayton and C. W. White (eds.), *Ion Implantation and Ion Beam Processing of Materials, Materials Research Society Symp. Proc.,* Vol. 27, Elsevier, New York, 1984, p. 621.

3 G. Dienel, A. Kolitsch and E. Richter, *Proc. 11th Plansee Semin.,* Metallwerk Plansee, Reutte, 1984, p. 757.

4 D. W. Oblas and V. K. Sarin, *Carbide Tool J., 18* (1986) 15.

5 S. B. Luyckx, Y. Cassuto and J. P. F. Sellschop, *Mater. Sci. Eng., 90* (1987) 339.

6 S. B. Luyckx, Y. Cassuto, T. E. Derry and J. P. F. Sellschop, *Nucl. Instrum. Methods B, 19–20* (1987) 158.

7 R. K. Viswanadham and J. D. Venables, *Metall. Trans. A, 8* (1977) 187.

8 J. H. Fremlin and N. A. Askouri, *Nature, London, 249* (1974) 137.

9 S. B. Luyckx and J. P. F. Sellschop, *J. Mater. Sci. Lett., 7* (1988) 265.

10 G. Dearnaley, *Proc. Int. Conf. on Science of Hard Materials, Wyoming, U.S.A.,* Plenum, New York, 1983, pp. 467–484.

11 J. Larsen-Basse and N. Devnani, *Proc. 2nd Int. Conf. on Science of Hard Materials, Rhodes,* Adam Hilger, 1986, pp. 883–895.

12 S. Fayeulle, D. Treheax, P. Guiraldeng, J. Dubois and G. Fantozzi, *J. Mater. Sci., 21* (1986) 1814–1818.

Adhesion at Elevated Temperatures of TiN Coatings on Cemented Carbides*

M. T. LAUGIER

Materials Research Centre, National Institute for Higher Education, Limerick (Ireland)

(Received November 9, 1987)

Abstract

Thin hard coatings for mechanical protection under tribological conditions must be both tough and adherent. Adhesion is a measure of spalling resistance and may be thought of as the toughness of the interface. The widely used scratch test provides a simple means of investigating coating failure and, although theoretical understanding is incomplete, a simple energy formulation of the scratch test is available. The adhesion of TiN protective coatings on cemented carbides is investigated by means of the scratch test at temperatures up to 1050 °C (typical of cutting tool applications). The adhesion energies are G (20 °C) = 69 J m^{-2} *and G (400 °C) = 75 J m*$^{-2}$.

1. Introduction

Hard coatings only a few microns in thickness can often significantly improve performance in wear, bearing and cutting applications. Coating failure mode will depend on the scale of contact which is determined by the application.

In bearing applications, performance may be determined by the resistance of the coating to hertzian fracture [1] which occurs under essentially elastic conditions; this will involve the coating toughness G_c and also the frictional properties (in the case of sliding contacts). In some cutting applications, coatings fail by spalling preceded by local plastic flow which may result from contact with small hard workpiece inclusions. Resistance to spalling is traditionally ascribed to "adhesion"; a better description might be in terms of an interfacial toughness parameter.

In wear applications, high hardness and low frictional properties are general requirements for good performance.

Indentation methods are well suited to investigations where contact size, load conditions and friction must be controlled and include tests of hardness, toughness, wear and "adhesion". This work looks for the first time at "adhesion" as determined by the scratch test at temperatures up to 1050 °C close to operating temperatures of TiN-coated cemented carbide cutting tools.

1.1. *Indentation fracture*

For a crack contained within the localized region of high strain beneath a loaded indenter, the energy required for extension will scale as c^2, where c is a characteristic dimension of the strain field, while the energy available in the field scales as c^3. As a result, there will be a critical dimension c_{crit} below which plastic deformation may occur, and above which crack extension occurs, leading to an indentation size effect [2].

1.2. *Fracture modes*

Indentation with spheres of radius greater than 2 mm generally leads to ring cracking under essentially elastic conditions. For smaller indenters, ring cracking does not occur; instead the material beneath the indenter deforms plastically. Cracks formed at this stage in the plastic zone of intense inhomogeneous staining will ultimately extend into the elastic region as the radial and lateral cracks familiar from the Vickers indentation in ceramics [3].

1.3. *Scratch test*

The scratch test in which an indenter (typically diamond of radius 0.1–0.2 mm) is loaded and translated across the coating, leading to coating removal, is traditionally thought of as a test of

*Paper presented at the 3rd International Conference on the Science of Hard Materials, Nassau, The Bahamas, November 9–13, 1987.

0921-5093/88/$3.50 © Elsevier Sequoia/Printed in The Netherlands

"adhesion" (the critical normal load is often quoted as a measure of "adhesion") [4]. When applied to hard coatings on hard substrates, coatings often fail by spalling with relatively little plastic deformation [5]. Under such conditions, it may be reasonable to write a simple Griffith-type energy balance giving the approximate result for interfacial toughness [6]:

$$G_{\text{interface}}=\frac{\sigma^2 h}{2E} \qquad (1)$$

where h is the coating thickness, E is Young's modulus of the coating and σ is the compressive stress in the coating ahead of the indenter.

In addition to the applied stress, there will also be thermal and intrinsic coating stresses. When, as in the case of TiN coatings on cemented carbide substrates, spalling occurs under largely elastic conditions [5], the applied stress resulting from the sliding indenter may be written

$$\sigma_{\text{appl}}=\frac{(10f-1)P^{1/3}(3E_{\text{substrate}}/4kR)^{2/3}}{4\pi} \qquad (2)$$

using the results of refs. 6 and 7, where f is the coefficient of friction between the indenter and the coating, R is the indenter radius and k is a parameter from hertzian theory involving the elastic modulus of the substrate and the indenter ($k=0.84$ for the cemented carbide substrate and diamond indenter; $E_{\text{substrate}}=6\times10^5$ MPa). The thermal stresses may also be estimated, when the deposition temperature is known, from the difference between thermal expansion coefficients of the coating and the substrate, but little information is available concerning the intrinsic stress component which results from non-equilibrium effects arising during the coating process. The intrinsic stress tends to increase with decreasing deposition temperature, increasing deposition rate and increasing impurity content and can promote loss of coating integrity [8].

The coating stress due to differential thermal contraction between coating and substrate arising on cooling is

$$\sigma_{\text{thermal}}=\frac{E\,\Delta\alpha\,\Delta T}{1-\nu} \qquad (3)$$

where $\Delta\alpha$ is the difference between the thermal expansion coefficients, ΔT is the temperature difference and ν is Poisson's ratio for the coating.

2. Experimental details

2.1. Materials

The TiN coatings tested were produced commercially by the chemical vapour deposition (CVD) technique on cemented carbide substrates (commercial TiN-coated cutting tips; substrate, WC–1wt.%Ta–10wt.%Co). The deposition temperature is in the region of 1000 °C. Figure 1 is a scanning electron micrograph of a typical fracture surface. Near the TiN–cemented carbide interface the CVD coating appears to consist of dense submicron grains; the grain size increases with a tendency to columnar morphology on approaching the outer TiN surface.

No interfacial voids or flaws were seen, in contrast with observations made on a TiC fracture surface [5]. The coating thickness was 6 μm.

The Vickers microhardness values (load, 0.1 kgf) for the TiN coatings were found to be close to 1900 at room temperature (approximate penetration depth, 1.5 μm); the room temperature substrate hardness was 1540. The elevated temperature (1050 °C) hardness (load, 5 kgf) for a dwell time of 10 s was 850.

2.2. Apparatus

Specimens were placed on an alumina pedestal fixed to a low friction trolley. The diamond indenter (Rockwell C; radius, 0.2 mm) was lowered into contact with the specimen via a single-beam balance; dead loading was provided vertically above the indenter. The low friction trolley was rigidly coupled to a driving motor via a polymer tube; a linear variable-differential transducer clamped to the tube had previously been calibrated against the applied load and was used to provide a measure of the horizontal force

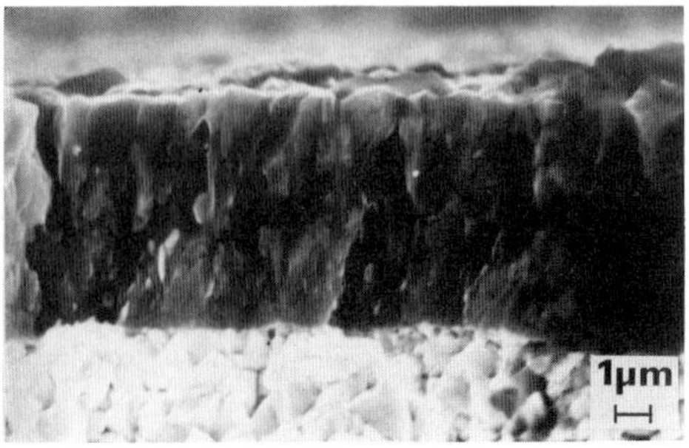

Fig. 1. Scanning electron micrograph of a typical TiN fracture surface.

and hence the friction coefficient of the diamond–TiN system. The pedestal was placed within a two-loop coil used for r.f. induction heating of the indenter–specimen pair. Testing was carried out in a nitrogen atmosphere; temperature measurement was by optical pyrometer and thermocouple.

3. Results

Room temperature testing confirmed previous experience with TiN coatings [9]; complete coating removal occurred under a load P of 15 kgf; very little spalling was apparent. Figure 2 is a scanning electron micrograph of a scratch track under a load of 15 kgf; evidence of limited coating plasticity may be seen. The apparent coefficient of friction was 0.2 but this contains a "ploughing" component. The true coefficient of friction was 0.15 as determined from low load (less than 1 kgf) measurements. Figure 3 shows a scratch track produced at 400 °C under a load of 10 kgf. Complete removal has occurred; again there is evidence of "smearing" of the coating, but there is also more coating "break-up" at the end of the scratch, suggesting greater brittleness. The apparent coefficient of friction was 0.3.

Figure 4 shows a scratch track produced under a load of 10 kgf at 1050 °C; although removal has not occurred along the track, spalling within the coating itself has occurred along the edges and ahead of the track. The light central region in Fig. 4 may be an indication that some degree of chemical reaction between the diamond indenter, the TiN coating and the test atmosphere (mostly nitrogen) has occurred during high temperature testing. Figure 5 shows the end of the track at higher magnification. In addition to the partial spalls already referred to, there is a region of complete coating removal; this may be a result of adhesion of the indenter to the coating, leading to coating detachment on raising the indenter at the end of the test. The apparent coefficient of friction was 0.6.

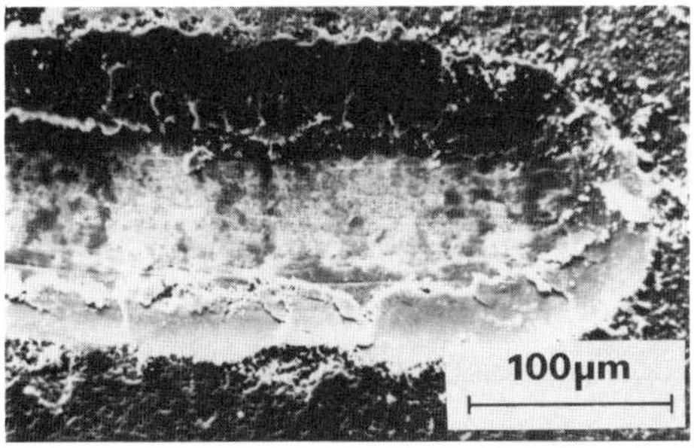

Fig. 2. Scanning electron micrograph of a scratch track obtained on a TiN coating at room temperature at a load P of 15 kgf.

Fig. 4. Scanning electron micrograph of a scratch track obtained on a TiN coating at 1050 °C at a load P of 10 kgf.

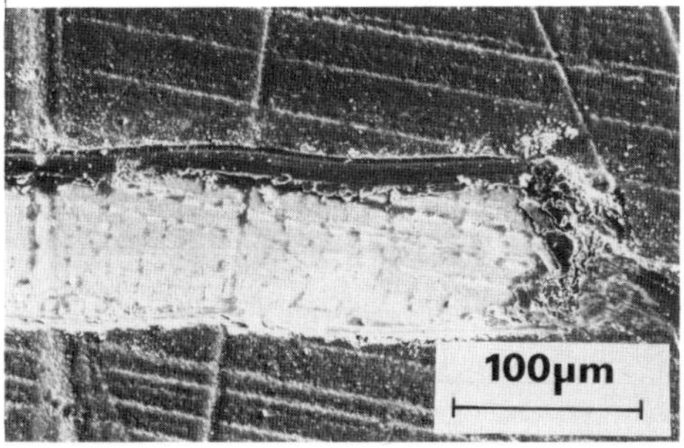

Fig. 3. Scanning electron micrograph of a scratch track obtained on a TiN coating at 400 °C at a load P of 10 kgf.

Fig. 5. Scanning electron micrograph at a higher magnification by the end of the scratch track in Fig. 4.

Cracking along the scratch track which had been observed previously [10] when TiN coatings were tested in air at 700 °C was not found under the present conditions.

4. Discussion

The scratch test adhesion theory described (Section 1.3) may be applicable to the materials tested at room temperature and at 400 °C, where evidence of brittle failure was found; the model may not be applicable to the results obtained at 1050 °C owing to significant substrate softening (a Vickers hardness (load, 5 kgf) of 850; see Section 2.1). The interfacial toughnesses G (adhesion energies) determined by direct application of eqns. (1)–(3) are $G(20\,°C) = 69$ J m^{-2} and $G(400\,°C) = 75$ J m^{-2} (the coefficients of thermal expansion used are 5×10^{-6} K^{-1} (WC–Co) and 9.3×10^{-6} K^{-1} (TiN), Young's modulus for the coating was taken as 2.5×10^5 MPa, and the true coefficient f of friction of 0.15 was used). These values are higher by a factor of 1.5 than results obtained previously on CVD TiN coatings under somewhat different conditions and appear to be a reflection of the better quality of the present coatings [5]. The reasonableness of these values provides some support for a simple elastic model even in the presence of limited coating plasticity; it is emphasized that coating removal does indeed occur by spalling, plasticity being limited to the contact region.

5. Conclusion

TiN coatings have a good scratch resistance at all temperatures up to 1050 °C; little evidence of spalling was found at 1050 °C but evidence of coating fracture was found at 400 °C. The adhesion energies G calculated from the simple elastic model are $G(20\,°C) = 69$ J m^{-2} and $G(400\,°C) = 75$ J m^{-2}.

References

1 M. T. Laugier, *J. Mater. Sci. Lett., 2* (1983) 419.
2 K. E. Puttick, *J. Phys. D, 13* (1980) 2249.
3 B. Lawn and R. Wilshaw, *J. Mater. Sci., 10* (1975) 1049.
4 J. Valli, U. Makela, A. Matthews and V. Murawa, *J. Vac. Sci. Technol. A, 3* (1985) 2411.
5 M. T. Laugier, *J. Mater. Sci., 21* (1986) 2269.
6 M. T. Laugier, *Thin Solid Films, 117* (1984) 243.
7 M. T. Laugier, *Thin Solid Films, 79* (1981) 289.
8 D. S. Campbell, in L. I. Maissel and R. Glang (eds.), *Handbook of Thin Film Technology,* McGraw-Hill, New York, 1970, Section 12.3.
9 M. T. Laugier, *J. Vac. Sci. Technol. A, 5* (1987) 67.
10 M. T. Laugier, *2nd Int. Conf. on Surface Engineering, Stratford on Avon, June 15–18, 1987.*

Materials Science and Engineering, A105/106 (1988) 517–523

Synthesis of Sintered Diamond with a High Electrical Resistivity and High Hardness*

MINORU AKAISHI, SHINOBU YAMAOKA, JUNZO TANAKA and TOSHIKAZU OHSAWA

National Institute for Research in Inorganic Materials, 1-1 Namiki, Sakura-Mura, Niihari-Gun, Ibaraki 305 (Japan)

OSAMU FUKUNAGA

Faculty of Engineering, Tokyo Institute of Technology, 2-12-1, Ohokayama, Meguro, Tokyo 152 (Japan)

(Received November 9, 1987)

Abstract

Diamond powder to which about 1 vol.% metal was homogeneously added was obtained by sputtering partially graphitized diamond powder which was vibrating by means of an ultrasonic generator. A well-sintered diamond with a fine-grained homogeneous microstructure was synthesized at 7.7 GPa and 2000 °C when diamond powder with 1–5 vol.% Co or Ni additive was used as the starting material. It has a high hardness and a high electrical resistivity.

The thermal resistance of the sintered diamond was strongly dependent on the metallic content and types of metal species. Sintered diamond with nickel additive was more easily graphitized at high temperatures than that with the cobalt additive.

1. Introduction

Most sintered diamonds available commercially are synthesized by sintering diamond powder as a laminate on a WC–Co substrate at a high temperature and a high pressure in the thermodynamically stable region of diamond. Details of the sintering procedure for such materials have been described in the *U.S. Patents* by Wentorf and Rocco [1, 2]. The hardness and electrical resistivity of these sintered diamonds at room temperature are 63.7–78.4 GPa and 3.5×10^{-5} Ω cm respectively [3, 4]. These values are lower than the values of about 100 GPa and 10^{18} Ω cm for natural single-crystal diamond. In particular, the lower electrical resistivity is due to a continuous network of metallic cobalt phase which is infiltrated from the WC–Co layer as a sintering aid. The cobalt content in the sintered diamond was calculated to be 5.57–9.42 vol.% from the reported density [3], showing that a fairly large amount of cobalt is contained in the commercially available diamonds. The large amount of metallic phase in the sintered diamond causes the physical properties of the mass sintered at a high temperature to deteriorate because of the catalytic graphitization of diamond. Chipping and microcracking also occur as a result of the thermal stress caused by the difference between thermal expansion coefficients of diamond and the additive metal [5, 6].

Therefore, sintered diamond with a small amount of metal is expected to have better physical properties than that with a large amount of metal. In this context, we have recently reported the synthesis of sintered diamond with a high hardness and a high electrical resistivity using a diamond powder mechanically mixed with 5 vol.% ultrafine cobalt powder [7]. Also, our previous work concluded that it was difficult to synthesize a well-sintered diamond with below 2.5 vol.% Co additive using this powder mixture. The reason is not clear, but it may be due to the inhomogeneous distribution of ultrafine cobalt powder in the mixture. Therefore, it is considered that a homogeneous powder mixture is important as a starting material for the synthesis of sintered diamonds with a small amount (about 1 vol.%) of metal additive.

In this paper the preparation of a homogeneous powder mixture of diamond and cobalt or nickel

*Paper presented at the 3rd International Conference on the Science of Hard Materials, Nassau, The Bahamas, November 9–13, 1987.

0921-5093/88/$3.50

© Elsevier Sequoia/Printed in The Netherlands

metal, its sintering behaviour at a high temperature and a high pressure and the properties (such as hardness, electrical resistivity and thermal resistance) of the sintered diamonds obtained are described.

2. Experimental procedures

2.1. *Starting materials*

Commercially available synthetic diamond powder (General Electric Co.; 2–4 μm), ultrafine cobalt or nickel powders (Vacuum Metallurgical Co. Ltd.; 30 nm) and cobalt or nickel plates (purity, 99.9%) were used as starting materials. The starting material for the sintering experiments was prepared by the following two methods.

(1) Powder mixtures of diamond and ultrafine cobalt or nickel below 5 vol.% were mixed in a vibratory mill using a chamber lined with polyacetal resin for 0.5 h and pressed uniaxially at 400 MPa into cylindrical compacts. These compacts were next treated at 500 °C for 2 h in a vacuum of 0.1 Pa before it was used for the sintering experiments.

(2) A small amount (about 1 vol.%) of cobalt or nickel was added by magnetron sputtering on the surface of diamond powder which was vibrating by means of an ultrasonic generator, as shown in Fig. 1. To prevent the diamond powder from forming agglomerates during sputtering, it was graphitized partially beforehand. The details are given in Section 2.2.

The starting materials and sintering conditions in the present study are shown in Table 1.

2.2. *Sintering procedure*

High pressure sintering was carried out using a modified belt-type high pressure apparatus with a bore diameter of 25 mm [8]. Samples were heated by means of an internal graphite heater with an outside diameter of 9.4 mm, an inside diameter of 8 mm and a length of 12 mm.

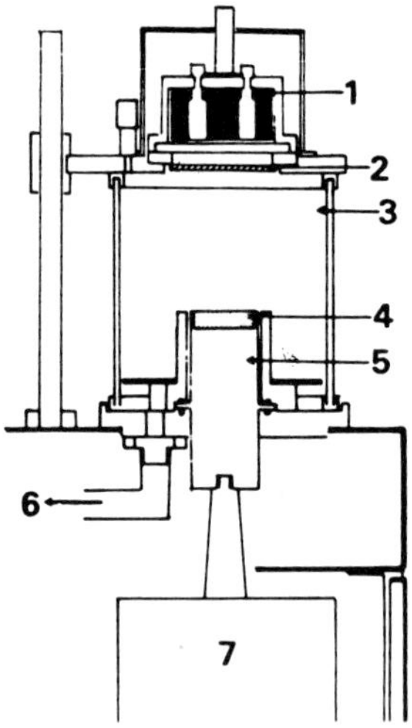

Fig. 1. Magnetron sputtering for metal film coating on partially graphitized diamond powder: 1, magnet; 2, target; 3, vacuum chamber; 4, diamond powder; 5, horn; 6, rotary pump; 7, ultrasonic generator.

The temperature was measured with a (Pt–6wt.%Rh)–(Pt–30wt.%Rh) thermocouple without correction of the effect of pressure on the electromotive force. In routine experiments, the temperature was estimated from an extrapolated relation between the power input and the temperature. The pressure was calibrated at room temperature using the known pressure-induced phase transitions of bismuth, thallium and barium. Pressure corrections at a high temperature were made by using the melting curve of silver and its extrapolation based on the data obtained in a piston-and-cylinder apparatus [9]. Samples are first subjected to pressure, and then the temperature is applied and held for 1 h. Sintering was performed at 1800–2000 °C and 7.7 GPa, since it was desired to keep the diamond powder within the thermodynamically stable region above the

TABLE 1 Starting materials and sintering conditions

Sample	*Amount of additive* (vol.%)	*Mixing*	*Sintering conditions*	
			Temperature (°C)	*Pressure* (GPa)
A	5 Co	Mechanical mixing	1800	7.7
B	5 Co	Mechanical mixing	1900	7.7
C	5 Co	Mechanical mixing	2000	7.7
D	2.5 Co	Mechanical mixing	2000	7.7
E	5 Ni	Mechanical mixing	2000	7.7
F	1.5 Co	Magnetron sputtering	2000	7.7
G	1 Ni	Magnetron sputtering	2000	7.7

diamond–Co or diamond–Ni eutectic temperature.

The samples were characterized by scanning electron microscopy (SEM) with an energy-dispersive X-ray analyser, atomic absorption analysis, optical microscopy, X-ray diffraction, Vickers hardness tests (1 kgf load) and thermal resistance and electrical resistivity measurements. The obtained sintered diamonds for use in the Vickers hardness tests and thermal resistance investigations were polished using a rotating cast iron scaife.

3. Results and discussion

3.1. Homogeneous addition of metals to the diamond powder

Comparatively homogeneous diamond–metal powder mixtures were obtained by mechanical mixing using an ultrafine metal powder as an additive metal. Figure 2(a) shows a backscattered electron image of mechanically mixed diamond–2.5 vol.% ultrafine cobalt powder mixture. The lighter area in this figure corresponds to cobalt particles. It is found that, although macroscopically the very fine cobalt particles appeared to be distributed homogeneously, microscopically they were distributed inhomogeneously.

Fig. 2. Backscattered electron images of powder compacts: (a) diamond–2.5vol.%Co powder mixture obtained by mechanical mixing; (b) diamond–1.5vol.%Co powder mixture obtained by magnetron sputtering.

To obtain a homogeneous diamond–metal powder mixture from both a macroscopic and a microscopic point of view, magnetron sputtering was adopted as shown in Fig. 1; it was intended to deposit a metal thin film on the diamond powder homogeneously.

However, many agglomerates were often formed in as-received fine diamond powders. This prevents the diamond powder from being coated homogeneously. However, no agglomerates were observed in partially graphitized diamond powders. Therefore, it is very effective to use partially graphitized diamond powder. For this purpose, commercially available diamond powder (2–4 μm) was treated at 1480 °C and 20 min in a vacuum of 6×10^{-3} Pa. After this heat treatment, partially graphitized diamond powder containing about 10 vol.% graphite was obtained and used for sputtering.

To obtain a homogeneous thin metal film coating on the diamond powder, a combination of magnetron sputtering with an ultrasonic generator which could vibrate the powder during the sputtering was adopted. After the partially graphitized diamond powder had been sputtered for 5 h, the cobalt content analysed by atomic absorption analysis was 1.5 vol.%. Figure 2(b) shows a backscattered electron image of the coated powder compact without a gold coating for SEM observation. It can be clearly seen from this figure that a homogeneous cobalt film coating was obtained on each diamond powder. A homogeneous nickel film coating was also obtained on the diamond powder.

3.2. Sintering behaviour of diamond with a cobalt or nickel additive of 1–5 vol.%

Diamond–metal powder mixtures prepared by the magnetron sputtering method and mechanical mixing method were sintered under the conditions shown in Table 1. To remove the zirconium foil used for wrapping, the surfaces of the sintered body were ground with a diamond wheel.

Of the mechanically mixed samples, sample A

was a weakly bonded sintered body, although no graphite was observed in it. From their grinding resistance, samples B–E were thought to be strongly bonded sintered bodies. Many cracks were observed in samples B and D. By contrast, no cracks were observed in samples C and E. The present results suggest that, under sintering conditions of 7.7 GPa and 2000 °C, a metal content of at least 5 vol.% was necessary to obtain a well-sintered body without cracks when mechanical mixing is used.

In contrast, when the diamond powder homogeneously coated with a thin metal film (about 1 vol.%) by magnetron sputtering was used as the starting material, a well-sintered body could be obtained. Samples F and G, for example, were sintered at 7.7 GPa and 2000 °C for 1 h. From the grinding resistance and optical observation of the sintered bodies, the samples obtained were thought to be strongly bonded sintered bodies.

The microstructures of the successfully obtained samples C, E, F and G were investigated using optical microscopy and SEM.

From these observations, no extreme grain growth was observed in the microstructure of sample C. As shown in Fig. 3(a), the average grain size of the sintered body was about 4 μm, which is similar to that of the starting powders. The cobalt distribution in the sintered body can be clearly seen in Fig. 3(b), where the lighter area corresponds to cobalt particles. The figure shows that cobalt was distributed discretely and homogeneously, and neither large cobalt pools nor continuous cobalt layers were seen in the sintered body. The cobalt content of the same sample calculated from the value measured by atomic absorption analysis was 3.5 vol.%, if voids and graphite are assumed to be free. This is lower than that of the starting materials, suggesting that some cobalt was "swept out" of the sintered body during the sintering process.

The above assumption is quite acceptable in this sample for the following reason. No trace of graphite was detected by Raman spectroscopy and X-ray diffraction. Because the former is particularly sensitive to the presence of graphite, it is concluded that no graphite was present. Voids can also be neglected because none could be seen in SEM observations and the density calculated from the sample volume obtained on this assumption (3.69 g cm^{-3}) and the density obtained by Archimedes' method (3.73 g cm^{-3}) were quite consistent within experimental error.

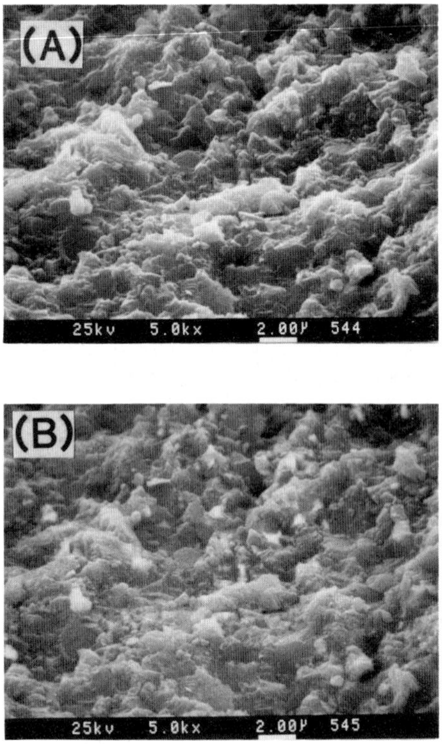

Fig. 3. SEM photographs of the fracture surface of sample C: (a) secondary electron image; (b) backscattered electron image of the same spot in (a).

The fracture surface of sample E with the mechanically added nickel was also investigated by SEM. From these observations the grain size and metal distribution were very similar to those of sample C. The nickel content of the sintered body calculated from the atomic absorption analysis was 4 vol.%, which is lower than that of the starting materials. It is also considered that some nickel was "swept out" of the sintered body during the sintering process.

It is very interesting to investigate the microstructure of samples F and G which have a much smaller amount of metal additive than do samples C and E. As shown in Fig. 4(a), no extreme grain growth was observed in the microstructure of sample F and the grain size was almost the same as that of the starting materials. The cobalt distribution in the sintered body can be clearly seen in Fig. 4(b). Cobalt with a fine grain size was distributed discretely and homogeneously, and no formation of a continuous cobalt network was seen.

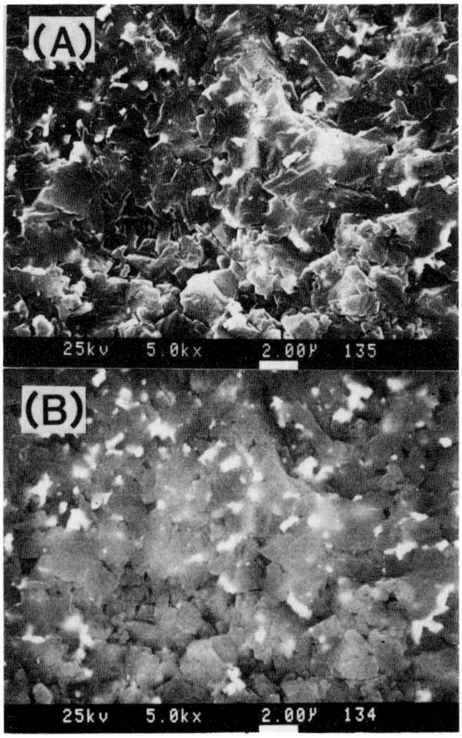

Fig. 4. SEM photographs of the fracture surface of sample F: (a) secondary electron image; (b) backscattered electron image of the same spot in (a).

The same microstructure was also seen in sample G.

3.3. Physical properties of the sintered diamonds

Sintered diamonds (samples C, E, F and G) for the hardness tests were polished using a rotating cast iron scaife. The hardnesses (1 kgf load) of the sintered diamonds were measured on the Vickers scale. Figure 5 is an optical micrograph of the indented surface of sample F. The hardness was calculated from a measurement of the diagonal of the indenter trace. The value was above 80 GPa on the Vickers scale. The hardnesses of samples C and G were almost the same as that of sample F, but the value of sample E was about 60 GPa.

As described above, samples C, E, F and G had high hardnesses, and cobalt or nickel in the sintered bodies was not present continuously. We then investigated the electrical resistivity of these sintered diamonds.

The electrical resistivities of samples C, E and F in the shape of a rectangular prism (1.2 mm × 2.1 mm × 2.3 mm) was measured by a two-probe d.c. method using an electrometer in the atmosphere. The electrical resistivity of sample C is higher than 10^8 Ω cm at room temperature and decreases with increasing temperature, as shown in Fig. 6. The temperature gradient of the resistivity is linear up to 117 °C. At 117 °C the slope changes and it becomes steeper above 117 °C. The activation energy calculated from the full line was 0.42 eV. The temperature dependences of the electrical resistivities of the other samples, E

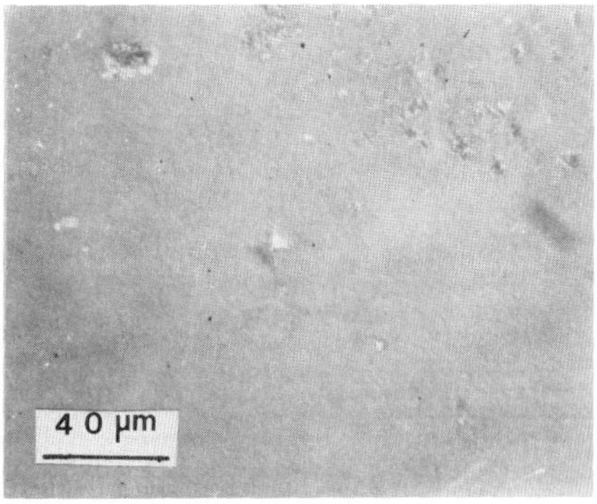

Fig. 5. Optical micrograph of the indented surface of sample F (differential interference microscopy).

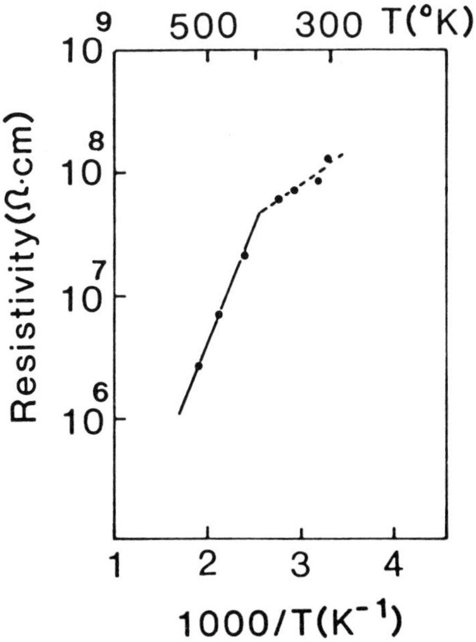

Fig. 6. Electrical resistivity plotted against reciprocal temperature for the sintered diamond sample C.

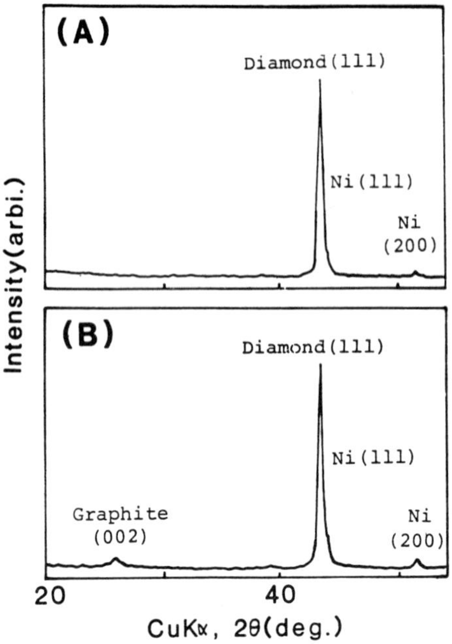

Fig. 7. X-ray diffraction patterns of the sintered diamond sample G: (a) before heat treatment; (b) after heat treatment at 1100 °C for 30 min in a vacuum of $(1-1.8)\times 10^{-3}$ Pa.

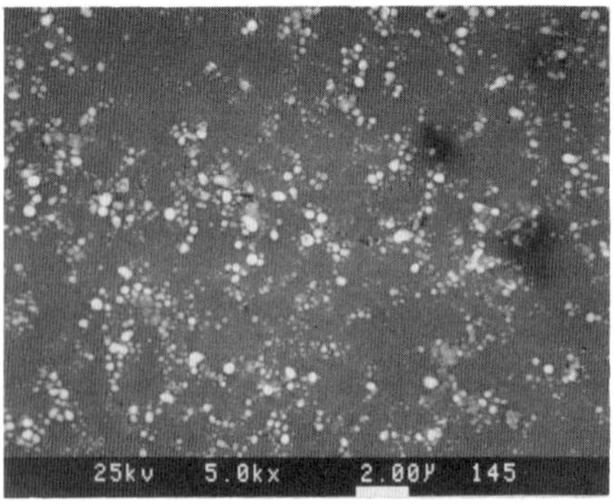

Fig. 8. Backscattered electron image of sintered diamond sample G after heat treatment.

and F, showed a similar tendency to that of sample C. From these results, it is found that the sintered diamonds obtained were insulators or semiconductors with a very low carrier concentration.

The thermal resistances of these sintered diamonds (samples C, E, F and G) were investigated by heat treatment at 1100 °C for 30 min in a vacuum of $(1-1.8)\times 10^{-3}$ Pa. Before the heat treatment, these sintered diamonds were polished using a rotating cast iron scaife. X-ray diffraction patterns were obtained before and after heat treatment of these sintered diamonds. Also, the surfaces of sintered diamonds after heat treatment were observed by SEM and optical microscopy.

No graphite was observed in the X-ray diffraction patterns of these sintered diamonds before heat treatment. After heat treatment, a small amount of graphite was detected in the sintered diamond with 1 vol.% Ni (sample G), as shown in Fig. 7. Also, a fairly large amount of graphite was observed in the X-ray diffraction pattern of sample E. The amount of graphite increased qualitatively on increase in the amount of nickel additive. In contrast, no graphite was detected in the sintered diamond samples C and F with cobalt additive.

From these results, it is considered that sintered diamonds with nickel additive are easily graphitized at high temperatures compared with those with cobalt additive.

The surfaces of sintered diamonds after heat treatment were observed by optical microscopy and SEM with an energy-dispersive X-ray analyser. Many cracks were observed in sample C, but only two cracks were observed in sample F. Some microcracks were observed in sample E. By contrast, no cracks were observed in sample G. From these results, it is found that cracking by heat treatment decreased on decrease in the amount of metal additive.

More detailed observation of the surfaces was made by SEM. Many tiny precipitates with a spherical shape were observed in all the samples examined (samples C, E, F and G). Figure 8 shows the backscattered electron image of sample G. Many tiny precipitates with a spherical shape can be clearly seen in this figure. Energy-dispersive X-ray analysis confirmed that these precipitates were composed of nickel or of nickel-rich phases.

4. Conclusion

In the mechanically mixed systems, the metal particles in the mixtures were distributed inhomogeneously on each diamond powder. However, diamond powders coated with a thin homogeneous metal film were obtained by magnetron sputtering partially graphitized diamond powders which were vibrating by means of an ultrasonic generator.

In the mechanically mixed systems, a metal content of at least 5 vol.% was necessary to obtain well-sintered diamonds without cracks at 7.7 GPa and 2000 °C. By contrast, a well-sintered diamond with about 1 vol.% metal was obtained under the same sintering conditions using the diamond powder coated with a homogeneous metal film by magnetron sputtering.

From microstructure observations on the sintered diamonds obtained from the starting powders by the mechanical mixing and magnetron sputtering methods, it could be clearly seen that the cobalt or nickel particles in the sintered bodies were not continuous. The Vickers hardnesses (samples C, F and G) and electrical resistivities (samples C, E and F) of the sintered diamonds were above 80 GPa and above 10^8 Ω cm respectively.

The thermal resistances of the sintered diamonds were strongly dependent on the metal content and types of metal species. The features were summarized as follows: sintered diamonds with a nickel additive were easily graphitized compared with those with a cobalt additive. Cracking on heat treatment is decreased on decrease in the amount of metal additive.

References

1 R. H. Wentorf, Jr., and W. A. Rocco, *U.S. Patent 3,745,623,* July 17, 1973.

2 R. H. Wentorf, Jr., and W. A. Rocco, *U.S. Patent 3,767,371,* October 23, 1973.

3 R. H. Wentorf, Jr., R. C. DeVries and F. P. Bundy, Sintered superhard materials, *Science, 208* (1980) 873–890.

4 P. D. Gigl, Strength of polycrystalline diamond compacts, in K. D. Timmerhaus and M. S. Barber (eds.), *High Pressure Science and Technology,* Vol. 1, Plenum, New York, 1979, pp. 914–922.

5 N. Iijima, H. Takeyama and M. Kashiwase, Study on machining performance of sintered diamond tool and its wear mechanism (2nd report) — mechano-physical wear mechanism of sintered diamond tool, *J. Jpn. Soc. Precis. Eng., 50* (1984) 1110–1116.

6 N. Iijima, H. Takeyama and M. Kashiwase, Study on machining performance of sintered diamond tool and its wear mechanism (2nd report) — thermo-chemical wear mechanism of sintered diamond tool, *J. Jpn. Soc. Precis. Eng., 50* (1984) 1469–1474.

7 M. Akaishi, S. Yamaoka, J. Tanaka, T. Ohsawa and O. Fukunaga, Synthesis of sintered diamond with high electrical resistivity and hardness, *J. Am. Ceram. Soc., 70* (1987) C237–239.

8 O. Fukunaga, S. Yamaoka, T. Endo, M. Akaishi and H. Kanda, Modification of belt-like high-pressure apparatus, in K. D. Timmerhaus and M. S. Barber (eds.), *High Pressure Science and Technology,* Vol. 1, Plenum, New York, 1979, pp. 846–852.

9 P. W. Mirwald, I. C. Getting and G. C. Kennedy, Low friction cell for piston cylinder high pressure apparatus, *J. Geophys. Res., 80* (1975) 1519–1525.

Characteristics of Thin Film Growth in the Synthesis of Diamond by Chemical Vapour Deposition and Application of the Thin Film Synthesis Technology for Tools*

N. KIKUCHI and T. KOMATSU

Central Research Institute, Mitsubishi Metal Corporation, Omiya, Saitama (Japan)

H. YOSHIMURA

Tokyo Plant, Mitsubishi Metal Corporation, Nishishinagawa, Tokyo (Japan)

(Received November 9, 1987)

Abstract

Diamond films were synthesized under various reaction conditions by hot-filament thermal chemical vapour deposition (CVD). Under reaction conditions of a low total pressure, or a high CH_4 concentration, or a long distance between the filament and the substrate, the surface morphologies of diamond films are of a very-fine-grained coagulated spherical structure. In these cases, strong peaks of amorphous carbon phases in laser Raman spectra were detected. From reflection high energy electron diffraction results and X-ray diffraction analysis, it was concluded that an amorphous carbon phase exists on the surfaces and also between the grain boundaries of diamond polycrystals. Cutting performance tests were carried out to investigate the possibility of utilizing CVD diamond films as a wear-resistant coating for cutting tools. It was also concluded that the CVD diamond coating on cemented-carbide inserts is an effective means of prolonging the life of the insert for turning non-ferrous metals, and, in particular, aluminium alloys and semisintered ceramics.

1. Introduction

Diamond is the hardest material known and has been widely used in many industrial applications such as grinding tools and polishing powders. Diamond is either synthesized by the ultrahigh pressure technique, which was developed in the early 1950s [1], in powder form or it is obtained in a block form by mining. Since the ultrahigh pressure technique of diamond synthesis is rather costly, more cost-effective methods have been investigated. Spitsyn and Derjaguin [2] and Eversole [3] showed that it is possible to produce diamond under a reduced pressure by the chemical vapour deposition (CVD) method in metastable thermodynamic states. In contrast, Aisenberg and Chabot [4] reported that a very hard, optically transmissible and high electrical resistance film can be obtained by the ion beam technique and they called this film a diamond-like carbon film. This result was reinvestigated by Spencer *et al.* [5] and they reported that this film consists of very fine diamond crystals.

Six years ago, Matsumoto *et al.* [6, 7] reported that thin diamond films can be produced either by a modified thermal technique or by a plasma CVD technique from a low pressure gas mixture of H_2 and CH_4. Since then, much work has been done on these techniques, *i.e.* studies on the reaction mechanisms [8, 9] and on new synthesis methods [10–12]. Advances in such synthesis techniques for thin diamond films will certainly widen the applications areas of diamond. CVD diamond films have been reported to have various morphologies depending on the reaction conditions [13, 14], but systematic studies on this subject appear to be scarce, and many applications of diamond films have been considered but there appear to be no reports which show the possible practical uses of these diamond films.

The purpose of the present work is to investigate the effects of CVD reaction conditions on

*Paper presented at the 3rd International Conference on the Science of Hard Materials, Nassau, The Bahamas, November 9–13, 1987.

0921-5093/88/$3.50 © Elsevier Sequoia/Printed in The Netherlands

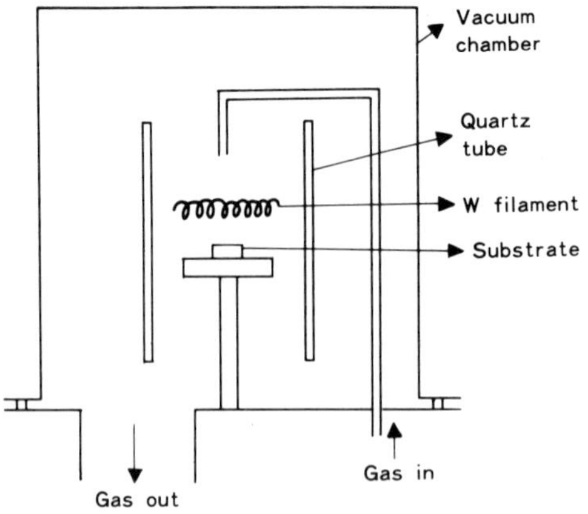

Fig. 1. Schematic picture of apparatus for CVD diamond coating.

morphologies and on crystal structures of diamond films and also the possibility of utilizing CVD diamond films as wear-resistant coatings for cutting tools.

2. Experimental details

Figure 1 shows a schematic picture of the hot-filament CVD diamond coating apparatus in which H_2 and CH_4 were used as source gases. The inner diameter of the quartz tube is 200 mm and the diameter of the substrate holder which is made from molybdenum is 70 mm. Three filaments were made by winding 0.3 mm tungsten wire into coils of 2 mm diameter with 50 turns and these were placed over the centre of the substrate holder. They were kept at the centre of the vacuum chamber which can be evacuated to 10^{-3} Torr before reaction. Hydrogen (purity, 99.99%) and CH_4 (purity, 99.9%) were used as the reactant gases.

Tungsten plates (10 mm × 10 mm × 1 mm) and cemented-carbide throw-away inserts (94wt.%WC–6wt.%Co; CIS type SPP422) were used as substrates. The tungsten plates were polished using grade 180 emery paper and then 3 μm diamond powders. The surfaces of the cemented-carbide inserts were finished on a no. 180 diamond grinding wheel. The substrates were placed on the holder and the chamber was evacuated; then the reaction gases were introduced into it at the established pressure for controlling the reaction. The reaction was initiated by heating the filament to 2000 °C. During the reaction the temperatures on surface of the substrates were kept at 750 °C by radiation from the filaments when the distance between the substrates and filaments was 30 mm. This temperature decreased to 650 °C when the distance between the substrates and filaments increased to 40 mm.

TABLE 1 Reaction conditions

Conditions	*Vacuum pressure* (Torr)	CH_4 (%)	*Distance* (mm)
A	3.5	1	30
B	7	1	30
C	7	0.7	30
D	7	1.4	30
E	7	1	40

The reaction pressure, concentration of CH_4 and the distance between the hot filaments and the substrate were varied, and these experimental conditions are shown in Table 1. Scanning electron microscopy (SEM) observations were made to investigate the surface morphologies of the diamond films. The crystal structures of the diamond films were examined by reflection high energy electron diffraction (RHEED), by X-ray diffraction and by laser Raman spectroscopy. Cutting performances for turning aluminium alloys and also semisintered SiC ceramic were tested.

3. Results and discussion

3.1. Observations of surface morphologies of chemically vapour-deposited diamond films

Figure 2 shows the surface morphologies of diamond films which were formed under several different reaction conditions. When the reaction pressure used was 3.5 Torr (Fig. 2(a)), the diamond crystals formed were very fine and were coagulated with each other to form a spherical structure. On increase in the reaction pressure to 7 Torr (Fig. 2(b)), the diamond crystals became larger and they were characterized by sharp crystal habits. This spherical structure was also observed when the CH_4 concentration was high (Fig. 2(d)) or when the distance between the hot filament and the substrate was increased (Fig. 2(e)). In the latter case, the reaction temperature decreased and the spherical structure developed. When the CH_4 concentration was decreased, the crystal size of the diamond became small (Fig. 2(c)). This means that the decreased deposition rate causes a reduction in crystal growth *i.e.* in Fig. 2(c).

Fig. 2. Surface morphologies of diamond films under (a) conditions A, (b) conditions B, (c) conditions C, (d) conditions D and (e) conditions E.

3.2. Results of laser Raman spectroscopy

Figure 3 shows the results of laser Raman spectroscopy. The broad peaks of amorphous carbon (1600 and 1350 cm^{-1}) were detected clearly in the Raman spectrum of low pressure products (Fig. 3(a)). As the reaction pressure was increased (Fig. 3(b)), the sharp peak of diamond (1333 cm^{-1}) and the broad peak of amorphous carbon at around 1500 cm^{-1} were seen. Similar results were obtained in both cases of increasing CH_4 concentration (Fig. 3(d)) and increasing the distance between the hot filament and the substrate (Fig. 3(e)). From these results, it can be inferred that amorphous carbon exists more in

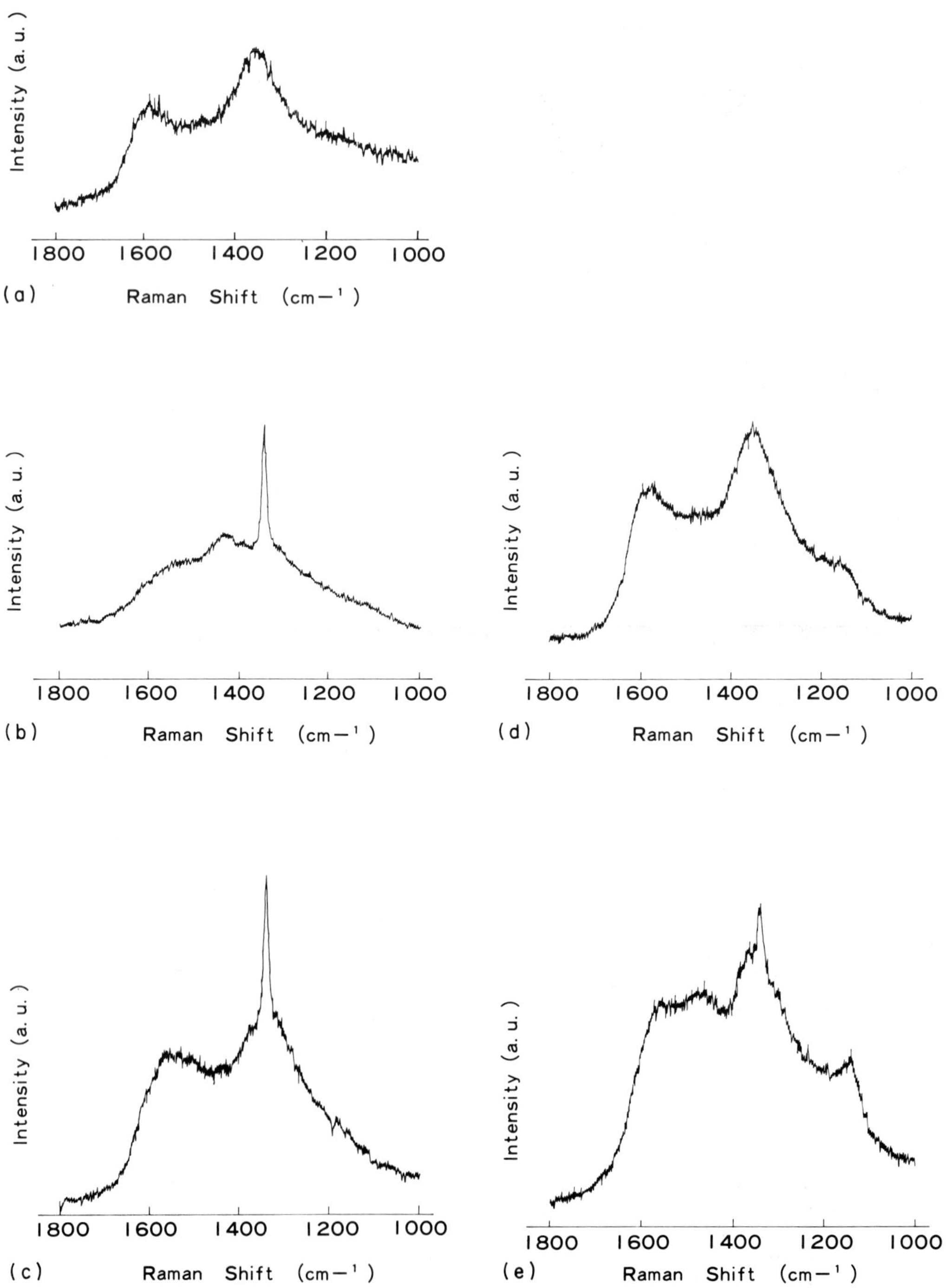

Fig. 3. Laser Raman spectra of diamond films under (a) conditions A, (b) conditions B, (c) conditions C, (d) conditions D and (e) conditions E.

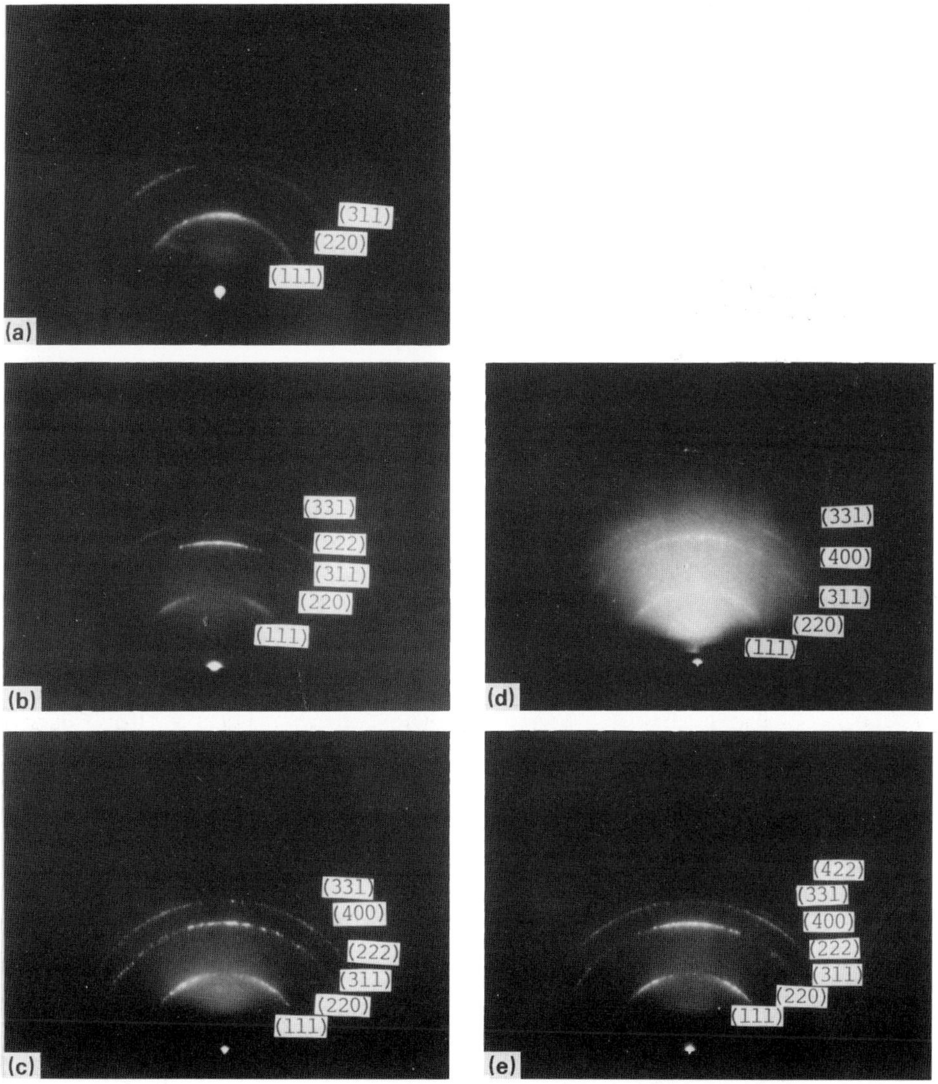

Fig. 4. RHEED patterns of diamond films under (a) conditions A, (b) conditions B, (c) conditions C, (d) conditions D and (e) conditions E.

the diamond films with a fine and coagulated spherical structure than in the diamond film with a sharp crystal habit.

3.3. Results of reflection high energy electron diffraction

Figure 4 shows the results of the RHEED study of coated diamond films. The diffraction lines of diamond crystals detected are sharp in all the experimental conditions studied. A halo pattern of amorphous carbon was detected in the case of the high CH_4 concentration (Fig. 4(d)). This is consistent with the results of the laser Raman spectroscopy study.

Interplanar spacings (d values) of the coated diamond were calculated from RHEED lines and

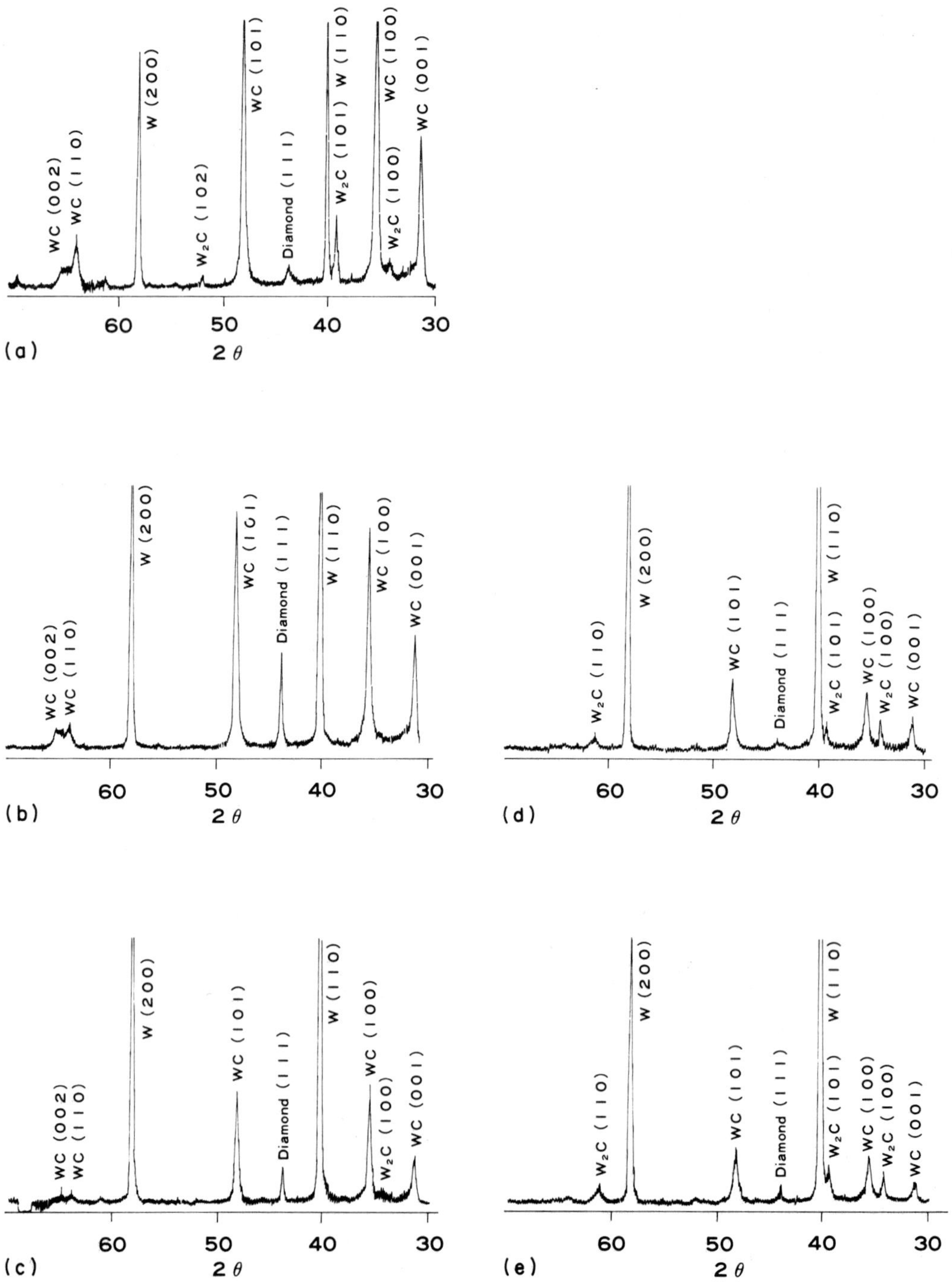

Fig. 5. X-ray diffraction patterns of diamond films under (a) conditions A, (b) conditions B, (c) conditions C, (d) conditions D and (e) conditions E.

are shown in Table 2. All *d* values are very similar and match those in *ASTM 6-675 Standard* [15] well. These results indicate that all diamond crystals produced under different experimental conditions of this study are identical and are pure diamond.

3.4. Results of X-ray diffraction

Figure 5 shows the results of the X-ray diffraction of coated diamond films. The diamond diffraction line (111) is clear and sharp in Figs. 5(b) and 5(c). In contrast, in Figs. 5(a), 5(d) and 5(e) the diamond diffraction lines (111) are broad and weak. These results are consistent with those of laser Raman spectroscopy.

All the substrates used were carbonized, and WC and W_2C were formed. W_2C can be seen more clearly in Figs. 5(a), 5(d) and 5(e) than in Figs. 5(b) and 5(c). In the case of carbonization of tungsten, W_2C is formed at first and then WC is formed. Therefore the results above suggest that carbon diffuses into the substrate faster in Figs. 5(b) and 5(c) than in Figs. 5(a), 5(d) and 5(e).

3.5. Summary of the results

The results of Sections 3.1–3.4 are summarized in Table 3. The spherical structure in a deposited diamond film was observed in the case of a low reaction pressure, or a high CH_4 concentration, or a long distance between filaments and substrates. In such cases, strong amorphous carbon peaks were detected in laser Raman spectra, and the X-ray diffraction lines of crystal diamond become broad and weak. However, the results of RHEED indicate that these films are constituted mainly of crystalline diamond particles. These results strongly suggest that the spherical structure of these diamond films consist of very fine diamond crystals and that the amorphous carbon exists mainly on the surfaces and between grain boundaries of the diamond polycrystalline film [16].

From these results, we can classify the diamond film formation processes into two types. One type is when the epitaxial growth of the diamond crystal occurs more easily than does nucleation. In this case, the morphology of the diamond film appears to have sharp crystal habits. The other is when the nucleation of diamond occurs on the surface of diamond more easily than does epitaxial growth. In this case, the surface morphology of diamond films has a spherical structure. These two reaction processes are drawn schematically in Fig. 6.

3.6. Results of cutting performance tests

Single-crystal and sintered diamonds have been used as cutting tools for the machining of non-ferrous metals or hard materials such as semisintered ceramics. We made cutting performance tests to estimate the characteristics of diamond-coated inserts. These coatings were made in the experimental condition (Fig. 6(b)) using the throw-away type of cemented-carbide inserts (CIS type SPP422) as substrates. Figure 7 is an SEM micrograph of a fracture surface of a diamond-coated cemented carbide. In this figure, it

TABLE 2 ***d* values from reflection high energy electron diffraction**

	d value (Å) for the following conditions					
	A	*B*	*C*	*D*	*E*	*ASTM Standard 6-675*
(111)	2.06	2.13	2.09	2.09	2.09	2.06
(220)	1.28	1.28	1.28	1.28	1.28	1.261
(311)	1.08	1.09	1.08	1.09	1.08	1.075

TABLE 3 Summary of results

Variation of reaction conditions	*Surface morphologies*	*Laser Raman spectra*	*RHEED*	*X-ray diffraction*
Reaction pressure high → low	Spherical structure increases	Peak of amorphous carbon increases	Diamond	Diamond decreases; W_2C increases
CH_4 concentration low → high	Spherical structure increases	Peak of amorphous carbon increases	Diamond Halo pattern intensity increases	Diamond decreases; W_2C increases
Distance between filament and substrate short → long	Spherical structure increases	Peak of amorphous carbon increases	Diamond	Diamond decreases; W_2 increases

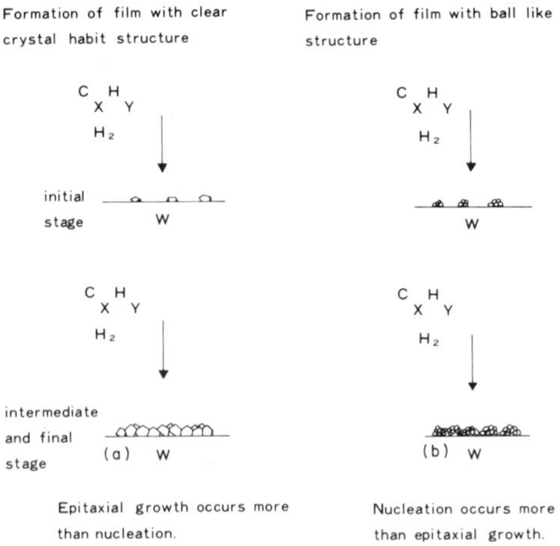

Fig. 6. Schematic drawing of the reaction processes of diamond film formation (a) with a clear crystal habit structure, in which epitaxial growth occurs more than nucleation and (b) with a ball-like structure, in which nucleation occurs more than epitaxial growth.

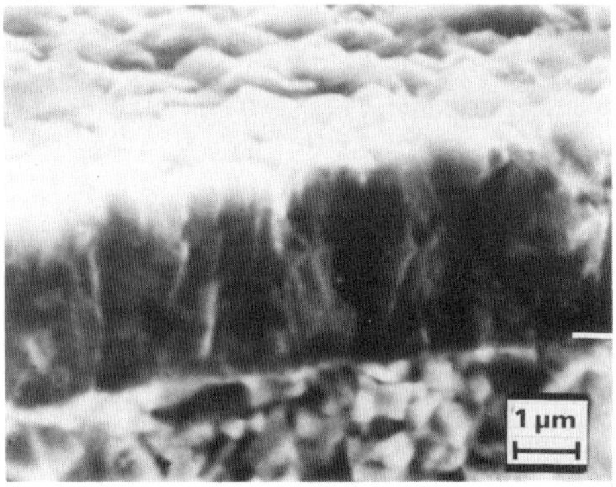

Fig. 7. Fracture surface of diamond-coated cemented carbide. The cemented carbide substrate lies below the line and the diamond layers is above the line.

is seen that diamond is formed on the cemented-carbide substrate surface with good cohesion.

Figure 8 consists of photographs of the cutting edges of a diamond-coated insert (Fig. 8(a)) and an uncoated insert (Fig. 8(b)) after the cutting test with 89wt.%Al–11wt.%Si alloy. Although a very large build-up of the welded aluminium alloy was observed on the edge of the uncoated insert, such a welding phenomenon was almost absent in the case of the diamond-coated insert.

Figure 9 shows the surface of the aluminium alloy after the cutting test. The surface of the work machined by the diamond-coated insert

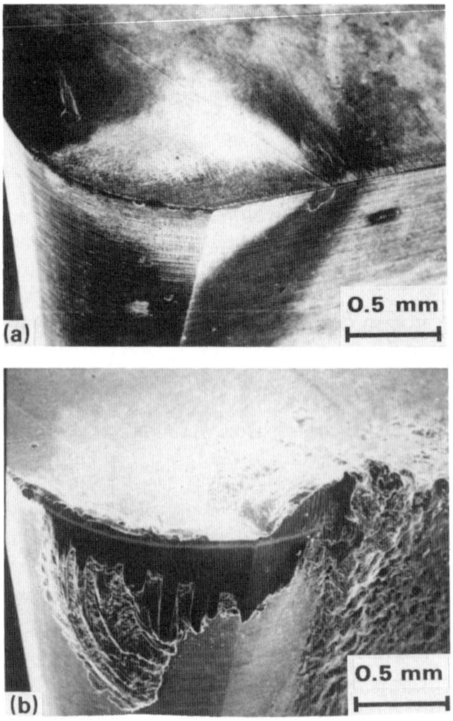

Fig. 8. Photographs of the cutting edges of (a) diamond-coated and (b) uncoated inserts (cutting of aluminium alloy 89at.%Al–11at.%Si; speed, 500 m min^{-1}; feed, 0.1 mm rev^{-1}; depth of cut, 1 mm; time, 5 min).

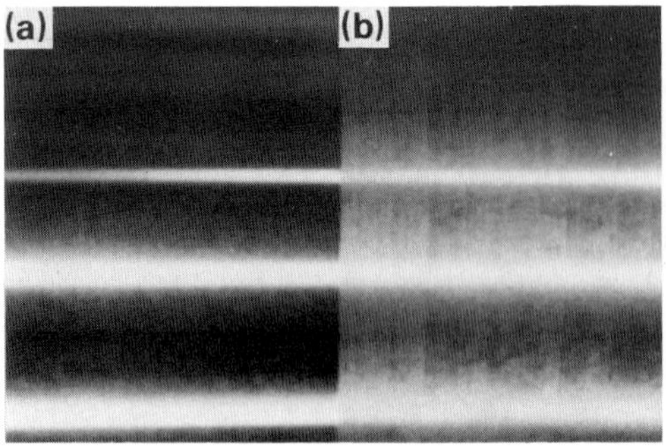

Fig. 9. Surface of aluminium alloy after cutting test with (a) diamond-coated and (b) uncoated inserts.

(Fig. 9(a)) is very smooth and lustrous. In contrast, the machined surface (Fig. 9(b)) is rough and much less lustrous in the case of the uncoated insert.

Figure 10 shows the cutting edges of a coated insert (Fig. 10(a)) and an uncoated insert (Fig.

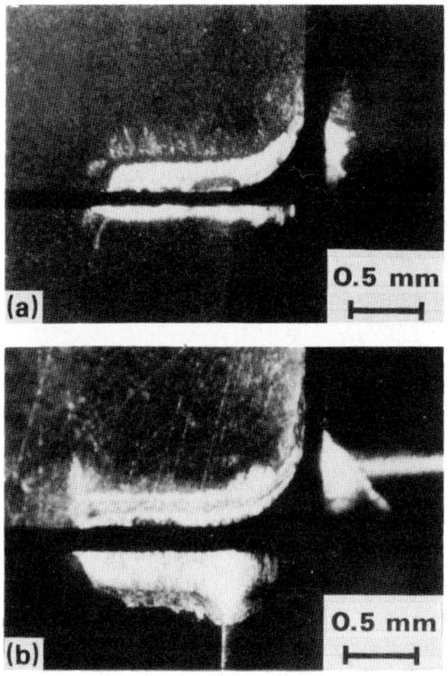

Fig. 10. Photographs of the cutting edges of (a) diamond-coated and (b) uncoated inserts respectively (cutting of high silicon alloy 81.5at.%Al–18.5at.%Si; speed, 250 m min^{-1}; feed, 0.1 mm rev^{-1}; depth of cut, 1.5 mm; time, 1 min).

10(b)) after cutting test of a high silicon 81.5wt.%Al–13.5wt.%Si alloy. This material is very hard and it is difficult to turn it. Although an uncoated insert was worn considerably in the short cutting time, the diamond-coated insert was worn very little.

Figure 11 shows the results of a cutting test of a semisintered SiC ceramic. Although an uncoated insert (Fig. 11(b)) shows very large wear in the short cutting time, *e.g.* 1 min, the diamond-coated insert (Fig. 11(a)) was worn very little in a long cutting time of about 15 min. These results show that CVD diamond coating of cemented-carbide inserts is an effective means of prolonging the life of the insert for turning non-ferrous metals and, in particular, aluminium alloys and semisintered ceramics.

4. Conclusion

The effects of CVD reaction conditions on the morphology and the crystal structure of diamond films were studied, and the possibility of utilizing a CVD diamond film as a wear-resistant coating for cutting tools was investigated; the following conclusions were obtained.

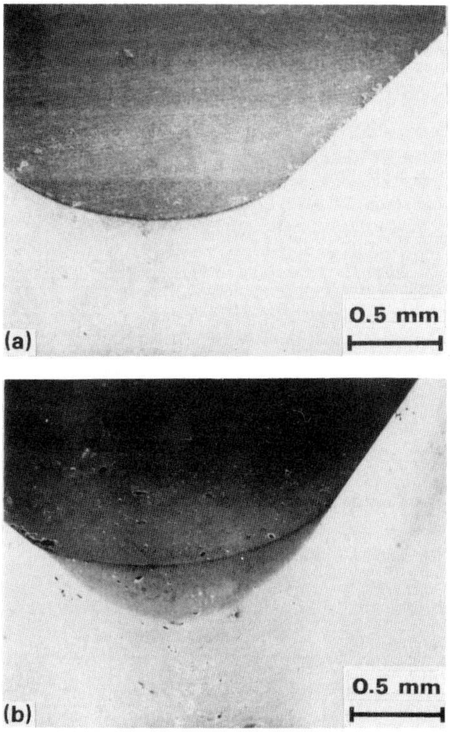

Fig. 11. Photographs of the cutting edges of (a) diamond-coated insert after 15 min and (b) uncoated insert after 1 min (cutting of semisintered SiC ceramic; speed, 40 m min^{-1}; feed, 0.05 mm rev^{-1}; depth of cut, 0.5 mm.

(1) The surface morphologies of CVD diamond films are much affected by the reaction conditions. Under reaction conditions of a low pressure, or a high CH_4 concentration, or a long distance between the filament and the substrate, a spherical structure was observed.

(2) Diamond films of spherical structure were found to be made of coagulated fine diamond crystals, and amorphous carbon phases were seen on surfaces and also between grain boundaries.

(3) CVD diamond coating on cemented-carbide inserts is an effective means of prolonging the life of the insert for turning non-ferrous metals and, in particular, aluminium alloys and semisintered ceramics.

References

1 F. Bundy, H. Hall, H. Strong and R. Wentorf, *Nature, London, 176*(1955) 51.

2 B. V. Spitsyn and B. V. Derjaguin, *U.S.S.R. Patent 339134*, 1980.
3 W. G. Eversole, *U.S. Patents 3,030,187 and 3,030,188*, 1962.
4 S. Aisenberg and R. Chabot, *J. Appl. Phys.*, *42* (1971) 2953.
5 E. G. Spencer, P. H. Schmibt, D. C. Joy and F. J. Sansalont, *Appl. Phys. Lett.*, *29*(1976) 118.
6 S. Matsumoto, Y. Sato, M. Kamo and N. Setaka, *Jpn. J. Appl. Phys.*, *21* (1982) L183.
7 S. Matsumoto, Y. Sato, M. Tsutsimi and N. Setaka, *J. Mater. Sci.*, *17*(1982) 3106.
8 N. Kikuchi, T. Komatsu and H. Yamashita, *Proc. 9th Symp. on Ion Sources and Applied Technology, Tokyo, 1985*, 1985, p. 233.
9 M. Tsuda, M. Nakajima and S. Oikawa, *Proc. 8th Int. Symp. on Plasma Chemistry, 1987*, 1987, p. 2453.
10 A. Sawabe and T. Inuzuka, *Appl. Phys. Lett.*, *46* (1985) 146.
11 S. Matsumoto, *J. Mater. Sci. Lett.*, *4*(1985) 600.
12 T. Komatsu and H. Yamashita, Y. Tamou and N. Kikuchi, *Proc. 8th Int. Symp. on Plasma Chemistry, 1987*, 1987, p. 2487.
13 Y. Mitsuda, Y. Kojima, T. Yoshida and K. Akashi, *J. Mater. Sci.*, *22*(1987) 1557.
14 S. Matsumoto, *J. Mater. Sci.*, *17*(1987) 3106.
15 *ASTM Stand. 6–675*, 1953.
16 N. Kikuchi, T. Komatsu and H. Yamashita, *Proc. 10th Symp. on Ion Sources and Applied Technology, Tokyo, 1986*, 1986, p. 359.

Growth of Diamond on Single Crystals of Tungsten Carbide in Microwave Plasma*

MUTSUKAZU KAMO, HITOSHI CHAWANYA†, TAKAHO TANAKA, YOICHIRO SATO and NOBUO SETAKA

National Institute for Research in Inorganic Materials, 1-1 Namiki, Tsukuba, Ibaraki 305 (Japan)

(Received November 12, 1987)

Abstract

Crystalline diamond was deposited on (0001), (10$\bar{1}$0) and (11$\bar{2}$0) faces of tungsten carbide (WC) by vapour deposition. Deposition was carried out in microwave plasma in a stream consisting of a mixture of methane and hydrogen gas. The deposits were studied by scanning electron microscopy and Raman spectroscopy. The nucleation density and the growth rate of diamond varied with the crystallographic plane orientation of the substrate. Acceleration of the decomposition of methane on the (10$\bar{1}$0) face is suggested. The growth rate of diamond increased but the nucleation density decreased with increasing substrate temperature.

1. Introduction

The hardness, high electrical resistivity, high thermal conductivity, optical transparency and chemical inertness make diamond a desirable material for hard coating, protection, insulation and heat transmission.

Diamond crystals have been synthesized from the gas phase by chemical transport using atomic hydrogen in a closed system [1]. We have reported that diamond can be grown from a gaseous mixture of hydrogen and hydrocarbons using a hot tungsten filament [2, 3] and microwave plasma [4, 5]. These techniques are suitable for coating applications because thin films of diamond can be deposited on various non-diamond materials such as silicon, silica and molybdenum.

Tungsten carbide (WC) has been used as the major constituent of cemented carbide cutting tools. Attempts are being made to coat the surface of these tools with diamond in order to achieve improved hardness and durability. Deposition of diamond on WC–Co alloy substrate by the hot-filament method has been reported [6]. The deposition behaviour on cemented carbides may be complicated by the composition of the substrate, existence of binder metal and various other factors.

In the present paper, we deal with the deposition of diamond in microwave plasma on single crystals of WC which had been grown by the floating-zone method. The nucleation-and-growth process in the early stage of the deposition was closely investigated and is discussed. The effects of the crystallographic plane of the substrate and of changes in the deposition parameters were also studied.

2. Experimental details

The deposition apparatus for diamond was the same as that reported in previous papers [4, 5]. The substrates were put on a silica glass plate and placed in a silica glass reaction tube 40 mm in diameter. After the reaction tube had been evacuated, a gaseous mixture of methane and hydrogen was introduced into the tube. The total gas flow rate was 100 standard cm^3 min^{-1}. The concentration of methane was varied in the range 0.5%–1%. The pressure in the tube was maintained at 5.3 kPa by differential evacuation. Microwaves (2.45 GHz) generated by a magnetron were transmitted to the tube through a set of waveguides, a power monitor and a tuner. The power of the microwave was varied in the range 300–450 W. The temperature of the substrate was measured with an optical pyrometer.

*Paper presented at the 3rd International Conference on the Science of Hard Materials, Nassau, The Bahamas, November 9–13, 1987.

†Present address: Tokyo Research Laboratory, Japan Synthetic Rubber Co. Ltd., 3-5-1 Higashiyurigaoka, Asao-ku, Kawasaki, Kanagawa 215, Japan.

0921-5093/88/$3.50

© Elsevier Sequoia/Printed in The Netherlands

Single crystals of WC were grown by the floating-zone method [7]. (0001), $(10\bar{1}0)$ and $(11\bar{2}0)$ wafers of WC, which were spark machined from the single-crystal rods, were used as the substrate for deposition. The surfaces of the wafers were polished successively using diamond paste of decreasing particle sizes (8–16, 0–3 and 0–0.5 μm) each for 1 h. The Si (111) substrate was used for comparison. The mirror-polished surface of commercially available silicon wafers was roughened using diamond powder of particle size 5–12 μm in order to increase the nucleation density. The substrates were heated owing to absorption of the microwave as well as to heat transfer from the plasma. The rise in the temperature was somewhat larger for WC substrates than for silicon substrates. Thus the temperature of the WC and silicon substrates were 900 °C and 880 °C respectively at a microwave power of 450 W, 870 °C and 860 °C respectively at 390 W, and 830 °C for both substrates at 300 W.

The same areas of the substrates were observed at appropriate time intervals during a deposition experiment using scanning electron microscopy (SEM), and the nucleation and growth of diamond crystals were followed.

Raman spectra were recorded by backscattering from the sample at room temperature. The 488.0 nm line of an argon ion laser was used for excitation.

3. Results and discussion

Figure 1 shows SEM images of diamond deposited on the surface of (0001) plane of a WC substrate at a methane concentration of 0.5% and a substrate temperature of 830 °C. The formation of diamond nuclei 0.04–0.3 μm in diameter was observed after deposition for 2 h (Fig. 1(a)). The number of nuclei and the particle size increased with increasing deposition time (Figs. 1(b)–(f)). However, the rate of nucleation began to decrease after deposition for 6 h. It is shown in the photographs that the crystals formed increased in size, gradually came into contact with each other, covered the surface of the substrate and finally formed a film. The crystals show well-defined habits and are predominantly multiply twinned particles consisting of 5 or 20 single-crystal units as described elsewhere [1].

The nucleation behaviours on the $(10\bar{1}0)$ and $(11\bar{2}0)$ faces of WC and (111) face of silicon were similar to that on the (0001) face. The nucleation was fast in the first stage, but the rate of the nucleation decreased after deposition for 6–10 h.

Figure 2 shows morphology of diamond deposited on four different substrates at the same conditions for 14 h. On the $(10\bar{1}0)$ face of WC the nucleation density is larger than that on (0001) face as shown in Figs. 2(a) and 2(b), but particles are smaller, indicating a lower growth rate of diamond on this substrate. The irregular shapes of the particles suggest that each particle consists of many small crystallites. When the deposit was grown to form a film, a square-shaped (100) face of diamond frequently appeared on the film. These characteristics of the deposit are similar to those observed on silicon substrate at a higher methane concentration [5]. On the $(11\bar{2}0)$ face, multiply twinned particles with well-defined habits were observed; these are similar to those on the (0001) plane. However, the nucleation density on this plane was somewhat higher. The nucleation density was higher on the silicon substrate than on any plane of the WC. The deposited particles showed well-defined habits, and many of them were of a spinel-type twin form.

The growth of diamond occurred in two stages on the silicon substrate. Particles of diamond 0.02–0.1 μm in diameter were generated and the number of the particles increased for the first 6 h of the deposition. Growth of the particles was not observed in this period. In the second stage, the nucleation rate decreased and the particles began to grow in size. However, on the WC the growth rate was observed to be nearly constant throughout deposition.

Table 1 summarizes the nucleation densities and lateral growth rates of the particle determined from the SEM photographs. The nucleation density, defined as the mean value of some measurement of a particle count per area, is lower on WC substrates than on silicon substrates and varies with the crystallographic plane orientation of the WC substrate. The nucleation density on $(10\bar{1}0)$ plane was about four times that on the (0001) plane. The growth rate of the particle on the WC substrate was higher on the crystal plane with lower nucleation densities.

Figure 3 shows the Raman spectra of diamond films deposited on four different substrates for a deposition time of 20 h. Only a Raman band of diamond at 1333 cm^{-1} is observed in the spectrum for a film on a silicon substrate. In the films on the (0001) and $(11\bar{2}0)$ faces of WC an additional broad band was observed at 1450–1460

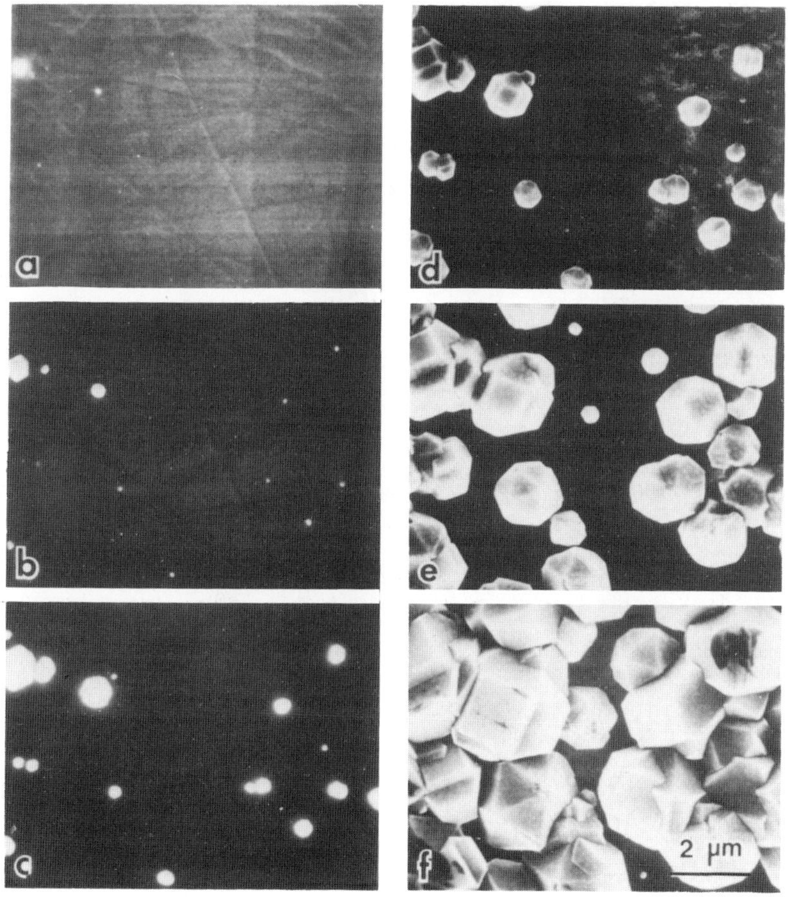

Fig. 1. Growth of diamond on WC(0001) substrates at 830 °C for various deposition times (concentration of methane, 0.5%): (a) 2 h; (b) 4 h; (c) 6 h; (d) 8 h; (e) 14 h; (f) 20 h.

cm^{-1}. A broad band was found at 1500 cm^{-1} with the film on the $(10\bar{1}0)$ face. These broad bands may be attributed to the presence of double bonds which form no periodic structure.

Figure 4 shows SEM images of diamond deposited for 5 h with the methane concentration raised to 1%. The nucleation density and growth rate increased as the methane concentration was increased. The film-like deposits on a silicon substrate were obtained after 5 h. The nucleation density increased by a factor of 1.2–2.6 on WC and by a factor of 3.6 on silicon, and the growth rate increased by a factor of 2 on WC and by a factor of 1.4 on silicon.

The morphologies of the particle became more irregular at the higher methane concentration, indicating that more imperfections were introduced into the diamond crystals. The deposits on the (0001) face have many irregularities on the crystal face of diamond. On the $(10\bar{1}0)$ face, spherical particles consisting of many small crystallites were formed. The deposits on the $(11\bar{2}0)$ face were aggregates of small crystallites.

The Raman spectra of the deposits shown in Fig. 4 again indicated the presence of a broad band at 1450–1500 cm^{-1}, which was observed at a methane concentration of 0.5%. They grew in intensity and shifted towards higher wavenum-

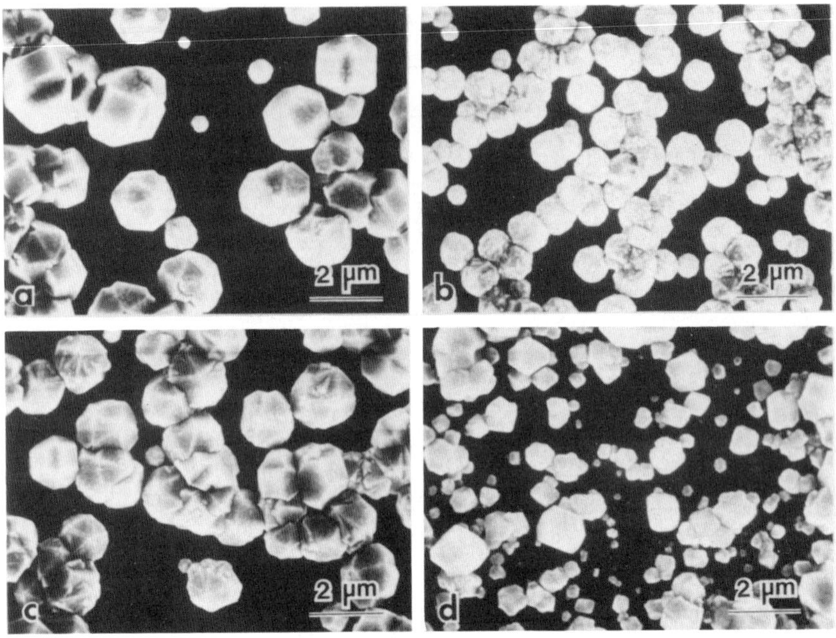

Fig. 2. Diamond deposited on different substrates at 830 °C after a deposition time of 14 h (concentration of methane, 0.5%): (a) WC(0001); (b) WC(10$\bar{1}$0); (c) WC(11$\bar{2}$0); (d) Si(111).

TABLE 1 Nucleation densities and growth rate of diamond on various substrates at 830 °C

Substrate	*Nucleation density* (10^8 cm^{-2})	*Growth rate* (μm h^{-1})
WC(0001)	0.30	0.089
WC(10$\bar{1}$0)	1.29	0.039
WC(11$\bar{2}$0)	0.43	0.067
Si(111)	2.80	0.063

bers. On the (10$\bar{1}$0) face, this band was especially large, and Raman bands of highly disordered graphite at 1360 and 1600 cm^{-1} were also observed.

Figure 5 shows SEM images of diamond deposited at higher substrate temperatures. The temperatures were 900 °C and 880 °C for WC and silicon substrates respectively. The particles deposited on WC were larger, indicating a higher growth rate at higher temperatures. In contrast, the nucleation densities were observed to be lower.

Diamond crystals with well-defined habits were deposited on the (10$\bar{1}$0) plane of WC, on which particles with irregular shapes were deposited at a lower temperature. At the higher temperature, the formation of multiply twinned particles predominated, and (100) faces appeared more frequently than did (111) faces. As the Raman spectra in Fig. 6 show, the relative inten-

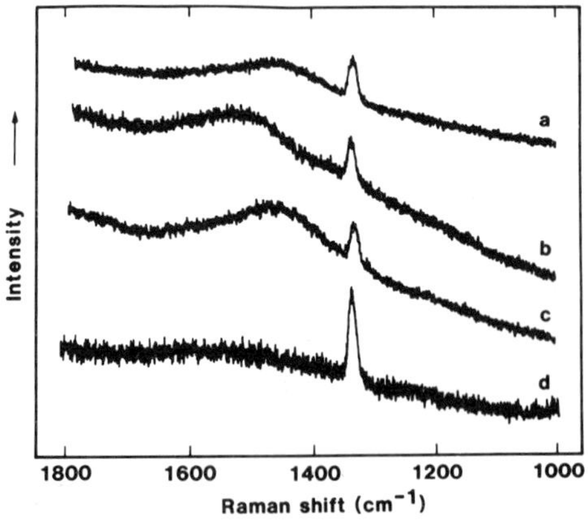

Fig. 3. Raman spectra of diamond films deposited on different substrates at 830 °C (concentration of methane, 0.5%): (a) WC(0001); (b) WC(10$\bar{1}$0); (c) WC(11$\bar{2}$0); (d) Si(111).

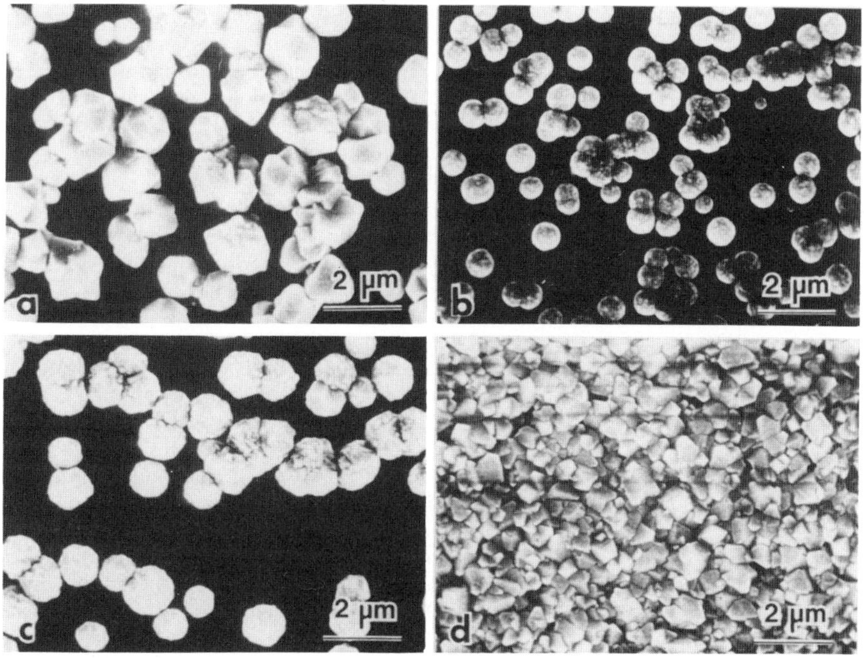

Fig. 4. Diamond deposited on different substrates at 830 °C for 5 h (concentration of methane, 1%): (a) WC(0001); (b) WC(10$\bar{1}$0); (c) WC(11$\bar{2}$0); (d) Si(111).

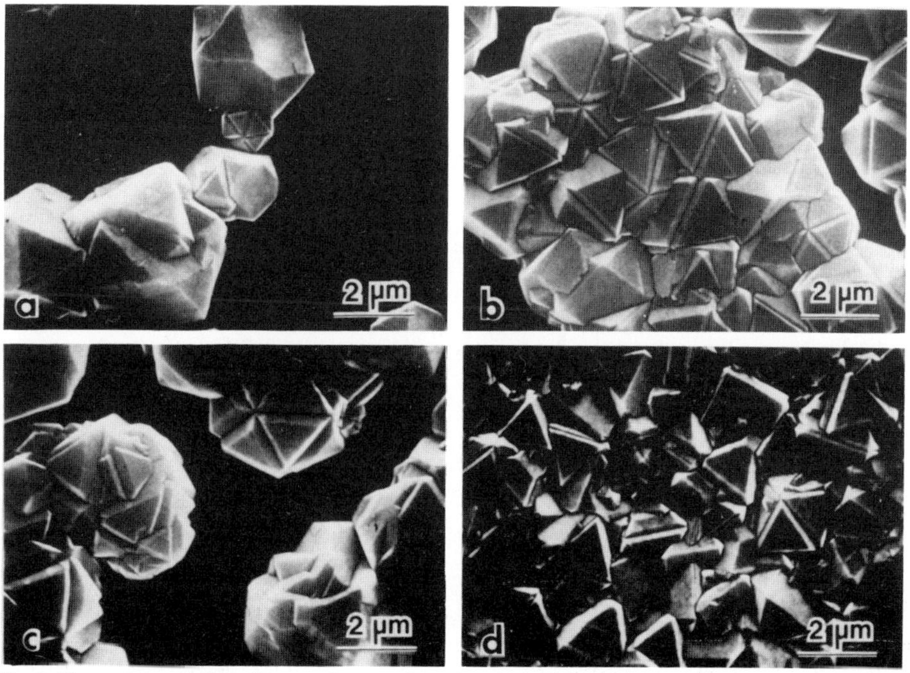

Fig. 5. Diamond deposited on different substrates after a deposition time of 9 h (concentration of methane, 0.5%): (a) WC(0001) deposited at 900 °C; (b) WC(10$\bar{1}$0) deposited at 900 °C; (c) WC(11$\bar{2}$0) deposited at 900 °C; (d) Si(111) deposited at 880 °C.

sity of the band at 1500 cm^{-1} with respect to the band for diamond is smaller on the (0001) face of WC and silicon substrates. This indicates that the diamond crystals grown on these substrates have a better crystalline quality. However, little morphological difference was noted from the SEM observations for all the substrates.

Figure 7 and Fig. 8 show the variation in the nucleation density and the growth rate respectively with substrate temperature. The temperature of the WC substrate was varied from 830 to 900 °C. No appreciable deposition was observed at 800 °C. The nucleation density on the WC decreased in the order $(10\bar{1}0)$, $(11\bar{2}0)$ and (0001) faces. The nucleation density decreased as the substrate temperature was raised. However, on silicon substrates, the nucleation density increased with increasing substrate temperature. In contrast, the growth rate increased with increasing temperature on all the substrates.

The morphology of diamond was also affected by the crystal face of the substrate. Diamond particles deposited on the $(10\bar{1}0)$ face have irregular shapes and are often aggregates of small crystallites, and Raman spectra revealed a broad band at about 1500 cm^{-1}. These features of the deposits have been observed in diamond crystals deposited on silicon at higher methane concentrations. Thus, it is suggested that the decomposition of methane is accelerated on the $(10\bar{1}0)$ face of WC, and that the WC$(10\bar{1}0)$ face acts as a catalyst [8]. The deposition features on the face indicate that this is possibly so.

A correlation was found between the nucleation density and the growth rate; the higher the nucleation density, the lower was the growth rate. The correlation holds not only at 830 °C, as

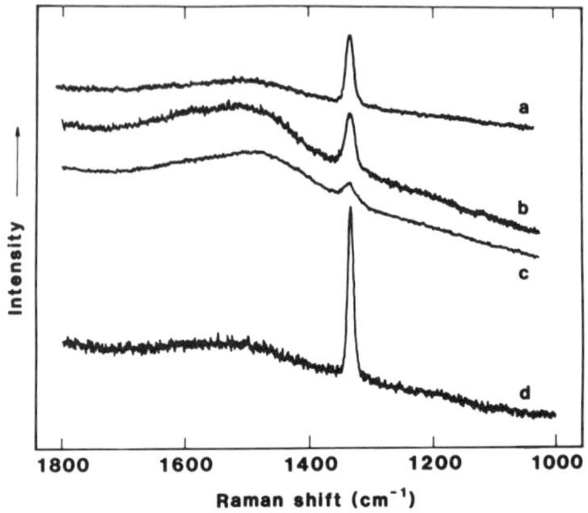

Fig. 6. Raman spectra of diamond films deposited on different substrates (concentration of methane, 0.5%): (a) WC(0001) deposited at 900 °C; (b) WC$(10\bar{1}0)$ deposited at 900 °C; (c) WC$(11\bar{2}0)$ deposited at 900 °C; (d) Si(111) deposited at 880 °C.

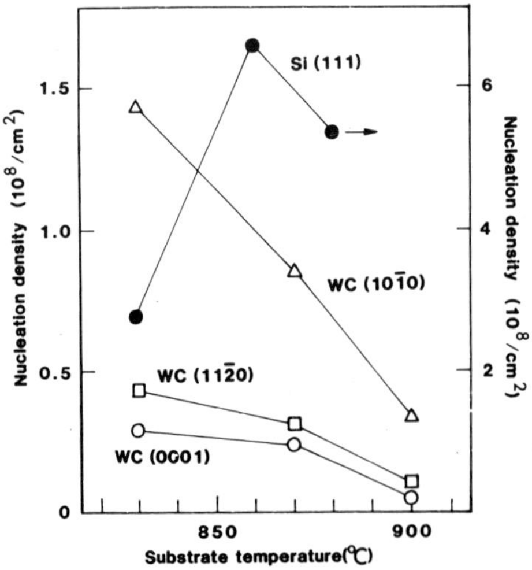

Fig. 7. Dependence of the nucleation density on the substrate temperature (concentration of methane, 0.5%).

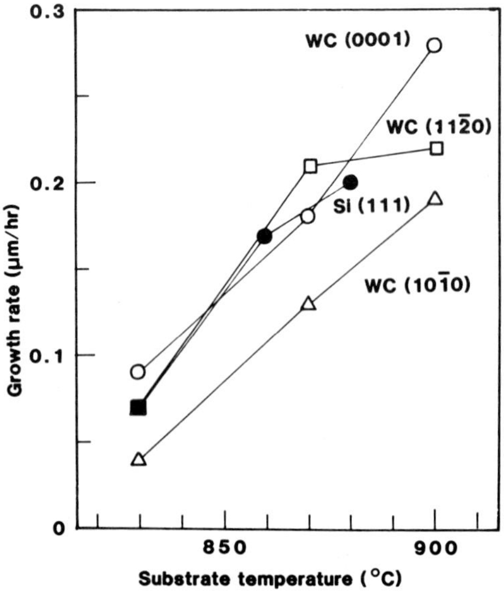

Fig. 8. Dependence of the growth rate on the substrate temperature (concentration of methane, 0.5%).

shown in Table 1, but also at higher temperatures, as seen from Figs. 7 and 8.

The results indicate that not only the direct deposition of carbon but also the diffusion of carbon deposited on the substrate surface contribute appreciably to the growth of diamond within the reaction duration of the present work.

As to the deposition on silicon at a substrate temperature of 830 °C and a methane concentration of 0.5%, the growth of diamond was suppressed for the first 6 h, while the number of nuclei increased. After that period, the nucleation rate decreased and the particles began to grow. At substrate temperatures of 860 and 880 °C the growth rate was also small in the early stage of deposition, although nucleation and growth occurred simultaneously. The smaller growth rate in the early stage may be perhaps due to the concurrent carbonization of the silicon substrate.

4. Conclusions

The variations in growth features have been noted for the three different faces of WC.

The nucleation density varied with the crystallographic plane orientation of the substrate. The growth rate of the particles is lower on the crystal face with the higher nucleation density.

The deposition features on the $(10\bar{1}0)$ face of WC resemble those for higher methane concentrations on a silicon substrate. Acceleration of the decomposition of methane on the $(10\bar{1}0)$ face is suggested.

The growth rate of diamond increased but the nucleation density decreased with increasing substrate temperature.

References

1 B. V. Spitsyn, L. L. Bouilov and B. V. Derjaguin, *J. Cryst. Growth, 52* (1981) 219.
2 S. Matsumoto, Y. Sato, M. Kamo and N. Setaka, *Jpn. J. Appl. Phys., 21* (1982) L183.
3 S. Matsumoto, Y. Sato, M. Tsutsumi and N. Setaka, *J. Mater. Sci., 17* (1982) 3106.
4 M. Kamo, Y. Sato, S. Matsumoto and N. Setaka, *J. Cryst. Growth, 62* (1983) 642.
5 M. Kamo, Y. Sato and N. Setaka, *J. Chem. Soc. Jpn., Chem. Ind. Chem., 1984* (1984) 1642.
6 H. Suzuki, H. Matsubara and N. Horie, *J. Jpn. Soc. Powder Powder Metall., 33* (1986) 262.
7 T. Tanaka, S. Otani and Y. Ishizawa, *J. Mater. Sci., 23* (1988) 665.
8 G. Schulz-Ekloff, D. Baresel and W. Sarholz, *J. Catal., 43* (1976) 353.

Preparation of a Diamond–Corundum Layer Composite Using Low Pressure Diamond Plasma-assisted Chemical Vapour Deposition*

R. BICHLER, J. PENG, R. HAUBNER and B. LUX

Institute for Chemical Technology of Inorganic Materials, Technical University Vienna, A-1060 Vienna (Austria)

(Received November 11, 1987)

Abstract

During the last few years, numerous publications and patents for the low pressure plasma-assisted chemical vapour deposition (CVD) of diamond and its application have been published. Recently, composites consisting of such diamond particles combined with other materials have been described in patents. In the present paper, we deal with a tool surface where low pressure diamonds are embedded in an Al_2O_3 surface layer produced by normal CVD methods. In this composite layer the good wear resistance and adhesion of the Al_2O_3 layer and the high hardness of the diamond could be successfully combined.

1. Introduction

Diamond deposition from the gas phase at a low pressure from a CH_4–H_2 gas mixture recently became possible using different gas activation methods [1–9]. Diamond composite particles [7] and other material combinations [8] between low pressure diamond particles and a hard material have also been described recently.

In the following paper a composite layer coating consisting of low pressure diamond particles formed by vapour phase synthesis and surrounded by a corundum matrix is described. The single-crystal diamond particles were first deposited on a tool substrate, which was then covered by an Al_2O_3 layer, produced by normal chemical vapour deposition (CVD) using the $AlCl_3$–H_2–CO_2 system [11–14]. The result was a composite layer combining the good wear and adhesion properties of Al_2O_3 layers with the useful properties of the superhard diamond particles.

*Paper presented at the 3rd International Conference on the Science of Hard Materials, Nassau, The Bahamas, November 9–13, 1987

0921-5093/88/$3.50

2. Experimental procedure

2.1. Diamond deposition

Cemented-carbide inserts for metal cutting pre-coated by CVD with a 6 μm TiC layer (hereafter called (WC–Co)/TiC) were used as substrates. Figure 1 shows a schematic illustration of the reactor used for the diamond deposition [10]. A tantalum filament [9], heated by an a.c. source, was placed above the substrate, which had been cleaned ultrasonically in acetone. After the reactor chamber had been flushed with pure hydrogen and the electric furnace heated to the selected deposition temperature, the tantalum-filament was switched on and the reaction gas was introduced for the chosen deposition time. The following deposition conditions were used: temperature of the tantalum-filament, 2000 °C; temperature of the substrate, 650–850 °C; gas

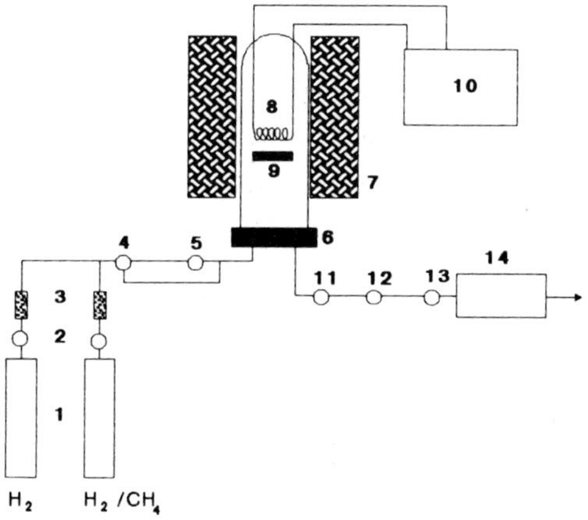

Fig. 1. Experimental arrangement of the reactor for diamond deposition: 1, gas supply; 2, reduction valve; 3, flow meter; 4, three-way stopcock; 5, needle valve; 6, reaction chamber; 7, furnace; 8, filament, 9, sample; 10, constant-power source; 11, vacuum gauge, 12, needle valve; 13, stopcock; 14, pump.

© Elsevier Sequoia/Printed in The Netherlands

pressure, 5 Torr; gas composition, 0.5%CH_4–H_2; total gas flow rate, 30 Nl h^{-1}; deposition time, 5–22 h. The substrate temperatures were monitored with a Ni–(Ni–Cr) thermocouple placed beneath the substrate. The filament temperature was measured by an optical pyrometer.

2.2. Al_2O_3 deposition

The Al_2O_3 CVD process [11–13] was carried out in the $AlCl_3$–H_2–CO_2 system in a hot-wall reactor at 1050 °C and 50 Torr. The flow rates in the reactor were as follows: H_2, 24.0 Nl h^{-1}; CO_2, 1.0 Nl h^{-1}; $AlCl_3$ 1.5 Nl h^{-1}. The deposition times were 2–5 h. Figure 2 shows the experimental set-up.

3. Experimental results

3.1. Diamond deposition

First a given number of isolated faceted single-crystal diamond crystals (polyhedral and cubo-octahedral) were deposited using different deposition conditions on a (WC–Co)/TiC substrate surface.

At a deposition temperature of 680 °C and a deposition time of 5 h, these crystals had a diameter of 5 μm. At a deposition time of 16 h, crystals 10–12 μm in diameter were obtained (Fig. 3(a)). At 800 °C the size of the deposited diamond particles ranged, depending on the deposition time, from 10 μm for 5 h to 30 μm for 22 h (Fig. 3(b)). With the various process parameters chosen (flow rate, CH_4-to-H_2 ratio etc.), additional diamond nucleation with increasing deposition time was observed only rarely. That is, with extended deposition time the initially formed single-crystal particles merely increased in size and the formation of a continuous diamond layer was prevented [10].

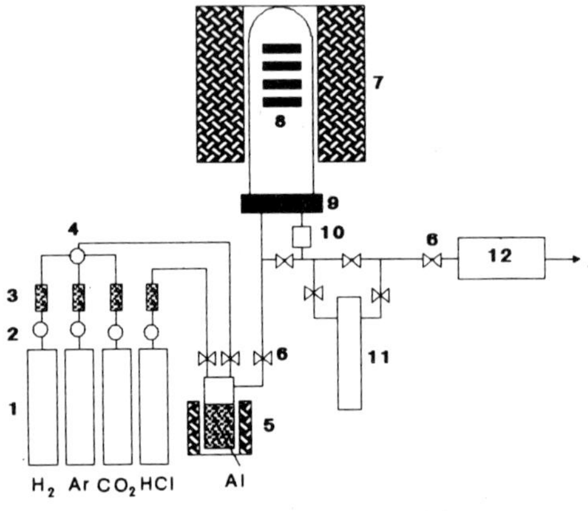

Fig. 2. Experimental arrangement for Al_2O_3 deposition: 1, gas supply; 2, reduction valve; 3, flow meter; 4, four-way stopcock; 5, evaporator; 6, stopcock; 7, furnace; 8, sample; 9, reaction chamber; 10, vacuum gauge; 11, cooling chamber; 12, pump.

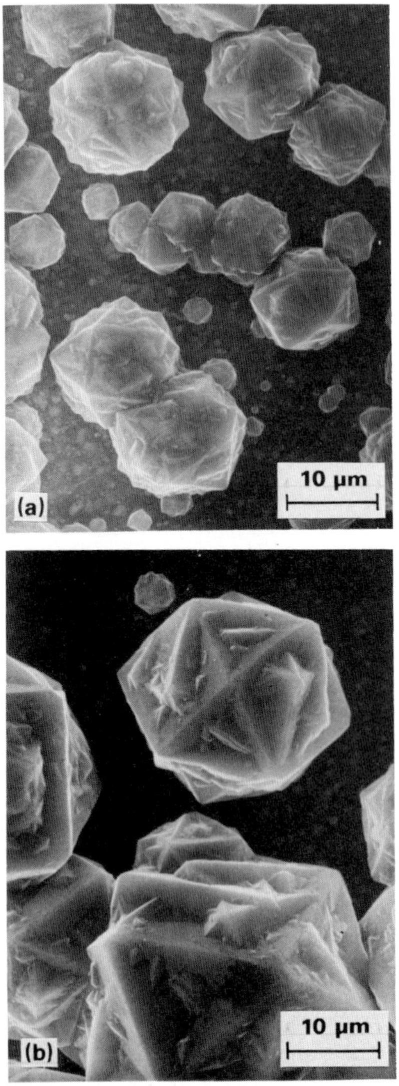

Fig. 3. Diamond deposition on the (WC–Co)/TiC substrate: (a) 680 °C for 16 h; (b) 800 °C for 22 h.

The desired average density of diamond nuclei on the substrate surface was, in the case described, about equal to or slightly less than one crystal per 100 μm^2. This value can be varied to suit the intended application. In order to create the right amount of diamond particles in the composite layer, their number may be increased considerably (*e.g.* deposition on Si–Al–O–N substrates).

3.2. Al_2O_3 deposition

In a second operation the previously formed single-crystal diamond particles and the entire substrate surface were covered with a thin Al_2O_3 layer. With the deposition parameters given in Section 2.2, a finely crystalline Al_2O_3 layer with a uniform thickness of 3 μm was formed on the TiC surface as well as on the diamonds. The morphology and the layer thickness of the Al_2O_3 depositions are the same on the diamond crystals and on the TiC, but the Al_2O_3 crystal size is slightly smaller on the diamond than on the TiC (1 μm on the diamond particles, and 2 μm on the TiC surface). Owing to differences between the thermal expansion of diamond and corundum, the Al_2O_3 layer on the diamond crystals cracked during cooling (Fig. 4), a phenomenon well known from industrial Al_2O_3 coatings on cemented-carbide inserts. Such cracks are not generally detrimental to the tool performance [13]. All photographs shown were obtained by scanning electron microscopy.

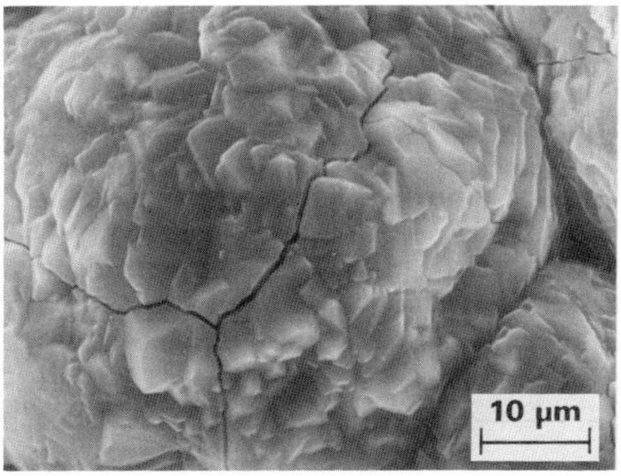

Fig. 4. Fracture in the Al_2O_3 layer deposited on low pressure diamond.

4. Discussion

The composite layer obtained combines the high hardness, wear resistance and other interesting properties of the corundum with those of the superhard diamond. An inherent weakness of some diamond layers formed by low pressure plasma-assisted CVD can be their poor adhesion on certain substrates [15–17]. This can be overcome by covering the isolated diamond particles with the strongly adherent Al_2O_3 CVD layer which acts as a matrix and a binder for the composite layer. The micrographs in Fig. 5 illustrate the excellent adherence obtained in most cases, even to the point that fracture of the small diamond particles occurred.

The detailed analyses of the fracture and the interfaces TiC–diamond–corundum revealed the following.

(1) The Al_2O_3 layer is stripped off (Fig. 5(b)).

(2) The main part of a diamond crystal broke off; the Al_2O_3 layer remained on the substrate (Fig. 5(c)).

(3) An Al_2O_3–TiC interface, showing a clean fracture through a diamond crystal, indicates very good adhesion (Fig. 5(d)).

These micrographs clearly show the excellent bonding between the diamond crystal and the TiC layer and the somewhat weaker adhesion between the diamond and the Al_2O_3 layer. This is probably due to the rather different chemical natures of diamond and Al_2O_3. However, other hard intermetallic components such as carbides, borides, nitrides, silicides or even other oxides exhibiting a better controllable chemical interface interaction with diamond can be easily applied by the same methods. Some of these have already been the subject of patent applications [8, 15–17].

Further experimental work is required to determine which material combinations give the optimal interface bonding between all components and what compromises are necessary for the expansion coefficients in order to minimize mechanical stresses. Further, the specific practical applications and the wear and friction behaviour of different material combinations have to be evaluated. It is of course also possible to make multiple use of the production methods described above, *e.g.* a second deposition of diamond particles on top of the covering hard layer, which would then be covered with another coating. Thicker layers and tailor-made combinations thus become easily feasible. Also, many other surface coating techniques such as physical

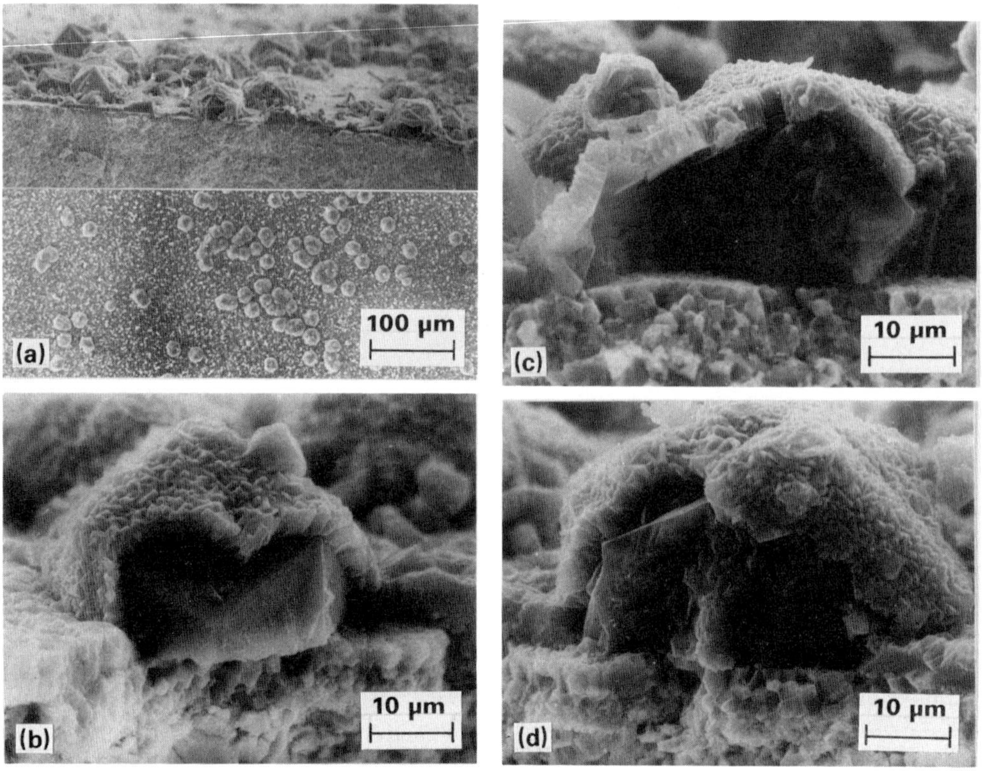

Fig. 5. Different behaviour of Al_2O_3-coated low pressure diamond crystals under stress: (a) general view; (b) pull-out of a diamond crystal; (c) the main part of the diamond is broken, whereas the Al_2O_3 layer remains; (d) clean fracture through the Al_2O_3 layer and a diamond crystal.

vapour deposition, chemical plating or electroplating could be used instead of CVD.

5. Conclusion

The experiments described above have shown that it is possible to produce composite layer coatings consisting of low pressure gas phase diamond particles embedded in a corundum matrix. Such layers could be of great technological interest as surface layers for wear parts, grinding tools and other appropriate applications. Instead of Al_2O_3, other hard or soft layers could be used as a matrix to form such composite coatings.

The advantage of these composite layers is twofold: an increase in the wear resistance of the Al_2O_3 CVD layer [12] and a considerable reduction in the stresses appearing at the interface between diamond and substrate, which usually occur on fully covered diamond-coated substrates. These stresses arise during cooling from the reaction temperature to room temperature and are the result of the considerable difference in thermal expansion between most substrates and the diamond.

Repeated application of the two deposition processes described allows the production of tailor-made and thicker wear-resistant layers.

Acknowledgments

The authors would like to thank Sandvik AB Coromant and Hard Material Division for financial support and Professor B. Aronsson, Dr. M. Sjöstrand and Dr. P. O. Snell for helpful discussions.

References

1 A. Sawabe and T. Inazuka, *Appl. Phys. Lett.*, *46* (1985) 146.

2 N. Kikuchi, T. Komatsu and H. Yamashita, *Proc. 9th Symp. on Ion Sources and Applied Technology, Tokyo, 1985*, 1985, p. 233.

3 N. Fujimori, T. Imani and A. Doi, in H. Oechsner (ed.), *Proc. 5th Int. Conf. on Ion and Plasma Assisted Techniques, Munich, 1985*, 1985, pp. 307–311.

4 M. Kamo, Y. Sato, S. Matsumoto and N. Setaka, *J. Cryst. Growth, 62* (1983) 642.

5 Y. Saito, S. Matsuda and S. Nogita, *J. Mater. Sci. Lett., 5* (1986) 565.

6 S. Matsumoto, in C. J. Timmermans (ed.), *Proc. 7th Int. Symp. on Plasma Chemistry, Eindhoven, July 1985*, A-1-10, pp. 79–84.

7 B. Lux, *Germ. Patent DE 3546 113*, 1985; *Germ. Patent EP 226898*, 1987.

8 H. Schachner, B. Lux, K. Stjernberg, A. Thelin and H. Tippmann, *Germ. Patent DE 3522583*, 1985.

9 H. Schachner, B. Lux, K. Stjernberg, A. Thelin and H. Tippmann, *Eur. Patent EP 166708*, 1984.

10 R. Bichler, R. Haubner and B. Lux, in R. Porat (ed.), *Proc. 6th Eur. Chemical Vapour Deposition Conf., Jerusalem, 1987*, 1987, p. 413.

11 B. Lux, C. Colombier, H. Altena and K. Stjernberg, *Proc. 2nd Int. Conf. on the Science of Hard Materials, Rhodes, September 23–28, .1986*, in *Inst. Phys. Conf. Ser., 75* (1986).

12 B. Lux and R. Haubner, in R. Porat (ed.), *Proc. 6th Eur. Chemical Vapour Deposition Conf., Jerusalem, 1987*, 1987, p. 51.

13 B. Lux, R. Haubner, H. Altena and C. Colombier, *Fischmeister/Jehn Hartstoffschichten zur Verschleißminderung Berichte 39 WEH-Semin. Bad Honef, 1986*, Informations Gesellschaft, Deutsche Gesellschaft für Metallkunde, Oberursel, 1986, pp. 71–112.

14 M. Sjöstrand and T. A. Thelin, *Met. Powder Rep., 41* (12) (1986).

15 N. Kikuchi, *59-166670 (A) Jpn. Patent Appl. 5840286*, 1984.

16 N. Kikuchi, *59-166671 (A) Jpn. Patent Appl. 5840287*, 1984.

17 N. Kikuchi, *59-170261 (A) Jpn. Patent Appl. 5841706*, 1984.

The Microstructure of Ultrahard Material Compacts Studied by Transmission Electron Microscopy*

J. C. WALMSLEY

H. H. Wills Physics Laboratory, University of Bristol, Tyndall Avenue, Bristol BS8 1TL (U.K.)

(Received November 9, 1987)

Abstract

Some microstructural features of ultrahard compacts consolidated at a high pressure have been illustrated with respect to three commercially available examples. The cobalt solvent–catalyst present in Syndite encourages a high degree of intergrowth resulting in a large internal area of strong direct bonding between diamond grains. This is responsible for most of the coherence of the material. In Amborite and Syndax 3 the area of contact between adjacent grains is not increased beyond that produced by physical compression. Bonding in the latter two compacts is dominated by the binders produced in the reaction of aluminium with BN and of silicon with C.

The materials described represent products with a successful combination of constituents and processing. Knowledge of the fine detail of microstructure is important in understanding why desirable properties are achieved and can hopefully contribute to the development of new products.

1. Introduction

Transmission electron microscopy has been applied to three commercial polycrystalline ultrahard compacts. They are De Beers Syndite, Syndax 3 and Amborite. Syndite is diamond based and has a catalytically active binder which encourages the conversion of graphite to diamond in suitable thermodynamic conditions. Amborite and Syndax 3 are cubic boron nitride (BN) based and diamond based respectively and have catalytically inactive binders. The regrowth caused by the binder in Syndite can be observed directly. No regrowth is observed in the other two compacts and most of the bonding is seen to be due to the binder phases.

Transmission electron microscopy (TEM) observations of the three materials discussed have been reported [1–5]. An overview of these results, concentrating on the aspects of the microstructure relating to the binder phases is given here.

The manufacture of ultrahard compacts has been reviewed in detail [6, 7]. They are marketed commercially by several manufacturers including De Beers Industrial Diamond Division, General Electric and Sumitomo Electric. Diamond or cubic BN grains are consolidated at a high pressure (60 kbar or higher) and a moderate temperature (above 1400 °C) adjacent to a source of an additive phase. The mean pressure in the mass of grains is within the range of diamond and cubic BN thermodynamic stability. Initially the load distribution is very uneven, being extremely high at points of direct contact between adjacent grains. The pressure is effectively zero at internal surfaces where rapid degradation to graphite or hexagonal BN takes place as the temperature is raised. When the additive melts and penetrates between the grains, it dissolves or reacts with the graphite and hexagonal BN and possibly the unconverted grain surfaces.

2. Specimen preparation

Roughly parallel-sided pieces of compact were spark eroded from the as-provided material, with maximum dimensions of 3 mm and a thickness of several hundred microns. One side was then polished flat on a cast-iron "scaife", charged with diamond powder. The slice was then turned over and glued to a polycrystalline diamond substrate and attached to a holder on which its thickness could be monitored by a depth gauge micrometer.

*Paper presented at the 3rd International Conference on the Science of Hard Materials, Nassau, The Bahamas, November 9–13, 1987.

0921-5093/88/$3.50

© Elsevier Sequoia/Printed in The Netherlands

Mechanical polishing on the "scaife" then took the thickness down to 50 μm or less. Final thinning to electron transparency was achieved by ion beam bombardment from both sides of the slice.

All ultrahard compact samples produced in this way show an uneven surface topography, largely owing to different thinning rates of the binder and the ultrahard grains. Consequently, it is necessary to study thick continuous sections together with thinner less-complete areas in order to build up an overall idea of the microstructure.

3. Plastic deformation

The mass of starting grains is densified quite considerably by the application of pressure. This is largely accommodated by plastic deformation which provides the most obvious features within the diamond and cubic BN grains. The result is a high dislocation density, often sufficiently high that individual dislocations cannot be discerned. In both diamond and cubic BN grains, microtwin lamellae are found, produced by the movement of partial dislocations across the grains. The plastic deformation of diamond grains has been the main subject of previous studies [1, 8, 9]. The rest of this paper will be devoted to observations which relate to the behaviour of the additive phase in each of the three materials examined.

4. Syndite

Syndite is a diamond-based compact consolidated in the presence of cobalt, an effective metal solvent–catalyst for the conversion of graphite to diamond. As the metal infiltrates between the diamond grains, it takes the graphite at grain surfaces into solution. When the intergranular pressure and temperature move into the diamond-stable region of the C–Co phase diagram, the dissolved carbon starts to precipitate out as diamond [7]. The result is a considerable amount of intergrowth between the diamond grains. The regrown diamond can be identified by its impurity content and by the fact that it has generally undergone less plastic deformation than the original material.

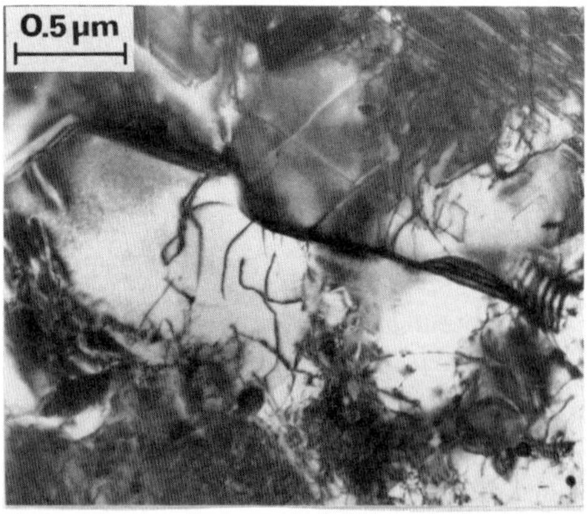

Fig. 1. A field containing a low dislocation density regrown diamond, lying on either side of the large-angle grain boundary which bisects the field.

These points are illustrated in Fig. 1. Here regrown diamond is seen to bridge a gap about 1.5 μm wide between two dislocation-rich original grains which lie in the upper right and the lower left parts of the field respectively. Both grains show good diffraction contrast. Diamond regrowth after the initial stages of heavy plastic deformation, characterized by a low dislocation density, has taken place on both grains. The regrowth fronts have met at the large-angle grain boundary which runs across the field from slightly above middle on the left to slightly below on the right.

In both grains the transition from deformed original to regrown crystal is marked by a high concentration of cobalt impurity. This is not surprising as the onset of regrowth must correspond to diamond growth at minimum sufficient pressure. The dark field image in Fig. 2, formed by a 200-type reflection from the lower grain, shows a band of strongly diffracting cobalt inclusions to be lying at the boundary between regrown and original material. The strong contrast arises because they are in parallel orientation with the matrix around them. The cobalt encountered in Syndite, both in the intergranular channels and enclosed in the grains, is of the high temperature cubic β form which can be retained at room tem-

Fig. 2. Dark field image, showing the same area as Fig. 1, demonstrating contrast from the oriented cobalt inclusions present in the lower grain.

Fig. 3. Diffraction pattern from a diamond matrix containing a cobalt impurity. The inner four spots (forming a square with vertical sides) are the 200-type reflections from the cobalt.

Fig. 4. A point of direct contact between two cubic BN grains in Amborite. On thinning, the two grains have separated slightly.

perature even after moderately rapid cooling. Figure 3 shows a selected area diffraction pattern from an area of diamond matrix containing a single cobalt inclusion. The matrix has been tilted so that it has [001] nearly parallel to the electron beam. There is no zero-layer double-diffraction route to produce intensity in the kinematically forbidden diamond 200-type reflections. The intensity present is due to the cobalt in whose structure 200-type reflections are allowed. All the cobalt encountered in Syndite is found to contain sufficient interstitial carbon for its lattice parameter to match that of diamond, within the accuracy of measurements taken from electron diffraction patterns; so cobalt reflections which coincide with allowed diamond reflections in Fig. 3 are masked. The impurity content, which includes graphite as well as cobalt, of regrown material is always greater than that of original material.

It is clear that the solvent–catalyst behaviour of the cobalt is responsible for strong direct bonding between the diamond grains in Syndite. When the metal is leached out, the coherence of the compact is reduced only slightly [7].

At normal pressures the presence of cobalt at internal diamond surfaces accelerates graphitization at temperatures above about 700 °C, leading to degradation of the compact. Even at temperatures below this, the large mismatch in thermal expansivity between the two components of the compact is undesirable. These problems can be avoided by the use of non-catalytic additives during compaction. The microstructure of two such materials is described below.

5. Amborite

Amborite is manufactured by compacting grains of cubic BN in the presence of aluminium. The metal reacts to form a mixture of AlN and AlB_2 which remain as a stable and catalytically inactive binder. The distribution of the two aluminium compounds in the material is quite characteristic.

Figure 4 shows a point of direct contact between two cubic BN grains. They have mutually indented, showing increased plastic deformation around the point of contact. There is no sign that any regrowth has taken place and the area of direct contact is that produced by mechanical compaction. This is quite a typical observation in the material. The total area of direct contact between cubic BN grains in Amborite is lower than that observed between the diamond grains in Syndite.

Figure 5 shows a roughly octahedral cubic BN facet which appears dark and occupies the upper left portion of the field; it has been tilted so that it is viewed edge on. A rind of AlN, which appears

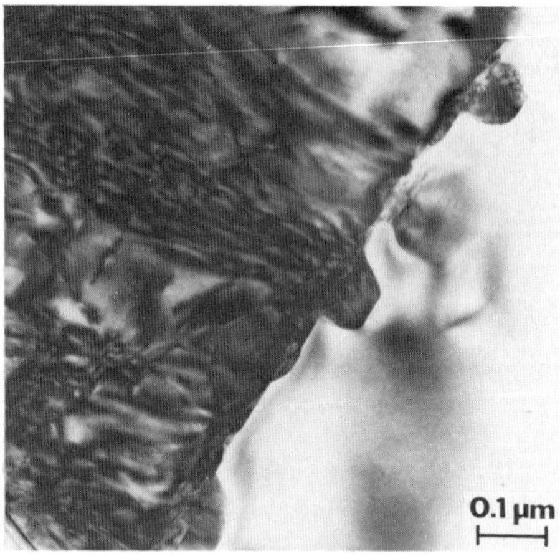

Fig. 5. The edge of a cubic BN grain in Amborite, with [110] parallel to the beam. The AlN rind runs diagonally across the field from lower left to top right. Outside this is the AlB_2.

Fig. 6. Selected area diffraction pattern from the central part of Fig. 4, showing the relative orientation of the phases present.

to be continuous and has developed into "nodules" at several points along its length, is present at the grain surface. Most of the AlN occurs as such a rind enclosing the exposed cubic BN surfaces and is invariably oriented with principal crystallographic directions parallel to those in the cubic BN. The relative orientation of cubic BN and AlN in Fig. 5 is demonstrated by the selected area diffraction pattern of Fig. 6. A basic cubic

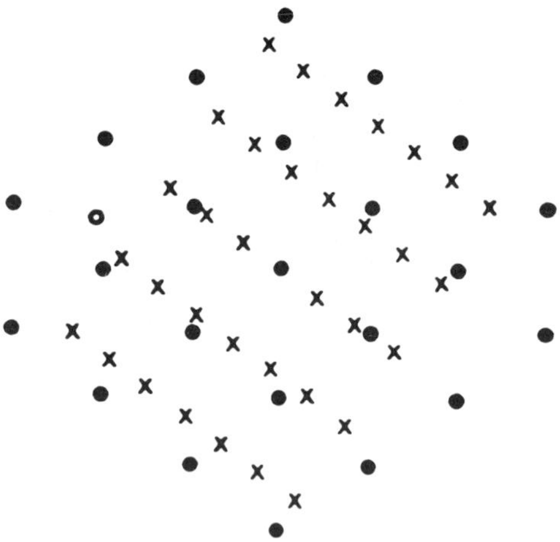

Fig. 7. Diagram identifying the spots present in Fig. 6: ●, cubic BN; ○, AlB_2; ×, AlN. The extra rows of spots present in Fig. 6 are due to double diffraction.

[110] BN and rectangular $[11\bar{2}0]$ AlN zone axis pattern are superimposed. Principal diffraction spots are identified in Fig. 7. The predominant orientation relationship between cubic BN and AlN, found at octahedral facets and most irregular surfaces, including those that deviate considerably from being octahedral, is that represented in Fig. 6, *i.e.* $(1\bar{1}1)_{BN}//(0001)_{AlN}$ and $[110]_{BN}//[11\bar{2}0]_{AlN}$. The prevalence of this orientation is plausible, given that the arrangement of nitrogen atoms in the basal plane of AlN is virtually identical with that in the cubic BN octahedral plane. On well-defined cube facets the orientation relationships $(001)_{BN}//(0001)_{AlN}$ and $[110]_{\text{cubic BN}}//[11\bar{2}0]_{AlN}$ were found to occur. AlB_2 forms after the AlN and occupies the channels between the cubic BN grains as a continuous network. The AlB_2 network largely avoids plastic deformation as is illustrated by the dislocation-free material, which appears light on the photograph and occupies the lower right-hand corner of Fig. 5.

6. Syndax 3

Syndax 3 is produced by compacting diamond grains in the presence of silicon. The two components react to produce a binder of β-SiC which is stable and catalytically inactive. The carbide is found to be in the form of a fine-grained, randomly oriented network which contains small

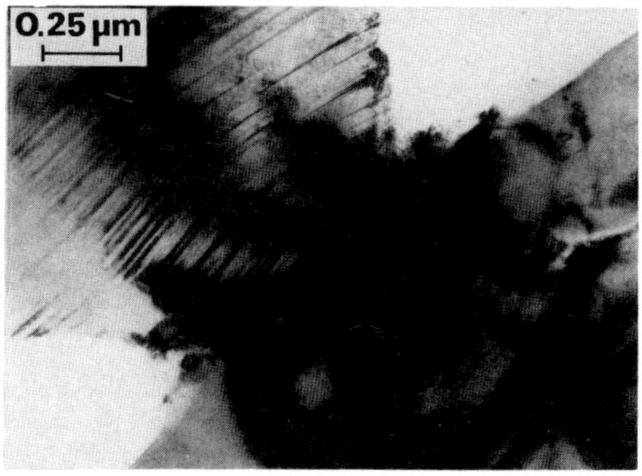

Fig. 8. A point of direct contact between two diamond grains in Syndax 3.

fragments of diamond and a small constant level of unreacted silicon. The survival of diamond fragments, presumed to derive from the large original grains during consolidation, and unreacted silicon, the latter containing an estimated 7 vol% of substitutional carbon, suggests that the reaction between silicon and graphite or diamond during manufacture is not as strong as that of aluminium with hexagonal or cubic BN in Amborite.

There is no evidence of diamond regrowth in Syndax 3 and again the total area of direct contact between adjacent grains is that produced by the physical compression of the compact. Figure 8 shows an area where most of the carbide has been removed from around the interface between two grains. The grains appear to have "welded" together early in the compaction process and silicon has been completely excluded.

7. Discussion

The differences in bonding between compacts manufactured with catalytic and non-catalytic binder phases are clearly illustrated by TEM. The coherence of Syndite is mostly due to the high degree of intergrowth and direct bonding between adjacent diamond grains. In Amborite and Syndax 3, most of the bonding must be attributed to the binder phases although they exhibit mechanical properties comparable with those of Syndite.

Undoubtedly the physical compression of the grains during ultrahard material compaction is important in producing a stable densified mass of grains. Weak points and areas of loosely packed grains are largely eliminated by fracture and the subsequent fine-scale local rearrangement that takes place. While direct bonding between neighbouring grains is observed for Syndax 3 and Amborite, the properties of compacts with a non-catalytic binder should be understood in terms of their composite nature.

Acknowledgments

The author would like to thank Professor A. R. Lang for continuous advice and help in the interpretation of results, and Dr. R. Caveney, De Beers Industrial Diamond Division (Pty) Ltd., for providing samples and encouraging the study.

Financial support to attend this meeting is gratefully acknowledged from De Beers Industrial Diamond Division (Pty), the Institute of Physics and Physics Department, University of Bristol. The work was carried out under a grant from the Science and Engineering Research Council.

Figures in Sections 4 and 5 are reproduced from the *Journal of Materials Science* courtesy of Chapman and Hall Ltd.

References

1 J. C. Walmsley and A. R. Lang, *J. Mater. Sci. Lett., 2* (1983) 785–788.
2 J. C. Walmsley and A. R. Lang, in P. Daniel (ed.), *Ultrahard Materials Application and Technology,* DeBeers Industrial Diamond Division, Ascot, 1984, pp. 11–21.
3 J. C. Walmsley and A. R. Lang, *J. Mater. Sci., 22* (1987) 4093–4102.
4 J. C. Walmsley and A. R. Lang, *J. Mater. Sci., 23* (1988) 1829–1834.
5 J. C. Walmsley and A. R. Lang, in C. G. Barrett (ed.), *Ultrahard Materials Application and Technology,* Vol. 4, DeBeers Industrial Diamond Division, Ascot, 1987, pp. 61–75.
6 R. H. Wentorf, R. C. DeVries and F. P. Bundy, *Science, 208* (1980) 873–880.
7 P. N. Tomlinson and R. J. Wedlake, in N. R. Comins and J. B. Clark (eds.), *Proc. Int. Conf. on Recent Developments in Speciality Steels and Hard Materials*, Pergamon, Oxford, 1982, pp. 173–184.
8 S. Yazu, T. Nishikawa and Y. Doi, in N. R. Comins and J. B. Clark (eds.), *Proc. Int. Conf. on Recent Developments in Speciality Steels and Hard Materials*, Pergamon, Oxford, 1982, pp. 449–456.
9 M. Lee, R. C. De Vries and E. F. Koch, *GE Rep. 84CRD241*, 1984 (General Electric).

Materials Science and Engineering, A105/106 (1988) 555-563

Demineralization and Microstructure of Carbonado*

J. P. DISMUKES, P. R. GAINES†, H. WITZKĖ, D. P. LETA, B. H. KEAR‡, S. K. BEHAL and S. B. RICE

Exxon Research and Engineering Company, Route 22 East, Clinton Township, Annandale, NJ 08801 (U.S.A.)

(Received November 9, 1987)

Abstract

The objectives of this study were to characterize the mineral phases, the diamond phase and the pore structure in the rare polycrystalline form of natural diamond, carbonado, and to determine whether the inorganic mineral content could be extracted by leaching treatments, without degrading the diamond matrix. The carbonado samples evaluated were found to contain more than two dozen impurity elements, in total concentrations varying from about 1.5 to 5 wt.%, not including oxygen. In addition to oxygen, the most abundant were sodium, potassium, calcium, strontium, barium, lead, aluminum, iron, lanthanum, cerium, yttrium, silicon, titanium, zirconium and phosphorus. Extraction of the mineral phases was accomplished by a sequential three-step leaching procedure, beginning with $HClO_4$ to remove all elements except potassium and silicon, followed by H_2O to remove potassium, and finishing with HF solution to remove silicon. The leaching rate is sample size dependent and appears to be diffusion limited. Imaging secondary ion mass spectrometry and scanning electron microscopy (SEM) characterization of fractured surfaces, comparing leached and unleached carbonado samples, showed that the minerals are contained in partially filled interconnected pores and channels and that these minerals are removed by the leaching treatment. SEM evaluation of the pure diamond phase remaining after leaching showed no evidence of foreign phases within the diamond grains or at grain boundaries. The low temperature stability of the minerals detected in the interconnecting void and channel structure suggest that these minerals were formed by a natural infiltration process after the initial formation of carbonado at elevated temperatures and pressures in the Earth.

1. Introduction

Nature has produced a porous, yet tough polycrystalline form of diamond, carbonado, which also contains a number of impurity elements in mineral phases [1-12]. The bulk density reported for carbonado varies from about 3.1 to 3.4 g cm^{-3}, compared with the density of pure diamond which is 3.511 g cm^{-3}. In spite of carbonado's porosity and mineral content, this material has been reported to exhibit superior transverse rupture strength to both natural single-crystal diamond and synthetic polycrystalline diamond composites [13]. Although carbonado is a rare material, mined annually in quantities of about 20 kg world wide, manufacturers nevertheless have found it advantageous over other diamond materials for use in special applications, such as in the most wear-prone regions of diamond matrix body bits [1, 14]. In fact, earlier studies of carbonado as a model material have stimulated the development of fully dense synthetic polycrystalline diamond composites containing cobalt or SiC as an interpenetrating second phase [15, 16].

Previous studies reported in the literature [1-12] of carbonado have dealt with some aspects of its occurrence, composition and microstructure. Rather than being found in Kimberlite diamond pipes, carbonado typically occurs in sedimentary deposits together with various minerals common to the Earth's crust, *e.g.* orthoclase, perovskite, zircon, rutile, olivine, ilmenite, magnetite and others. Annual world-wide production from Brazil, Venezuela, South Africa and the Soviet Union is of the order of 100 000 carats or 20 kg. Trueb and coworkers [1, 7] and Mori-

*Paper presented at the 3rd International Conference on the Science of Hard Materials, Nassau, The Bahamas, November 9-13, 1987.

†Present address: Union Carbide Corporation, P.O. Box 670, River Road, Bound Brook, NJ 08805, U.S.A.

‡Present address: Rutgers University, Center for Materials Synthesis, New Brunswick, NJ 08901, U.S.A.

0921-5093/88/$3.50

© Elsevier Sequoia/Printed in The Netherlands

TABLE 1 Sequential leaching procedures tested for demineralization of carbonado

Step	*Procedure 1*	*Procedure 2*	*Procedure 3*	*Procedure 4*
1	Aqua regia, 24 h reflux (IA and IB, IIA and IIB, Al, IIIB, VIIIB and rare earths)[a]	Aqua regia–HF–H_2SO_4, 72 h reflux (all)[a]	Aqua regia, 24 h reflux (IA and IB, IIA and IIB, Al, IIIB, VIIIB and rare earths)[a]	$HClO_4$ (72%), 24 h reflux (IA and IB, IIA and IIB, Al, IIIB, VIIIB and rare earths)[a]
2	$KHSO_4$ fusion, 1 h, 600 °C (IVB, VB)[a]	—	HF–H_2SO_4, 8 h reflux (IVB, VB, Si)[a]	H_2O, 24 h reflux (K (as $KClO_4$))[a]
3	HF–H_2SO_4, 8 h reflux (Si)[a]	—	—	HF (40%), 24 h reflux

[a]The concepts for each leaching procedure were designed on the basis of the known chemical behavior of mineral compounds and modified on the basis of test results.

yoshi *et al.* [2] have conducted investigations of the diamond microstructure by scanning electron microscopy (SEM) and transmission electron microscopy (TEM). Trueb and Butterman [1] have identified impurity elements and mineral phases such as quartz and aluminum silicates in carbonado samples and have discussed the reaction of hydrofluoric acid (HF) with exposed silica or silicate minerals on the surface of carbonado samples. However, none of the publications that we have found deals with systematic attempts to remove the mineral phases by leaching or to characterize the microporosity of carbonado. The objectives of this effort have been to characterize further the mineral phases, pore structure and diamond microstructure in carbonado, and in particular to determine whether the inorganic mineral content could be removed by leaching treatments, without degrading the diamond phase.

2. Experimental procedures

Samples of carbonado used in these studies were obtained from Dubbeldee Diamond Corporation, Anco Diamond Abrasives Corporation and Reed Rock Bit Company. The first issue addressed was to determine whether leaching procedures could be developed which would remove the mineral content from carbonado. This of course depends on the reactivity of the leaching reagents, the nature of the mineral phases to be attacked and the type of pores (*i.e.* open or closed) within the carbonado. The second issue was to determine whether a leaching procedure for complete mineral removal could be developed which did not significantly attack the diamond matrix. Since diamond is attacked by oxygen and some oxidizing molten salts at

TABLE 2 Matrix of demineralization tests on carbonado samples

Demineralization procedure[a]	*Carbonado samples*						
	A	*B*	*C*	*D*	*E*	*F*	*G*
Aqua regia, step 1; $KHSO_4$, step 2; HF–H_2SO_4, step 3	×						
Aqua regia–HF–H_2SO_4, one step	×	×					
Aqua regia, step 1; HF–H_2SO_4, step 2		×	×	×	×		
$HClO_4$, step 1; H_2O, step 2; HF, step 3					×	×	×

[a]Refer to Table 1.

temperatures above about 300 °C, most of the effort was devoted to strong inorganic acids.

Table 1 gives the sequential leaching procedures tested for carbonado demineralization, showing the chemicals used, typical reflux times and temperatures, and the elemental groups designed to be attacked, based on the known reactivity [17, 18] of mineral types possibly associated with carbonado. Table 2 shows the matrix of four different demineralization procedures evaluated on eight different carbonado samples, A–G, of mass ranging from about 30 to 100 mg. All acids used were Baker Ultrex purity grade, and the $KHSO_4$ was Baker reagent grade. The aqua regia, $HClO_4$ and H_2O treatments were all performed under reflux conditions using Kjeldahl digestion apparatus equipped with 30 ml Kjeldahl flasks. All acid treatments involving HF were performed in a Teflon round-bottomed flask equipped with a Teflon condenser. After completion of each of the demineralization sequences in Table 1, all samples were heated with deionized–distilled water in an open beaker for 1 h followed by drying at 110 °C for 1 h in a vacuum drying oven. The initial sample masses as well as

TABLE 3 Elements detected in carbonado by inductively coupled plasma emission spectroscopy, X-ray fluorescence and secondary-ion mass spectrometry

Subgroup	*Group I*	*Group II*	*Group III*	*Group IV*	*Group V*	*Group VI*	*Group VII*	*Group VIII*
A	Na, K	Mg, Ca, Sr, Ba	B, Al	C, Si, Pb	P	O, S	Cl	
B	Cu	Zn, Hg	Sc, Y, RE[a]	Ti, Zr	V	Cr	Mn	Fe, Ni

[a]RE ≡ La, Ce, Pr etc.

the masses after drying were obtained using a digital electronic five-place analytical balance.

The elemental impurity analyses of leachate solutions were performed by inductively coupled plasma emission spectroscopy (ICPES) using a JACO model 1100 direct-reading, inductively coupled spectrometer, capable of detecting the elements in Table 3 to an accuracy of about 5% relative. The analyses of non-HF-containing acid demineralization solutions were obtained after dilution with water to contain about 3.5 vol.% acid. For HF-containing leachate solutions, the HF was first removed by fuming in the presence of excess H_2SO_4, before diluting to about 3.5 vol.% acid.

Investigation of the solid carbonado samples, both before and after leaching, was conducted by X-ray diffraction (XRD), X-ray fluorescence (XRF), SEM and imaging secondary-ion mass spectrometry (imaging SIMS). A Philips APD 3600 diffractometer with computer-controlled scanning was used for the XRD analysis. An ISI-5540 scanning electron microscope equipped with a PGT System-4 energy-dispersive X-ray fluorescence system or a JEOL JSM-35C system similarly equipped was used for SEM studies. A Kevex 0700 energy-dispersive XRF spectrometer operated under a helium atmosphere was used to obtain the XRF data. The imaging SIMS analyses were performed using a Cameca IMS-3f instrument modified as previously described [19], which has capabilities for argon, oxygen and cesium mass-filtered primary-ion bombardment, positive and negative secondary-ion collection and a high resolution mass spectrometer operating in an ion-imaging mode. Most of the analyses in the present study were conducted employing a 10.5 keV O_2^+ primary-ion beam for the generation of positive secondary-ion images. This modified imaging SIMS system also includes the addition of an ultrahigh gain video camera system capable of single-ion detection, as well as digital imaging-processing capabilities and optical laser disk storage of ion images.

3. Results and discussion

3.1. *Elemental analysis of carbonado samples*

The impurity elements detected in carbonado are indicated in Table 3, arranged by group and subgroup of the periodic table. Imaging SIMS analysis of fractured carbonado surfaces provided initial qualitative indication of presence of all these elements with the exception of copper, zinc, mercury, manganese, nickel and sulfur, which were present in trace quantities. These trace impurities, in addition to the major impurities, were detected by the XRF analyses on the carbonado samples and by the ICPES analyses on leachate solutions. In addition, the SIMS images on fractured carbonado surfaces showed that the detected elements were present as heterogeneous oxide mineral phases on pore walls rather than homogeneously distributed and that the diamond phase was thus relatively pure.

Quantitative analyses of the major impurity elements in the eight carbonado samples, A–G, are shown in Table 4, obtained by ICPES on the leachate solutions. The data indicate not only that the total elemental concentrations are sample dependent ranging from about 1 to 5 wt.% but also that the ratio of specific elements such as aluminum and iron or of elemental groups such as group IIA and the rare earths are also variable. Of all the elements, iron is the most variable, ranging from 0.07 to 3.2 wt.%. This could indicate that the impurity concentrations are quite variable within the larger specimens from which the samples analyzed were taken or that the impurity concentrations of carbonado samples from different locations in the Earth are widely different.

3.2. *Effectiveness of the demineralization procedures*

Our initial premise in devising and testing leaching procedures for demineralization was that only high temperature forms of oxide minerals would be present as a result of the high temperatures and pressures of diamond synthesis. In

TABLE 4 Elemental concentrations in carbonado samples

Element	*Amount* (wt.%) *for the following samples*						
	A	*B*	*C*	*D*	*E*	*F*	*G*
Na + K	0.21	NA	NA	NA	0.20	0.19	0.21
Ca	0.02	0.08	0.35	0.22	0.05	0.02	0.08
Sr + Ba	NA	0.64	0.14	0.28	0.20	0.04	0.62
Pb	0.01	0.13	0.05	NA	0.14	0.12	0.10
Al	0.25	0.47	1.00	0.70	0.76	0.57	2.14
Fe	1.60	0.35	3.20	0.27	0.07	0.12	0.20
Y	0.01	0.01	0.02	NA	0.11	0.01	0.08
La + Ce	NA	NA	NA	0.10	0.50	0.08	0.53
Si	0.25	0.05	0.05	NA	0.05	NA	NA
Ti	0.07	0.02	0.01	0.03	0.03	0.03	0.08
Zr	0.01	0.04	0.09	NA	0.02	0.02	0.09
P	NA	NA	NA	NA	NA	0.07	0.72
Other[a]	0.04	0.03	0.02	0.02	0.03	0.03	0.04
Total	2.47	1.82	4.93	1.62	1.96	1.30	4.89

NA, no analysis.
[a]Elements from Table 3 at less than 0.01 wt.%.

procedure 1 of Table 1, aqua regia was designed to attack mineral salts, molten $KHSO_4$ was designed to attack acid-insoluble forms of oxides, *e.g.* TiO_2 as brookite, and the HF–H_2SO_4mixture was designed to remove silicon if it were present as SiO_2. For sample A in Table 4, aqua regia extraction for 72 h removed more than 95% of most of the elements, but removed less than 20% of the titanium and zirconium. Hence silicon is believed to be present as silicates rather than as SiO_2 found in some samples by Trueb and Butterman [1]. The $KHSO_4$ fusion was ineffective in removing the remaining impurities, strongly attacked the diamond and contaminated it with small platinum particles. The final HF-H_2SO_4 treatment for 8 h removed all traces of silicon and the remaining titanium and zirconium. Although SEM examination of fractured surfaces of carbonado sample A showed no remaining minerals, procedure 1 was abandoned because of the attack of carbonado by $KHSO_4$.

Procedure 2 was next tested in the hope of simplifying the leaching sequences of procedure 1 to one step, based on published reports that this three-component mixture is very effective in dissolution of minerals [17, 18]. However, the group IIA and rare earth elements were only partially removed, probably owing to the insolubility of their fluorides in highly acid solutions. Modification of this treatment to a two-step procedure 3 using aqua regia for 72 h followed by HF–H_2SO_4 for 8 h removed the elements only partially and in variable amounts from carbonado samples B–E.

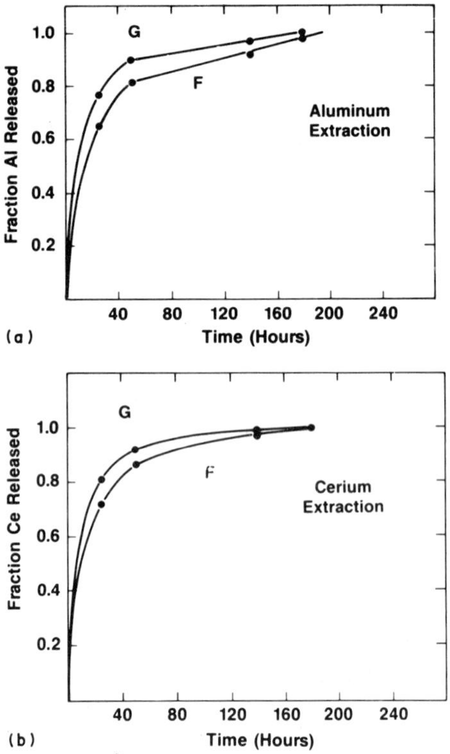

Fig. 1. Time dependence of elemental extraction from carbonado samples F and G by $HClO_4$: (a) fraction of aluminum removed; (b) fraction of cerium removed.

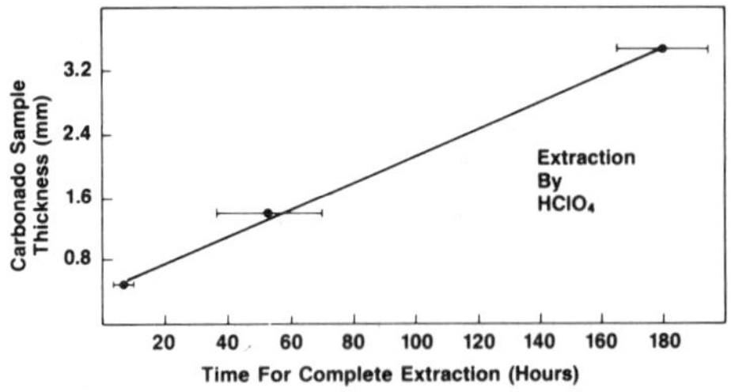

Fig. 2. Sample thickness *vs.* time for complete extraction of carbonado samples F and G by $HClO_4$.

We therefore concluded that leaching procedures involving aqua regia alone or in combination do not extract the minerals from carbonado.

Boiling of carbonado samples in concentrated H_2SO_4 showed that this treatment opened the pore and channel structure in carbonado, although it gave incomplete mineral removal because of the precipitation of insoluble sulfates of the group IIA elements. This provided the clue for the development of the very successful procedure 4, in which $HClO_4$ was used to dissolve all of the elements with the exception of potassium, which was removed in a second H_2O-rinse step, and silicon, which was removed in a final step with HF. We conclude that perchloric acid is effective for several reasons, namely that, like sulfuric acid, it is a strong dehydrating agent which can attack and break up the structure of metal silicates, that all its salts except that of potassium are soluble and that it is a powerful oxidizing agent. After demineralization by this procedure, the only elements which could be detected in carbonado samples by XRF were microgram traces of calcium and iron at the 50 wt.ppm level. SIMS analysis of such specimens showed the presence of calcium and iron uniformly distributed in the diamond phase. SEM analysis, however, did not reveal second-phase particles. Therefore, it is possible that calcium and iron are occluded in closed nanoscale pores or as nanoscale carbide precipitates, which are not connected to the porous network containing the oxide mineral phases.

3.3. *Rate of mineral extraction from carbonado*

Since extremely long times ranging from less than 1 day to about 2 weeks were found to be necessary for extraction of minerals from carbonado, both the rate and the mechanism of mineral removal were of interest in establishing an effective procedure for its complete demineralization. Conceptually the demineralization process may be separated into three steps: (1) acid transport through the porous structure of carbonado to the mineral phases, (2) mineral attack and dissolution and (3) transport of the dissolved mineral elements back out through the porous structure to the bulk solution. Steps (1) and (3) are mass transport limited, whereas step (2) is limited by reaction kinetics.

For the successful $HClO_4$ extraction, the fraction of impurity elements removed was moni-

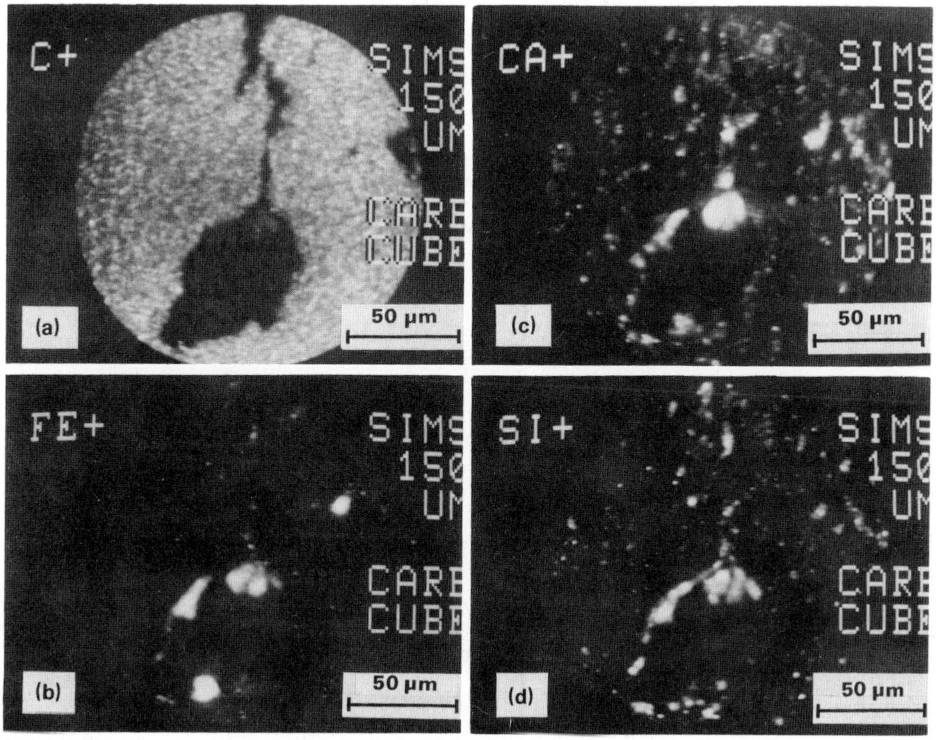

Fig. 3. SIMS images of carbonado sample C showing a deep pore with a channel connecting to another pore and mineral phases partially covering pore and channel wall: (a) C^+ image highlighting the diamond phase; (b), (c), (d) positive-ion images indicating oxide minerals containing (b) Fe^+, (c) Ca^+ and (d) Si^+.

tored as a function of time in order to determine the rate. Figure 1 shows rate curves for extraction of aluminum and cerium, respectively, from carbonado samples F and G, indicating that for samples of about 3 mm in average dimension the time for complete extraction approaches 200 h. Figure 2 shows that the time for complete extraction depends linearly on the carbonado sample thickness dimension. Similar behavior was observed for the other impurity elements. These results suggest that diffusion through the pore structure of carbonado is the rate-limiting step in mineral dissolution. Other factors are only important in the case of partial removal of minerals, where reprecipitation of fluorides or possible non-reactivity of silicates with aqua regia play a role.

3.4. Microstructure of carbonado

Figures 3 and 4 show SIMS micrographs of two different regions of carbonado sample C, before demineralization. The C^+ image in Fig. 3 highlights a deep pore 60 μm in diameter in the diamond phase, connected by a narrow channel to another pore. The other positive-ion images in Fig. 3 show mineral phases containing iron, calcium and silicon coating the pore and channel walls. The C^+ image in Fig. 4 shows two adjacent shallow pores with channels leading from them. The other positive-ion images in Fig. 4 show the mineral phases on the bottom of the pores, containing potassium, aluminum and silicon. Although not shown in the figures, titanium was also observed in smaller quantities as isolated particles. SIMS images of other carbonado samples, such as sample G, have also shown a close association of group IIA elements, the rare earths and phosphorus in the mineral phase.

Figure 5 compares backscatter mode SEM photographs of carbonado sample F fractured before demineralization, with those obtained on fracturing after demineralization. Typical frac-

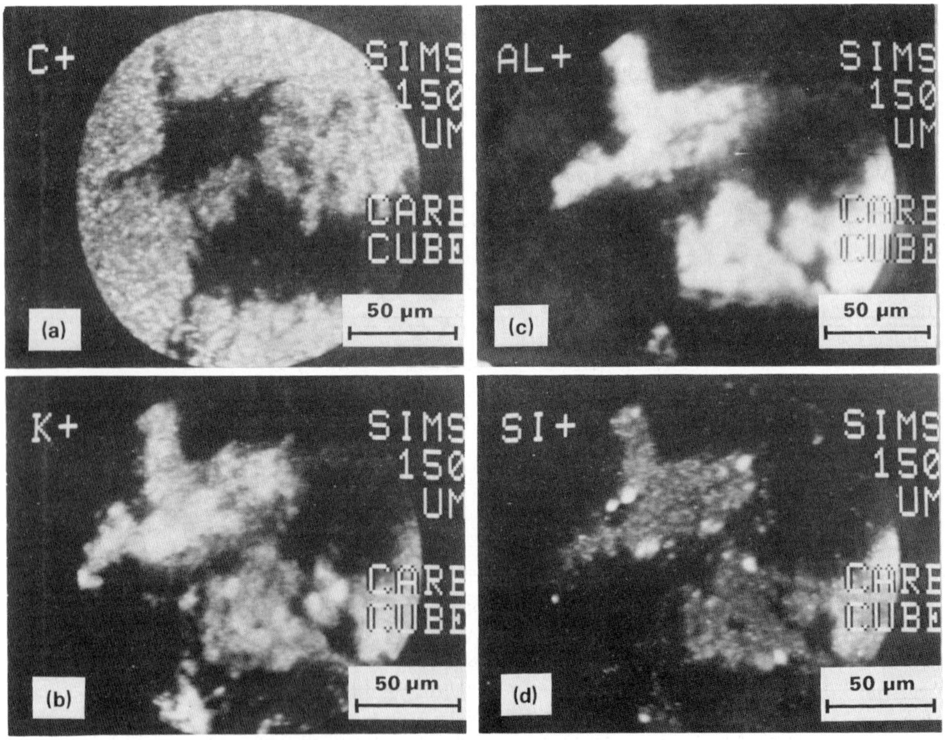

Fig. 4. SIMS images of carbonado sample C showing adjacent shallow pores with channels leading from them, and mineral phases covering the bottom of the pore: (a) C^+ image highlighting the diamond phase; (b), (c), (d) positive-ion images showing the mineral phases containing (b) K^+, (c) Al^+ and (d) Si^+.

Fig. 5. Backscatter mode SEM photographs of surfaces of carbonado sample F fractured (a), (b) before and (c), (d) after demineralization (a) revealing pores 100–300 μm in size with mineral phases coating the pore wall, (b) of the mineral phase coating, revealing mineral particles several microns in size, (c) revealing the fractured diamond surface and several pores completely free of mineral phases and (d) showing faceted diamond grains 1–5 μm in size at a pore wall.

tured surfaces (*e.g.* Fig. 5(a)) before demineralization show pores 100–300 μm in size partially filled with mineral phases of less than 5 μm particle size (Fig. 5(b)) which in this case were found by spot energy-dispersive X-ray analysis (EDXA) examination to contain silicon, phosphorus, aluminum, iron, calcium and the rare earths. In other samples, different types of mineral were observed within the pores, *e.g.* sample C in Table 2, which contains primarily iron oxide. Typical fractured surfaces after demineralization using the $HClO_4$ method show no evidence of minerals, verifying the effectiveness of this demineralization procedure, whereas some residual minerals were observed when using the other demineralization procedures tested. Several pore and channel features are revealed in Fig. 5(c) after leaching of the mineral phases. A closer view of one of the pore walls (Fig. 5(d)) reveals a number of faceted diamond grains ranging in size from about 1 to 10 μm. This is the same grain size range observed by Trueb and Butterman [1] in their microscopic characterization of carbonado samples.

Figure 6 compares secondary-mode SEM photographs of carbonado sample G fractured after demineralization, showing large channels 10–20 μm in size in Figs. 6(a) and 6(b) and smaller channels about 1 μm in size in Figs. 6(c) and 6(d). Although there may have been slight etching of the diamond phase by the perchloric acid, we believe that the micron-size holes, seen in Fig. 6(d), are not etch pits but are small channels forming a part of the porous network within carbonado. This interconnected network provides access for diffusion of the demineralizing acid to attack and remove the mineral phases.

X-ray diffractomer analysis of various carbonado samples showed the strong lines of the diamond phase in the intensity ratios expected for a non-oriented polycrystalline specimen. In addition, this X-ray analysis showed weak lines which are not due to the diamond matrix. These lines coincide with the strongest lines of several known mineral phases, including Fe_2O_3, Fe_3O_4, kaolinite [$Al_2(OH)_4Si_5$] [20] and florencite [21]. Florencite can be represented as substituted

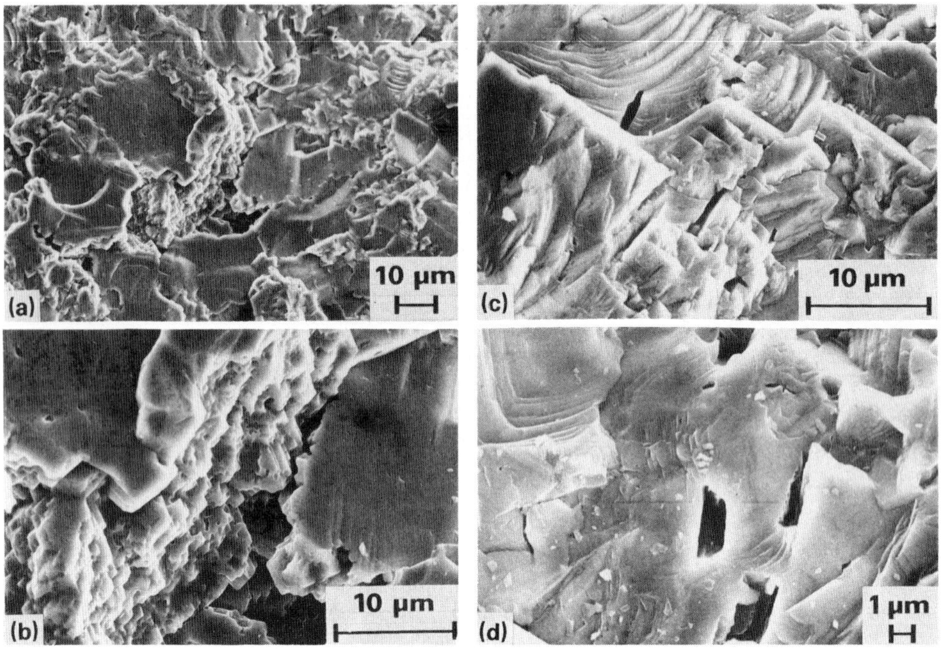

Fig. 6. Secondary-mode SEM photographs of fractured surfaces of carbonado sample G after demineralization (a) showing channels 10–20 μm in size, (b) showing enlargement of one channel in (a), (c) showing micron-sized channels and (d) showing micron-sized channels enlarged.

plumbogummite group mineral [22] of variable composition between $RAl_3(PO_4)_2(OH)_6$ and $MAl_3(PO_4)_2(OH)_5(H_2O)$, where R is a trivalent rare earth element and M is a divalent element such as calcium, strontium, barium and lead. Although only weak lines of this phase were observed, the analyses for carbonado, *e.g.* samples F and G in Table 4, and spot EDXA scans are in agreement with this interpretation of the XRD data.

4. Conclusions

On the basis of the complete leaching of oxide minerals from carbonado and on microscopic observations by imaging SIMS and SEM analyses, the oxide minerals in carbonado are contained in a porous network of large (100–300 μm) pores interconnected by medium-sized (10–20 μm) as well as micron-size channels. A successful leaching procedure was developed based on $HClO_4$ to remove all elements except potassium, which precipitates as $KClO_4$ to be removed by a water rinse, and silicon which is removed by a final HF treatment. A time of up to 200 h is required for demineralization of 100 mg samples, and the rate of demineralization appears to be diffusion limited. The impurity analyses by ICPES on leachate solutions show total impurities, not including oxygen, in the range 1–5 wt.%. The low temperature stability of the mineral phases detected by XRD suggest that these minerals were formed by a natural infiltration process after the initial formation of carbonado at elevated temperatures and pressures in the Earth [10].

Acknowledgments

The authors thank M. Zygmund for carrying out the demineralization experiments, C. V. Schenck and M. A. Modrick for the XRF data, and G. B. Ansell and P. J. Kwiatek for the XRD data.

References

1 L. F. Trueb and W. C. Butterman, Carbonado: Microstructural Study, *Am. Mineral.*, *54* (3–4) (1969)

412–425.
2 Y. Moriyoshi, M. Kamo, N. Setaka and Y. Sato, The microstructure of natural polycrystal diamond, carbonado and ballas, *J. Mater. Sci., 18* (1) (1983) 217–224.
3 Y. L. Orlov and F. V. Kaminskii, Carbonado with lonsdalite—a new polycrystalline aggregate of diamond, *Dokl. Akad. Nauk S.S.S.R., 259* (2) (1981) 459–461.
4 D. N. Robinson, The characteristics of natural diamond and their interpretation, *Miner. Sci. Eng., 10* (2) (1978) 55–72.
5 E. V. Frantsesson and F. V. Kaminskii, Carbonado, a diamond variety of nonkimberlite genesis, *Dokl. Akad. Nauk S.S.S.R., 219* (1) (1974) 187–189.
6 T. D. Varfolomeeva, A. V. Gulyutin and L. N. Dzhavadov, Impurities in samples of natural carbonado, *Dokl. Akad. Nauk S.S.S.R., 202* (1) (1972) 185–186.
7 L. F. Trueb and E. C. DeWys, Carbonado: natural polycrystalline diamond, *Science, 165* (1969) 799–802.
8 R. S. Clark, D. E. Appleman and D. R. Ross, An Antarctic iron meteorite contains preterrestrial impact-produced diamond and lonsdalite, *Nature, London, 291* (1981) 396–398.
9 J. Wilks and E. M. Wilks, Remarks on the classification of natural polycrystalline diamond, *Ind. Diamond Rev., 39* (1979) 156–161.
10 J. V. Smith and J. B. Dawson, Carbonado: diamond aggregates from early impacts of crustal rocks, *Geology, 13* (1985) 342–343.
11 C. Jaynes, Natural polycrystalline diamond, *Ind. Diamond Rev., 38* (1) (1978) 14–23.
12 Yu. A. Klynev, Distortive features of structure of natural polycrystalline diamond, *Dokl. Akad. Nauk S.S.S.R., 240* (1978) 1104–1107.
13 Diamond in new form finds uses in product designs, *Des. Eng.* (1976) 35–36.
14 M. Thompson, Reed Rock Bit Company, personal communication, 1984.
15 W. C. Maurer, Stratapax bits, in *Advanced Drilling Techniques*, Petroleum Publishing Company, Tulsa, OK, 1979, Chapter 22, pp. 717–779.
16 P. M. Tomlinson, N. J. Pipkin, A. Lammer and K. P. Burnand, High performance drilling—Syndax 3 shows versatility, *Ind. Diamond Rev., 45* (1985) 299–304.
17 W. J. French and S. E. Adams, *Anal. Chim. Acta, 62* (1973) 324–328.
18 W. E. Kline and H. S. Fogler, *Chem. Eng. Sci., 36* (1981) 871–884.
19 D. P. Leta, *Springer Series in Chemical Physics,* Vol. SIMS5, *Secondary Ion Mass Spectroscopy,* Springer, Berlin, 1986, pp. 232–234.
20 *Powder Diffraction File*, Joint Committee on Powder Diffraction Standards, International Center for Diffraction Data, Swarthmore, PA, 1984, Card 29-1488.
21 *Powder Diffraction File,* Joint Committee on Powder Diffraction Standards, International Center for Diffraction Data, Swarthmore, PA, 1984, Card 15-320.
22 G. W. Brindley and G. Brown (eds.), *Crystal Structures of Clay Minerals and Their X-ray Identification, Mineralogical Society Monogr. 5,* Mineralogical Society, London, 1980.

The Dynamometer as a Qualitative Gauge in Diamond Polishing*

Y. YARNITSKY†

Israel Institute of Technology, Haifa (Israel)

J. P. F. SELLSCHOP, M. REBAK and S. BARTOLUCCI LUYCKX

Wits Council for Scientific and Industrial Research, Schonland Research Centre, University of the Witwatersrand, Johannesburg (South Africa)

(Received November 9, 1987)

Abstract

A dynamometer is described which can be used to determine efficiently and rapidly the orientation of diamonds and the optimal polishing direction on diamond faces. The dynamometer has proved useful in diamond polishing research as shown, for example, by experiments on the effects of temperature and scaife velocity on diamond polishing. The results of these experiments are consistent with the theory that the abrasion of diamond is essentially a process of microfracture.

1. Introduction

Diamonds are polished by holding the stone in contact with a rotating metal wheel or "scaife", charged with diamond powder. Diamonds are anisotropic, *i.e.* different crystallographic faces present different resistances to polishing and, on one and the same face, polishing is easier in one direction than in others. Diamond polishers usually find the most suitable polishing directions by trial and error and the orientation of diamonds is usually assessed by optical goniometers [1]. In this paper, we present a more efficient method for establishing orientation and polishing directions and illustrate applications of the method to the study of fundamental problems in the field of abrasion and polishing of diamonds.

The method is based on the use of a dynamometer, which measures the frictional force (or drag) between diamond and scaife.

2. Apparatus and experiments

The dynamometer consists of a horizontal arm (Fig. 1) with a spindle able to rotate about a vertical axis. One end of this arm is pivoted; the other is fitted with a vertical spindle. At the lower end of the spindle the diamond is mounted, while the upper end is connected to an electric motor which enables the spindle to rotate continuously and to be set with precision in selected angular positions.

The horizontal arm is constrained to a "zero" position by a leaf spring. When the frictional forces between diamond and scaife change, the arm deflects and the deflections are measured by strain gauges. Two strain gauges are glued onto opposite sides of the leaf spring to increase sensitivity and to compensate for temperature changes.

The leaf spring can be calibrated to calculate the frictional force. In the present work, however, only qualitative results are presented. Frictional force values vary from diamond to diamond depending on the geometry [2] and orientation [1] of the abrader and vary with the applied load [3]. Thus, quantitative results involve a large number of parameters. Quantitative results are beyond the scope of the present paper (as stated in the title of the paper) but will be the aim of future research.

The response of the strain gauge system is instantaneous; thus frictional forces can be measured continuously while rotating the spindle relatively to the scaife. There is no need to remove the stone to measure the parameters under investigation, which is a very considerable advantage.

By connecting a plotter to the strain gauge system, it is possible to record the frictional force *vs.*

---

*Paper presented at the 3rd International Conference on the Science of Hard Materials, Nassau, The Bahamas, November 9–13, 1987.

†Visiting Professor at the University of the Witwatersrand, Johannesburg, South Africa.

0921-5093/88/$3.50 © Elsevier Sequoia/Printed in The Netherlands

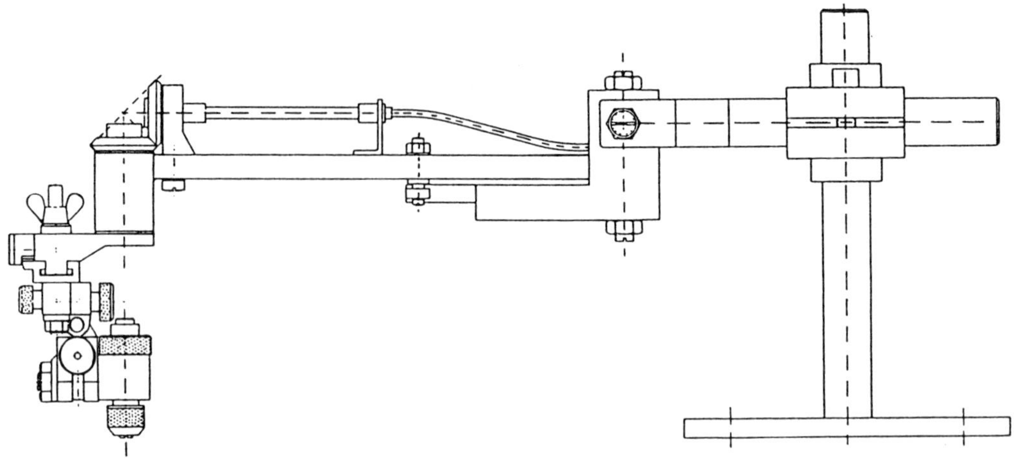

Fig. 1. Schematic representation of the dynamometer described in the text.

the angle through which the diamond has been rotated. The graph obtained is called a "tribogram".

During some of the experiments described below, the amount of material removed in polishing was measured with a Sylvac 25 height gauge, which is able to detect height variations as small as 10^{-4} mm.

Up to now, the dynamometer has been used as follows: to establish the most favourable polishing directions on various diamond planes; to study the effect of temperature on polishing; to study the effect on polishing of the velocity of the scaife.

In order to establish the most favourable polishing directions, the diamond was mounted in the dynamometer with the face to be tested perpendicular to the axis of the spindle. During polishing, the diamond was rotated and the frictional forces were recorded throughout the rotation. The peaks in the tribograms obtained represent the maxima of frictional force and correspond to favourable polishing directions, as was found by measuring the amount of material removed in polishing.

The effect of temperature on polishing was investigated by heating either the diamond or the scaife. The temperature was measured by means of a thermocouple attached to the protruding diamond, near the plane being polished.

The effect of the velocity of the scaife was investigated by increasing the velocity of the scaife from 0 to 3000 rev min^{-1} while keeping the diamond at a constant distance from the centre of the scaife. The friction and the material removed were measured at intervals of 500 rev min^{-1}, after polishing for 1 min at each interval.

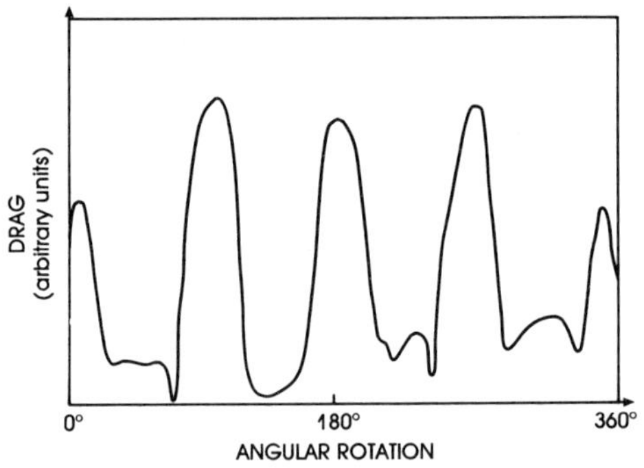

Fig. 2. Typical tribogram from a (100) diamond face. The peaks correspond to ⟨100⟩ directions.

3. Results

Figures 2–4 are characteristic tribograms from (100), (110) and (111) diamond faces. The tribogram in Fig. 2 exhibits four peaks within a 360° range, the one in Fig. 3 two peaks and the one in Fig. 4 six peaks. The peaks are maxima of frictional force and agree with the peaks in the coefficient of friction graphs that were obtained by previous investigators by sliding a diamond stylus on diamond surfaces (see for example ref. 4). The experiments with a diamond stylus showed that the frictional anisotropy of (100) faces exhibits

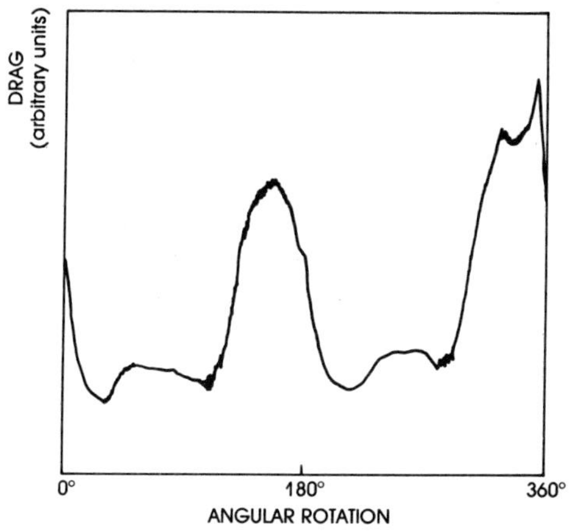

Fig. 3. Typical tribogram from a (100) diamond face. The two higher peaks correspond to ⟨100⟩ directions and the two lower peaks to ⟨110⟩ directions.

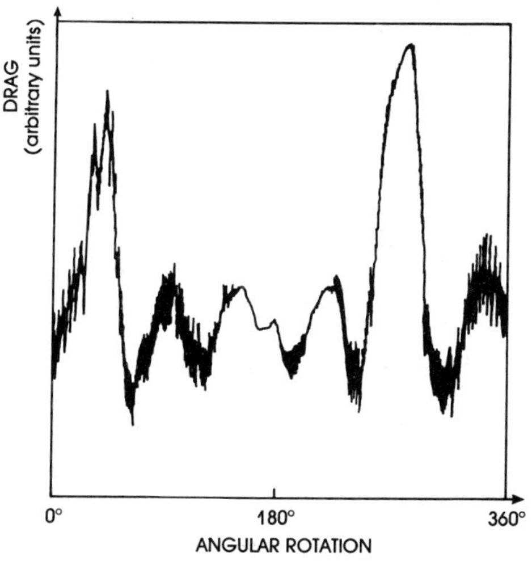

Fig. 4. Typical tribogram from a (111) diamond face.

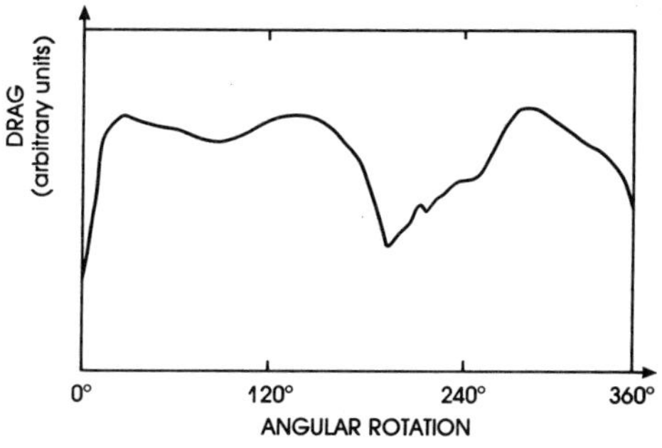

Fig. 5. Tribogram from a diamond face at about 10° to a (100) face.

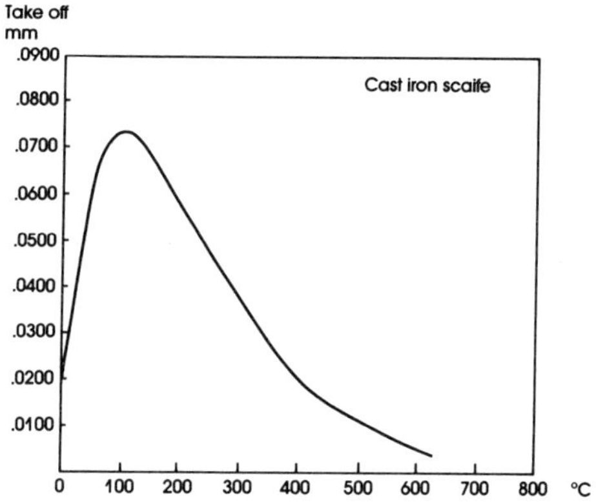

Fig. 6. Dependence of amount of material removed in polishing (which is proportional to the frictional force) on the temperature of the scaife.

fourfold symmetry and (110) faces exhibit lowest friction along ⟨110⟩ directions (small peaks in Fig. 3) and highest along ⟨100⟩ directions (large peaks in Fig. 3). According to the previous experiments [4], (111) faces have triad symmetry (in agreement with Fig. 4) but exhibit low friction in all directions.

Tribograms have also been obtained from planes deviating from low index planes by several degrees. In general, such planes produce tribograms exhibiting peaks broader or in larger number than for low index planes. An example is given in Fig. 5.

The effect of temperature on polishing was investigated by heating the diamond or by heating the scaife. When the diamond was heated, no effective increase in material removal was observed. When the scaife was heated, the material removal was found to increase up to a maximum and then to decrease. Again, increases in material removal were found to correspond to increases in friction. The maximum material removal occurred at about 145 °C (Fig. 6) and the anisotropy was retained above room temperature.

The effect of scaife velocity was tested at velocities ranging from 500 to 3000 rev min^{-1}, keeping the diamond at constant distance from the centre of the scaife and applying to the diamond 1.45 or 2.25 kgf loads. As shown in Fig. 7,

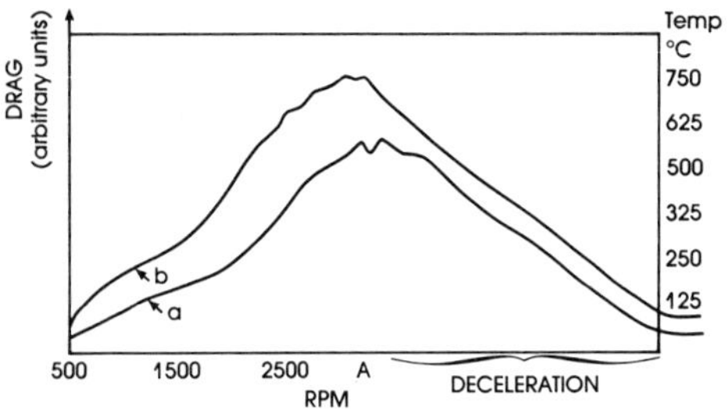

Fig. 7. Dependence of the frictional force on the velocity of the scaife at a load of (a) 1.45 kgf and (b) 2.25 kgf.

the frictional force increased with increasing velocity and with increasing applied load.

4. Discussion

The anisotropy in polishing characteristics of diamond planes was confirmed by the present experiments. The anisotropy was previously studied by sliding a diamond stylus on diamond surfaces [4] and the results were extrapolated to represent industrial polishing conditions. By contrast, the experimental conditions of the present work are almost identical with industrial conditions; thus they complement previous work and justify previous extrapolations.

When comparing the present tribograms with the coefficient of friction graphs reported by other researchers (see for example ref. 4), it can be seen that the tribograms are more irregular. For example, the two higher peaks in Fig. 3 have different heights although they represent friction on the same (110) diamond face and in equivalent ⟨100⟩ directions, whilst the corresponding coefficient of friction *vs.* direction graphs found in the literature exhibit peaks of equal heights in equivalent directions [4].

The irregularity of the tribograms, however, can be accounted for by the results of the experiments using a stylus. Those experiments, in fact, showed that the values of friction depend not only on the surface being abraded but also on the geometry [2] and the orientation [1] of the diamond stylus and on the contact pressure [3]. In the present case, where the diamond stylus was replaced by randomly oriented and variously shaped grit particles, fluctuations in the friction values are to be expected. While rotating the diamond from one direction to an equivalent one, some microchipping occurs of both the surface being abraded and the grit particles, which changes the geometry of the abrasion couple and the contact pressure and causes the friction values to fluctuate continuously throughout the experiment.

The fluctuations in friction values, however, do not prevent the identification of low index planes, since these are characterized by tribograms exhibiting narrow peaks, the identification of the plane being polished, since this is based just on the number of peaks, and the identification of "good" polishing directions, since these coincide with clearly recognizable peaks.

For an interpretation of the anisotropy in polishing characteristics the reader is referred to the review paper by Wilks and Wilks [5] according to which abrasion of diamonds occurs by microcleavage. If it is assumed that diamonds prefer to fracture along cleavage planes, we can account for all the observed directions of hard and easy abrasion.

The present results from the study of the effect of temperature on polishing (Fig. 6) can also be interpreted in terms of the theory that polishing proceeds by microchipping. In these experiments, both the abraded diamond surface and the abrading scaife surface were rough at the microscale level; the abraded surface presented the typical asperities of an unpolished diamond surface and the scaife was studded with grit, *i.e.* protruding diamonds of irregular shapes. At room temperature, therefore, only the highest points of each surface came into contact and only at these contacts could microchipping occur. When the temperature was increased, the yield stress of the scaife material decreased and so the most protruding grit particles, which were subjected to very high stresses, sunk into the scaife until the height of their exposed part equalled the height of less protruding particles. The applied load was now spread over a larger number of particles; so the stress on each particle was lower. When the stress on each particle became insufficient to push the particle into the scaife, the sinking stopped. At this stage the number of contacts between abraded diamond and abrading scaife was larger than at room temperature, microchipping increased and so did material removal. At temperatures higher than about 145 °C, however, the rate of material removal decreased because further softening of the scaife caused the grit to sink completely into the scaife, reducing the diamond–grit contacts to zero.

By contrast, when the diamond was heated, the asperities on the diamond surface remained equal in number and approximately equal in size because the yield stress of diamond does not decrease appreciably below at least 1900 K [6]. Thus the number of diamond–grit contacts remained unchanged and so did the rate of microchipping and material removal.

The results from the study of the effects on polishing of the velocity of the scaife at two different loads further support the theory that the polishing of diamond proceeds by microchipping. The friction increased with increasing velocity of the scaife (Fig. 7) because the amount of debris produced by microchipping increases with increasing number of revolutions. The debris build up ahead of the abrading particles increasing the friction, as was shown by Samuels and Wilks [7] when sliding a diamond stylus repeatedly on a diamond surface.

When increasing the load from 1.45 to 2.25 kgf, the friction increased because the amount of fragmentation of both diamond surface and grit particles increased and so did the amount of debris ahead of the abrading grit.

5. Conclusions

The dynamometer described in this paper has been used to show that the anisotropic polishing characteristics of diamond established by sliding a stylus on a diamond surface are observed also under industrial conditions, *i.e.* when the stylus is replaced by a scaife charged with diamond powder. The dynamometer has also been used to study the effects of temperature and scaife velocity on the polishing of diamond. The results are consistent with the theory that the polishing of diamond proceeds by microchipping and indicate that the parameter controlling the rate of polishing is the number of contacts between the diamond surface and the grit embedded in the scaife.

Acknowledgments

The authors acknowledge with appreciation the support of the Foundation for Research Development and the Diamond Research Laboratory for the provision of materials.

References

1 E. M. Wilks and J. Wilks, *Diamond Conf., Abstracts, Cambridge, 1983*, De Beers Industrial Diamond Division, London, 1983, p. 7.
2 F. U. Hillebrecht and J. Wilks, *Diamond Conf., Abstracts, Cambridge, 1979*, De Beers Industrial Diamond Division, London, 1979, p. 33.
3 D. Tabor, *Proc. R. Soc. London, Ser. A, 373* (1981) 405.
4 D. Tabor, in J. E. Field (ed.), *The Properties of Diamond*, Academic Press, London, 1979, p. 383.
5 J. Wilks and E. M. Wilks, in J. E. Field (ed.), *The Properties of Diamond*, Academic Press, London, 1979, p. 351.
6 T. Evans, in J. Burls (ed.), *Science and Technology of Industrial Diamonds*, Vol. 1, Industrial Diamond Information Bureau, London, 1967, p. 105.
7 B. Samuels and J. Wilks, *Diamond Conf. Abstracts, Reading, 1985*, De Beers Industrial Diamond Division, London, 1985, p. 9.

Size Separation for Hard Microparticles*

CAO TIANSHOU

East China Institute of Metallurgy, Maanshan, Anhui (China)

(Received November 9, 1987)

Abstract

It is always necessary to reduce the particle sizes in raw materials (whether ceramic composites or superhard materials) to 100 μm and to separate the microparticles in order to make the sizes uniform. Generally, microparticles are classified by sedimentation or overflow. The key to the technology is to use a reliable formula for calculation.

According to the principles of particle dynamics, a theoretical formula is developed in this paper. This new formula is suitable for use over a wider range than Stokes' law can be applied to because it conforms to the law of motion of multi-sized angular particles, whereas Stokes' law can be applied only to small spherical particles.

1. Introduction

The size separation of hard microparticles is essential in the development of new materials and in the improvement of hard materials in order to meet the technological challenge of the future. The sizes of hard microparticles affect the properties of raw materials in many important ways. For example, the particle size determines the sintering time and shrinkage of ceramic composites, the activity of chemical catalysts (the specific surface area should be determined from size distribution data) and the abrasive properties of micropowders of superhard materials. Superconductive materials of the future are also strongly affected by the size of the powder used.

Fluid classification is a process for size separation of microparticles. A wide variety of equipment is available, ranging in capacity from many tons per hour for the larger units to a few grams per hour for laboratory machines.

2. Newton's law of resistance and Stokes' theory

If a particle moves relative to the fluid in which it is suspended, the force opposing the motion is known as the drag force. Knowledge of the magnitude of this force is essential if the particle motion is to be studied. Conventionally, the drag force R is expressed, according to Newton, by

$$R = \frac{\rho_f U_{Newton}^2}{2} \frac{\pi D_s^2}{4} F_d \qquad (1)$$

where U is the velocity of the particle relative to the fluid, ρ_f is the fluid density, D_s is the particle diameter (hence $\pi D_s^2/4$ is the area of the particle projected in the direction of the motion) and F_d is the drag coefficient.

Newton assumed that the drag force is due to the inertia of the fluid and that F_d would then be constant. However, experiments show that F_d is generally a function of the particle Reynolds' number Re:

$$Re = \frac{\rho_f U D_s}{M} \qquad (2)$$

where M is the fluid viscosity.

At low Reynolds' numbers (Re < 0.2), Stokes assumed that the drag force is due to viscous forces within the fluid and the relationship between F_d and Re becomes

$$F_d = \frac{24}{Re} \qquad (3)$$

Elimination of F_d and Re in eqns. (1)–(3) gives Stokes' law:

$$R = 3\pi M D_s U_{Stokes} \qquad (4)$$

A single particle settling in a gravity field is subjected primarily to a drag force, a gravity force and buoyancy; Newton's equation of motion becomes

$$m\frac{dU}{dt} - \left(mg - mg\frac{\rho_f}{\rho_s} - R\right) = 0 \qquad (5)$$

*Paper presented at the 3rd International Conference on the Science of Hard Materials, Nassau, The Bahamas, November 9–13, 1987.

0921-5093/88/$3.50 © Elsevier Sequoia/Printed in The Netherlands

where m is the particle mass, g the acceleration due to gravity and ρ_s the density of the solid. With increasing time, $dU/dt \to 0$. The particle mass m is given by

$$m = \frac{\pi}{6} D_s^3 \rho_s \tag{6}$$

Substituting this in eqn. (5) and with $dU/dt = 0$, we find that

$$\frac{\pi}{6}(\rho_s - \rho_f) g D_s^3 - R = 0 \tag{7}$$

For Re > 1000, elimination of R between eqns. (1) and (7) gives

$$U_{\text{Newton}} = \left(\frac{4}{3} g \frac{\rho_s - \rho_f}{\rho_f} \frac{D_s}{F_d} \right)^{1/2} \tag{8}$$

For Re < 0.2, elimination of R between eqns. (4) and (7) gives

$$U_{\text{Stokes}} = \frac{g D_s^2}{18M} (\rho_s - \rho_f) \tag{9}$$

3. Calculation of settling velocity for intermediate particle sizes

Experience has shown that, when the solids are greater than 1 mm in diameter or when Reynolds' number for the system is greater than 1000, $F_d = 0.44$ [1], eqn. (8) is applicable under the conditions stated.

When the particle size is smaller than 20 μm or when Reynolds' number for the system is less than 0.2, the free-settling velocity of a sphere is given by eqn. (9).

For intermediate particle sizes, the sizes most common in hard materials (20–200 μm), F_d is neither constant nor equal to 24/Re. Generally, for this intermediate region, graphical methods exist by which the velocity can be estimated (ref. 2, p. 227). Unfortunately, this graphical method cannot be used for an automatic process of classification.

Attempts at theoretical solutions of the relationship between F_d and Re have been made. Oseen [3] partially allowed for inertial effects and obtained

$$F_d = \frac{24(1 + 3\,\text{Re}/16)}{\text{Re}} \tag{10}$$

Proudman and Pearson [4] obtained, as a first-order correction,

$$F_d = \frac{24\{1 + 3\,\text{Re}/16 + 9\,\text{Re}^2 \ln(\text{Re}/20)\}}{\text{Re}} \tag{11}$$

The experimental relationship between Reynolds' number and the drag coefficient is shown graphically in Fig. 1.

In the intermediate region, the drag force on a sphere is given by [5]

$$R_{\text{int}} = 3\pi(M + C) D_s U_{\text{int}} \tag{12}$$

where M and C are the fractions of the drag forces in Stokes' and Newton's equations respectively.

Comparing this with Stokes' equation (eqn. (9)) gives

$$U_{\text{int}} = \frac{D_s^2(\rho_s - \rho_f) g}{18(M + C)} \tag{13}$$

Considering the work of Oseen and of Proudman and Pearson, let

$$F_d = \frac{24}{\text{Re}} + \frac{9b}{2} \tag{14}$$

where the wake flow coefficient b is a function of Reynolds' number.

In the intermediate region, with increasing Reynolds' number, the boundary layer begins to detach itself from the rear of the particle and to form vortices as in Fig. 2.

At very high Reynolds' numbers, substituting $9b/2$ in eqn. (14) for F_d in eqn. (8), gives

$$U_{\text{int}} = \left\{ \frac{8g(\rho_s - \rho_f) D_s}{27 \rho_f b} \right\}^{1/2} \tag{15}$$

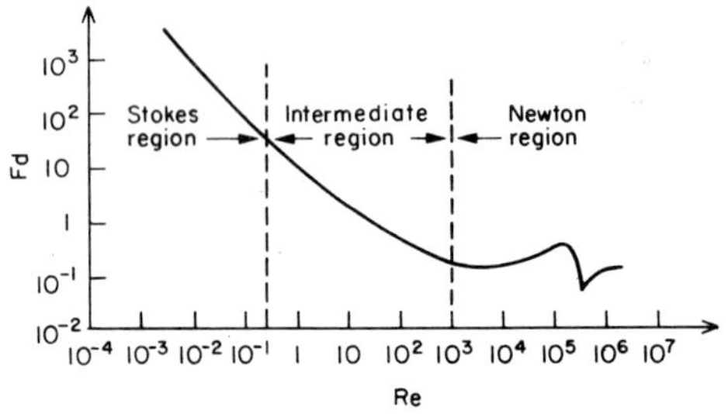

Fig. 1. Experimental relationship between Re and F_d.

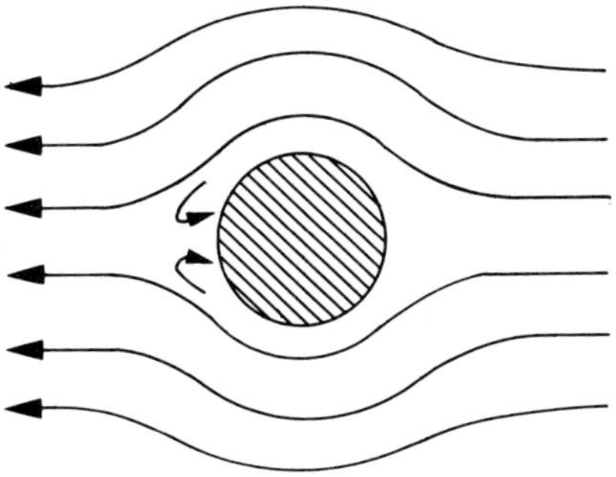

Fig. 2. Formation of vortices behind a spherical particle.

It is shown that, when the drag force used by Stokes is negligible at very high Reynolds' numbers, eqn. (13) becomes

$$U_{int} = \frac{gD_s^2(\rho_s - \rho_f)}{18C} \quad (16)$$

Comparing eqn. (15) with eqn. (16) gives

$$C = \left\{ \frac{bg\rho_f D_s^2(\rho_s - \rho_f)}{96} \right\}^{1/2} \quad (17)$$

Substituting this into eqn. (13) gives

$$U_c = \frac{D_s^2(\rho_s - \rho_f)g}{\{(3D_s/2)^3 b\rho_f(\rho_s - \rho_f)g\}^{1/2} + 18M} \quad (18)$$

4. Discussion on the application of the equation

The new equation, eqn. (18), can be used wherever eqns. (8) and (9) have been used in the past. For Re < 0.2, when the spherical particle is very small, eqn. (18) becomes eqn. (9). For Re > 1000, when the spherical particle is very large, eqn. (18) becomes eqn. (8). It can be seen that the new equation is an intermediate equation between Stokes' equation and Newton's equation. The new equation is applicable to a wide range of particle sizes.

The wake flow coefficient b does not have a wide range of numerical values. For the Stokes region, comparing eqn. (10) with eqn. (14) gives $b = 1$. For the Newton region, Re > 1000 and $F_d = 0.44$; substituting these into eqn. (14) gives $b = 0.1$. It can be seen that $0.1 \leqslant b \leqslant 1$ and hence $0.32 \leqslant b^{1/2} \leqslant 1$.

It is interesting at this point to consider this range of numerical values. Reynolds' number Re has a wide range of numerical values (from 0 to 1000); the wake flow coefficient b is just the opposite to Re. It has distinct advantages for integrating Newton's equation and Stokes' equation.

Experiments show that b is near unity for $D_s < 100$ μm. For $D_s > 20$ μm, the free-settling velocity of a spherical particle calculated from eqn. (18) is smaller than that from eqn. (9) as shown in Table 1.

TABLE 1 Ratios of U_c to U_{Stokes}

D_s (μm)	U_c/U_{Stokes}
1	1.000
5	0.995
10	0.985
20	0.960
40	0.893
50	0.860
100	0.687
150	0.534
200	0.427
300	0.287

TABLE 2 Viscosity of water at room temperature

Temperature (°C)	*Experimental value* (N s m^{-2})	*Value from eqn. (19)* (N s m^{-2})
0	0.0017921	0.0017921
5	0.0015188	0.0015172
10	0.0013077	0.0013056
15	0.0011404	0.0011387
20	0.0010500	0.0010044
25	0.0008937	0.0008944
30	0.0008007	0.0008030

5. Effects of temperature

For the fluid at different temperatures the present author [6] has given an expression for the drag coefficient instead of referring to relevant tables:

$$M = \frac{M_0}{(1 + eT)^n} \quad (19)$$

where M_0 (N s m^{-2}) is the viscosity of the fluid at 0 °C, e the free-space coefficient of the fluid, T (°C) the temperature of the fluid and n an exponential coefficient.

If the fluid is water at room temperature, experiments give the following values: $M_0 = 17.921 \times 10^{-4}$ N s m^{-2}; $e = 0.023$; $n = 1.53$. The viscosity of water at room temperature is given in Table 2.

6. Effects of particle shapes

Perry [7] introduced the work of Pettyjohn and Christiansen in which they present a correlation which allows for the effect of particle shapes on the free-settling velocity for isometric particles. In the Stokes region, the terminal velocity is given by

$$U_{\text{Stokes}} = k\frac{D_s^2(\rho_s - \rho_f)}{18M}g \tag{20}$$

where k ($=0.843 \log(w/0.065)$) is the shape factor and w is the sphericity (which ranges from 0.906 to 0.670). The sphericity w can be determined from the specific surface area:

$$w = \frac{S_p}{S_s}$$

where S_p is the surface area of a sphere having the same volume as the particle and S_s is the surface area of the particle. Therefore

$$S_s = \frac{S_p}{w} = \frac{\pi D_s^2}{w}$$

The volume-specific surface area S_u is given by

$$S_u = \frac{S_s}{V} = \frac{\pi D_s^2/w}{\pi D_s^3/6} = \frac{6}{wD_s}$$

where V is the volume of the particle. The average volume-specific surface area $\overline{S_u}$ is given by

$$\overline{S_u} = \sum_i S_u X_i \Big/ \sum_i X_i = \frac{6}{w}\left\{\sum_i (X_i/D_i) \Big/ \sum_i X_i\right\} = \frac{6}{wD_h}$$

where

$$D_h = \sum_i X_i \Big/ \sum_i (X_i/D_i)$$

is called the harmonic mean of the diameters. Hence

$$w = \frac{6}{\overline{S_u} D_h}$$

The parameters in this formula can be measured with instruments.

For example, let us calculate the shape factor of SiC micropowder, given the following data: $\overline{S_u} = 3.69 \times 10^5 \text{ m}^{-1}$; $D_h = 22.7 \times 10^{-6}$ m. Hence

$$w = \frac{6}{3.69 \times 10^5 \times 22.7 \times 10^{-6}} = 0.717$$

$$k = 0.843 \log\left(\frac{0.717}{0.065}\right) = 0.879$$

Asymmetric particles, such as ellipsoids and discs, do not generally fall vertically. A set of identical particles can have a range of terminal velocities. This range is fairly limited, however. The ratio of maximum to minimum velocities is less than 2 to 1, even with a length-to-diameter ratio of 10 to 1 (ref. 2, p. 231).

If r is the ratio of length to diameter, $a = r^{0.3}$; this effect can be estimated from

$$U = \frac{U_{\text{Stokes}}}{a} \tag{21}$$

where a is the instability.

7. Concentration effects

Steinour [8], who compared previous tests on uniform spheres, pointed out that a portion of the liquid suspension medium is carried downwards with angular emery particles during their fall whether the suspensions are flocculated or not. Therefore the concentration effect is greater for non-spherical and angular particles than for spherical particles.

Steinour's final equation was

$$U = U_{\text{Stokes}}\, p^2 \times 10^{-1.82(1-p)} \tag{22}$$

where the porosity $p = 1 - c$ (c is the concentration of particles by volume).

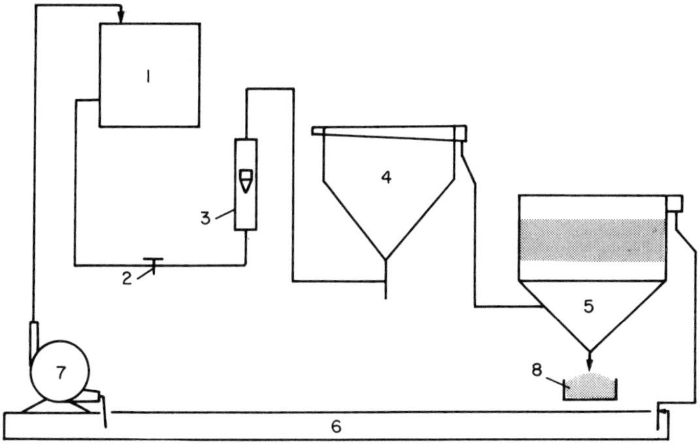

Fig. 3. Operating principle of the single-stage overflow classifier.

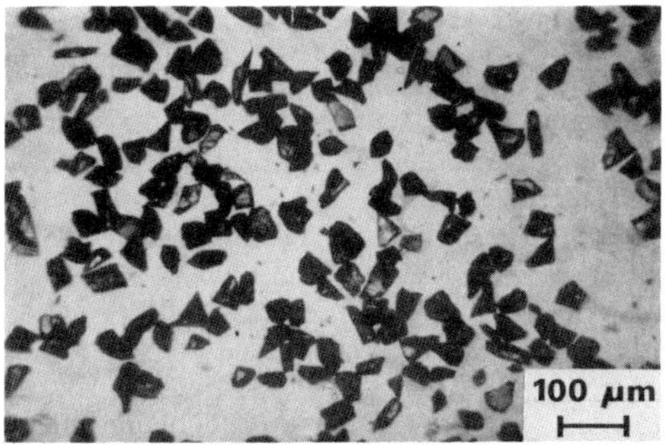

Fig. 4. Microphotograph of SiC micropowder.

Selim [9], in later investigations, developed a model for predicting the sedimentation velocity in suspensions of multisized non-flocculating solids, in which the retarding effect of the smaller particles on the settling velocities of the larger particles is taken into account.

8. Size separation

Fluid classification is a process of size separation of microparticles. The single-stage overflow classifier is a process of grading particles by means of an upward current of fluid, usually water. The process is therefore the reverse of gravity sedimentation. The parameter by which particles are classified is the speed at which they fall.

From eqns. (18)–(22),

$$U = \frac{kD_s^2(\rho_s - \rho_f)gp^2 10^{-1.82(1-p)}}{a[\{(3D_s/2)^3 b\rho_f(\rho_s - \rho_f)g\}^{1/2} + 18M_0/(1+eT)^n]} \qquad (23)$$

All the parameters in the formula can be measured with instruments, and the industrial process can conveniently be controlled with a microcomputer.

The operating principle of the single-stage overflow classifier is shown in Fig. 3. Figure 4 is a microphotograph of SiC micropowder.

9. Conclusions

Basic on the principles of particle dynamics, according to the boundary layer theory and the linear additional law, a settling velocity formula that is adapted to a wide range of particle diameters is introduced in the paper. The latest model (eqn. (23)) included the effect of the fluid temperature, the particle shape and the suspension concentration on the settling velocity. The simple model is very important for atomic control in the working process and it is also quite useful for measuring the particle diameter. To date, it has been successfully made use of in micropowder production.

Acknowledgments

This work was supported by the Science and Technology Committee of Maanshan City (China).

I would like to thank my coworker Zhang Qiujin and Professor Sun Duanfang, reference librarian of East China Institute of Metallurgy.

References

1 L. Svarovsky, Sedimentation, *Encyclopedia of Chemical Techniques*, Vol. 20, 3rd edn., Wiley, New York, 1980, p. 561.
2 T. Allen, *Particle Size Measurement*, 3rd edn., Chapman and Hall Ltd., New York, 1981.
3 C. W. Oseen, *Neuere Methoden und Ergebrisse in der Hydrodynamik,* Akademische, Leipzig, 1927.
4 I. Proudman and J. R. A. Pearson, Expansions at small Reynolds number for the flow past a sphere and a circular cylinder, *J. Fluid Mech., 2* (1957) 237.
5 C. Tianshou, Free-settling velocity of multi-angular particles, *J. East China Inst. Met., 3* (1) (1986) 109–117.
6 C. Tianshou, *J. Abrasives Grinding,* (5) (1987) (in Chinese).
7 R. H. Perry, *Particle Dynamics, Chemical Engineering Handbook*, McGraw-Hill, New York, 1973, pp. 5-61–5-65.
8 H. H. Steinour, Rate of sedimentation, *Ind. Eng. Chem., 36* (9) (1944) 840.
9 M. S. Selim, Sedimentation of multisized particles in concentrated suspensions, *AIChE. J., 29* (6) (1983) 1029–1038.

Author Index

© Elsevier Sequoia/Printed in the Netherlands

Subject Index

© Elsevier Sequoia/Printed in the Netherlands